Nanotechnology in Agriculture, Energy and Environment

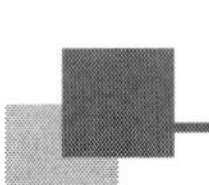

The Editors

Dr. A. Lakshmanan is the Head of the Department of Nano Science and Technology, TNAU, Coimbatore and involved in Undergraduate and Post graduate teaching since 1995. Underwent post-doctoral training on Nano composites in waste water treatment at Monterry Tech, Mexico and on SWAT modelling at Texas A&M university, USA.He has expertise on Environmental nanotechnology and currently working on nano systems in pollution management and disease control. He has more than 50 research publications and authored four books, guided 12 Post graduate and 3 Ph.D students and mentoring two PDFs. Obtained 18 externally funded projects with a total out lay of around Rs. 13 crores to address issues pertaining to Bio & Nano remediation, Climate change adaptation and Biosafety of engineered nano devices. Involved in the preparation of policy manuals on Climate change Adaptation in Agriculture and Water sectors for the states of Tamil Nadu and Telangana in coordination with respective governments.

Dr. K. S. Subramanian is presently the Director of Research, Tamil Nadu Agricultural University, Coimbatore and NABARD Chair Professor at the Department of Nano Science & Technology. He is with the University for more than 32 years involved in teaching, research and extension activities of the University. He did his Ph.D. at the University of Ottawa, Canada, utilizing one of the most prestigious Canadian Common wealth Scholarships (1993-1998). He has got expertise in the fields of nanotechnology in agriculture and soil biology, generated more than Rs. 25 crores of research grants to build infrastructure and human resource. He is the Founder Head of the Department of Nano Science & Technology and evolved the research and educational frame works in the sphere of fascinating field of Nano-science and nanotechnology suiting to agricultural sciences. He is one of the core committee members constituted by DBT, GOI, New Delhi for the regulation of nano-agri products in the country. Dr. K. S. Subramanian published over 200 research papers in peer-reviewed journals and six books on Nanotechnology in Agriculture, participated as a lead speaker in various national and international conferences, visited several countries. He received more than 12 awards and distinctions to his credentials.

Nanotechnology in Agriculture, Energy and Environment

— *Editors* —

A. Lakshmanan

K.S. Subramanian

Tamil Nadu Agricultural University
Coimbatore – 641 003, Tamil Nadu

2020

Daya Publishing House®

A Division of

Astral International Pvt. Ltd.

New Delhi – 110 002

ISBN 978-93-89719-33-8

Published by : **Daya Publishing House®**
A Division of
Astral International Pvt. Ltd.
– ISO 9001:2015 Certified Company –
4736/23, Ansari Road, Darya Ganj
New Delhi-110 002
Ph. 011-43549197, 23278134
E-mail: info@astralint.com
Website: www.astralint.com

Digitally Printed at : **Replika Press Pvt. Ltd.**

M.S. Swaminathan
Founder Chairman, M.S.Swaminathan Research Foundation, Ex-Member of Parliament (Rajya Sabha)

Foreword

Nanotechnology has touched all realms of life including agriculture and the success is widely embraced by the scientific community. The unique properties of materials at nano scale make them suitable candidates for the design and development of novel tools in support of smart agriculture. Nanotechnology has the potential to enhance the agricultural productivity, managing environmental pollution and maintaining ecosystem health. Considering the challenges we face due to the growing global population and climate change, the application of nanotechnologies as well as the introduction of nanomaterials in agriculture can potentially contribute to address the issue of food and nutritional security. Nanotechnology can intervene the agricultural sector in many ways. The use efficiency of fertilizers and pesticides can be enhanced through Nano scale carriers and compounds and by reducing the amount to be applied without compromising the productivity. Nanotechnologies can also have an impact on the reduction of waste through contributing to a more efficient production system and efficient reuse of waste.

I am happy to note that the Tamil Nadu Agricultural University is the first State Agricultural University in India to establish an exclusive Department of Nano science and Technology a decade back with a view to initiate research in the cross-cutting areas. The Nano center at TNAU is effectively involved in teaching, research and dissemination of Nano inventions to various stakeholders. It is heartening to note that the Nano Centre of Tamil Nadu Agricultural University is releasing a book on "Nanotechnology in Agriculture, Energy and Environment" to commemorate the ten years of excellence. This book presents a comprehensive overview of new and emerging nanotechnologies that have potential impacts on

3rd Cross Road, Taramani Institutional Area, Chennai (Madras) - 600 113, India
Phone: +91-44-2254 2790, 2254 1698, Fax: +91-44-2254 1319
E-mail: founder@mssrf.res.in, swami@mssrf. res.in

agriculture, energy, food and environmental sectors. The book covers important subjects that are relevant to the topic and the chapters are written in a lucid language with clarity. The important topics such as nano agro inputs, nano biosensors, nanotechnologies for food processing, packaging and storage, crop improvement, pollution management, water treatment and plant protection are covered with case studies that would heighten the readers interest. The one among the major challenges for nano products is bio safety concerns and it is interesting to note that biosafety, toxicology and guidelines for commercializing nano inputs and products have been discussed elaborately in this book. Climate change is a global concern and agriculture not only impacts climate change by contributing greenhouse gas emission but also affected by changing climate. Role of nanotechnologies in minimizing greenhouse gas emission and promoting climate resilient agriculture has been discussed with case studies. The chapter on future perspectives of nano technological interventions in agriculture would trigger the interest of students to initiate research in such emerging fields to find solutions to unanswered field problems. The authors have wisely linked nanotechnologies with smart farming and artificial intelligence that would certainly help market driven nano research in agriculture. The editors, Dr.A. Lakshmanan and Dr.K.S.Subramanian have excellent track record in Nano research and have led the group of proficient authors to bring out this book that offers a concrete, useful and in-depth insight on Nano Agriculture to a variety of readers in particular the student community. This book is lucidly written, well structured and immensely practical and should be in every library. I congratulate the authors and editors for their efforts and wish them success in their endeavors.

M.S. Swaminathan

TAMIL NADU AGRICULTURAL UNIVERSITY

Prof. N. Kumar, Ph.D., F.H.S.I.,
Vice Chancellor

Message

Nanotechnology is one of the world's fastest-growing fields and nanotechnology research and development has been growing virtually across all scientific disciplines and industries. Although nanotechnology deals with very small molecules to create useful materials, its impact on business seems to be significantly huge. Nanotechnologies are now widely considered to have the potentials to bring benefits in agriculture and allied sectors.

The Department of Nano Science and Technology, Tamil Nadu Agricultural University is involved in teaching, research and dissemination of nano inventions to various stakeholders. Collaborative research works are going on with institutes of national and international reputation to find solutions to various field problems.

This book on Nanotechnology in Agriculture, Energy and Environment is a compilation of basic and applied aspects of nanotechnology and this book can serveas an effective reading material for students and research scholars. It is interesting to note that many chapters of this book are dealing with nano agricultural inputs and the findings of the department are included in the chapters that would help the scientific community and students to understand better on the impact of nanotechnology in agriculture. The book has case studies in many chapters and hence readers can gain a broader perspective of nano interventions in agriculture, energy and environment sectors. Climate change is considered as a major challenge to sustain the productivity in agriculture and it is interesting to note that various strategies to manage the impacts of climate change through nanotechnological interventions have been explained in this book. Similarly special emphasis has been given to enlighten the readers on various nano concepts to minimize the post harvest losses in agriculture through advanced food packages and preservatives. As nanotechnology is a fast evolving science, it is imperative to carryout need based research. The future perspectives of nanotechnology in agriculture has been elaborated well in the book that would help the students to plan their research programme in tune with farmers, need. It is my hope and expectation that this book will providean effective learning experience and referenced resource for researchers and students.

I congratulate all the contributors for explaining various topics in a lucid language and my special appreciation to the editors Dr.A.Lakshmanan and Dr.K.S.Subramanian for the adept compilation of various topics in an orderly manner.

(N.KUMAR)

TAMIL NADU AGRICULTURAL UNIVERSITY

Dr. R. Santhi, Ph.D.
Director
Directorate of Natural Resource Management

Message

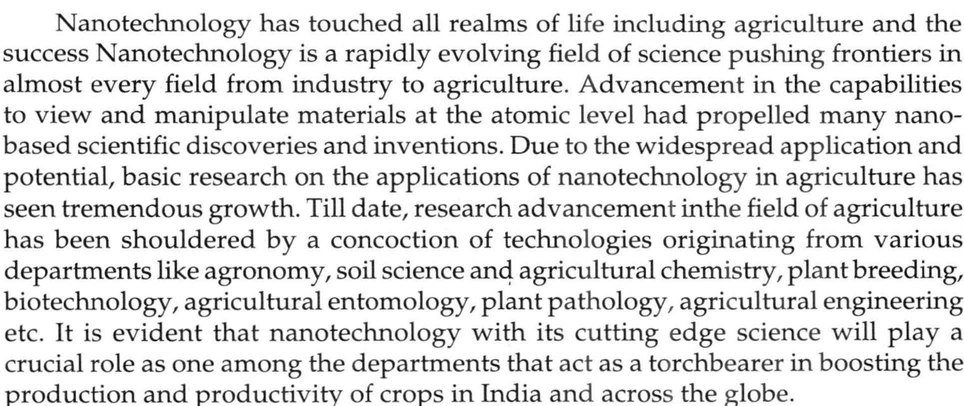

Nanotechnology has touched all realms of life including agriculture and the success Nanotechnology is a rapidly evolving field of science pushing frontiers in almost every field from industry to agriculture. Advancement in the capabilities to view and manipulate materials at the atomic level had propelled many nano-based scientific discoveries and inventions. Due to the widespread application and potential, basic research on the applications of nanotechnology in agriculture has seen tremendous growth. Till date, research advancement inthe field of agriculture has been shouldered by a concoction of technologies originating from various departments like agronomy, soil science and agricultural chemistry, plant breeding, biotechnology, agricultural entomology, plant pathology, agricultural engineering etc. It is evident that nanotechnology with its cutting edge science will play a crucial role as one among the departments that act as a torchbearer in boosting the production and productivity of crops in India and across the globe.

Tamil Nadu Agricultural University, an oldest and renowned institution was quick to see the possibilities and potential of the role of nanotechnology in transforming Indian agriculture for the better and undertook laborious task of establishing a dedicated centre for agricultural nanotechnology under the wings of Directorate of Natural Resource Management and channelled considerable funding for the development and establishment of the department to pursue cutting edge research in the field of agricultural nanotechnology.

After a decade long journey of learning and relearning the nuances and in-depth science involved in nanotechnology research, the department has been able to not only keep abreast with the constantly evolving field but also to establish a well-structured curriculum for PG and doctoral students and to develop many nano-based products that meets the dire requirement of farmers and agricultural allied sectors.

This book entitled "Nanotechnology in Agriculture, Energy and Environment" is truly an astounding compilation of chapters enlightening diverse concepts and research emerging in the field of agricultural nanotechnology. The chapters in the book cover various basic fields like nanophysics, nanochemistry and applied fields like nano-based agri-inputs, hydrogels, biosensors, smart packaging, nano remediation, nanotechnology in genome engineering and nano in post-harvest technology. The editors and authors of the book chapters have meticulously framed the chapters and its content to meet the intellectual needs of both beginners (UG, PG and Doctoral) and advanced avid readers (researchers and scientists).

I congratulate the editors and contributors of the book for presenting the advanced concepts and research in the simplest form possible in a lucid language. I hope this book is widely read and fulfil the knowledge requirement of numerous budding minds who seek to learn and grow and to carve out a space of their own in nanotechnological research.

Best Wishes

(R. Santhi)

Preface

Nanotechnology is one of the fastest growing field of science capable of developing processes and products that are precise to deliver inputs or facilitate early detection as it involves atom-by-atom by manipulation. This has been widely exploited in the sphere of medicine, energy, environment and rarely in agricultural sciences. In the past one decade, agricultural scientists across the globe began to work and published papers in various dimensions in agricultural sciences encompassing nano-agri-inputs, nano-agri food systems, early detection of pests and diseases and safety of nano-materials in various trophic levels in the agri-ecosystems.

The book on '**Nanotechnology in Agriculture, Energy and Environment"** is a compilation of basic and applied aspects to serve as a reading material to students and young researchers in nanotechnology involved in agriculture, food technology, energy, environment and ecosystem health. This book encompasses twelve units and thirty three chapters covering synthesis of nano-materials, their basic properties, structure and characterization, applications in various fields besides biosafety. The Department of Nano Science & Technology, Tamil Nadu Agricultural University, Coimbatore, one of the early birds in the niche area of agricultural nanotechnology, is carrying multi-disciplinary research team who have shared their knowledge and experience in addressing various issues.

The unit on **nano-agri inputs** carries nano-fertilizers, nano-technology for moisture conservation, weed management, seed invigouration and plant protection. This provides insights of nano particles mediated smart delivery of functional molecules in soil and plant systems. This chapters provides base information to the readers for further applied research and product commercialization. Delivering genome editing tools to plants is a major challenge and the efficiency of genome editing can be improved through nanotechnological interventions. The unit on Nano Biotechnology elucidates the role of nanomaterials as probes and sensors besides plant genome engineering.

The unit on **Nano Food Systems** is a blend of basic and applied aspects of nano concepts aiming to minimize the post-harvest losses, enhance the product shelf life and advanced nano packaging. The importance of using biopolymers and natural ingredients to develop various nano formulations is discussed with case studies in this unit that would promote the concept of green nanotechnology among readers and young researchers.

Nanotechnology has huge impact in **renewable energy** generation, pollution management and ecosystem health. The units 9 and 10 are dealing with energy and environmental nanotechnology in which various advanced nano systems such as Halloysite Nanotube (HNT), Graphene oxide, chitosan nanofibrils, carbon nanodots have been discussed elaborately. As these nano-systems are derived from natural sources, their role in pollution management is highly significant and would help the readers to concentrate on eco-friendly management of environmental issues.

Ensuring food security in tune with **changing climate** would be a big challenge in the coming years and climate adaptive breeding using novel genome editing tools would be the right choice to invest and explore. The chapter on Harnessing Nanotechnology to Manage Greenhouse Gases would provide valuable lead towards future research on climate adaptive breeding. Nanotechnological interventions could certainly make huge positive impacts on agriculture in the coming years. However, there are few concerns on the commercialization and field application of nano products and the major apprehension is on the biosafety of nano products. Hence, four chapters have been devoted to deal with impact of nanoparticles on ecosystem, nanotoxicity and bio safety guidelines. These chapters would educate the readers on biosafety protocols and guidelines for safe technology outreach.

Many problems in agriculture, energy and environmental sectors can be better handled with the intervention of nanotechnology. Hence, the last unit of the book deals with the **Emerging Opportunities** of Nanotechnology in Agriculture and Allied Sectors. Plant nanobionics to enhance basic functions in plant systems, Internet of NanoThings in Agriculture, Nanotechnology and Artificial Intelligence, and nano barcoding are some of the fascinating areas that could transform the agriculture in the coming years and would trigger the interest of students and researchers to initiate investigations in such emerging areas to find solutions to many field problems in agriculture.

This book can help teachers, students and researchers to develop a clear vision and better understanding of nanotechnological interventions in agriculture, besides providing valuable lead for future research. It is our hope that this book helps to create a firm foundation of Nanotechnology in Agriculture, Energy and Environment that anyone can build on in any way they wish.

We thank all the contributors for providing valuable resources and we are indebted to **Dr. N. Kumar**, Vice Chancellor (TNAU) for the motivating message and guidance and **Dr. M.S. Swaminathan** for the inspiring foreword and the blessings. Our special thanks to M/s. Astral International Pvt. Ltd., New Delhi for their commitment and interest in publishing this book in a short span of time.

"Oh God, Thy sea is so great and my boat is so small"

A. Lakshmanan

K.S. Subramanian

Contents

Unit 3: Principles and Application of Major Nanotechnological Instruments and Software

Unit 4: Synthesis of Nanomaterials

Unit 5: Nano Inputs in Agriculture

Unit 6: Nano Biotechnology

List of Contributors

R. Abinayaa
Post Doctoral Fellow
Department of Nano Science and Technology
Tamil Nadu Agricultural University
Coimbatore – 641 003
e-mail: nstabhi@gmail.com

Badal Kumar Biswal
Research Scholar
Department of Nano Science and Technology
Tamil Nadu Agricultural University
Coimbatore – 641 003
e-mail: 2017801301@tnau.ac.in

Dr. C.R. Chinnamuthu
Professor and Head
Department of Agronomy
Tamil Nadu Agricultural University
Coimbatore - 641 003
e-mail: crchinnamuthu@yahoo.com

P. Divyabharathi
Research Scholar
Department of Nano Science and Technology
Tamil Nadu Agricultural University
Coimbatore – 641 003
e-mail: divyabhrtipandian@gmail.com

Dr. K. Elango
Research Fellow
Department of Entomology
Tamil Nadu Agricultural University
Coimbatore – 641 003
e-mail: elaento@gmail.com

Dr. V. Geethalakshmi
Director (Crop Management)
Tamil Nadu Agricultural University
Coimbatore – 641 003
e-mail: geetha@tnau.ac.in

K. Geetha Thanuja
Research scholar
Department of Agricultural Microbiology
Tamil Nadu Agricultural University
Coimbatore – 641 003
e-mail: k.geethanu@gmail.com

K. Girija
Assistant Professor
Department of Physics
Dr. N. G. P. Arts and Science Collage
Coimbatore – 641 048
e-mail: kgirija.ngp@gmail.com

Golla Gowtham
Department of Nano Science and Technology
Tamil Nadu Agricultural University
Coimbatore – 641 003
e-mail: goutham221@icloud.com

Dr. R. Gowtham
Research scholar
Agro Climate Research Centre
Tamil Nadu Agricultural University
Coimbatore – 641 003
gowtham.acrc@gmail.com

Dr. K. Govindaraju
Scientist E/Associate Professor
Centre for Ocean Research,
Sathyabama Institute of Science and Technology
Chennai – 600 119
e-mail: govindtu@gmail.com

Dr. S. Haripriya
Assistant Professor (Horticulture)
Department of Nano Science and Technology
Tamil Nadu Agricultural University
Coimbatore – 641 003
e-mail: haripriya.s@tnau.ac.in

Dr. D. Jeya Sundara Sharmila
Assistant Professor (Physics)
Department of Nano Science and Technology
Tamil Nadu Agricultural University
Coimbatore – 641 003
e-mail: djssharmila@gmail.com

Dr. M. Kannan
Assistant Professor (Agrl Entomology)
Department of plant protection
TNAU Horticultural Collage and Research Institute
Periyakulum – 625604
e-mail: kanento@gmail.com

Dr. V. Karthika
Post Doctoral Fellow
Department of Nano Science and Technology
Tamil Nadu Agricultural University
Coimbatore – 641 003
e-mail: karthikaphd@gmail.com

Dr V.P. Krithika
Research Fellow
Department of Nano Science and Technology
Tamil Nadu Agricultural University
Coimbatore – 641 003
e-mail: krithika1706@gmail.com

M.S. Kumuthan
Research Scholar
Department of Nano Science and Technology
Tamil Nadu Agricultural University
Coimbatore – 641 003
e-mail: kumuthan93@gmail.com

Dr. A. Lakshmanan
Professor & Head
Department of Nano Science and Technology
Tamil Nadu Agricultural University
Coimbatore – 641 003
e-mail: microlaxman@gmail.com

M. Latha
Research Scholar
Department of Nano Science and Technology
Tamil Nadu Agricultural University
Coimbatore – 641 003
e-mail: latha.evergreen13@gmail.com

Dr. S. Marimuthu
Assistant Professor (Agronomy)
Department of Nano Science and Technology
Tamil Nadu Agricultural University
Coimbatore – 641 003
e-mail: Sm20@tnau.ac.in

J. Mohanraj
Research Scholar
Department of Nano Science and Technology
Tamil Nadu Agricultural University
Coimbatore – 641 003
e-mail: mohanbscagri@gmail.com

Dr. N. Natarajan
Professor (Retired)
Suki Illam, Sri Krishna Layout
Anandha Nagar
Coimbatore - 641 041
e-mail: divkavnn@gmail.com

Paladugu Krishnadev
Research Scholar
Department of Nano Science and Technology
Tamil Nadu Agricultural University
Coimbatore – 641 003
e-mail: kisnadevp5@gmail.com

D. Pradeep
Research Scholar
Department of Nano Science and Technology
Tamil Nadu Agricultural University
Coimbatore – 641 003
e-mail: pradeepkutty007@gmail.com

Dr. M. Praghadeesh
Research Fellow
Department of Nano Science and Technology
Tamil Nadu Agricultural University
Coimbatore – 641 003
e-mail: praghadeesh.m@gmail.com

Dr. K. Raja
Assistant Professor (Seed Science & Technology)
Department of Nano Science and Technology
Tamil Nadu Agricultural University
Coimbatore – 641 003
e-mail: rajaksst@gmail.com

Dr. S.K. Rajkishore
Assistant Professor (Environmental Sciences)
O/o the Dean (SPGS)
Tamil Nadu Agricultural University
Coimbatore – 641003
e-mail: rajkishoresk@gmail.com

Dr.V. Ramamoorthi
Research Fellow
Department of Nano Science and Technology
Tamil Nadu Agricultural University
Coimbatore – 641 003
e-mail: vasudevanramamoorthi@gmail.com

Dr. Pon. Sathya Moorthy
Assistant Professor (Physics)
Department of Nano Science and Technology
Tamil Nadu Agricultural University
Coimbatore – 641 003
e-mail: sathyanindia2020@gmail.com

S. Shanmugapriya
Research Scholar
Department of Nano Science and Technology
Tamil Nadu Agricultural University
Coimbatore – 641 003
e-mail: shan.vks1412@gmail.com

Dr. C. Sharmila Rahale
Assistant Professor (SS&AC)
Department of Nano Science and Technology
Tamil Nadu Agricultural University
Coimbatore – 641003
e-mail: sharmicj@yahoo.com

L. Sivashankari
Research Fellow
Department of Nano Science and Technology
Tamil Nadu Agricultural University
Coimbatore – 641 003
e-mail: sivashankari.lr@gmail.com

Dr. S. Srivignesh
Assistant Professor
Department of Horticulture and Floriculture
School of Life Sciences
Central University of Tamil Nadu
Thiruvarur-610 005
e-mail: srivignesh07@gmail.com

Dr. K.S. Subramanian
Director of Research and NABARD Chair Professor
Tamil Nadu Agricultural University
Coimbatore – 641 003
e-mail: kssubra2001@rediffmail.com

N. Swathika
Research Scholar
Department of Nano Science and Technology
Tamil Nadu Agricultural University
Coimbatore – 641 003
e-mail: swathikanataraj97@gmail.com

Dr. S. Thirumalairajan
Ramalingaswami Re-entry Faculty Fellow
Department of Nano Science and Technology
Tamil Nadu Agricultural University
Coimbatore – 641 003
e-mail: sthirumalairajan@gmail.com

Dr. N. Viji
Research Fellow
Department of Nano Science and Technology
Tamil Nadu Agricultural University
Coimbatore – 641 003
e-mail: mathumitha08@gmail.com

Chapter 1

Physical Concepts and Structural Properties of Nanomaterials

Pon. Sathya Moorthy, D. Jeya Sundara Sharmila and Badal Kumar Biswal

Department of Nano Science and Technology, Tamil Nadu Agricultural University, Coimbatore – 641 003, Tamil Nadu
e-mail: sathyanindia2020@gmail.com

When the world renowned Theoretical Physicist Dr. Richard Feynman delivered a classic lecture on December 29th, 1959 at the annual meeting of the American Physical Society at the California Institute of Technology (Caltech), he excellently opened a technical introduction to nanotechnology by declaring: There's Plenty of Room at the Bottom: An Invitation to Enter a New Field of Physics. Ever since, "scientists are trying to understand the unique physics at the nanoscale level that would be at the heart of present and future nanotechnology. Nanotechnology has begun to seep into the national and international consciousness and awareness and is being spoken about as a revolutionizing technology that will change everything from basic building materials, computers, space travels, health care, agriculture, energy and environment. The term "nanotechnology" was in use as early as 1974, which was defined by Professor Norio Taniguchi in a paper titled "On the Basic Concept of Nanotechnology" (Taniguchi, 1974). Here nanotechnology is referred to as the ability to engineer materials precisely at the nanometer level. Similarly Eric Drexler used "nanotechnology" interchangeably with "molecular technology" (now referred to as "molecular manufacturing") in his book "Engines of Creation" (Drexler, 1986). Drexler placed his nanotechnology focus on molecular machining saying that the "new technology will handle individual atoms and molecules with control and precision." These definitions clearly distinguish the difference between nanotechnology and bulk technology. In principle, Nanotechnology is the application of Nano Science. In turn, Nano Science is the science that relates to objects with at least one dimension between 1–100 nm. The ability to work with atoms at nanoscale–at the molecular level–with atomically precise design, promises to open up a lot of areas of technological development" (Ashrafi, 2011).

Quantum Mechanical Concepts for Nanomaterials

The amazing accomplishment of quantum hypothesis in providing both qualitative and quantitative explanation of photoelectric effect, blackbody radiation, emission of line spectra and Compton effect coupled with sufficient rationalization of the phenomenon of interference, diffraction and polarization of electro-magnetic theory established beyond doubt the dual aspect of radiation (wave aspect and particle aspect). It was in this background, that the French Scientist Louis de Broglie in 1924 extended the idea of dual nature of radiation to matter itself.

Wave – Particle Duality and deBroglie Wavelength

The concept of matter waves was proposed by de Broglie that "a material particle of mass 'm' moving with a velocity 'v' should have an associated wave length 'λ' called the de Broglie wave length, given by

$$\lambda = h/\text{momentum} = h/mv = h/p$$

where h is the Plank's constant.

Schrodinger Wave Equation and Particle in a Box of a Free Electron

The free electron theory assumes that the nuclei stay fixed on their lattice sites surrounded by the inner or core electrons whilst the outer or valence electrons that can travel freely are called free electrons. Quantum mechanical descriptions of the outer electrons are depicted by just considering one of these electrons in the well-known "particle in a box".

In the case of minimum a one dimensional nanomaterials, the electron is confined to a line of length a (the length of the nanomaterials), which we shall call the x-axis. Because we are ignoring the cores, there is nothing for the electron to interact with and so it experiences zero potential within the solid. The Schrödinger equation for the electron is

$$\frac{-h}{2m_e}\frac{d^2\Psi}{dx^2} = (E - V)\Psi \tag{1}$$

where is Planck's constant divided by 2π, m_e is the mass of the electron, V is the electrical potential, Ψ is the wave function of the electron and E is the energy of an electron with that wave function. When V=0, the solutions to this equation are simple sine or cosine functions, the equation (1) forms:

$$\Psi = \sin\sqrt{\frac{2m_e E}{h^2}}x \tag{2}$$

Differentiating equation (2) gives:

$$\frac{d\Psi}{dx} = \sqrt{\frac{2m_e E}{h^2}} \cos \sqrt{\frac{2m_e E}{h^2}} x \qquad (3)$$

Differentiating equation (2) twice, we will get,

$$\frac{d^2\Psi}{dx^2} = -\frac{2m_e E}{h^2} \sin \sqrt{\frac{2m_e E}{h^2}} x = -\left(2m_e \frac{E}{h^2}\right)\Psi$$

$$\Rightarrow \frac{d^2\Psi}{dx^2} = -\frac{2m_e E}{h^2}\Psi \qquad (4)$$

Multiplying equation (4) with $-\left(\frac{h^2}{2m_e}\right)$ gives equation (1) when V=0.

Energy Quantization

The electron is not allowed outside the box and to ensure this we put the potential to infinity outside the box. Since the electron cannot have infinite energy, the wave function must be zero outside the box and since it cannot be discontinuous, it must be zero at the boundaries of the box. If we take the sine wave solution, then this is zero at x=0. To be zero at x=a as well, there must be a whole number of half waves in the box. Sine functions have a value of zero at angles of nπ radians where n is an integer and so $a\sqrt{\frac{2m_e E}{h^2}} = n\pi$. The energy thus quantised, $E = \frac{n^2 h^2}{8 m_e a^2}$ with quantum number n.

For three dimensions, the bulk meterial can be taken as a rectangular box a x b x c, the equation is given by:

$$E = \frac{h^2}{8m_e}\left(\frac{n_a^2}{a^2} + \frac{n_b^2}{b^2} + \frac{n_c^2}{c^2}\right) \qquad (5)$$

Each set of quantum numbers n_a, n_b, n_c will give rise to an energy level. However, in three dimensions, many combinations of n_a, n_b, and n_c exist which will give the same energy, whereas for the one-dimensional model there were only two levels of each energy (n and –n).The number of states with the same energy is known as the degeneracy.

Density of States

The equation for density of states is given by:

$$N(E)dE = \sqrt{(2m_e)^3 E} \times (V / 2\pi^2 h^3) dE \qquad (6)$$

The density of states increases with increasing energy–the higher the energy, the more states there are in the interval dE. In metals, the valence electrons fill up the states from the lowest energy up with paired spins. For sodium, for example, each atom contributes one 3s electron and the electrons from all the atoms in the crystal occupy the levels until all the electrons are used up. The highest occupied level is called the Fermi level.

Structural Properties of Nanomaterials

Solids

Particle theory is the basic model to explain the properties and behaviors in three states namely solids, liquids and gasses shown in the Figure 1.1. A solid is a state of matter characterized by particles arranged such that their shape and volume are relatively stable. They generally have geometric shapes and flat faces. The reason a solid has a rigid shape is that the atoms or molecules are tightly connected via chemical bonds. The bonding may produce either a regular lattice or an amorphous shape (S.O. Pillai, 2010).

Figure 1.1. Three different States of Matters.
Courtesy: https://www.quora.com/What-are-the-three-states-of-matter-and-their-properties

Classification of Solids

Based on the atomic arrangement, binding energy, physical and chemical a properties. Solids can be further classified in to (i) Crystalline, (ii) Amorphous and (iii) Poly crystalline solids as illustrated in the Figure 1.2.

The arrangement of particles in a crystalline solid is in a very orderly fashion is called as crystals. The spaces between the atoms are very less due to high intermolecular forces. This results in crystals having high melting and boiling points. The intermolecular force is also uniform throughout the structure. Crystals

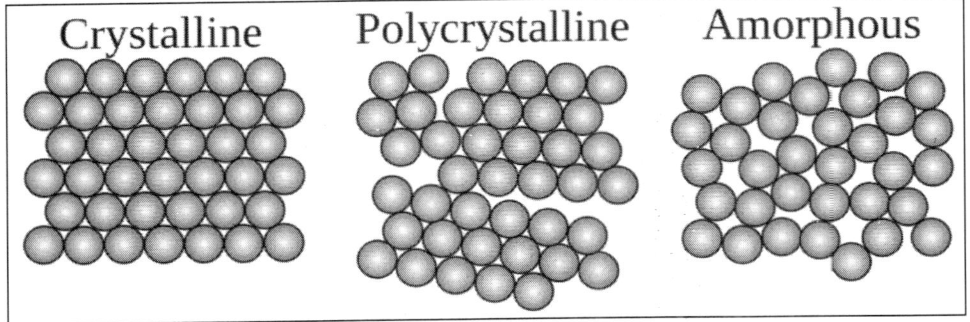

Figure 1.2. Classificatiom of Solid Based on the Atomic Arrangement.
Courtesy: https://www.toppr.com/guides/chemistry/the-solid-state/crystalline-and-amorphous-solids/

have a long-range order, which means the arrangement of atoms is repeated over a great distance. Polycrystalline materials are solids that are composed of many crystallites of varying size and orientation. The variation in direction can be random or directed, possibly due to growth and processing conditions. Almost all common metals, and many ceramics are polycrystalline. Powder grains can themselves be composed of smaller polycrystalline grains. Amorphous solids are rigid structures but they lack a well-defined shape. They do not have a geometric shape. The most common example of an amorphous solid is Glass. Gels, plastics, various polymers, wax, ect. (S.O. Pillai, 2010).

Lattice Points and Space Lattice

It is very convenient to imagine points in space about which these atoms are located such points in space are called as lattice points (S.O. Pillai, 2010 and Murugeshan and Kiruthiga Sivaprasath, 2014). Crystal lattice refers to the geometry of a set of points in space. Schematic representation of lattice points is shown in the Figure 1.3. Space lattice id defined as a regular, indefinitely repeated array of points in three dimensions in which the points lie at the intersections of three sets of parallel equidistant planes and shown in Figure 1.4.

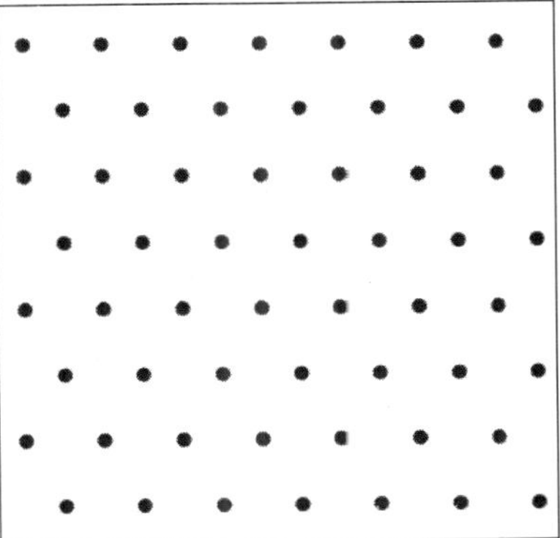

Figure 1.3. Lattice of a Acrystal.
Courtesy: https://en.wikipedia.org/wiki/Lattice_(group)

Basis

The unit assembly of atoms or molecules identical in composition, arrangement and orientation is called the basis or the pattern (S.O. Pillai, 2010 and Murugeshan and Kiruthiga Sivaprasath, 2014)). Repetition of basis with correct periodicity in all direction gives the actual crystal structure as shown in the Figure 1.5.

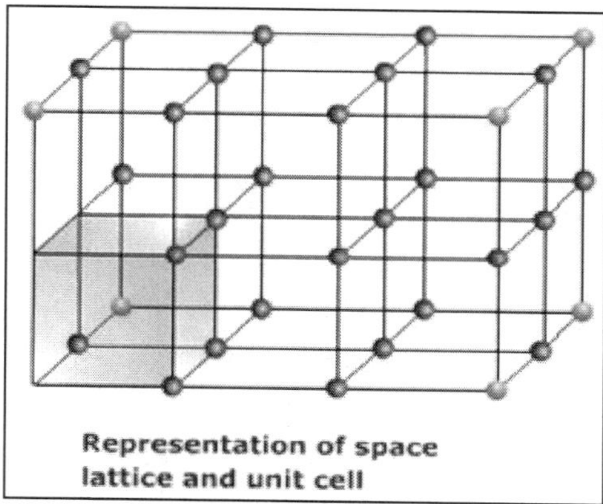

Figure 1.4. Representation of Space Lattice.
Courtesy: https://sseaimes.wordpress.com/tag/lattice-structure/

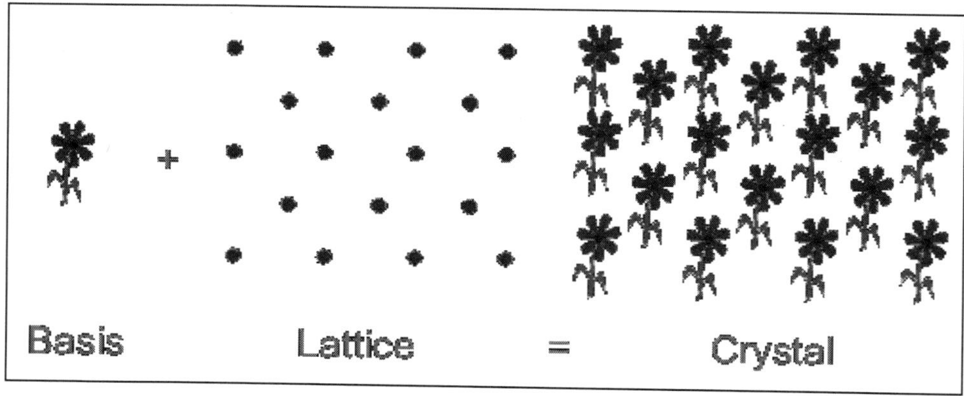

Figure 1.5. Schematic Representation of Basis and Lattice Forming Crystal Structure.
Courtesy: http://britneyspears.ac/physics/crystals/wcrystals.htm

Unit Cell and Primitive Cell

Unit cell is the structural and functional unit of a crystal. Unit cell is the smallest group of atoms that has the overall symmetry of a crystal, and from which the entire lattice can be built up by repetition in three dimensions. Therefore, unit cells are the repeating units of crystal lattices. Unit cell may have more than one lattice per cell. Primitive cell can be a unit cell but unit cell cannot be a primitive cell. Unit cell may be base centered, body centered and face centered system. A unit cell is described using lattice parameters and lattice points. Lattice parameters are lengths between edges of unit cell (given by a, b and c symbols) and angles of the unit cell (given by α, β and γ symbols). Lattice points are atoms, molecules or ions from which the lattice is made of (S.O. Pillai, 2010, Murugeshan and Kiruthiga Sivaprasath, 2014).

Unit cell has a geometry known as parallele piped (a 3D figure formed from 6 parallelograms). This geometry is described by the six lattice parameters (mentioned above). The positions of lattice points are given by fractional coordinates denoted by x_i, y_i and z_i, which are measured from a reference point. According to Auguste Bravais (1850), there are 14 types of lattices, known as Bravais lattices. The unit cells of these Bravais lattices are shown in the Figure 1.6. The names of unit cells (1-14) in the above image are given below. (Here, P refers to "primitive centering",

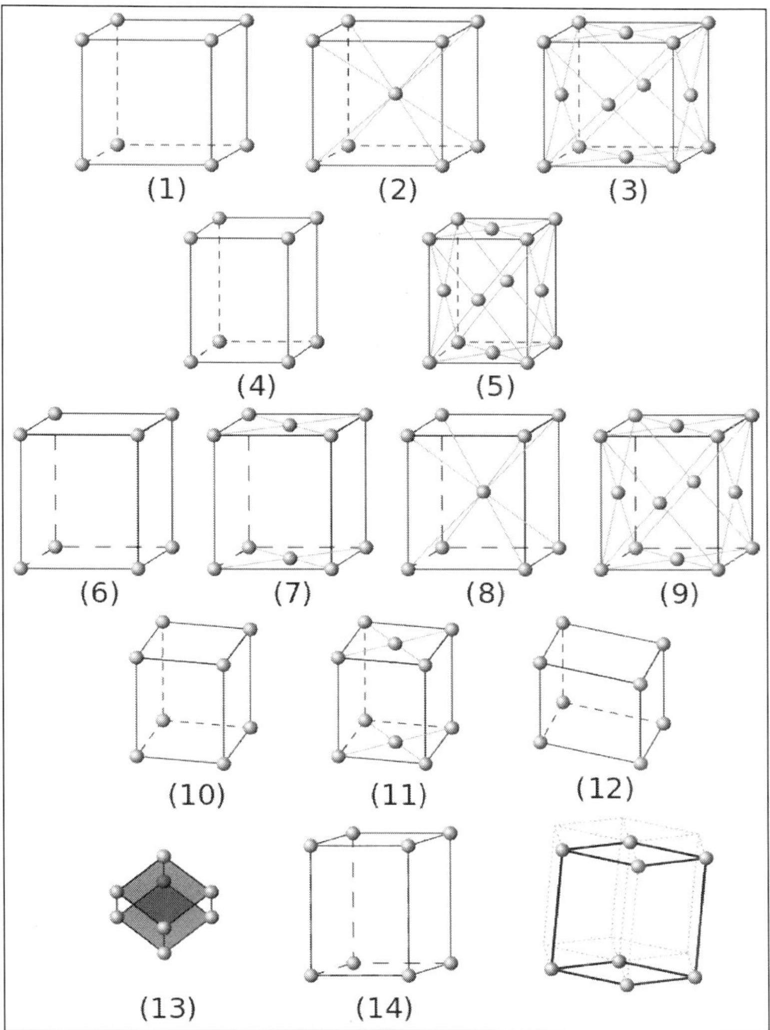

Figure 1.6. Schematic Representation 14 Bravaise Lattice of Unit Cells.

1. Cubic P; 2. Cubic I; 3. Cubic F; 4. Tetragonal P; 5. Tetragonal I; 6. Orthorhombic P; 7. Orthorhombic C; 8. Orthorhombic I; 9. Orthorhombic F; 10. Monoclinic P; 11. Monoclinic C; 12. Triclinic; 13. Rhomboedral; 14. Hexagonal.

Courtesy: https://www.differencebetween.com

C refers to "centered on a single phase" and I refers to "body centered" whereas F refers to "face centered").

Primitive unit cells contain only one lattice point, which is made up from the lattice points at each of the corners. There are two types of primitive cells: two-dimensional primitive cells and three-dimensional primitive cells. The two-dimensional primitive cells are parallelograms. This means, there can be orthogonal angles, equal lengths or both in these two-dimensional primitive cells. Types of primitive cells are in the Figure 1.7.

Parallelogram (Monoclinic)	
Rhombus (Orthorhombic)	
Rectangle (Orthorhombic)	
Square (Tetragonal)	

Figure 1.7. Schematic Representation 2D Primitive Cells.
Courtesy: https://www.differencebetween.com

A three-dimensional primitive cell is known to be parallelepiped (a 3D figure formed from 6 parallelograms). It has orthogonal angles, equal lengths or both. Types of three-dimensional primitive cells are listed below.

1. Parallelepiped (Triclinic)
2. Oblique rhombic prism (Monoclinic)
3. Oblique rectangular prism (Monoclinic)
4. Right rhombic prism (Orthorhombic)
5. Rectangular cuboid (Orthorhombic)
6. Square cuboid (Tetragonal)
7. Trigonal trapezohedran (Rhombohedral)
8. Cube (Cubic)

Crystal Systems

The seven crystal systems are discussed below:

There are thirty two classes of crystal systems based on the geometrical considerations (*i.e.*, symmetry and internal structure). But, it is a common practice to divide all the crystal systems into seven groups or basic systems. These seven basic crystal systems are distinguished from one another by the angles between the three axes and the intercepts of the faces along them. The basic crystal systems are:

1. Cubic (Isometric)
2. Tetragonal
3. Orthorhombic
4. Monoclinic
5. Triclinic
6. Trigonal (Rhombohedral)
7. Hexagonal

The seven crystal systems are discussed below:

Cubic Crystal Systems

$a = b = c, \alpha = \beta = \gamma = 90^{o}$

The crystal axes are perpendicular to one another, and the repetitive interval is the same along all the three axes. Cubic lattices maybe simple, body- centered or face- centered.

Tetragonal Crystal Systems

$a = b \neq c, \alpha = \beta = \gamma = 90^{o}$

The crystal axes are perpendicular to one another. The repetitive intervals along two axes are the same, but the interval along the third axis is different. Tetragonal lattices maybe simple or body -centered.

Orthorhombic Crystal Systems

$a \neq b \neq c, \alpha = \beta = \gamma = 90^{o}$

The crystal axes are perpendicular to one another, but the repetitive intervals are different along the three axes. Orthorhombic lattices maybe simple, body- centered or face- centered.

Monoclinic Crystal Systems

$a \neq b \neq c, \alpha = \beta = 90^{o} \neq \gamma$

Two of the crystal axes are perpendicular to each other, but the third is obliquely inclined. The repetitive intervals are different along all the three axes. Monoclinic lattices maybe simple or base- centered.

Triclinic Crystal Systems

$a \neq b \neq c, \alpha \neq \beta \neq \gamma \neq 90^{o}$

None of the crystal axes is perpendicular to any of the others, and the repetitive intervals are different along all the three axes.

Trigonal (Sometimes called rhombohedral) Crystal Systems

$a = b = c, \alpha = \beta = \gamma \neq 90^o$

The three axes are equal in length and are equally inclined to each other at an angle other than

Hexagonal Crystal Systems

$a = b \neq c, \alpha = \beta = \gamma = \gamma = 90^o$

Two of the crystal axes are apart while the third is perpendicular to both of them. The repetitive intervals are the same along the axes that are apart, but the interval along the third axis is different. The seven systems and their properties are given below in the Figure 1.8.

Sr. No.	Crystal System	Axial length of Unit Cell	Inter axial angles	Number of Lattice in the system
1	Cubic	$a = b = c$	$\alpha = \beta = \gamma = 90^0$	3
2	Tetragonal	$a = b \neq c$	$\alpha = \beta = \gamma = 90^0$	2
3	Orthorhombic	$a \neq b \neq c$	$\alpha = \beta = \gamma = 90^0$	4
4	Monoclinic	$a \neq b \neq c$	$\alpha = \beta = 90^0 \neq \gamma$	2
5	Triclinic	$a \neq b \neq c$	$\alpha \neq \beta \neq \gamma \neq 90^0$	1
6	Trigonal	$a = b = c$	$\alpha = \beta = \gamma < 120^0, \neq 90^0$	1
7	Hexagonal	$a = b \neq c$	$\alpha = \beta = 90^0, and\ \gamma = 120^0$	1

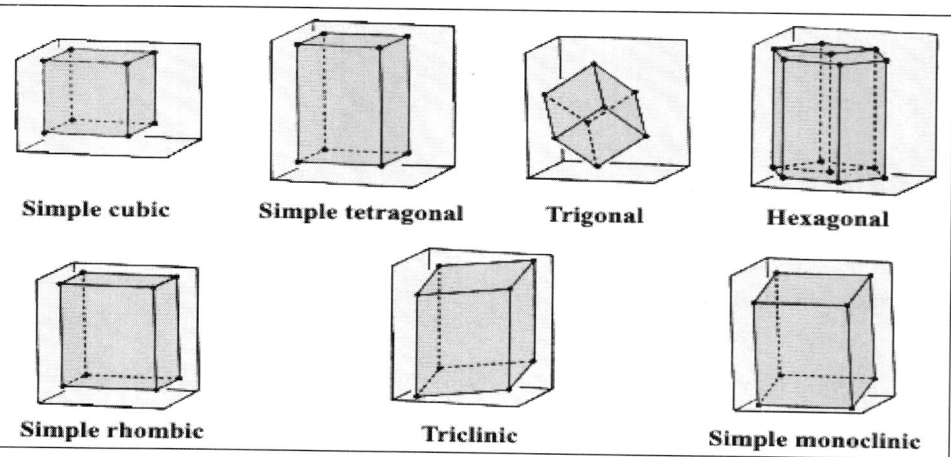

Figure 1.8. Diagrammatic Representation of Seven Crystal Systems
Courtesy: https://www.tankonyvtar.hu/en/tartalom/tamop425/0038_foldrajz_mineralogy_Da/ch01s03.html

A Symmetry operation is an operation that can be performed either physically or imaginatively that results in no change in the appearance of an object. Again it is emphasized that in crystals, the symmetry is internal, that is it is an ordered geometrical arrangement of atoms and molecules on the crystal lattice. But, since the internal symmetry is reflected in the external form of perfect crystals, we are going to concentrate on external symmetry, because this is what we can observe. There are 3 types of symmetry operations: rotation, reflection, and inversion.

Rotation Symmetry

There are two basic types of rotational symmetry operations. *Proper* rotations move an object, but do not change the handedness of the object. *Improper* rotations include a proper rotation as well as a component that inverts the handedness of the object. A body is said to possess rotational symmetry about an axis if after rotation of the body about this axis by some angle $\tilde{O}_n$, the body appears as it did prior to rotation *i.e.* the body is left invariant as a result of rotation. The axis of symmetry alone are possible as a line such that the crystal assumes a congruent position for every rotation of [360/n]° (B.D. Cullity and S.R.Stock, 2001, S.O. Pillai, 2010). The value of n determines the fold of the axis. Only one, two, three, four and six fold rotation axes of symmetry alone are possible in a single crystal lattice. Rotation symmetry of order six is represented in Figure 1.9.

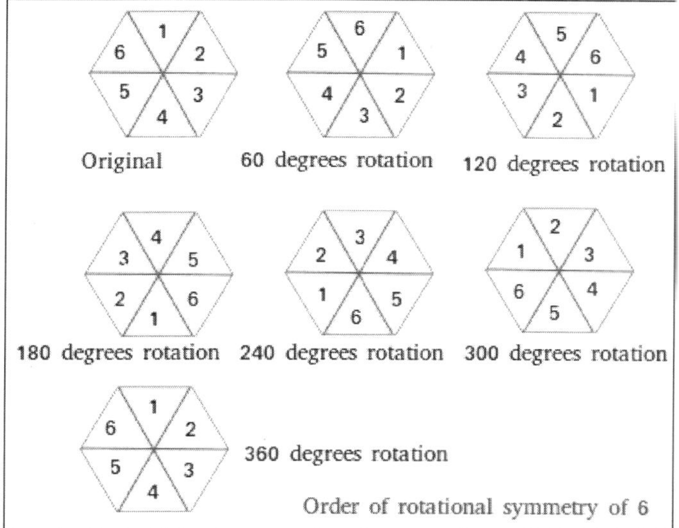

Figure 1.9. Rotation Symmetry of Order of Six.
Courtesy: https://www.basic-mathematics.com/order-of-rotational-symmetry.html

Reflection Symmetry

A crystal is said to possess reflection symmetry about a plane if it is left unchanged in every way after being reflected by the plane. Highly regular crystals may be bilaterally symmetrical about several planes cutting them in different directions. There may be several planes of symmetry within the crystal shown in the Figure 1.10.

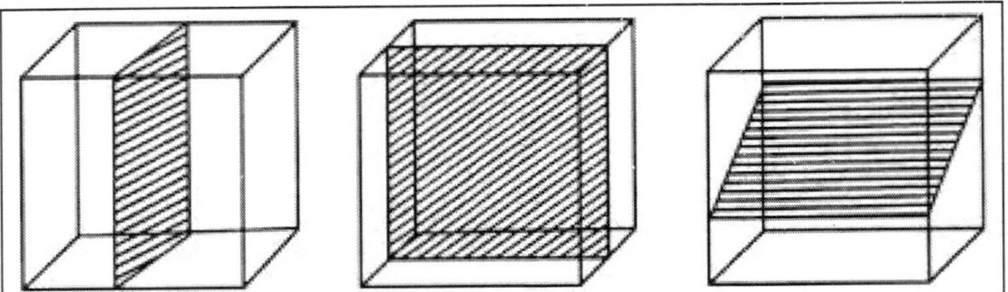

Figure 1.10. Schematic Representation Planes of Symmetry within the Crystals.
Courtesy: http://www.engineeringenotes.com/engineering/crystal-symmetry-in-solids-materials-engineering/33707

Centre of Inversion

A cube possesses three such pairs of parallel and opposite faces. Hence a cube is said to show a centre of symmetry, *i.e.*, the body centre of the cube is a centre of symmetry. This centre lies at equal distances from various symmetrical positions (B.D. Cullity and S.R.Stock, 2001, S.O. Pillai, 2010). Centre of symmetry is also known as inversion centre. Cube exhibiting centre of symmetry is shown in the Figure 1.11.

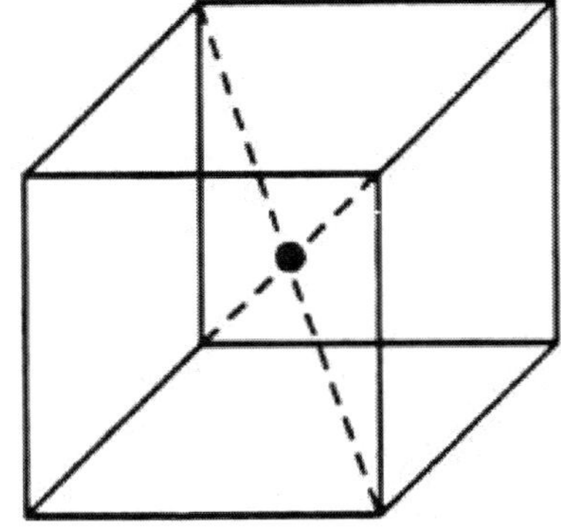

Figure 1.11. Crystal Cube Exhibiting Centre of Inversion Symmetry.
Courtesy: http://www.engineeringenotes.com/engineering/crystal-symmetry-in-solids-materials-engineering/33707

Imperfection in Crystals

Imperfections or defects: Any deviation from the perfect atomic arrangement in a crystal is said to contain imperfections or defects. Adding alloying elements to a metal is one way of introducing a crystal defect. Crystal imperfections have strong influence upon many properties of crystals, such as strength, electrical conductivity and hysteresis loss of ferromagnets. Thus some important properties of crystals are controlled by as much as by imperfections and by the nature of the host crystals.

The conductivity of some semiconductors is due entirely to trace amount of chemical impurities.

1. Color, luminescence of many crystals arise from impurities and imperfections
2. Atomic diffusion may be accelerated enormously by impurities or imperfections
3. Mechanical and plastic properties are usually controlled by imperfections

Imperfections in crystalline solids are normally classified according to their dimension as follows

1. Point imperfections (Zero dimensional defects)
2. Line imperfections (One dimensional defects)
3. Plane or surface imperfections (Two dimensional defects)
4. Volume imperfections (Three dimensional defects)

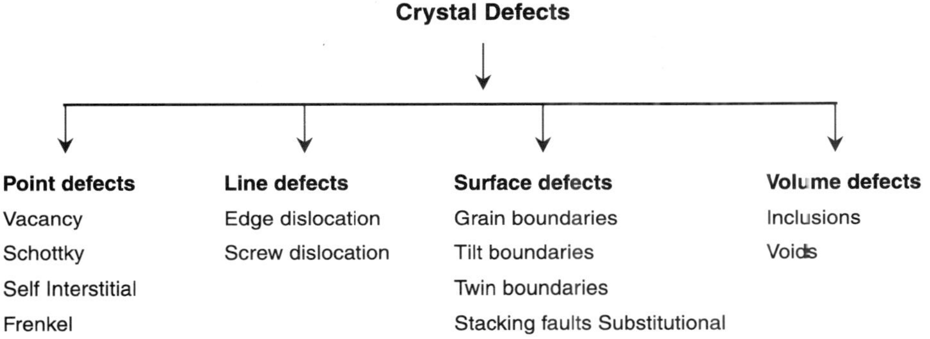

(1) Point Defects

As the name indicates, these defects are at some points in the crystal. So, these are also called Zero dimensional defects. The point defects are divided into three

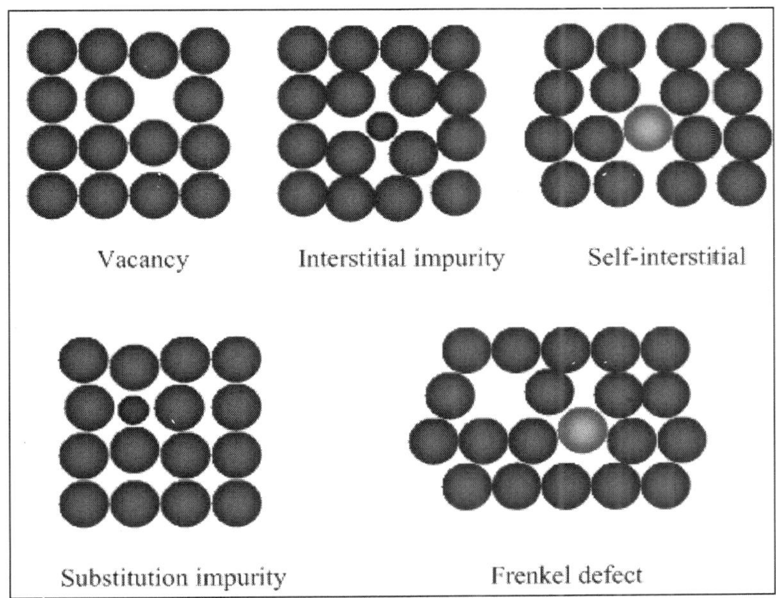

Figure 1.12. Point Defects in Crystals.
Courtesy: https://www.askiitians.com/iit-jee-solid-state/imperfections-in-solids-and-defects-in-crystals/

categories: (a) lattice site defects; this includes vacancies (Schottky defect) and interstitialcies (Frenkel defect), (b) compositional defects; this includes substitutional impurity and interstitial impurity (c) electronic defects (S.O. Pillai, 2010). These efects include below.

Lattice site defects: In this type of defects, some atoms may not be present in their regular atomic sites. They are:

Vacancies: At a lattice point, one or two or three atoms are missed, and this is referred to as single or double or triple vacancies, respectively. The vacancies are formed due to the imperfect packing during crystallization or due to thermal vibrations at high temperature and shown in Figure 1.12.

Scottky defect: In ionic crystals, if a cation vacancy exists, then in the very nearby place an anion vacancy also exists. *i.e.*, usually an anion and cation pair moved to the surface of the crystal, so that charge neutrality is maintained in the vacancy region and shown in Figure 1.12.

This is known as Schottky defect. Crystals such as NaCl, KCl, KBr, *etc.* show Schottky defect.

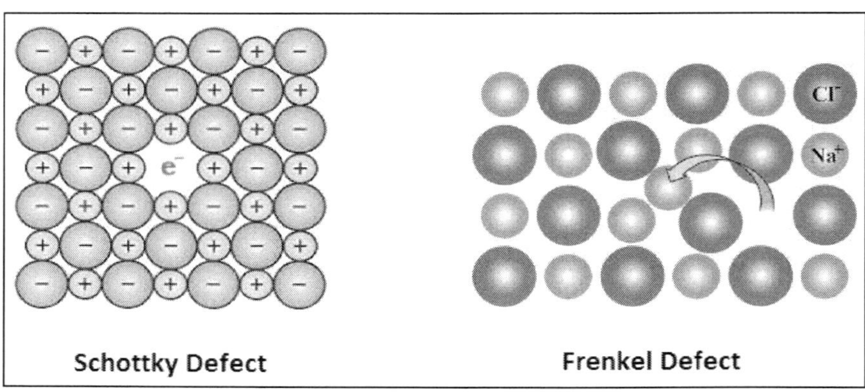

Figure 1.13. Schematic Representation of (a) Schottky and (b) Frenkel Defect.
Courtesy: https://en.wikipedia.org/wiki/Schottky_defect; https://en.wikipedia.org/wiki/Frenkel_defect

Interstitial defect: If an atom is moved to an interstitial space in the crystal, then the defect is known as interstitial defect.

Frenkel defect: In ionic crystals, if a cation (positive ion) moves to an interstitial space, then a vacancy is formed in its atomic position. Here, charge neutrality is maintained in the defective region. This type of defect is known as Frenkel defect and shown in Figure 1.12. Crystals such as CaF_2, AgBr, AgI, *etc.* show Frenkel defect.

Compositional defect: The presence of impurity atoms in the crystal leads to compositional defects is shown in Figure 1.12. Impurity atoms are present at the sites of regular parent atoms or in the interstitial spaces. These defects are described below.

Substitutional defect: During the crystallization few foreign atoms occupy the regular parental atom sites is shown in Figure 1.12. For *e.g* in extrinsic semiconductors either third or fifth group atoms occupy the sites of silicon or germanium atoms.

Interstitial impurity defect: The spaces between the parental atoms in a crystal are known as interstitial spaces. Small sized (lower atomic number) atoms, such as hydrogen. *etc.* may fit into these interstitial spaces. These atoms are known as interstitial atoms and the defects formed due to the presence of the interstitial atoms are known as interstitial defect is shown in Figure 1.12.

Electronic defects: Non uniformity of charge or energy distribution in the crystal is referred to as electronic defect.

(2) Line Defects

If a crystal plane ends somewhere in the crystal, then along the edges of that incomplete plane produces defect in a crystal called line defect. The line defect is of two types: they are (i) edge dislocation and (ii) screw dislocation (S.O. Pillai, 2010). These are described below.

(i) Edge Dislocation

This defect occurs due to the termination of the edge of an atomic plane within the crystal instead of passing all the way through. The way in which an edge dislocation moves under the shear and helps in the deformation of a crystal is shown in Figure 1.14.

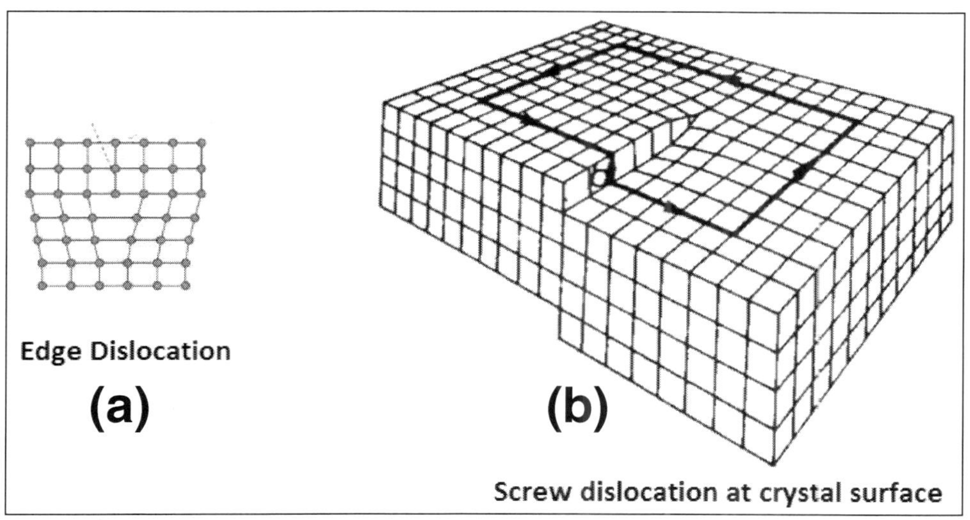

Figures 1.14 (a) and (b). Schematic Representation of (a) Edge Dislocation and (b) Screw Dislocation.
Courtesy: Researchgate.net/publication/270761961

Thus because of the shear, the dislocation moves across the crystal with the net result that the top half of the crystal gets displaced one lattice distance with respect to the bottom half of the crystal.

(ii) Screw Dislocation

To understand it, imagine a cut in a crystal with the particles to the left of the cut pushed up through the distance of one unit cell, as shown in Figure. The unit cells now make a continuous spiral around the end of the cut, the screw axis. It means that the atomic planes around the screw axis (dislocation line) form a spiral or helical ramp. Such a displacement of atoms is known as screw dislocation. The existence of screw dislocation would help in the further growth of the crystal because atoms can easily settle down on the step. However, when the lower terrace of the crystal is completely covered, further growth ceases. Just like the edge dislocations, the screw dislocations can also move under the influence of appropriate shear force and thus help in the deformation.

3. Surface Defects

Surface imperfections of a structural nature arise from a change in the stacking of atomic planes across a boundary. The changes may be one of the orientation or of the stacking sequence of the planes (S.O. Pillai, 2010).

(i) Grain Boundaries

Grain boundaries are those surface imperfections which separate crystals of different orientation in a poly-crystalline aggregation during nucleation or crystallization as shown in the Figure 1.15. The shape of a grain is usually influenced by the presence of surrounding grains. The boundary atoms in two randomly oriented grains, therefore, cannot have a perfect complement of the surrounding atoms. As a result, a region of transition exists in which the atomic packing is imperfect.

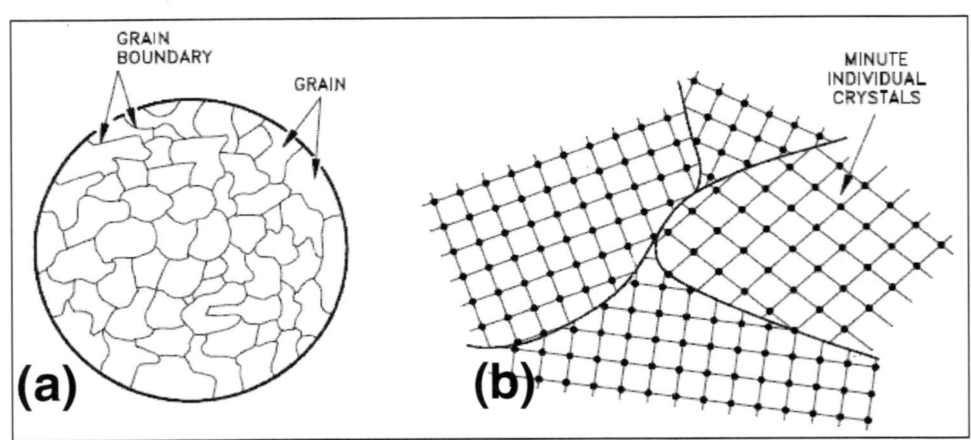

Figure 1.15. Schematic Representation of (a) Grain, Grain Boundary and (b) Grain Slip.

Courtesy: http://www.amie.nbcafe.in/phpbb/write-short-note-on-slip, grain-structure-and-grain-boundary–t2521.html

(ii) Stacking Fault

A stacking fault is a surface imperfection that results from the stacking of one atomic plane out of sequence on another while the lattice on either side of the fault is perfect.

For example, the stacking sequence in an ideal face centered cubic (FCC) crystal may be described as ABC, ABC,. A stacking fault might change the sequence to ABC, ABA, BCA.

The stacking fault in this case is due to the **A** plane of atoms after the second **B** and may be described as a very thin region of (HCP) stacking in a (FCC) crystal. Such stacking faults may occur during crystal growth or may result from the separation of two particular dislocations.

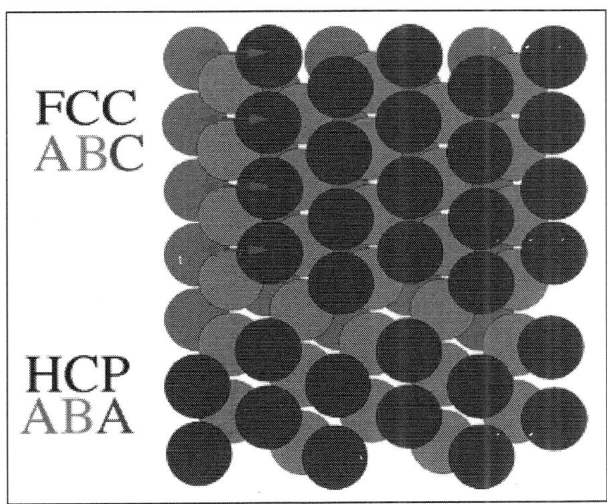

Figure 1.16. Comparison of FCC and HCP Lattices, Explaining the Formation of Stacking Faults in Close-Packed Crystals.
Courtesy: https://en.wikipedia.org/wiki/Stacking_fault

(iii) Tilt Boundary

A small angle grain boundary is formed when the edge dislocations are aligned in the manner as shown in the figure. This type is called a tilt boundary; the angle of misorientation, is also indicated in the Figure 1.17.

(iv) Twin Boundary

These are the boundaries in the grains at which the atomic arrangements on one side of the boundary is the mirror image of the atoms on the other side.It may be produced by the shear formation. The region between the pair of boundaries is called the twinned region is shown in the Figure 1.18.

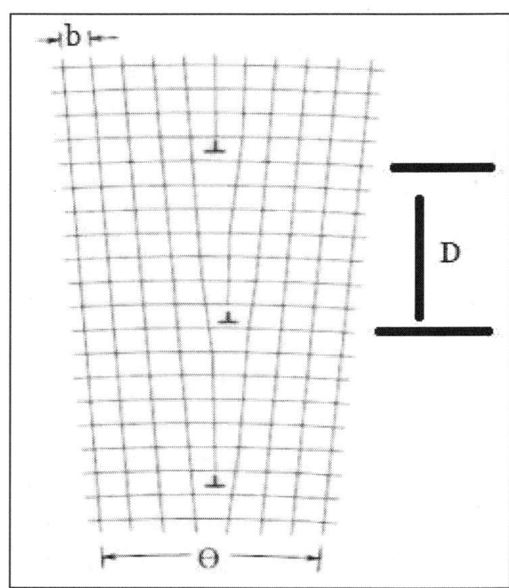

Figure 1.17. Comparison of fcc and hcp Lattices, Explaining the Formation of Stacking Faults in Close-Packed Crystals.
Courtesy: https://www.smartway2study.com/2016/03/tilt-boundary.html

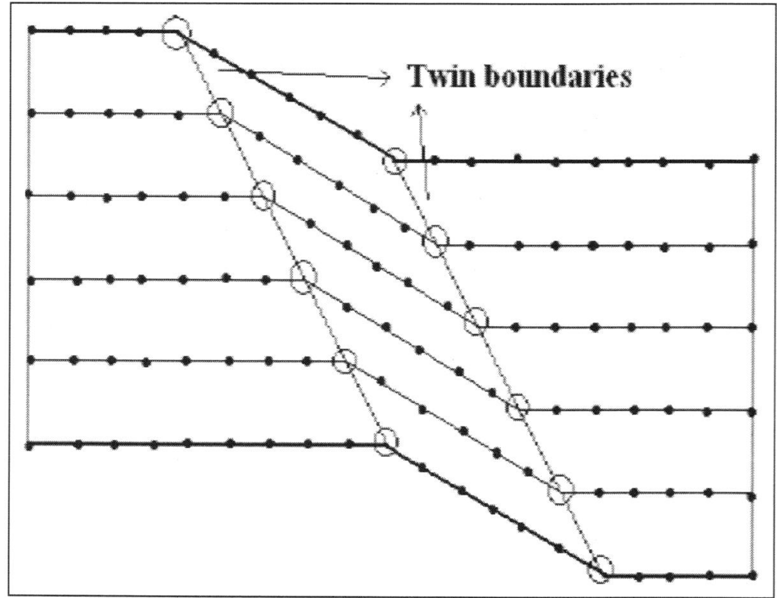

Figure 1.18. Schematic Representation of Twin Boundary.
Courtesy: https://nptel.ac.in/content/storage2/courses/112108150/pdf/Lecture_Notes/MLN_03.pdf

(4) Volume Defects

A crystal structure deviating from ideality by having two or more chemical species in one or more crystal sites. Three-dimensional macroscopic defects are called bulk defects. They generally occur on a much larger scale than the microscopic defects. These macroscopic defects generally are introduced into a material during refinement from its raw state or during fabrication processes. These include cracks, pores, foreign inclusions, and other phases. The working and forging of metals can cause cracks that act as stress concentrators and weaken the material. Any welding or joining defects may also be classified as bulk defects (S.O. Pillai, 2010).

1. Three-dimensional macroscopic or bulk defects, such as pores, cracks, or inclusions.
2. Voids – small regions where there are no atoms, and which can be thought of as clusters of vacancies.
3. Impurities can cluster together to form small regions of a different phase. These are often called precipitates.

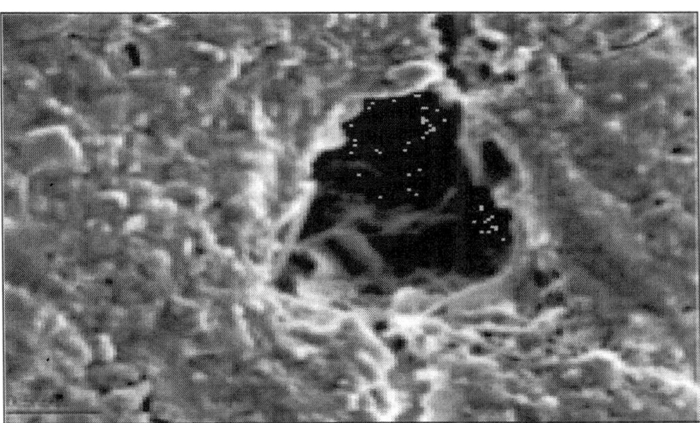

Figure 1.19. SEM Image of Void Defect in Crystal.
Courtesy: https://www.nded.org/EducationResources/CommunityCollege/Materials/Structure/bulk_defects.html

Bragg's Law

Definition

Bragg's Law derived by the English physicists Sir W.H. Bragg and his son Sir W.L. Bragg in 1913 to explain why the cleavage faces of crystals appear to reflect X-ray beams at certain angles of incidence (theta, θ). The variable d is the distance between atomic layers in a crystal, and the variable lambda λ is the wavelength of the incident X-ray beam (see applet); n is an integer. Bragg's Law refers to the simple equation (B.D. Cullity and S.R.Stock, 2001, S.O. Pillai, 2010)

$$n\lambda = 2d \sin \theta$$

Deriving Bragg's Law

Bragg's Law can easily be derived by considering the conditions necessary to make the phases of the beams coincide when the incident angle equals and reflecting angle. The rays of the incident beam are always in phase and parallel up to the point at which the top beam strikes the top layer at atom z Figure 1.20. The second beam continues to the next layer where it is scattered by atom B. The second beam must travel the extra distance AB + BC if the two beams are to continue traveling adjacent and parallel. This extra distance must be an integral (n) multiple of the wavelength (λ) for the phases of the two beams to be the same:

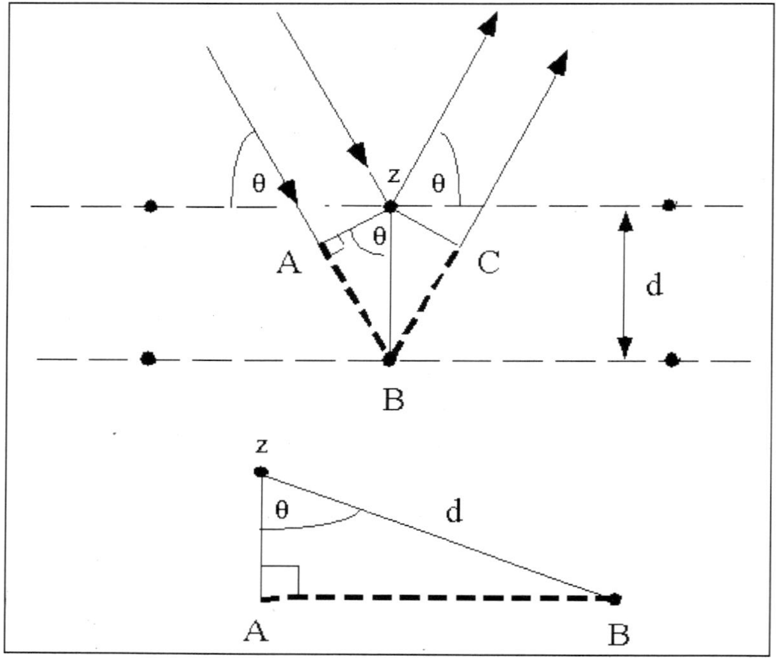

Figure 1.20. Schematic Representation of Bragg's Law.

λ = AB +BC

Deriving Bragg's Law using the reflection geometry and applying trigonometry. The lower beam must travel the extra distance (AB + BC) to continue traveling parallel and adjacent to the top beam. Recognizing d as the hypotenuse of the right triangle Abz, we can use trigonometry to relate d and θ to the distance (AB + BC). The distance AB is opposite θ so,

AB = d sinθ, BC= d sinθ

Because AB = BC d sinθ

$n\lambda = 2AB$

$n\lambda = 2\, d \sin\theta$

and Bragg's Law has been derived. The location of the surface does not change the derivation of Bragg's Law.

Scherrer Equation

The Scherrer equation is a formula that relates the size of sub micrometre crystallites, in a solid to the broadening of a peak in a diffraction pattern. If is often referred to, incorrectly, as a formula for particle size measurement or analysis. It is named after scientist Paul Scherrer. It is used in the determination of size of crystals in the form of powder (https://en.wikipedia.org/wiki/Scherrer_equation).

The Scherrer equation can be written as: $\tau = K\lambda/\beta \text{Cos}\theta$

where:

τ is the mean size of the ordered (crystalline) domains, which may be smaller or equal to the grain size, which may be smaller or equal to the particle size

K is a dimensionless shape factor, with a value close to unity. The shape factor has a typical value of about 0.9, but varies with the actual shape of the crystallite

λ is the X-ray wavelength

β is the line broadening at half the maximum intensity (FWHM), after subtracting the instrumental line broadening, in radians. This quantity is also sometimes denoted as Δ or 2 θ.

θ is the Bragg angle.

The Scherrer equation is limited to nano-scale crystallites, or more-strictly, the coherently scattering domain size, which can be smaller than the crystallite size (due to factors mentioned below). It is not applicable to grains larger than about 0.1 to 0.2 μm, which precludes those observed in most metallographic and ceramographic microstructures.

It is important to realize that the Scherrer formula provides a lower bound on the coherently scattering domain size, referred to here as the crystallite size for readability. The reason for this is that a variety of factors can contribute to the width of a diffraction peak besides instrumental effects and crystallite size; the most important of these are usually inhomogeneous strain and crystal lattice imperfections. The following sources of peak broadening are listed in reference:[3] dislocations, stacking faults, twinning, microstresses, grain boundaries, sub-boundaries, coherency strain, chemical heterogeneities, and crystallite smallness. If all of these other contributions to the peak width, including instrumental broadening, were zero, then the peak width would be determined solely by the crystallite size and the Scherrer formula would apply. If the other contributions to the width are non-zero, then the crystallite size can be larger than that predicted by the Scherrer formula, with the "extra" peak width coming from the other factors. The concept of crystallinity can be used to collectively describe the effect of crystal size and imperfections on peak broadening.

Although "particle size" is often used in reference to crystallite size, this term should not be used in association with the Scherrer method because particles are often agglomerations of many crystallites, and XRD gives no information on the particle size. Other techniques, such as seiving, image analysis, or visible light scattering do directly measure particle size. The crystallite size can be thought of as a lower limit of particle size.

REFERENCES

1. Arthur Beiser, (1997)., Concept of modern physics by Tata McGraw-Hill Publishing Company Ltd. New Delhi.
2. Ashrafi, A. (2011). "Quantum Confinement: An Ultimate Physics of Nanostructures" In: Encyclopedia of Semiconductor Nanotechnology Ed. Umar, A., 5: 1–67.
3. Taniguchi, N.,(1974). Japan Society of Precision Engineering, Tokyo, Japan.
4. Drexler, K. E. (1986). "Engines of Creation." p. 298, Anchor Books, New York.
5. S. O. Pillai. (2010) *"Solid State Physics"* New Age International (p) Limited, New Delhi.
6. R. Murugeshan and Er. Kiruthiga Sivaprasath. (2014). Modern Physics by S. Chand and Company PVT. Ltd. Seventeenth edition. New Delhi.
7. Elements of X-ray Diffraction. By B.D. Cullity and S.R.Stock Pearson; 3 edition, 2001.
8. https: //www.quora.com/What-are-the-three-states-of-matter-and-their-properties
9. http: //britneyspears.ac/physics/crystals/wcrystals.html
10. http: //www.engineeringenotes.com/engineering/crystal-symmetry-in-solids-materials-engineering/33707
11. http: //www.engineeringenotes.com/engineering/crystal-symmetry-in-solids-materials-engineering/33707
12. https: //en.wikipedia.org/wiki/Lattice_(group)
13. https: //sseaimes.wordpress.com/tag/lattice-structure/
14. https: //www.basic-mathematics.com/order-of-rotational-symmetry.html
15. https: //www.differencebetween.com
16. https: //www.differencebetween.com
17. https: //www.toppr.com/guides/chemistry/the-solid-state/crystalline-and-amorphous-solids/

Chapter 2

Magnetic, Electrical and Optical Properties of Nanostructured Materials

S. Thirumalairajan[1] and K. Girija[2]

[1]Ramalingaswami Re-entry Faculty Fellow, Department of Nano Science and Technology, Tamil Nadu Agricultural University (TNAU), Coimbatore – 641003
e-mail: sthirumalairajan@gmail.com
[2]Assistant Professor, Department of Physics, Dr. N.G.P. Arts and Science College, Coimbatore – 641 048, Tamil Nadu

Magnetic Properties

Introduction

Earlier the researcher believed that the material properties can be altered only by varying the chemical composition. But later it was found that the material properties can be tuned by different shape and controlled size of the material at nanoscale without changing the chemical composition. The transition from bulk materials to nanoscale lead to a number of changes in their physical properties especially magnetic, electrical and optical properties. Magnetic materials play an important role in the advancement of research in the field of nanoscience and technology. From the scientific point of view, when a material is placed within a magnetic field, the magnetic forces of the electrons will be affected. This effect is known as Faraday's Law of Magnetic Induction. From technological point of view, the magnetic based nanostructures instruct considerable larger storage density due to a phenomenal increase in the number of bits stored in a unit area. The magnetic properties in nanostructures are governed by a number of factors such as shape, composition, size, topology, surface morphology, anisotropy, *etc.*, They are invariably used in potential appliances such as, power generation, digital and analogue data storage, medical applications like magnetic therapy, magnetic resonance imaging (MRI) and drug delivery, sensors and actuators, scientific instruments, *etc.* This section, deals with classification and properties of magnetic materials, magnetic materials analysis technique, important parameters of magnetic analysis, magnetic anisotropy, magnetic domains, shape dependent properties of different perovskite nanostructures and commercial applications of nanostructure magnetic materials.

Classification of Magnetic Materials

Most materials come under the five kinds of magnetism classified as follows and as shown in Figure 2.1.

- ☆ **Diamagnetic** : Characterized by atoms having no inherent magnetic moments.
- ☆ **Paramagnetic** : Characterized by non-interacting atomic moments.
- ☆ **Ferromagnetic, antiferromagnetic and ferrimagnetic:** Characterized by interacting magnetic moments.

Diamagnetic materials have a weak, negative susceptibility to magnetic fields. Diamagnetic materials are slightly repelled by magnetic field and the material does not retain the magnetic properties when the external field is removed. In diamagnetic

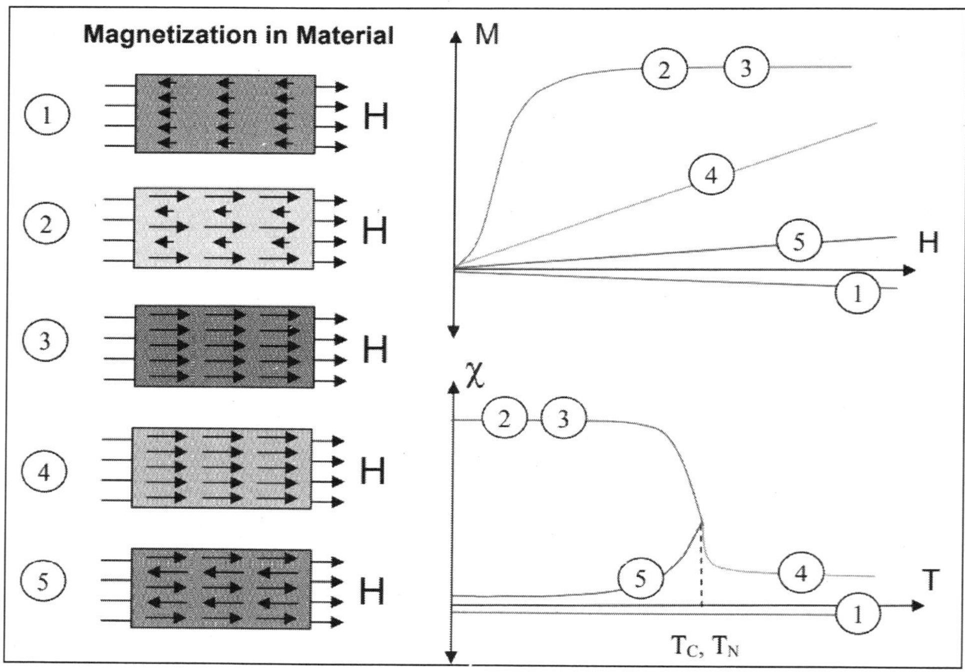

Figure 2.1. Schematic Representation of Spin Orientation in Classification of Magnetic Materials. (1) Diamagnetic (2) Ferrimagnetic (3) Ferromagnetic (4) Paramagnetic (5) Antiferromagnetic materials.
Courtesy: GREG VIALLE, Georgia Institute of Technology, 2009.

materials all the electrons are paired so there is no permanent net magnetic moment per atom. This property arise from the realignment of electron paths under the influence of an external magnetic field. Most elements in the periodic table, including copper, silver, and gold, are diamagnetic.

Paramagnetic materials have a small, positive susceptibility to magnetic field. These materials are slightly attracted by a magnetic field and the material does not retain the magnetic properties when the external field is removed. Paramagnetic properties are due to the presence of some unpaired electrons, and also from the realignment of the electron paths caused by the external magnetic field. Paramagnetic materials include magnesium, molybdenum and tantalum.

Ferromagnetic materials have a large positive susceptibility to an external magnetic field. They exhibit a strong attraction to magnetic field and are able to retain their magnetic properties after the external field has been removed. Ferromagnetic materials have some unpaired electrons so their atoms have a net magnetic moment. They get their strong magnetic properties due to the presence of magnetic domains. In these domains, large numbers of atom moments (10^{12} to 10^{15}) are aligned parallel so that the magnetic force within the domain is strong. When a ferromagnetic material is in the unmagnetized state, the domains are randomly organized and the net magnetic field for the part as a whole is zero. When a magnetizing force is applied, the domains become aligned to produce a strong magnetic field within the part. Iron, nickel, and cobalt are examples of ferromagnetic materials. Components with these materials are commonly inspected using the magnetic particle method.

Depending on the details of the crystal structure (which determines the size and sign of the exchange integral), exchange energy is minimum when electronic spins are aligned parallel or anti-parallel. This type of magnetic ordering is called **ferrimagnetism**. The magnetic structure is composed of two magnetic sublattices (called A and B) separated by oxygen. The exchange interactions are mediated by the oxygen anions. When this happens the interaction is called indirect super exchange interaction. In ferrimagnets, the magnetic moments of A and B sublattices are not equal and results in a net magnetic moment. Ferrimagnetism is therefore similar to ferromagnetism. Magnetite is a well-known ferrimagnetic material with unequal sublattice corresponding to Fe^{3+} and Fe^{2+} site. If the A and B sublattice moments are exactly equal but opposite, the net moment is zero. This type of magnetic ordering is called **antiferromagnetism**.

Magnetic Materials Analysis Technique

Vibrating Sample Magnetometer (VSM): VSM is a scientific instrument which measures the magnetic properties of materials in an external magnetic field, invented in 1955 by Simon Foner at Lincoln Laboratory, MIT. A VSM converts the dipole field of the sample into an AC electrical signal. The sample is placed inside a uniform magnetic field and then vibrated sinusoidally. Creates a corresponding variation of the magnetic flux in pickup coils placed nearby, inducing a sinusoidal voltage. The amplitude of the AC voltage is proportional to the sample magnetic moment and measured by a lock-in amplifier. The VSM can be used to measure the magnetic properties of prepared samples with an applied magnetic field. The M-H loop (hysteresis loop) can be used to derive the essential magnetic parameters, namely, saturation magnetization, remanent magnetization and coercivity. The magnetic characterization (diamagnetic, paramagnetic, ferromagnetic or antiferromagnetic) of any material is best described in terms of their hysteresis loop. In Figure 2.2, magnetic properties of various size and shaped nanostructures were analyzed using vibrating sample magnetometer (VSM-EG and G, PARC) with an applied magnetic field of 20 kOe at room temperature.

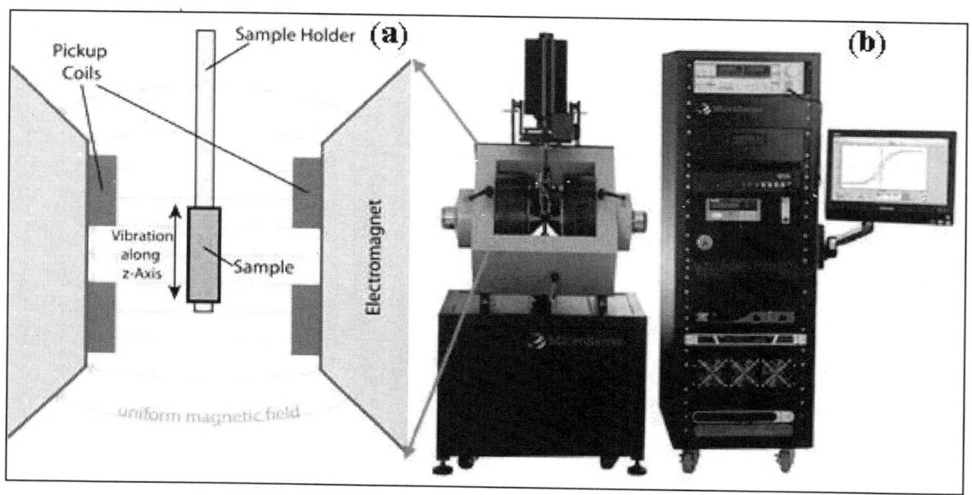

Figure 2.2(a). Function of Sample Analysis and (b) Vibration Sample Magnetometer (VSM) with Electromagnet.
Courtesy: Quantum Design, Europe, 2018

Important Parameters of Magnetic Materials Properties

M-H Hysterias Loop

A great deal of information can be learned about the magnetic properties of a material by studying its hysteresis loop (Kodama R.H. *et al.*, 1997). The hysteresis loop shows the relationship between the induced magnetic flux density **(B)** and the magnetizing force **(H).** It is often referred to as the B-H loop. An example hysteresis loop is shown in Figure 2.3. The loop is generated by measuring the magnetic flux of a ferromagnetic material while the magnetizing force is changed. A ferromagnetic material that has never been previously magnetized or has been thoroughly demagnetized will follow the dashed line as **H** is increased.

As the line demonstrates, greater the amount of current applied **(H^+)**, stronger the magnetic field in the component **(B^+)**. At point 'a' almost all of the magnetic domains are aligned and an additional increase in the magnetizing force will produce very little increase in magnetic flux. The material has reached the point of **magnetic saturation**. When reduced to zero, the curve will move from point 'a' to point 'b'. At this point, it can be seen that some magnetic flux remains in the material even though the magnetizing force is zero. This is referred to as the point of retentivity

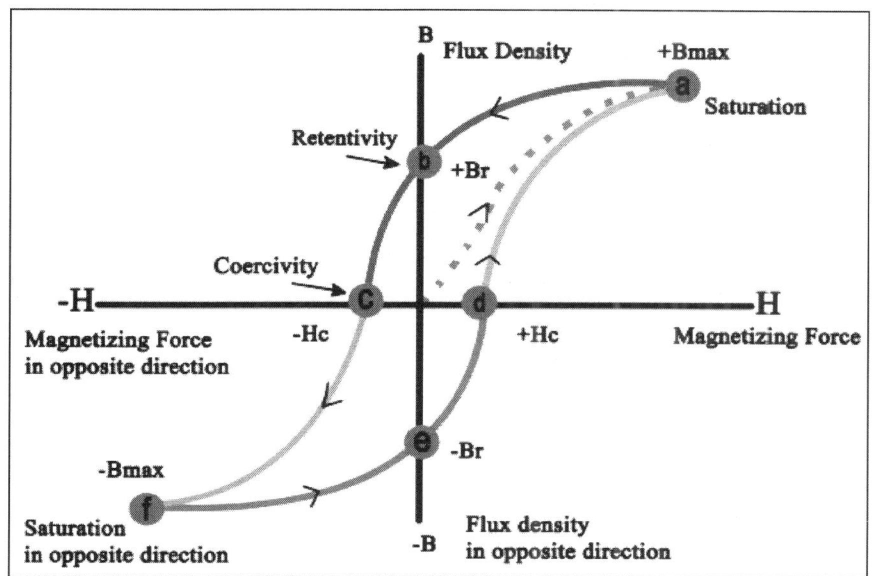

Figure 2.3. Schematic Illustration of Magnetic Hysteresis Loop.
Courtesy: Electrical Academic 2010.

on the graph and indicates the remanence or level of residual magnetism in the material. (Some of the magnetic domains remain aligned but some have lost their alignment). As the magnetizing force is reversed, the curve moves to point 'c', where the flux has been reduced to zero. This is called the point of **coercivity** on the curve. (The reversed magnetizing force has flipped enough of the domains so that the net flux within the material is zero.) The force required to remove the residual magnetism from the material is called the coercive force or coercivity of the material. As the magnetizing force is increased in the negative direction, the material will again become magnetically saturated but in the opposite direction (point "d"). Reducing **H** to zero brings the curve to point 'e'. It will have a level of residual magnetism equal to that achieved in the other direction. Increasing **H,** back in the positive direction will return **B** to zero. Notice that the curve did not return to the origin of the graph because some force is required to remove the residual magnetism. It will take a different path from point 'f' back to the saturation point were completed (Winkler E., *et al.*, 2005).

From the hysteresis loop, a number of primary magnetic properties of a material can be determined. But three major magnetic parameters that can be determined from the hysteresis loop are as follows:

Saturation magnetization (Ms): The maximum magnetization of a ferromagnetic material, when all the magnetic moment of the samples is oriented along the applied field direction.

Remenant magnetization (Mr): It is a measure of the magnetization retained by the sample when the external applied field is brought to zero.

Coercive Force: It is a measure of reverse magnetic field which must be applied to a magnetic material to make the magnetization zero.

Shape of Hysteresis Loop

The shape of the hysteresis loop tells a great deal about the material being magnetized. The hysteresis curves of two different materials are shown in the graph.

Relative to other materials, a material with a wider hysteresis loop has:

- ☆ Lower Permeability
- ☆ Higher Retentivity
- ☆ Higher Coercivity
- ☆ Higher Reluctance
- ☆ Higher Residual Magnetism

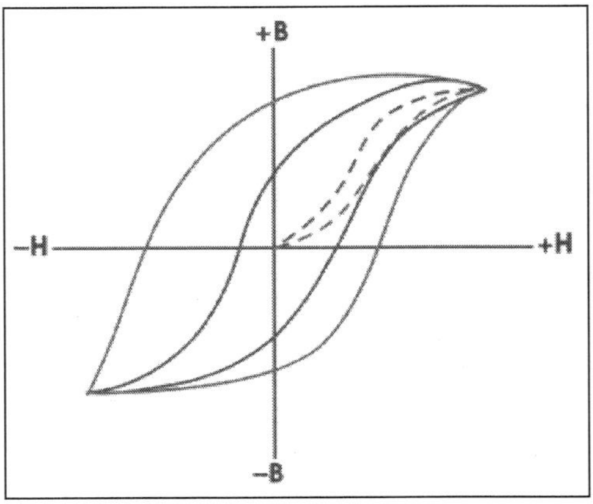

Relative to other materials, a material with the narrower hysteresis loop has:

- ☆ Higher Permeability
- ☆ Lower Retentivity
- ☆ Lower Coercivity
- ☆ Lower Reluctance
- ☆ Lower Residual Magnetism.

In magnetic particle testing, the level of residual magnetism is important. Residual magnetic fields are affected by the permeability, which can be related to the carbon content and alloying of the material. A component with high carbon content will have low permeability and will retain more magnetic flux than a material with low carbon content.

The shape of the M-H loop and the most important magnetic parameters, which determine the applicability of a magnetic material in various applications, depend mainly on two factors,

- ☆ Magnetic anisotropy
- ☆ Types of magnetic domain

Magnetic Anisotropy

The dependence of magnetic properties on a preferred direction (z-axis) is called magnetic anisotropic. The magnetic anisotropy is generally associated with ferromagnetic materials. The anisotropy energy creates barriers to free rotation of the magnetization within the magnetic crystal. This leads to different magnetic behaviour of a same material with different size, shape and crystal structure. There are many causes for anisotropy energy. The most important are

- ☆ **Magnetocrystalline anisotropy- crystal structure**
- ☆ **Shape anisotropy – grain shape**

Magnetocrystalline Anisotropy – Crystal Structure

Magnetocrystalline anisotropy is an intrinsic property which depends on the crystal orientation in a sample (Mazumder, R., *et al.*, 2007). In the magnetic field, the magnetization reaches saturation in different field. Magnetocrystalline anisotropy energy is the energy necessary to deflect the magnetic moment in a single crystal from the easy to the hard direction. When magnetized and demagnetized is along the easy axis, coercitivty value is lowered. On contrary, the hysteresis loop shows a high coercivity value which when magnetized along the hard axis.

Shape Anisotropy - Grain Shape

Shape of a material/grain is another source of magnetic anisotropy in ferromagnetic materials. For example, consider a long thin needle shaped grain. The anisotropy energy will be less if the magnetization is along the long axis than along one of the short axis. This produces an easy axis of magnetization along the long axis. A sphere, on the other hand, has to shape anisotropy (Folven, E *et al.*, 2010). Enhanced anisotropy induces large magnetic coercivity, where the magnetic spins are preferentially aligned along the long axis and their reversal in the opposite direction requires higher energies than that for spheres.

Types of Magnetic Domain

The existence of domains is hinted by the observation that some magnetic properties in particular, coercivity and remanence vary greatly with grain size. The magnetic behavior can be subdivided on the basis of grain size into four ranges:

- ☆ **SPM: superparamagnetic**
- ☆ **SD: single domain**
- ☆ **PSD: pseudo-single domain**
- ☆ **MD: multidomain**

The maximum coercivity for a given material occurs within its SD range. For larger grain, coercivity decreases as the grain subdivides into domains. For smaller grain, coercivity again decreases, but this time due to the randomizing effects of thermal energy. Domains constitute a fundamental concept in magnetism. A ferro or ferrimagnetic material may be generally defined as one that possesses a spontaneous magnetization (Ms) dependent on temperature, but only slightly dependent on applied field. As the grain size decreases, a critical size will be reached where the grain can no longer accommodate a wall. Below this critical size, the grain contains a single domain (SD). An SD grain is uniformly magnetized to its saturation magnetization. Changing the magnetization of a MD grain is an energetically easy process, which needs translation of the domain wall. Hence it can be accomplished in relatively low fields. Thus MD grains are magnetically soft with low values of coercivities and remancence. However, the only way to change the magnetization of a SD grain is to rotate the magnetization, an energetically difficult process. Thus,

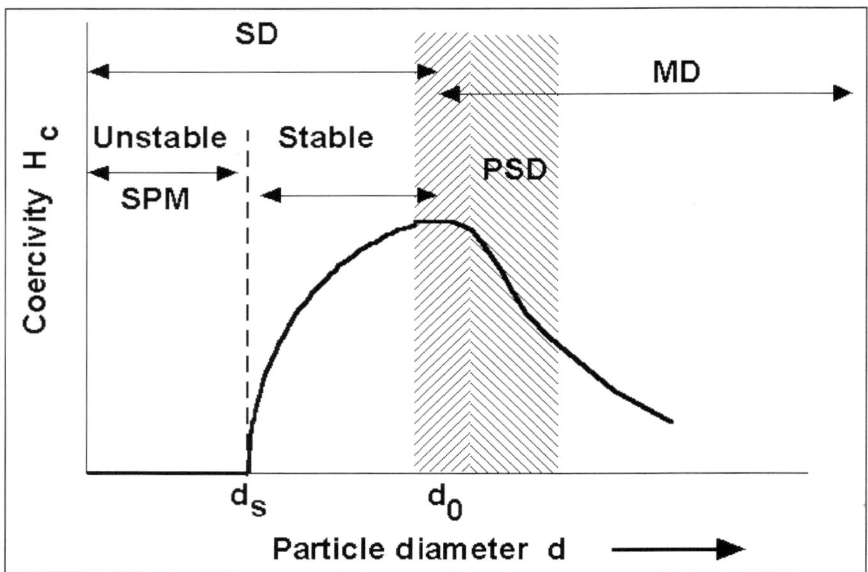

Figure 2.4. Variation of Intrinsic Coercivity with Particle Size.

SD grains are magnetically hard and have high coercivities and remanence. The critical size for SD behaviour depend on several factors including, the saturation magnetization and the shape of the grain (Kriener, M., *et al.*, 2004).

The distinction between SD and MD is straightforward. However, small MD grains exhibit a mixture of SD like (high remanence) and MD like (low coercivity) behaviour. This state is called pseudo-single domain (PSD), which occurs in smaller particles with few multi-domain grains. A particle size continues to decrease with the SD range, another critical threshold is reached, at which remanence and coercivity reaches zero. When this happens, the grain becomes superparamagnetic (SPM). An SD particle approaching the SPM size has a uniform magnetization directed along the easy axis of magnetization. If the volume is small enough, the thermal energy will be sufficient to overcome the anisotropy energy separating the magnetization states and cause a spontaneous reveal of magnetization. Figure 2.4, shows the variation of coercivity with particle diameter (equivalent to particle size) and the corresponding domain type demarcation as SD, MD, SPM and PSD.

The coercivity of small particle has a striking dependence on their size, as the particle size is reduced. It is typically found that the coercivity increases, goes through a maximum and then tends towards zero. The maximum coercivity for a given material occurs within its PSD range, which is characterized by the presence of multidomains in smaller particles. For larger grain size, coercivity decreases as the grain subdivides into domains. For grain size lesser than the PSD range, the particles are purely SD. For very small grain size, coercivity again decreases, because of the randomizing effect of thermal energy as seen in superparamagnetisim. The PSD range thus marks an upper limit to the SD range and lower limit to the MD range.

Magnetic Properties of different Morphologies of $LaFeO_3$ Nanostructures

The magnetic behaviour of the different morphologies are different for $LaFeO_3$ nanostructures and studied by measuring the magnetic hysteresis using VSM at room temperature, maximum applied magnetic field 20 KOe. $LaFeO_3$ is known to be antiferromagnetic with weak ferromagnetism (Huang, S., *et al.*, 2006). The net magnetization of $LaFeO_3$ should be very small because of the antiferromagnetic order of the Fe^{3+} spins. Only a slight canting of the adjacent Fe^{3+} spins produced weak ferromagnetism. The spontaneous magnetization of $LaFeO_3$ bulk crystals was only ~0.1 emu/g at room temperature (Kimura *et al.*, 2003)

The shape of hysteresis loop showing the small spontaneous magnetization were characteristics of weak ferromagnetism. All the morphologies exhibit slight hysteresis features, suggesting the weak ferromangnetic behaviour at room temperature. The M-H loop of the six different morphologies of $LaFeO_3$ are shown in Figure 2.5, the hysteresis loops doesn't get saturation even up to the maximum applied magnetic field of 20 T. It indicates the presence of large anisotropy and uncompensated surface spins, *i.e* the presence of more surface atoms in the nanomaterials. The important magnetic parameters like saturation magnetization, remanent magnetization and coercivity were calculated from the hysteresis loop and are tabulated in Table 2.1, for all the $LaFeO_3$ nanostructures. Also the ratio of saturation and remanent magnetizations M_r/M_s was calculated for each morphology to determine the type of magnetic domain present in the prepared nanostructures.

On the basis of the proposed criteria [18], the nature of domain in a ferromagnetic material of a particular size that can be determined, by calculating the Mr/Ms value from the hysteresis loop. According to hypothesis, single domain (SD) particle have a M_r/M_s value large than 0.5, while the pseudo-single-domain (PSD) particle have a value between 0.1 to 0.5 and multidomain (MD) particle have a value lower than 0.1. Figure 2.5 a-d shows the M-H loop recorded for the different nanostructures like nanocubes, flowers, dendrites and nanorods respectively. Among all the prepared nanostructures, the nanocubes were found to exhibit a high magnetization and remanenance. The M_r/M_s values of 0.0.09891 were calculated from the M-H loop of the nanocubes suggests the presence of multidomain (MD) behaviour. Similarly, the M_r/M_s value calculated for the M-H loop for the other nanostructures like flowers, dendrites and nanorods shows the values of 0.1837, 0.2096 and 0.2137, respectively again suggesting the presence of all nanostructure pseudo-single domains (PSD). The variation in M_r/M_s value can be related to the average particle size and morphology which is related to the use of surfactant in the reaction,

The M-H loop of the sphere like morphologies of $LaFeO_3$ nanostructures are shown in Figure 2.5 (e and f). The M_r/M_s value of 0.0231 and 0.0062 are estimated for microspheres and nanospheres respectively, and it suggest the formation of multidomain. The presence of MD is also reflected in the coercivity value, as shown in Table 2.1.

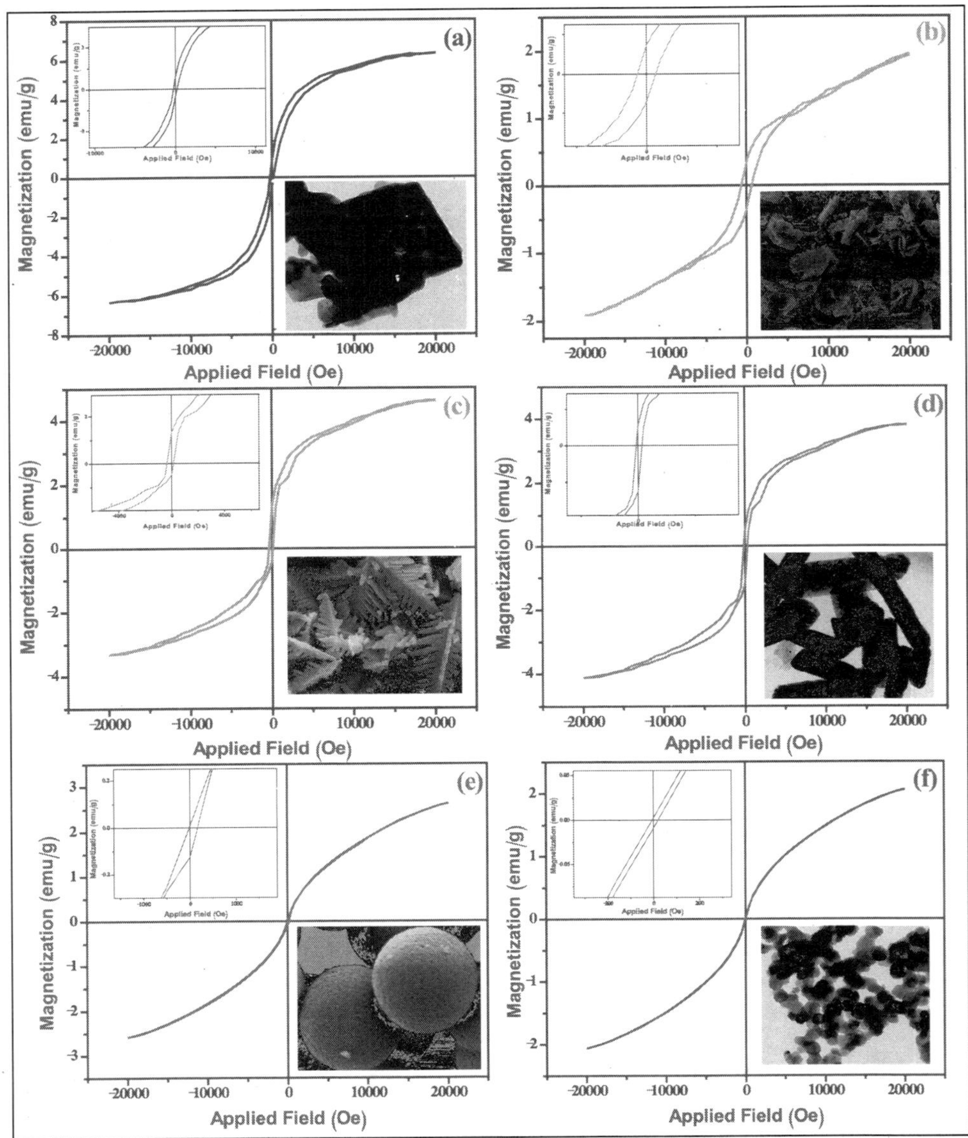

Figure 2.5. Room Temperature M-H loop of the different $LaFeO_3$ Nanostructures and Inset (left) shows the Magnified View of the M-H Loop at Low Applied Magnetic Field (Right) the Corresponding Nanostructures, (S. Thirumalairajan *et al.*, 2016).

Table 2.1. Magnetic Parameters of the different Morphologies of $LaFeO_3$ Nanostructures

Nanostructures	*Saturation Magnetization (M_s) (emu/g)*	*Remanent Magnetization (M_r) (emu/g)*	*Coercivity (Oe)*	*M_r/M_s*
Nanocubes	6.47	0.640	226	0.09891
Flowers	1.924	0.353	431	0.1837
Dendrites	4.03	0.845	205	0.2096
Nanorods	3.86	0.825	198	0.2137
Microspheres	2.60	0.062	85	0.0231
Nanospheres	1.02	0.006	14	0.0062

The large coercivity value for the nanocubes, flowers, dendrites and nanorods might be due to the combined effect of:

- ✰ Shape anisotropic induced due to the deviation from micro and nanospheres
- ✰ Small particles in PSD range as reflected from the M_r/M_s value calculated from the M-H
- ✰ Surfactant used in the preparation of these nanostructures, which might have resulted in OH group and associated surface spin orientation.

The prepared $LaFeO_3$ nanostructures are weak ferromagnetic and multi disperse system consisting of the SD and MD particles. The magnetization of the sample is considered as the sum of two terms:

$$M(H) - M^fH + M^{sp}(H) \tag{1}$$

Here, $M^f(H)$ is the contribution of ferromagnetic (f) nanoparticles (multiple domains and $M^{sp}(H)$ is the contribution from the superparamagnetic (sp) nanoparticles (single domain). Further, the effect of size dispersion is presented in a real system. The magnetization of superparamagnetic nanoparticles can be represented by the expression (Fonseca, F., *et al.*, 2005)

$$M^{sp}(H) = M^{sp}(\infty)^{\Sigma}_{j} f(m_j) L\left[\frac{m_j H}{K_E T}\right] \tag{2}$$

For particle with sphere shape, the distribution of particle size $f(D)$ is given by the expression (Lee, S.J *et al.*, 2004).

$$f(D) = \frac{1}{\sqrt{2\pi}\sigma D} \exp\left(-\frac{\ln(D/\overline{D})^2}{2\sigma^2}\right) \tag{3}$$

Where, σ is the standard deviation and $\overline{D}$ is the average particle size. $f(m_j)$ can be calculated from D. Figure 2.5 shows the curve of the $LaFeO_3$ nanostructures with fitting results in Langevin function (Lee, S.J *et al.*, 2004). The results indicate that the magnetization of $LaFeO_3$ nanopetals is higher than all other $LaFeO_3$ morphologies.

The magnetic moment of the antiferromagnetic $LaFeO_3$ nanoparticles originate from the spin uncompensation at the surface and due to the canting of Fe-Fe antiferromagnetic order. As the particle size decreases, an increasing fraction of atoms lie at or near the surface and then the spin uncompensation increases. Hence, the moment of an antiferromagnetic particle is expected to increase significantly with decreasing particle size. The magnetic properties are intimately tied to the Fe-O-Fe bonding of the FeO_6 octahedra (Vijayanand, S., *et al.*, 2009), which can be attributed to the larger size of $LaFeO_3$ nanopetals than the other nanostructures. The results demonstrate that the $LaFeO_3$ nanostructures exhibit weak ferromagnetism which corresponds with its bulk form (Lee, S.J *et al.*, 2004). So, we deduce that the magnetic behaviour of the $LaFeO_3$ nanostructures originate from their shape and finite size.

Commercial Application of Nanostructure Magnetic Materials

They are invariably used in potential applications in magnetic materials such as, power generation, digital and analogue data storage, medical applications like magnetic therapy, magnetic resonance imaging (MRI) and drug delivery, sensors and actuators, scientific instruments, *etc.*

Electrical Properties

Electrical transport is a complex property of materials. Ever since the discovery of the electron by J. J. Thomson in 1897, the conductivity in materials has been widely studied. Later Sommerfeld modified Drude's theory, mainly by replacing the Maxwell-Boltzmann electron velocity distribution with the Fermi-Dirac distribution (Malk, M., *et al.*, 2014). The equilibrium free-electron theory of Sommerfeld was extended by Bloch theory to be valid for a system in the band structure of the metal. This theory, referred to as the semiclassical approach to electron dynamics, describes the dynamics of electrons in an applied electric and magnetic field, neglecting the presence of electron scattering (Dwyer, R.P., *et al.*, 2017).

The electrical properties like conductivity analysis, dielectric constant, dielectric loss, electric dipole moment and polarisation are very important parameter of electrical materials. Measurements of the dielectric properties of a material, which are characterized by its complex values are complicated task and special kind of instrumentation are needed. The impedance and four probe methods has been widely used to study the dielectric behaviour of the crystalline, polycrystalline ceramic and nanostructure materials. In general, the dielectric properties of ferroelectromagnetic materials arise due to intra-grain, inter-grain and other electrode effects. The motion of charges could take place by charge displacement, dipole reorientation, space charge formation *etc.* (Dwyer, R.P., *et al.*, 2017). In order to understand the electrical properties of a prepared nanostructure sample, grain, grain boundary and electrode contributions must be separated out. Complex impedance analysis has emerged as a very powerful tool for separating out electrical properties. This section, deals with the dielectric materials, dielectric parameter and properties, dielectric constant, electric dipole moment, polarization, conductance impedance spectroscopy measurements technique and also electrical properties of nanoscale materials in commercial applications.

Important Parameters of Electrical Properties

Ohm's Law

When an electric potential *V* is applied across a material, a current of magnitude *I* flows. In most metals, at low values of *V*, the current is proportional to *V*, according to Ohm's law:

$$I = V/R$$

where *R* is the electrical resistance. *R* depends on the intrinsic *resistivity r* of the material and on the geometry (length *l* and area *A* through which the current passes).

$$R = rl/A$$

Electrical Conductivity

The *electrical conductivity* is the inverse of the resistivity: $s = 1/r$.

The electric field in the material is $E=V/l$, Ohm's law can then be expressed in terms of the current density $j = I/A$ as:

The electric field in the material is $E=V/l$, Ohm's law can then be expressed in terms of the current density $j = I/A$ as:

$$j = s\,E$$

The conductivity is one of the properties of materials that varies most widely, from 10^7 (W-m) typical of metals to 10^{-20} (W-m) for good electrical insulators. Semiconductors have conductivities in the range 10^{-6} to 10^4 (W-m) (Iqbal, M.J., *et al.*, 2008).

Dielectric Materials

Dielectric materials are materials having no free electrons to conduct. The main difference from insulators is that it can store energy. All the insulators in electrical circuits are dielectrics. The change of dielectric material with and without electric field is represented in Figure 2.6. A dielectric material is a substance with poor conductor of electricity, but an efficient supporter of electrostatic field s. If the flow of current between opposite electric charge poles is kept to a minimum while the electrostatic lines of flux are not impeded or interrupted, an electrostatic field can store energy. This property is useful in capacitor s, especially at radio frequencies. Dielectric materials are also used in the construction of radio-frequency transmission lines. In practice, most dielectric materials are solid. Examples include porcelain (ceramic), mica, glass, plastics, and the different metal oxides. Some liquids and gases can serve as good dielectric materials. Dry air is an excellent dielectric, and is used in variable capacitors and some types of transmission lines. Distilled water is a fair dielectric. A vacuum is an exceptionally efficient dielectric.

The **dielectric material** which is used in capacitors fulfils the following functions.

- ✰ Decreases the useful electric field between the capacitor plates.
- ✰ Boosts the capacitance of the capacitor plate structure.

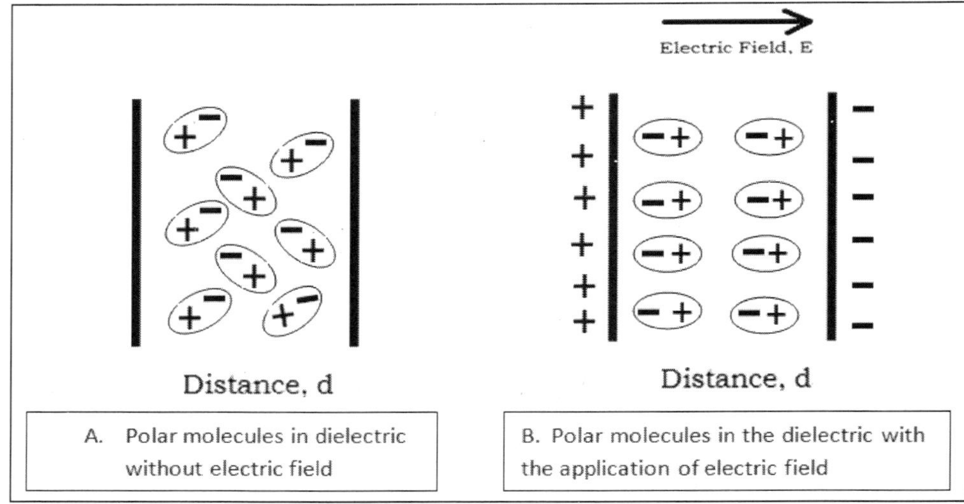

Figure 2.6. Polarized Dielectric Materials.

- ✰ Keep the conducting plates from coming in contact.
- ✰ Reduce the possibility of shorting out by sparking during high voltages.

As said above, the presence of dielectric material will decrease the electric field produced by the charge density. The effective electric field which is produced by the charge density can be given by:

$$E_{eff} = E - E_{polar} = \frac{\sigma}{\kappa\varepsilon_0}$$

σ: Charge density in A/m^2

κ: Dielectric constant.

ε_0: Permittivity of free space in F/m = 8.854 x 10^{-12} F/m.

The effective field is decreased by polarisation of the dielectric by a factor 'κ'. It is called the dielectric constant of the material. For a vacuum, it is equal to 1.

The capacitance of the normal capacitor;

$$C = \frac{\varepsilon_0 A}{d}$$

The capacitance of the capacitor which is filled with the dielectric will be increased by the factor called the dielectric constant, κ which is given by

$$C = \frac{\kappa\varepsilon_0 A}{d}$$

'κ' is defined as the relative permittivity of a **dielectric material** and it is an important parameter in deciding the capacitor characteristics.

Dielectric Constant

The dielectric constant is the ratio of the permittivity of a substance to the permittivity of free space. It is an expression of the extent to which a material concentrates electric flux, and is the electrical equivalent of relative magnetic permeability as shown in Figure 2.7. As the dielectric constant increases, the electric flux density increases, if all other factors remain unchanged. This enables objects of a given size, such as sets of metal plates, to hold their electric charge for long periods of time, and/or to hold large quantities of charge. Materials with high dielectric constants are useful in the manufacture of high-value capacitors. A high dielectric constant, in and of itself, is not necessarily desirable. Generally, substances with high dielectric constants break down more easily when subjected to intense electric fields, than materials with low dielectric constants. For example, dry air has a low dielectric constant, but it makes an excellent dielectric material for capacitors used in high-power radio-frequency (RF) transmitters. Even if air does undergo dielectric breakdown (a condition in which the dielectric suddenly begins to conduct current), the breakdown is not permanent. When the excessive electric field is removed, air returns to its normal dielectric state. Solid dielectric substances such as polyethylene or glass, however, can sustain permanent damage.

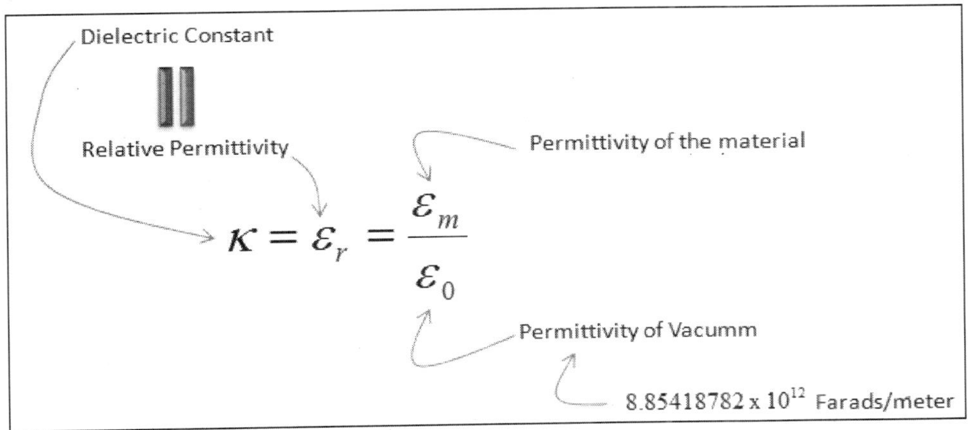

Figure 2.7. Schematic Expression for Dielectric Constant.
Courtesy: European passive components Institute, 2014.

Dielectric Polarization and their Types

When an electric field is applied to a capacitor, the dielectric material (or electric insulator) become polarized, such that the negative charges in the material orient themselves towards the positive electrode and the positive charges shift towards the negative electrode. Since charges are not free to move in an insulator, the polarization effect that opposes the applied field draws charges onto the electrodes, thus storing energy in the capacitor (Coìrdoba P., *et al.*, 2015)

The more easily a material be polarized, the greater the amount of charge can be stored in the capacitor. This ability to store energy in an electric field is referred to as the dielectric constant K, or relative permittivity ε_r. The degree of polarization P is related to the dielectric constant K and the electric field strength E as follows.

$$P = \varepsilon_0(K-1) E$$

where ε_0 is a physical constant known as the vacuum permittivity.

The total polarization of a dielectric comes the sum of four sources of charge displacement: electronic displacement P_e, ionic displacement P_i, orientation of permanent dipoles P_d, and space charge displacement P_s.

$$P_t = P_e + P_i + P_d + P_s$$

The polarization in the dielectric material occurs when it is subjected to the external field and it is classified according to their origin as shown in Figure 2.8.

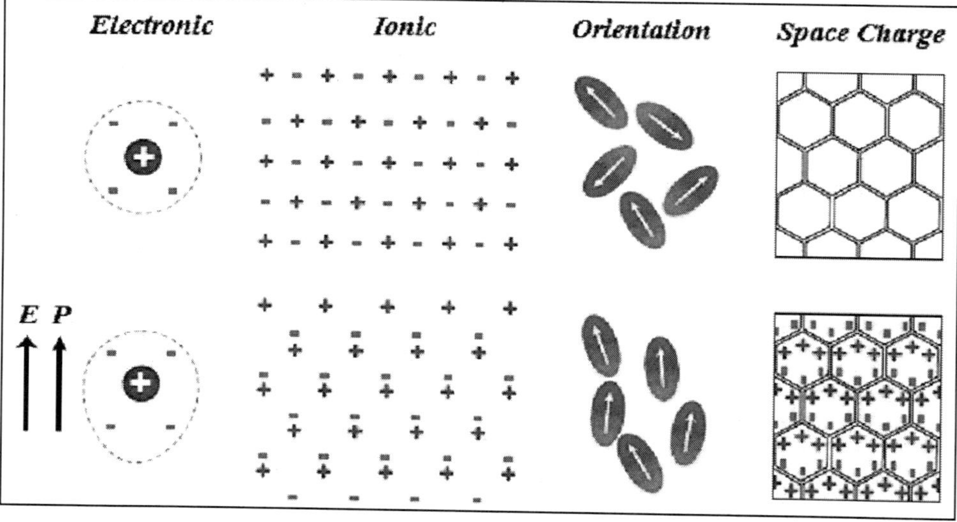

Figure 2.8. Four Types of Polarization Mechanisms.
Courtesy: Muhammad Usman, 2015.

Types of polarizations are:

1. Electronic polarization
2. Atomic or ionic polarization
3. Dipolar polarization
4. Interface or space charge polarization

Electronic Polarization

This effect occurs in all atoms under the application of an electric field. The nucleus of the atom and the center of its electron cloud shift away from each other, creating a tiny dipole with very small polarization effect.

Ionic Polarization

In ionic solids such as ceramic materials, the ions are symmetrically arranged in a crystal lattice with a net zero polarization. Once an electric field is applied, the cations and anions are attracted to opposite directions. This creates a relatively large ionic displacement (compared to electronic displacement), which can give rise to high dielectric constants in ceramics popularly used in capacitors.

Dipole (or Orientation) Polarization

Certain solids have permanent molecular dipoles that, under an electric field, rotate themselves in the direction of the applied field, creating a net average dipole moment per molecule. Dipole orientation is more common in polymers since their atomic structure permits reorientation.

Space Charge (or Interfacial) Polarization

In ceramics, this phenomenon arises from extraneous charges that come from contaminants or irregular geometry in the interfaces of the polycrystalline solids. These charges are partly mobile and migrate under an applied field, causing this extrinsic type of polarization.

Dielectric Loss

Interestingly, each type of polarization has a different time response capability to an applied field frequency, which means the net effective polarization to the dielectric constant is frequency dependent as shown in Figure 2.9:

- ☆ Electronic displacement is very rapid, so this polarization occurs at frequencies up to 10^{17}
- ☆ Ionic polarization is a bit slower and occurs at frequencies up to 10^{13}
- ☆ Dipole polarization occurs at frequencies less than 10^{10}
- ☆ Space charge polarization is the slowest and occurs at less than 10^{4}

Therefore, the dielectric constant (and therefore the capacitance value) always decreases with increased frequency, since the polarization mechanisms become less effective. The frequency at which a dielectric is used has an important effect on the polarization mechanisms, notably the relaxation time displayed by the material when following field reversals in an alternating circuit.

- ☆ Case 1: If the relaxation time for polarization is much longer and slower than the field reversals, the ions cannot follow the field at all and losses are small.
- ☆ Case 2: If the relaxation time is much faster than the field reversals, the polarizing processes can easily follow the field frequency and losses are small.
- ☆ Case 3: If the relaxation time and field frequency are the same, the icons can follow the field but are limited by their lag, thus generating the highest loss with frequency.

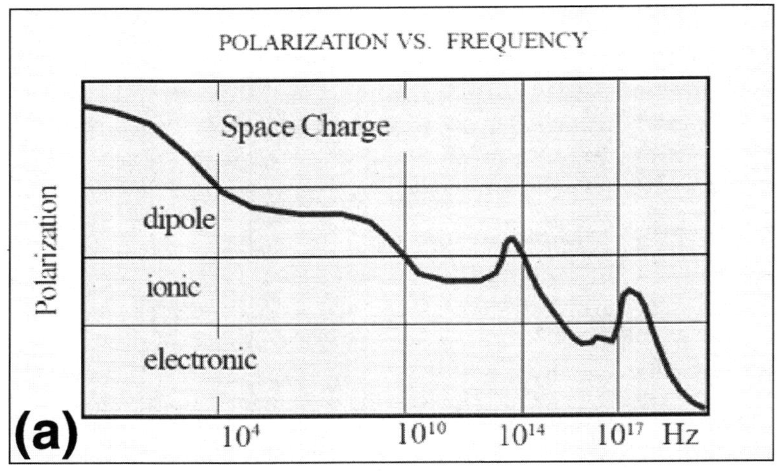

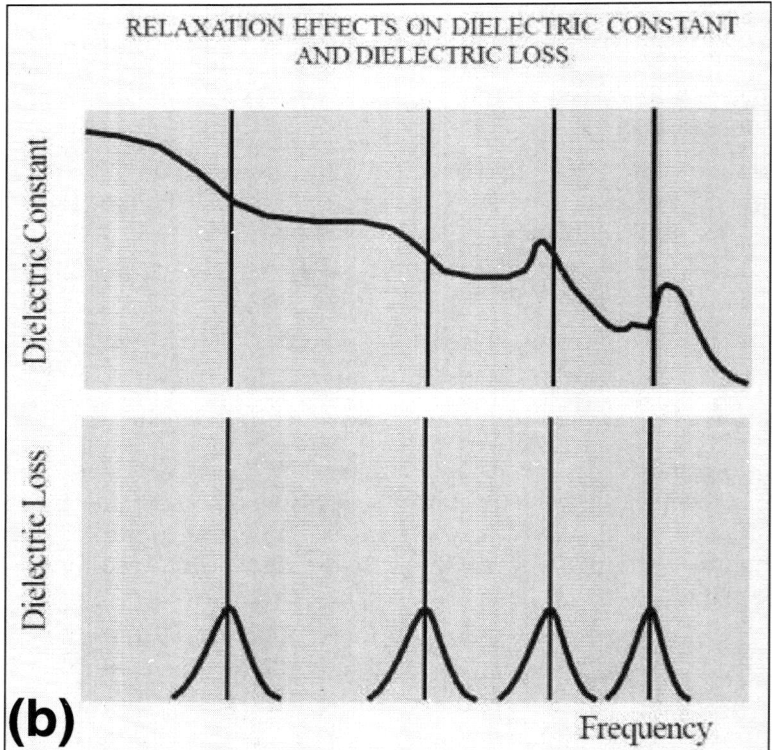

Figure 2.9. (a) Effect of Frequency on Polarization Mechanisms, (b) Changes in Dielectric Constant and Dielectric Loss Caused by Frequency.
Courtesy: Advanced Centre of Research in High Energy Materials, (ACRHEM), India, 2015.

Therefore, dielectric losses are highest at the frequency where the applied field has the same period of the relaxation process. Ceramic dielectric formulations always show a range of relaxation times over the frequency spectrum, since these materials consist of polycrystalline matter. In high frequency applications, this parameter is often known as the Q factor, which is the reciprocal of the loss tangent: Q =1/(tan δ) (Wu, J., *et al.*, 2002). Hopefully, above discussion provides better understanding of the dielectric polarization and how it affects your specific application.

Electrical Properties Analysis using Impedance Spectroscopy

Impedance Spectroscopy (IS): Impedance spectroscopy is a flexible tool for simultaneous electrical and dielectric characterization of materials. This powerful technique has been widely used to characterize the dielectric behaviour of single crystal, polycrystalline, amorphous ceramic and nanostructure materials. The experimental data contains three variables, real, imaginary component of the electrical response and frequency together with four inter related electrical response formalism: impedance (Z*), admittance (Y*), relative permittivity (ε*) and electrical modulus (M*). Consequently, data can be presented in many formats; in practice, it is often found that different formats may highlight different electrical characteristics of a sample. It is the most commonly used experimental technique to analyze the dynamics of the ionic movement in solids. The general approach is to apply a monochromatic (single frequency) voltage to the sample and measure the corresponding response current as the frequency is scanned. In this section, Solartron-SI1260 impedance/grain phase analyzer with frequency range 100 μHz to 10 MHz, fully automated with a personal computer was used to study the electrical transport properties of the samples as shown in Figure 2.10. Following discussion with the impedance measurements and also dielectric properties of nanostructure $LaFeO_3$ with various morphology.

Different morphologies of $LaFeO_3$ nanostructured powder was made into pellets of 8 mm diameter and ~ 2 mm thickness for the impedance measurements. AC impedance spectroscopic measurements have been carried out using a Hewlett–Packard HP 4284A model Precision LCR Meter over the frequency range from 1 MHz to 20 Hz in the temperature domain 300 – 340 K. The electrical conductivity value was estimated using the relation $\sigma = t/(R_b A)$, where t is the thickness of the specimen, A the area covered by the silver electrodes in contact with the specimen and R_b is the bulk resistance of the material derived from the intercept of the complex impedance plot *i.e.*, Nyquist plot on the real axis. The equivalent circuit software developed by Boukamp has been effectively used to extract the accurate value of grain boundary from the intercept on the real axis at the high frequency end of the relevant Nyquist plot.

Electrical Impedance Analysis of different $LaFeO_3$ Nanostructures

The electrical impedance and dielectric behavior of the nanostructured materials depend on the chemical composition, shape and size effect. When nanostructure materials are sintered under special conditions, the valance state of ions change, and the individual cations thus formed in the sample lead to high conductivity (Macdonald, J.R., *et al.*, 2005). When such a material is cooled in an oxygen atmosphere, it transforms to a material with high resistivity over the constituent grains. Studies on the effect of temperature, shape, size and frequency on the

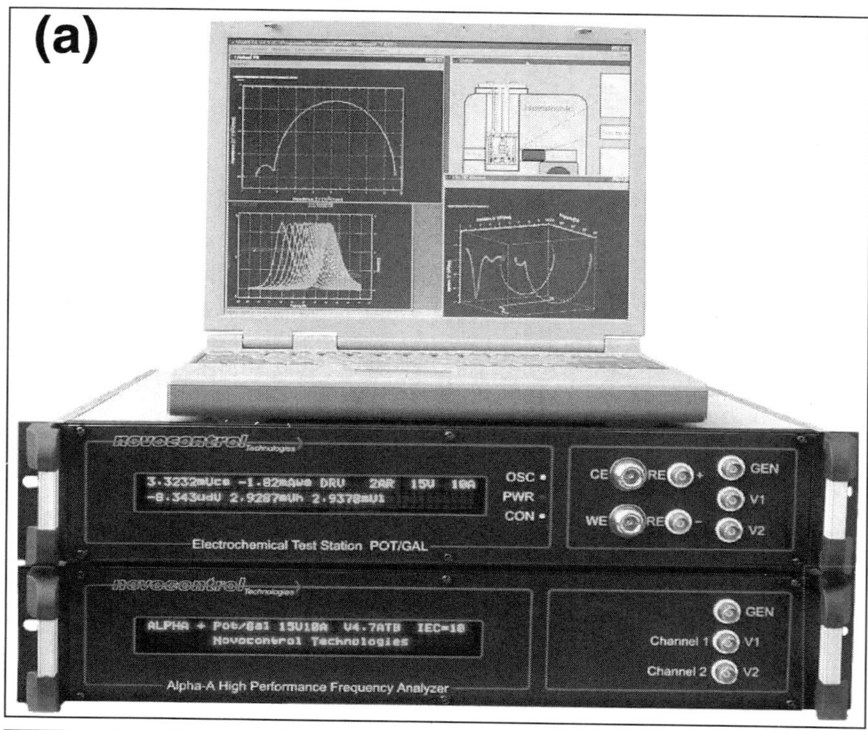

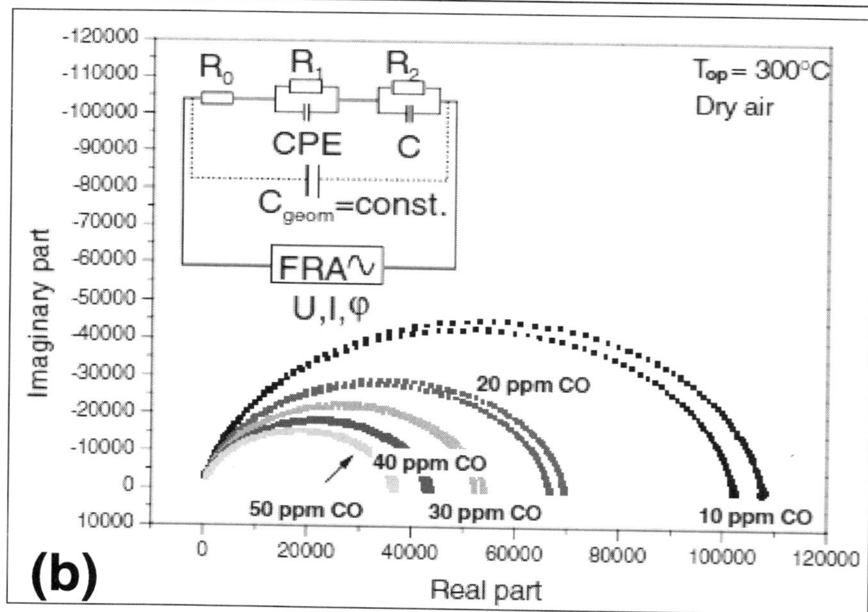

Figure 2.10. (a) Photograph of Impendence Spectroscopy, (b) Cole – Cole Plot of Impedance Spectrum.

Courtesy: Novo control Technology, Germany, 2019.

dielectric behavior and electrical impedance offer valuable information about the conduction phenomenon based on the localized electric charge in the nanostructured ceramics. The frequency dependence of the electrical impedance was determined in the frequency range between 20 and 10^6 Hz during the temperature plateau for nanostructured $LaFeO_3$ nanopetals, dendrites, nanorods and nanospheres. In Figure 2.10a, the complex Cole-Cole (or Nyquist) plot Im(Z) versus Re(Z)) consists of a well resolved semicircle whose center is located below the real axes, *i.e.* a non Debye behavior, and a spike at low frequencies for a selected ceramic of $LaFeO_3$ nanopetals at room temperature. The spike is an evidence of the electrode polarization effect during our measurements Jonscher *et al.*, 1992.

Besides, these behaviors were found for all $LaFeO_3$ ceramics with different morphologies. In order to describe the complex impedance of materials, we employed a phenomenological model based on two serial (R_1//CPE_1) and (R_2//CPE_2) elements associated with two distinct contributions to the impedance response: one at low frequency and one at high frequency. The total electrical impedance is given by:

$$Z = \left[\frac{1}{R_1} + \frac{1}{Z_{CPE,1}} \right]^{-1} + \left[\frac{1}{R_2} + \frac{1}{Z_{CPE,2}} \right]^{-1} \tag{1}$$

In both cases, CPE (constant phase element) denotes the pseudo capacitance defined by Jonscher *et al.*, 1992. The impedance of CPE is expressed as follows (Jorcin *et al.*, 2006).

$$Z_{CPE} = \frac{1}{Q(j\omega)^{\alpha}} \tag{2}$$

where Q is a CPE constant in $F \cdot s^{\alpha-1}$, $\omega = 2\pi f$ is the angular frequency (f being the frequency in Hz), α is the dimensionless CPE parameter ranging from 0 up to 1, and $j = \sqrt{-1}$ is the imaginary unit. The equivalent circuit based on the impedance data for the nanopetals sample is shown in Figure 2.11a inset. Probably, the occurrence of one semicircle indicates that the grain and grain boundary are overlapped owing to the similar relaxation frequency of both contributions or the grain boundary conduction is almost negligible. Therefore, the parameter R_1 denotes the total resistance of the dc-resistances R_b and R_{gb} associated to the grain and grain boundary. The parameters of Eq. (1) were obtained for three different temperatures (300, 310, and 320 K) for all the nanostructures by analyzing the data using the non-linear least squares (NLLS) fitting procedure, as illustrated in Figure 2.11b for the ceramic $LaFeO_3$ nanopetals at room temperature. Then, the total conductivity (σ_{DC}) is the sum of the two conductivities of both grain and grain boundary ($\sigma_{DC} = \sigma_g + \sigma_{gb}$). From Figure 2.11c, we note that the total resistance of the $LaFeO_3$ nanopetals ceramic decreases with the increase in temperature suggesting that the total conductivity is a thermally activated quantity. These behaviors were also found for all ceramics with different morphologies.

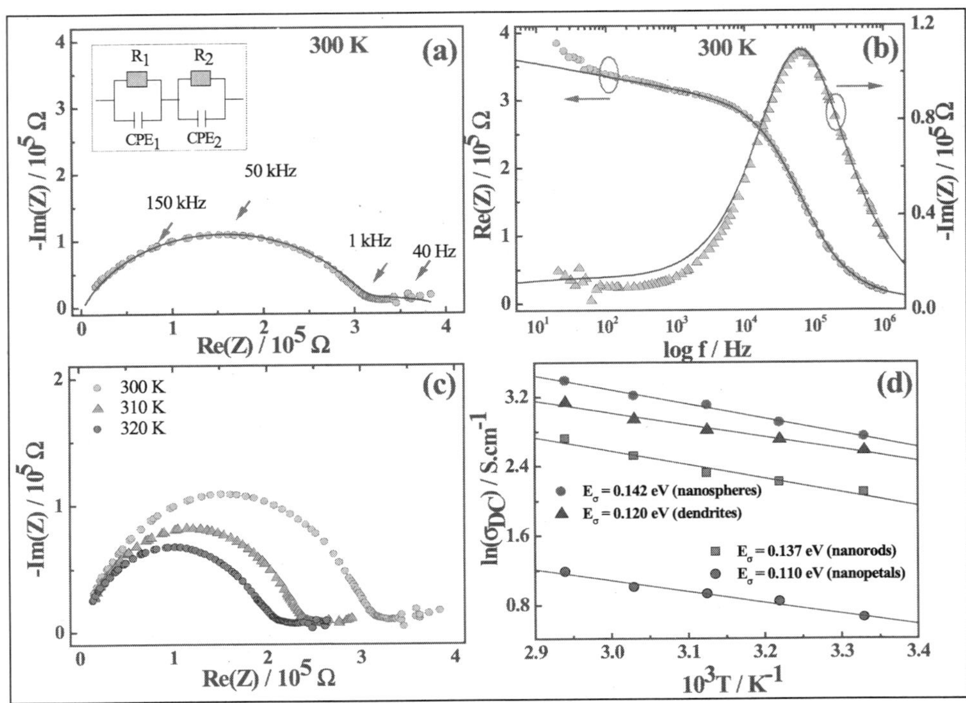

Figure 2.11. Impedance Spectra for $LaFeO_3$ Nanopetals (a) Room Temperature, (b) Real Part Re(Z*) and the Opposite of the Imaginary Part–Im(Z*) as a Function of Frequency at RT (c) Impedance Spectra at different Temperatures and (d) Arrhenius Plot of the Conductivity for different $LaFeO_3$ Nanostructures (S. Thirumalairajan, *et al.*, 2016).

From Figure 2.11d, the activation energies for the total conductivity are found to be in the range 0.110, 0.120, 0.137 and 0.142 eV for $LaFeO_3$ nanopetals, dendrites, nanorods and nanospheres. The values of activation energies are characteristic of the electron hopping between Fe^{2+} and Fe^{3+} ions in *B* sites (electronic conductivity). Therefore, the conduction due to anion vacancies may be responsible for the conductivity behavior with higher activation energy (~ 0.142 eV) in the high temperature region for nanospheres with grain size (50 nm). Probably, the different activation energy values may be explained based on the variation of the zero-point energy induced by the powder morphology. Therefore, one can conclude that the size and morphology of the $LaFeO_3$ nanoparticles changed the vibrational frequencies associated with iron and oxygen motions in the octahedral cage, resulting in the variation of activation energy values during the electronic conductivity process. The value of activation energy exhibited by nanospheres is characteristic to the diffusion of oxygen vacancies. The conduction mechanism of $LaFeO_3$ nanostructures can be explained on the basis of hopping of the charge carriers between Fe^{2+} and Fe^{3+} on the octahedral sites. The compositional dependence on ac conductivity is also depicted. It can be evidenced that the magnitude of conductivity increases with decrease in grain size. The maximum conductivity was obtained for nanospheres, which could be attributed to the increased active carrier concentration with decreased grain size.

Optical Properties

One of the most fascinating and useful aspects of nanomaterials is their optical properties. Applications based on optical properties of nanomaterials include optical detector, laser, sensor, imaging, phosphor, display, solar cell, photocatalysis, photo electrochemistry and biomedicine. Many of the underlying principles are similar in these different technological applications that span a variety of traditional disciplines including chemistry, physics, biology, medicine, materials science, engineering, electrical and agricultural science. It is essential for students in these various science and engineering fields to have a basic understanding of the fundamental optical properties and related spectroscopic techniques. This section deals with, band theory of solids, semiconductors, classification of semiconductors, Intrinsic and Extrinsic semiconductors, size and shape dependent optical properties and applications of nanostructure optical materials.

Band Theory of Solids

A band gap is the distance between the valence band of electrons and the conduction band. Essentially, the band gap represents the minimum energy that is required to excite an electron up to a state in the conduction band where it can participate in conduction (Eustis, *et al.*, 2006). The lower energy level is the valence band, and thus if a gap exists between this level and the higher energy conduction band, energy should be input for electrons to become free. The size and existence of this band gap allows one to visualize the difference between conductors, semiconductors, insulators (Kuznetsov, *et al.*, 2016). These distances can be seen in diagrams known as band diagrams, shown in Figure 2.12.

Valence Band

The energy band involving the energy levels of valence electrons is known as the valence band. It is the highest occupied energy band. When compared with insulators, the bandgap in semiconductors is smaller. It allows the electrons in the valence band to jump into the conduction band on receiving any external energy.

Conduction Band

It is the lowest unoccupied band that includes the energy levels of positive (holes) or negative (free electrons) charge carriers. It has conducting electrons resulting in the flow of current. The conduction band possess high energy level and are generally empty. The conduction band in semiconductors accepts the electrons from the valence band.

Semiconductors

In conductors, the valence band overlaps with the conduction band. This overlap causes the valence electrons to be essentially free to move into the conduction

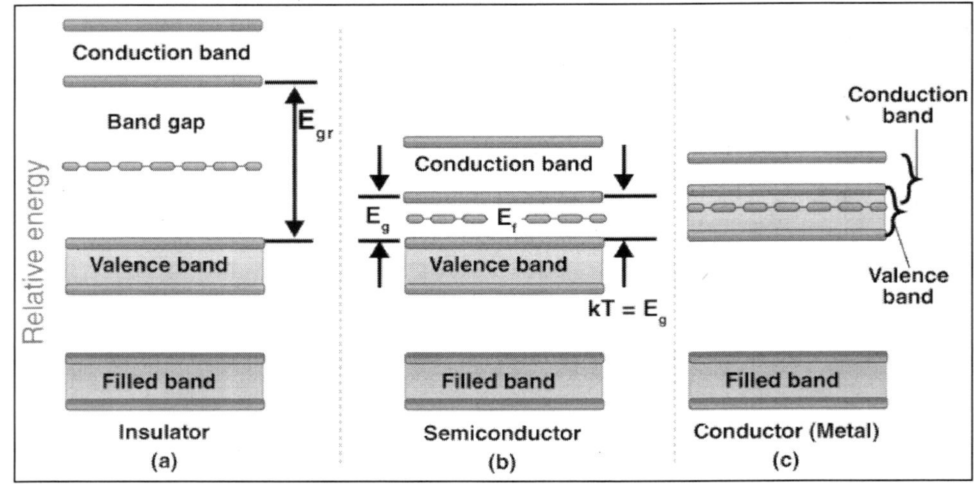

Figure 2.12. Band Gap Diagram showing the different Sizes of Band Gaps for Conductors, Semiconductors and Insulators (*Source*: Gold hydric, 2009).

band and participate in conduction. Since it is not a full overlap, only a fraction of the valence electrons can move through the material, but this is still enough to make conductors conductive (Roldan, M.V., *et al.*, 2008). Semiconductors are the materials which have a conductivity between conductors (generally metals) and non-conductors or insulators (such ceramics). Semiconductors can be compounds such as gallium arsenide or pure elements, germanium or silicon. Physics explains the theories, properties and mathematical approach governing semiconductors.

Properties of Semiconductors

Semiconductors can conduct electricity under preferable conditions or circumstances. This unique property makes it an excellent material to conduct electricity in a controlled manner as required. Unlike conductors, the charge carriers in semiconductors arise only because of external energy (thermal agitation). It causes a certain number of valence electrons to cross the energy gap and jump into the conduction band, leaving an equal amount of unoccupied energy states, *i.e.* holes. Conduction due to electrons and holes are equally important.

- **Resistivity:** 10^{-5} to 10^{6} Ωm
- **Conductivity:** 10^{5} to 10^{-6} mho/m
- **Temperature coefficient of resistance:** Negative
- **Current Flow:** Due to electrons and holes

Some other important Properties of Semiconductors are:

1. Semiconductor acts like an insulator at Zero Kelvin. On increasing the temperature, it works as a conductor.

2. Due to their exceptional electrical properties, semiconductors can be modified by doping to make semiconductor devices suitable for energy conversion, switches, and amplifiers.
3. Lesser power losses.
4. Semiconductors are smaller in size and possess less weight.
5. Their resistivity is higher than conductors but lesser than insulators.
6. The resistance of semiconductor materials decreases with the increase in temperature and vice-versa.

Types of Semiconductors

- **Intrinsic Semiconductor**
- **Extrinsic Semiconductor**

Intrinsic Semiconductor

An intrinsic type of semiconductor material is made to be very pure chemically. It is made up of only a single type of element. Germanium (Ge) and Silicon (Si) are the most common type of intrinsic semiconductor elements. They have four valence electrons (tetravalent). They are bound to the atom by covalent bond at absolute zero temperature. When the temperature rises, due to collisions, few electrons are unbounded and become free to move through the lattice, thus creating an absence in its original position (hole). These free electrons and holes contribute to the conduction of electricity in the semiconductor. The negative and positive charge carriers are equal in number as shown in Figure 2.13(I).

Extrinsic Semiconductor

The conductivity of semiconductors can be greatly improved by introducing a small number of suitable replacement atoms called impurities. The process of adding impurity atoms to the pure semiconductor is called doping. Usually, only 1 atom in 10^7 is replaced by a dopant atom in the doped semiconductor. An extrinsic semiconductor can be classified into:

- ***n*-type Semiconductor**
- ***p*-type Semiconductor**

n-Type Semiconductor

- Mainly due to electrons
- Entirely neutral
- $I = I_h$ and $n_h >> n_e$
- Majority charge carrier are Electrons and Minority charge carrier are Holes.

When a pure semiconductor (Silicon or Germanium) is doped by pentavalent impurity (P, As, Sb) then, four electrons out of five valence electrons bonds with the four electrons of Ge or Si. The fifth electron of the dopant is set free. Thus the impurity atom donates a free electron for conduction in the lattice and is called

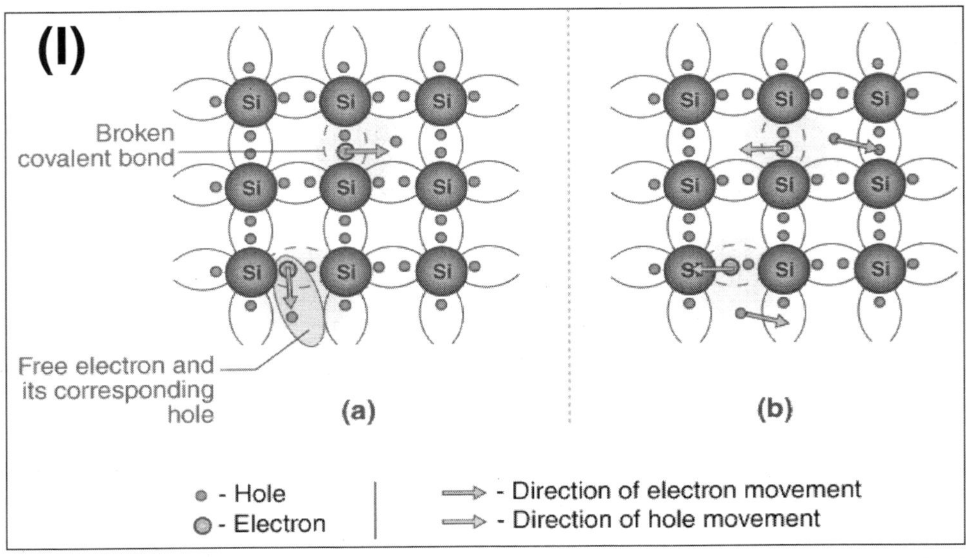

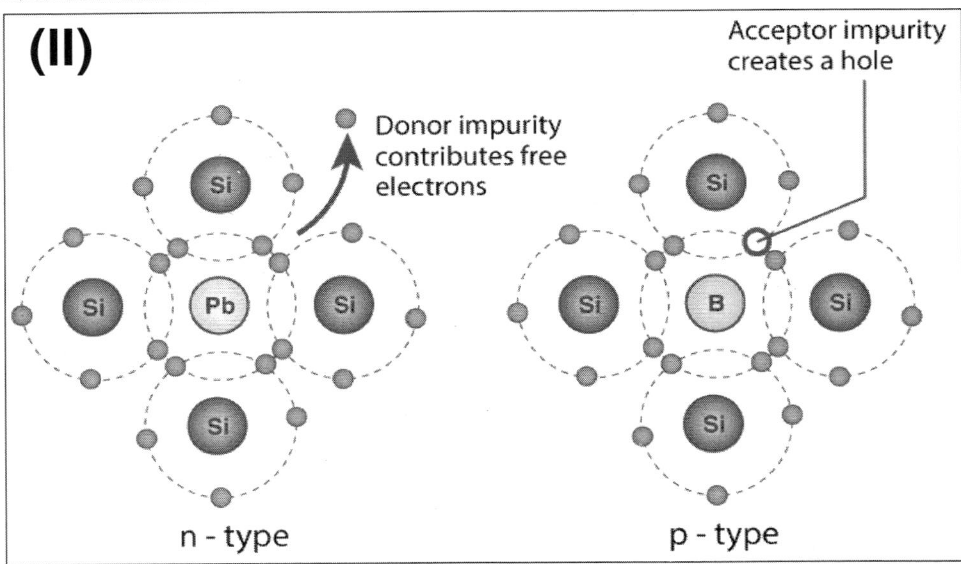

Figure 2.13. (I) Conduction Mechanism in Case of Intrinsic Semiconductors (a) In absence of electric field (b) In presence of electric Field, (II) Classification of Extrinsic Semiconductor.

'**Donar**'. Since the number of free electron increases by the addition of an impurity, the negative charge carriers increase. Hence it is called n-type semiconductor as seen in Figure 2.13(II). Crystal as a whole is neutral, but the donor atom becomes an immobile positive ion. As conduction is due to a large number of free electrons, the electrons as n-type semiconductor are the majority carriers and holes are the minority carriers.

p-Type Semiconductor

- ☆ Mainly due to holes
- ☆ Entirely neutral
- ☆ $I = I_h$ and $n_h >> n_e$ and Majority charge carrier are Holes and Minority charge carrier are Electrons.

Table 2.2. Difference between Intrinsic and Extrinsic Semiconductors

Intrinsic Semiconductor	*Extrinsic Semiconductor*
Pure semiconductor	Impure semiconductor
Density of electrons is equal to the density of holes	Density of electrons is not equal to the density of holes
Electrical conductivity is low	Electrical conductivity is high
Depend on temperature only	Dependence on temperature as well as on the amount of impurity
No impurities	Trivalent impurity, pentavalent impurity

When a pure semiconductor is doped with a trivalent impurity (B, Al, In, Ga) then, the three valence electrons of the impurity bonds with three of the four valence electrons of the semiconductor. This leaves an absence of electron (hole) in the impurity. These impurity atoms which are ready to accept bonded electrons are called '**Acceptors**'. With the increase in the number of impurities, holes (the positive charge carriers) are increased. Hence, it is called p-type semiconductor. Crystal as a whole is neutral, but the acceptors become an immobile negative ion. As conduction is due to a large number of holes, the holes in the *p*-type semiconductor are majority carriers and electrons are minority carriers (Grangvist, C.G., *et al.*, 1977).

Size and Shape Dependent Optical Properties

One of the most fascinating and useful aspects of nanomaterials is their optical properties. Applications based on optical properties of nanomaterials include optical detectors, sensors, imaging, phosphors, solar cell, photocatalysis, photoelectron chemistry display and biomedicine.

The optical properties of perovskite nanomaterials depend on parameters such as feature size, shape, surface characteristics, and other variables including doping and interaction with the surrounding environment or other nanostructures. Likewise, shape causes have dramatic influence on the optical properties of metals, perovskite mixed metal nanostructures. Figure 2.14 exemplifies the difference in the optical properties of metal and semiconductor nanoparticles. In the CdSe semiconducting nanoparticles, a simple change in size alters the optical properties of the nanoparticles. When metal nanoparticles are enlarged, their optical properties

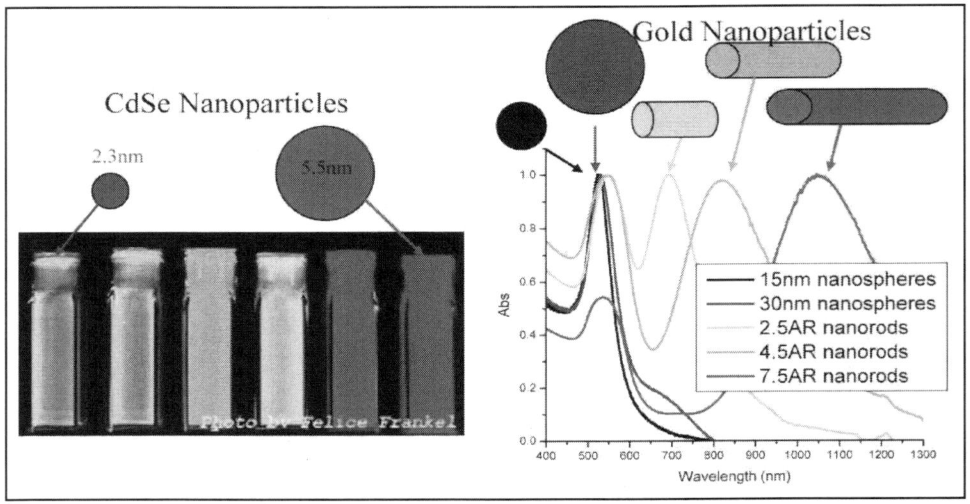

Figure 2.14. Fluorescence Emission of CdSe Quantum Dots of Various Sizes and Absorption Spectra of Various Sizes and Shapes of Gold Nanoparticles (Eustis *et al.*, 2006).

change only slightly as observed for the different samples of gold nanospheres in Figure 2.16. However, when anisotropy is added to these nanoparticles, such as growth of nanorods, the optical properties of nanoparticles change dramatically.

Shape Dependent Optical Properties of Nanostructure Materials

Compounds with the perovskite structure have been used as model systems for spectroscopic studies in the infrared, visible, and ultraviolet region. Perovskite oxides (ABO_3) are particularly suited since the effect of magnetic ordering of the transition metal ion in the B-site can be studied in addition to that of the rare earth ion A. The absorption spectra either for the host ion of the perovskite structure or after incorporation of a low concentration "probe" into a diamagnetic lattice can be studied. In the former case of magnetically concentrated systems, a broad spectrum is obtained, with line widths typically of 0.5-5 cm^{-1}, while in the latter a sharp spectrum is obtained, with line widths of about 0.1 cm^{-1}. The optical properties of many ABO_3 compounds, in which B is a transition metal ion with an empty or partially filled d-shell, have been explored. For the $LaFeO_3$ perovskite containing an empty d-shell transition ion (Fe^{3+}), the emission spectrum shows luminescence which is independent of sample history and contamination and shows a band maximum at about 550 nm, with low quenching temperature (35 K) and a small Stokes shift (7000 cm-1). The Fe octahedral sharing corners in $LaFeO_3$ are favorable for energy band formation. The broadening of the energy levels of the Fe octahedron results in a considerably lower energy for the first absorption transition, and the excited state of the Fe becomes mobile. This situation has been described due to the coexistence of free and delocalized (self-trapped) excitons. The optical transitions involved were shown to be of the metal to- metal charge-transfer type, *e.g.*, Fe^{3+}, Cr^{3+} et., which are responsible for the brownish colour of the composition.

The identification of the nature and structure of the exposed sites on the perovskite surfaces as well as the possible involvement of these sites in catalytic reactions have frequently been studied by the adsorption of suitable probe molecules. Both the equilibrium and the kinetics of adsorption and, in some cases, the desorption of molecules from the surface were analyzed. Optical properties of the perovskite oxide materials were investigated by UV-Vis diffuse reflectance spectroscopy as shown in Figure 2.15. The energy band features of perovskite oxide materials are considered the key factor in determining its photocatalytic activity. It can be observed from Figure 2.15 that the six different morphologies of $LaFeO_3$ nanostructures have similar and strong visible light absorption ability. The absorption cut-off wavelength in the range 580 to 620 nm indicates that the catalyst exhibits a typical optical absorption of perovskite oxide from UV to 620 nm. In perovskite-type oxides the strong absorption edge can principally be attributed to the electronic transition from the valance band to conduction band, O2p Fe 3d. The diffuse reflectance spectrum of $LaFeO_3$ allows us to calculate the optical band gap based on the absorption edge using the formula, Band gap: $E_g = 1240/\lambda_g$ (Yang, H.G., *et al.*, 2004).

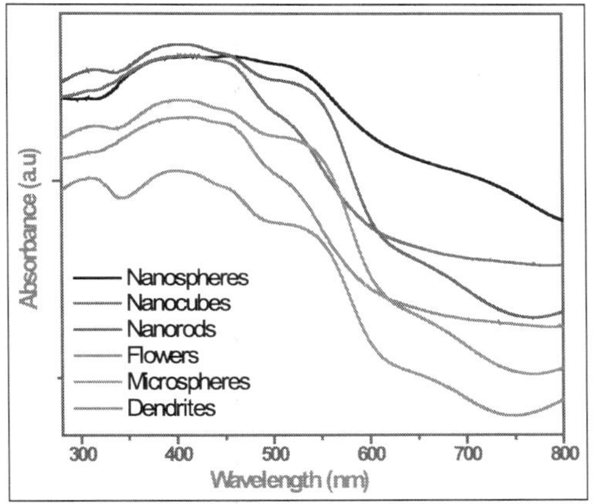

Nanostructures	*Cut-off Wavelength (nm)*	*Bandgap Energy (eV)*
Nanocubes	582	2.14
Flowers	591	2.10
Dendrites	601	2.07
Nanorods	609	2.05
Microspheres	614	2.03
Nanospheres	619	2.00

Figure 2.15. UV-Vis DRS of different $LaFeO_3$ Nanostructures and Table 2.3 Representative Corresponding band gap energy (eV).

From the absorption measurements, the onset absorption wavelength and the corresponding bandgap energies of $LaFeO_3$ nanostructures such as nanocubes, flowers, dendrites, nanorods, microspheres and nanospheres are 2 - 2.27 eV, respectively as shown in Table 2.3. It has been reported in literature that the photocatalytic activity of catalysts depended on many factors, such as crystallinity, surface properties, absorption and optical properties. Under visible light irradiation, the photocatalyst absorbs light above 400 nm, which helps in the formation of photogenerated electron-hole pairs in semiconductors (Ding, J., *et al.*, 2010). Therefore, $LaFeO_3$ nanostructures possess suitable band gap that can be activated by visible light for photocatalytic decomposition of organic contaminants.

Application of Nanostructured Optical Materials

The optical materials, components and systems are used to enhance existing capabilities and enable new capabilities. The major products include IR countermeasures, conformal optics and directed energy weapons, components for maritime platforms, components for submarines, components for IR simulators, and components to enable compact EO/IR imaging systems

REFERENCES

1. Coìrdoba-Torres, P., Mesquita, T. J., and Nogueira, R. P. (2015). Relationship between the origin of constant-phase element behavior in electrochemical impedance spectroscopy and electrode surface structure. *The Journal of Physical Chemistry C, 119*(8), 4136-4147.
2. Ding, J., Lü, X., Shu, H., Xie, J., and Zhang, H. (2010). Microwave-assisted synthesis of perovskite $ReFeO_3$ photocatalyst. *Materials Science and Engineering: B, 171*(3), 31-34.
3. Dwyer, R. P., Smieska, L. M., Tirmzi, A. M., and Marohn, J. A. (2017). Vector electric field measurement via position-modulated Kelvin probe force microscopy. *Applied Physics Letters, 111*(17), 173106.
4. Eustis, S., and El-Sayed, M. A. (2006). Why gold nanoparticles are more precious than pretty gold: Its enhancement of the radiative and nonradiative properties of nanocrystals of different shapes. *Chemical society reviews, 35*(3), 209-217.
5. Folven, E., Tybell, T., Scholl, A., Young, A., Retterer, S. T., Takamura, Y., and Grepstad, J. K. (2010). Antiferromagnetic domain reconfiguration in embedded LaFeO3 thin film nanostructures. *Nano letters, 10*(11), 4578-4583.
6. Fonseca, F. C., Ferlauto, A. S., Alvarez, F., Goya, G. F., and Jardim, R. F. (2005). Morphological and magnetic properties of carbon–nickel nanocomposite thin films. *Journal of Applied Physics, 97*(4), 044313.
7. Granqvist, C. G., and Hunderi, O. (1977). Optical properties of ultrafine gold particles. *Physical Review B, 16*(8), 3513.
8. Huang, S., Ruan, K., Lv, Z., Zhuang, L., Wei, P., Wu, H., and Cao, L. (2006). Magnetic and transport properties in layered $Nd_{1-x}Sr_{1+x}$ CoO 4. *Physical Review B, 73*(9), 094431.
9. Iqbal, M. J., and Ahmad, Z. (2008). Electrical and dielectric properties of lithium manganate nanomaterials doped with rare-earth elements. *Journal of Power Sources, 179*(2), 763-769.
10. Jonscher, A. K. (1992). The universal dielectric response and its physical significance. *IEEE Transactions on Electrical Insulation, 27*(3), 407-423.
11. Jorcin, J. B., Orazem, M. E., Pébère, N., and Tribollet, B. (2006). CPE analysis by local electrochemical impedance spectroscopy. *Electrochimica Acta, 51*(8-9), 1473-1479.

12. Kimura, T., Goto, T., Shintani, H., Ishizaka, K., Arima, T. H., and Tokura, Y. (2003). Magnetic control of ferroelectric polarization. *nature, 426*(6962), 55.

13. Kodama, R. H., Makhlouf, S. A., and Berkowitz, A. E. (1997). Finite size effects in antiferromagnetic NiO nanoparticles. *Physical Review Letters, 79*(7), 1393.

14. Kriener, M., Zobel, C., Reichl, A., and Lorenz, T. (2004). Structure, magnetization, and resistivity of La 1– x M x CoO 3 (M= Ca, Sr, and Ba). *Physical Review B, 69*(9), 094417.

15. Kuznetsov, A. I., Miroshnichenko, A. E., Brongersma, M. L., and Luk'yanchuk, B. (2016). Optically resonant dielectric nanostructure. *Science, 354*(6314), aag2472.

16. Lee, S. J., Jeong, J. R., Shin, S. C., Kim, J. C., and Kim, J. D. (2004). Synthesis and characterization of superparamagnetic maghemite nanoparticles prepared by coprecipitation technique. *Journal of Magnetism and Magnetic Materials, 282*, 147-150.

17. Louis, L., Gemeiner, P., Ponomareva, I., Bellaiche, L., Geneste, G., Ma, Dkhil, B. (2010). Low-symmetry phases in ferroelectric nanowires. *Nano letters, 10*(4), 1177-1183.

18. Macdonald, J. R., and Barsoukov, E. (2005). Impedance spectroscopy: theory, experiment, and applications. *History, 1*(8), 1-13.

19. Malki, M., Hoo, C. M., and Schneider, H. (2014). Electrical conductivity of mullite ceramics. *Journal of the American Ceramic Society, 97*(6), 1923-1930.

20. Mazumder, R., Sujatha Devi, P., Bhattacharya, D., Choudhury, P., Sen, A., and Raja, M. (2007). Ferromagnetism in nanoscale BiFeO3. *Applied Physics Letters, 91*(6), 062510.

21. Roldán, M. V., Scaffardi, L. B., de Sanctis, O., and Pellegri, N. (2008). Optical properties and extinction spectroscopy to characterize the synthesis of amine capped silver nanoparticles. *Materials Chemistry and Physics, 112*(3), 984-990.

22. Thirumalairajan, S., Girija, K., and Mastelaro, V. R. (2015). Investigation on magnetic and electric properties of morphologically different perovskite $LaFeO_3$ nanostructures *Journal of Materials Science: Materials in Electronics, 26*(11), 8652-8662.

23. Vijayanand, S., and Joy, P. A. (2009). Origin of high room temperature ferromagnetic moment of nanocrystalline multiferroic $BiFeO_3$. *Applied physics letters, 94*(18), 182507.

24. Winkler, E., Zysler, R. D., Mansilla, M. V., and Fiorani, D. (2005). Surface anisotropy effects in NiO nanoparticles. *Physical Review B, 72*(13), 132409.

25. Wu, J., Nan, C. W., Lin, Y., and Deng, Y. (2002). Giant dielectric permittivity observed in Li and Ti doped NiO. *Physical review letters, 89*(21), 217601.

26. Yang, H. G., and Zeng, H. C. (2004). Preparation of hollow anatase TiO_2 nanospheres via Ostwald ripening. *The Journal of Physical Chemistry B, 108*(11), 3492-3495.

Chapter 3

Mechanical and Thermal Properties of Nanomaterials

Pon. Sathya Moorthy

Department of Nano Science and Technology, Tamil Nadu Agricultural University, Coimbatore – 641 003, Tamil Nadu

e-mail: sathyanindia2020@gmail.com

Introduction

The material with the dimensions in the range of 0.1 nm to 100 nm and tolerance of 0.1 nm or specific surface area > 60 m^2/gm is known as nanomaterials. When a bulk material is brought to nano scale by any of the method of synthesis the change in property of the material with bulk material is observed obviously. Distinct property of the material is due (i) reduction in size of the material to nano scale, (ii) The particle in nanoregime exhibits a high surface to volume ratio as shown in Figure 3.1 and Table 3.1. Change in property is observed in (i) Chemical properties – reactivity, catalysis, *etc.* (ii) Thermal properties – Melting temperature, Lattice

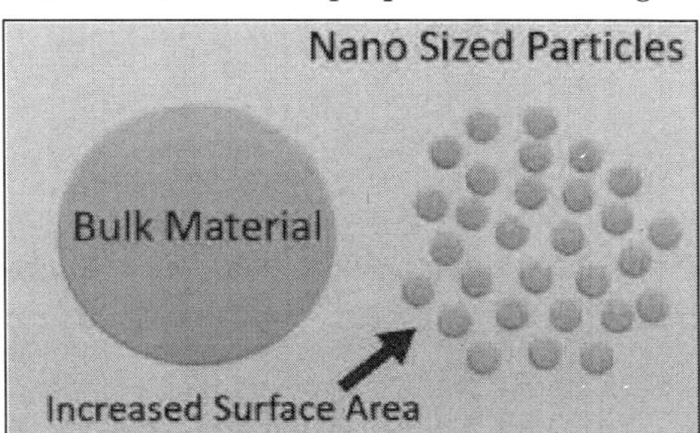

Figure 3.1. Schematic Diagram of Nano Particles Exhibiting Increased Surface Area.
Courtesy: J. Appl. Emerg. Sci., 2017, 7(1), 12-21.

constant *etc.* (iii) Mechanical properties – Hardness, Fracture strength, Density *etc.*, Optical properties – absorption and scattering of light, *etc.* (iv) Electrical properties – tunnelling current, *etc.* (6) Magnetic properties – Superparamagnetic effect, Giant Magnetic Resonance, *etc.*

Table 3.1. Showing the Size of the Cube and its Equivalent Collective Surface Area

Size of Cube	*Number of Cube*	*Collective Surface Area*
1M	1	$6M^2$
0.1 M	10	$60M^2$
0.01M	10^6	$600M^2$
0.001M	10^9	$6000M^2$
10^{-9}M	10^{27}	$6000KM^2$

Mechanical Properties of Nanomaterials

Fullerene was discovered in 1985 by Sir Harold W. Kroto of the United Kingdom and by Richard E. Smalley and Robert F. Curl, Jr., of the United States. LASER was used to vaporize graphite rods in an atmosphere of helium gas, to obtained cage like molecules composed of 60 carbon atoms (C) joined together by single and double bonds to form a hollow sphere with 12 pentagonal and 20 hexagonal faces–a design that resembles a football, or soccer ball as shown in the Figure 3.2. Each carbon atom is bonded to three others and is sp^2 hybridized. Sir Harold W. Kroto, Richard E. Smalley and Robert F. Curl was awarded noble prize in the year 1996 for the discovery of new allotrope of carbon called as fullerene. Diameter of the C60 molecule is $7Å = 7.0 \times 10^{-10}$m. The different properties of graphite and diamond are a consequence of the different way the carbon atoms in the materials are bonded together. Other than the crystalline nature of carbon namely diamond, and amorphous nature of carbon namely graphite, third new form of carbon with remarkable electrical conductivity and mechanical strength is named as fullerene. 1 dimensional fullerenes or carbon nano tubes were identified by SumioLijimaof Japan in the year 1991.

Some nanomaterials have inherent exceptional mechanical strength. One such material is carbon nanotubes: these are extremely small tubes having the same honeycomb structure of graphene, but with different properties compared to graphite and diamond. They can be single-walled or multi-walled as illustrated in the Figure 3.3.

Carbon nanotubes are 100 times stronger than steel but six times lighter. The solids consists of tiny tubes made up of numerous concentric "graphene" cylinders, each cylinder wall consisting of a sheet of carbon atoms arranged in hexagonal rings. High-resolution transmission electron microscopy later revealed that these multiwalled carbon nanotubes (MWNTs) are seamless and that the spacings between adjacent layers is about 0.34 nm , close to the spacing observed between sheets of graphite. The number of concentric cylinders in a given tube ranged from 3 to 50 nm, and the ends were generally capped by fuilerene domes that included pentagonal

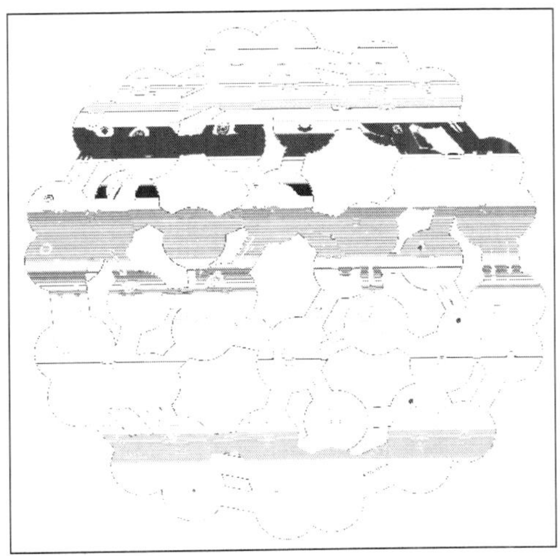

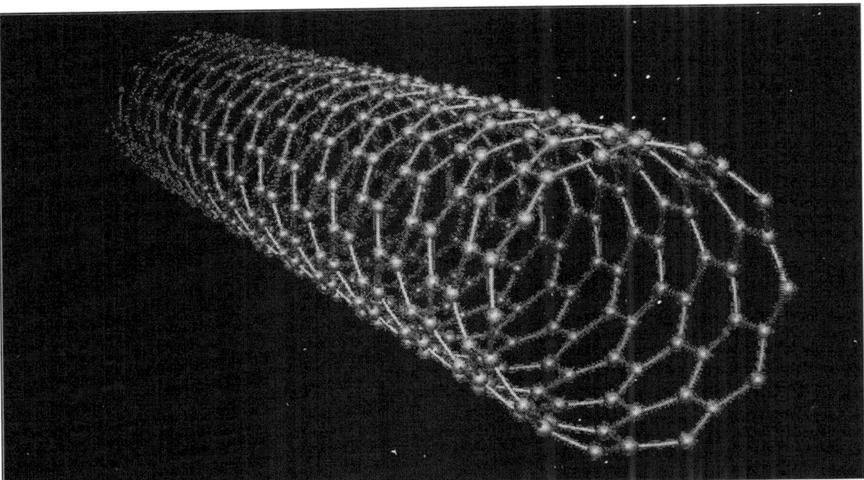

Figure 3.2. Schematic Representation of Fullerene and Carbon Nanotubes.
Courtesy: http://www.polifab.polimi.it/seminar-on-carbon-nanotubes/
https://www.cheaptubes.com/product/carbon-fullerenes-c60/

rings (B.C. Yadav and Ritesh Kumar, 2008). It was soon shown that single-walled nanotubes (SWNTs) could be produced by this method if a cobalt-nickel catalyst was used. In 1996 a group led by Smalley produced SWNTs in high purity by laser vaporization of carbon impregnated with cobalt and nickel. TEM image of SWNT and MWNT are represented in Figure 3.4. These nanotubes are essentially elongated fullerenes. Physical properties of C60 (fullerene) is Density : 1.65 g cm-3, Standard heat of formation : 9.08 k cal mol^{-1}, Index of refraction : 2.2 (600nm), Boiling point : Sublimes at 800K, Resistivity : 1014 ohms m^{-1}, Vapor density : N/A, Crystal form Hexagonal cubic, Vapor pressure : 5 x 10^{-6} torr at room temperature : 8 x 10^{-4}torr at

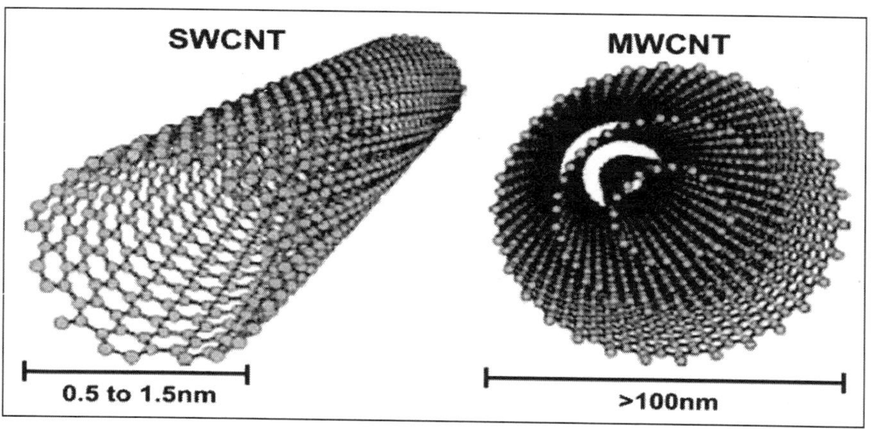

Figure 3.3. Graphene and Carbon Nanotubes as Single Wall Carbon Nanotube (SWCNT) and Multi-wall Carbon Nanotube (MWCNT) Structures.
Courtesy: Frontiers in Systems Neuroscience, 2014, 8(91): 91.

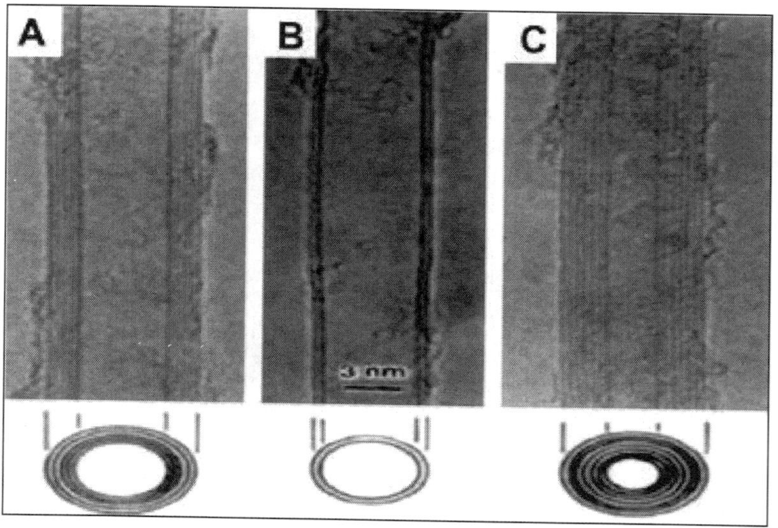

Figure 3.4. TEM Image of SWNT and MWNT.

800K, Organoleptic properties : Appearance, Bucky ball soot : Very finely divided black powder, Fullerite : Brown/black powder, C60 : Black solid, Odour : Odourless.

Comparison of tensile strength, and Young's modulus of SWNT and MWNT are listed in the Table 3.2 (Ayesha Kausar *et al.*, 2017)

Young'smodulus and the ultimate tensile strength of carbon-based nanostructures can reach as high as 670GPa and 70GPa,respectively, which are much superior to those of previously reported carbon-based nanostructures (Ayesha Kausar *et al.*, 2017).

Table 3.2. Comparison of Tensile Strength, and Young's Modulus of SWNT and MWNT

Material	Young's modulus (GPa)	Tensile strength (GPa)
Epoxy	3.500	0.005
SWCNT	1054.000	150.000
MWCNT	1200.000	150.000
Steel	208.000	0.400
Wood	16.000	0.008

SWCNT, single-walled carbon nanotube; MWCNT, multi-walled carbon nanotube.

Young's Modulus

Thomas young a scientist had explained the concept Young's modulus in the year 1769. It is a measure of stiffness of the solid material. Elastic property of the material is explained using the stress and strain. Stress is force per unit area causing the deformation. Strain is defined as change in dimension to its original dimension of the material. Young's modulus or modulus of elasticity of the material is defined as the ratio of stress to strain is constant provided the material is within elastic limit this is also called as Hook's law. The schematic representation of young's modulus with respect to stress and strain are represented in the Figure 3.5.

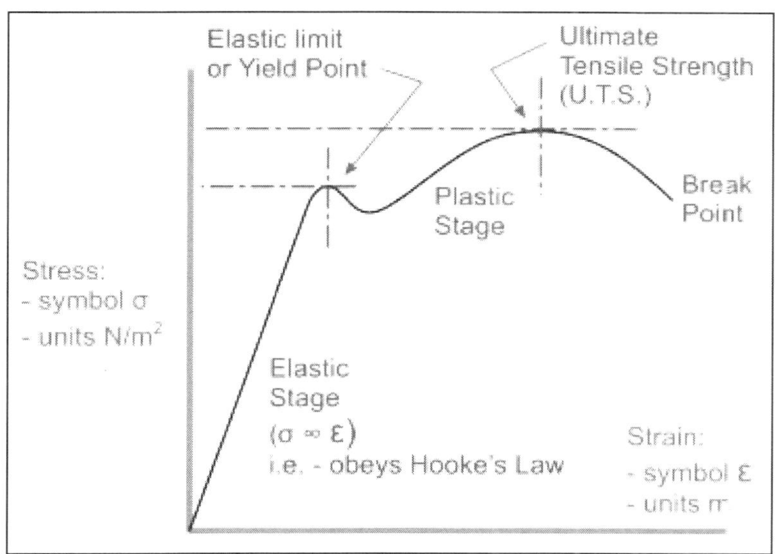

Figure 3.5. Schematic Representation of Young's Modulus or Modulus of Elasticity.
Courtesy: http://wiki.dtonline.org/index.php/Young per cent 27s_Modulus

If an external force is applied to the material due to the application of the external force there is a change in dimension is observed either due to compression or expansion of the material. When the same external force is withdrawn the material will set back to its original dimension until the material is in elastic limit. If the force is continued to be applied beyond this phase, the material reaches a point at which it stays permanently stretched. This is the Yield Point or Elastic Limit and beyond this, during its Plastic Stage, the material will continue to remain permanently deformed until eventually it breaks. Much can be determined about a material from its Stress/Strain Curve - for example ductile materials will exhibit an extended plastic stage whereas materials would have little or none at all. In short:

- A stiffer material will exhibit a steeper slope
- A brittle material will lack a plastic region
- A stronger material will show a higher ultimate tension strength

The shape of the Stress/Strain curve therefore changes but the area under it remains the same because the work done to fracture is constant.

Nano scale materials exhibit the higher elasticity due to absence of defects or because of unstable or mobile dislocations. Due to the unstable or absence of dislocations the grain of nano crystal remains with enhanced internal and external perfection. The crystal may contain imperfections at the grain boundaries where the lattice orientation changes beyond 10 nm. Smaller the grain size the imperfections are observed ant the interfaces whereas in larger crystals the combination of sliding and intra-grain dislocations activity is observed. Hardness is the measure of resistance of the material to the various kinds of permanent shape change when a compressive force is applied. Fracture toughness is a property of materials resistance to crack or fracture. The Hall–P*etc*h relationship tells us that we could achieve strength in materials that is as high as their own theoretical strength by reducing grain size. Indeed, their strength continues to increase with decreasing grain size to approximately 20–30 nm. Indeed, the peak yield strength of pure nanostructured copper with grain size approaching approximately 20 nm can reach as high as 800–900 MPa compared to 200 MPa for coarse-grained copper. But decreasing grain size beyond 20 nm reverses the H–P effect: in other words the material start to soften instead of further strengthening is known as Inverse Hall Petcheffect.

$$\text{Yield strength} = \sigma_{TS} = \sigma_0 + \frac{K_{TS}}{\sqrt{d}}$$

In general, nanostructured metals are characterized as having very high strength with poor ductility. In other words, the strength increase trades off with ductility in nanostructured metallic materials obviously demonstrated in Figure 3.6.

$$\text{Hardness} = H = H_0 + \frac{K_H}{\sqrt{d}}$$

Where, K_{TS} is the strengthening coefficient (a constant unique to each material), σ_o is a materials constant for the starting stress for dislocation movement (or the resistance of the lattice to dislocation motion), d is the grain diameter, and σ_y is the yield stress. Yield strength and Hardness of the materials are inversely proportional to the grain size. As the grain size increases, the decrease in yield strength and hardness of the materials decreases.

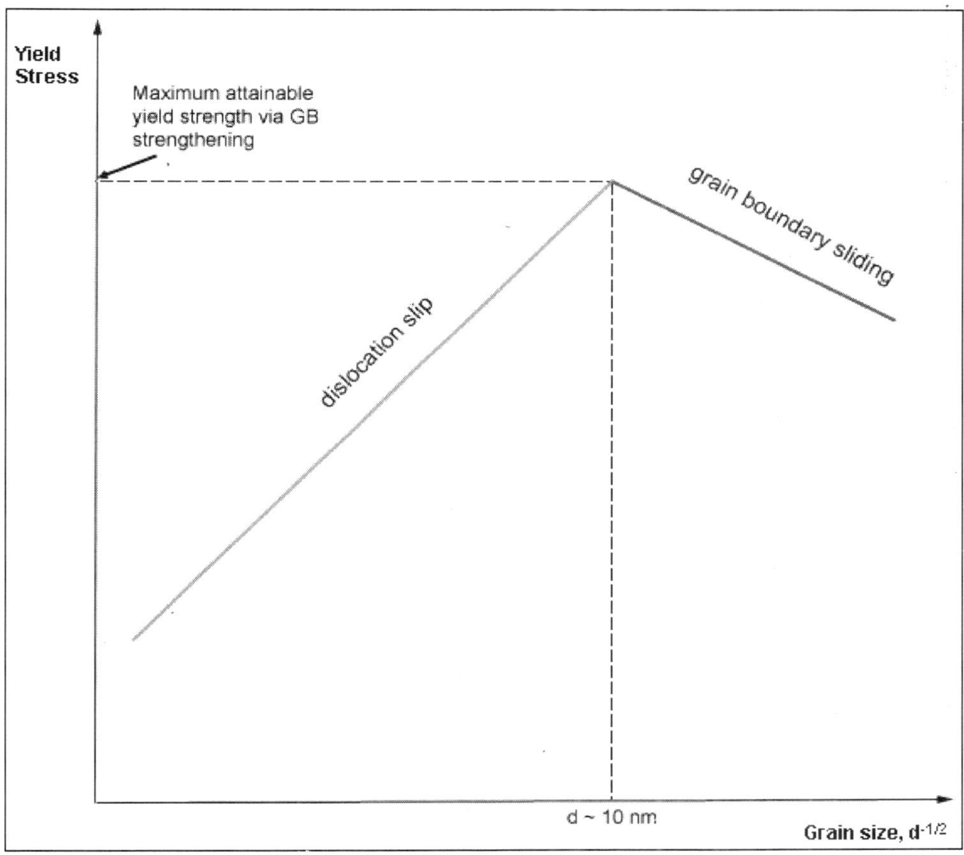

Figure 3.6. Schematic Representation of Hall-Petch Strengthening Limit.
Courtesy: https://en.wikipedia.org/wiki/Grain_boundary_strengthening

Thermal Properties

Soon after the discovery of carbon nanotubes (CNTs) in 1991 by Sumio Lijima, numerous researches has been initiated in order to study its inherent property of the material. Many scientists reported the physical, chemical, mechanical and electric properties of both Single walled nano tubes (SWNTs) and Multiwalled nanotubes (MWNTs). Thermal property of the CNTs remains unexplored till 2006 and thermal conductivity of MWCNTs exhibit longitudinal thermal conductivity of 3000W/m K and 2000W/m K for SWCNTs for a single nanotube at room temperature (Kim *et al.*, 2001, Yu *et al.*, 2005). Thermal conductivity of SWNTs and MWNTs is comparable to diamond and higher than graphite and carbon fibers, as well as aspect ratio usually in the order 10^3 (Guthy *et al.*, 2007). SWCNTs are significantly smaller in diameter compared to MWCNTs and the thermal properties may differ significantly. Similar to other non-metallic materials, the transport of thermal energy in CNTs is assumed to occur *via* a phonon conduction mechanism. The phonon conduction in nanotubes is influenced by several processes such as the number of phonon active modes, the boundary surface scattering, the length of the free path for the phonons and inelastic Umklapp-scattering (an anharmonic phonon–phonon or electron–phonon scattering process). The thermal conductivity of CNTs depends on atomic arrangement (*i.e.*, how the sheets of graphite are "rolled"), the diameter and length of the tubes, the number of structural defects and the morphology, as well as on the presence of impurities as shown in Figure 3.7. SWCNTs are significantly smaller in diameter compared to MWCNTs and the thermal properties may differ significantly. The atomic structure of nanotubes is described in terms of the tube chirality, or helicity, defined by the chiral vector (ch = na1 + ma2) and the chiral angle. Thus, the structure of CNT can be specified by (n, m), for arm chair (n = m), zig–zag (n = 0 or m = 0), or chiral (all others).

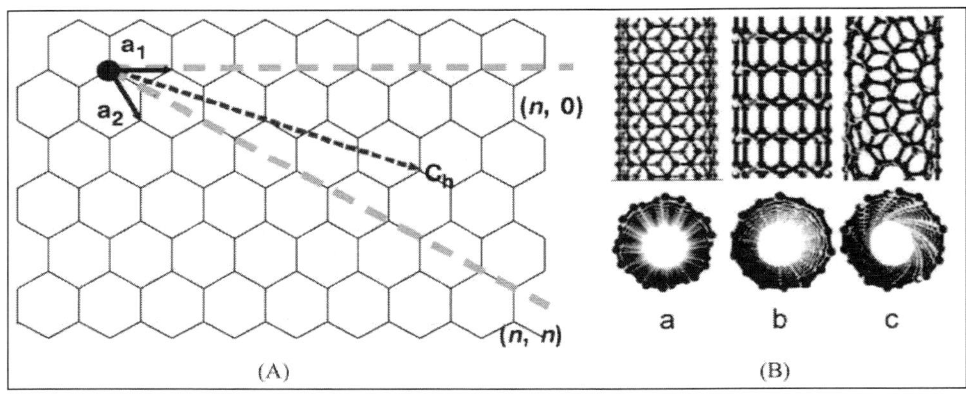

Figure 3.7. Schematic Model of (a) An armchair nanotube, (b) A zig-zag nanotube, and (c) a chiral nanotube
Courtesy: Polymer Science 36 (2011) 914–944.

Mechanism of heat conduction may occur due to phonons or electrons movement critically depends on the band gaps of materials. Whereas, heat transfer mechanism of CNTs is strongly depends on the chirality, which determines the size of their band gaps and electronic properties. The largest band gap (on the order of 1.5 eV) is found with chiral nanotubes exhibiting semiconductor nature and the conduction mainly governed by the phonon (Odom *et al.*, 2000). In case of armchair nanotubes(n = m), the band gap is considerably smaller in the case of armchair nanotubes exhibiting metallic nature (Ando, 2004). Thus, the electronic contribution to the thermal conductivity will be significant in metallic CNTs with a small band gap.

The phonon mean free paths are thought to be relatively long in nanotubes, 500nm for MWNTs and even longer for SWNTs. It is well known that CNTs are characterized by a large aspect ratio and a huge surface area. Diameter and length are two key parameters to describe CNTs and directly affect the thermal conductivity of both CNTs and composites containing CNTs.

The thermal conductivity should be higher for SWCNTs with smaller diameters. The thermal conductivity at 300 K was approximately inversely proportional to the diameter of SWCNT. Fujii*et al.* (2005) measured the thermal conductivity of a single MWCNT using a suspended sample-attached T-type nano sensor. The thermal conductivity of MWCNT at room temperature increased as its diameter decreased, *i.e.*, the thermal conductivity increased as the number of walls decreased, varying from about 500W/m K for an outer diameter of 28 nm to 2069W/m K for a 10 nm diameter (Cao *et al.*, 2004).

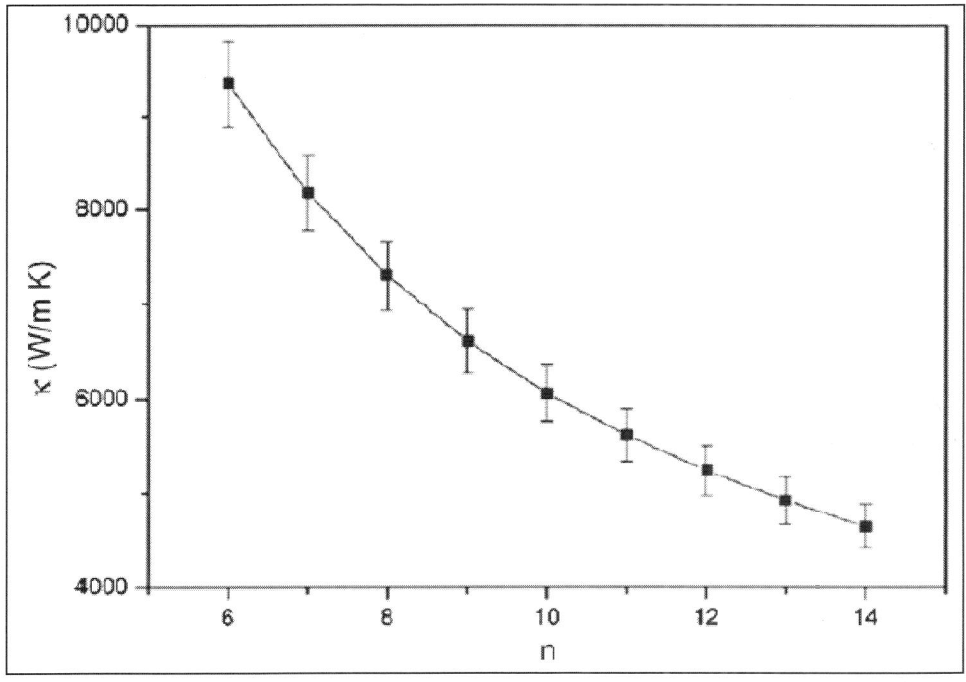

Figure 3.8. Thermal Conductivity of (n, 0) SWCNTs (n = 6–14) at 300 K.
Courtesy: Polymer Science 36 (2011) 914–944

Simulation results predict an effect of CNT length on its intrinsic thermal conductivity. By the molecular dynamics method, the thermal conductivity of SWCNTs with chiral indices (5, 5) and (10, 10) was found to increase as shown in the Figure 3.9.

The length dependence of the thermal conductivity at room temperature was also revealed for SWCNTs with different chiral indices: (5, 5), (10, 0), (7, 7), (10, 10),(17, 0), (15, 15) by using reverse non-equilibrium molecular dynamics simulations

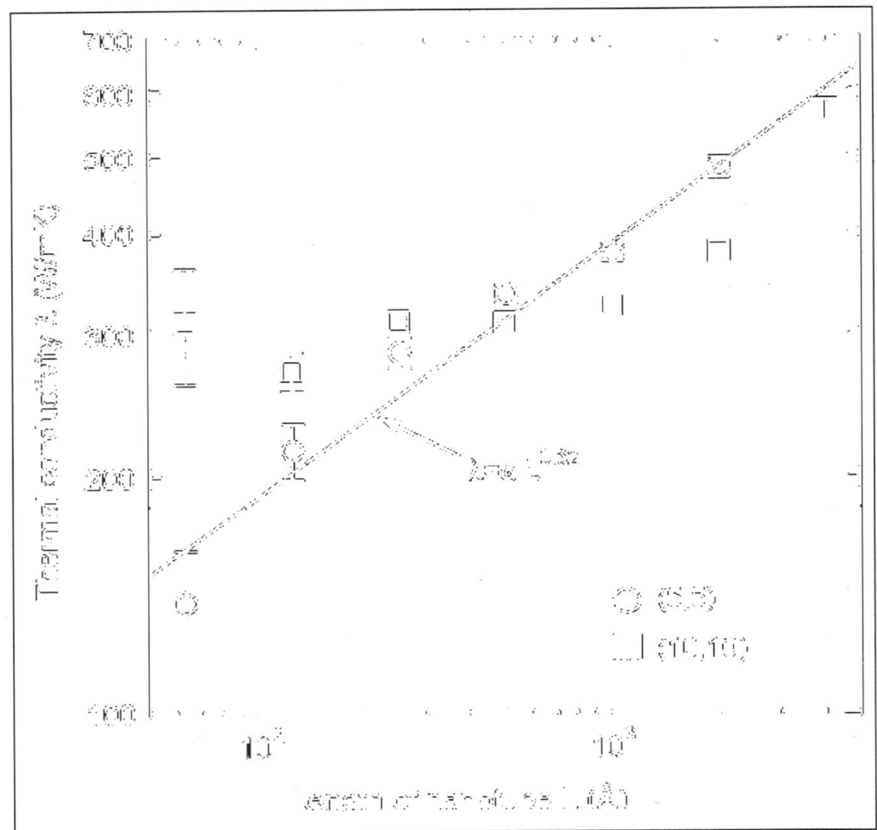

Figure 3.9. Dependence of Thermal Conductivity on Length of CNT at 300K.
Courtesy: Polymer Science 36 (2011) 914–944.

(Figure 3.10). For length parameters(L) from 5 to 350 nm, the calculated thermal conductivity increased with increasing tube length, with values between 0.54 (100nm< L < 350 nm)and 0.77 (L < 25 nm). These phenomena were explained by the variable ratio between the phonon mean free path and the CNT length. Based on these results, the thermal conductivity is expected to become constant when the tube length is much longer than the mean free path of the energy-carrying phonons.

It has been well established both experimentally and theoretically that the melting temperature (*T*) of nanoparticles depends on the particle size. When a bulk material is brought to nano scale the change in melting point of the materials is observed due to the breakage of bonds and shortening of lattice parameters as shown in the Figure 3.11 and Table 3.3.

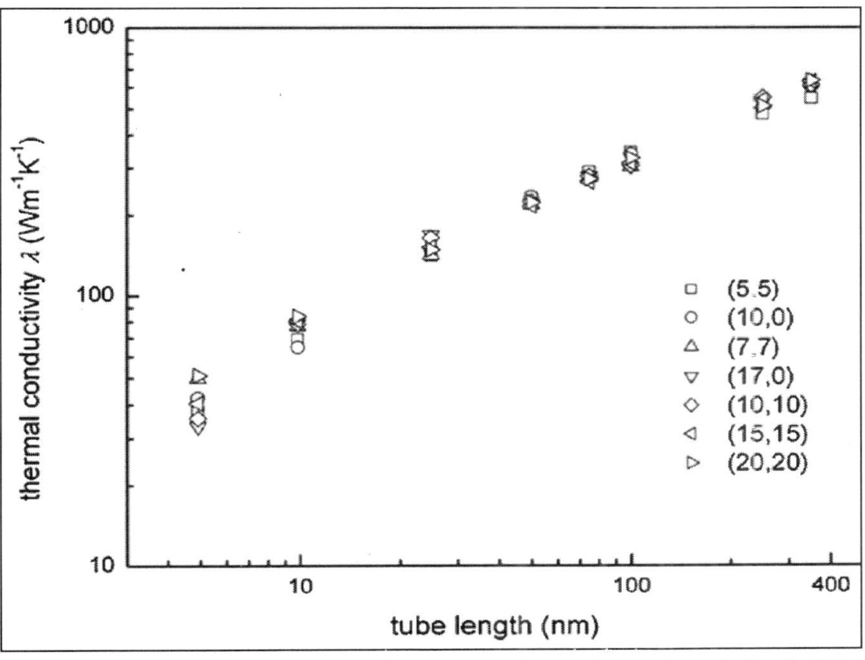

Figure 3.10. Thermal Conductivity versus Tube Length of (5, 5), (10, 0), (7, 7), (10, 10), (17, 0), (15, 15), and (20, 20) SWCNTs at 300K in a Double-Logarithmic Scale.

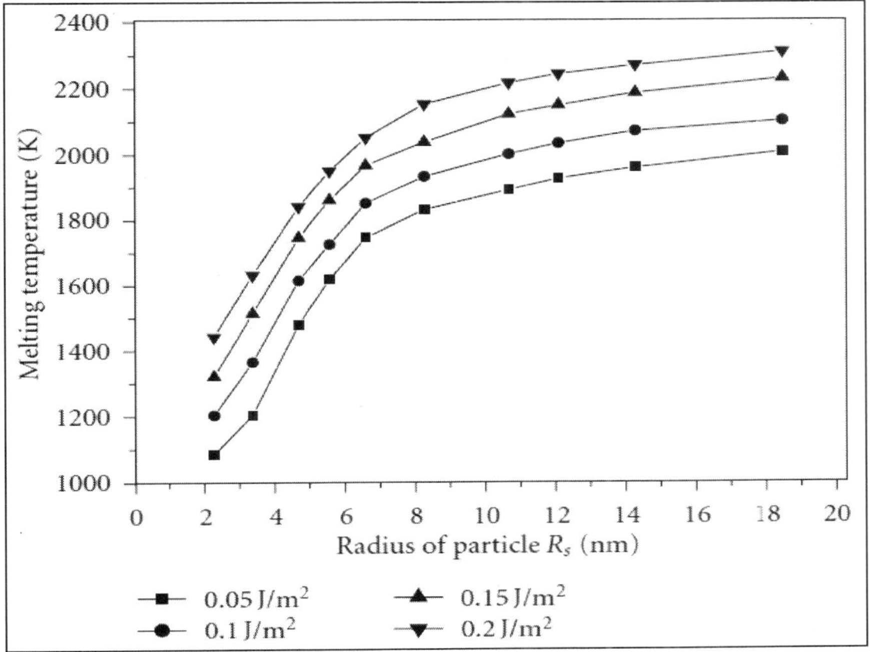

Figure 3.11. Variation of Melting Point of GaN Spherical Nanoparticles against the Size of the Particles.

Table 3.3. Variation of Melting Point of GaN Spherical Nanoparticles against the Size of the Particles

S. no.	R_s (nm)	$1/R_s$ (nm)$^{-1}$	T_{GN} (K)			
			$\gamma_{lb} = 0.05$ J/m^2	$\gamma_{lb} = 0.10$ J/m^2	$\gamma_{lb} = 0.15$ J/m^2	$\gamma_{lb} = 0.20$ J/m^2
1	2.28	0.439	1085.07	1203.18	1321.29	1443.12
2	3.38	0.296	1204.39	1365.48	1514.15	1632.27
3	4.69	0.213	1480.29	1615.33	1747.07	1841.65
4	5.58	0.179	1618.64	1726.84	1861.88	1949.85
5	6.61	0.151	1747.07	1851.97	1966.78	2051.33
6	8.27	0.121	1831.74	1932.92	2037.81	2152.92
7	10.68	0.094	1892.44	2000.64	2122.06	2216.63
8	12.10	0.083	1926.31	2034.11	2149.32	2243.89
9	14.29	0.069	1959.76	2071.26	2186.48	2270.73
10	18.48	0.054	2007.25	2101.82	2230.26	2311.21

Courtesy: Journal of Nanomaterials, Volume 2012, Article ID 415797, 11 pages.

REFERENCES

1. Ando, T. (2003). Quantum anomalies in carbon nanotubes. *Solid state communications, 127*(2), 69-78.
2. Ando, T. (2004). Carbon nanotubes and exotic transport properties. *Physica E: Low-dimensional Systems and Nanostructures, 22*(1-3), 656-661.
3. Antoniammal,P., and Arivuoli, D. (2011). A thermo dynamical modelfor the shape and size effect on melting of boron carbide nanoparticles. *Journal of Nanoscience and Nanotechnology,*11, 1–8.
4. Biercuk, M. J., Llaguno, M. C., Radosavljevic, M., Hyun, J. K., Johnson, A. T., and Fischer, J. E. (2002). Carbon nanotube composites for thermal management. *Applied physics letters, 80*(15), 2767-2769.
5. Buffat, P.H., and Borel, J.P. (1976). Size effect on the melting temperatureof gold particles.*Physical Review A,* 13(6), 2287–2298.
6. Choi, E. S., Brooks, J. S., Eaton, D. L., Al-Haik, M. S., Hussaini, M. Y., Garmestani, H. and Dahmen, K. (2003). Enhancement of thermal and electrical properties of carbon nanotube polymer composites by magnetic field processing. *Journal of Applied physics, 94*(9), 6034-6039.
7. Choi, S. U. S., Zhang, Z. G., Yu, W., Lockwood, F. E., and Grulke, E. A. (2001). Anomalous thermal conductivity enhancement in nanotube suspensions. *Applied physics letters, 79*(14), 2252-2254.
8. Ercolessi, F., Andreoni, W., and Tosatti, E. (1991). Melting of smallgold particles: Mechanism and size effects. *Physical ReviewLetters,* 66(7), 911–914.
9. Fischer, J. E. (2006). Carbon nanotubes: structure and properties. In *Carbon nanomaterials* (pp. 49-83). CRC Press.
10. Grujicic, M., Cao, G., and Gersten, B. (2004). Atomic-scale computations of the lattice contribution to thermal conductivity of single-walled carbon nanotubes. *Materials Science and Engineering: B, 107*(2), 204-216.

11. Guthy, C., Du, F., Brand, S., Winey, K. I., and Fischer, J. E. (2007). Thermal conductivity of single-walled carbon nanotube/PMMA nanocomposites. *Journal of heat transfer*, *129*(8), 1096-1099.
12. Hepplestone, S. P., Ciavarella, A. M., Janke, C., and Srivastava, G. P. (2006). Size and temperature dependence of the specific heat capacity of carbon nanotubes. *Surface science*, *600*(18), 3633-3636
13. Hone, J., Whitney, M., Piskoti, C., and Zettl, A. (1999). Thermal conductivity of single-walled carbon nanotubes. *Physical Review B*, *59*(4), R2514
14. *http: //wiki.dtonline.org/index.php/Young per cent 27s_Modulus*
15. http: //www.polifab.polimi.it/seminar-on-carbon-nanotubes/
16. https: //en.wikipedia.org/wiki/Grain_boundary_strengthening
17. https: //www.cheaptubes.com/product/carbon-fullerenes-c60/
18. Iijima, S. (1991). Helical microtubules of graphitic carbon. *nature*, *354*(6348), 56.
19. Kashchiev, D., and van Rosmalen, G. M. (2003). Nucleation in solutions revisited.*Crystal Research and Technology*, 38 (7), 555–574.
20. Kim, H. S., Chae, Y. S., Park, B. H., Yoon, J. S., Kang, M., and Jin, H. J. (2008). Thermal and electrical conductivity of poly (L-lactide)/multiwalled carbon nanotube nanocomposites. *Current Applied Physics*, *8*(6), 803-806.
21. Kim, P., Shi, L., Majumdar, A., and McEuen, P. L. (2001). Thermal transport measurements of individual multiwalled nanotubes. *Physical review letters*, *87*(21), 215502.
22. Lewis, L. J. P., Jensen, P., and Barrat, J. L. Melting, freezing,and coalescence of gold nanoclusters. *Physical Review B*, 56(4), 2248–2257.
23. Mensah, S. Y., Allotey, F. K. A., Nkrumah, G., and Mensah, N. G. (2004). High electron thermal conductivity of chiral carbon nanotubes. *Physica E: Low-dimensional Systems and Nanostructures*, *23*(1-2), 152-158.
24. Nan, C. W., Shi, Z., and Lin, Y. (2003). A simple model for thermal conductivity of carbon nanotube-based composites. *Chemical Physics Letters*, *375*(5-6), 666-669.
25. Odom, T. W., Huang, J. L., Kim, P., and Lieber, C. M. (1998). Atomic structure and electronic properties of single-walled carbon nanotubes. *Nature*, *391*(6662), 62.
26. Odom, T. W., Huang, J. L., Kim, P., and Lieber, C. M. (2000). Structure and electronic properties of carbon nanotubes.
27. Pawlow, P. (1909). The dependency of the melting point on the surface energy of a solid body. *Zeitschriftf ̈urPhysikalischeChemie*, 65(5), 545–548.
28. Ribeiro, B., Botelho, E. C., Costa, M. L., and Bandeira, C. F. (2017). Carbon nanotube buckypaper reinforced polymer composites: A review. *Polímeros*, *27*(3), 247-255.
29. Ruoff, R. S., and Lorents, D. C. (1995). Mechanical and thermal properties of carbon nanotubes. *Carbon*, *33*(7), 925-930.

30. R.Murugeshan and Er. KiruthigaSivaprasath -Modern physics,(2014) S.Chand and Company pvt. ltd., Ram Nagar, New Delhi – 110055.
31. S.O.Pillai, Solid state physics(1999).New Age International (P) Ltd., New Delhi.
32. Stroscio, M. A., Dutta, M., Kahn, D., and Kim, K. W. (2001). Continuum model of optical phonons in a nanotube. *Superlattices and microstructures, 29*(6), 405-409.
33. Thostenson, E. T., Ren, Z., and Chou, T. W. (2001). Advances in the science and technology of carbon nanotubes and their composites: a review. *Composites science and technology, 61*(13), 1899-1912.
34. Wen, D., and Ding, Y. (2004). Effective thermal conductivity of aqueous suspensions of carbon nanotubes (carbon nanotube nanofluids). *Journal of Thermophysics and Heat Transfer, 18*(4), 481-485.
35. Wronski, C. R. M. (1967). The size dependence of the melting point of small particles of tin. *British Journal of Applied Physics*, 18(12), 1731–1737.
36. Xie, H., Cai, A., and Wang, X. (2007). Thermal diffusivity and conductivity of multiwalled carbon nanotube arrays. *Physics Letters A, 369*(1-2), 120-123.
37. Yamamoto, T., Watanabe, S., and Watanabe, K. (2004). Universal features of quantized thermal conductance of carbon nanotubes. *Physical review letters, 92*(7), 075502.
38. Yu, C., Shi, L., Yao, Z., Li, D., and Majumdar, A. (2005). Thermal conductance and thermopower of an individual single-wall carbon nanotube. *Nano letters, 5*(9), 1842-1846.
39. Zhang, W., Zhu, Z., Wang, F., Wang, T., Sun, L., and Wang, Z. (2004). Chirality dependence of the thermal conductivity of carbon nanotubes. *Nanotechnology, 15*(8), 936.

Chapter 4

Electronic Structure and Surface Chemistry of Nanomaterials

D. Jeya Sundara Sharmila

Department of Nano Science and Technology, Tamil Nadu Agricultural University, Coimbatore – 641 003, Tamil Nadu
e-mail: djssharmila@gmail.com

Introduction to Functional Groups, Bonds and Electronic Structure of Nanomaterials

A functional group is an atom or a group of atoms with characteristic chemical and physical properties. It is the reactive part of the molecule. Most organic compounds have C–C and C–H bonds. However, many organic molecules possess other structural features such as, heteroatoms that are atoms other than carbon or hydrogen and π bonds are the most common bonds occur in C–C and C–O double bonds. These structural features distinguish one organic molecule from another. They determine a molecule's geometry, physical properties, and reactivity, and comprise what is called a functional group. Heteroatoms and π bonds confer reactivity on a particular molecule. Heteroatoms have lone pairs and create electron-deficient sites on carbon. π Bonds are easily broken in chemical reactions. A π bond makes a molecule a base and a nucleophile. The C–C and C–H bonds are important as they form the carbon backbone or skeleton to which the functional group is attached. Hydrocarbons are compounds made up of only the elements carbon and hydrogen (Tables 4.1 and 4.2). They may be aliphatic or aromatic. Aromatic hydrocarbons are so named because many of the earliest known aromatic compounds had strong characteristic odors. The simplest aromatic hydrocarbon is benzene. The six-membered ring and three π bonds of benzene comprise a single functional group. When a benzene ring is bonded to another group, it is called a phenyl group.

Table 4.1. Basic Hydrocarbon Functional Groups

Type of compound	General structure	Example	Functional group
Alkane	R–H	CH_3CH_3	--
Alkene	>C=C<	$H_2C=CH_2$	double bond
Alkyne	–C≡C–	H–C≡C–H	triple bond
Aromatic compound	benzene ring	benzene ring	phenyl group

Compounds Containing the C=O Group is called a "carbonyl group". The polar C–O bond makes the carbonyl carbon an electrophile, while the lone pairs on O allow it to react as a nucleophile and base. The carbonyl group also contains a π bond that is more easily broken than a C–O σ bond. A functional group determines the properties of a molecule, intra molecular bonding, shape, type and strength of intermolecular forces, physical properties, chemical reactions and reactivity.

Intra-Molecular Bonding

A molecule is a group of atoms held together by energy and the holding force is called a chemical bond. There are three kinds of chemical bonds:

1. Ionic bonds
2. Covalent bonds
3. Hydrogen bonds

Ionic Bonds

Ionic bonds form between atoms by the transfer of one or more electrons. Ionic compounds form crystals instead of individual molecules. Example: NaCl (sodium chloride). The key properties of Ionic bonds are

- Strong: But not as strong as covalent bonds
- *Not* directional: They are not formed between particular ions in the compound

Ionic bond between positively charged sodium atom and negatively charged chloride atom (Figure 4.1).

Table 4.2. Organic Compounds with different Functional Groups

Type of compound	General structure	Example	3-D structure	Functional group
Alkyl halide	R–X: (X = F, Cl, Br, I)	CH_3–Br:		–X halo group
Alcohol	R–OH	CH_3–OH		–OH hydroxy group
Ether	R–O–R	CH_3–O–CH_3		–OR alkoxy group
Amine	R–NH_2 or R_2NH or R_3N	CH_3–NH_2		–NH_2 amino group
Thiol	R–SH	CH_3–SH		–SH mercapto group
Sulfide	R–S–R	CH_3–S–CH_3		–SR alkylthio group
Aldehyde	:O: ‖ R–C–H	:O: ‖ CH_3–C–H		C=O carbonyl group
Ketone	:O: ‖ R–C–R	:O: ‖ CH_3–C–CH_3		C=O carbonyl group
Carboxylic acid	:O: ‖ R–C–OH	:O: ‖ CH_3–C–OH		–COOH carboxy group
Ester	:O: ‖ R–C–OR	:O: ‖ CH_3–C–OCH_3		–COOR
Amide	:O: ‖ R–C–N–H (or R), H (or R)	:O: ‖ CH_3–C–NH_2		–$CONH_2$, –CONHR, or –$CONR_2$
Acid chloride	:O: ‖ R–C–Cl:	:O: ‖ CH_3–C–Cl:		–COCl

Covalent Bonds

Covalent bonds are formed by the sharing of two or more electrons (Figure 4.2). Electron sharing produces covalent bonds in molecules and the key properties of Covalent bonds are:

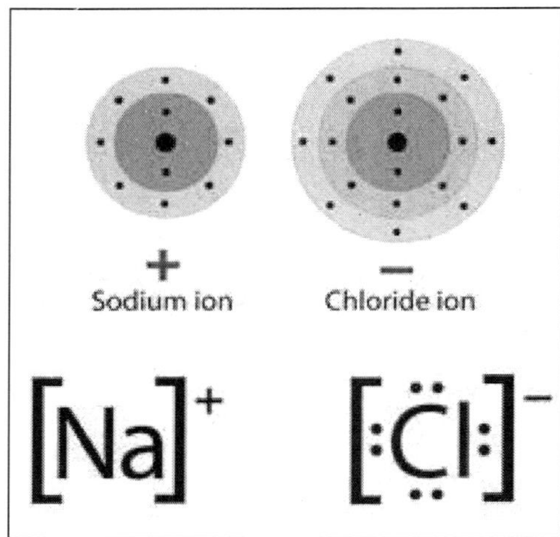

Figure 4.1. Ionic Bond Formation in NaCl.
Courtesy: http://people.fmarion.edu/tbarbeau/physio_chemical_supplements.htm.

- ✰ Strong: The strength increases with the number of shared electrons
- ✰ *Very* directional: They are formed between two specific atoms

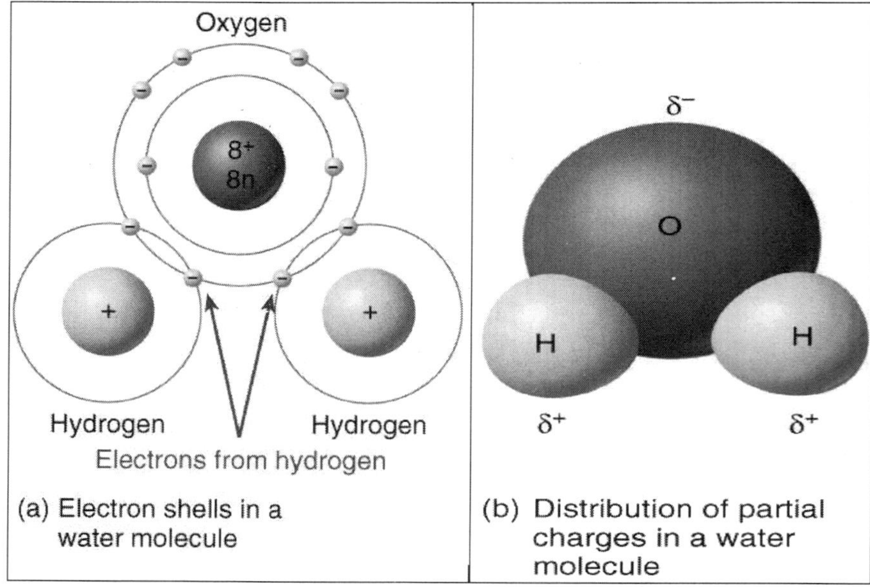

Figure 4.2: Water Molecules contain Two Covalent Bonds.
Courtesy: http://people.fmarion.edu/tbarbeau/physio_chemical_supplements.htm.

Electrons shared equally between atoms produce nonpolar molecules. Electrons shared unequally produces polar molecules. Atoms with six or seven valence shell electrons are electronegative whereas atoms with one or two valence shell electrons are electropositive.

Hydrogen Bonds

Hydrogen Bonds are formed by the attraction of opposite partial electric charges between two polar molecules. They are too weak to bind atoms together and common in dipoles such as water (Figure 4.3). They are responsible for surface tension in water and important as intramolecular bonds, giving the molecule a three-dimensional shape. The Hydrogen bond potential is

$$E_{Hbond} = \Sigma\, A\, r_{ij}^{-12} - C\, r_{ij}^{-10}$$

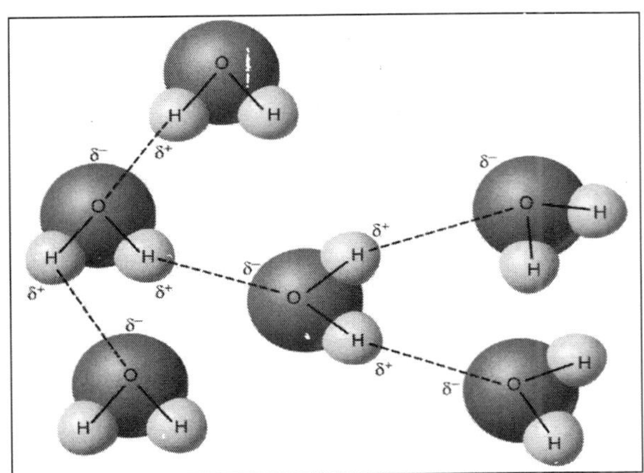

Figure 4.3. Hydrogen Bonding in Water Molecules.
Courtesy: http://people.fmarion.edu/tbarbeau/physio_chemical_supplements.htm

Chemical Reactions occur when chemical bonds are formed, rearranged, or broken and are written in symbolic form using chemical equations.

Patterns of Chemical Reactions

- Combination reactions: Synthesis reactions which always involve bond formation

$$\mathbf{A + B \rightarrow AB}$$

- Decomposition reactions: Molecules are broken down into smaller molecules

$$\mathbf{AB \rightarrow A + B}$$

- Exchange reactions: Bonds are both made and broken

$$\mathbf{AB + C \rightarrow AC + B}$$

- All chemical reactions are theoretically reversible

$$A + B \rightarrow AB$$
$$AB \rightarrow A + B$$

If neither a forward nor reverse reaction is dominant, chemical equilibrium is reached.

Oxidation-Reduction (Redox) Reactions

Reactants losing electrons are electron donors and are oxidized. Reactants taking up electrons are electron acceptors and become reduced. Generally, the atom that is reduced contains the most energy.

Reaction Type	*Energy Change and Heat in Reaction*
Endothermic	Heat absorbed
Exothermic	Heat released

Energy Flow in Chemical Reactions: Exergonic reactions are the reactions that release energy and Endergonic reactions are the reactions whose products contain more potential energy than did its reactants. The activation energy is the minimum energy needed for a reaction to take place. When a collision provides energy equal to or greater than the activation energy, product can form.

Factors Influencing Rate of Chemical Reactions

- Temperature – chemical reactions proceed quicker at higher temperatures
- Particle size – the smaller the particle the faster the chemical reaction
- Concentration – higher reacting particle concentrations produce faster reactions
- Catalysts – increase the rate of a reaction without being chemically changed
- Enzymes – biological catalysts

Influence of Functional Groups on Reactivity

Functional groups create reactive sites in molecules. Electron-rich sites react with electron poor sites. All functional groups contain a heteroatom, a π bond or both, and these features create electron-deficient (or electrophilic) sites and electron-rich (or nucleophilic) sites in a molecule. Molecules react at these sites

Intermolecular Forces

Intermolecular forces are interactions that exist between molecules. Functional groups determine the type and strength of these interactions. There are several types of intermolecular interactions. Ionic compounds contain oppositely charged particles held together by extremely strong electrostatic inter-actions. These ionic inter-actions are much stronger than the intermolecular forces present between covalent molecules. Covalent compounds are composed of discrete molecules. The nature of the forces between molecules depends on the functional group present.

van der Waals Forces

van der Waals forces are also known as London forces. They are weak interactions caused by momentary changes in electron density in a molecule. They are the only attractive forces present in nonpolar compounds.

$$\text{Lennard-Jones potential } E_{vdW} = \Sigma\, 4\, \varepsilon_{ij}\, ((\sigma_{ij}/r_{ij})^{12} - (\sigma_{ij}/r_{ij})^{6})$$

All compounds exhibit van der Waals forces. The surface area of a molecule determines the strength of the van der Waals interactions between molecules (Figure 4.4). The larger the surface area, the larger the attractive force between two molecules, and the stronger the intermolecular forces. van der Waals forces are also affected by polarizability. Polarizability is a measure of how the electron cloud around an atom responds to changes in its electronic environment.

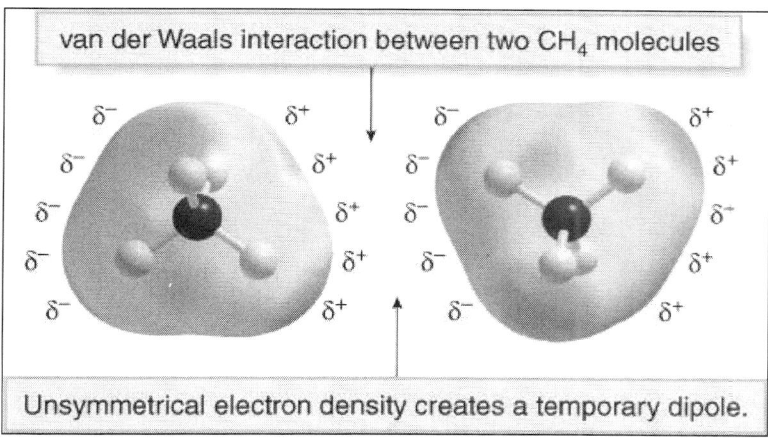

Figure 4.4: van der Waals interactions.
Courtesy: Klein, Organic Chemistry 3e.

Electrostatic Interactions and Dipole-Dipole Interactions

Dipole–dipole interactions are the attractive forces between the permanent dipoles of two polar molecules. The dipoles in adjacent molecules (Figure 4.5) align so that the partial positive and partial negative charges are in close proximity. These attractive forces caused by permanent dipoles are much stronger than weak van der Waals forces.

Electrostatic Potential

$$E_{es} = \Sigma\, Q_i\, Q_j/r_{ij}$$

Hydrogen Bonding

Hydrogen bonding typically occurs when a hydrogen atom bonded to O, N, or F, is electrostatically attracted to a lone pair of electrons on an O, N, or F atom in another molecule (Figure 4.6).

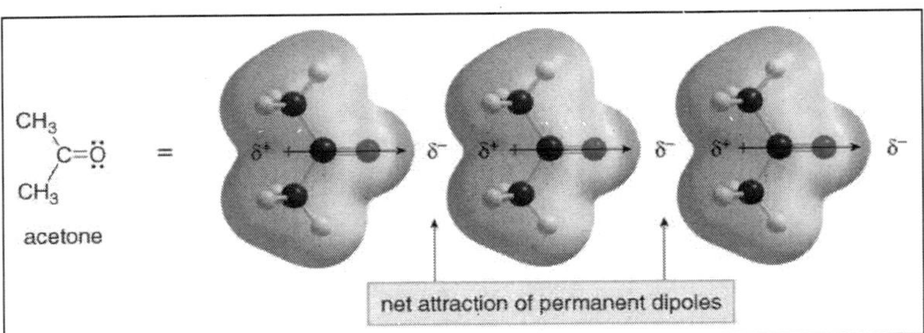

Figure 4.5. Dipole Interactions in Acetone.
Courtesy: Klein, Organic Chemistry 3e.

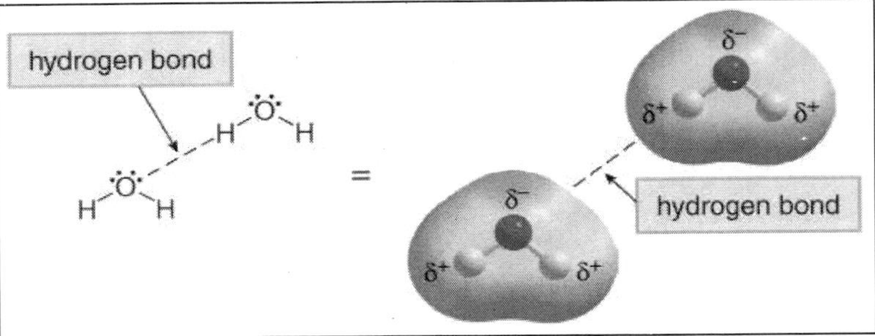

Figure 4.6. Hydrogen Bond and Electrostatic Attraction.
Courtesy: Klein, Organic Chemistry 3e.

The strength of intermolecular forces enhances, as the polarity of an organic molecule increases.

Physical Properties–Boiling Point and Melting point

The boiling point of a compound is the temperature at which liquid molecules are converted into gas. In boiling, energy is needed to overcome the attractive forces in the more ordered liquid state. The stronger the intermolecular forces, the higher the boiling point. The melting point is the temperature at which a solid is converted to its liquid phase. In melting, energy is needed to overcome the attractive forces in the more ordered crystalline solid. The stronger the intermolecular forces, the higher the melting point. Given the same functional group, the more symmetrical the compound, the higher is the melting point. The larger the surface area, there is higher the boiling point and the more polarizable the atoms also result in the higher the boiling point. The nonpolar part of a molecule that is not attracted to H_2O is said to be hydrophobic. The polar part of a molecule that can be hydrogen bond to H_2O is said to be hydrophilic.

Electronic Structure – Density of States (DOS) Properties as a Function of Nano-structures

Nanostructured materials derive their special properties from having one or more dimensions made small compared to a length scale critical to the physics of the process.

Classification: 0-Dimensional, 1-D, 2-D nanostructures and 3-D

Each band in Density of States (DOS) has a width that reflects the interaction between atoms, with a band gap between the conduction and the valence bands that reflects the original separation of the bonding ad antibonding states. Density of State (DOS = dn/dE) is a function of dimensionality (0D, 1D and 2D) nanomaterials and 3D is for bulk particles (Figure 4.7).

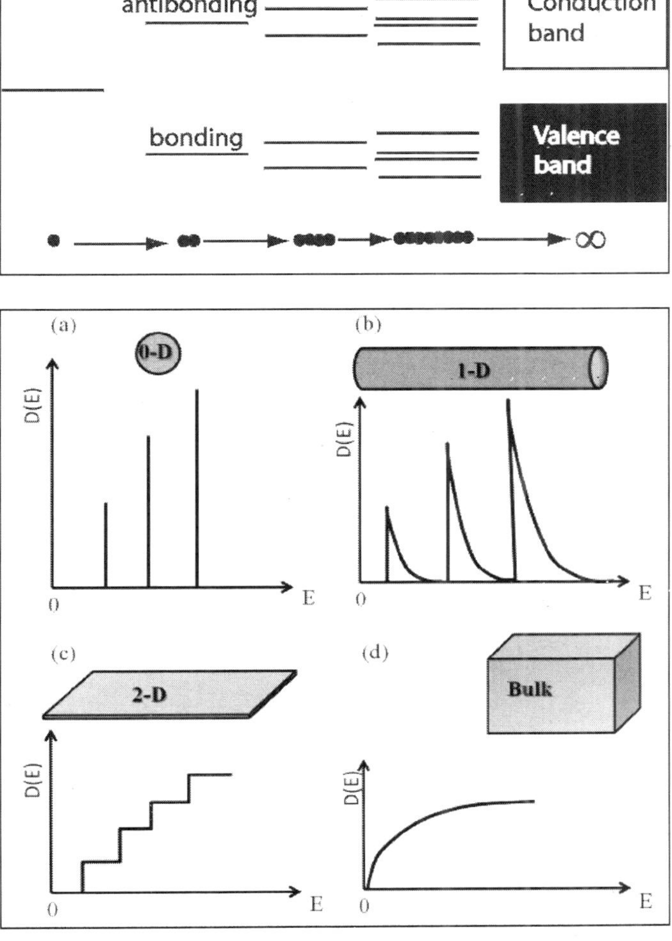

Figure 4.7. Density of States (DOS) as a Function of Dimensionality of Nanomaterials.
Courtesy: Amin, G. (2012)

Size Quantization and Quantum Confinement in Nanostructures

For the 0-D Nanomaterials, the electrons are fully confined in all the 3-Directions and for

E.g.: Quantum Dots ⇒ Electrons can easily move in 0 Dimensions! (Figure 4.8).

For the 1-D and 2-D Nanomaterials, the electron are confinement and delocalization co-exist.

Hence, for the 1-D Nanomaterials, the electrons are confined in 2-Directions and for *E.g.*: Quantum Wires ⇒ Electrons can easily move in 1 Dimension!

However, for the 2-D Nanomaterials, the electrons are confined in 1-Direction and for *E.g.*: Quantum Walls (thin films) ⇒ Electrons can easily move in 2 Dimensions!

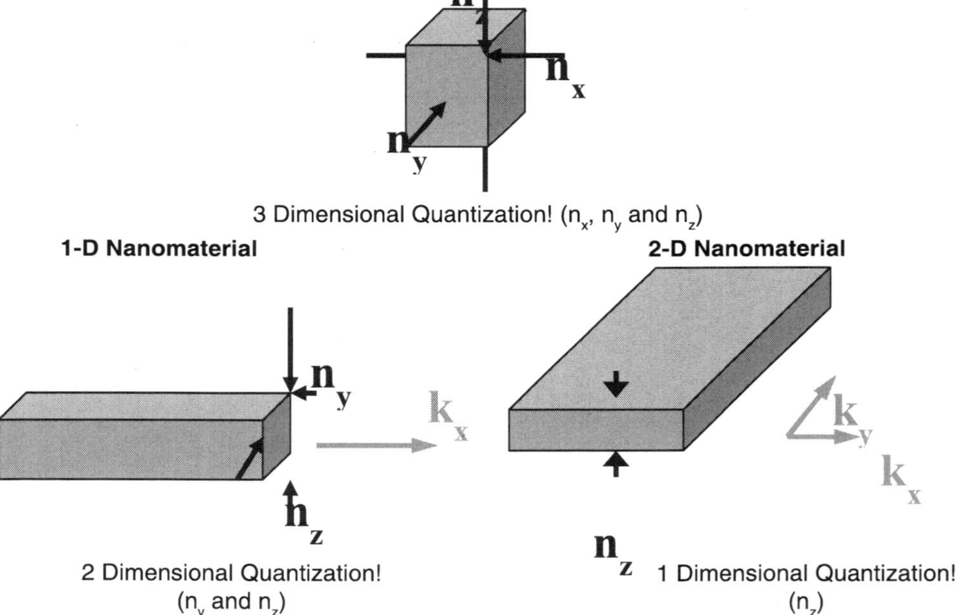

Figure 4.8. Quantum Confinement in Nanostructures.

Each further confinement direction changes a continuous k component to a discrete component characterized by a quantum number n.

Electronic Energy Levels of Nanostructures

$$(0-D) \qquad E_n = \left[\frac{\pi^2 h^2}{2mL^2}\right](n_x^2 + n_y^2 + n_z^2)$$

(1 – D) $$E_n = \left[\frac{\pi^2 h^2}{2mL^2}\right](n_x^2 + n_y^2)$$

(2 – D) $$E_n = \left[\frac{\pi^2 h^2}{2mL^2}\right](n_x^2)$$

Where $\hbar$ equals h divided by 2π; h is Planck's constant; m is the mass of the electron; L is the width (confinement) of the deep potential well; n_x, n_y and n_z are the principal quantum numbers in the three dimensions x,y and z respectively.

In the size dependance of DoS, the k-space is considered filled with an uniform grid of points each separated in units of 2π/L along any axis. The volume of k-space occupied by each point is $(2\pi/L)^3$.

Colloids and Surface Chemistry of Nanomaterials

Dispersed Systems

Dispersed systems consist of particulate matter (dispersed phase), distributed throughout a continuous phase (dispersion medium). They are classified according to the particle diameter of the dispersed material.

Molecular dispersions (less than 1 nm): Particles are invisible in electron microscope and can pass through semipermeable membranes and filter paper. The particles do not settle down on standing and undergo rapid diffusion *e.g.* ordinary ions, glucose.

Colloidal dispersions (1 nm - o.5 μm): Particles not resolved by ordinary microscope and can be detected by electron microscope. They pass through filter paper but do not pass through semipermeable membrane. Particles made to settle by centrifugation and diffuse very slowly *e.g.* colloidal silver sols, naural and synthetic polymers.

Coarse dispersions (> 0.5 μm): Particles are visible under ordinary microscope and do not pass through filter paper or semipermeable membrane. Particles will settle down under gravity and do not diffuse *e.g.* emulsions, suspensions, red blood cells.

Size and Shape of Nano-Colloids

Particles lying in the nano-colloidal size have large surface area when compared with the surface area of an equal volume of larger particles. Specific surface is the surface area per unit weight or volume of material. The possession of large specific surface results in as catalyst only when found in colloidal form due to large surface area which adsorb reactant on their surface *e.g.* platinium. The colour of colloidal dispersion is related to the size of the paticles *e.g.* red gold sol takes a blue colour when the particles increase in size. The shape of colloidal particles (Figure 4.9) in dispersion is important as the more extended the particle the greater its specific surface area and the greater the attractive force between the particles of the dispersed

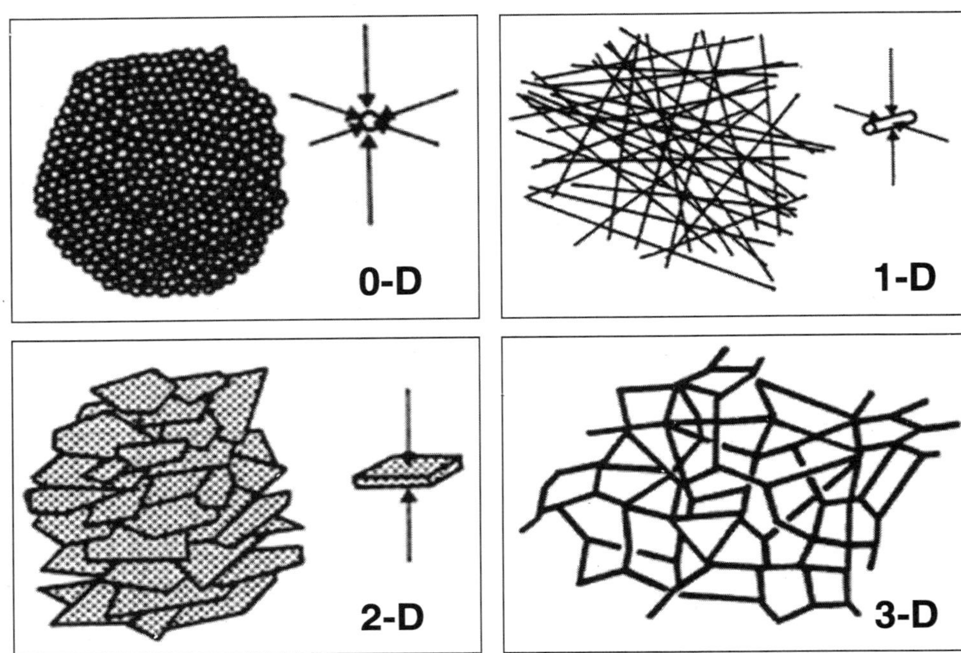

Figure 4.9. Shape of Colloidal Nanosystem.
Courtesy: https://www.slideshare.net/zoraizhaiderzoraizhaider/lecture-colloids.

phase and the dispersion medium. Flow, sedimentation and osmotic pressure of the colloidal system affected by the shape of colloidal particles.

Purification of Colloidal Solutions

When a colloidal solution is prepared is often contains certain electrolytes which tend to destabilize it. The following methods are used for purification:

Dialysis

Semipermeable cellophane membrane prevent the passage of colloidal particles, yet allow the passage of small molecules or electrolytes (Figure 4.10).

Electro-dialysis

In the dialysis unit, the movement of ions across the membrane can be speeded up by applying an electric current through the electrodes induced in the solution. The dialysis membrane allows small particles (ions) to pass through but the colloidal size particles do not pass through the membrane (Figure 4.10).

According to the states of dispersed phase and dispersion medium, the colloidal system can be segregated as given in Table 4.3.

According to the nature of interaction between dispersed phase and dispersion medium, the colloidal system can be classified as follows:

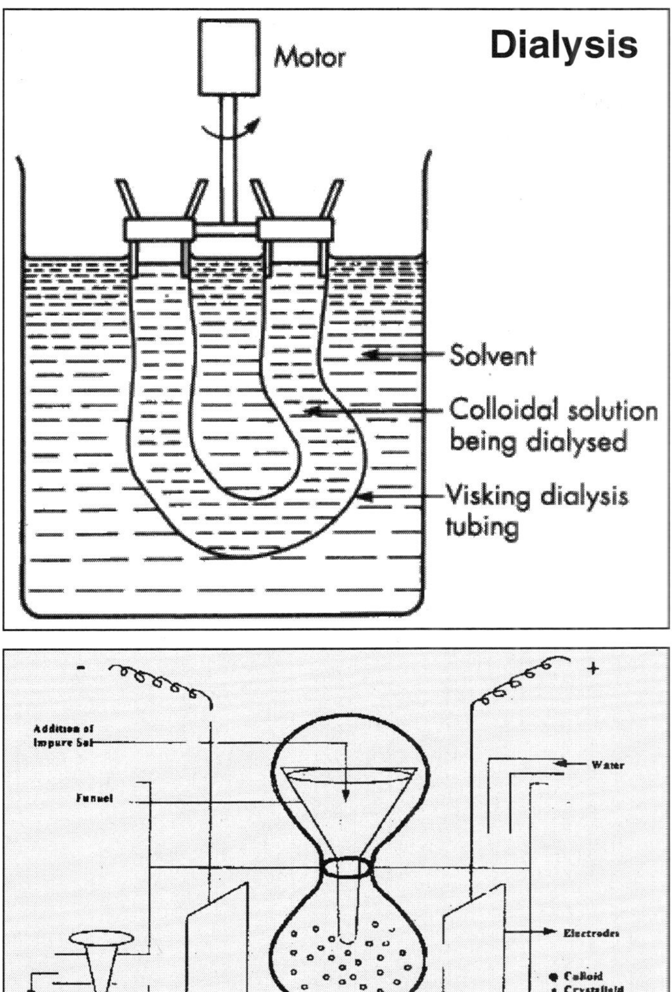

Figure 4.10. Purification of Colloids.
Courtesy: Shaw, D. J., 1992.

Lyophilic Colloids (Solvent Attracting) (Solvent Loving)

The particles in a lyophilic system have a great affinity for the solvent. If water is the dispersing medium, it is often known as a hydrosol or hydrophilic that will be readily solvated (combined chemically or physically, with the solvent) and dispersed, even at high concentrations and more viscid. Examples of lyophilic sols include sols of gum, gelatin, starch, proteins and certain polymers (rubber) in organic solvents.

Table 4.3. Classification of Colloids According to the Original States of their Constituent Parts.

Dispersed Phase	*Dispersion Medium*	*Common Name*	*Examples*
Solid	Solid	Solid Sol	Coloured gemstones, coloured glasses
Solid	Liquid	Sol	Paints, muddy water, gold sol, starch sol. arsenous sulphide sol
Solid	Gas	Aerosol	Smoke, dust
Liquid	Solid	Gel	Gellies, cheese
Liquid	Liquid	Emulsion	Milk, cod liver oil
Liquid	Gas	Liquid alcosol	Mist, fog, cloud
Gas	Solid	Solid foam	Foam rubber, pumice stone
Gas	Liquid	Foam	Froth, whipped cream

Lyophobic Colloids (Solvent Repelling) (Solvent Hating)

The particles resist solvation and dispersion in the solvent. The concentration of particles is usually relatively low and less viscid. These colloids are easily precipitated on the addition of small amounts of electrolytes, by heating or by shaking. These are less stable as the particles surrounded only with a layer of positive or negative charge. Once precipitated, it is not easy to reconstitute the sol by simple mixing with the dispersion medium. Hence, these sols are called irreversible sols. Examples of lyophobic sols include sols of metals and their insoluble compounds like sulphides and oxides *e.g.* gold in water.

Association/Amphiphilic Colloids

Certain molecules are termed amphiphiles (Figure 4.11) or surface active agents, characterized by two regions of opposing solution affinities within the same molecule.

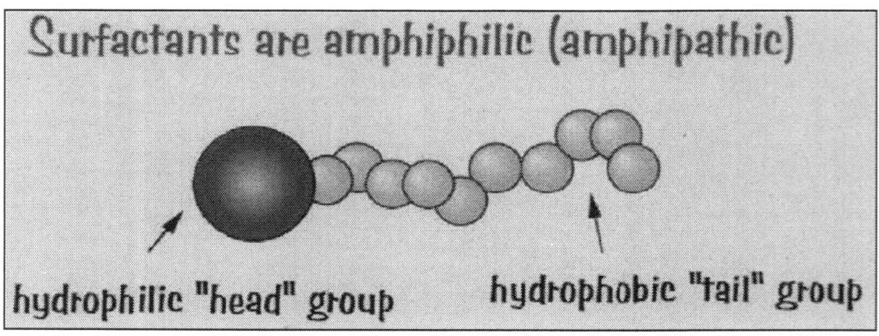

Figure 4.11. Amphiphilic Surfactant.
Courtesy: https://www.slideshare.net/

At low concentration, the amphiphiles exist separately (sub-colloidal size) and at high concentration, they form aggregates or micelles (50 or more monomers in colloidal size) (Figure 4.12).

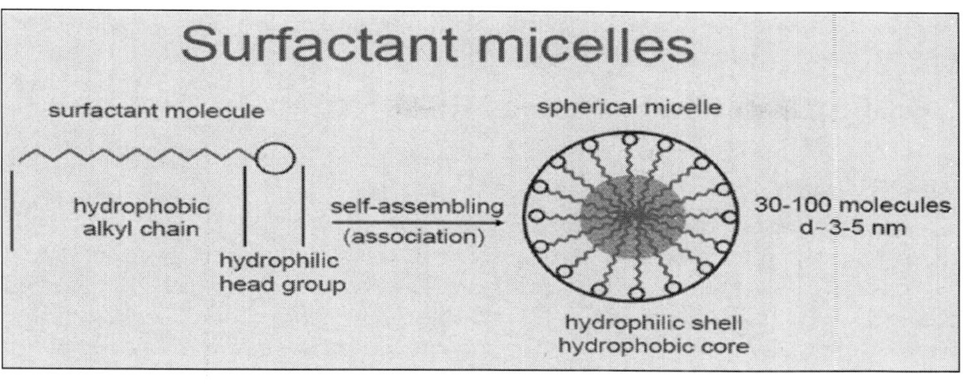

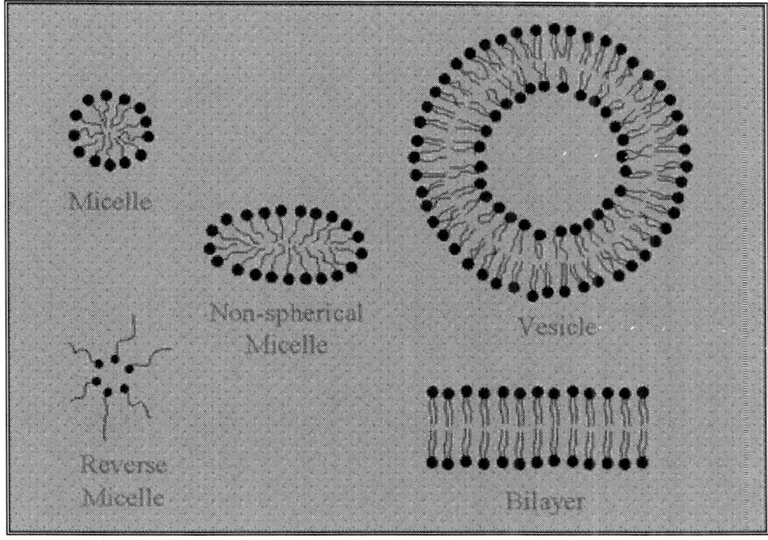

Association Colloids

Figure 4.12. Micelles and Association Colloids.
Courtesy: https://www.slideshare.net/

Critical Micelle Concentration (C.M.C)

The concentration at which micelle forms. The phenomenon of micelle formation can be explained as follows:

- Below C.M.C: amphiphiles are adsorbed at the air/water interface
- As amphiphile concentration is raised: both the inter-phase and bulk phase become saturated with monomers (C.M.C)
- Any further amphiphile added in excess: amphiphiles aggregate to form micelles

The formation of association colloids is spontaneous, provided the concentration of amphiphile in solution exceeds C.M.C.

Classification of Surfactants

"The compounds termed surfactants are amphiphiles and surface-active agents that can be classified as given in **Table 4.4**.

Anionic Surfactants are dissociated in water in an amphiphilic anion, and a cation, which is in general an alcaline metal (Na+, K+) or a quaternary ammonium. They are the most commonly used surfactants. They include alkylbenzene sulfonates (detergents), (fatty acid) soaps, lauryl sulfate (foaming agent), di-alkyl sulfosuccinate (wetting agent), lignosulfonates (dispersants) *etc*.... Anionic surfactants account for about 50 per cent of the world production.

Table 4.4: Classification of Surfactants

Class	Examples	Structures
Anionic		
	Na stearate	$CH_3(CH_2)_{16}COO^-Na^+$
	Na dodecyl sulfate	$CH_3(CH_2)_{11}SO_4^-Na^+$
	Na dodecyl benzene sulfonate	$CH_3(CH_2)_{11}C_6H_4SO_3^-Na^+$
Cationic		
	Laurylamine hydrochloride	$CH_3(CH_2)_{11}NH_3^+Cl^-$
	Trimethyl dodecylammonium chloride	$C_{12}H_{25}N^+(CH_3)_3Cl^-$
	Cetyl trimethylammonium bromide	$CH_3(CH_2)_{15}N^+(CH_3)_3Br^-$
Non-ionic		
	Polyoxyethylene alcohol	$C_nH_{2n+1}(OCH_2CH_2)_mOH$
	Alkylphenol ethoxylate	$C_9H_{19}–C_6H_4–(OCH_2CH_2)_nOH$
	Polysorbate 80 $w+x+y+z=20$ $R=(C_{17}H_{33})COO$	$HO(C_2H_4O)_w$, $(OC_2H_4)_xOH$, $CH(OC_2H_4)_yOH$, $CH_2(OC_2H_4)_zR$
	Propylene oxide-modified polymethylsiloxane (EO = ethyleneoxy, PO = propyleneoxy)	$(CH_3)_3SiO((CH_3)_2SiO)_x(CH_3SiO)_ySi(CH_3)_3$ $CH_2CH_2CH_2O(EO)_m(PO)_nH$
Zwitterionic	Dodecyl betaine	$C_{12}H_{25}N^+(CH_3)_2CH_2COO^-$
	Lauramidopropyl betaine	$C_{11}H_{23}CONH(CH_2)_3N^+(CH_3)_2CH_2COO^-$
	Cocoamido-2-hydroxypropyl sulfobetaine	$C_nH_{2n+1}CONH(CH_2)_3N^+(CH_3)_2CH_2CH(OH)CH_2SO_3^-$

Nonionic Surfactants come as a close second with about 45 per cent of the overall industrial production. They do not ionize in aqueous solution, because their hydrophilic group is of a nondissociable type, such as alcohol, phenol, ether, ester, or amide. A large proportion of these nonionic surfactants are made hydrophilic by the presence of a polyethylene glycol chain, obtained by the polycondensation of ethylene oxide. They are called polyethoxylated nonionics.

In the past decade glucoside (sugar based) head groups, have been introduced in the market, because of their low toxicity. As far as the lipophilic group is concerned, it is often of the alkyl or alkylbenzene type, the former coming from fatty acids of natural origin. The polycondensation of propylene oxide produce a polyether which (in oposition to polyethylene oxide) is slightly hydrophobic. This polyether chain is used as the lipophilic group in the so-called polyEOpolyPO block copolymers, which are most often included in a different class, *e.g.* polymeric surfactants, to be dealt with later.

Cationic Surfactants are dissociated in water into an amphiphilic cation and an anion, most often of the halogen type. A very large proportion of this class corresponds to nitrogen compounds such as fatty amine salts and quaternary ammoniums, with one or several long chain of the alkyl type, often coming from natural fatty acids. These surfactants are in general more expensive than anionics, because of the high pressure hydrogenation reaction to be carried out during their synthesis. As a consequence, they are only used in two cases in which there is no cheaper substitute, *i.e.* (1) as bactericide, (2) as positively charged substance which is able to adsorb on negatively charged substrates to produce antistatic and hydrophobant effect, often of great commercial importance such as in corrosion inhibition.

When a single surfactant molecule exhibit both anionic and cationic dissociations it is called **amphoteric** or **zwitterionic**. This is the case of synthetic products like betaines or sulfobetaines and natural substances such as amino acids and phospholipids" (Salager, 2002).

Gemini Surfactant

"The term Gemini surfactant was coined by Menger and has become accepted in the surfactant literature for describing dimeric surfactants, that is, surfactant molecules that have two hydrophilic (chiefly ionic) groups and two tails per surfactant molecule. These twin parts of the surfactants are linked by a spacer group of varying length (most commonly a methylene spacer or an oxyethylene spacer). A block diagram of a typical Gemini surfactant is shown in Figure 4.13.

Gemini surfactants possess a number of superior properties when compared to conventional single-headed, single-tailed surfactants, the Geminis tending to exhibit lower CMC values (by about an order of magnitude), increased surface activity and lower surface tension at the CMC, enhanced solution properties such as hard-water tolerance, superior wetting times, and lower Krafft points.

The Krafft Point

The solubilities of micelle-forming surfactants show a strong increase above a certain temperature, termed the Krafft point (Figure 4.13). Below the Krafft point the solubility of the surfactant is too low for micellization so solubility alone determines the surfactant monomer concentration. Above the Krafft point, a relatively large amount of surfactant can be dispersed in micelles and solubility increases greatly" (Schramm *et al.*, 2003).

Surface Energy

"Surface energy, γ, by definition, is the energy required to create a unit area of "new" surface.

$$\gamma = \left(\frac{\partial G}{\partial A}\right)_{n_i,T,P}$$

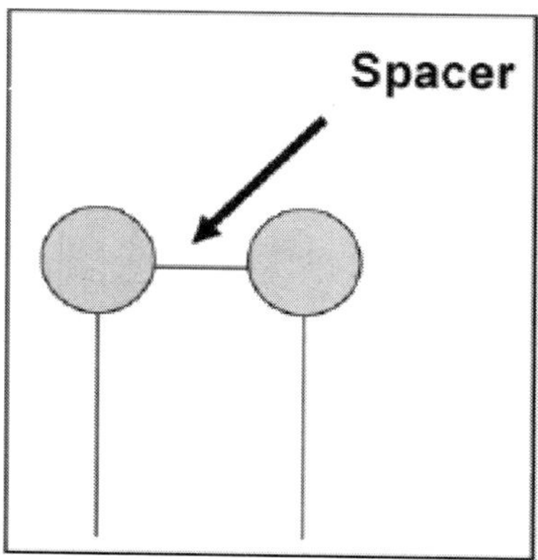

Gemini Surfactant Illustration

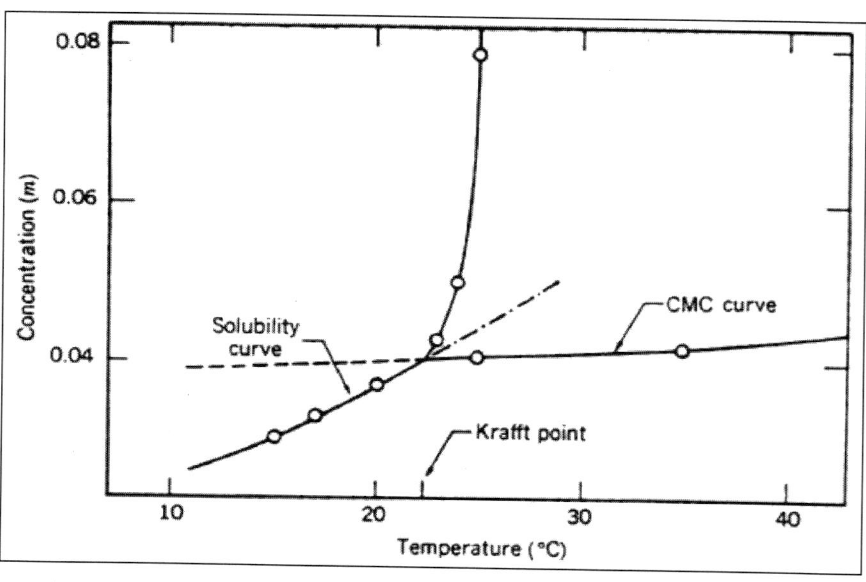

Krafft Point

Figure 4.13. Gemini Surfactant and Krafft Point.
Courtesy: Schramm *et al.*, (2003) *Annu. Rep. Prog. Chem.*, Sect. C, 99, 3–48.

where A is the surface area, G is the Gibbs free energy.

Therefore, the surface energy can also is given as

$$\gamma = \frac{1}{2} N_b \varepsilon \rho_a$$

where N_b is the number of broken bonds, ε is the half of the bond strength, ρ_a is the surface atomic density, the number of atoms per unit area on the new surface.

For a given surface with a fixed surface area, the surface energy can be reduced for achieving stability through (i) surface relaxation, the surface atoms or ions shift inwardly which occur more readily in liquid phase than in solid surface due to rigid structure in solids, (ii) surface restructuring through combining surface dangling bonds into strained new chemical bonds, (iii) surface adsorption through chemical or physical adsorption of terminal chemical species onto the surface by forming chemical bonds or weak attraction forces such as electrostatic or van der Waals forces, and (iv) composition segregation or impurity enrichment on the surface through solid-state diffusion.

Reduction of Overall Surface Energy of Nanostructures

At the overall system level, mechanisms for the reduction of overall surface energy include (i) combining individual nanostructures together to form large structures s_o as to reduce the overall surface area, if large enough activation is available for such a process to proceed, and (ii) agglomeration of individual nanostructures without altering the individual nanostructures.

Specific mechanisms of combining individual nanostructures into large structures include (i) sintering, in which individual structures merge together and (ii) Ostwald ripening, in which relatively large structures grow at the expense of smaller ones (Figure 4.14). (a) Sintering is to combine individual particles to a bulk with solid interfaces to connect each other and (b) Ostwald ripening is to merge smaller particles into a larger particle. Both processes reduce the solid-gas surface area.

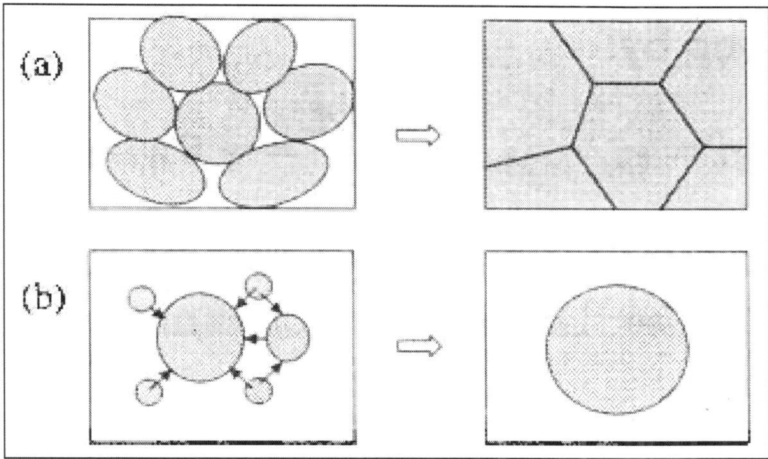

Figure 4.14. Mechanism of Sintering and Ostwald Ripening.
Courtesy: G. Cao, 2004.

Chemical Potential as a Function of Surface Curvature

The Young-Laplace equation describes the chemical potential of an atom in a spherical surface with respect to a flat reference surface.

If any curved surface described by two principal radii of curvature, R_1 and R_2, so we have:

$$\Delta\mu = \gamma\Omega\left(\frac{1}{R_1} + \frac{1}{R_2}\right)$$

Ostwald Ripening Process in Nanomaterials

In the Ostwald ripening processing (Figure 4.15), smaller particle has a larger solubility or vapor pressure due to its larger curvature, whereas the larger particle possesses a smaller solubility or vapor pressure. To maintain the local concentration equilibrium, smaller particle would dissolve into the surrounding medium; solute at proximity of smaller particle diffuses away; solute at proximity of larger particle would deposit. The process would continue till disappearance of the smaller particle.

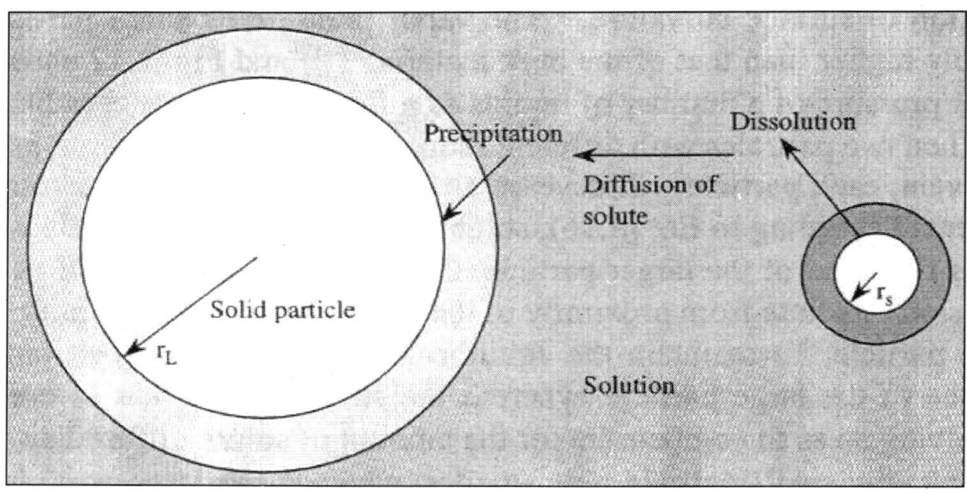

Figure 4.15: Ostwald Ripening Process
Courtesy: G. Cao, 2004.

This phenomenon of Ostwald ripening occurs also in the forms of solid-state diffusion and of evaporation-condensation. Depending on the process and applications, Ostwald ripening can have either positive or negative influence on the resulting materials. Ostwald ripening can either widen or narrow the size distribution, depending on the control of the process conditions. The reduction of overall surface energy is the driving force for the surface restructuring through sintering and Ostwald ripening. These are the reduction mechanisms for individual surface, individual nanostructures and the overall system. The system can have another mechanism to reduce the overall surface energy, in addition to sintering and Ostwald ripening. This is agglomeration. When small nanostructures form agglomerates, it is very difficult to disperse them. In nanostructure fabrication and processing, it is very important to overcome the huge total surface energy to create the desired nanostructures. It is equally important to prevent the nanostructures from agglomeration. As the dimension of nanostructured materials reduces, van der Waals attraction force between nanostructured materials becomes increasingly important.

Electrostatic Stabilization and Surface Charge Density of Nanoparticles

When a nano-scale solid emerges in a polar solvent or an electrolyte solution, a surface charge will be developed through one or more of the following mechanisms:

1. Preferential adsorption of ions
2. Dissociation of surface charged species
3. Isomorphic substitution of ions
4. Accumulation or depletion of electrons at the surface
5. Physical adsorption of charged species onto the surface.

For a given solid surface in a given liquid medium, a fixed surface electrical charge density or electrode potential, *E*, will be established, which is given by the Nernst equation:

$$E = E_0 + \frac{R_g T}{n_i F} \ln a_i$$

where ***E***, is the standard electrode potential when the concentration of ions is unity, n_i is the valence state of ions, a_i is the activity of ions, R_g is the gas constant and ***T*** is temperature, and ***F*** is the Faraday's constant.

Electric potential at the proximity of Nanoparticle Surface: Electric double layer

Such inhomogeneous distributions of ions in the proximity of the nanoparticle solid surface lead to the formation of so-called double layer structure, that consists of two layers, Stern layer and Gouy layer (also called diffuse double layer), and the two layers are separated by the Helmholtz plane. Between the solid surface and the Helmholtz plane is the Stern layer, where the electric potential drops linearly through the tightly bound layer of solvent and counter ions. Beyond the Helmholtz plane until the counter ions reach average concentration in the solvent is the Gouy layer or diffuse double layer. Schematic illustrating electrical double layer structure is given in (Figure 4.16) and the electric potential near the solid surface with both Stern and Gouy layers indicated. Surface charge is assumed to be positive.

Interactions between the Surface of Nanoparticles: DLVO Theory

The total interaction between two particles, which are electrostatic stabilized, is the combination of van der Waals attraction (Φ_A) and electrostatic repulsion (Φ_B):

$$\Phi = \Phi_A + \Phi_B$$

The electrostatic stabilization of nanoparticles in a suspension is successfully described by the DLVO theory, named after Derjaguin, Landau, Venwey and Overbeek. The interaction between two nanoparticles in a suspension is considered

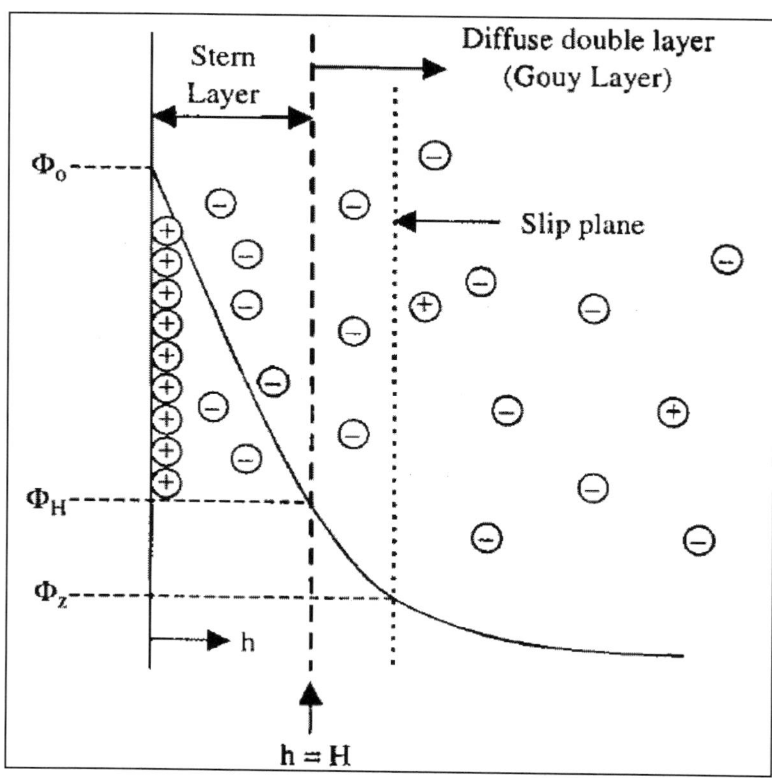

Figure 4.16. Electric Double Layer.
Courtesy: G. Cao, 2004.

as the combination of van der Waals attraction potential and the electric repulsion potential. There are some important assumptions in DLVO theory, such as follows: Infinite flat solid surface, Uniform surface charge density, No redistribution of surface charge, *i.e.* the surface electric potential remains constant, No change of concentration profiles of both counter ions and surface charge determining ions, *i.e.* the electric potential remains unchanged, and Solvent exerts influences via dielectric constant only, *i.e.* no chemical reactions between the particles and solvent. Depiction of DLVO potential is shown in (Figure 4.17) in which the van der Waals attraction potential (V_A), electric repulsion potential (**V_R**), and the combination of the two opposite potentials as a function of distance from the surface of a spherical particle besides the conditions for the occurrence of electrostatic repulsion between two nanoparticles.

When two particles are far apart or the distance between the surfaces of two particles is larger than the combined thickness of two electric double layers of two particles, there would be no overlap of diffusion double layers, and thus there would be no interaction between two particles. However, when two particles move closer and the two electric double layers overlap, a repulsion force is developed. As the distance reduces, the repulsion increases and reaches the maximum when the

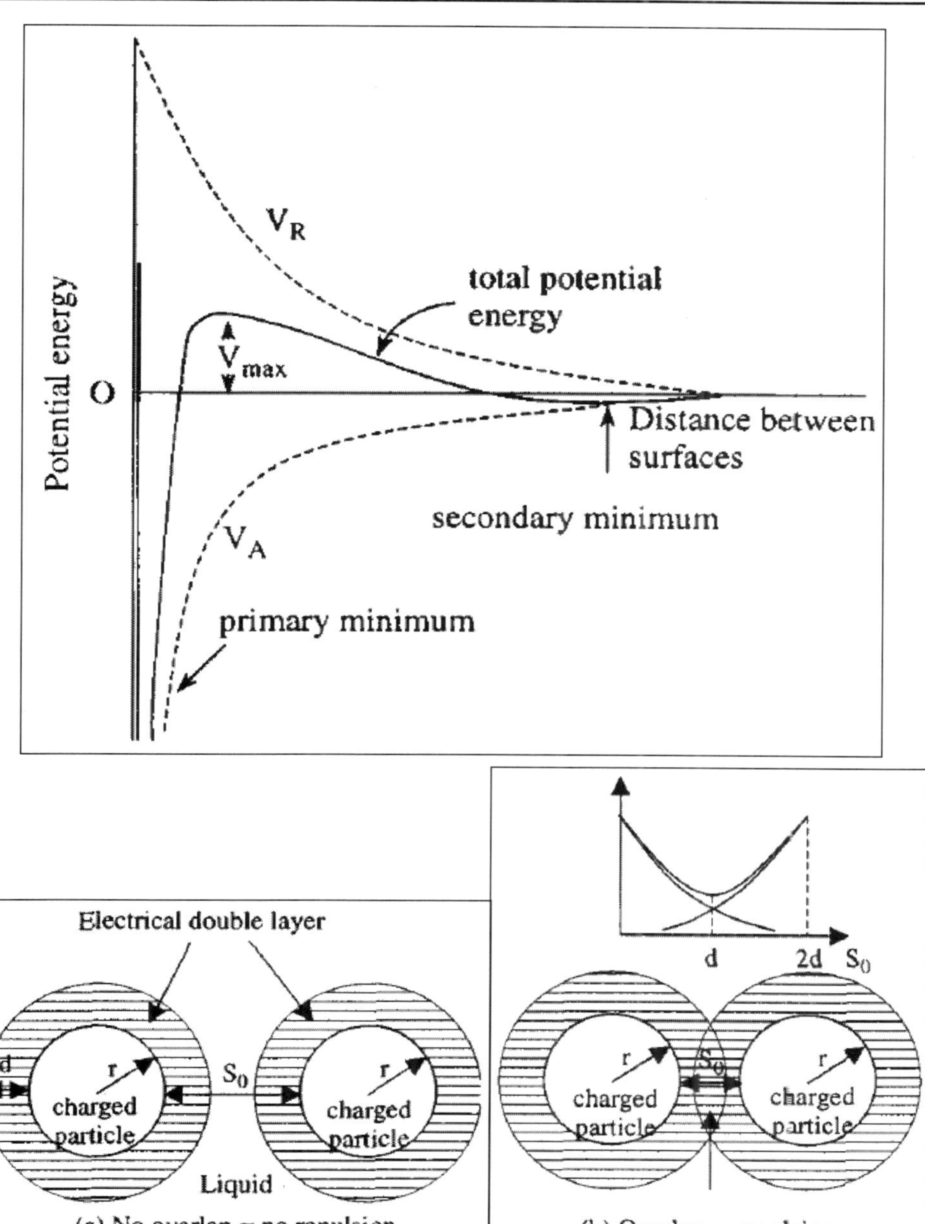

Figure 4.17: DLVO Potential as a Function of Distance between the Surfaces of Nanoparticles.
Courtesy: G. Cao, 2004.

distance between two particle surfaces equals the distance between the repulsive barrier and the surface.

Although many important assumptions of the DLVO theory are not satisfied in the really colloidal systems consists of nanostructures and nanomaterials, in which small particles are dispersed in a diffusive medium, the DLVO theory is still valid and has been widely applied in practice, as far as the following conditions are met:

1. Dispersion is very dilute, so that the charge density and distribution on each particle surface and the electric potential in the proximity next to each particle surface are not interfered by other particles.
2. No other force is present besides van der Waals force and electrostatic potential, *i.e.* the gravity is negligible or the particle is significantly small (Nano-scale), and there exist no other forces, such as magnetic field.
3. Geometry of particles is relatively simple, so that the surface properties are the same over the entire particle surface, and, thus surface charge density and distribution as well as the electric potential in the surrounding medium are the same.
4. The double layer is purely diffusive, so that the distributions of counter ions and charge determining ions are determined by all three forces: electrostatic force, entropic dispersion and Brownian motion" (Cao, 2004).

REFERENCES

1. Amin, G. (2012). ZnO and CuO Nanostructures: Low Temperature Growth, Characterization, their Optoelectronic and Sensing Applications. Ph.D. Thesis, Linköping University, Sweden.
2. Cao, G. (2004). Nanostructures and Nanomaterials: Synthesis, Properties, and Applications. Imperial College Press Heinemann – Elsevier Oxford.
3. Klein, D. R. (2016). Organic Chemistry, 3rd Edition, Wiley.
4. Salager, J.L. (2002). Surfactants Types and Uses. FIRP Booklet # E300-A, Universidad de los Andes, Colombia.
5. Schramm, L.L., Stasiuk, E. N. and Marangoni, D. G. (2003). Surfactants and their applications. Annu. Rep. Prog. Chem., Sect. C, 99, 3–48
6. Shaw. D. J. (1992). Introduction to Colloid and Surface Chemistry. 4th Edn., by Butterworth.

Chapter 5

Solid State, Polymer and Electrochemistry of Nanomaterials

Badal Kumar Biswal, Paladugu Krishnadev and D. Jeya Sundara Sharmila

Department of Nano Science and Technology, Tamil Nadu Agricultural University, Coimbatore – 641 003, Tamil Nadu

Molecular Orbital Theory for Nano-crystals

In particular, Schrodinger wave equation is applied to atomic level materials whereas a small crystal could contain 10^{16} number of atoms. In such situations, an approximation often used for smaller molecules is that combining atomic wave functions can form the molecular wave functions. This linear combination of atomic orbitals (LCAO) approach can also be applied to solids.

As the number of atoms increases, the number of levels increases, but the spread of energies appears to increase only slowly and is levelling off for long chains. Extrapolating to crystal size increase, there would be a very large number of levels within a comparatively small range of energies. A series of hydrogen atoms is a very simple and an artificial model; as an estimate of the energy separation of the levels. A metal crystal might contain 10^{16} atoms and the range of energies is only 10^{-19}J. The average separation between levels would thus be only 10^{-35} J. The lowest energy levels in the hydrogen atom are separated by energies of the order of 10^{-18} J so that the energy separation in a crystal is minute. The separation is in fact so small that, as in the free electron model, the set of levels as a continuous range of energies. Such a continuous range of allowed energies is known as an energy band.

For solids the LCAO equation modelled as:

$$\Psi(k) = \Sigma_n \, e^{ik.r_n} a_{nk} \Phi_n$$

where k is the wave vector, r is the position vector of a nucleus in the lattice, and a_{nk} is the coefficient of the orbital on the nth atom for a wave vector of k.

Crystalline Solids

A crystal is a solid form of material where the atoms, molecules, or ions pack together to form a regular repeating array.

Arrangement of Crystal Structures

The atoms in a crystal can be arranged in layer by layer by following alternate layer in a close packing system such as ABABABAB....n. this type of configuration is known as hexagonal close packing (hcp). Further, a third layer can also be added (C), known as cubic close packing (ccp) with a structure of ABC ABC ABC....n (Figure 5.1).

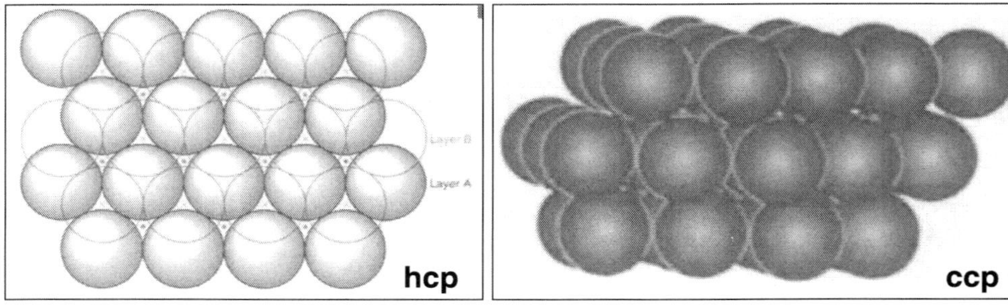

Figure 5.1. Packing of Crystal Structure.

Close-packing represents the most efficient use of space when packing identical spheres. The spheres occupy 74 per cent of the volume: the packing efficiency is said to be 74 per cent. Each sphere in the structure is surrounded by twelve equidistant neighbours-six in the same layer, three in the layer above and three in the layer below: the coordination number of an atom in a close-packed structure is thus 12.Two different types of space are contained within a close-packed structure: the first we will consider is called an octahedral hole where 6 spheres surround the hole. The other one is known as tetrahedral hole which is surrounded by 4 atoms. Octahedral hole in a close-packed array of spheres of radius r is 0.414r. For a tetrahedral hole, the radius is 0.225r.

Some metals do not adopt a close-packed structure but have a slightly less efficient packing method: this is the body-centred cubic structure (bcc) in which the coordination number has dropped from twelve to eight and the packing efficiency is 68 per cent, compared with 74 per cent for close-packing.

Entropy and Enthalpy

Energy is required to form a defect in a crystal and formation of defects is always an endothermic process. It may seem surprising that defects exist in crystals at all, and yet they do even at low temperatures, albeit in very small concentrations. The reason for this is that although it costs energy to form defects, there is a commensurate gain in entropy. The enthalpy of formation of the defects is thus balanced by the gain in entropy such that, at equilibrium, the overall change in free energy of the crystal due to the defect formation is zero according to the equation:

$$\Delta G = \Delta H - T\Delta S$$

The number of Schottky defects in a crystal of composition MX is given by:

$$n_S = N\exp\left(\frac{-\Delta H_S}{2KT}\right)$$

The Boltzmann formula tells us about the entropy of a system which is given by:

S = k lnW, where W= number of ways in distributing n defects, k= Boltzmann constant = 1.380662 x 10^{-23} J K^{-1}.

The change in entropy due to introducing the defects into perfect crystal is given by,

$$\Delta S = 2\text{k}\ln\left(\frac{N!}{(N-n_S)!n_S!}\right)$$

Raising the temperature introduces more defects. This happens because defect formation is an endothermic process and Le Chatelier's principle tells us that increasing the temperature of an endothermic reaction will favour the products-in this case defects. Second, if it were possible to decrease the enthalpy of formation of a defect, ΔH_s or ΔH_F, this would also increase the proportion of defects present.

Ionic Conductivity

One of the most important aspects of point defects is that they make it possible for atoms or ions to move through the structure. If a crystal structure were perfect, it would be difficult to envisage how the movement of atoms, either diffusion through the lattice or ionic conductivity (ion transport under the influence of an external electric field) could take place. Two possible mechanisms for the movement of ions through a lattice are possible, an ion hops or jumps from its normal position on the lattice to a neighbouring equivalent but vacant site. This is called the vacancy mechanism. The interstitial mechanism, where an interstitial ion jumps or hops to an adjacent equivalent site. These simple pictures of movement in an ionic lattice are known as the hopping model, and ignore more complicated cooperative motions.

Ionic conductivity, σ is defined by,

$$\sigma = nZe\mu$$

where n is the number of charge carriers per unit volume, Ze is their charge (expressed as a multiple of the charge on an electron, e=1.602 189 x 10–19 C), and μ is their mobility, which is a measure of the drift velocity in a constant electric field.

NaCl Crystal

In case of NaCl, which contains Schottky defects, the Na+ ions are the ones which move because they are the smaller; however, even these meet quite a lot of resistance. The direct route is clearly going to be very unlikely because it leads directly between two Cl^- ions, which in a close-packed structure such as this, are going to be very close together. The other pathway first passes through a one of the triangular faces of the octahedron (point 1), then through one of the tetrahedral holes (point 2), and finally through another triangular face (point 3), before finally arriving at the vacant octahedral site (Figure 5.2). The coordination of the Na+ ion will change from 6→3→4→3→6 as it jumps from one site to the other. While there is clearly going to be an energy barrier to this happening, it will not be as large as for the direct path where the Na^+ ion becomes 2-coordinate. The energy of the ion is the same at the beginning and end of the jump; the energy required to make the jump, E_a, is known as the activation energy for the jump. This means that the temperature dependence of the mobility of the ions can be expressed by an Arrhenius equation:

$$\mu \propto \mu_0 \exp\left(\frac{-E_a}{KT}\right)$$

where μ_0 is a proportionality constant known as a pre-exponential factor. μ_0 depends on several factors: the number of times per second that the ion attempts the move, v, called the attempt frequency (this is a frequency of vibration of the lattice of the order of 10^{12}–10^{13} Hz); the distance moved by the ion; and the size of the external field.

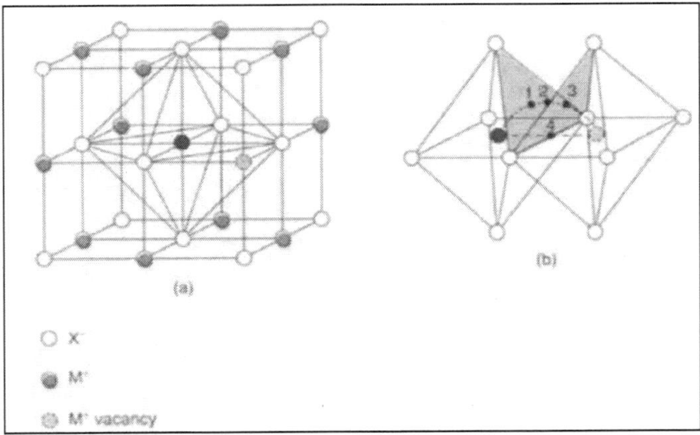

Figure 5.2. NaCl Structure.

If the external field is small (up to about 300 V cm^{-1}), a temperature dependence of 1/T is introduced into the pre-exponential factor" (Smart and Moore, 2016)

Polymer Chemistry

The word "polymer" is derived from the Greek word 'poly' and 'meros', which defines as a long molecule composed of repeating units connected by covalent bonds. Monomer is a repeated unit in a polymer and the reactant for the polymerization reaction. Some of the scientist working in this area prefer to choose the word "macromolecule," or large molecule, instead of polymer. Others usually prefer to use as naturally derived polymers, or biopolymers, and synthetic polymers were studied in different courses. In other way these large molecules are simply called as a "giant molecules." However, these principles apply to all polymers similarly. The major differences between all polymers (plastics, fibers, and elastomers or rubbers) are determined primarily by the intermolecular and intramolecular forces between the molecules and within the individual molecule, respectively. The functional groups are present and it is based on their size, allowing a cumulation of these forces. Among all the polymers, protein based polymer is widely used and a rich source of content of amino acids and boost energy as well. During ancient time people use to degrade or depolymerize the protein in meat by the process of aging and cooking, and the egg albumin are generally denatured by heating or addition of vinegar to the eggs. In modern life of humans, people learned how to process the dye, and weave the proteinaceous fibers from natural sources of wool and silk into useful one as well as carbohydrate rich fibers from flax and cotton. During 'Aztecs' a South American civilizations, they used natural rubber (NR) (*Hevea brasiliensis*) for making elastic articles and for waterproofing fabrics. There has always been an abundance availability of natural fibers and elastomers, but few plastics. The crude plastic art are usually employed in tanning the protein from the animal skins to make and convert into leather and in heat-formed tortoise shells.

Classification of Polymers

Polymer is applicable to almost of all materials with high molecular weight. These materials exist in countless forms and numbers because of types of atoms present in their molecules. Polymers basically possess with different kind of properties such as chemical structures, physical properties, mechanical behavior, thermal characteristics, *etc.* and are clearly classified one by one in different ways, as now discussed.

Natural and Synthetic Polymers

Based on their source, polymers can be grouped into two *i.e.*, natural or synthetic. Those polymer materials which are derived from natural resources are called natural polymers. Most common examples are: silk, cotton, rubber and wool. Chemical modifications of naturally derived polymers such as cellulose rayon, cellophane, and leather are followed in fact. Processing of polymers from low molecular weight compounds are called synthetic polymers. Typical examples are polyethylene, PVC, nylon and terylene.

Organic and Inorganic Polymers

A polymer consist of backbone chain which are essentially made up of carbon atoms is termed an organic polymer. The individual atoms are attached to the side valencies of backbone of carbon atoms (hydrogen, oxygen, nitrogen, *etc.*). The majority of synthetic polymers are organic and they are very extensively studied. In fact, the number and variety of organic polymers are so large that when we refer to 'polymers', we normally mean organic polymers. The molecules of inorganic polymers, on the other hand, generally contain no carbon atom in their chain backbone. Typical examples of inorganic polymers are glass and silicone rubber.

Thermoplastic and Thermosetting Polymers

During course of polymer heating it becomes soften and can be converted or moulded into desirable shape that they can retain on cooling. This process of heating, reshaping and retaining the same on cooling step can be repeated twice. Such polymers, that often on heating and stiffen on cooling, are termed 'thermoplastics'. Most common thermoplastics examples are polyethylene, PVC, nylon and sealing. During heating of polymer it undergo some chemical changes and convert themselves into an infusible mass. They are alike the yolk of the egg, which on heating sets into a mass, and, once set, cannot be reshaped. Such polymers, that become an infusible and insoluble mass on heating, are called 'thermosetting' polymers.

Plastics, Elastomers, Fibres and Liquid Resins

Based on form a polymer can be classified as plastic, elastomer, fibre or liquid resin. When the polymer is shaped into hard and tough utility articles by the application of heat and pressure, it is used as a 'plastic'. Most common examples are polystyrene, PVC and polymethyl methacrylate. When the rubber is vulcanized into rubbery products exhibiting with good strength and elongation, polymers are used as 'elastomers'. Some of the common examples are natural rubber, synthetic rubber, silicone rubber. Usually the length of the filament-like materials is at least 100 times its diameter, polymers which are said to have been converted into 'fibres'. Examples are nylon and terylene. Polymers are used as potting compounds, adhesives, sealants, *etc.* which are generally in a liquid form are described as liquid resins. Epoxy adhesives and polysulphide sealants are commercially available typical examples.

Polymer Structure–Property Relationships

Polymer properties are not only related to their chemical nature, but also attribute to each and individual factors such as extent and distribution of long polymer chain length, distribution of crystallinity, and amount of additives, such as reinforcing agents, fillers, and plasticizers. These factors generally influence all essential polymeric properties to certain case including flammability, hardness, weatherability, biological response, chemical stability, comfort, flex life, appearance, moisture retention, dyeability, softening point, and electrical properties. List of selected property-structure relationships are given in Table 5.1.

Crystalline and Amorphous Combinations

Most polymers consist of both crystalline and amorphous domains. In crystals such as spherulites, where (Figure 5.3a–c), the regions are less ordered with folded crystalline lamellae, by approximating amorphous regions. This combination of crystalline and amorphous regions is important for the formation of materials that have both good strength (contributed largely by the crystalline regions) and some flexibility or "softness" (derived from the amorphous portions). Figure 5.3b contains a space-filled model for PE chains (a total of about 400 units with 5 branches, one longer and four shorter).

Table 5.1. Some of the Selected Property-Structure Relationships

Glass Transition Temperature	*Solubility*	*Crystallinity*
Increases with the presence of: ❒ bulky pendant groups ❒ stiffening groups as 1,4-phenylene ❒ chain symmetry ❒ polar groups ❒ cross-linking Decreases with the presence of: ❒ dditives like plasticizers ❒ flexible main chain groups ❒ nonpolar groups ❒ dissymmetry	Favored by: ❒ Longer chain lengths ❒ Low interchain forces ❒ Disorder and dissymmetry ❒ Increased temperature ❒ Compatible solvent	Favored by: ❒ High interchain forces ❒ Regular structure; high symmetry ❒ Decrease in volume ❒ Increased stress ❒ Slow cooling from melt ❒ Homogenous chain length

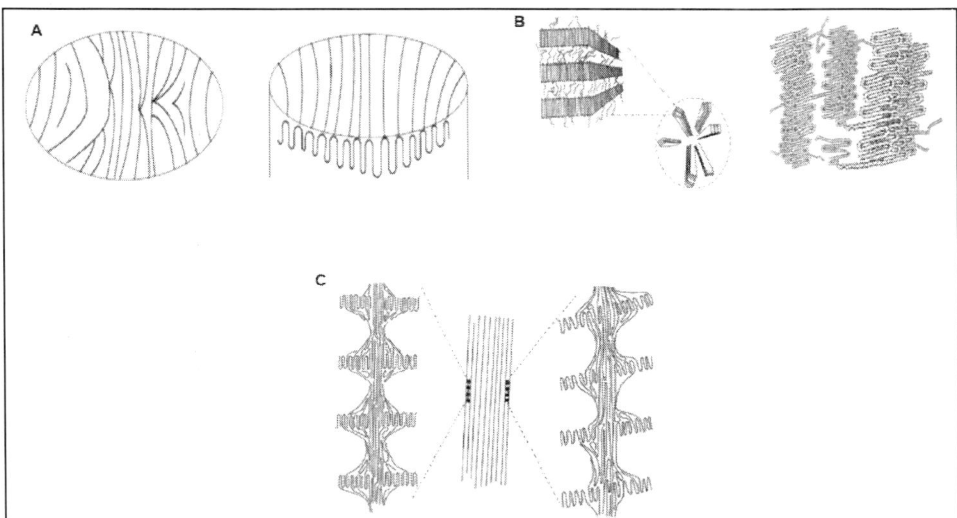

Figure 5.3. (a) Structure of a spherulite from the bulk; (b) Spherulite structure showing the molecular-level lamellar chain-folded platelets and tie and frayed chain arrangements; (c) Crystalline polymer structures formed under applied tension including flow conditions.
Courtesy: Carraher Jr, 2003.

Rheology of Polymeric Materials

The name rheology is given by Bingham, who some call as the father of modern rheology. The prefix rheo is derived from the Greek *rheos*, meaning current of flow. The study of deformation and flow of materials is called as Rheology. The main two branches of rheology are of mechanics-fluid and solid. Low molecular weight solids and liquids exhibit their characteristics flow behavior according to Hooke's equation or Newton's equation, which are now discussed as Hooke's equation.

Hooke's Equation

The behavior of an ideal solid obeying Hooke's law can be represented by elongation versus time plot in Figures 5.4 a and b. Application of a load at a certain point of time t_1 to the solid material elongates the material by a certain length and the extended length remains as long as the load is retained. We remove the load at time t_2 and the material goes back to its original length instantaneously. This is the deformation behavior of an ideal solid. On other words, an ideal solid elongates instantaneously upon loading and retracts instantaneously upon unloading. This behavior is called *elasticity or elastic deformation* and it time-independent. The stress-strain (or force elongation) plot for such a deformation will be as represented in Figures 5.5 a and b.

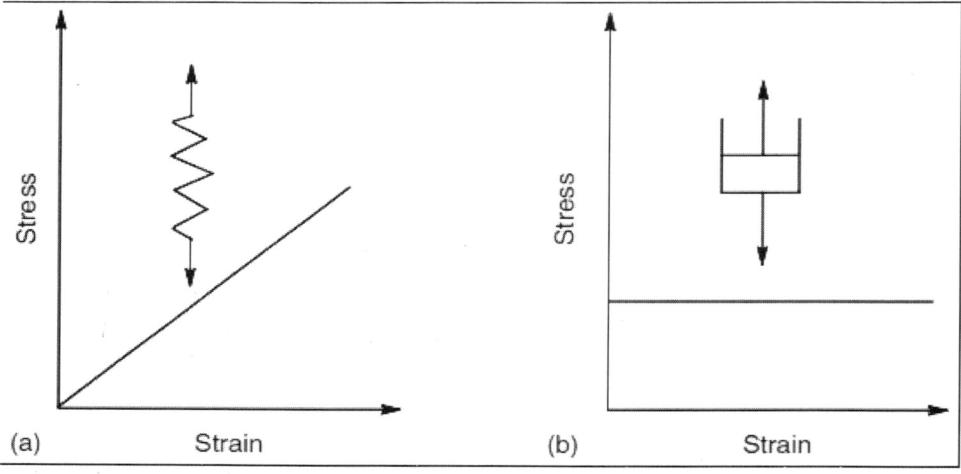

Figure 5.4. (a) Stress–strain plots for a Hookean spring where E is the slope, (b): Newtonian dashpot where S is a constant.
Courtesy: Carraher Jr, 2003.

From the diagrams, it can be seen that the applied stress (*S*) is proportional to the resultant strain or elongation (*E*) and is independent of the rate of strain (*dE/dt*). This is *Hookes's law*, and is represented as follows:

$$S = GE$$

where *G* is the modulus of elasticity, a quantity that is representative of the material resistance to deformation. Differentiation of Eqn. (1) gives

Now, elongation is given by the stress divided by modulus:

$$E = \frac{S}{G}$$

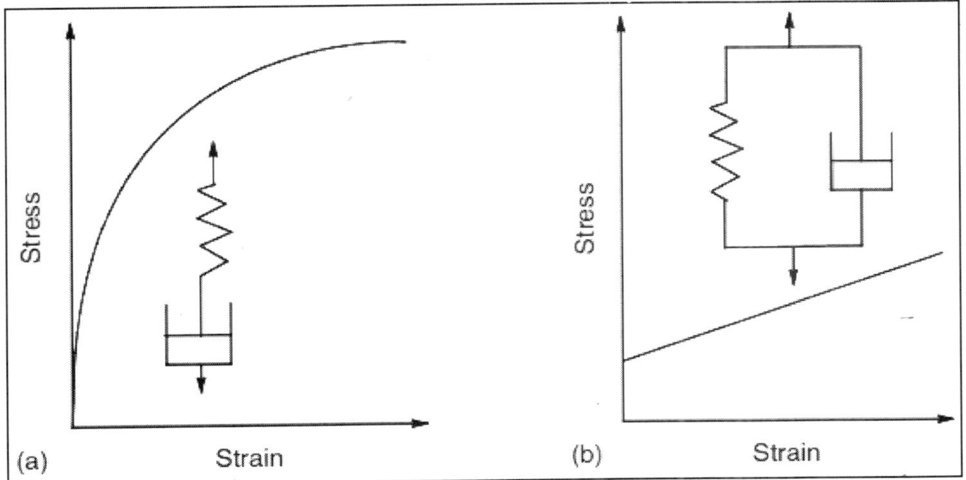

Figure 5.5. (a): Stress–strain plot for stress relaxation for the Maxwell model; (b): Voigt–Kelvin model.
Courtesy: Carraher Jr, 2003.

Thus, the larger the modulus, the more rigid will be material. The behavior of an ideal solid can be compared to that or a steel spring which elongates readily upon loading and goes back to its original shape upon unloading.

Typical Stress – Strain Behaviour

The stress-strain behavior typically consists of three main polymer groupings Figure 5.6. Mostly nondestructive physical test are evaluated. The three types of mechanical stress measures will be considered Figures 5.7.a, b and c. The stress to strain ratio is called as *Young's modulus*. This ratio usually called as elasticity of modulus and tensile modulus. It is generally calculated by simply dividing stress by strain.

Young's modulus= stress (Pa)/strain (mm/mm)

The main reasons for studying rheology (stress –strain relationships) are:

1. To determine the suitability of materials to serve specific applications; and
2. To relate the results to polymer structure and form. Studying structure–property relationships allows a better understanding of the observed results on a molecular level resulting in a more knowledgeable approach to the design of materials.

Polymer Nanocomposites

Polymers are most commonly reinforced by various nano sized filler materials to alleviate some of their limitations and besides broadening their applications The usage of nanoscale fillers in composites will to improve mechanical and physical

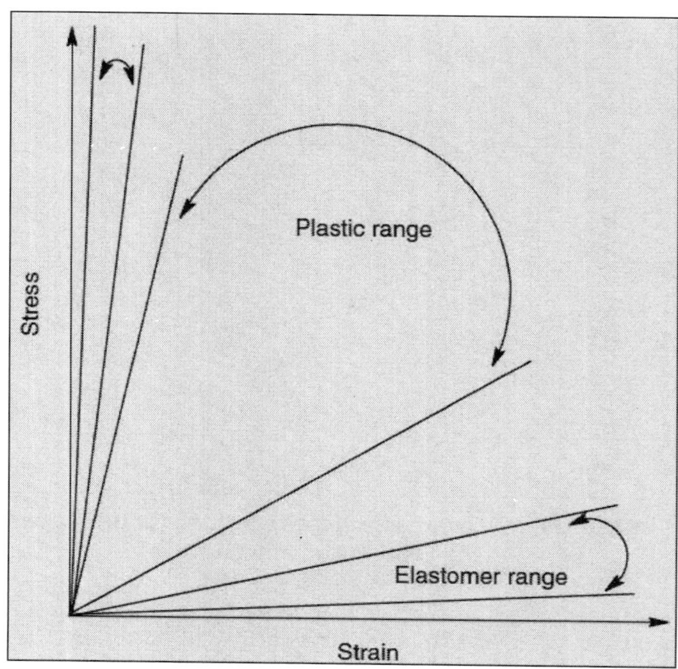

Figure 5.6. Typical stress–Strain Behavior for Fibers, Plastics, and Elastomers.
Courtesy: Carraher Jr, 2003.

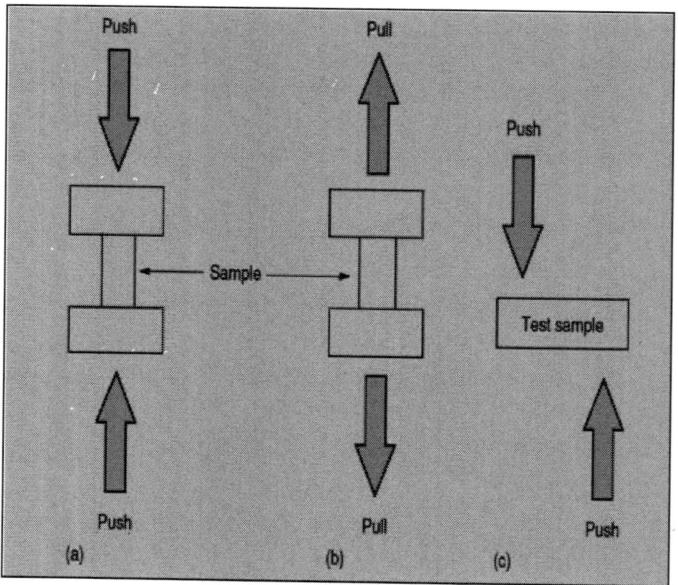

Figure 5.7. Major Types of Stress Tests: Compressive (a) Pulling Stress or Tensile Strength (b) and Shear Stress (c).

properties (Reynaud *et al.*, 1999; Young *et al.*, 2012). These nanoscale fillers have a characteristic length scale of nanometers and vary from isotropic to anisotropic sheet or needle-like morphologies. Nanoscience and nanotechnology always provides with unique opportunities to create a alternative composite material with combinations of polymer matrix and nanoscale fillers to get a desirable polymer nanocomposites for different application purposes. Polymer nanocomposites are categorized into three major types based on their dimensions. In the first type, the two dimensional (2D) nanoscale fillers layered silicate (Albdiry *et al.*, 2013), graphene (Young *et al.*, 2012) or MXene (Li *et al.*, 2018; Tu *et al.*, 2018) usually measured in the range of 1-100 nm thick to 100-1000 nm long like polymer sheets. In the second type, two dimensions nanoscale fillers are nanofibers or nanotubes, *e.g.*, carbon nanofibers and nanotubes (Calvert, 1999) or halloysite nanotubes are typically in nanometer scale (Liu *et al.*, 2014). These are used as a reinforcing nanofillers materials to obtain materials with good properties. The third type is the nanocomposites film which contains one of the nanoscale fillers with three dimensions in range of nanometers. Some of the examples are semiconductor nanoclusters (Herron and Thorn, 1998), silica (Reynaud *et al.*, 1999; Mark, 1996) and quantum dots (Huang *et al.*, 2016). Figure 5.8. Shows the various types of nanofillers

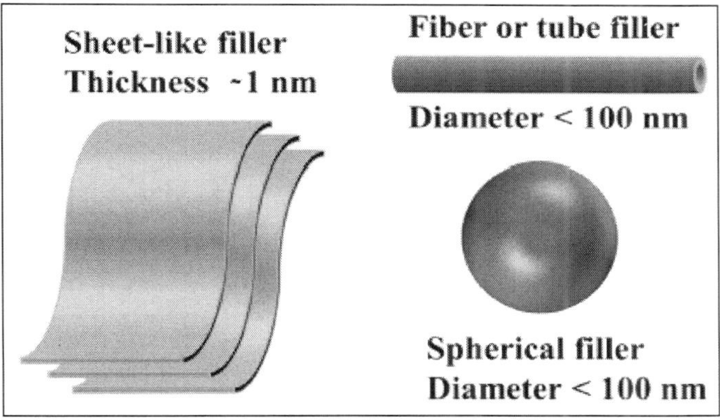

Figure 5.8. The Three Major Types of Nanoscale Fillers.

Processing of Polymer Nanocomposites

Good quality of polymer nanocomposite with high performance can be prepared by using the various processing techniques *viz.*, melt mixing, solution mixing, in situ polymerization, in-situ particle processing and other approaches (Coleman *et al.*, 2011; Mayerberger *et al.*, 2017; Petrovic *et al.*, 2015; Yang *et al.*, 2015). The components of polymer nanocomposites consist of (polymer matrix, organic cation, and layered silicate) are shown in Figure 5.7. During the processing, if the polymer is not intercalating between the silicate sheets, a phase separated micro-composite is obtained as shown (Figure 5.9a). The other two types of polymer nanocomposites can be achieved when the clay is intercalated or exfoliated, followed by the uniformly dispersed silicate layers in a continuous polymer medium are successfully obtained as exfoliated structure. For intercalated and exfoliated polymer are shown in (Figure 5.9b and c).

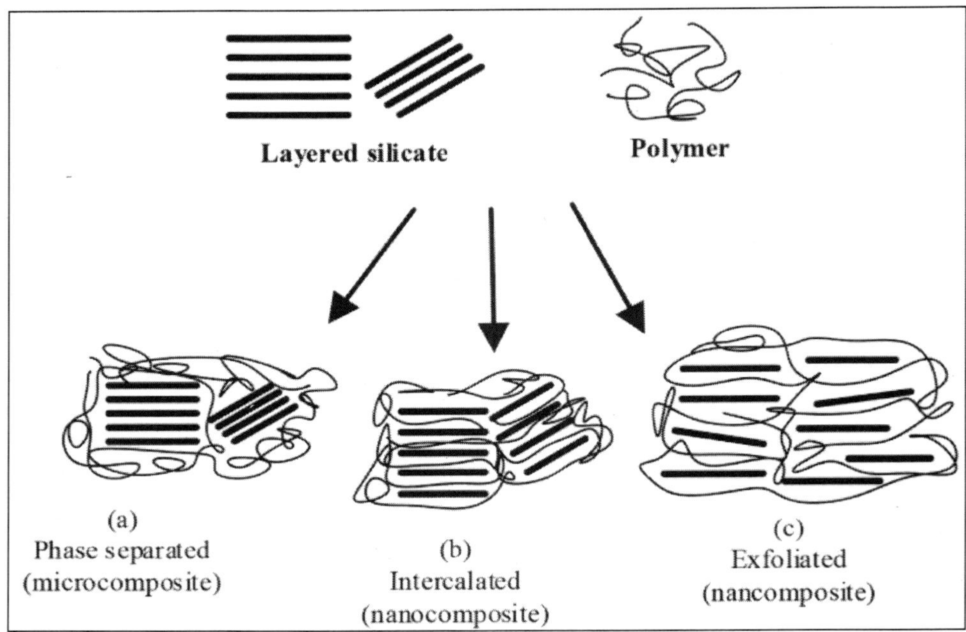

Figure 5.9: Schematic Illustration of Three different Types of Nanocomposites Arising from the Interaction of Layered Silicates and Polymers (a) Phase separated polymer micro-composite; (b) intercalated polymer nanocomposite; and (c) exfoliated polymer nanocomposite.
Courtesy: Alexandre and Dubois, 2000.

Plasticizers in Polymer Nanocomposites

According to the IUPAC (International Union of Pure and Applied Chemistry) council plasticizer is defined as "a substance or material incorporated in a polymer or elastomer material to enhance composite workability, flexibility, or distensibility"(Vieira *et al.*, 2011). Other properties are optical clarity, degree of crystallinity, electric conductivity, fire behaviour and resistance to biological degradation, amongst other physical properties. (Florjan´czyk and Florjan´czyk, 2007). In this scenario, during the processing of polymers few drops to milliliter of plasticizer molecules are incorporated with polymer chains, but the primary bonds are not chemically attached and therefore they are lost through evaporation, extraction, or migration. (Frados, 1976)

Plasticizers are generally classified as primary and secondary (Krauskopf, 1993). At high concentration of polymer the plasticizer is soluble and it said to be called as a primary plasticizer. These particular plasticizers are commonly used as either main element or as the sole plasticizer. The secondary plasticizers have limited compatibility and lower gelation capacity with respect to the polymer, which are typically blended along with the primary plasticizers, and overall improves the product properties or reduce the cost (Chanda and Roy, 1986).

Electrochemistry

Fundamentals of Electrode Processes

"Electrochemistry is the branch of chemistry concerned with the interrelation of electrical and chemical effects. The field of electrochemistry encompasses a huge array of different phenomena (*e.g.*, electrophoresis and corrosion), devices (electrochromic displays, electro analytical sensors, batteries, and fuel cells), and technologies (the electroplating of metals and the large-scale production of aluminium and chlorine).

Electrochemical Cells

In electrochemical systems, the processes and factors that affect the transport of charge across the interface between chemical phases, for example, between an electronic conductor (an electrode) and an ionic conductor (an electrolyte). Charge is transported through the electrode by the movement of electrons (and holes). Typical electrode materials include solid metals (*e.g.*, Pt, Au), liquid metals (Hg, amalgams), carbon (graphite), and semiconductors (indium-tin oxide, Si). In the electrolyte phase, charge is carried by the movement of ions. The most frequently used electrolytes are liquid solutions containing ionic species, such as, H^+, Na^+, Cl^-, in either water or a non- aqueous solvent. To be useful in an electrochemical cell, the solvent/electrolyte system must be of sufficiently low resistance (*i.e.*, sufficiently conductive) for the electrochemical experiment envisioned. Less conventional electrolytes include fused salts (*e.g.*, molten NaCl-KCl eutectic) and ionically conductive polymers (*e.g.*, Nation, polyethylene oxide-$LiClO_4$). Solid electrolytes also exist (*e.g.*, sodium j8-alumina, where charge is carried by mobile sodium ions that move between the aluminium oxide sheets). The interfaces at which chemical reaction takes place through electrode are known as electrochemical cells.

The overall chemical reaction taking place in a cell is made up of two independent half-reactions, which describe the real chemical changes at the two electrodes. Each half-reaction (and, consequently, the chemical composition of the system near the electrodes) responds to the interfacial potential difference at the corresponding electrode. Most of the time, one is interested in only one of these reactions, and the electrode at which it occurs is called the working (or indicator) electrode. To focus on it, one standardizes the other half of the cell by using an electrode (called a reference electrode) made up of phases having essentially constant composition.

The internationally accepted primary reference is the standard hydrogen electrode (SHE), or normal hydrogen electrode (NHE). Potentials are often measured and quoted with respect to reference electrodes other than the NHE, which is not very convenient from an experimental standpoint. A common reference electrode is the saturated calomel electrode (SCE). Since the reference electrode has a constant makeup, its potential is fixed. Therefore, any changes in the cell are ascribable to the working electrode.

Observing or control the potential of the working electrode with respect to the reference is equivalent to observing or controlling the energy of the electrons within the working electrode. By driving the electrode to more negative potentials (*e.g.*, by connecting a battery or power supply to the cell with its negative side attached to the working electrode), the energy of the electrons is raised. They can reach a level high enough to transfer into vacant electronic states on species in the electrolyte. In that case, a flow of electrons from electrode to solution (a reduction current) occurs. Similarly, the energy of the electrons can be lowered by imposing a more positive potential, and at some point electrons on solutes in the electrolyte will find a more favourable energy on the electrode and will transfer there. Their flow, from solution to electrode, is an oxidation current. The critical potentials at which these processes occur are related to the standard potentials, E^o, for the specific chemical substances in the system.

When one plots the current as a function of the potential, one obtains a current potential (i -E) curve. Such curves can be quite informative about the nature of the solution and the electrodes and about the reactions that occur at the interfaces. It is possible to calculate the open-circuit potential from thermodynamic data, that is, from the standard potentials of the half-reactions involved at both electrodes via the Nernst equation. The key point is that a true equilibrium is established, because a pair of redox forms linked by a given half-reaction (*i.e.*, a redox couple) is present at each electrode.

Ideal Polarized Electrode

An electrode at which no charge transfer can occur across the metal-solution interface, regardless of the potential imposed by an outside source of voltage, is called an ideal polarized (or ideal polarizable) electrode (IPE). For example, a mercury electrode in contact with a deaerated potassium chloride solution approaches the behaviour of an IPE over a potential range about 2 V wide. At sufficiently positive potentials, the mercury can oxidize in a charge-transfer reaction.

Electrochemical Thermodynamics

Reversibility

Since thermodynamics can strictly encompass only systems at equilibrium, the concept of reversibility is important in treating real processes thermodynamically. Reversibility is of 3 types: chemical reversibility, thermodynamic reversibility and practical reversibility. If the cell reaction reverses itself just by reversing the cell current, it is termed as chemically reversible. A process is thermodynamically reversible when an infinitesimal reversal in a driving force causes it to reverse direction. Obviously, this cannot happen unless the system feels only an infinitesimal driving force at any time; hence it must essentially be always at equilibrium. A reversible path between two states of the system is therefore one that connects a continuous series of equilibrium states. Traversing it would require an infinite length of time. A cell that is chemically irreversible cannot behave reversibly in a thermodynamic sense. Further, Nernst equation used to provide a linkage between electrode potential E and the concentration of participants in the electrode process. If a system follows Nernst equation it is said to be thermodynamically or electrodynamically reversible.

$$E = E^0 + \frac{RT}{NF} \ln \frac{C_O}{C_R}$$

Free Energy and Cell emf

The potential difference is therefore always the equilibrium (open-circuit) value. Since the extent of reaction is supposed to be small enough that all activities remain constant, the potential also remains constant. Then, the energy dissipated in R is given by:

$$|\Delta G| = nF|E|$$

where n is the number of electrons passed per atom of zinc reacted (or the number of moles of electrons per mole of Zn reacted), and F is the charge on a mole of electrons, which is about 96,500 C.

A positive emf gives spontaneous reaction, which will give a formula as,

$$\Delta G^O = -nFE^0_{rxn}$$

where E^0_{rxn} is known as standard emf of the cell reaction.

The equilibrium constant of the reaction given by:

$$RT \ln K_{rxn} = -\Delta G^o = nFE^0_{rxn}$$

Liquid Junction Potential

The potential differences in equilibrium electrochemical systems can be treated exactly by thermodynamics. However, many real cells are never at equilibrium, because they feature different electrolytes around the two electrodes. There is somewhere an interface between the two solutions, and at that point, mass transport processes work to mix the solutes. Unless the solutions are the same initially, the liquid junction will not be at equilibrium, because net flows of mass occur continuously across it.

For example in (Figure 5.10),

$$Cu/Zn/Zn^{2+}/Cu^{2+}/Cu'$$

$$\alpha \qquad \beta$$

The overall cell potential at null current is then,

$$E = (\phi^{Cu} - \phi^{\beta}) - (\phi^{Cu} - \phi^{\alpha}) + (\phi^{\beta} - \phi^{\alpha})$$

There are 3 types of liquid junction potential (Figure 5.11),

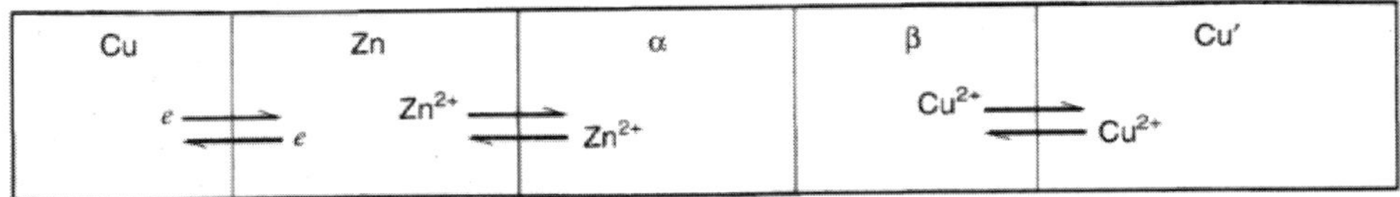

Figure 5.10. Liquid Junction Potential Example.

1. Two solutions of the same electrolyte at different concentrations
2. Two solutions at the same concentration with different electrolytes having an ion in common
3. Two solutions not satisfying conditions 1 or 2

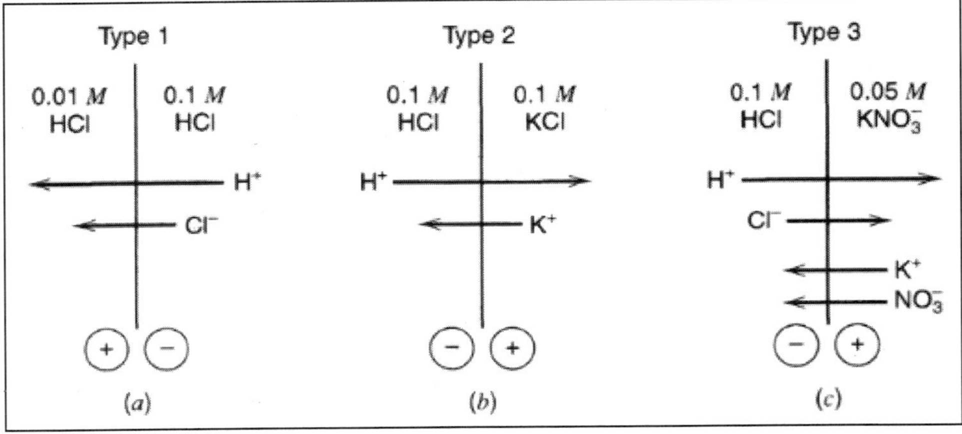

Figure 5.11. Types of Liquid Junction Potential.

Conductance, Transference Numbers, and Mobility

When an electric current flows in an electrochemical cell, the current is carried in solution by the movement of ions. The fractions of the current carried by H^+ and Cl^- are called their transference numbers.

$$t^+ + t^- = 1$$

Transference numbers are determined by the details of ionic conduction, which are understood mainly through measurements of either the resistance to current flow in solution or its reciprocal, the conductance, L. The value of L for a segment of solution immersed in an electric field is directly proportional to the cross-sectional area perpendicular to the field vector and is inversely proportional to the length of the segment along the field. The proportionality constant is the conductivity, ê, which is an intrinsic property of the solution:

$$L = KA/1$$

Since the passage of current through the solution is accomplished by the independent movement of different species, ê is the sum of contributions from all ionic species, i. It is intuitive that each component of ê is proportional to the concentration of the ion, the magnitude of its charge |Zi|, and some index of its migration velocity.

That index is the mobility, u_i, which is the limiting velocity of the ion in an electricfield of unit strength. Mobility usually carries dimensions of cm^2V^{-1}/cm^{-1}). When a field of strength ξ is applied to an ion, it will accelerate under the force imposed by the field until the frictional drag exactly counter balances the electric force. Then, the ion continues its motion at that terminal velocity.

The proportionality factor relating an individual ionic conductivity to charge, mobility, and concentration turns out to be the Faraday constant; thus

$$\kappa = F\sum_i |z_i| u_i C_i$$

Electro-Capillary Phenomena

Let us now consider a specific chemical system in which a mercury surface contacts an aqueous KCl solution. The potential of the mercury is controlled with respect to a reference electrode having no liquid junction with the test solution. Suppose also that the aqueous phase contains a neutral species, M, that might be interfacially active. For example, the cell could be

Cu'Ag/AgCl/K^+, Ñl$^-$, M/Hg/Ni/Cu

Applying Gibbs adsorption isotherm,

$$-d\gamma = \left(\Gamma_{Hg} d\bar{\mu}_{Hg} + \Gamma d\bar{\mu}_e^{Hg}\right) + \left(\Gamma_K + d\bar{\mu}_K + \Gamma_{Cl^-} d\bar{\mu}_{Cl^-}\right) + \left(\Gamma_M + d\bar{\mu}_M + \Gamma_{H_2O} d\bar{\mu}_{H_2O}\right)$$

In which condition applies,

$$\bar{\mu}_e^{Hg} = \bar{\mu}_e^{Cu}$$

$$\bar{\mu}_{KCl} = \mu_{KCl} = \bar{\mu}_{K^+} + \bar{\mu}_{Cl^-}$$

$$\bar{\mu}_{H_2O} = \mu_{H_2O}$$

$$\bar{\mu}_M = \mu_M$$

Further recognising $d\bar{\mu}_{Hg} = d\bar{\mu}^0_{Hg} = 0$ and from reference interface, $d\bar{\mu}_{AgCl} = d\bar{\mu}_{Ag} = 0$, we get

$$-d\gamma = \Gamma_s d\bar{\mu}_s^{Cu} - [\Gamma_{K^+} - \Gamma_{Cl^-}] d\bar{\mu}_s^{Cu'} + \Gamma_{K^+} d\mu_{KCl} + \Gamma_M d\mu_M + \Gamma_{H2O} d\mu_{H2O}$$

The excess charge density on the metallic side of the interface is

$$\sigma^M = -F\Gamma_s$$

An equal, but opposite charge density resides on the solution side:

$$\sigma^S = -\sigma^M = F(\Gamma_{K^+} - \Gamma_{Cl^-})$$

In addition,

$$d\bar{\mu}_{e}^{Cu} - d\bar{\mu}_{e}^{Cu'} = -Fd(\phi^{Cu} - \phi^{Cu'}) = -FdE_{-}$$

$$\Rightarrow -d\gamma = \sigma^{M} dE_{-} + d\mu_{KCl} + \Gamma_{M} d\mu_{M} + \Gamma_{H2O} d\mu_{H2O}$$

Again, according to Gibbs-Duhem relation for the aqueous phase,

$$\sum_{i} X_i \, d\mu_i = 0$$

Hence, $X_{H2O} \, d\mu_{H2O} + X_{KCl} d\mu_{KCl} + X_{M} d\mu_{M} = 0$, which gives,

$$-d\gamma = \sigma^{M} dE_{-} + \left[\Gamma_{K^+} - \frac{X_{KCl}}{X_{H2O}}\Gamma_{H2O}\right] d\mu_{KCl} + \left[\Gamma_{M} - \frac{X_{M}}{X_{H2O}}\Gamma_{H2O}\right] d\mu_{M}$$

which is known as electrocapillary equation.

Zeta Potential

Zeta potential is a scientific term for electro-kinetic potential in colloidal dispersions. In the colloidal chemistry literature, it is usually denoted using the Greek letter zeta (delu-123), hence ζ-potential. The usual units are volts (V) or millivolts (mV). From a theoretical viewpoint, the zeta potential is the electric potential in the interfacial double layer (DL) at the location of the slipping plane relative to a point in the bulk fluid away from the interface. In other words, zeta potential is the potential difference between the dispersion medium and the stationary layer of fluid attached to the dispersed particle.

The zeta potential is caused by the net electrical charge contained within the region bounded by the slipping plane, and also depends on the location of that plane. Thus, it is widely used for quantification of the magnitude of the charge. However, zeta potential is not equal to the Stern potential or electric surface potential in the double layer, because these are defined at different locations. Such assumptions of equality should be applied with caution. Nevertheless, zeta potential is often the only available path for characterization of double-layer properties (Figure 5.12).

The zeta potential is a key indicator of the stability of colloidal dispersions. The magnitude of the zeta potential indicates the degree of electrostatic repulsion between adjacent, similarly charged particles in a dispersion. For molecules and particles that are small enough, a high zeta potential will confer stability, *i.e.*, the solution or dispersion will resist aggregation. When the potential is small, attractive forces may exceed this repulsion and the dispersion may break and flocculate. So, colloids with high zeta potential (negative or positive) are electrically stabilized while colloids with low zeta potentials tend to coagulate or flocculate.

Let us consider the application of an electric field along a glass capillary tube filled with an electrolyte solution: The walls of the capillary tube will be charged at most pH values, because of protonation/deprotonation equilibria involving surface Si-OH groups. At a pH above about 3, the surface charge will be negative.When an

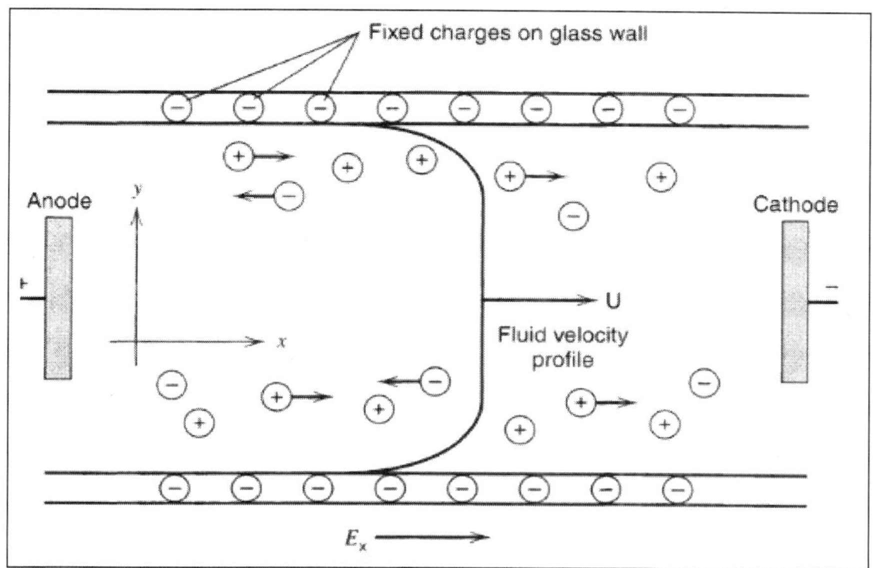

Figure 5.12. Calculating Electro-Capillary Phenomenon and Zeta Potential.

electric field is applied along the capillary axis, this excess charge (in the form of electrolyte cations) will move toward the cathode and the net viscous drag on the solvent will induce convection of the solution. In most cases, the radius of the tube will be large compared to the thickness of the diffuse layer, so that curvature can be neglected, and the flow considered as taking place in a planar channel.

It can be expressed mathematically as,

$$\eta_s \frac{\partial^2 v}{\partial y^2} = -\rho_E \xi_x = \varepsilon\varepsilon_0 \frac{\partial^2 \phi}{\partial y^2} \xi_x$$

where the charge density has been obtained from the Poisson equation,

$$\rho_E = -\varepsilon\varepsilon_0 \frac{\partial^2 \phi}{\partial y^2}$$

Integrating, first equation, it will give,

$$\eta_s \left(\frac{\partial v}{\partial y}\right) = \varepsilon\varepsilon_0 \left(\frac{\partial \phi}{\partial y}\right) \xi_x$$

A second integration from a position near the wall, where v = 0 and ϕ = g, to the middle of the channel, where v = U and ϕ = 0, gives the Helmholtz-Smoluchowski equation,

$$U = -\varepsilon\varepsilon_0 \zeta \xi_x / \eta_s$$

The parameter ξ is called the zeta potential, which is the potential in a not-very-well-defined position in the diffuse layer called the shear plane and U is the electroosmoticsolution velocity past a plane charged surface.

Cyclic Voltammetry

Cyclic voltammetry (CV) is a type of potentiodynamic electrochemical measurement. In a cyclic voltammetry experiment, the working electrode potential is ramped linearly versus time. Unlike in linear sweep voltammetry, after the set potential is reached in a CV experiment, the working electrode's potential is ramped in the opposite direction to return to the initial potential. These cycles of ramps in potential may be repeated as many times as needed. The current at the working electrode is plotted versus the applied voltage (that is, the working electrode's potential) to give the cyclic voltammogram trace. Cyclic voltammetry is generally used to study the electrochemical properties of an analyte in solution or of a molecule that is adsorbed onto the electrode.

The shape of the curve on reversal depends on the switching potential, E_{λ}, or how far beyond the cathodic peak the scan is allowed to proceed before reversal. Cyclic voltammetry, a three electrode electrochemical cell, is based on varying the applied potential at a working electrode in both forward and reverse directions (at some scan rate) while observing the current. This technique involves a linear and a cyclic variation of electrode potential between the working and reference electrodes within a potential window by measuring the current that flows between working and counter electrodes. Cyclic voltammetry (CV) is a potentio-dynamic electrochemical measurement in which working electrode's potential is ramped in the opposite direction to return to its initial potential. These cycles may be repeated many times. Cyclic voltammogram gives the plot of current versus applied potential.

In CV technique, the electrode potential increases linearly with time in cyclic manner. Scan rate is given by the rate of voltage variation over time during each cycle. The voltage is applied between the working electrode and reference electrode and current is measured between working electrode and counter electrode. Cyclic voltammetry consists of variation of working electrode's potential using a triangular wave form. In the initial forward scan between t_0 to t_1, the potential is increased. The current (cathodic) also increases over time because of the analyte species. During this period, the reduction takes place. At some point the reduction potential of the analyte reaches (at t_1), the current decreases as the analyte concentration is lowered. During the reverse scan (from t_1 to t_2) the reduced analyte starts to re-oxidize which gives rise to a current (anodic) of reverse polarity to before. The oxidation and reduction peak will be more similar if the reaction is more reversible. Hence cyclic voltammetry data provides the information about the redox potentials and electrochemical reaction rates. In this way the potential is swept from E_1 to E_2 at a given scan rate. Single or multiple cycles can be performed. In a given potential range, the potentiostat gives the resulting current according to voltage Figure 5.13.

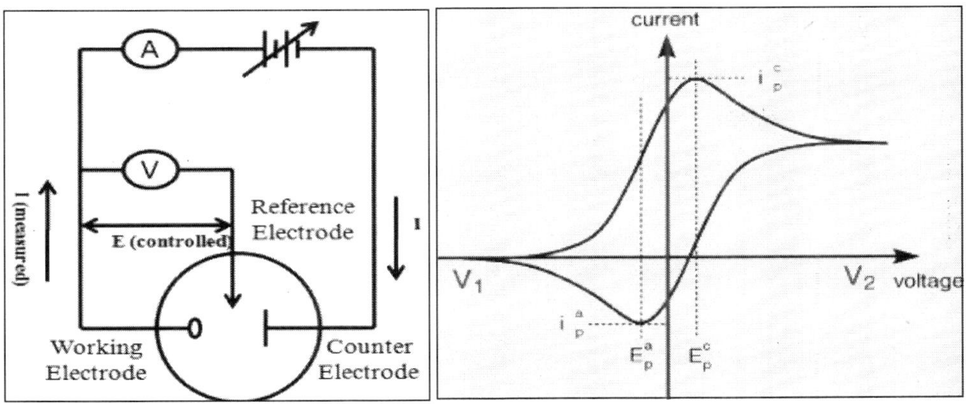

Figure 5.13. Principle and Voltammogram in Cyclic Volammetry.

Analysis of Voltammogram

Cyclic voltammetry is a widely used technique in electrochemistry. It gives the basic information about a capacitive electrochemical cell which includes:

1. Capacitance,
2. Voltage window,
3. Cycle life.

Peak potentials (E_{pc}, E_{pa}) and peak currents (i_{pc} and i_{pa}) of cathodic and anodic peaks are the important parameters in a cyclic voltammogram

The peak current i_p is given by,

$$i_p = (2.69 \times 10^5) n^{3/2} A D^{1/2} V^{1/2} C$$

The midpoint potential of the two peaks in the voltammogram is given by:

$$E_{mid\ point} = \frac{(E_{p,cathodic}) + (E_{p,anodic})}{2} = E_0' + \frac{RT}{NF} \ln\left(\frac{D_R^{1/2}}{D_o^{1/2}}\right)$$

where, E_o' isredox potential and D_o and D_R are the diffusion coefficients for the oxidized and reduced halves of that couple.

Separation between the two peaks of the voltammogram is given by,

$$\Delta E_p = \left| E_{p,anodic} - E_{p,cathodic} \right|$$

The charge storage capacity can be quantitatively evaluated by the expression,

$$specific\ capacitance = \frac{Voltametric\ Charge}{(Potential\ window \times Mass)}$$

Voltammetric charge is given by sum of anodic and cathodic voltammetric charges. It is given by integral area of CV curve/scan rate. Hence the specific capacity is give by,

$$C = \int_{E_1}^{E_2} i(E)dE/2(E_2 - E_1)\, mv$$

Instrument Set-up for Cyclic Voltammetry

Cyclic voltammetry experiments are carried out on a solution in a cell fitted with electrodes. The electrochemical cell, where the voltammetric experiment is carried out, consists of a working electrode, are reference electrode, and usually a counter (auxiliary) electrode Figure 5.14.

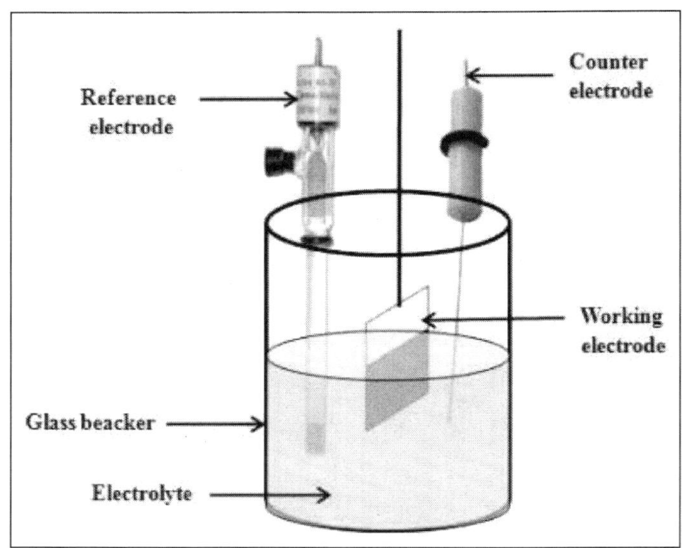

Figure 5.14: Cyclic Voltammetry Set-up.

The working electrode is the electrode where the potential is controlled and the current is measured. This is the electrode at which the electrochemical phenomena (reduction or oxidation) being investigated are taking place. The working electrode serves as a surface on which the electrochemical reaction takes place.

The reference electrode is used to measure the working electrode potential. It should have a constant electrochemical potential as long as no current flows through it. Commonly used reference electrodes are the silver-silver chloride electrode (Ag/AgCl) or the Saturated Calomel electrode (SCE).

The counter or auxiliary electrode which serves as a source or sink for electrons so that current can be passed from the external circuit through the cell. Simply, it is a conductor that completes the cell circuit. The current that flows into the solution via the working electrode leaves the solution via the counter electrode. The counter electrode should be a non-reactive, high surface area electrode such as platinum or carbon.

These three electrodes are immersed in an electrolyte. The collection of the electrodes, the solution, and the container holding the solution are referred to as an electrochemical cell. Electrolyte provides the sufficient conductivity. The solvent, electrolyte and the material used for working electrode, all these factors determine the potential range that can be accessed during the cyclic voltammetry experiment.

Applications of Cyclic voltammetry

Cyclic voltammetry is used in various fields.

- ✰ CV has become an important and widely used electro-analytical technique in many areas of chemistry. It is often used to study a variety of redox processes, to determine the stability of reaction products, the presence of intermediates in redox reactions, reaction.
- ✰ Analytical chemists routinely use voltammetric techniques for the quantitative determination of a variety of dissolved inorganic and organic substances.
- ✰ Inorganic, physical, and biological chemists widely use voltammetric techniques for a variety of purposes, including fundamental studies of oxidation and reduction processes in various media, adsorption processes on surfaces, electron transfer and reaction mechanisms, kinetics of electron transfer processes, and transport, speciation, and thermodynamic properties of solvated species.

Electro-Chemical Impedance Spectroscopy

Consider Ohm's law, which describes the relationship of voltage to a direct current passingthrough a resistor: $E = IR$

Impedance is, very simply, extends the concept of resistance to an alternating current circuit, and generally represented as Z.

$$E = IZ$$

With applying Oscillating voltage,

$$E(t) = \|E\| \sin(\omega t)$$

$$I(t) = |I| \sin(\omega t + \theta)$$

Hence,

$$Z = \frac{E(t)}{I(t)} = \frac{|E| \sin(\omega t)}{|I| \sin(\omega t + \theta)} = |Z| \frac{\sin(\omega t)}{\sin(\omega t + \theta)}$$

Further, according to Euler's formula,

$$e^{jx} = \cos(x) + j \sin(x)$$

Applying Euler's formula,

$$Z = |Z|e^{j\theta}$$

$$\Rightarrow E = IZ = I|Z|e^{j\theta}$$

We can represent any Z on an Argand diagram, as in the graph below Figure 5.15. This is the basis for the Nyquist plot, which is the plot of the real and imaginary parts of the impedance that you'll come across most often. An impedance measurement for a single frequency is a single point on a Nyquist plot. An impedance spectrum is therefore a series of points, where each point is a different frequency. On the Nyquist Plot the impedance can be represented as a vector (arrow) of length |Z|. The angle between this vector and the X-axis, commonly called the "phase angle", is φ (=arg Z).Nyquist Plots have one major shortcoming. When you look at any data point on the plot, you cannot tell what frequency was used to record that point.

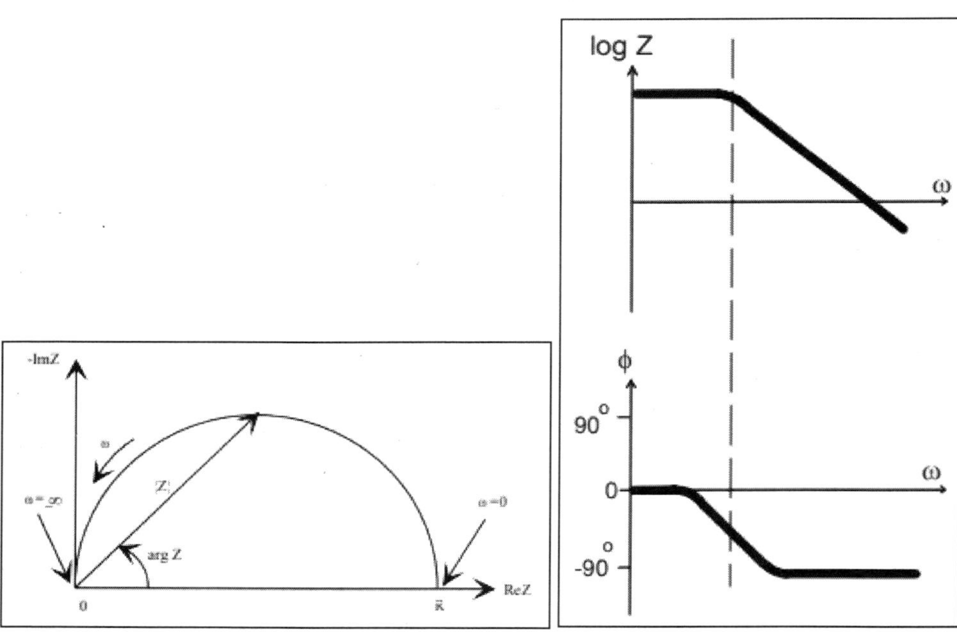

Figure 5.15. Nyquist and Bode Plot.

Another popular presentation method is the Bode Plot. The impedance is plotted with log frequency on the X-axis and both the absolute values of the impedance ($|Z|=Z_0$) and the phase-shift on the Y-axis" (Bard *et al.*, 1980).

REFERENCES

1. Albdiry, M. T., Yousif, B. F., Ku, H., and Lau, K. T. (2013). A critical review on the manufacturing processes in relation to the properties of nanoclay/polymer composites. *Journal of composite materials*, *47*(9), 1093-1115.

2. Alexandre, M., and Dubois, P. (2000). Polymer-layered silicate nanocomposites: preparation, properties and uses of a new class of materials. *Materials Science and Engineering: R: Reports, 28*(1-2), 1-63.
3. Bard, A. J., Faulkner, L. R., Leddy, J., and Zoski, C. G. (1980). *Electrochemical methods: fundamentals and applications* (Vol. 2). New York: wiley.Figure 4. Typical stress–strain behavior for fibers, plastics, and elastomers.
4. Bia³ecka-Florjañczyk, E., and Florjañczyk, Z. (2007). Solubility of plasticizers, polymers and environmental pollution. In *Thermodynamics, Solubility and Environmental Issues* (pp. 397-408). Elsevier.
5. Calvert, P. (1999). Nanotube composites: a recipe for strength. *Nature, 399*(6733), 210.
6. Carraher Jr, C. E. (2003). *Seymour/Carraher's polymer chemistry*. CRC Press.
7. Chanda, M., and Roy, S. K. (1986). Plastic polymers handbook.
8. Coleman, J. N., Lotya, M., O'Neill, A., Bergin, S. D., King, P. J., Khan, U., and Shvets, I. V. (2011). Two-dimensional nanosheets produced by liquid exfoliation of layered materials. *Science, 331*(6017), 568-571.
9. Fu, S., Sun, Z., Huang, P., Li, Y., and Hu, N. (2019). Some basic aspects of polymer nanocomposites: A critical review. *Nanomaterials Science, 1*(1), 2-30.
10. Gowariker, V. R., Viswanathan, N. V., and Sreedhar, J. (1986). *Polymer science*. New Age International.
11. Herron, N., and Thorn, D. L. (1998). Nanoparticles: uses and relationships to molecular cluster compounds. *Advanced Materials, 10*(15), 1173-1184.
12. Huang, P., Shi, H. Q., Fu, S. Y., Xiao, H. M., Hu, N., and Li, Y. Q. (2016). Greatly decreased redshift and largely enhanced refractive index of mono-dispersed ZnO-QD/silicone nanocomposites. *Journal of Materials Chemistry C, 4*(37), 8663-8669.
13. Krauskopf LG. Monomeric plasticizers. In: Wickson EJ, editor. New York: Wiley; 1993.
14. Li, X., Wang, C., Cao, Y., and Wang, G. (2018). Functional MXene materials: progress of their applications. *Chemistry–An Asian Journal, 13*(19), 2742-2757.
15. Liu, M., Jia, Z., Jia, D., and Zhou, C. (2014). Recent advance in research on halloysite nanotubes-polymer nanocomposite. *Progress in polymer science, 39*(8), 1498-1525.
16. Mark, J. E. (1996). Ceramic reinforced polymers and polymer modified ceramics. *Polymer Engineering and Science, 36*(24), 2905-2920.
17. Mayerberger, E. A., Urbanek, O., McDaniel, R. M., Street, R. M., Barsoum, M. W., and Schauer, C. L. (2017). Preparation and characterization of polymer Ti3C2Tx (MXene) composite nanofibers produced via electrospinning. *Journal of Applied Polymer Science, 134*(37), 45295.

18. Paul, D. R., and Robeson, L. M. (2008). Polymer nanotechnology: nanocomposites. *Polymer, 49*(15), 3187-3204.

19. Petroviæ, Z. S., Javni, I., Waddon, A., and Bánhegyi, G. (2000). Structure and properties of polyurethane–silica nanocomposites. *Journal of Applied Polymer Science, 76*(2), 133-151.

20. Reynaud, E., Gauthier, C., and Perez, J. (1999). Nanophases in polymers. *Revue de Metallurgie, 96*(2), 169-176.

21. Seymour, R.B. and Carraher, C.E. 1984. Structure–Property Relationships in Polymers. Plenum, New York.

22. Smart, L. E., and Moore, E. A. (2016). *Solid state chemistry: an introduction*. CRC press.

23. Tu, S., Jiang, Q., Zhang, X., and Alshareef, H. N. (2018). Large dielectric constant enhancement in MXene percolative polymer composites. *ACS nano, 12*(4), 3369-3377.

24. Vieira, M. G. A., da Silva, M. A., dos Santos, L. O., and Beppu, M. M. (2011). Natural-based plasticizers and biopolymer films: A review. *European Polymer Journal, 47*(3), 254-263.

25. Yang, F., Ou, Y., and Yu, Z. (1998). Polyamide 6/silica nanocomposites prepared by in situ polymerization. *Journal of Applied Polymer Science, 69*(2), 355-361.

26. Young, R. J., Kinloch, I. A., Gong, L., and Novoselov, K. S. (2012). The mechanics of graphene nanocomposites: a review. *Composites Science and Technology, 72*(12), 1459-1476.

Chapter 6

Electron and Probe Microscopy for Nanometrology

K. Raja[1], N. Natarajan[2], C. Sharmila Rahale[1], S. Shanmugapriya[1] and J.Mohanraj[1]

[1]Department of Nano Science and Technology,
[2]Retd. Professor, Department of Seed Science and Technology,
Tamil Nadu Agricultural University, Coimbatore – 641 003, Tamil Nadu

Microscopes are optical instruments used to examine micro-scale structures of plants, animals and material samples that are too small to be seen by the naked eye. Owing to the advent of advancement in sciences, scientists have started to examine the specimens at nano or atomic scales for which conventional optical microscopes may be of least importance. Due to limitations of the light microscope on the magnification of 1000X or the maximum of 2000X and resolution of 200 nm, advanced microscopic techniques *i.e.* Electron Microscopes were developed. The limitation in the use of optical light microspore had been reached during the early 20th century, and there was a scientific desire to see the fine details of the ultrastructures of cells such as nucleus, mitochondria. Golgi bodies, endoplasmic reticulum, *etc.*). This required 10,000X plus magnification which was just not possible using light microscopes. Electron Microscopes are scientific instruments that use a beam of highly energetic electrons to examine objects on a very fine scale. This examination can yield information about the topography (surface features of an object), morphology (shape and size of the particles making up the object), composition (the elements and compounds that the object is composed of and the relative amounts of them) and crystallographic information (how the atoms are arranged in the object).

Scanning Electron Microscope

The first prototype electron microscope with 400X was developed in 1931 by the physicist Ernst Ruska and the electrical engineer Max Knoll. This was the first apparatus to demonstrate the principle of electron microscopy. The first Scanning Electron Microscope (SEM) was introduced in 1942 with the first commercial instruments around 1965. Its late development was due to the electronics involved in "scanning" by the beam of electrons across the sample. The SEM is an essential tool for a large part of the scientific community with applications ranging from archeology and geology to biology and material science. It can obtain three-dimensional images as well as elemental analysis and crystal structure. The versatility of the SEM stems from its ability to detect a wide array of signals with secondary electrons, backscattered electrons, and characteristic x-rays being some of the most common. The SEM also benefits from the ability to image and characterize a diverse field of specimens ranging from the organic world of plants animals and insects to inorganic materials such as crystals and polymers

Principle

A high energy beam of electrons interacts with the sample and producing signals (Secondary electrons, Backscattered electrons, and characteristic X-ray, *etc.*,) which reflect the information about the sample's surface topography and elemental compositions

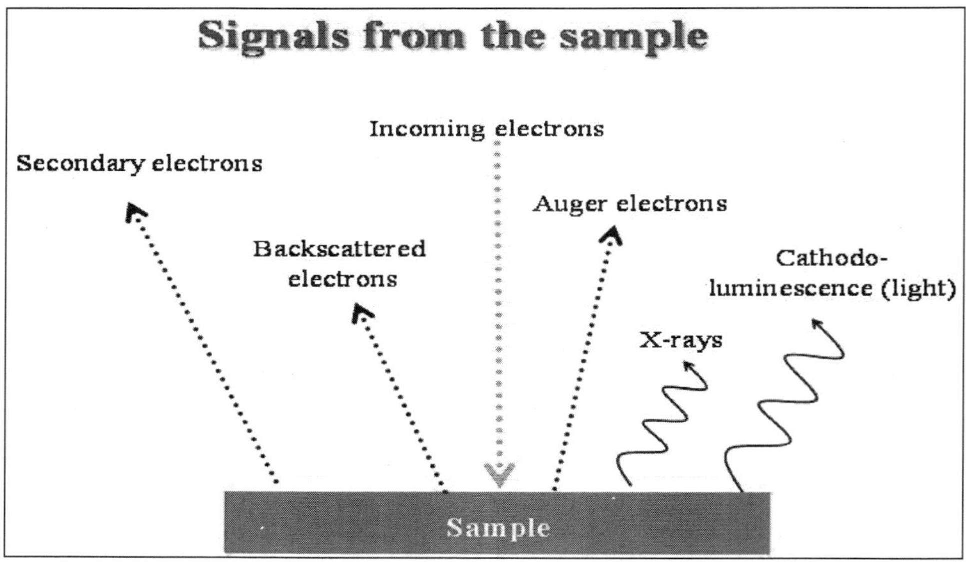

Figure 6.1. Interaction of Electrons with the Sample.

Components

The SEM instrument consists of two main components, electron column and the electronic console system (Figure 6.2).

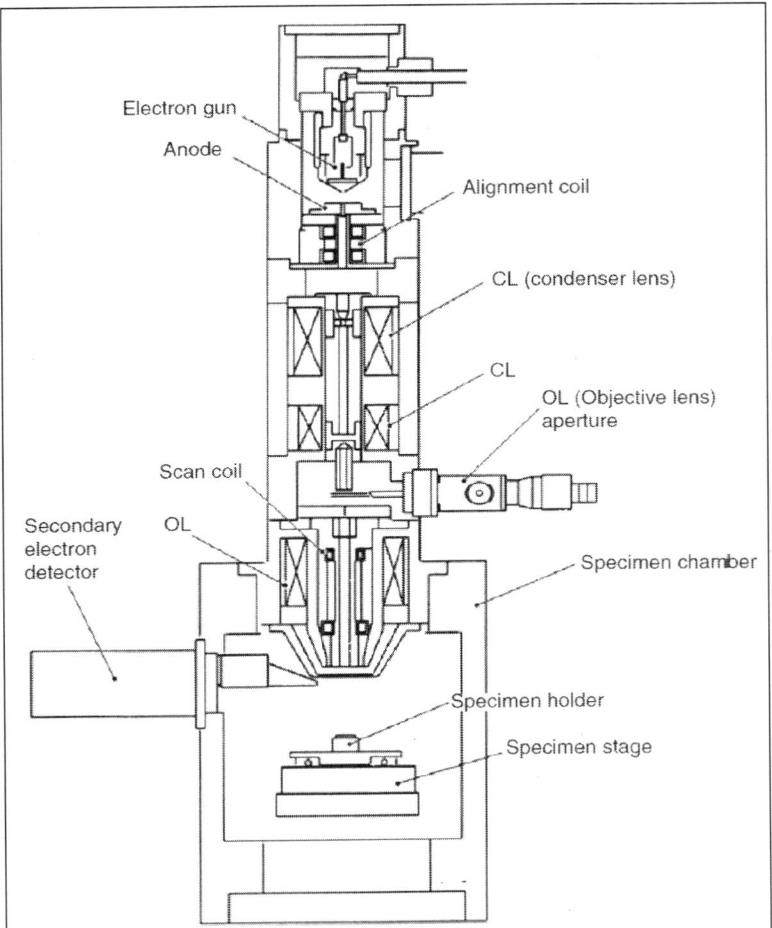

Figure 6.2. Schematic Diagram of Scanning Electron Microscope (JSM-5410).
Courtesy: JEOL, USA.

Electronic Console

The electronic console provides control knobs and switches that allow for instrument adjustments such as filament current, accelerating voltage, focus, magnification, brightness, and contrast. The recent model electron microscopes use a computer system in conjunction with the electronic console making it unnecessary to have a bulky console that houses control knobs, Cathode Ray Tubes (CRT), and an image capture device. All of the primary controls are accessed through the computer system using the mouse and keyboard. The image that is produced by the SEM is usually viewed on the computer monitor. The image that is captured can be saved in digital format or printed directly.

Electron Column

The electron column is a vacuum filled in a chamber wherein the electron beam is generated, focused on a single beam with a small diameter, and scanned across the surface of a specimen by electromagnetic deflection coils. Electron column consists of an electron gun, electromagnetic condenser lenses, apertures, objective lenses, scanning deflection coils, detectors, and specimen chamber. The secondary electron detector is located above the sample stage inside the specimen chamber (Lower portion of the chamber). Specimens are mounted and secured onto the stage which is controlled by a goniometer. The manual stage controls are found on the front side of the specimen chamber and allow for x-y-z movement, 360 rotation and 90 (Figure 6.3).

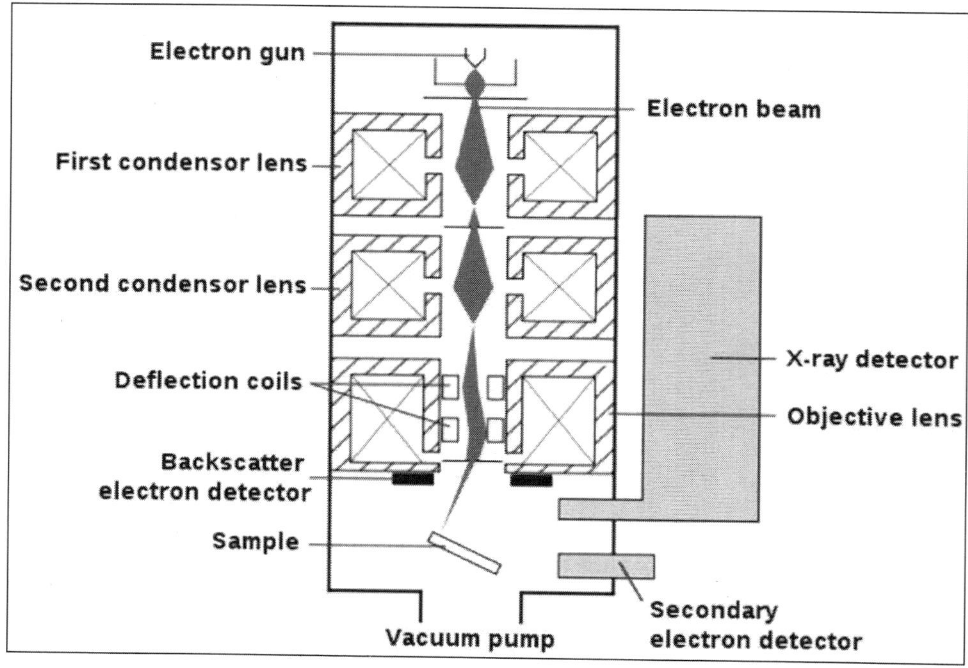

Figure 6.3. Scanning Electron Microscope Column.

Electron Gun

The electron gun generates the electron beam in electron microscopy. It has three components: 1) a filament or cathode made of Tungsten wire or Lanthanum hexaboride (LaB6) crystal, 2) a grid cap (Wehnelt Cylinder) that controls the flow of electrons (bias), and 3) a positively charged anode plate that attracts and accelerates the electrons down the column to the specimen (Figure 6.4). There are 2 classes of emission sources, thermionic emitters, and field emitters.

Old model or the first generation SEM system has tungsten or lanthanum hexaboride (LaB6) filaments, whereas the modern SEM mostly the field emission

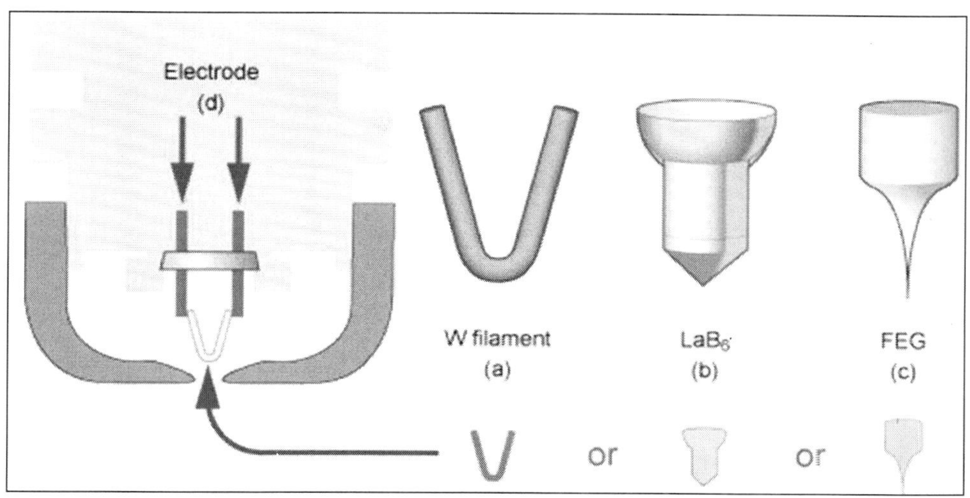

Figure 6.4. Filament Types of the Electron Gun.

sources are used, which provide enhanced current and lower energy dispersion. While selecting the emitters, the lifetime of the emitters is to be considered.

Thermionic Emission

Thermal emission uses heat to excite the electrons off the filament. The filament current control knob controls the heating of the filament. As the filament current is increased, the filament temperature will also increase. As the temperature increases, the number of electrons emitted by the filament will increase up to a point (Figure 6.5). Beyond this point, an increase in filament current will have no increase in emitted electrons. Continued increases in filament current will only increase the filament temperature, hastening filament burn out.

This is why filament current should be carefully adjusted to maximize the number of emitted electrons while the filament temperature is no hotter than necessary. This is called the saturation point. Exceeding the required filament temperature for the saturation point will decrease the life of the filament significantly. Adjusting the filament current below the saturation point will result in an unstable beam current and a poor image. Proper adjustment of filament current control can extend the life of the filament depending on filament type.

There are two types of thermal emission filaments. Tungsten filaments that are hair pen shape (Figure 6.4) or some variation of the hair pen, LaB6 filaments that are a single lanthanum hexaboride crystal ground to a fine point. a) *Tungsten filaments*: These are relatively inexpensive and can be operated at lower vacuums (approximately 10^{-4} Torr). Disadvantages of tungsten filaments include short life (approximately 100 hr. even in a high vacuum) and low electron emission (inadequate for high-resolution work in an SEM). b) *LaB6 filaments*: These filaments have higher electron emission, better resolution and a longer life span (approximately 2000 hr.) than tungsten filaments. The disadvantages of the LaB6 filaments are cost is much

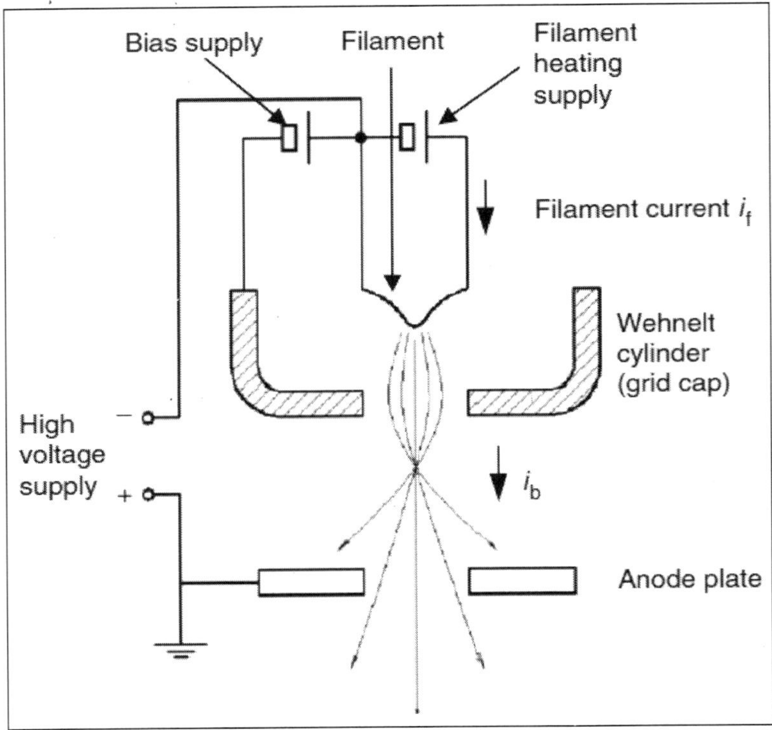

Figure 6.5. Schematic of Thermionic Emitters (Weilie Zhou *et al.*, 2006).

higher (approximately $1000) and a higher vacuum is needed (approximately 10^{-6} Torr). Otherwise, the life of the filament is too short to be economical.

Field Emission

Field emission sources (cold cathode) do not heat the filament material. The electrons are drawn from field emission gun by placing the filament at a huge electrical potential gradient, so large that the work function of the material is overcome, and electrons are drawn off of the filament. The filament consists of a fine pointed tungsten wire (ideally one atom across at the tip) that is kept at a high negative charge along with the rest of the cathode. The anode is kept at a positive charge. The large difference in potential on a small point pulls the electrons off the filament. Field emission sources utilize two anode plates placed below the gun assembly. Associated with the first anode is the extraction voltage (Figure 6.6).

The extraction voltage is usually in the range of 3-5 kilovolts and is the amount of voltage necessary to draw electrons from the source. The second anode has the accelerating voltage associated with it. The accelerating voltage determines the velocity at which the electrons travel down the column. Both of these anodes act as electrostatic lenses, focusing the beam into a small initial crossover. The advantages to this system are that the emission is cool, the emitted beam is much smaller in

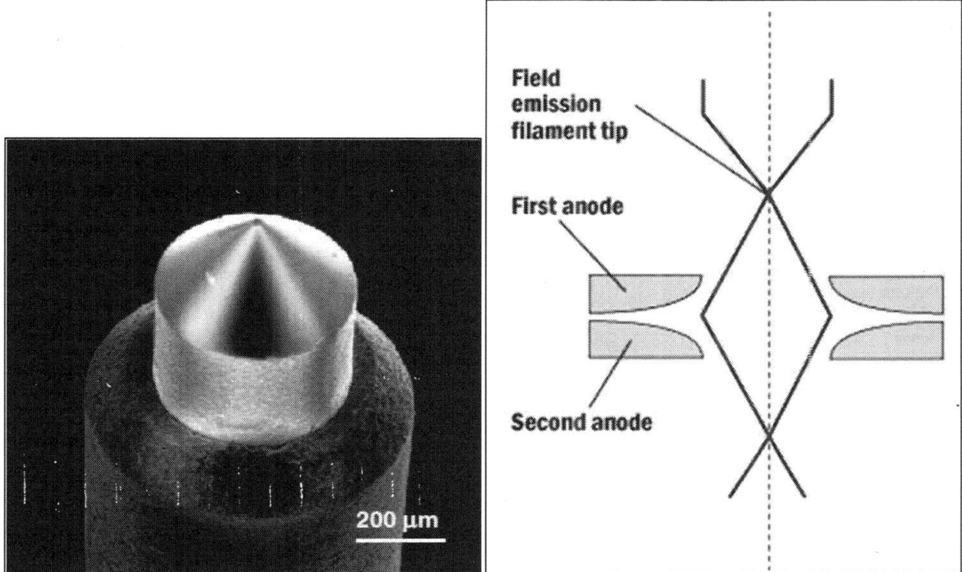

Figure 6.6. SEM Image and Schematic of Field Emission Gun (FEG).

diameter (better resolution), and filament life is generally longer. The disadvantages are a much higher vacuum (approximately 10-7 Torr) and a cleaner microscope (special pumps not oil diffusion) is needed for operation. Another disadvantage of field emission is long acquisition times during X-ray analysis. Field emission gun will not generate many X-rays since low beam currents and the small beam diameters are used.

Electron Beam Manipulations

Electrons in motion are affected only by electrostatic fields and magnetic fields. In the electron gun, electrons are controlled by an electrostatic field, while throughout the rest of the SEM the electrons are controlled by magnetic lenses. Electrostatic lenses are formed when negative and positive fields are near each other. This is the situation between the gun and the anode.

Electromagnetic Lenses

These lenses are used in the column. Electromagnetic lenses create a circular magnetic field that demagnifies (condense) the electron beam as it passes through. The strength of the lenses can be changed by varying the current supplied to the lens. Changing the lens current changes the focal length of the lens. There are two types of lens sets:

The Condenser Lens

This is at the top and the objective lens at the bottom. Each does a different job. The condenser lens convergences the cone of the electron beam to a spot below it (the electron beam is essentially focused down to 1000 times its original size), before the cone flares out again and is converged back again by the objective lens and down onto the sample. This initial convergence can be at different heights, that is, close to the lens, or further away. The closer it is to the lens, the smaller the spot diameter at the point of convergence. The further away, the larger the diameter of this point. So the condenser lens current controls this initial spot size and is referred to as the spot size control. The diameter of this initial convergence (also called a cross-over point) affects the final diameter of the spot the beam makes on the sample. Normally two condenser lenses are used in SEM to converge the beam and pass through a focal point (Figure 6.7).

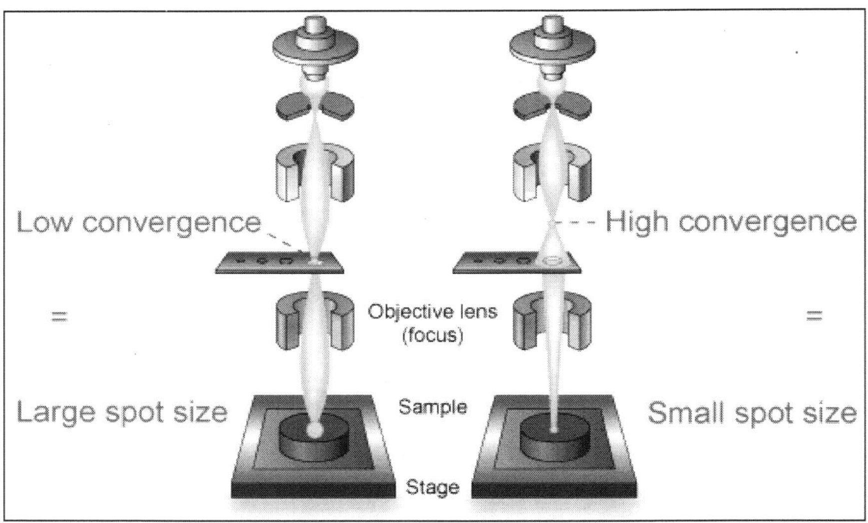

Figure 6.7. Electron Focussing by Condenser Lenses.

Objective Lens

The objective lens also has some influence over the diameter of the spot size of the electron beam on the specimen surface. But its main role is in focusing the beam onto the sample. A focused beam produces a smaller spot on the surface than an under or over-focused beam (Figure 6.8b). There are three designs of objectives lenses *viz.*, asymmetric pinhole lens, which has fewer aberrations, symmetric immersion lens in which small specimen can be observed with small lens aberrations and the third one is snorkel lens where the magnetic field extends up to the specimen providing small lens aberrations and large specimen (Figure 6.8a)

There are some problems inherent in electromagnetic lenses. They are spherical aberration, chromatic aberration, diffraction, and astigmatism. It is possible to correct for these problems, once one understands them.

Spherical Aberrations

These aberrations are the inability of the lens to image central and peripheral portions of the electron beam at the same focal point (Figure 6.9). The strength of the

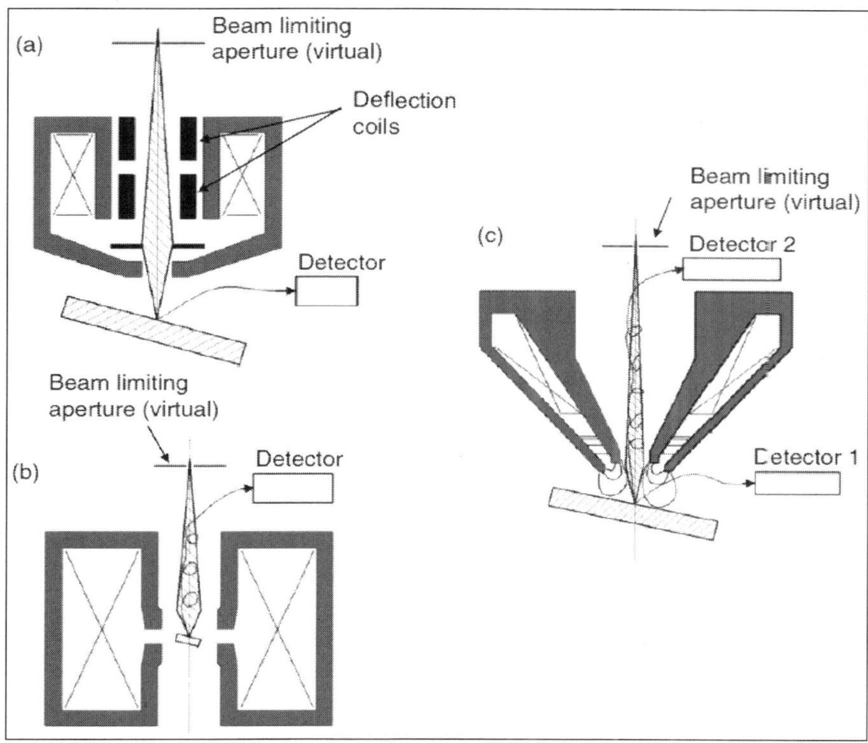

Figure 6.8a. Electron Focussing by Objective Lenses.

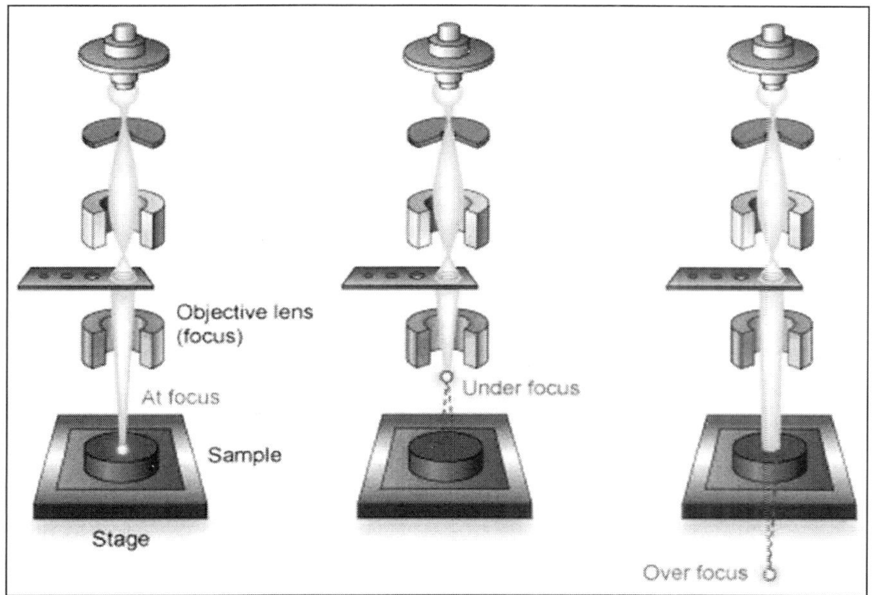

Figures 6.8b. Objective Lenses. a) asymmetric, b) Symmetric, c) Snorkel
Courtesy: Goldstein *et al.*, 2003.

magnetic field is the strongest near the surface of the lens. Therefore, the electrons that travel through a lens close to its surface will have their paths altered more than an electron which travels through the center of the lens and results in a loss of electrons from the beam. Manufacturing techniques are getting better, but the resolution is still limited by the lens with the worst spherical aberration.

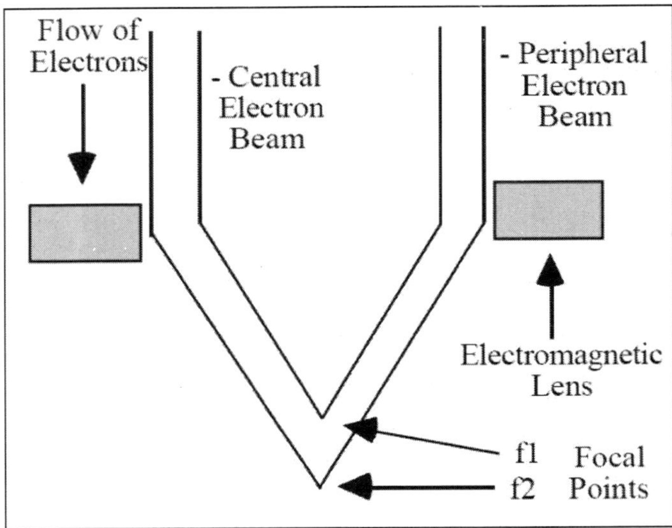

Figure 6.9. Effect of Spherical Aberration on the Focal Point of the Beam.

Chromatic Aberration

These aberrations relates to the varying energy of the electrons that comprise the beam. Not all electrons generated at the gun assembly have the same energy. Electrons with different energies have different wavelengths. A magnetic field will have more of an influence on a longer wavelength electron. Thus, due to a variance in the energies of the electrons in the beam, it is not focused on a discrete focal point.

Diffraction

It is of most importance at the final probe forming lens. Diffraction is caused by the wavelengths of the electron being out of phase. Thus, a lens will focus electrons of different phase to a different point depending upon the position of an electron in its wavelength when it passed through the lens. To correct diffraction, the electrons would have to be monochromatic and coherent. If these conditions were met, the lens could focus the stream of electrons to a point rather than to a disk of confusion.

Astigmatism

It occurs because of manufacturing imperfections within the electromagnetic lens. It is extremely difficult to manufacture an electromagnetic lens that forms a perfectly even magnetic field. Variances in the strength of the magnetic field around the circumference of the lens cause an elliptical focal spot. This problem can be overcome by segmenting the lens into many parts and being able to adjust the excitation to each part. In this manner, weak portions of the lens can be strengthened.

Apertures

Along with the electromagnetic lenses, metal aperture strips can be used to refine the beam. It is common for the modern SEM to be equipped with a variable condenser and objective lens apertures. These variable aperture strips will have a variety of pinhole sizes to choose from (Figure 6.10).

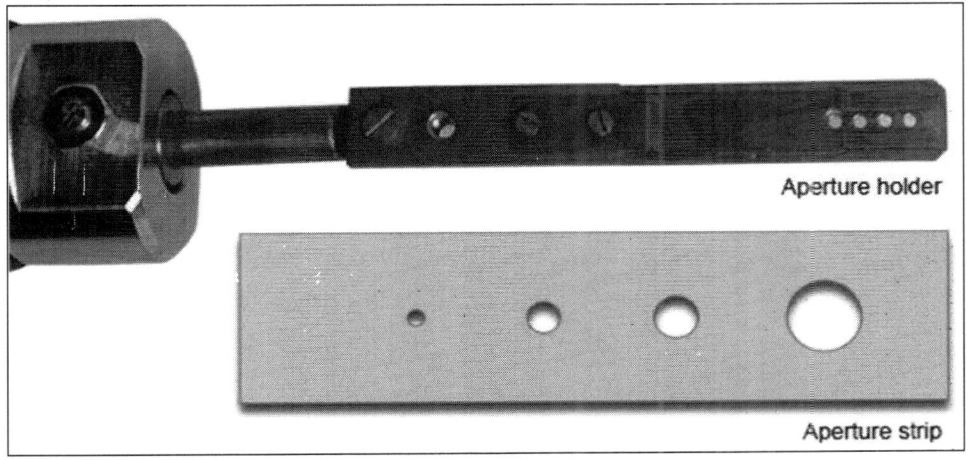

Figure 6.10. Aperture Holder and Aperture Strip.

To generalize, small objective aperture sizes will produce images with good resolution, good depth of field and minimal charging. Depth of field refers to the ability to have a large change in specimen topography appear to remain in focus. SEMs allow for the specimen to be translated along the x, y and z axes (Figure 6.11).

The z-axis is defined as the distance from the final lens to the surface of the specimen. This distance is also referred to as the "working distance." Long working distances and small apertures provide an image that appears to be in focus over a large change in z. An ideal specimen would be conductive. If a less than ideal (non-conductive) specimen is attempted to be viewed, beam electrons can build up on the surface of the specimen, resulting in the phenomenon known as charging.

Scanning/Raster Coil

After the objective aperture, the last manipulation of the beam prior to impingement upon the sample occurs in the final lens (Objective lens). The objective lens is the heart of the SEM and gives the instrument its name. Within the final lens are the raster coils. These coils raster or scan the focused electron beam over the surface of the specimen. Hence, the name is scanning electron microscope. The raster coils scan the focused electron beam across the specimen much as one would read the page. We start at the top, read across the page to the end of the first

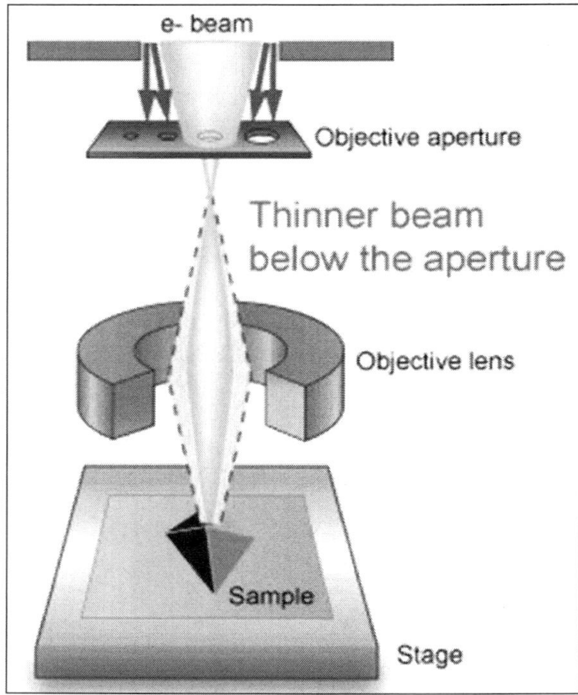

Figure 6.11. Effect of Objective Aperture on the Spot Size of the Beam on the Sample.

line, drop down a line and back to the left and repeat. The scan of the raster coils is synchronized with the scan of the viewing screen (Figure 6.12). The raster coils are used to change the magnification. To increase magnification, the coils can be made to scan a shorter line on the specimen. Since the size of the viewing Cathode Ray Tubes is fixed, the information generated by a shorter scan on the specimen must be enlarged to fill the viewing CRT. This is how magnification is changed. The stigmator or astigmatism corrector is located in the objective lens and uses a magnetic field in order to reduce aberrations of the electron beam. The electron beam should have a circular cross-section when it strikes the specimen however it is usually elliptical thus the stigmator acts to control this problem.

Detection of Signals and Detectors

The signals produced during the electron beam-specimen interactions (SE, BSE, and X-rays) are picked up by detectors. The detectors are specifically designed to specialize in and detect one or more types of signals.

Secondary Electron Detector

In SEM, topographical information is mainly provided by secondary electrons that are produced by the interaction of the beam with the specimen. "Everhart-Thornley" detector is used to detect the secondary electrons emitted from the specimen. The low energy secondary electrons are emitted from the sample in

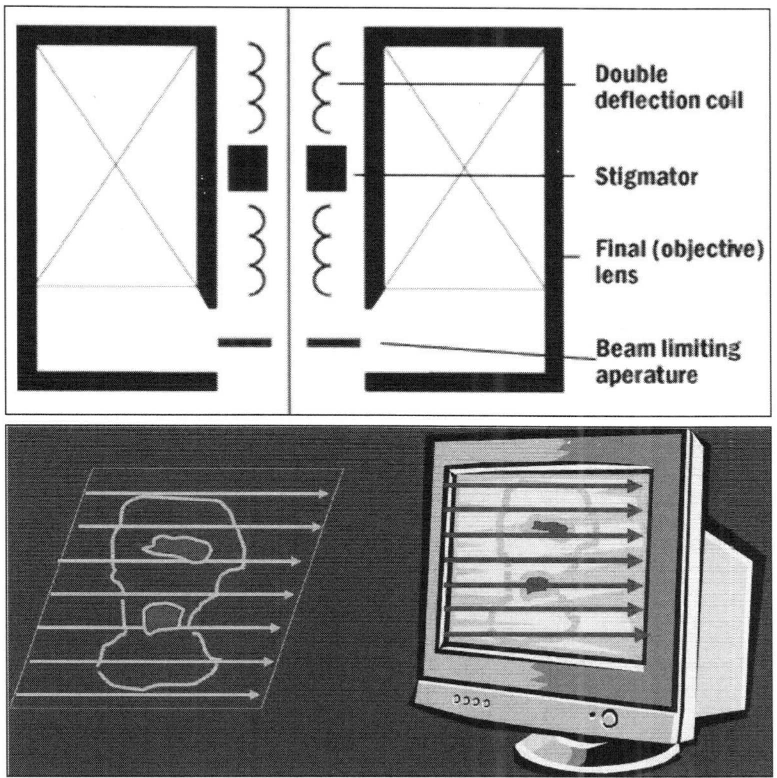

Figure 6.12. Raster Coil and Raster Scanning Over the Surface of the Sample.

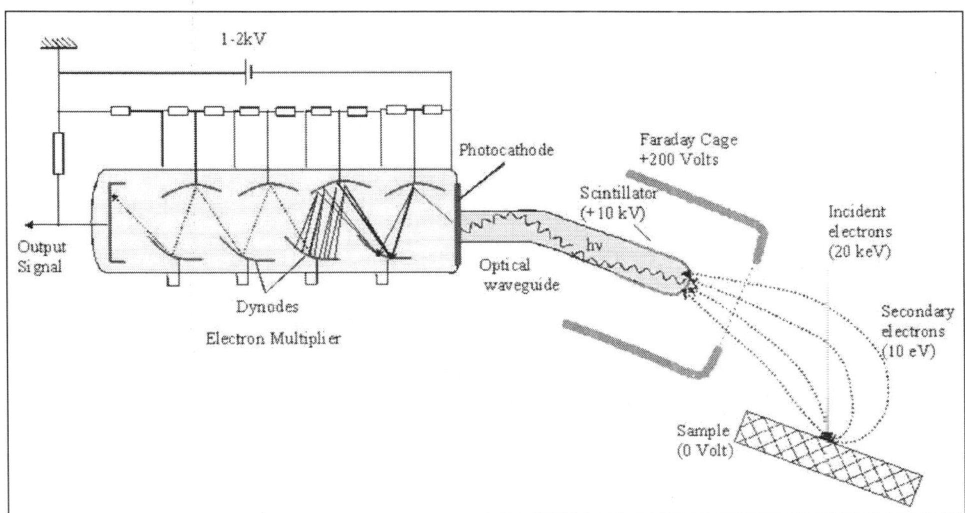

Figure 6.12a. Secondary Electron Detector (Everhart-Thornley) Detector.

all directions and possess relatively low energies. They are initially gathered by a charged collector grid (or Faraday cage), which can be biased from -50 to +300 V (Figure 6.13).

There are three types of ET detector a) conventional (below-lens) scanning electron microscope with below-lens ET detector (Figure 6.13a), b) condenser/ objective (in-lens) scanning electron microscope with above-lens ET detector (Figure 6.13b); and (c) high-resolution (near-lens) scanning electron microscope with above-lens ET detector (Figure 6.13c)

This draws the secondary electrons towards the scintillator, which is a thin plastic disk coated with a short-persistence phosphor that is highly efficient at converting the energy contained in the electrons into ultraviolet light photons (4000

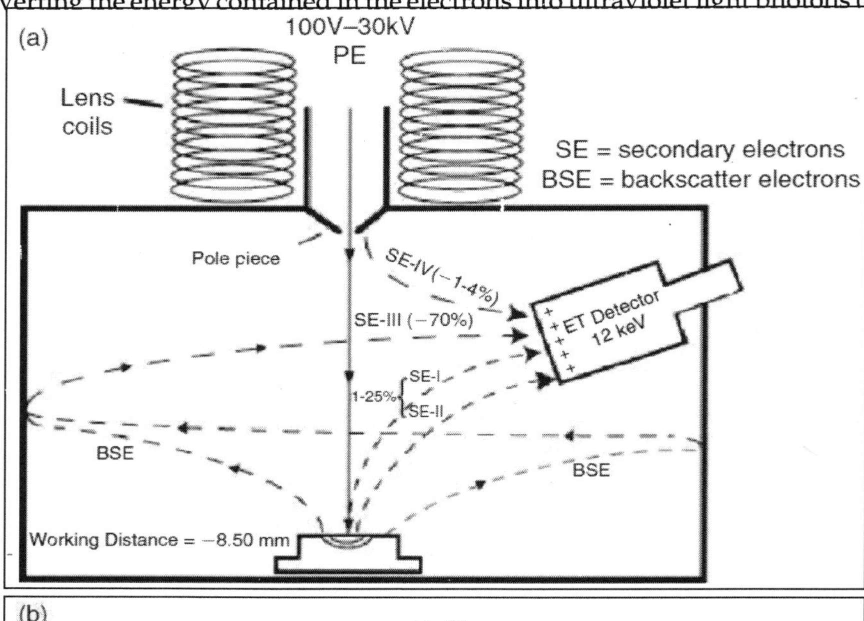

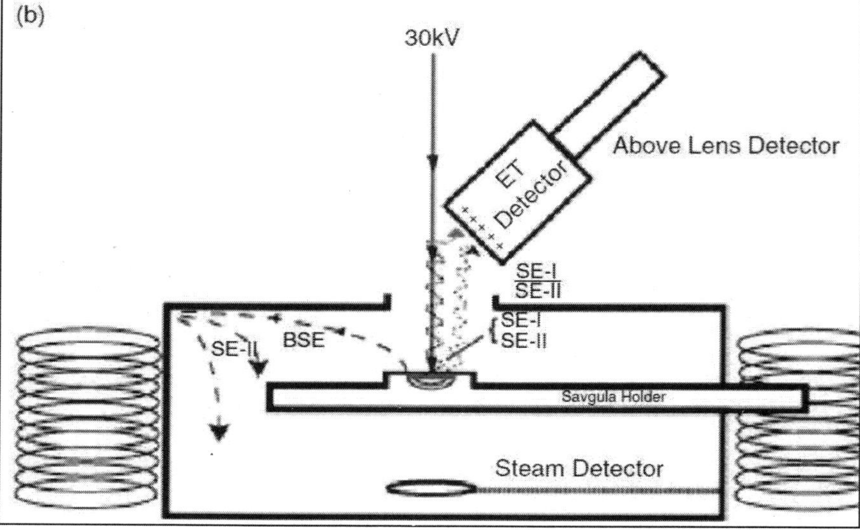

Figure 6.13a-b. Three Types of Secondary Electron Detectors.
Courtesy: Weilie Zhou *et al.*, 2006.

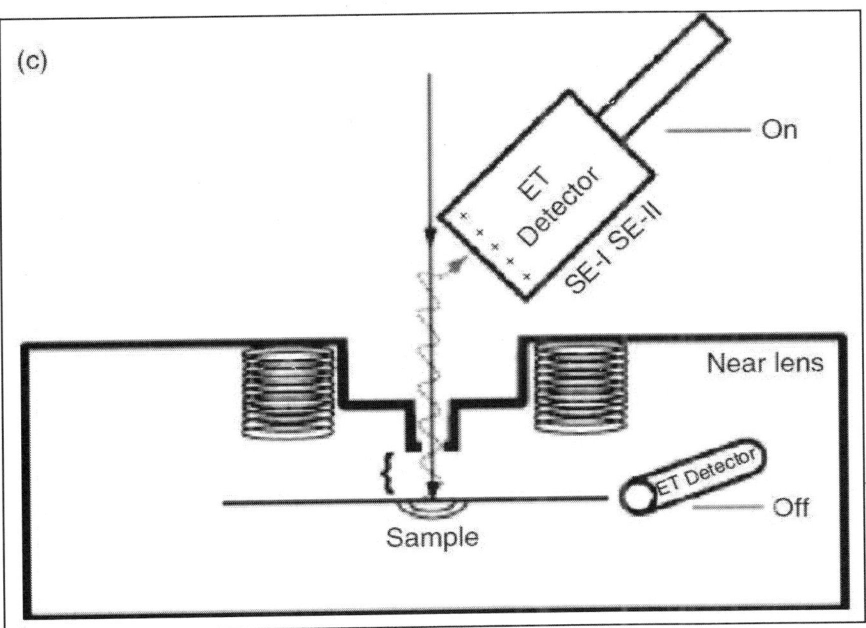

Figure 6.13c. Three Types of Secondary Electron Detectors.
Courtesy: Weilie Zhou *et al.*, 2006.

Å). The response time of the phosphor is fast and permits high-resolution scanning. The outer layer of the scintillator is coated with a thin layer (10-50 nm) of aluminum, positively biased at approximately 10 keV, which accelerates the electrons to the scintillator surface. The charged collector grid, in addition to collecting secondary electrons from the sample, helps to alleviate some of the negative effects of the scintillator aluminum layer bias which can actually distort the incident beam. The aluminum layer also acts as a mirror to reflect the photons produced in the phosphor layer down the light pipe and out through the specimen chamber wall. The photoelectrons strike a photocathode that converts the UV photons back into electrons. The electron multiplier consists of a series of dynodes, each held at a more positive voltage than the previous one. The tube interior is a vacuum. Electrons are accelerated towards the first dynode, strike it, and emit more low energy electrons. These, in turn, are accelerated towards the second dynode, and so forth. The dynodes are coated with a material with high SE yield so at least two (sometimes as many as ten) secondary electrons are emitted for each photoelectron. The result is a cascade of electrons that eventually strike the anode; a single photon produces ~106 final electrons.

The number of cascade electrons produced by the PMT depends on the voltage applied across the cathode and anode of the PMT. One can increase the gain by increasing the voltage to the PMT (this is essentially what is accomplished when adjusting the contrast). The amplified electrical signal is sent to further electrical amplifiers, which increase the electrical signal thus increasing brightness. The topographical aspects of a secondary electron image depend on how many electrons actually reach the detector. When the incident electron beam intersects the edges of topographically high portions of a sample at lower angles, it puts more energy into the volume of secondary electron production. Thus, high points produce more secondary electrons, generating a larger signal. Faces oriented towards the detector also generate more secondary electrons. Secondary electrons that are prevented from reaching the detector do not contribute to the final image and these areas will appear as shadows or darker in contrast to those regions that have a clear electron path to the detector. Using the signal of secondary electrons image resolution less than 0.5 nm is possible.

Backscattered Electron Detector (Solid State Detector)

The backscattered electrons are usually measured with a solid-state detector located on the primary beam pole piece. The detector consists of a diode with a thin gold conductor across the front surface. Backscattered electrons have sufficient energy to pass through the front surface and produce electron-hole pairs that produce a current in the diode. The size of the current (signal) is proportional to the number of electrons hitting the PN junctions (Figure 6.14).

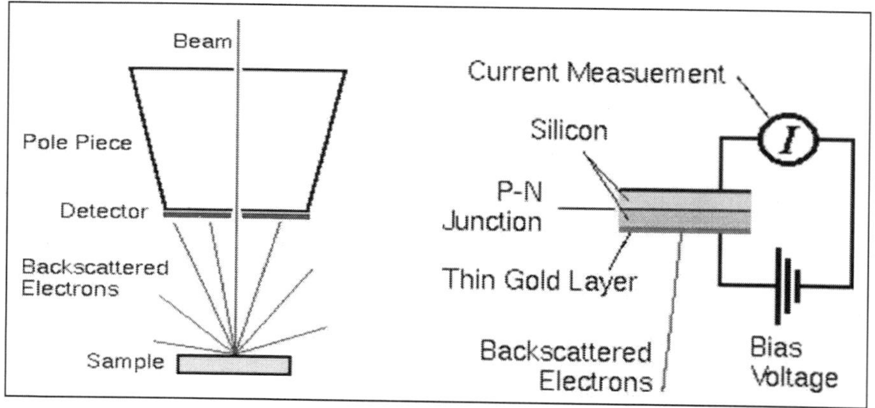

Figure 6.14. Backscattered Electron Detector (Solid State Detector).

Larger atoms (with a greater atomic number, Z) have a higher probability of producing an elastic collision because of their greater cross-sectional area. Consequently, the number of backscattered electrons (BSE) reaching a BSE detector is proportional to the mean atomic number of the sample. Thus, a "brighter" BSE intensity correlates with greater average Z in the sample, and "dark" areas have lower average Z. BSE images are very helpful for obtaining high-resolution compositional maps of a sample and for quickly distinguishing different phases. They are often used in conjunction with spot probe analyses by either EDS or WDS methods. Figure 6.15 shows the Backscatter image of a composite (polished cement fragment) in which low atomic weight particles appear dark and high atomic weight particles are white.

Energy-Dispersive X-Ray Spectroscopy (EDS)

Interaction of an electron beam with a sample target produces a variety of emissions, including x-rays. An energy-dispersive (EDS) detector is used in SEM to separate the characteristic x-rays of different elements into an energy spectrum, and EDS system software is used to analyze the energy spectrum in order to determine the abundance of specific elements. EDS can be used to find the chemical composition of materials down to a spot size of a few microns, and to create element composition maps over a much broader raster area. Together, these capabilities provide fundamental compositional information for a wide variety of materials.

EDS systems include a sensitive x-ray detector, a liquid nitrogen dewar for cooling, and software to collect and analyze energy spectra. The detector is mounted in the sample chamber of the main instrument at the end of a long arm, which is itself cooled by liquid nitrogen. The most common detectors are made of Si(Li) crystals that operate at low voltages to improve sensitivity, but recent advances in detector technology make available so-called "silicon drift detectors" that operate at higher count rates without liquid nitrogen cooling (Figure 6.15).

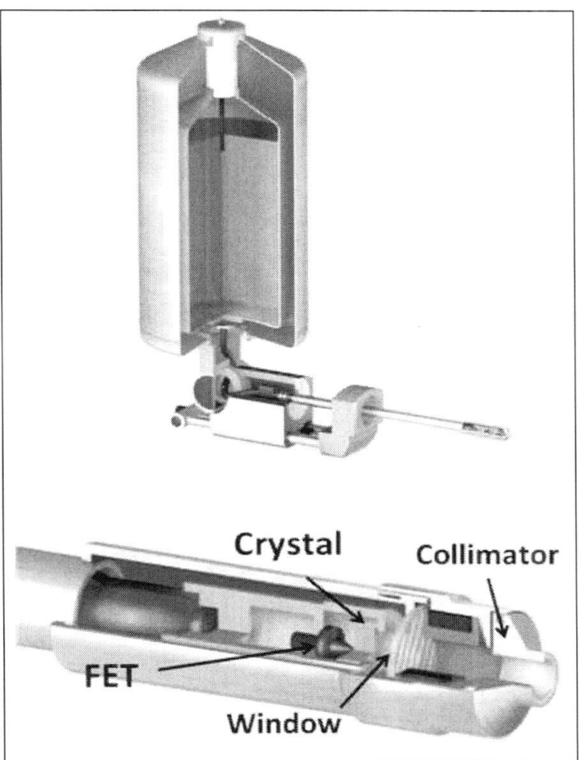

Figure 6.15. Cut-away Diagram showing the Construction of a Typical EDS Detector.

An EDS detector contains a crystal that absorbs the energy of incoming X-rays by ionization, yielding free electrons in the crystal that become conductive and produce an electrical charge bias. The x-ray absorption thus converts the energy of individual X-rays into electrical voltages of proportional size; the electrical pulses correspond to the characteristic x-rays of the element.

Vacuum System

All electron microscopes are high-vacuum instruments. The environment within the column is an extremely important part of the neutron microscope. Without sufficient vacuum in the SEM, the electron beam cannot be generated nor controlled. High vacuum pressure is required in SEM for a variety of reasons. First, the current that passes through the filament causes the filament to reach temperatures around 2700K. A hot tungsten filament oxidizes and burns out in the presence of air at atmospheric pressure. Secondly, the ability of the column optics to operate properly requires a fairly clean, dust-free environment. Third, air particles and dust inside the column can interfere and block the electrons before they ever reach the specimen in the sample chamber A basic requirement for the general operation of the SEM is the control and operation of the vacuum system. In order to provide adequate vacuum pressure inside the column, a vacuum system consisting of two or more pumps is typically present.

Most SEM's are designed to operate in three separate vacuum modes, high vacuum (10 to 10 Torr), low vacuum (0.1 to 20 Torr) and ESEM (0.1 to 20 Torr). To pump from atmospheric pressure down to 10 Torr, two classes of pumps are used: a low vacuum pump (atmosphere down to 10) and a high vacuum pump (10 down to 10 or greater depending on the type of pump). There will be one or more of each class on SEM.

In the low vacuum and ESEM mode of operation, the pressure inside the specimen chamber operates at a lower pressure than that inside the actual column which is always at a high vacuum. The high vacuum mode is typically where most SEMs operate. In this mode, the highest resolutions and magnifications can be achieved although, it is not suitable for all specimens. Generally, anything that is conductive and has a high density works well in high vacuum mode. Specimens that are non-conductive or have a low density are more suitable for the low vacuum or ESEM mode. The ESEM mode of operation is necessary for wet, non-conductive samples.

Electron Beam-Specimen Interactions

When the electron beam strikes the surface of the specimen, the primary (beam) electrons interact in either an elastic or inelastic fashion with atoms of the specimen

Elastic scattering events are those where the primary electron comes close to the nucleus alters the path of the primary electron with a minimal loss in electron velocity. The signal generated by elastic scattering is known as Backscattered electrons. Inelastic events are those where the primary beam electron has a collision with a nucleus or electron of an atom of the specimen. The primary electron undergoes a change in direction, as well as transferring energy to the specimen. The types of signals generated by inelastic events are: Auger electron, secondary electron, characteristic x-rays, light, *etc.*

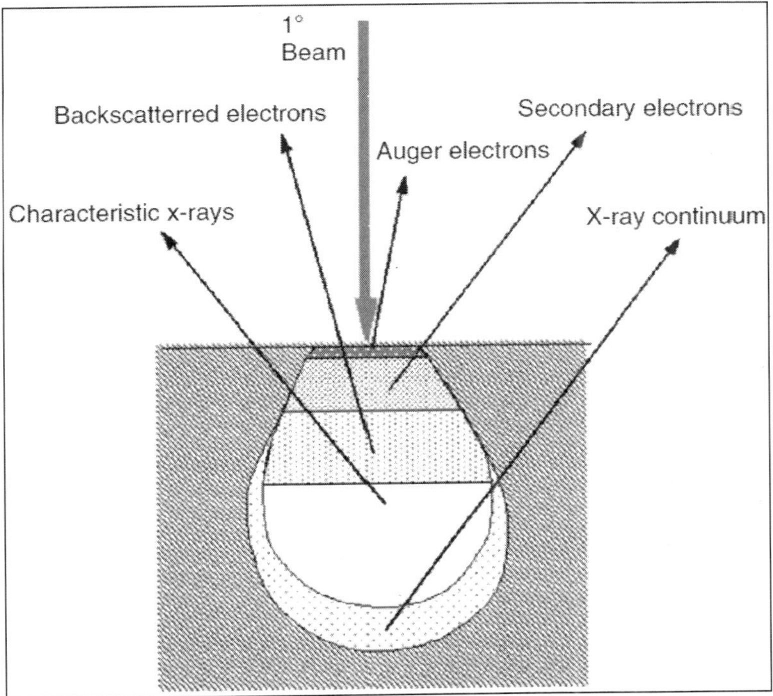

Figure 6.16. Scattering Effects Produced by Electron Bombardment of a Material.
Courtesy: Weilie Zhou *et al.*, 2006.

Backscattered Electrons

A primary electron whose path is altered enough to have it leave the specimen is referred to as a backscattered electron. Since Backscattered electrons are the original beam electrons they have a high energy level, near that of the gun voltage. Backscattered electrons have an energy range from 50eV (electron volts) up to the accelerating potential the SEM is being operated at (Figure 6.17).

The amount of backscattered electrons emitted from a specimen is dependent upon the atomic number of the specimen. This is known as the backscattered coefficient, and as a general rule, the number of backscattered electrons emitted from a specimen increases with an increasing atomic number. Also, increasing the beam current will excite more backscattered electrons from the specimen. With proper detector, the image formed from the backscattered electrons emitted from a specimen will show regions of atomic number inhomogeneity.

Auger Electron

An incoming electron can ionize an atom in the material, and leave a 'hole' (Figure 6.18a). If this electron is from an inner shell of the atom, the internal rearrangement of electrons may take place, resulting in an electron in an outer shell falling into the hole (Figure 6.18b). The energy emitted by this process may

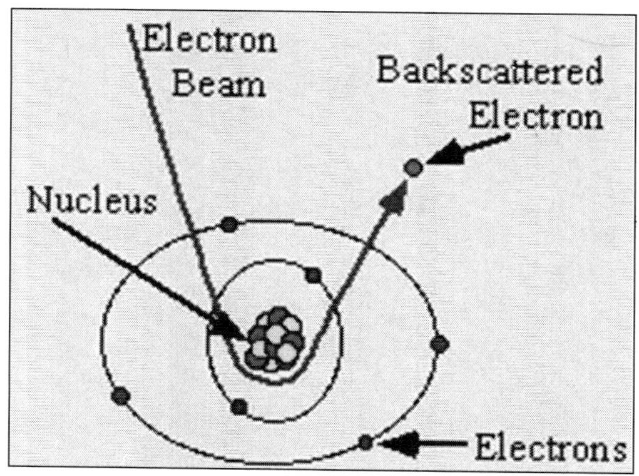

Figure 6.17. Backscattered Electron.

either be released as an X-ray or used to eject another less tightly bound electron (Figure 6.18c).

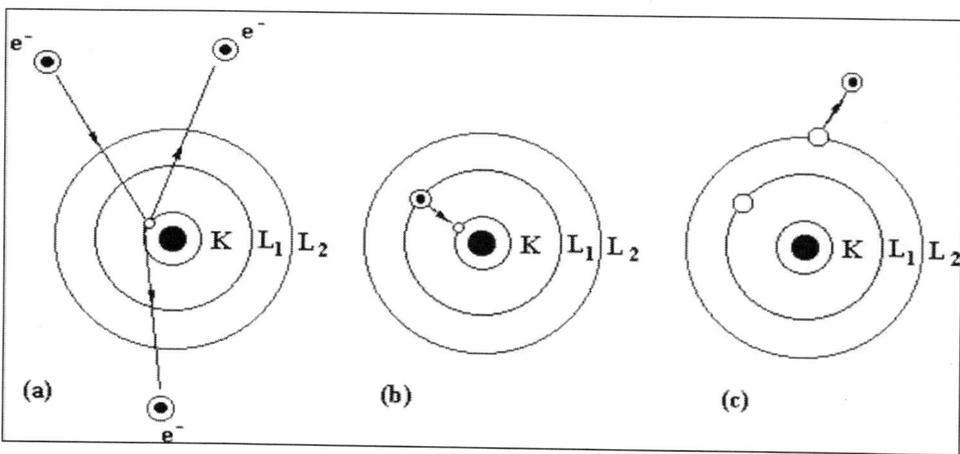

Figure 6.18a-c. Auger Electron.

The latter process is known as Auger emission and the electron known as the Auger electron. The energy of the Auger electron is characteristic of the material. As Auger electrons have low kinetic energy, only those within 5 nm of the surface can contribute to observable emission. Thus the Auger process is useful in providing surface chemical information.

Secondary Electrons

A secondary electron is an electron of an atom of the specimen that has been knocked from its orbit. Secondary electrons can also be generated by other secondary electrons. Secondary electrons have a low energy level of only a few electron volts, thus, they can only be detected when they are dislodged near the surface of the reaction vessel (Figure 6.19).

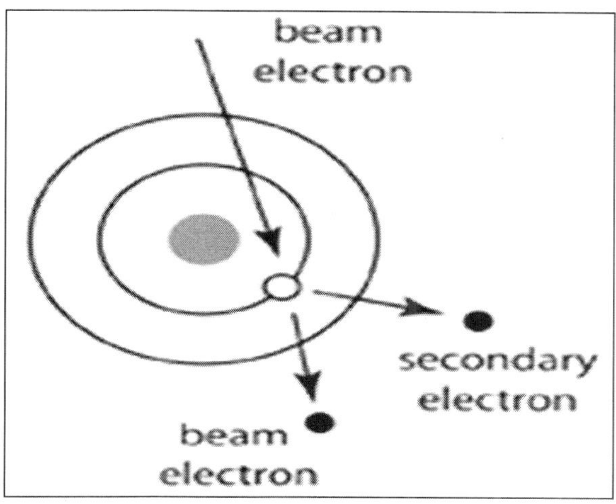

Figure 6.19. Secondary Electron.

The energy range for secondary electrons states that the electron must have energy greater than the 0eV but less than 50eV. The energy range also produces what is referred to as the edge effect. Edges of the specimen provide greater amounts of surface area from which the weak secondary electrons can escape. Thus, at the edges of features on the specimen, more secondary electrons are emitted. These areas appear brighter in the image. Two of the foremost reasons for operating in the secondary electron imaging mode are to obtain topographical information and high resolution. An excellent feature of imaging in the secondary mode is that the contrast and soft shadows of the image closely resemble that of a specimen illuminated with light. Thus, image interpretation is easier because the images appear more familiar. Another advantage of viewing a specimen in the secondary mode is that the primary beam electrons may give rise to several secondary electrons by multiple scattering events, potentially increasing the signal. With an increased signal and shallow escape depth of secondary electrons, the spatial resolution in the second mode can be greater than in the other modes.

The group of secondary electrons (SE) can be divided further. There are SEI, SEII, SEIII, and SEIV. These electrons are classified according to how they are generated. An SEI is an electron that is generated at the point of primary beam impingement on the surface of the specimen. Thus, it carries the highest resolution information. An SEII is an electron that is generated when a backscattered electron leaves the surface of the specimen. Due to the energy of backscattered electrons, this SEII could leave the surface of the specimen microns away from the primary beam impingement site. SEIIs hurt the resolution of the image, but add greatly to overall image brightness. A SE III is an electron released when an energetic backscattered electron strikes the interior of the specimen chamber, causing a SE to be released.

SEIVs are formed when the primary beam strikes an aperture within the electron column. SEIIIs and SEIVs contribute noise to the image. By understanding the signal formation, the specimen can be properly prepared for analysis (Figure 6.20).

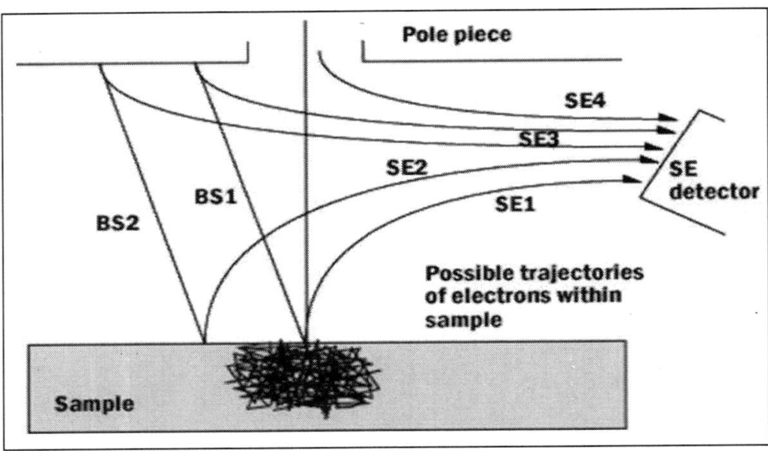

Figure 6.20. Trajectories of Electrons within Sample X-rays.

X-rays

When the incident beam bounces through the sample creating secondary electrons, it leaves thousands of the sample atoms with holes in the electron shells where the secondary electrons used to be. If these "holes" are in inner shells, the atoms are not in a stable state. To stabilize the atoms, electrons from outer shells will drop into the inner shells, however, because the outer shells are at a higher energy state, to do this the atom must lose some energy. It does this in the form of X-rays. Elemental Composition can be obtained in the X-ray mode through EDS (Energy dispersive x-ray spectroscopy) analysis of x-ray signals because the X-ray generated has a wavelength and energy characteristic of the elemental atom from which it originated.

Image Formation

When the primary electron beam interacts with the sample, the energy exchange between the electron beam and the sample takes place which results in the reflection of high-energy electrons by elastic scattering, emission of secondary electrons by inelastic scattering and the emission of electromagnetic radiation, each of which can be detected by specialized detectors. Electronic amplifiers of various types are used to amplify the signals, which are displayed as variations in brightness on a computer monitor (or, for vintage models, on a cathode ray tube). Each pixel of computer video memory is synchronized with the position of the beam on the specimen in the microscope, and the resulting image is, therefore, a distribution map of the intensity of the signal being emitted from the scanned area of the specimen. In older microscopes, images may be captured by photography from a high-resolution cathode ray tube, but in modern machines image is saved to a computer data storage.

Operation of Scanning Electron Microscope

In SEM, a source of electrons is focused in a vacuum into a fine probe that is rastered over the surface of the specimen. The electron beam passes through scan coils and objective lenses that deflect horizontally and vertically so that the beam scans the surface of the sample.

As the electrons penetrate the surface, a number of interactions occur that can result in the emission of electrons or photons from or through the surface. A reasonable fraction of the electrons emitted can be collected by appropriate detectors, and the output can be used to modulate the brightness of a cathode ray tube (CRT) whose x-and y-inputs are driven in synchronism with the x-y voltages rastering the electron beam. In this way an image is produced on the CRT; every point that the beam strikes on the sample is mapped directly onto a corresponding point on the screen. As a result, the magnification system is simple and linear magnification is calculated by the equation:

$$M=L/l \quad (1)$$

Where L is the raster's length of the CRT monitor and l the raster's length on the surface of the sample.

SEM works on a voltage between 2 to 50kV and its beam diameter that scans the specimen is 5nm-2μm. The principle images produced in SEM are of three types: Secondary electron images, backscattered electron images, and elemental X-ray maps. Secondary and backscattered electrons are conventionally separated according to their energies. When the energy of the emitted electron is less than about 50eV, it is referred to as a secondary electron and backscattered electrons are considered to be the electrons that exit the specimen with an energy greater than 50eV. Detectors of each type of electrons are placed in the microscope in proper positions to collect them.

Advantages and Disadvantages

Electrons in scanning electron microscopy penetrate into the sample within a small depth, so that it is suitable for surface topology, for every kind of samples (metals, ceramics, glass, dust, hair, teeth, bones, minerals, wood, paper, plastics, polymers, *etc.*). It can also be used for the chemical composition of the sample's surface since the brightness of the image formed by backscattered electrons is increasing with the atomic number of the elements. This means that regions of the sample consisting of light elements (low atomic numbers) appear dark on the screen and heavy elements appear bright. Backscattered are used to form diffraction images, called EBSD, that describe the crystallographic structure of the sample. In SEM, X-rays are collected to contribute to Energy Dispersive X-Ray Analysis (EDX or EDS), which is used to the topography of the chemical composition of the sample.

Consequently, SEM is only used for surface images and both resolution and crystallographic information are limited (because they're only referred to the surface). Other constraints are first that the samples must be conductive, so non-conductive materials are carbon-coated and secondly, that materials with an atomic number smaller than the carbon are not detected with SEM.

Environmental SEM (ESEM)

The major growth of SEMs is in the development of specialized instruments. Environmental SEM uses differential pumping to permit the observation of specimens at low-pressure gaseous environments (*e.g.* 1-50 Torr), at high relative humidity (up to 100 per cent) and at higher pressures. This was made possible by the development of a secondary-electron detector capable of operating in the presence of water vapor and by the use of pressure-limiting apertures with differential pumping in the path of the electron beam to separate the vacuum region (around the gun and lenses) from the sample chamber. In this type of SEM, there's no need for conductive coating, the secondary electron detector operates in the presence of water vapor, and in the microscope's column, there are pressure-limiting apertures. The ESEM is ideal for non-metallic surfaces, such as biological materials, plastics, and elastomers

Techniques for High-Resolution Imaging

Scanning microscopes, with the advancement of electronics and the introduction of new techniques, are becoming much easier to use these days. Anybody can acquire relatively good images just after a short amount of training. Problems like unclear image and difficulties in acquiring particular features encountered while operating the microscope can be resolved by determining the exact cause by focusing on the following techniques for successful imaging.

Image Disturbances

Image disturbances, while imaging specimens may be due to defects with the instrument, inexperienced operator, improper sample preparation, external influences such as instrument vibration. Disturbances can be examined and corrected using a systematic approach based on their appearance as indicated in Table 6.1.

Table 6.1. Image Disturbances and their Possible Causes

Lack of Sharpness	❒ Improper accelerating voltage setting. ❒ Instability of gun emission caused by insufficient heating of the filament ❒ Improper electron probe diameter ❒ Improper setting and incorrect centering of objective aperture. ❒ Insufficient astigmatism correction ❒ Improper focal depth ❒ Too large magnification ❒ Specimen charge up and magnetization ❒ Defocus of the camera system

Low image quality	❒ Improper accelerating voltage setting. ❒ Improper probe current setting ❒ Incorrect astigmatism correction ❒ Noise caused by excessive photomultiplier (PMT) gain ❒ Improper contrast and brightness ❒ The improper specimen preparation process ❒ The improper positional relation between specimen and cetector
Noises	❒ Instability of accelerating voltage and gun emission ❒ Discharge of detector ❒ Charge up of specimen surface ❒ An external stray magnetic field ❒ Mechanical vibration
Image distortion and deformation	❒ Specimen charge up ❒ An external stray magnetic field ❒ Electron beam damage ❒ Deformation of the specimen itself during its preparation ❒ Image drift caused by column interior charge up specimen drift on heating and cooling stages

Effect of Accelerating Voltage

Choosing the right accelerating voltage is critical for obtaining a good clear image however the most suitable voltage level depends mostly on the type of material being examined. The more conductive the material the better it will behave under higher voltages. Higher voltages (15-30 Kv) generally allow for high resolution

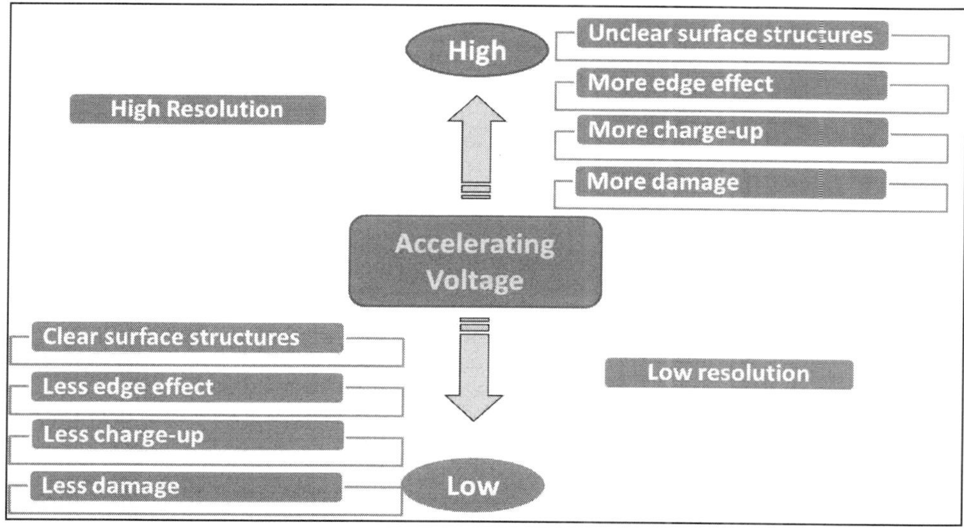

Figure 6.21: Effect of accelerating voltage on Image quality

at high magnifications although, this can damage the specimen very quickly if it is not highly conductive. Thus, when imaging polymers and ceramics it is more suitable to use voltages below 10 kV.

Effect of Working Distance

Image quality is influenced by the choice of working distance and spot size besides accelerating voltage. Ideally, a working distance of 10mm should be used for allowing for a good depth of field while maintaining good resolution. In most cases, the working distance is reduced to achieve better resolution, especially when using lower accelerating voltages.

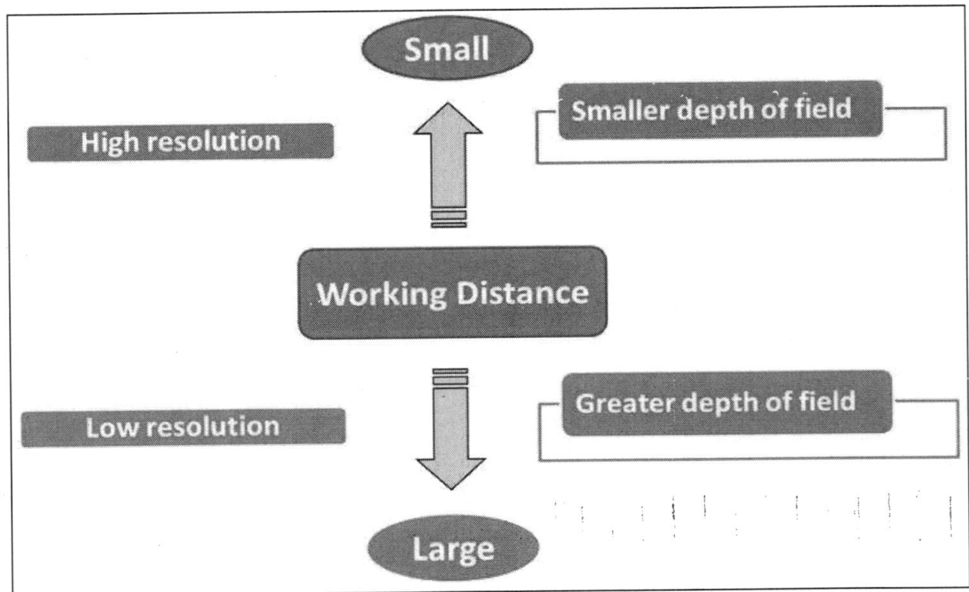

Figure 6.22. Effect of accelerating voltage on Image quality.

Effect of Spot Size

Size basically restricts the beam current and thereby influencing brightness and contrast compensations. Smaller spot sizes facilitating higher resolution and a greater depth of field, require higher brightness and contrast levels which may be a limiting factor (Figure 6.23).

Sample Preparation for SEM

The objective of specimen preparation is to manipulate the specimen as little as possible insuring that the sample is of appropriate size, stable in the vacuum, electrically conductive, and possessing characteristics resembling the natural state. The sample selected must be representative of the entire group if data is to be extrapolated to the entire population. Since most metallic samples meet these conditions, they require little or no preparation. Many other materials such as

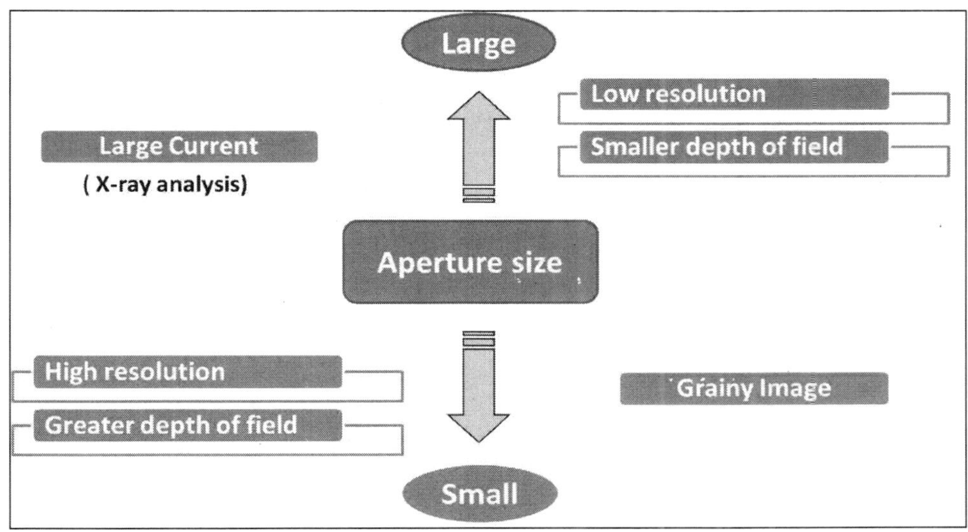

Figure 6.23. Effect of Spot size/Aperture size on Image resolution.

ceramics, plastics, and minerals need only to be coated with a conducting metal. Most biological samples need to go through several treatments to meet the above requirements.

Biological Sample

Natural topographies of most biological specimens will not be retained in a high vacuum. Thus samples must be carefully prepared for examination in the microscope. The softer and more hydrated the sample is, the more care needs to be taken in its preparation. The specimen when collected should be stabilized immediately. The two most commonly used methods of stabilization are freezing and chemical fixation. Both methods are used to prevent natural changes in the sample and to increase the specimen's resistance to deformation by subsequent preparatory manipulations.

Preservation of the sample for a vacuum environment usually means that the sample must be dry. With special equipment, frozen specimens may be placed directly in the specimen chamber and kept frozen while viewing. Freeze-drying is an alternative method to dehydrating a sample but must be performed under rigid conditions in order to prevent ice crystal damage. For biological specimens, the method of preparation is very similar to the methods employed for TEM. Primary fixation is routinely carried out making use of aldehydes. After primary fixation, a secondary fixation using osmium tetroxide is performed. This stabilizes some fats and lipids within the sample. Dehydration through a graded series of ethanol removes the water from the sample. To remove ethanol from the sample, it can either be critical point dried or chemically dried with hexamethyldisilazane (HMDS).

Critical Point Drying

Preventing the gas/liquid interface is the main goal of critical point drying. Whenever a liquid evaporates into the gaseous phase, large surface tension phenomena occur. These surface tensions can harm fine surface details on the surface of the sample. Since the SEM is a surface imaging technique, it is desirable to avoid such artifacts. For critical point drying, the samples are loaded into the chamber of a critical point drying apparatus. This chamber is surrounded by a water jacket that allows the operator to vary the temperature of the unit. Initially, the chamber is cooled below 13 degrees Celsius. At this temperature, liquid carbon dioxide will flow into the chamber and remain in the liquid state. Ethanol is denser than liquid carbon dioxide and will flow out of the specimen, to the bottom of the chamber. There are ports in the bottom of the chamber that allow the operator to purge away the ethanol. When all of the ethanol has been purged away, the level of liquid carbon dioxide in the chamber is lowered (prevents high pressures from being attained), and the chamber is heated above 31 degrees Celsius. The "critical point" of carbon dioxide is approximately 31 degrees Celsius and 1070 lbs/square inch. At this pressure and temperature, carbon dioxide wants to exist as both a liquid and a gas. In this state, the carbon dioxide can be purged off, preventing the gas/liquid interface. Critical point drying usually provides the best sample preservation, but can take up to 3 hours to perform.

Material Science Samples

For material science samples the specimen preparation must initially involve desiccation. The specimen must have all volatile materials removed from it so as to not degrade the vacuum of the SEM. Once dry, the preparation may proceed and may take many forms. A few would be ion milling, heating/baking, epoxy embedment, ashing and *etc.*hing (acid/base). The information desired from the specimen will dictate which preparation approach to take. For example, the best quantitative X-ray information is collected from a specimen with a flat surface. This is due to the possibility of surface protrusions absorbing X-rays before they reach the X-ray detector. This is known as shadowing. To prevent this, specimens for quantitative X-ray microanalysis should be polished flat. Once the specimen is prepared, it next needs to be mounted.

After drying, the specimen is mounted on a specimen holder (stub). Usually, stubs are aluminum pegs with a flat surface. Sample holders are available for 0.4",0.5",1" and 1.25" diameter cylindrical mounts (Figure 6.24). The specimen is attached to the stub with conductive double-sided adhesive carbon tape.

For most metallic specimens and large nonporous biological specimens, this method is adequate. Remember that if the sample surface is flush with the top of the holder, the working distance (selected by using the "Z" control) will be corrected. If the sample is taller than the top of the holder, the working distance will be reduced. Importantly, tall samples can damage the EDS detector or the objective lens pole piece if moved too high. For most non-metallic specimens, it is necessary to make them conductive by coating them with metal. If the specimen is conductive, it might not require coating. Metallic coatings can be used to boost the secondary

Figure 6.24. Stubs for Mounting Samples.

electron emission of the specimen as well as rendering it conductive. Aluminium, gold, platinum, chromium, tungsten, tantalum, and palladium are common metals used to coat specimens.

Sputter Coating

The most common method of specimen coating is sputter coating. Sputter coating takes place in a vacuum chamber. The specimen to be coated is loaded upon the anode. A vacuum is generated. Prior to coating, the vacuum is compromised with an inert gas (usually argon). When a high tension is applied to the cathode where the metal source resides, the argon gas molecules are attracted to the cathode. The ionized argon strikes the metallic target, knocking loose metal grains, which are attracted to the anode (Figure 6.25).

Figure 6.25. Sputter Coater.

Due to the directional randomness of the argon/target collisions, an omnidirectional coating is achieved upon the specimen. The sputtering method is relatively easy and inexpensive. High vacuums are not required and the amount of gold or other metal used is comparatively small. Once prepared, the sample should be stored in a vacuum desiccator. This prevents hydration to atmospheric humidity levels and reduces oxidation of the metal coating. Most specimens can be stored indefinitely with little appreciable degradation.

Applications of Scanning Electron Microscopy

Topography

The surface features of an object or "how it looks", its texture; direct relation between these features and materials properties (hardness, reflectivity *etc.*) ***Morphology:*** The shape and size of the particles making up the object; direct relation between these structures and materials properties (ductility, strength, reactivity *etc.*) ***Composition:*** The elements and compounds that the object is composed of and the relative amounts of them; a direct relationship between composition and materials properties (melting point, reactivity, hardness *etc.*) **Crystallographic Information**: How the atoms are arranged in the object; direct relation between these arrangements and materials properties (conductivity, electrical properties, strength *etc.*)

Transmission Electron Microscope

The transmission electron microscope (TEM) operates on the same basic principles as the light microscope but it uses a beam of electrons as a source instead of light and used to observe the features of very small specimens. TEMs use electrons as "light source" and their much lower wavelength makes it possible to get a resolution a thousand times better than with a light microscope. The physicist Ernst Ruska and Electrical engineer Max Sknol developed the first electron microscope, a TEM in 1931, and has evolved over the years to become a common technique that is used globally in science and engineering to look at micro and nanoparticle, TEM can be used to observe particles at a much higher magnification and resolution than can be achieved with a light microscope because wavelength of an electron is much shorter than that of a photon. It also provides higher resolution images than a scanning electron microscope, which can only be used to scan and view the surface of a sample. Using TEM, scientists can be used to view specimens to the atomic level, which is less than 1nm. TEMs produce two-dimensional images, being used in a wide range of educational, science and industry applications for analyzing the structure, crystallization, and morphology of a specimen.

Principle

A high energy beam of electrons produced and focused towards the specimen by electromagnetic lenses where the electrons transmitted through a very thin sample, and the interactions between the electrons and the atoms will be used to observe the features such as the crystal structure and features in the specimen.

Components of TEM

Transmission electron microscope has an electron gun which contains filament made up of lanthanum hexaboride (LaB_6) or tungsten. Electron gun produces the electron beam (primary electrons), which are gathered by an anode and condensed by the condenser system. The condenser system has the electromagnetic lenses which focus the beam on the sample. The important part of the TEM is the image-producing system, consisting of the objective lens, movable specimen stage, intermediate and projector lenses. The objective lens and the specimen holder/stage system is the heart of the TEM. This critical region usually extends over a distance of approximately 10mm at the center of the TEM. Herein all of the beam and specimen interactions take place and the two fundamental TEM operations occur namely, the creation of the various images and diffraction patterns that is subsequently magnified for viewing and recording. Clearly, the objective lens is the most important lens in a TEM because its quality determines the quality of all the information about the specimen. The third component system is the image recording system contains a fluorescent screen for viewing and focusing the image and a charge-coupled device (CCD). CCD is a metal-insulator silicon devise that stores the charge generated by light or electron beam. CCD arrays consist of several megapixels which are individual capacitors electrically isolated from each other through the creation of potential wells under each CCD cell so that they can accumulate charge in proportion to the incident beam. The image recording system converts the electron image into some form of perceptible to the human eye (Figure 6.26). Another component of the TEM is the vacuum system, which is essential to ensure electrons do not collide with gas atoms. A low vacuum is first achieved using either a rotary pump or diaphragm pumps

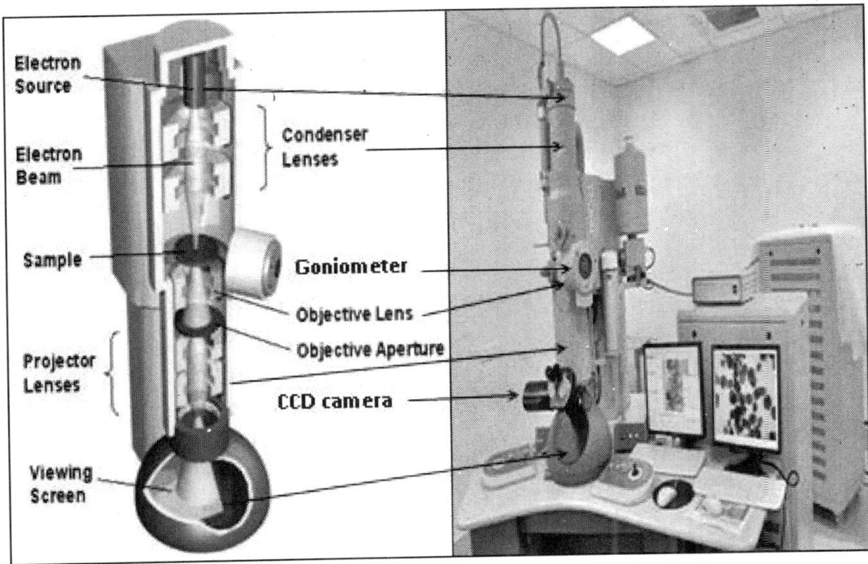

Figure 6.26. Structure of the Transmission Electron Microscope (TEM).
Courtesy: Kannan, 2018.

which enable a low enough pressure for the operation of a diffusion pump, which then achieves a vacuum level that is high enough for operations. High voltage TEMS require particularly high vacuum levels and a third vacuum system may be used.

Electron Gun

The electron source is sharply pointed rod-shaped lanthanum hexaboride. The filament is surrounded by a control grid called as Wehnelt cylinder, with a central aperture arranged on the axis of the column; the apex of the cathode is arranged to lie at or just above or below this aperture. The cathode and control grid is at a negative potential equal to the desired accelerating voltage and is insulated from the rest of the instrument. The final electrode of the electron gun is the anode, which takes the form of a disk with an axial hole. Electrons leave the cathode and accelerate toward the anode. The control and alignment of the electron gun are critical in ensuring satisfactory operation.

Condenser Lenses System

The intensity and angular aperture of the beam are controlled by the condenser lens system between the gun and the specimen. A single lens may be used to converge the beam onto the object, but, more commonly, a double condenser is employed. In this, the first lens is strong and produces a reduced image of the source, which is then imaged by the second lens onto the object. The use of a small spot size minimizes disturbances in the specimen due to heating and irradiation

Objective Lenses and Projector Lenses

The specimen grid is carried in a smallholder in a movable specimen stage. The objective lens is usually of short focal length (1–5 mm) and produces a real intermediate image that is further magnified by the projector lens or lenses. A single projector lens may provide a range of magnification of 5:1, and by the use of interchangeable pole pieces in the projector, a wider range of magnifications may be obtained. Modern instruments employ two projector lenses (one called the intermediate lens) to permit a greater range of magnification and to provide a greater overall magnification without a commensurate increase in the physical length of the column of the microscope. For image stability and brightness, the microscope is often operated to give a final magnification of 1,000–250,000X on the screen. If a higher final magnification is required, it may be obtained by photographic or digital enlargement. The quality of the final image in the electron microscope depends largely upon the accuracy of the various mechanical and electrical adjustments with which the various lenses are aligned to one another and to the illuminating system. The lenses require power supplies of a high degree of stability; for the highest standard of the resolution, electronic stabilization to better than one part in a million is necessary. The control of a modern electron microscope is carried out by a computer, and dedicated software is readily available.

Image Translation System (Image recording system)

TEM provides pieces of information in the form of variations of electron intensity in the image. The electron image is monochromatic and must be made visible to the eye either by allowing the electrons to fall on a fluorescent screen fitted at the base of the microscope column or by capturing the image digitally for display on a computer monitor. Computerized images are stored in a format such as TIFF or JPEG.

TEM Operation

- ☆ The electron gun emits electron when the filament is applied with high voltage current by electrical heating. The gun accelerates the electrons to extremely high speeds using electromagnetic coils and voltages of up to several million volts. The electron beam is focused into a thin, small beam by a condenser lens, which has a high aperture that eliminates high angle electrons. Having reached their highest speed, the electrons zoom through the ultra-thin specimen (50nm size, semitransparent for electrons and carries information about the structure of the specimen) loaded on a grid (Figure 6.27) inserted in the path and manipulated by goniometer and parts of the beam are transmitted depending on how transparent the sample is to electrons.
- ☆ The beam transmitted through an objective aperture and projected by the projector lens after corrected by intermediate lenses on the fluorescence screen. At the bottom of the microscope, the unscattered electrons hit a fluorescent screen, which gives rise to a "shadow image" of the specimen with its different parts displayed in varied darkness according to their density. The image is observed with the help of optical binocular attached to its viewing window.
- ☆ The spatial variation in the "image" is then magnified by a series of magnetic lenses until it is recorded by hitting a fluorescent screen or light-sensitive sensor such as a CCD (charge-coupled device) camera fitted either in the side or bottom of the photographic plate. TEM electron scattering rather than differences in absorbance produce contrast in the image. Scattering results from an interaction between specimen atoms and electrons of the illuminating beam.

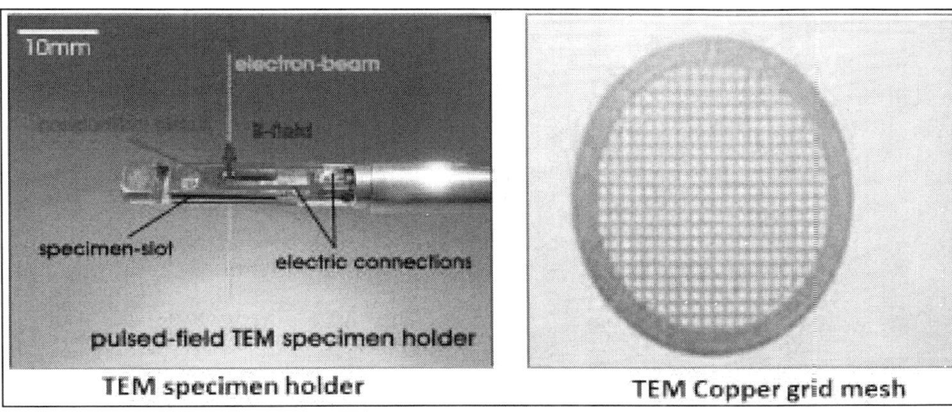

Figure 6.27. TEM Specimen Holder and Copper Grid.

☆ The negatively charged electron clouds around atomic nuclei scatter electrons by repelling them. These effects increase as electrons pass closer to specimen atoms. Nuclei of high atomic number, as in atom of heavy metals such as lead and uranium, cause the widest scattering. Transmission electron microscopes produce two-dimensional, black and white images. The TEM can easily resolve structures such as ribosome, microtubules, microfilaments and large molecules such as proteins. Even images of an individual heavy metal atom have been produced under the special operating conditions.

Preparation of Specimen for TEM

Biological Samples

In a TEM, the specimen must be of such a low density, it should allow electrons to travel through the tissue. There are different ways to prepare materials for that purpose. Thin slices of the specimen from a piece of tissue can be cut either by fixing it in plastic or working with it as frozen material. Another way to prepare the specimen is to isolate it and study a solution of for example viruses or molecules in the TEM. The specimen can be stained in different ways and use markers to locate specific things in the tissue. It can be stained with heavy metals like uranium and lead, which scatters electrons well and improves the contrast in the microscope.

Sections of Embedded Material

Biological material contains large quantities of water. Since the TEM works in a vacuum, the water must be removed. To avoid disruption as a result of the loss of water, preserve the tissue with different fixatives. These cross-link molecules with each other and trap them together as stable structures. The tissue is then dehydrated in alcohol or acetone. After that, the specimen can be embedded in plastic that polymerizes into a solid hard plastic block. The block is cut into thin sections by a diamond knife in an instrument called ultramicrotome. Each section is only 50-100 nm thick. The thin sections of samples are placed on a copper grid and stained with heavy metals. The slice of tissue can now be studied under the electron beam.

Whole Mounts

Small or very thin objects can be examined directly by mounting them onto a support film and introducing them directly into the electron beam. Contrast is provided by heavy metal precipitation in one of three ways.

1. **Positive staining**: The object is chemically stained with a solution of the metal salt and appears dark on a bright background.
2. **Negative staining**: The object remains unstained but is embedded in a dried film of the heavy metal salt. The specimen appears light on a dark background. This method of visualization has been used extensively in the study of virus particles but is also useful for cell fractions (*e.g.* coated vesicles).

3. **Shadowing**: A thin layer of heavy metal atoms is deposited on the specimen by evaporation in a vacuum chamber. Shadowing from one direction only produces a pseudo-three-dimensional image. Rotary shadowing, where the specimen is uniformly coated with heavy metal, is used to visualize nucleic acids and proteins.

Ultrathin Sectioning

The most popular technique for examining biological materials is to embed the material under study in plastic and cut ultrathin sections that can be examined in a TEM. The material is stabilized by chemical fixation (usually with aldehydes such as formaldehyde or glutaraldehyde), contrasted with solutions of heavy metal salts (osmium tetroxide and uranyl acetate), dehydrated in ethanol or acetone, and embedded in plastic (epoxy resin). Ultrathin sections (50 nm) cut with glass or diamond knives using an ultramicrotome (Figure 6.28) are floating on the water, transferred to specimen support grids and examined in the TEM. Often the sections are further contrasted with uranyl acetate and lead citrate prior to examination in the microscope.

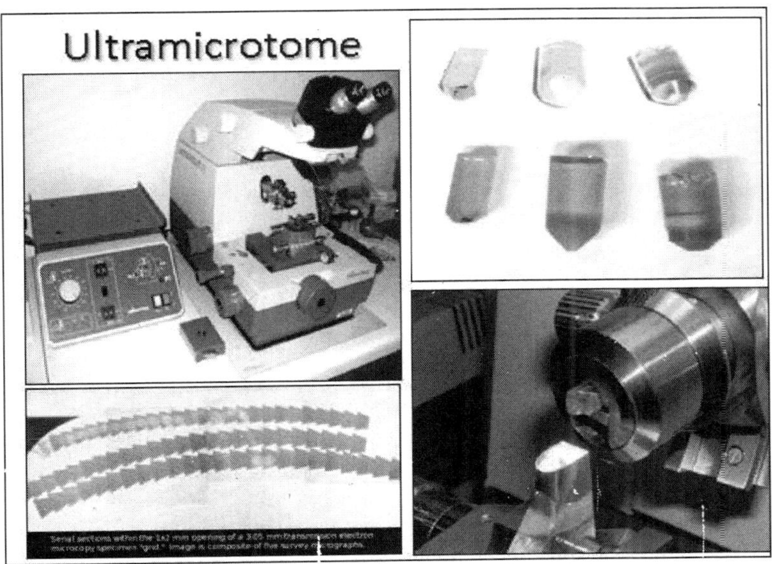

Figure 6.28. Ultramicrotome Sectioning.
Courtesy: Kannan, 2018.

In some cases, macromolecules can be specifically labeled prior to embedding and sectioning. For example, the location of some enzymes can be visualized by incubating the tissue with a substrate whose reaction with the enzyme leads to the local deposition of electron opaque material. Alternatively, antibodies can be coupled to such enzymes, and the electron opaque reaction product is used to localize the antigens recognized by the antibodies. Some embedding resins (*e.g.* Lowicryl resins and LR White resin) have been designed to enable antibodies and electron opaque markers (such as colloidal gold particles) to be applied to the ultrathin sections. In this way, subcellular antigens recognized by the antibodies can be localized with the TEM.

Another sectioning technique that is increasing in popularity is cryosectioning (the sectioning of vitrified, frozen material). After chemical fixation, the tissue is immersed in cryo-protectant (usually sucrose) and then quickly frozen in liquid nitrogen. The cryo-protectant allows the biological material to be frozen without the formation of ice crystals, which would damage ultrastructure. This type of freezing, or vitrification, is possible in the absence of cryo-protectants but is technically demanding. Sections cut from the vitrified block can be thawed and incubated with antibodies specific to subcellular antigens. Electron opaque markers allow the antibodies to be seen in the TEM.

Colloidal gold coupled to protein A (a protein from bacterial cell walls which binds to the Fc portion of some immunoglobulins) has been used extensively in recent years to localize antibodies on resin and frozen sections of biological materials. The ability to produce homogeneous populations of colloidal gold with different particle sizes has enabled researchers to use these probes to colocalize different structures in the same section.

Cryofixation

It is possible to freeze biological material fast enough to vitrify the water present inside the cells. Vitrification of water occurs when the freezing has occurred so fast that ice crystals have no time to form. Vitrified biological material can be sectioned at low temperatures. Thin films of vitrified water and sections of vitrified material can be examined in transmission electron microscopes that are equipped with specimen stages that can be kept cold.

Rapid Freezing Methods

There are seven main rapid freezing methods presently available. They are:

1. Immersion freezing - the specimen is plunged into the cryogen.
2. Slam (or metal mirror) freezing - the specimen is impacted onto a polished metal surface cooled with liquid nitrogen or helium.
3. Cold block freezing - two cold, polished metal blocks attached to the jaws of a pair of pliers squeeze-freeze the specimen.
4. Spray freezing - a fine spray of the sample in a liquid suspension is shot into the cryogen (usually liquid propane).
5. Jet freezing - a jet of liquid cryogen is sprayed onto the specimen.
6. High-pressure freezing - freezing the specimen at high pressure to subcool the water.
7. Excision freezing - a cold needle is plunged into the specimen, simultaneously freezing and dissecting the sample.

Freeze-Fracture followed by Freeze Etch and Replication

If for some reason, the object to be studied cannot be examined in the TEM, then a thin replica can be made. This is usually made by evaporating a thin layer of heavy metal (usually platinum) onto the specimen and then coating this with a thin layer of carbon. The object and the replica are separated either by floating off the replica or by digesting away the object. There are four basic steps to follow

1. The specimen is frozen (often without regard to vitrification).
2. The specimen is fractured, while still frozen, under vacuum.
3. The fractured specimen can then be *etc.*hed by leaving it frozen and under vacuum. Depending on the time of exposure, more or less water sublimes from the specimen (freeze-drying).
4. A replica of the fractured surface is made which is then examined in the electron microscope.

A recent modification of this method employs rapid freezing achieved by slamming cells against a copper block cooled to -269°C with liquid helium. If these frozen cells are then exposed to extensive freeze-drying (deep *etc.*hing), very impressive images of the internal structures of cells are uncovered.

Table 6.2. Sample Preparation for Transmission Electron Microscopy (General Schedule for Plant Tissue)

	Chemical	*Temperature*	*Time*	*Repetitions*
Primary fixation	2.5 per cent glutaraldehyde in buffer	room or 0-4°C	2-4 hours or microwave*	1
Wash	buffer	room or 0-4°C	30 minutes	3-5
Secondary fixation	1-4 per cent osmium tetroxide in buffer	room or 0-4°C	2-4 hours	1
Wash	buffer or distilled water**	room or 0-4°C	30 minutes	3-5
en bloc staining (optional)***	0.5 per cent uranyl acetate	0-4°C	overnight	1
Wash after *en bloc* staining	distilled water	room or 0-4°C	10-15 minutes	2
Dehydration	25 per cent ethanol 50 per cent ethanol 70-75 per cent ethanol 90-95 per cent ethanol 100 per cent ethanol Transition solvent if embedding resin is not miscible with ethanol	room or 0-4°C	20 minutes 20 minutes 20 minutes 20 minutes 30 minutes	1 1 1 1 2

	Chemical	*Temperature*	*Time*	*Repetitions*
Infiltration	1 part resin/2 parts solvent 1 part resin/1 part solvent (optional) 2 parts resin/1 part solvent 100 per cent resin	room room room room	1 hour-overnight 1 hour-overnight 1 hour-overnight 1 hour	1 1 1 1
Embedding	Place in 100 per cent resin in a suitable container			1
Degassing (optional)	Place in vacuum desiccator or vacuum oven	room-60° C	3-30 minutes	
Polymerization		60-70° C	> 8 hours	

* See instructions for microwave fixation: **Wash in water if *en bloc* staining is used.*** If omitting *en bloc* staining, proceed to dehydration at end of washes

Courtesy: Chinnamuthu, 2013.

Advantages

- ✰ TEMs offer the most powerful magnification, potentially over one million times or more
- ✰ TEMs have a wide range of applications and can be utilized in a variety of different scientific, educational and industrial fields
- ✰ TEMs provide information on the element and compound structure
- ✰ Images are high-quality and detailed
- ✰ TEMs are able to yield information of surface features, shape, size and structure
- ✰ They are easy to operate with proper training

Disadvantages

- ✰ TEMs are large and very expensive
- ✰ Laborious sample preparation
- ✰ Potential artifacts from sample preparation
- ✰ Operation and analysis requires special training
- ✰ Samples are limited to those that are electron transparent, able to tolerate the vacuum chamber and small enough to fit in the chamber
- ✰ TEMs require special housing and maintenance
- ✰ Images are black and white
- ✰ Electron microscopes are sensitive to vibration and electromagnetic fields and must be housed in an area that isolates them from possible exposure.
- ✰ A Transmission Electron Microscope requires constant upkeep including maintaining voltage, currents to the electromagnetic coils and cooling water

Applications

- ✰ A Transmission Electron Microscope is ideal for a number of different fields such as life sciences, nanotechnology, medical, biological and material research, forensic analysis, gemology, and metallurgy as well as industry and education.
- ✰ TEMs provide topographical, morphological, compositional and crystalline information.
- ✰ The images allow researchers to view samples on a molecular level, making it possible to analyze structure and texture.
- ✰ This information is useful in the study of crystals and metals but also has industrial applications.
- ✰ TEMs can be used in semiconductor analysis and production and the manufacturing of computers and silicon chips.
- ✰ Technology companies use TEMs to identify flaws, fractures, and damages to micro-sized objects; this data can help fix problems and/or help to make a more durable, efficient product.
- ✰ Colleges and universities can utilize TEMs for research and studies.
- ✰ Although electron microscopes require specialized training, students can assist professors and learn TEM techniques.
- ✰ Students will have the opportunity to observe a nano-sized world in incredible depth and detail

Atomic Force Microscope (A probe based Microscopy)

The Atomic Force Microscope (AFM) is being used to solve processing and materials problems in a wide range of technologies affecting the electronics, telecommunications, biological, chemical, automotive, aerospace, and energy industries. The materials being investigating include thin and thick film coatings, ceramics, composites, glasses, synthetic and biological membranes, metals, polymers, and semiconductors. The AFM is being applied to studies of phenomena such as abrasion, adhesion, cleaning, corrosion, *etc.* hing, friction, lubrication, plating, and polishing. By using AFM one can not only image the surface in atomic resolution but also measure the force at the nano-newton scale. The publications related to the AFM are growing speedily since its birth.

The first AFM was made by meticulously gluing a tiny shard of diamond onto one end of a tiny strip of gold foil. In the fall of 1985, Gerd Binnig and Christoph Gerber used the cantilever to examine insulating surfaces. A small hook at the end of the cantilever was pressed against the surface while the sample was scanned beneath the tip. The force between tip and sample was measured by tracking the deflection of the cantilever. This was done by monitoring the tunneling current tot a second tip positioned above the cantilever. They could delineate lateral features as small as 300 Å. The force microscope emerged in this way. In fact, without the breakthrough in tip manufacture, the AFM probably would have remained a curiosity in many research groups. It was Albrecht, a fresh graduate student, who fabricated the first silicon microcantilever and measured the atomic structure of boron nitride. Today the tip-cantilever assembly typically is microfabricated from Si or Si_3N_4. The era of AFM came finally when the Zurich group released the image of a silicon (111) 7X7 pattern. The world of surface science knew that a new tool for surface microscope was at hand. After several years the microcantilevers have been perfected, and the instrument has been embraced by scientists and technologists.

The force between the tip and the sample surface is very small, usually, less than 10^{-9} N. How to monitor such small forces is another story. The detection system does not measure force directly. It senses the deflection of the microcantilever. The detecting systems for monitoring the deflection fall into several categories. The first device introduced by Binnig was a tunneling tip placed above the metalized surface of the cantilever. This is a sensitive system where a change in the spacing of 1 Å between tip and cantilever changes the tunneling current by an order of magnitude. It is straight forward to measure deflections smaller than 0.01 Å. Subsequent systems were based on optical techniques. The interferometer is the most sensitive of the optical methods, but it is somewhat more complicated than the beam-bounce method which was introduced by Meyer and Amer. The beam-bounce method is now widely used as a result of the excellent work by Alexander and colleagues. In this system, an optical beam is reflected from the mirrored surface on the backside of the cantilever onto a position-sensitive photodetector. In this arrangement, a small deflection of the cantilever will tilt the reflected beam and change the position of the beam on the photodetector. A third optical system introduced by Sarid uses the cantilever as one of the mirrors in the cavity of a diode laser. The motion of the cantilever has a strong effect on the laser output, and this is exploited as a motion detector.

According to the interaction of the tip and the sample surface, the AFM can be classified as repulsive or Contact mode and attractive or Noncontact mode. Now the Tapping mode shows a prosperous future to image the micro-world.

Operation Principle of AFM

The AFM and light microscope amplify the image of the sample but the major difference between them is that the former does not use visible light; instead, it uses a cantilever made from silicon or silicon nitride having a very low spring constant to image a sample. At one end of the cantilever, a very sharp tip (around 100– 200 nm long and 20–60 nm radius of curvature) is fabricated using semiconductor processing techniques. The cantilever scans above the surface of the sample by progressively moving backward and forward across the surface. A piezo-electric crystal raises or lowers the cantilever to maintain a constant bending of the cantilever. The force exerted on the tip varies with the difference in the surface height and thus leads to the bending of the cantilever. A laser beam gets constantly reflected from the top of the cantilever towards a position-sensitive photodetector consisting of four side by- side photodiodes (Figure 6.29). This laser beam detects the bend occurring in the cantilever and calculates the actual position of the cantilever. Thus, AFM records a three-dimensional image of the surface topography of the sample under a constant applied force (as low as nano Newton range), which provides a maximum resolution

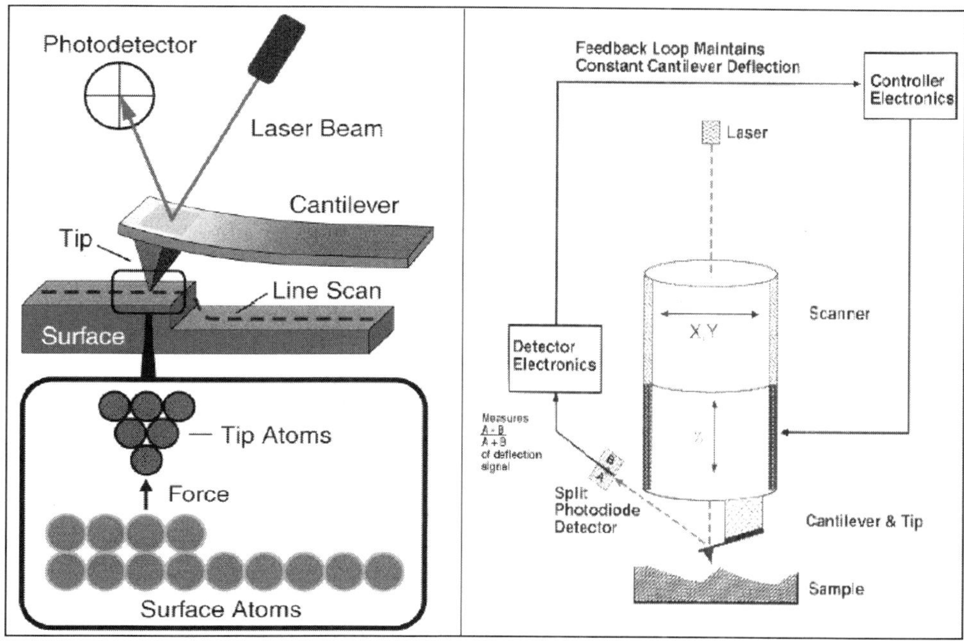

Figure 6.29. Schematic Diagram Showing the Operating Principles of the AFM.
Courtesy: Kannan, 2013.

image without causing any damage to the sample surface. As the tip scans the surface of the sample, the force between the tip and the sample varies. This change in force is sensed by the tip attached to the flexible cantilever. The amount of force between the probe and the sample is dependent on the spring constant of the cantilever and the distance between the probe and the sample surface. According to Hooke's Law, this force can be described as

$$F = -k\,\Delta x$$

Where F is the force, k is the spring constant and x is the cantilever deflection.

The interactions between the tip of the cantilever and the sample surfaces are regulated by different types of forces, *viz.*, van der Waals forces, capillary, and adhesive forces, and double-layer forces. Among these, van der Waals force is most commonly associated with AFM. In the contact region, the distance between the tip of the cantilever and the sample surface is less than a few angstroms (Å). Thus the probe experiences repulsive van der Waals force. In the non-contact region, the tip is several tens to hundreds of angstroms away from the sample surface and hence experiences an attractive van der Waals force. Various scanning modes operate in different regions of the van der Waals force vs distance curve as represented in Figure 6.30. The contact mode operates in the repulsive region and the non-contact mode operates in the non-repulsive region of the curve, while the intermittent or tapping mode fluctuates between the two.

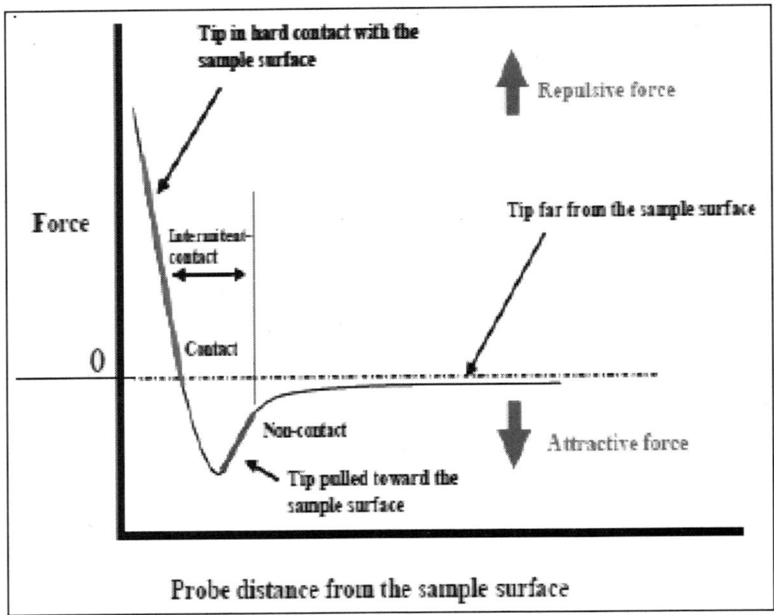

Figure 6.30. Graphical Representation Force Distance Relationship.
Courtesy: Kannan, 2013.

AFM Resolution

The concept of resolution in AFM is different from radiation-based microscopies because AFM imaging is a three-dimensional imaging technique. The ability to distinguish two separate points on an image is the standard by which lateral resolution is usually defined. There is clearly an important distinction between images resolved by wave optics and scanning probe techniques. The former is limited by diffraction, and later primarily by apical probe geometry and sample geometry. Usually, the width of a DNA molecule is loosely used as a measure of resolution because it has a known diameter of 2.0 nm in the B form. Some of the best values for AFM imaging are 3.0 nm quoted form DNA in propanol. Unfortunately, this definition of resolution can be misleading because the sample height clearly affects this value. Indeed; many authors have seen that it is the radius of curvature that significantly influences the resolving ability of the AFM. Images of DNA made by the sharper tip have shown dramatic improvements in resolution widths (Figure 6.31). Even greater improvements in resolution have been attained with tapping mode but contact imaging still is capable of high-resolution imaging.

Vibration Isolation

In order to obtain good AFM results, the vibration isolation platform is needed. The vibration isolation consists of a large mass attached to bungy cords firmly anchored to the building. (Notice that the resonance frequency only depends on the stretch of the Bungy cord). Damping of the oscillation is believed to result

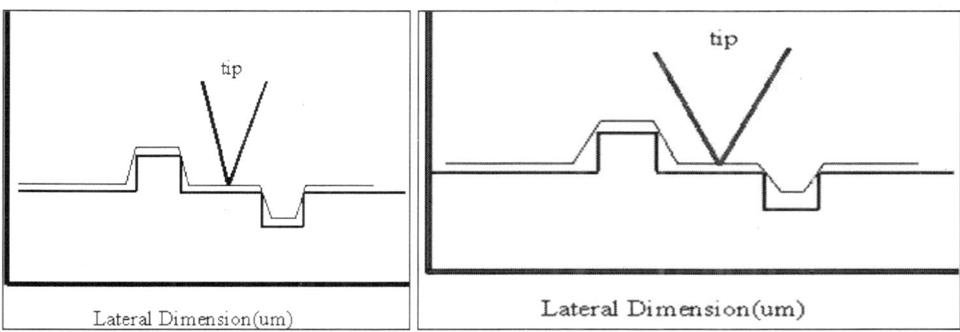

Figure 6.31. Dramatic Improvements in Resolution Widths.

from rubbing the rubber fibers inside of the Bungy cord against the outside lining material. Between the low resonance frequency of the Bungy cord system and the high resonance frequency of the microscope hardware itself (> 10 kHz), the AFM effectively comprises a bandpass filter. This allows the microscopists to safely image their samples in the intermediate range of about 1 - 100 Hz and obtain atomic resolution.

Operating Modes of AFM

Contact Mode

Contact mode or repulsive mode is the simplest mode of operation for AFM. This mode of operation involves sideways scanning by the cantilever tip over the sample surface. The tip makes soft physical contact with the sample and the scanner gently traces the tip across the sample. However, this mode is also associated with disadvantages, *viz.*, general integrity of the delicate biological samples can be lost. This drawback prevents the use of contact mode in studies pertaining to biological substances. In addition, contact mode yields very low resolution with large soft samples. As biological substances are very soft and delicate, the vertical and shear forces exerted by the tip can damage the sample. Therefore, contact mode is not ideal for biological systems. This lacuna can be overcome by applying tapping mode in the fluid that reduces the shear forces and minimizes the damage to biological samples.

Tapping Mode

Tapping mode or the intermittent contact AFM is the most preferred operating mode for high-resolution topographic imaging of subcellular structures and soft and delicate biological samples.

Tapping Mode in the Air

In this operating mode, a small piezoelectric crystal mounted in the multimode AFM tip holder makes the cantilever oscillate up and down at or slightly below its resonance frequency. The tip oscillates vertically, alternately contacts the surface and lifts off. The amplitude of this oscillation typically ranges from 20 nm to 100 nm. The oscillating tip lightly touches or 'taps' on the sample surface during scanning.

When the tip comes close to the sample surface, forces like van der Waals forces, dipole-dipole interactions, electrostatic forces, *etc.*, act on the cantilever and lead to a decrease in the amplitude of oscillation. Thus, the image is obtained by imaging the force of the oscillating contacts of the cantilever tip with the sample surface.

Tapping Mode in Fluids

Tapping mode operation in an aqueous medium is a very useful tool for biologists because the samples are in a state that closely resembles the *in vivo* environment as compared to dehydrated samples.

Non-Contact Mode

This mode is preferentially used in the study of chromatin dynamics. The Non-Contact AFM (or NC-AFM) operates with increased tip-sample separation without being in contact with the samples under normal imaging conditions. The cantilever oscillates above the sample surface with a small amplitude at a frequency larger than its resonance frequency. This results in increased sensitivity in comparison with tapping mode AFM. Outstanding spatial resolution can be achieved by performing NC-AFM in an ultra-high vacuum (UVH). In biological studies tapping mode and non-contact modes are most extensively used.

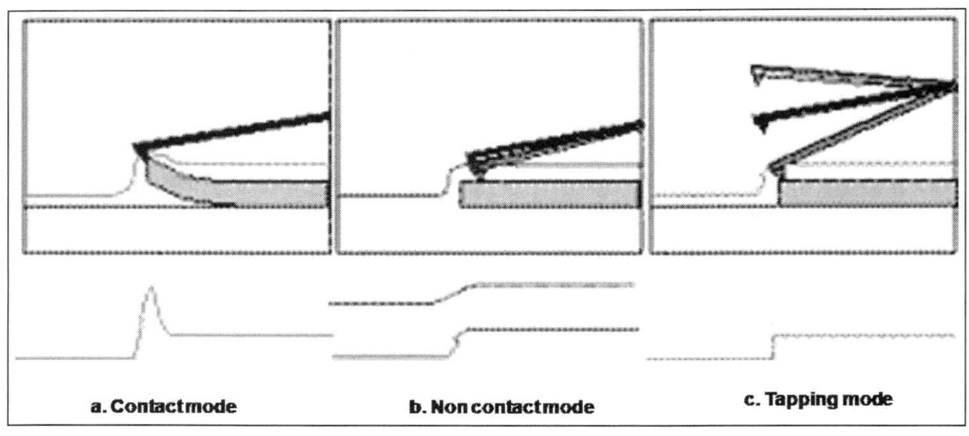

Figure 6.32. Common AFM Modes.

Applications of AFM

Compound Semiconductor

Imaging compound semiconductor surfaces on the atomic scale.

Electronic Materials

AFM tip is used to write and erase nanoscale **electronic** structures.

Data Storage

AFM is nowadays used for data storage due to space and cost constraints.

Life Science

AFM measures surface structures of various biological substances, and is capable of visualizing biological objects from living cells down to single-molecule levels.

Pharmaceuticals

In pharmaceutical industries, AFM is used for drug crystallization study, particle characterization and tablet coatings.

Semiconductor

AFM is used in characterizing semiconductors.

Optics

AFM is used in optics for the metrological measurement since the optical profilers are unable to image transparent specimens.

Polymers

Polymers are generally insulators. High-resolution images of the polymers are very difficult to obtain using SEM/TEM, as the samples need to be coated with a conductive layer. AFM, however, can generate images of polymers without sample preparation.

Limitation of AFM

- The AFM can be used to study a wide variety of samples (*i.e.* plastic, metals, glasses, semiconductors, biological samples such as the walls of cells, bacteria).
- Unlike STM or scanning electron microscopy it does not require a conductive sample. However, there are limitations in achieving atomic resolution.
- The physical probe used in AFM imaging is not ideally sharp.
- As a consequence, an AFM image does not reflect the true sample topography but rather represents the interaction of the probe with the sample surface (tip convolution).
- As a consequence, an AFM image does not reflect the true sample topography but rather represents the interaction of the probe with the sample surface (tip convolution).

Advantages of AFM over conventional microscopy techniques

- In contact AFM electrostatic and/or surface tension forces from the adsorbed gas layer pull the scanning tip toward the surface. It can damage samples and distort image data. Therefore, contact mode imaging is heavily influenced by frictional and adhesive forces compared to non-contact or tapping mode.

- ☆ Non-contact imaging generally provides low resolution and can also be hampered by the contaminant layer which can interfere with oscillation.
- ☆ Tapping Mode AFM was developed as a method to achieve high resolution without inducing destructive frictional forces both in air and fluid. With the Tapping Mode technique, the very soft and fragile samples can be imaged successfully. Also, incorporated with *Phase Imaging*, the tapping mode AFM can be used to analyze the components of the membrane

AFM in Life Sciences

The AFM has great promise for measuring the surface structure of biological material. For example, the AFM is the only microscopy technique that allows making nanometer-scale images with the sample submerged in liquid.

Cells

An AFM can readily measure images of cells both in ambient air and submerged in a liquid. The cells must rigidly adhere to a surface for successful imaging. With the AFM it is possible to measure the mechanical activity of a cell by simply placing the probe on the surface of the cell and monitoring the motion of the AFM cantilever. An advantage of the AFM for imaging cells is that the cell does not have to be coated, and in fact can still be alive when imaged.

Bio-Molecules

AFM images of bio-molecules such as DNA are easily measured as long as the bio-molecules are directly attached to a surface. Activated substrates are available for creating DNA images with an AFM. There are many examples of AFM DNA images as well as many images of other types of bio-molecules such as proteins. The AFM is capable of systematically removing genetic material from DNA or chromosomes by scratching the surface of the material.

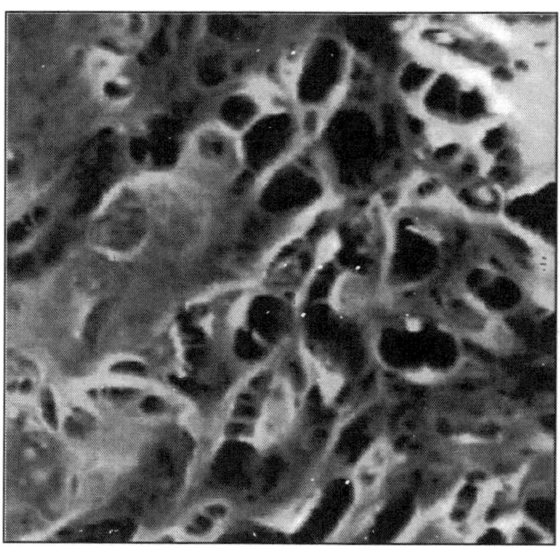

Figure 6.33. Breast Cancer Cells 78 x 78 microns.

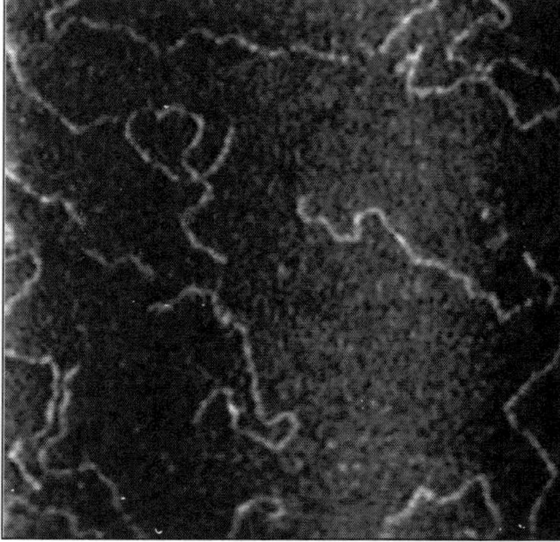

Figure 6.34. DNA Image 1.2u x 1.2u.

Conclusion

The Advancement in microscopic techniques offers the scientific community with powerful high-end tools to characterize the surface morphology, composition, cryptographic information and ultrastructural information of a variety of samples at the atomic scale with high resolution and magnification. Recent advances in electron and probe-based microscopy support the successful imaging of real-time images of different biochemical reactions, living cells of biological samples and the imaging of magnetic microstructures.

REFERENCES

1. Binnig, G., and Quate, C. F. (1986). Ch. Gerber–" Atomic force microscope. *Phys. Rev. Lett, 56*, 9.
2. Cheney, B. (2007). Introduction to scanning electron microscopy. *Materials Engineering department San Jose State University.*
3. Chinnamuthu C.R. and Brindha.K. (2013). Sample Preparation and Characterization of Biological and Material Samples Using TEM. Proceeding on Nanotechnology and Plant Disease Management (NPDM), Published by Department of Nano Science and Technology, Tamil Nadu Agricultural University (TNAU), Coimbatore. Pp. 88-96
4. FEI. The Quanta 200 User's Operation Manual 2nd ed. (2004).
5. Fultz, B., and Howe, J. M. (2012). *Transmission electron microscopy and diffractometry of materials*. Springer Science and Business Media.
6. Goldstein, J. (Ed.). (2012). *Practical scanning electron microscopy: electron and ion microprobe analysis*. Springer Science and Business Media.
7. Goldstein, J. I., Newbury, D. E., Michael, J. R., Ritchie, N. W., Scott, J. H. J., and Joy, D. C. (2017). *Scanning electron microscopy and X-ray microanalysis*. Springer.
8. JEOL. Guide to scanning electron Microscopy [Online]. Available at http: // www.jeol.com/sem/docs/sem_guide/tbcontd.html
9. Kannan M. 2018. Transmission Electron Microscope – Principle, Components and Applications. A Textbook on Fundamentals and Applications of Nanotechnology. Edited by Edited by: Subramanian K S, Janavi, G.J. Marimuthu S. Kannan. M. Raja. K. Haripriya. S. Jeaya Sundra Sharmila D. and Pon.Sathya Moorthy. Daya Publishing House, Astral International Private Limited, Delhi. Pages 33-43. ISBN 978935124932-0. pp. 93 – 102.
10. Kannan. M and Marimuthu. S. 2013. Atomic Force Microscopy - Basic Principle. Proceeding on Nanotechnology and Plant Disease Management (NPDM), Published by Department of Nano Science and Technology, Tamil Nadu Agricultural University (TNAU), Coimbatore. pp. 97-102
11. Loretto, M. H. (2012). *Electron beam analysis of materials*. Springer Science and Business Media.

12. Lyman, C. E., Newbury, D. E., Goldstein, J., Williams, D. B., Romig Jr, A. D., Armstrong, J. and Peters, K. R. (2012). *Scanning electron microscopy, X-ray microanalysis, and analytical electron microscopy: a laboratory workbook*. Springer Science and Business Media.

13. Marti, O., Elings, V., Haugan, M., Bracker, C. E., Schneir, J., Drake, B. and Weisenhorn, A. L. (1988). Scanning probe microscopy of biological samples and other surfaces. *Journal of microscopy, 152*(3), 803-809.

14. Notebook, S. P. M. T. (2000). Version 3.0. *Digital Instruments, Veeco Metrology Group*.

15. Transmission Electron Microscopy of Materials, Gareth Thomas and Michael J. Goringe, A Wiley-Interscience Publication, USA, 1979.

16. Transmission Electron Microscopy, Williams. D.B. and Carter. C.B. Plenum Press, New York, 1996.

17. Watt, I. M. (1997). *The principles and practice of electron microscopy*. Cambridge University Press.

18. www.matter.org.uk

19. Yao, N., and Wang, Z. L. (Eds.). (2005). *Handbook of microscopy for nanotechnology* (p. 745). Boston: Kluwer academic publishers.

20. Zhou, W., Apkarian, R., Wang, Z. L., and Joy, D. (2006). Fundamentals of scanning electron microscopy (SEM). In *Scanning microscopy for nanotechnology* (pp. 1-40). Springer, New York, NY.

Chapter 7

Characterization of Nanomaterials using Diffraction and Spectroscopy

C. Sharmila Rahale, K. Raja, N. Viji, M.S. Kumuthan and M. Latha

Department of Nano Science and Technology, Tamil Nadu Agricultural University, Coimbatore – 641 003, Tamil Nadu

Nanomaterials behave differently as the size changes with respect to the bulk. But measuring this nano dimension is not a very easy task. Although research is going on to synthesise nanostructured and nanophasic materials, characterizing these nano sized materials is also an emerging field posing lot of challenges to scientists and technonologists. It is necessary to characterize physical, structural and optical properties of a material to qualify as nanomaterial. Various characterization techniques are used to know the characteristics of the nanomaterials. This chapter describes the instruments utilized for various measurements towards the characterization particularly of diffraction and spectroscopy of the synthesized nanocrystals.

I. X-Ray Diffraction

X-ray powder diffraction (XRD) is a rapid analytical technique primarily used for phase identification of a crystalline material and can provide information on unit cell dimensions. The analyzed material is finely ground, homogenized, and average bulk composition is determined. About 95 per cent of solid materials are crystalline in nature, where atoms are arranged with symmetric repeats (Boddolla, 2018). When X-rays interact with a crystalline substance, it results in characteristic diffraction pattern. The same substance always gives the same pattern; and in a mixture of substances, each produces its pattern independently of others. The X-ray diffraction pattern of particular substance is like a fingerprint of the substance, which helps to identify the unknown substances. Similarly, X-diffraction is also useful in determination of phase of the materials, crystallite size and lattice stresses.

Principle

English physicists Sir W.H. Bragg and his son Sir W.L. Bragg developed a relationship in 1913 to explain why the cleavage faces of crystals appear to reflect X-ray beams at certain angles of incidence (theta, θ). The variable d is the distance between atomic layers in a crystal, and the variable lambda λ is the wavelength of the incident X-ray beam; n is an integer. This observation is an example of X-ray wave interference, commonly known as X-ray diffraction (XRD), and was direct evidence for the periodic atomic structure of crystals postulated for several centuries.

$$n\lambda = 2d \sin \theta$$

λ = Wavelength of the x-ray

Θ = Scattering angle

n = Integer representing the order of the diffraction peak.

d = Inter-plane distance

Although Bragg's law was used to explain the interference pattern of X-rays scattered by crystals, diffraction has been developed to study the structure of all states of matter with any beam, *e.g.*, ions, electrons, neutrons, and protons, with a wavelength similar to the distance between the atomic or molecular structures of interest.

Deriving Bragg's Law

Bragg's Law can easily be derived by considering the conditions necessary to make the phases of the beams coincide when the incident angle equals and reflecting angle. The rays of the incident beam are always in phase and parallel up to the point at which the top beam strikes the top layer at atom z. The second beam continues to the next layer where it is scattered by atom B. The second beam must travel the extra distance AB + BC if the two beams are to continue traveling adjacent and parallel. This extra distance must be an integral (n) multiple of the wavelength (λ) for the phases of the two beams to be the same:

$$\mathbf{n\,\lambda = AB + BC}$$

we can use trigonometry to relate d and to the distance (AB + BC). The distance AB is opposite Θ so, AB = dsinΘ

Because AB = BC, n λ = 2AB

$$\text{So, } n\lambda = 2d\sin\Theta$$

How X-rays are Produced

X rays are produced in device called X- ray tube, which consists of cathode and anode. The cathode acts as the source of electrons, which are accelerated to strike the anode (Grounded) and the cathode, is maintained at a high negative potential with respect to the anode. The anode acts as source of producing X rays

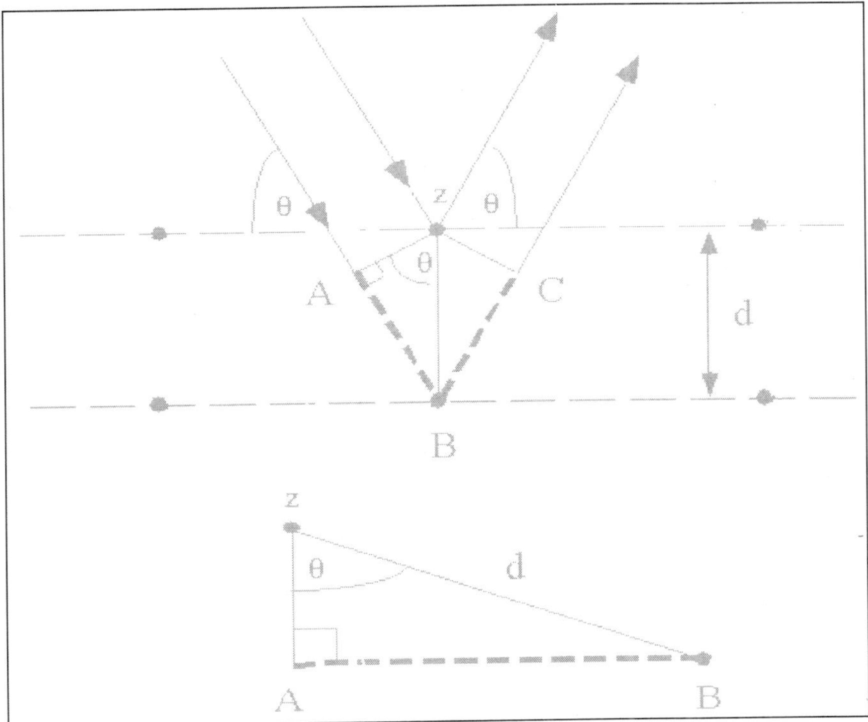

Figure 7.1. Derivation of Braggs' Law when the diffracted Beam Interfere Constructively.

(Figure 7.2). The high voltage maintained across cathode and anode, some tens of thousands of volts, rapidly draws the electrons to the anode, which they strike with very high velocity. X-rays are produced at the point of impact over anode and radiate in all directions. When the target material of the X-ray tube is bombarded with fast moving electrons from the cathode filament, two types of X-ray spectra are produced, which are continuous and characteristic X-rays.

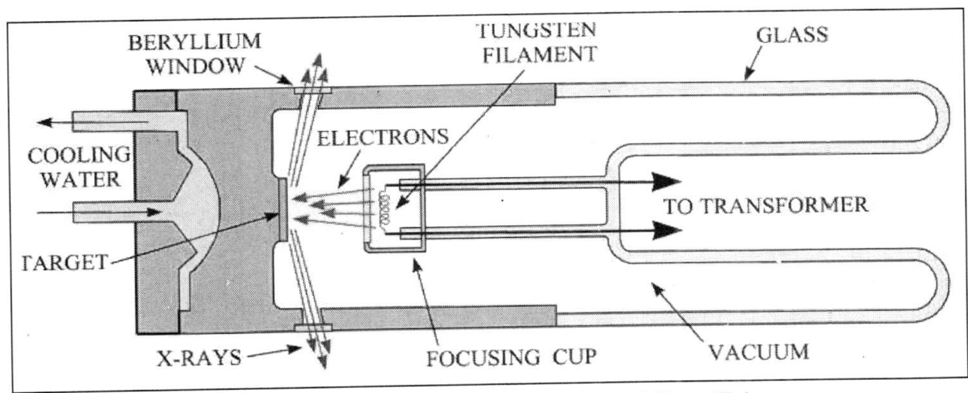

Figure 7.2. Schematic Cross Section of X-ray Tube.

When the target material of the X-ray tube is bombarded with fast moving electrons from the cathode filament, two types of X-ray spectra are produced,

1. Continuous X-rays
2. Characteristic X-rays

Continuous X-rays are created when free moving electrons electromagnetically interact with nuclei, whereas characteristic X-rays are formed during the electron transition processes that occur when an inner shell electron is released from an atom.

Continuous X-rays

A few fast moving electrons penetrate deep into the interior of the atoms of the target material and are attracted towards the nuclei by the attractive forces of their nuclei. Due to these forces, the electrons get deflected from their original path. As a result of this, the electrons are decelerated, and hence energy of the electron decreases continuously. This loss of energy during retardation is given off in the form of X-rays of continuously varying wavelength. This is known as continuum or continuous x rays. This is also known as Bremsstrahlung radiation (Braking or retarding Radiation), Brems – to brake and Strahlung-radiation. In this case, the radiation released will have energy in the range of energy of X rays.

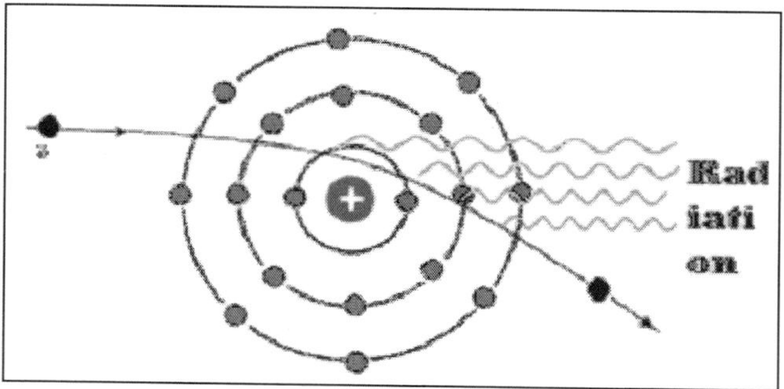

Figure 7.3. Production of Continuous X-rays.

Characteristic X-rays

Few of the fast moving electrons having velocity of about $(1/10)^{th}$ of the velocity of light may penetrate the surface atoms of the target materials and knock out the tightly bound electrons even from the inner most shells (like K, L shells) of the atom.

when the fast moving electrons knock off one electron from K-Shell and the vacancy is filled by the nearby electron from the L shell. During this transition, the energy difference is radiated in the form of X-rays of very small wave length. This corresponds to $K\alpha$ - line of the series. Suppose, the electron from M shell jumps to the K shell, it gives out $K\beta$ line and so on. If an electron jumps from the M-Shell to the vacant state in L-Shell, it contributes $L\alpha$ line and if the vacancy in L-Shell is filled up by an electron of N shell, it contributes $L\beta$ and so on. The frequency of radiation depends upon the target material. The X-ray spectra consists of sharp lines and is the characteristic of target material. Hence this spectra is known as characteristic spectra.

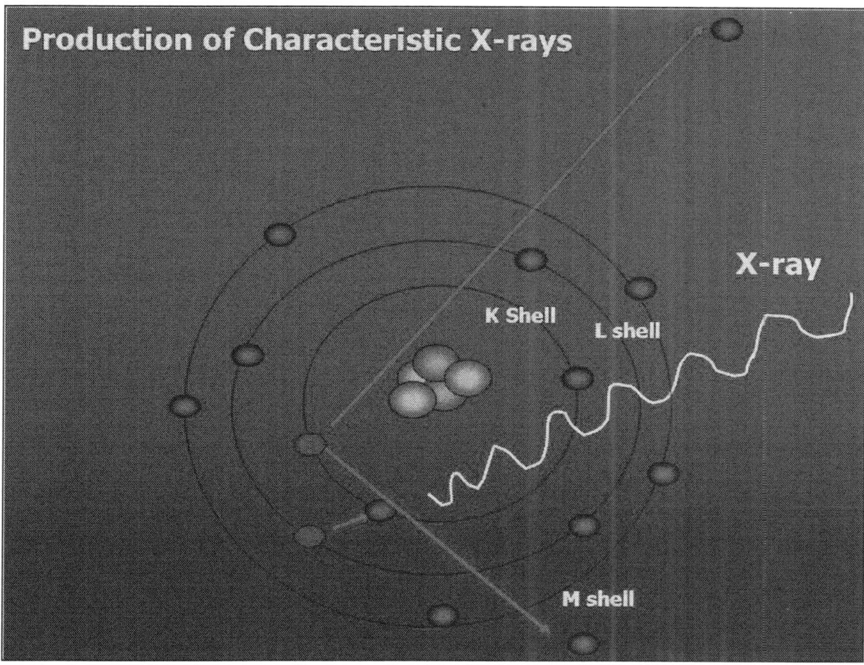

Figure 7.4. Production of Characteristic X-rays.

When X-rays are scattered from a crystalline solid they can constructively interfere, producing a diffracted beam. Interference occurs among the waves scattered by the atoms when crystalline solids are exposed to X-rays. There are two types of interference depending on how the waves overlap one another (Figure 7.5).

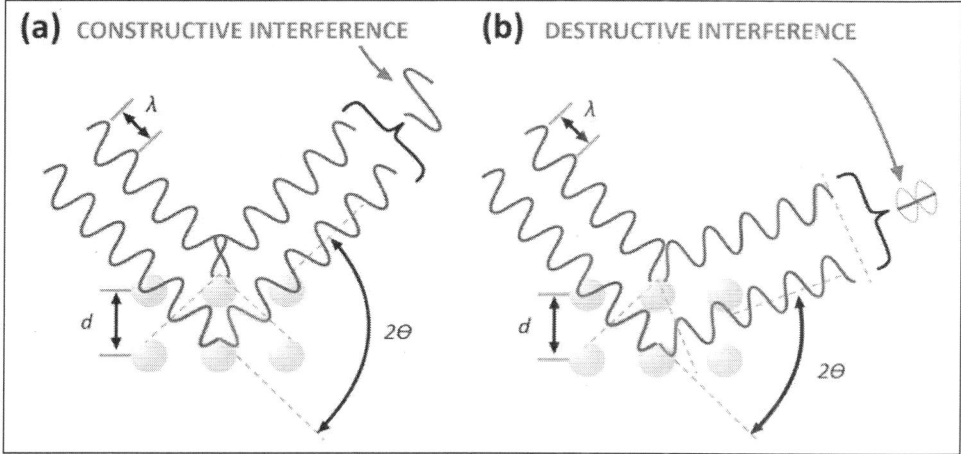

Figure 7.5. Constructive and Destructive interference

Constructive interference is when the crests or the troughs of the waves interfere with the other. Therefore their amplitudes of are added to find the total amplitude Constructive interference occurs when the waves are moving in phase with each other. Destructive interference is when a crest and a trough interfere with each other. This results in a lower amplitude than those of the original waves. Destructive interference occurs when the waves are out of phase. This constructive interference results in diffraction patterns.

X Ray Diffractometer

X-ray diffraction is based on constructive interference of monochromatic X-rays and a crystalline sample. A cathode ray tube, filtered to produce monochromatic radiation, collimated to concentrate, and directed toward the sample, generates X-rays. The interaction of the incident rays with the sample produces constructive interference (and a diffracted ray) when conditions satisfy Bragg's Law ($n\lambda=2d \sin \theta$). The diffracted X-rays are then detected, processed and counted. All possible diffraction directions of the lattice in the sample are attained by scanning the sample through a range of 2r angles. Conversion of the diffraction peaks to d spacing allows identification of the mineral because each mineral has a set of unique d spacing.

The instrument used for X ray diffraction studies is X-ray diffractometer, which consist of three basic components, an X-ray tube; a sample holder, and an X-ray detector. X-rays are generated in a cathode ray tube (Tungsten) by heating a filament to produce electrons, which are accelerated to impinge a target (metal node. Normally copper) by applying very high voltage. Cobalt and Molybdenum are used as anodes in X ray diffraction technique. When electrons have sufficient energy to dislodge inner shell K electrons of the target metal and characteristic X-rays are produced. Crystal monochrometers are required to produce monochromatic X-rays needed for diffraction. The fine-grained powder under investigation are placed at the designated place in glass plate. Gentle pressing flattens the sample and avoids

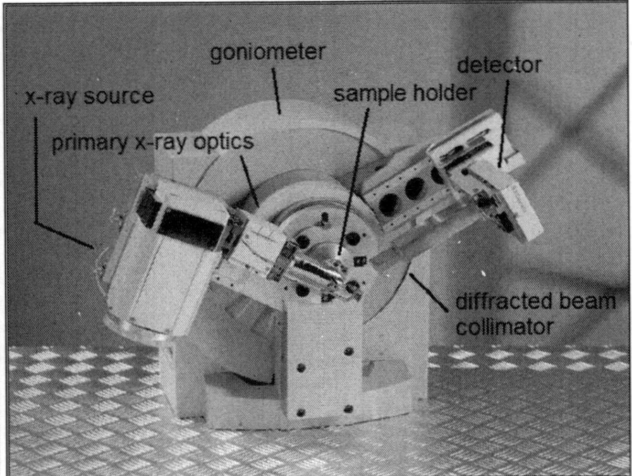

Figure 7.6. X-ray Diffractometer.

air pockets in the sample. The glass plate along with sample are kept in the sample holder. The X-rays are collimated and directed onto the sample.

The geometry of an X-ray diffractometer is such that the x ray tube rotates in the path of the collimated X-ray beam at an angle θ while the X-ray detector, mounted on an arm to collect the diffracted X-rays, rotates at an angle of 2θ, which is known as Bragg-Brentano configuration. The mechanism used to maintain the angle and rotate the X ray tube and detector is termed as Goniometer (Figure 7.6). The intensity of diffracted X- rays is continuously recorded as the sample and detector rotate through respective angles normally from 5° to 90°. When the geometry of the incident X-rays impinging the sample satisfies the Braggs' equation, constructive interference occurs results in peak intensity (maxima in the spectra). The diffracted X-rays are then detected, processed and counted

After a scan of the sample the X-ray intensity are plotted against the angle θ to produce a chart showing diffraction peak corresponding the angle 2θ which are converted to d spacing, using the Bragg equation. Conversion of the diffraction peaks to d spacing allows identification of the sample under investigation because each mineral has a set of unique d spacing. The comparison of spectra from the sample are compared with standard reference patterns of International Centre for Crystal Diffraction (ICCD) who have collected data diffraction peak and 2θ for thousands of crystalline substances for identifying the sample.

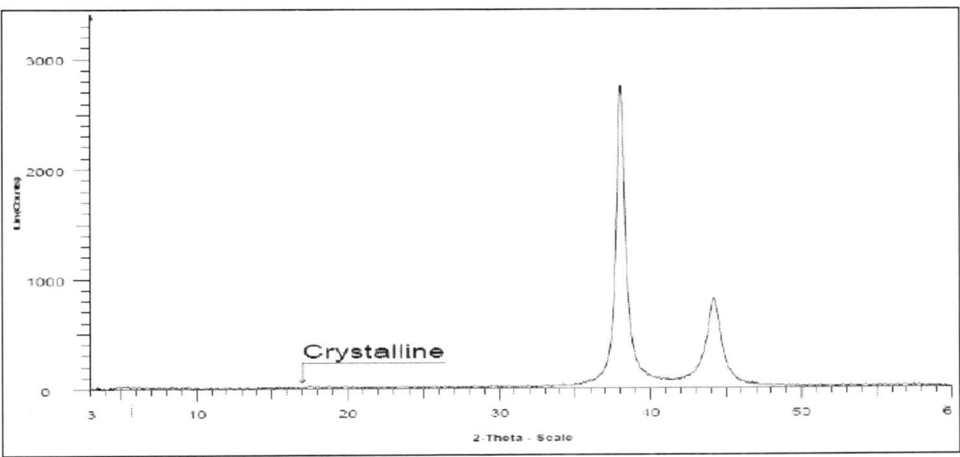

Figure 7.7. X-ray Diffraction Pattern of Crystalline Sample with Sharp Intense Peaks.

Sample Preparation

Sample Preparation Proper sample preparation is one of the most important requirements in the analysis of powder samples by X-ray diffraction. This statement is especially true for soils and clays that contain finely divided colloids, which are poor reflectors of X-rays, as well as other types of materials such as iron oxide coatings and organic materials that make characterization by XRD more difficult. Sample preparation includes not only the right sample treatments to remove

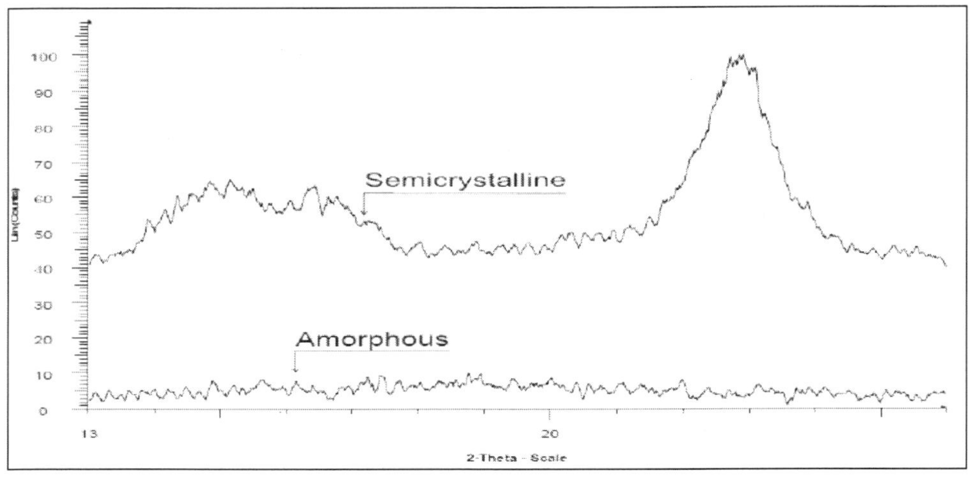

Figure 7.8: X-ray Diffraction Pattern of Semi-crystalline and Amorphous Samples.

undesirable substances, but also appropriate techniques to obtain desirable particle size, orientation, thickness, and other parameters. Several excellent books are available that deal with appropriate sample preparation techniques for clays and soils (Bish and Post, 1989; Iyengar *et al.*, 1997). Analysis of powders by XRD requires that they are extremely fine grained to achieve good signal-to-noise ratio (and avoid fluctuation in intensity), avoid spottiness, and minimize preferred orientation. Reduction of powders to fine particles also ensures enough particle participation in the diffraction process. The recommended size range is around 1–5 mm (Cullity, 1978; Klug and Alexander, 1974), especially if quantification of various phases is desired. For routine qualitative evaluation of mineral components, the samples are usually ground to pass through a 325-mesh sieve (45 mm). Grinding is accomplished either through hand grinding or in a mechanical grinder. The effects of excessive grinding include lattice distortion and possible formation of an amorphous layer (Beilby layer) outside the grains. There are two types of mounts normally employed depending on the nature of crystallite orientation required. Random mounts are preferred when identification of phases in a specimen is required. In this type of mount, particles ground to 1–5 mm are packed to a flat surface onto a sample holder to assume different orientations and ensure reflections from various planes (hkl). Oriented mounts are used when analyzing clay minerals, which rarely show strong diffraction effects from Bragg planes other than the (001). In general, these are prepared by making a slurry of the sample with distilled water. The water is then allowed to evaporate until the slurry is smeared into a sample holder (a glass slide or ceramic tile) (Figure 7.9).

Important Factors in Sample Preparation are

- ✰ Sample properties also influence the quality of a powder pattern by either reducing intensities or distorting intensities.

Figure 7.9. Glass Plate Sample Holder.

- Preferred orientation or texture: Texture means that the powder particles do not have an arbitrary shape but a strongly regular anisotropic shape, typically platelets or needles. On preparation these are then preferably oriented along the sample surface massively changing the peak intensities. Several techniques may be employed to minimize this effect:
- The most efficient way is to form a slurry in a highly viscous liquid such as nail varnish. In such a liquid, the random orientation is retained on drying.
- Alternatively, the anisotropic particle shape can be reduced by grinding in a ball mill. This should be done with great care as excessive grinding can easily break down the particle size to nano meter size and lead to amorphization. It is recommended to try the effect of subsequent 5 min grinding intervals to optimize the process on respective samples. In the case of coatings or thin films preferred orientation is often a desired effect. In this case Rietveld refinement can be used to determine the degree of texture.
- Crystallite size and strain: the broadness of a diffraction peak corresponds to the mean crystallite size in a reciprocal manner. The smaller the average crystallite size, the broader the reflections and the lower the absolute intensities. This effect becomes visible below an average crystallite size of less than 200 nm. Related to crystallite size broadening is strain broadening. Strain broadening occurs due to the presence of defects in crystals. Such strain can be introduced by substitution of constituting atoms but also by special thermal treatment. It is possible to distinguish strain broadening from size broadening as the angle dependence is much larger than in the latter case (He, 2009).
- Sample preparation height: Rotating sample holders improve the measurement statistics and thus provide the best results. However, they are not available for all machines. The most serious error during sample

preparation is to fill the sample holder too high or too low. Both result in a significant shift of peak positions, which can make the interpretation difficult.

Applications of X-ray Diffraction

- It has widespread applications in the fields of geology, pharmaceuticals, materials, polymers, environmental and forensic investigations.
- Identification of unknown crystalline substances and determination of crystallite size.
- Measurement of strain, or micro-strain effects (imperfections) in bulk and thin-film samples.
- Quantification of preferred orientation (texture) in thin films, multi-layer stacks, and manufactured parts.
- Determination of the ratio of crystalline to amorphous material in bulk materials and thin-film samples.
- Analyzing films as thin as 50 angstroms for texture and phase behaviors.
- Determining strain and composition in epitaxial thin films and, surface offcut in single crystal materials.
- Measuring residual stress in bulk metals and ceramics.

Advantages

- X ray diffraction is a non-destructive technique.
- Reliable and powerful tool for the identification of crystalline minerals.
- Simple sample preparation.
- X ray diffraction is faster method and deriving results are simple and straight forward.

Limitations

- X ray diffraction is highly suitable for identifying single phase material.
- Standard library is essential for interpreting results.
- X ray diffraction requires sample grain size of 2µ.
- Detection limit for phase identification in the sample is ~ 2 to 5 per cent.
- Identification of amorphous materials is difficult through X ray diffraction

X-ray diffraction (XRD) is an analytical technique used to characterize crystalline phases of a wide variety of materials, typically for mineralogical analysis and identification of unknown materials. Powder diffraction data are fundamentally derived by the atomic and molecular arrangements explained by the physics of crystallography. Minerals are the building blocks of the solid Earth. Some minerals are readily recognized by their distinctive colors or crystal forms, but in most cases, powder X-ray diffraction is the primary and most definitive method used to identify minerals. The high flux and density of X-rays produced at synchrotrons provide the microelectronics industry with a powerful probe of the structure and behavior of a wide array of solid materials that are being developed for use in devices of the future.

II. Fourier Transform-Infrared Spectroscopy (FTIR)

Fourier Transform-Infrared Spectroscopy (FTIR) is an analytical technique used to identify organic (and in some cases inorganic) materials. This technique measures the absorption of infrared radiation by the sample material versus wavelength. The infrared absorption bands identify molecular components and structures.

Infrared spectroscopy is one of the most important analytical technique in the field of advanced science. Infrared spectroscopy is a technique based on the vibrations of the atoms of a molecule. The molecule must possess a specific feature, *i.e.* an electric dipole moment which changes during the application of IR radiation. The dipole moment of an IR active, heteronuclear diatomic molecule changes its bond by stretching and bending. Whereas, infrared-inactive' molecule is a homonuclear diatomic molecule because its dipole moment remains zero no matter how long the bond. The interactions of infrared radiation with matter may be understood in terms of changes in molecular dipoles associated with vibrations and rotations. Liquids, solutions, pastes, powders, films, fibers, gases and surfaces which posses electric dipole moment can be examined through this technique. IR active molecules have unique IR absorption spectra that may be used as a "fingerprint" for identification.

Principle

When a material is irradiated with infrared radiation, absorbed IR radiation usually excites molecules into a higher vibrational state. The wavelength of light absorbed by a particular molecule is a function of the energy difference between the at-rest and excited vibrational states. The wavelengths that are absorbed by the sample are characteristic of its molecular structure.

Components of FT-IR

1. IR Light Sources

Instruments for measuring infrared absorption all require a source of continuous infrared radiation and a sensitive infrared transducer, or detector.

Infrared sources consist of an inert solid that is electrically heated to a temperature between 1,500 and 2,200 K. The heated material will then emit infra red radiation.The types of IR light sources as follows:

The Nernst Glower

The Nernst glower is constructed of rare earth oxides in the form of a hollow cylinder. Platinum leads at the ends of the cylinder permit the passage of electricity. Nernst glowers are fragile. They have a large negative temperature coefficient of electrical resistance and must be preheated to be conductive.

The Globar Source

A globar is a rod of silicon carbide (5 mm diameter, 50 mm long) which is electrically heated to about 1,500 K. Water cooling of the electrical contacts is needed to prevent arcing. The spectral output is comparable with the Nernst glower, except at short wavelengths (less than 5 mm) where it's output becomes larger.

The Carbon Dioxide Laser

A tunable carbon dioxide laser is used as an infrared source for monitoring certain atmospheric pollutants and for determining absorbing species in aqueous solutions.

The Mercury Arc

For the far infrared region of the spectrum (> 50 m), none of the thermal sources just described provides sufficient radiant power for convenient detection. Here, a high-pressure mercury arc is used. This device consists of a quartz-jacketed tube containing mercury vapor at a pressure greater than one atmosphere.

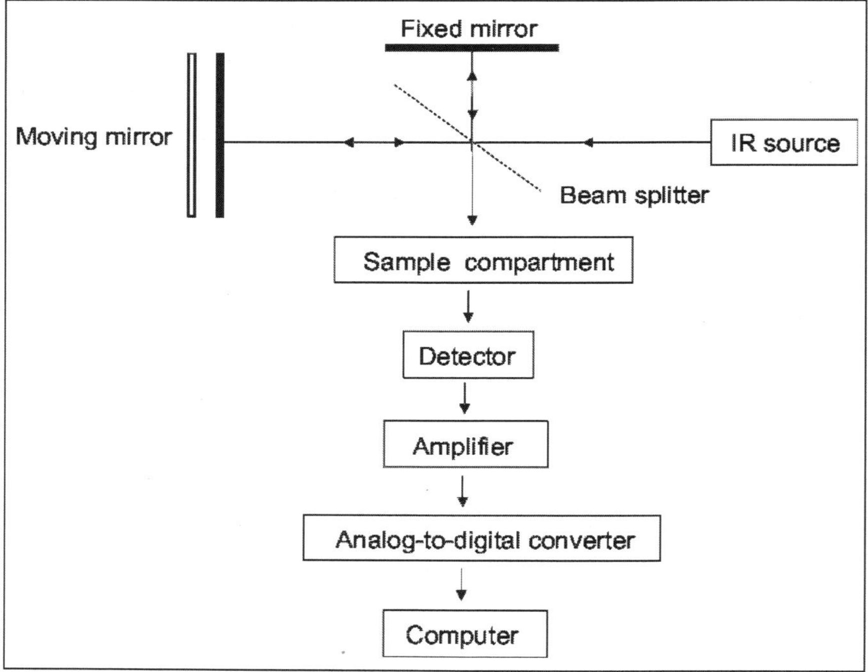

Figure 7.10. Components of FT-IR.

2. Sample Compartment

To hold the samples to be characterized.

3. Optical Components Used in Spectroscopy

Entrance Slit: Purpose is to provide rectangular optical image.

Collimating Mirror or lens: Purpose is to produce parallel beams of radiation, it overcomes diffraction.

Prism or Grating: Disperses radiation into its component wavelengths.

Focusing Mirror or lens: Reforms image from slit onto focal plane.

Exit Slit: Isolates Spectral Band.

4. Detector

To detect the absorbed or transmitted light from the sample.

The detectors can be classified into two categories. They are thermal detectors and Quantum detectors.

Thermal Detectors

Thermal detectors can be used over a wide range of wavelengths and they operate at room temperature. Their main disadvantages are slow response time and lower sensivity relative to other types of detectors. *e.g.* Thermocouple, Thermopile, Bolometer, Pneumatic detector (Golay) and Pyroelectric detector.

Thermocouple

In these detectors, the signal originates from a potential difference caused by heating a junction of the unlike metals by the infrared beam, while the other junction is kept at constant temperature

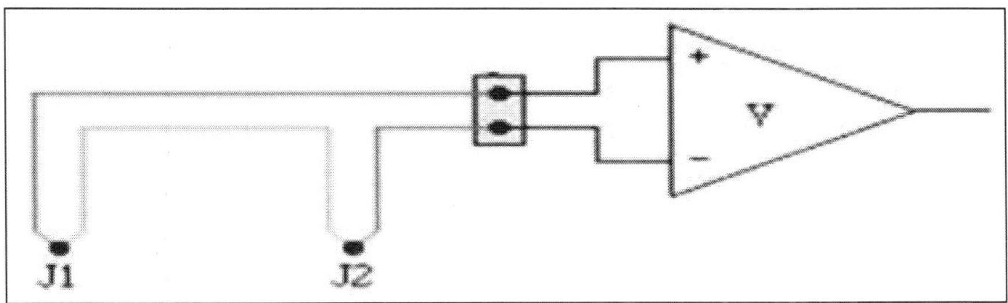

Figure 7.11. Schematic Diagram of Thermocouple Detector.

Thermopiles

It is possible to increase the output voltage by connecting several thermocouples in series. This arrangement is referred to as Thermopiles.

Bolometer

It provides an electrical signal as a result of the variation in resistance of a conductor with temperature.

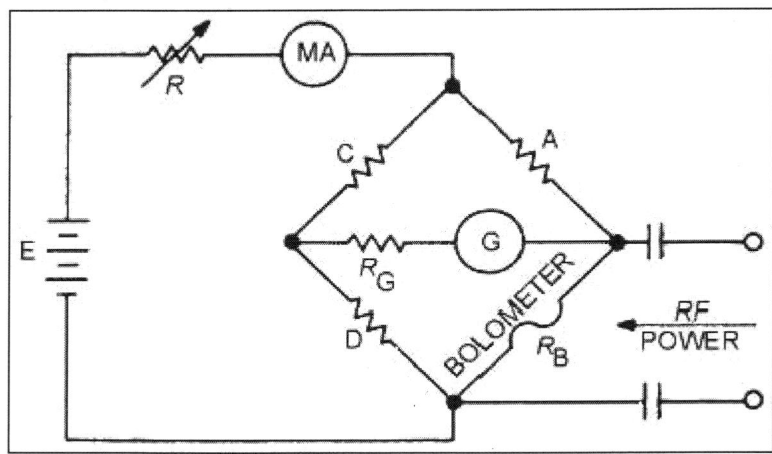

Figure 7.12. Schematic Diagram of Bolomter Detector.

Pneumatic Detector (Golay)

It measures the intensity of IR radiation by the expansion of a gas filled in its chamber, upon heating.

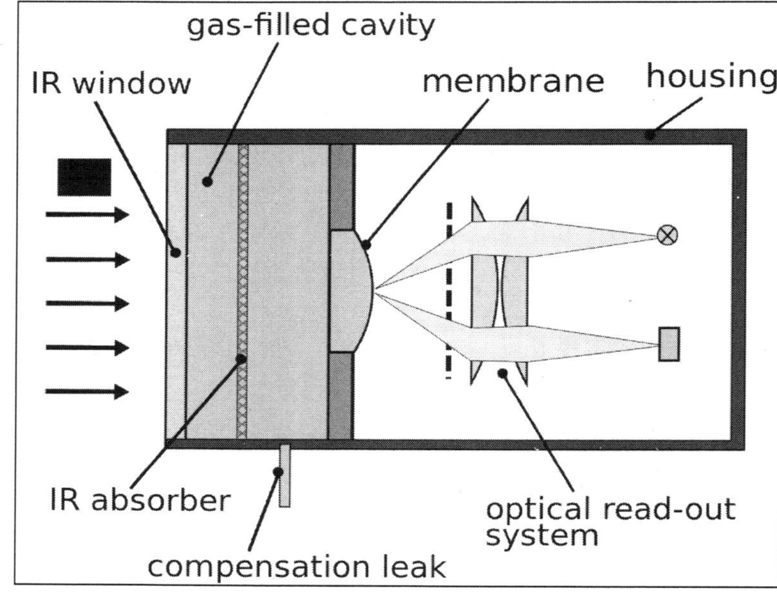

Figure 7.13. Schematic Diagram of Pneumatic Detector.

Pyroelectric Detectors

Pyroelectric detectors consists of a pyroelectric material which is an insulator with special thermal and electric properties. Triglycine sulphate is the most common material for pyroelectric infrared detectors. Unlike other thermal detectors the pyroelectric effect depends on the rate of change of the detector temperature rather than on the temperature itself. This allows the pyroelectric detector to operate with a much faster response time and makes these detectors the choice for Fourier transform spectrometers where rapid response is essential. Commonly used crystal material for pyroelectric detector is Tri Glycine Sulphate (TGS)

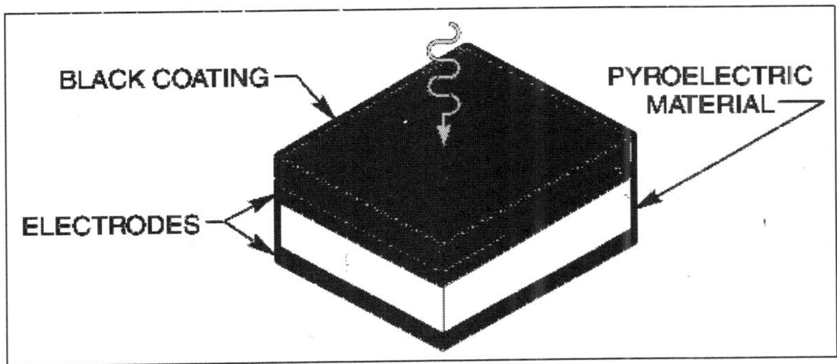

Figure 7.14. Schematic Diagram of Pyroelectric Detector.

Quantum Detectors

In **Quantum Detectors,** responsivity is wavelength dependent. These detectors need Cooling for operation. Advantages are fast response and high detectivity **Quantum Detectors are classified into two types.**

1. Intrinsic type (*e.g.*): Solid State photo detectors
2. Extrinsic type (*e.g.*): Photo conducting detectors

Properties	*Solid State Photo Detectors*	*Photo Conducting Detectors*
Principle	Photo electric effect	Electrical resistors, which decrease in resistance in relation to the intensity of light striking there surface
Material	Cadmium-Mercury Telluride(CMT) or Indium Antimonide (InSb)	Semiconductors like Lead sulphide or Lead telluride
Sensitivity	2-6 μ (InSb) 10-12 μ (CMT)	3.5 μ (Lead sulphide) 6 μ (Lead telluride)

Photo Conducting Detectors

Photoconducting detectors are the most sensitive detectors. They rely on interactions between photons and a semiconductor. The detector consists of a thin film of a semiconductor material such as lead sulphide, mercury cadmium telluride or indium antimonide deposited on a non conducting glass surface and sealed into an evacuated envelope to protect the semiconductor from the atmosphere. The lead sulphide detector is used for the near-infrared region of the spectrum. For mid- and far-infrared radiation the mercury cadmium telluride detector is used. It must be cooled with liquid nitrogen to minimize disturbances.

5. Amplifier

Amplifier is an electronic device that can increase the power of a signal.

6. ADC

Analog to digital converter is an electronic system which converts an analog signal, into a digital signal.

7. Recorder

A device that records and stores the data for future analysis

The Michelson Interferometer

An FTIR is typically based on The Michelson Interferometer Experimental Setup; an example is shown in Figure 7.15. The interferometer consists of a beam splitter, a fixed mirror, and a mirror that translates back and forth, very precisely. The beam splitter is made of a special material that transmits half of the radiation striking it and reflects the other half. Radiation from the source strikes the beam splitter and separates into two beams. One beam is transmitted through the beam

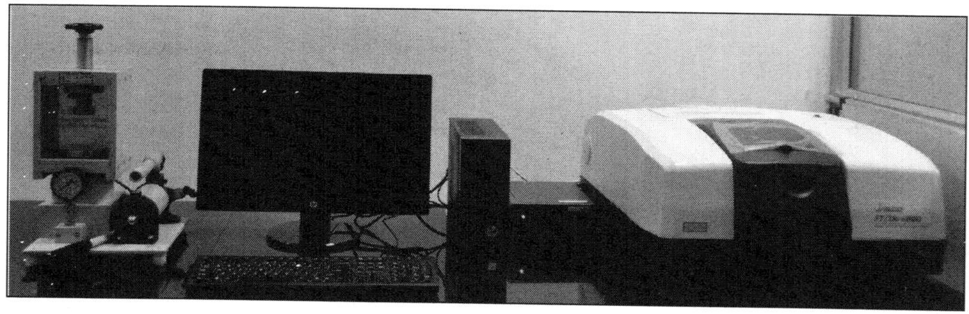

Figure 7.15a. FT-IR (Jasco - 6800).

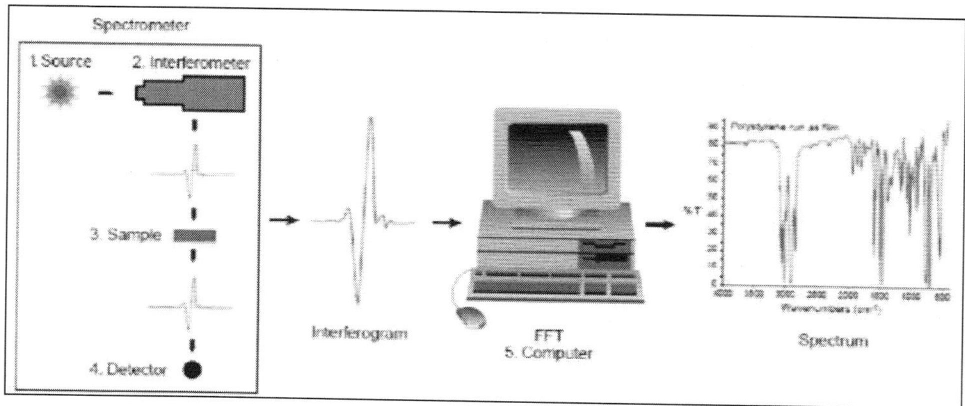

Figure 7.15b. Schematic Diagram of FT-IR Spectroscopy.

splitter to the fixed mirror and the second is reflected off the beam splitter to the moving mirror. The fixed and moving mirrors reflect the radiation back to the beam splitter. Again, half of this reflected radiation is transmitted and half is reflected at the beam splitter, resulting in one beam passing to the detector and the second back to the source.

The interferogram has to be measured from zero path difference to a maximum length that depends on the resolution required. In practice the scan can be on either side of zero resulting in a double-sided interferogram. Mechanical design limitations may mean that for the highest resolution the scan runs to the maximum OPD on one side of zero only.

The interferogram is converted to a spectrum by Fourier transformation. This requires it to be stored in digital form as a series of values at equal intervals of the path difference between the two beams. To measure the path difference a laser beam is sent through the interferometer, generating a sinusoidal signal where the separation between successive maxima is equal to the wavelength of the laser (typically a 633 nm HeNe laser is used). This can trigger an analog-to-digital converter to measure the IR signal each time the laser signal passes through zero. Alternatively, the laser and IR signals can be measured synchronously at smaller intervals with the IR signal at points corresponding to the laser signal zero crossing being determined by interpolation. This approach allows the use of analog-to-digital converters that are more accurate and precise than converters that can be triggered, resulting in lower noise (Smith 1968)

The result of Fourier transformation is a spectrum of the signal at a series of discrete wavelengths. The range of wavelengths that can be used in the calculation is limited by the separation of the data points in the interferogram. The shortest wavelength that can be recognized is twice the separation between these data points. For example, with one point per wavelength of a HeNe reference laser at 0.633 μm (15800 cm^{-1}) the shortest wavelength would be 1.266 μm (7900 cm^{-1}). Because of aliasing any energy at shorter wavelengths would be interpreted as coming from longer wavelengths and so has to be minimized optically or electronically. The spectral resolution, *i.e.* the separation between wavelengths that can be distinguished, is determined by the maximum OPD. The wavelengths used in calculating the Fourier transform are such that an exact number of wavelengths fit into the length of the interferogram from zero to the maximum OPD as this makes their contributions orthogonal. This results in a spectrum with points separated by equal frequency intervals (Table 7.1).

Assessment of Infrared Spectrum

Step 1

Overall Spectrum Appearance Is the spectrum simple? Are there only a few main characteristic absorption bands, say less than five?

If the spectrum is simple, as defined, then the compound may be a low-molecular-weight organic or inorganic compound, such as a simple salt of a common molecular ion (carbonate, sulfate, nitrate, ammonium, *etc.*) or a covalent species

Table 7.1. Sample and Sampling Techniques

Sample Type	*Sample Cell*	*Technique*	*Procedure*
Solid	NaCl	Mull	Sample slurry with nujol Mulls Technique Sample is crushed in agate mortar and pestle and mixed and ground together with nujol to a thick paste. A spot is applied between two NaCl plates and mounted in path of IR beam of instrument and a spectrum is recorded.
Solid	KBr	Pressed pellet	Sample pressed in Die Dried sample film The sample is dissolved in a solvent which is non-aqueous, non-reacting, non IR absorbing in the range to be measured. A drop of this solution is placed on NaCl plate and the solvent evaporated to dryness leaving a thin sample film.
Solid	NaCl plates	Thin film	Sample dries on plate Pressed pellet One milligram sample to 100 milligram KBr is finely ground and compressed into a transparent disc using an evacuable die and hydraulic press. Another disc is similarly made with pure dry KBr which is used as reference.
Solid	ATR crystal	Evanescent wave	Direct into IR beam ATR Evanescent wave This is an ATR crystal accessory used with IR spectrophotometer which allows both solid and liquid samples to be analyzed in their natural state without further sample preparation. Liquid samples are simply poured over the crystal and solid samples are clamped onto it. IR lights passed through it, an evanescent wave is formed which is collected and measured by the detector.
Liquid	NaCl Liquid cell	Neat sample	Fill Liquid cell Liquid sample cell Cell consist of two high quality pressed NaCl windows sealed and separated by a thin Teflon gasket and wetted with mercury. Cell thickness for most liquids is usually 0.01 – 0.05mm.
Gas	NaCl or KBr gas cell	Gas expansion	Fill evacuated cell

(chloroform, dichloromethane, methanol, water, *etc.*). In this case, the physical properties are important – if the material is a crystalline solid, then it is most likely a salt; if it is a mobile, possibly volatile, liquid, then the latter probably applies. Note that there are always exceptions, *e.g.* bromoform and iodoform, simple covalent compounds, are crystalline solids (both melt at low temperatures, whereas salts do not). One other class of compounds worth mentioning here is simple polymers, such as polyethylene and polytetrafluoroethylene, both of which exhibit very simple spectra when obtained from thin films. Are some of the bands broad? Hydrogen bonding possibly exists and can be exhibited in hydrates, water solutions, alcohols, ammonium compounds, amino compounds, *etc.* Note that some inorganics may exhibit a combination of both broad and very narrow bands, even for certain hydrated species. These are often related to the crystalline structure of the compound and the symmetry of certain aspects of the molecular/ionic structure (Coates, 2000)

Step 2

Testing for Organics and Hydrocarbons – Absorptions in the Region 3200–2700 cm^{-1}

Well-defined and characteristic absorptions in this region are normally characteristic of carbon- and hydrogen containing species, and are assigned to various forms of CH stretching. Are there absorptions above 3000 cm^{-1}. If yes, then the compound is likely to be unsaturated (contains CDC) or aromatic. If isolated absorptions occur at 3010 and/or 3040 cm^{-1}, then the absorbing species is mostly simple olefinic unsaturation. Is the main absorption below 3000 cm^{-1}. If so, the compound is probably aliphatic. If the main absorptions are approximately 2935 and 2860 cm1, and there are also absorptions at 1470 and 720 cm^{-1}, then the compound probably contains a long linear aliphatic chain.

Step 3

Testing for Hydroxy or Amino Groups – Absorptions in the Region 3650–3250 cm^{-1}

Hydroxy or amino groups mainly dominate this region, both giving rise to very characteristic band profiles. The presence or absence of hydrogen bonding is well delineated in this region. One other characteristic vibration that is observed in this region, which is unlikely to be confused with any other functional group, is the CH stretch of a terminal alkyne (acetylenic compound). This exhibits a relatively narrow absosorption at 3300 cm^{-1}, and is associated with a second (CC) band at 2200 cm^{-1}. Is the main absorption band in the area broad? If so, the compound is probably a hydroxy or amino group. Knowledge of the presence of nitrogen or a nitrogenous base (pH or titration) helps to support an amino compound. Are there additional moderate to intense bands in the ranges 1600–1300, 1200–1000 and 800–600 cm^{-1} If so, the compound is likely to be a simple hydroxy compound. Remember that the simple hydrogen-bonded OH absorption of a hydroxy (alcohol) function has a very characteristic shape. If the feature is relatively sharp and occurs between 3670 and 3550 cm^{-1}, the compound probably contains a non-hydrogen-bonded hydroxy group, often an alcohol or phenol with a sterically hindered OH group. Note also that this spectral feature is also exhibited by certain inorganics and minerals, and is indicative of a "free" OH group, either on the surface, or embedded within a crystal lattice, and free from interactions with other ions or groups.

Step 4

Testing for Carbonyl Compounds – Absorptions in the Region 1850–1650 cm^{-1}.

A major band in this region usually indicates the presence of a CDO group (carbonyl compound). Note that certain CDN (imino) and NDN (azo) groups may interfere, example compounds being purines. Is the absorption at the high end of the range, *e.g.* 1775 cm^{-1} or above? If yes, the compound probably contains a reactive carbonyl, such as an anhydride, acid halide (acyl halide) or a-halogenated carbonyl, or a strained-ring carbonyl, such as a lactone, or an organic carbonate. Is the absorption in the middle of the range, 1750–1700 cm^{-1} If yes, the compound is probably a simple carbonyl compound, such as a ketone, an aldehyde, an ester, or a carboxylic acid. Is the absorption at the low end of the range, *e.g.* below 1700 cm^{-1} If yes, the compound is probably an amide or a carboxylate (carboxylic acid salt). Note that conjugation with another carbonyl group, or a double bond or aromatic ring, will lower the carbonyl absorption by 30–50 cm^{-1}. Therefore, conjugated aldehydes, ketones, esters, and carboxylic acids may fall into this lower end category. If conjugation with a double bond is present, then a second strong absorption should be observed nearby, between 1650 and 1600 cm^{-1}. In the case of the involvement of an aromatic ring, it is important to look for evidence of the characteristic aromatic absorptions.

Step 5

Testing for Unsaturation–Weak to Moderate Absorption in the Region 1670–1620 cm^{-1}

Note that this is not a particularly definitive diagnostic without the support of other corroborating absorptions. A relatively narrow, weak-to-moderate absorption, normally centred around 1650 cm1, is indicative of olefinic unsaturation (CDC). A lowering of this frequency, accompanied by intensification of the band, is characteristic of conjugation with another double bond structure, such as CDC, CDO or an aromatic ring. The test for unsaturation normally involves a check for CH peaks above 3000 cm1, and typically as a single, or possibly a pair, of absorptions (dependent on configuration of the double bond) 43085 cm^{-1} and/o 43025 cm^{-1} – not to be confused with the multiple band situation for aromatics. Are there single or multiple bands between 1000 and 880 cm^{-1} (CH out-of-plane bending)? If so, this, in combination with the 1650 and 3010 and/or 3040 cm^{-1} bands, can be used as an assignment of unsaturation. This is either terminal (or vinyl $CHDCH_2$), with a pair of bands, normally located about 990 cm^{-1} and close to 900 cm1, or trans unsaturation (CHDCH), with a band centered around 965–960 cm^{-1}. A single band around 890 cm1, in conjunction with the other unsaturation absorptions, can indicate a vinylidene (or pendant, $CDCH_2$) type of olefinic double bond. Note that a cis configuration (CHDCH) is sometimes hard to characterize, because the normally characteristic CH out-of-plane bending is broad, with only a moderate absorption, and occurs in a region (around 700 cm^{-1}) where it can overlap with absorptions from other structural components, such as long-chain aliphatics, aromatics and hydrogen compounds.

Step 6

Testing for Aromatics – Well-defined Absorptions in the Region 1615–1495 cm^{-1}

One, but typically two, sets of bands in the region, with one set around 1600 cm^{-1} and the other around 1500 cm^{-1} are consistent with an aromatic compound. Sometimes the absorptions occur as single bands, at other times one or more of the band sets may appear split, as an asymmetric doublet. This assignment is usually confirmed by the presence of other bands in the spectrum. Are there a series of weak-to-moderate absorptions in the region 3150–3000 cm^{-1}, which can be assigned to aromatic CH stretching? These usually support the presence of the 1600/1500 cm^{-1} aromatic ring bands.

Step 7

Testing for Multiple Bonding (Often with a Bond Order of 2 or Higher) – Absorption in the Region 2300–1990 cm^{-1}

Multiply bonded nitrogen compounds and cumulated multiple bond compounds, such as cyanides (nitriles), cyanates, isocyanates, thiocyanates, and diazo compounds, exhibit characteristic absorptions in the region 2300–1990 cm^{-1}. Cyanide absorptions vary from weak to moderate to strong, depending on the nature of other substituents in the molecule. Conjugation, including substitution on an aromatic ring, modifies the intensity of this absorption. If the absorption band is extremely strong, the group is more likely to be a three-center functionality, such as a cyanate, an isocyanate, or a thiocyanate. Acetylenes and alkyne substituents also exhibit a weak to -moderate absorption within the same spectral region. If the acetylenic function is internal (medial), then the absorption can be extremely weak. However, a terminal alkyne group (C a"C) band may have moderate intensity, and this is supported by a characteristic C-H stretching absorption around 3310 cm^{-1}. A relatively low-frequency C-H bending vibration around 630 cm1 also confirms the terminal alkyne vibration. Metal carbonyls have unique bonding, where the bond order between the carbon and oxygen is between two and three, and where the additional electrons are provided by dp back-bonding from the accompanying transition metal atom. The multiply bonded CO group provides an extremely intense absorption band close to 2000 cm^{-1} (typically between 2100 and 1900 cm1), the actual position of the band(s) and the complexity of the bands being dependent on the structure of the compound. Note that an isolated weak-to-moderate intensity absorption above the specified region, and typically between 2700 and 2400 cm^{-1}, is not normally associated with a multiply bonded compound. These absorptions are usually from hydride vibrations, such as from silanes (SiH), thiols and sulfides (SH), phosphines (P-H), arsines (As-H), boranes (B-H), *etc.* Knowledge of the existence of the accompanying heteroatom helps in the assignment of the X-H bonds in these compounds.

Case Study

PVA nanofiber was developed using electrospinning method and it was loaded with the fungicide Tebuconazole. The loading pattern was confirmed by the following FT-IR spectrum (Figure 7.16).

The pure component of tebuconazole showed transmittance peaks at wavenumbers of 678 cm^{-1} (C-Cl stretch), 1087 cm^{-1} (C-O stretch), 1489 cm^{-1} (C=N stretch), 1508 cm^{-1} (N-N stretch), 1732 cm^{-1} (C=O stretch), 2915 cm^{-1} (C-H stretch) and 3287 cm^{-1} (O-H stretch). Similarly, the pure PVA registered the transmittance peaks at wavenumbers of 1091 cm^{-1} (C-O stretch), 1732 cm^{-1}(C=O stretch), 2910 cm^{-1}(C-H stretch) and 3306cm^{-1}(O-H stretch). FTIR spectra of tebuconazole loaded PVA nanofibre had the transmittance peaks at wavenumbers of 1038 cm^{-1} (C-O stretch), 1487 cm^{-1} (C=N stretch), 1506 cm^{-1} (NN stretch), 1734 cm^{-1} (C=O stretch),

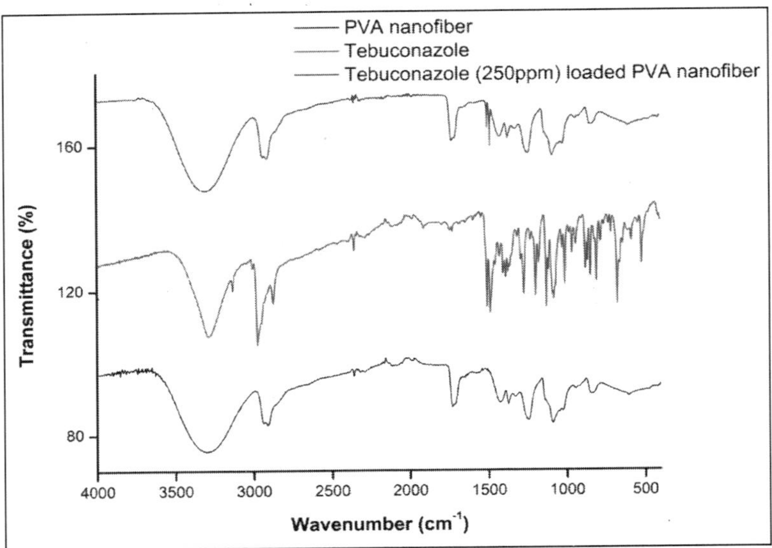

Figure 7.16. FTIR Spectra of Tebuconazole Loaded Nanofibres.

2914 cm^{-1} (C-H stretch) and 3320 cm^{-1} (O-H stretch). The stretching of functional groups C=N and N-N was found absent in PVA nanofibre, while presence was noted in the tebuconazole loaded nanofibers (Latha, 2019).

Applications

FT-IR spectroscopy used to….

a. Identification of polymers and polymer blends.
b. Indirect verification of trace organic contaminants on surfaces.
c. Routine qualitative and quantitative FTIR Analysis.
d. Thin film analysis.
e. Analysis of adhesives, coatings and adhesion promoters or coupling agents.
f. Small visible particle chemical analysis.
g. Analysis of stains and surface blemishes remnant from cleaning and degreasing processes combined with optical microscopy.
h. Analysis of resins, composite materials and release films.
i. Solvent extractions of leachable or contaminants, plasticisers, mould release agents and weak boundary layers coupled with XPS surface chemical analysis techniques.
j. Identification of rubbers and filled rubbers.
k. Determination of degrees of crystallinity in polymers.
   Comparative chain lengths in organics.
l. Measure air pollutants
m. Monitoring heavy metal from wastewater

Advantages

a. Feglet or multiplex advantage arises because all spectral elements are measured simultaneously, so the spectrum can be obtained quickly.

b. Jacquinet or throughput advantage arise due to the absence of slits which attenuates the IR light. So the optical throughput of the spectrometer is enhanced

c. Connes advantage arise due to the frequency scale of spectrum is known accurately.

All the above advantages reduce the signal to noise ratio of the spectra.

Disadvantages

a. FTIR instrument do not measure the spectra directly computer for transforming interferogram to spectra.

b. FTIR cannot be used for **homonuclear** diatomic molecules.

FTIR is one of method of infrared spectroscopy used to obtain as infrared spectrum of solid, liquid and gas. This infrared spectroscopy method is used to identify organic, polymeric, and some inorganic materials. Also this technique is used to assess the sample purity and composition identification.

III. UV–Vis Spectrometer

Ultraviolet–visible spectroscopy (UV–Vis) is an absorption or reflectance spectroscopy, in the ultraviolet- visible spectral ranges. It uses light in the visible and adjacent ranges near-UV and near-infrared region. It is used for the qualitative and quantitative analysis of different compounds such as metal, organic compounds and biological molecules. It is mainly for liquid samples to determine the concentration of the solution at a fixed wavelength (Skoog *et al.*, 2007).

Principle

The working mechanism of UV-Vis spectrometer is based on the Beer–Lambert law. The law states that the absorbance is directly proportional to the concentration of the absorbing compound in the solution and the path length *i.e.* The absorbance changes with the concentration. So for a fixed path length, UV-Vis spectrometer is used to determine the concentration of the absorber in the solution.

$$A = \log_{10} \frac{I_o}{I} = \varepsilon / C$$

A = Absorption of incident light

K″ = Molar Absorptivity

l = Path length of the light

c = Concentration of the sample

Io = Intensity of light entering the sample

I = Intensity of light emerging out of the sample

Key Components

1. **Light source**: A source which is used generates a broad band of electromagnetic radiation. Two lamps are fixed in the instrument, such as halogen lamp (UV light source) and deuterium lamp (Visible light source). Single beam and double beam spectrometers are available, where single beam is very easy to use but the fluctuation due to radiation source is eliminated in double beam spectrometer.
2. **Collimator**: Quartz or fused silica is used as a lens material. Reflecting mirrors are there to reflect and focus the light beam. To minimize the light loss, the mirrors are aluminized.
3. **Monochromator**: It is a dispersion device which is used to select a particular wavelength. It consists of an entrance slit, filters, dispersing device (prism) and an exit slit. The generated polychromatic radiation enters into the monochromator through the entrance slit. Then after reaching the collimating lens, the beam is collimated. The collimated beam strikes the dispersing element at a specific angle. Then the dispersing device is split the beam into its component wavelengths. Finally, only a particular wavelength radiation leaves the monochromator through the exit cell.
4. **Beam Splitter :** A beam splitter is an optical device which can split an incident light beam into two or more beams
5. **Sample cell:** Quartz cuvette are used for sample handling. The cuvettes for the sample and reference solution must be transparent to allow the radiation to pass through.
6. **Detectors**: It is device which converts light energy into electrical signals, are displayed in the monitor. A photomultiplier tube is commonly used as detector in UV-Vis spectroscopy. The tube consists of a photo-emissive cathode, an anode and several dynodes. The emitted electrons are being accelerated and the resulting current is amplified. The transmitted radiation used to determine the intensity of radiation absorbed by the sample.
7. **Differential amplifier:** Electronic amplifier that amplifies the difference between two input voltages but suppresses any voltage common to the two inputs

Working Mechanism

When the sample absorbs certain amount of energy from light in the UV-Vis region, the electrons get excited and move from the ground state to various excited states. The excited electrons may return to the ground state by vibrational transition. The absorbed energy released as heat in the sample. The absorbed energy from the sample is measured using a photo detector. Finally, a spectrum is generated which is a graphical representation of amount of light adsorbed or transmitted versus wavelength.

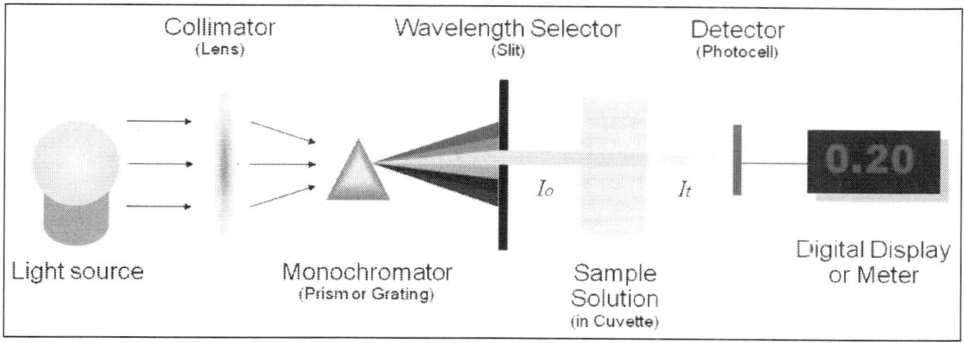

Figure 7.17. Schematic Diagram of UV-VIS Spectrometer.

Three modes are available with the instrument.

1. Spectral scan mode
2. Wavelength scan mode
3. Kinetic mode

Qualitative information like identification of substances, binding conditions and purity is obtained in spectral scan mode. Wavelength scan and kinetic modes are used to obtain quantitative information like concentration. In spectral scan mode, the absorbance maxima (λ^{max}) is measured for the identification of the compounds. Whereas in wavelength mode, optical density of the absorber is measures and concentration is calculated using a standard calibration curve.

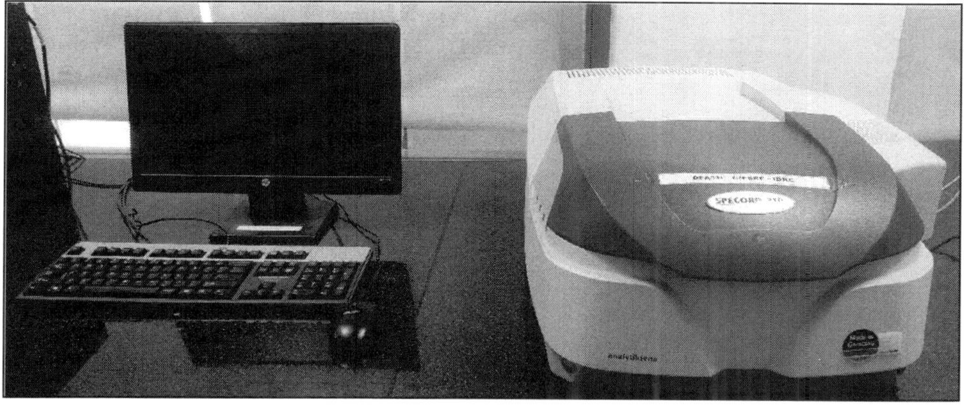

Figure 7.18. UV-VIS Spectrometer (SPECORD 210 plus- Analytic Jena).

Case Study

Silver nanoparticles were synthesized and used for degradation studies. The maximum absorbance of silver nanoparticles was found by UV-Vis spectrometer and the synthesis of silver nanoparticle was confirmed (Viji, 2016). Silver nanoparticles are known to exhibit a UV–Visible absorption maximum in the range of 400–500 nm due to its property. A narrow absorption band was observed at 435 nm which is a characteristic peak of mono dispersed silver nanoparticles.

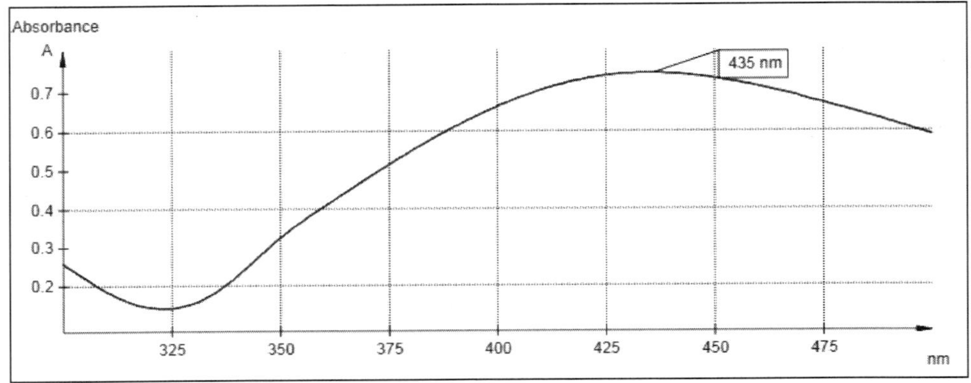

Figure 7.19. UV Spectra of Silver Nano Particles.

Applications

- ☆ Detection of impurities
- ☆ Structure elucidation of organic compounds
- ☆ Quantitative analysis of compounds
- ☆ Chemical kinetics
- ☆ Detection of functional group
- ☆ Molecular weight determination
- ☆ Estimation of concentration of sample.

Advantage

- ☆ Due to, quick analysis takes less time and easy operating procedure it is widely used in all the field of research

Disadvantage

- ☆ Scattering of light due to particulate in the sample
- ☆ Cannot be used in higher concentration of the sample.

UV-visible spectroscopy is a valid, simple and cost effective method for determining the concentration of absorbing species if applied to pure compounds, and used with the appropriate standard curve. A standard curve relating absorbance to concentration can be developed for any compound, and used to determine the concentration of samples containing the same compound. The analysis should be done at a wavelength with maximum absorption, and located in relatively flat region of the spectra so that absorbance will be high and constant in a narrow range around the chosen wavelength. The optimal wavelength should ensure good absorbance of the analyte and low absorbance by other species in the solution. This wavelength will allow valid absorption measurements to be made on analyte samples that contain mixtures of material

IV. Particle Size Analyzer

Particle size influences many properties of particulate materials and is a valuable indicator of quality and performance. This is true for powders, suspensions, emulsions, and aerosols. The size and shape of powders influences flow and compaction properties. Larger, more spherical particles will typically flow more easily than smaller or high aspect ratio particles. Smaller particles dissolve more quickly and lead to higher suspension viscosities than larger ones. Smaller droplet sizes and higher surface charge (zeta potential) will typically improve suspension and emulsion stability. Powder or droplets in the range of 2-5μm aerosolize better and will penetrate into lungs deeper than larger sizes. For these and many other reasons it is important to measure and control the particle size distribution of many products. Particle size analysis is an indispensable step in material analysis and material quality control. The size of the particles of specific material significantly affect the stability, appearance, fluidity, and chemical reaction characteristics of the material.

Particle Size Measurement

Particle size is the most important physical property of particulate samples and as a critical parameter which decides the fate of many products. It directly influences the material properties *viz.*, reactivity or dissolution rate, stability in suspension, efficacy of delivery, texture and feel, appearance, flow ability and handling, viscosity, packing density and porosity

There are wide ranges of particle size characterization techniques such as laser diffraction analysis; dynamic light scattering techniques and image analyzing are available. Each technique has it relative strength and limitations. In this chapter the principle, components and application of dynamic light scattering technique is described for measuring the particle size and distribution of sample

1. Dynamic Light Scattering

The method of dynamic light scattering (DLS) is the most common measurement technique for particle size analysis in the nanometer range.

Dynamic light scattering (DLS) is based on the Brownian motion of dispersed particles. When particles are dispersed in a liquid they move randomly in all directions. The principle of Brownian motion is that particles are constantly colliding with solvent molecules. These collisions cause a certain amount of energy to be transferred, which induces particle movement. The energy transfer is more or less constant and therefore has a greater effect on smaller particles. As a result, smaller particles are moving at higher speeds than larger particles. The relation between the speed of the particles and the particle size is given by the Stokes-Einstein equation.

Stokes Einstein: Relating Particle Size to Particle Motion

$D = kT/3\pi\lambda Dv$ *where,*

D is the diffusion coefficient

k is the Boltzmann's constant

λ is the solvent viscosity

T is the temperature

Dv is the particle diameter

The speed of the particles is given by the translational diffusion coefficient *D*. Further, the equation includes the viscosity of the dispersant and the temperature because both parameters directly influence particle movement. A basic requirement for the Stokes-Einstein equation is that the movement of the particles needs to be solely based on Brownian motion. If there is sedimentation, there is no random movement, which would lead to inaccurate results. Therefore, the onset of sedimentation indicates the upper size limit for DLS measurements. In contrast, the lower size limit is defined by the signal-to-noise ratio. Small particles do not scatter much light, which leads to an insufficient measurement signal.

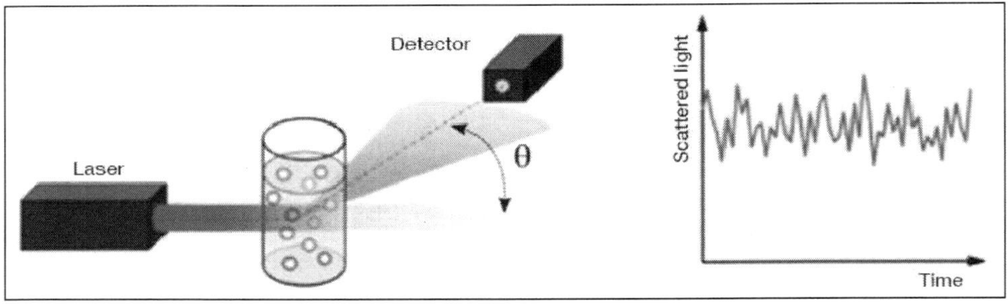

Figure 7.20. Schematic Diagram of DLS.

The basic setup of a DLS instrument is shown in Figure 7.20. A single frequency laser is directed to the sample contained in a cuvette. If there are particles in the sample, the incident laser light gets scattered in all directions. The scattered light is detected at a certain angle over time and this signal is used to determine the diffusion coefficient and the particle size by the Stokes-Einstein equation. The incident laser light is usually attenuated by a gray filter which is placed between the laser and the cuvette. The filter settings are either automatically adjusted by the instrument or can be set manually by the user. When turbid samples are measured the detector would not be able to process the amount of photons. Therefore, the laser light is attenuated to receive a sufficient but processable signal at the detector.

When light strikes moving or vibrating particles in the liquid medium, the frequency of the light exhibits a Doppler shift. It increases when the particle is moving towards the incident light and decreases when the particle is moving away. If a laser beam illuminates a cuvette containing particles dispersed in a liquid medium, and the speckle is focused on the screen and freezed for a moment, the similar speckle pattern will be obtained as show in the Figure 7.21. The dark and bright spots are appeared in the speckle, where dark spaces are due to the phase additions of the scattered light are mutually destructive and nullify each other. The bright spots in the speckle pattern are where the light scattered from the particles arrives with the same phase and interfere constructively to form a bright patch. The particles are under constant motion in Brownian movement and hence a speckle pattern is also changing accordingly, and forming new patterns. The rate of intensity fluctuations between the two phases of interferences depends on the size of the particles. The small particles cause the intensity to fluctuate more rapidly than the large ones.

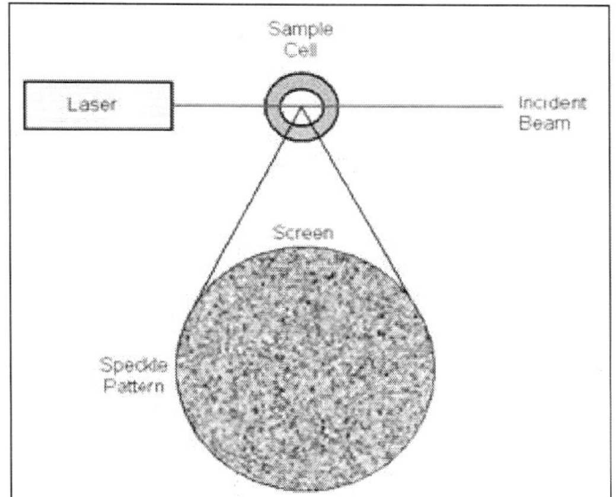

Figure 7.21. Speckle Pattern.

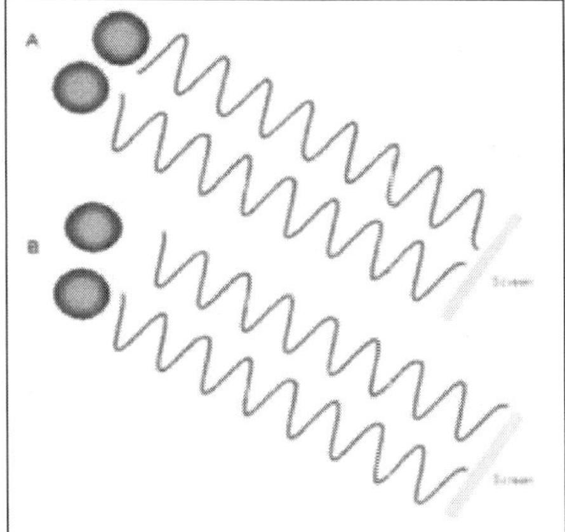

Figure 7.22. Destructive and Constructive Phases of Scattered Light.

The speckle pattern will be stationary both in speckle size and position because the whole system is stationary. The dark spaces are where the phase additions of the scattered light are mutually destructive and cancel each other out (Figure 7.23A). The bright blobs of light in the speckle pattern are where the light scattered from the particles arrives with the same phase and interfere constructively to form a bright patch (Figure 7.23B). For a system of particles undergoing Brownian motion, a speckle pattern is observed where the position of each speckle is seen to be in constant motion. This is because the phase addition from the moving particles is constantly evolving and forming new patterns. The rate at which these intensity fluctuations occur will depend on the size of the particles. Figure 7.23 schematically illustrates typical intensity fluctuations arising from a dispersion of large particles and a dispersion of small particles. The small particles cause the intensity to fluctuate more rapidly than the large ones.

A top view of the optical setup for DLS is shown below.

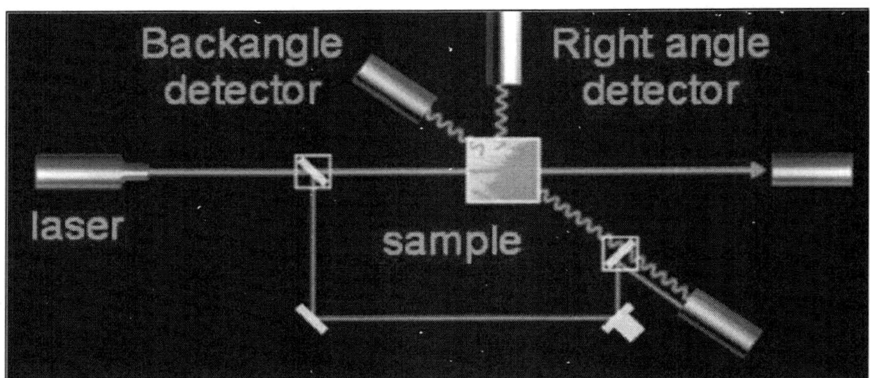

Figure 7.23A. Optical Setup of Dynamic Light Scattering.

Light from the laser light source illuminates the sample in the cell. The scattered light signal is collected with one of two detectors, either at a 90 degree (right angle) or 173 degree (back angle) scattering angle. The provision of both detectors allows more flexibility in choosing measurement conditions. Particles can be dispersed in a variety of liquids. Only liquid refractive index and viscosity needs to be known for interpreting the measurement results. The obtained optical signal shows random changes due to the randomly changing relative position of the particles. This is shown schematically in the graph below.

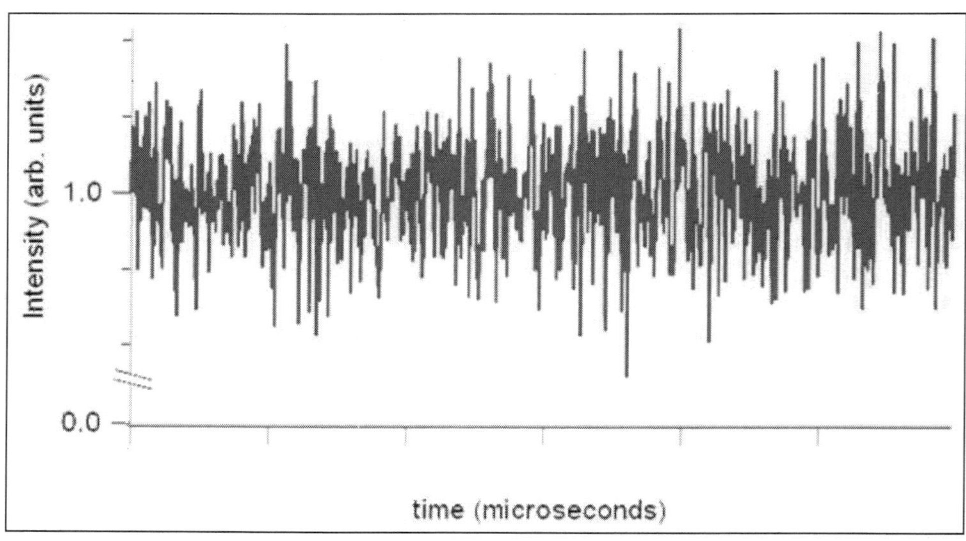

Figure 7.23B. Optical Signal from a Nanoparticle Sample on a Microsecond Timescale.

The "noise" is actually due to particle motion and will be used to extract the particle size. In contrast to laser diffraction, DLS measurements are typically made at a single angle, although data obtained at several angles can be useful. In addition, the technique is completely noninvasive; the particle motion continues whether or not it is being probed by DLS.

Zeta Potential of Nanomaterials

Zeta Potential analysis is a technique for determining the surface charge of nanoparticles in solution (colloids). Nanoparticles have a surface charge that attracts a thin layer of ions of opposite charge to the nanoparticle surface. This double layer of ions travels with the nanoparticle as it diffuses throughout the solution. The electric potential at the boundary of the double layer is known as the Zeta potential of the particles and has values that typically range from +100 mV to -100 mV. The magnitude of the zeta potential is predictive of the colloidal stability. Nanoparticles with Zeta Potential values greater than +25 mV or less than -25 mV typically have high degrees of stability. Dispersions with a low zeta potential value will eventually aggregate due to Van Der Waal inter-particle attractions.

Manipulating zeta potential is a method of enhancing suspension stability for formulation work, or speeding particle flocculation in applications such as water treatment. The stability behaviour of the colloid particles are depicted in the Table 7.2.

Table 7.2. The Stability Behavior of Colloidal Particles

Zeta Potential	*Stability*
0 to ±5	Rapid coagulation or flocculation
From ±10 to ±30	Incipient instability
From ±30 to ±40	Moderate stability
From ±40 to ±60	Good stability
More than ±61	Excellent stability

Electrophoretic Light Scattering Technique

A laser beam passes through the electrophoresis cell, irradiates the particles dispersed in it, and is scattered by the particles. The scattered light is detected by a photo-multiplier after passing through two pinholes. There are two types of optical systems: heterodyne and fringe. Ware and Flygare developed a heterodyne-type ELS instrument, that was the first instrument of this type. In a fringe optics ELS instrument, a laser beam is divided into two beams. Those cross inside the electrophresis cell at a fixed angle to produce a fringe pattern. The scattered light from the particles, which migrates inside the fringe, is intensity-modulated. The frequency shifts from both types of optics obey the same equations. The observed spectra resemble each other. Oka *et al.*, developed an ELS instrument of heterodyne-type optics that is now available commercially. If the frequencies of the intersecting laser beams are the same then it is not possible to resolve the direction of the motion of the migrating particles. Instead, only the magnitude of the velocity (*i.e.*, the speed) can be determined. Hence, the sign of the zeta potential cannot be ascertained. This limitation can be overcome by shifting the frequency of one of the beams relative to the other. Such shifting may be referred to as frequency modulation or, more colloquially, just modulation. Modulators used in ELS may include piezo-actuated mirrors or acousto-optic modulators. This modulation scheme is employed by the heterodyne light scattering method, too.

Advantages

- Particle size measurements using DLS can be achieved within one or two minutes and the measurements were found to be accurate, reliable and provided unbiased information on shape (flexible coil, rods, cylinders, spheres) of macromolecules (DNA, RNA, polysaccharides, proteins, and viruses).
- Measurement of particle concentration can be achieved calibration free and the mean size measurement only requires the knowledge of viscosity of the liquid.
- Samples having different concentrations such as simple or no sample preparation, high concentration and turbid samples can be analysed directly using DLS.
- This technology is also suitable for molecular weight determination and size measurements of molecules in the range of 10μm to less than 1 nm and having molecular weight less than 1000 Da can be determined.
- DLS' s major advantage is that very less amount of sample (< 3μL) is essential for analysis.
- DLS is found to be suitable for protein- RNA interaction studies and modified DLS in-line process systems are found to be suitable.
- It is possible to conduct experiments on wide range of sample buffer and wide temperature range using DLS technique.

Limitations

- Measurements obtained by DLS are altered by variations in temperature and solvent viscosity as it is highly sensitive to it. Therefore, both temperature and solvent viscosity must be kept constant for reliable results.
- DLS cannot differentiate molecules that are closely related (*e.g.*, monomer and dimer) since it is a low resolution method.
- DLS must be used on highly dilute solutions of food particles. It is difficult to dilute many food particles as it may result in variation of structural properties, for instance complex particles may dissociate. So, it is essential to ensure that the particles under study in a DLS experiment must retain their structures which are native to the food.
- Presence of large aggregates even in a small amount will adversely affect the measurements since scattering intensity is proportional to 6th

power of the size of macromolecules (I α d6, where d is a diameter of macromolecule). Therefore, proper cleaning of the sample-holding cuvette prior to measurement is essential.

- ☆ The signals from DLS vary with changes in size and concentration of macromolecules and so the concentration range is to be optimized prior to measurement for attaining reliable result

Applications

- ☆ Particle size analyzer is a non-destructive and eco-friendly method has wide application in the fields of cosmetics, ceramics, drug delivery, food and beverages, nanotechnology characterization, cement, nutraceuticals, pharmaceuticals aerosols, agriculture, pigments and inks, paper, polymer and plastics *etc.*, In all theses area, the particle size analyzer analysis the particle size for describing the quality of product and zeta potential to address the stability of materials
- ☆ DLS technique is primarily used for the size determination for standardizing the protocol to synthesize nanoparticles.
- ☆ Major application of this technology is in milk and other oil in water emulsions since milk contains large number of colloidal particles which has the ability to scatter light well and also the structure of milk is least affected due to dilution of milk.
- ☆ A modified version of DLS, DWS (Diffusion Wave Spectroscopy) can be used for concentrated suspensions which is having similar apparatus as that of DLS except that back scattered light and incident light are conducted through fibre optics.
- ☆ DLS has been used as a promising technology for conducting studies on homogeneity of RNA, proteins and their complexes. Multi-Angle Dynamic Light Scattering (MADLS®), a modified DLS technique, has the ability to improve the resolution of DLS and the size results obtained were found to be angular independent so that accuracy of DLS technique is improved.
- ☆ DLS has been widely used to determine the size distribution of casein micelles in bovine milk and also for molecular weight determination of molecules

Conclusion

DLS is a non-destructive technique that provides quick, precise, accurate and reproducible quality check for interaction studies during bio molecular preparation and the analysis requires very less amount of sample. Although DLS possess some minor limitations, it is gaining popularity in academic and industrial laboratories as it provides reliable results in a short period of time.

REFERENCES

1. Bish, D. L.; Post, J. E., eds. (1989). Modern Powder Diffraction: Reviews in Mineralogy, vol. 20; Mineralogical Society of America: Chantilly.

2. Bodolla S and S. Thodeti.(2018). A review on Characterization techniques of Nanomaterials. *International Journal of Engineering, Science and Mathematics,* 7(1): 16-175
3. Coates J. (2000). Interpretation of Infrared Spectra, A Practical Approach. *Encyclopedia of Analytical Chemistry.* R.A. Meyers (Ed.). pp. 10815–10837
4. https: / /www.atascientific.com.au / basic-principles-of-particle-size-analysis /
5. https: / /www.cif.iastate.edu / sites / default / files / uploads / Other_Inst / Particle per cent 20Size / Particle per cent 20Characterization per cent 20Guide.pdf
6. https: / /www.horiba.com / fileadmin / uploads / Scientific / eMag / PSA / Guidebook / pdf / PSA_Guidebook.pdf
7. https: / /www.innovatechlabs.com / newsroom / 672 / stuff-works-ftir-analysis
8. https: / /www.shimadzu.com / an / ftir / support / ftirtalk / talk8 / intro.html
9. https: / /www.sigmaaldrich.com / technical-documents / articles / biology / ir-spectrum-table.html https: / /teaching.shu.ac.uk / hwb / chemistry / tutorials / molspec / uvvisab3.htm.Metha and Akul (2011). Principle. *PharmaXChange.info.*
10. https: / /www2.chemistry.msu.edu / courses / cem434 / Principles per cent 20of per cent 20IR per cent 20spectroscopy.
11. Iyengar, S. S.; Buhrke, V. E.; Jenkins, R.; Smith, D. K. (1997).Sample Preparation for Clays in Preparation of Specimens for X-ray Fluorescence and X-ray Diffraction Analysis;Wiley-VCH: New York, 1997.
12. Klug, H. P.; Alexander, L. E. (1974).X-ray Diffraction Procedures; Wiley: New York, 1974.
13. Latha, M, Raja, K, Subramanian, K. S, Karthikeyan, M. and Lakshmanan, A. (2019). Fabrication and characterization of tebuconazole loaded PVA nanofibre. *International Journal of Agriculture Sciences, 11*(10): 8514-8517.
14. Skoog, Douglas A.; Holler, F. James; Crouch, Stanley R. (2007). *Principles of Instrumental Analysis* (6th ed.). Belmont, CA: Thomson Brooks / Cole. 169 -173.
15. Smith, D. R, Morgan, R. L. and Loewenstein, E. V. (1968). Comparison of the Radiance of Far-Infrared Sources. *Journal of the Optical Society of America. 58* (3): 433– 434. doi: 10.1364 / JOSA.58.000433
16. VACullity, B. D.(1989).Elements of X-ray Diffraction, 2nd ed.; Addison-Wesley: Menlo Park, CA,
17. Viji, N, Chinnamuthu, C. R and Chinnusamy, C. (2016). Depriving the purple nutsedge by degrading the starch present in the tubers using immobilized amylase on the surface of organically synthesized silver nanoparticles. *Green farming, 7*(1), 107-110.

Chapter 8

Software to Model Nanostructures

D. Jeya Sundara Sharmila and V.P. Krithika*

Department of Nano Science and Technology, Tamil Nadu Agricultural University
Coimbatore – 641 003, Tamil Nadu
**e-mail: djssharmila@gmail.com*

Modeling of Nanostructures

"Computational modeling of nanomaterials is a potent theoretical technique to explore their fundamental properties. The modeling and simulation of nanomaterials has attracted great interest due to its high efficiency on accurate prediction of fundamental properties of nanomaterials, which is crucial to various prospective applications such as structural manufacturing, functionalization of nanomaterials, and computational design. Recently, the development of cluster parallel computer offers powerful computational capabilities to calculate the fundamental properties of complex nanostructured systems and to simulate the physiochemical processes. Recent advances in computational modeling of physical chemistry properties of nanomaterials not only could give us better insight into the underlying mechanisms of physical and chemical reaction processes, but also facilitate the practical applications of nanostructured systems. Many application areas in nanotechnology are related to effects occurring at the junctions, interfaces and surfaces of nanomaterials (Figure 8.1). It is critical to be able to accurately model such phenomena from quantum theory using powerful software system" (Zeng *et al.*, 2015).

"Quantum Wise" – ATK and Virtual Nano Lab

QuantumWise offers a software platform for atomistic simulations of nanoscale systems with the following major modules: Atomistix Tool Kit (ATK), Virtual Nano Lab (VNL) (Figure 8.2) and Nano Language.

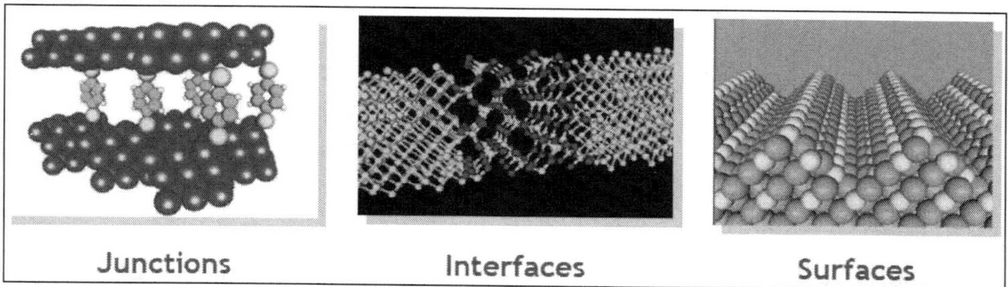

Figure 8.1. Modeling of Junctions, Interfaces and Surfaces of Nanomaterials.
Courtesy: QuantumWise - ATK Virtual Nano Lab Academic.

Atomistix Tool Kit (ATK) is a state-of-the-art DFT (Density Functional Theory) engine for electronic structure and transport calculations of nanostructures.

Virtual Nano Lab (VNL) is a modern GUI platform for setup and analysis of the Nanosystem.

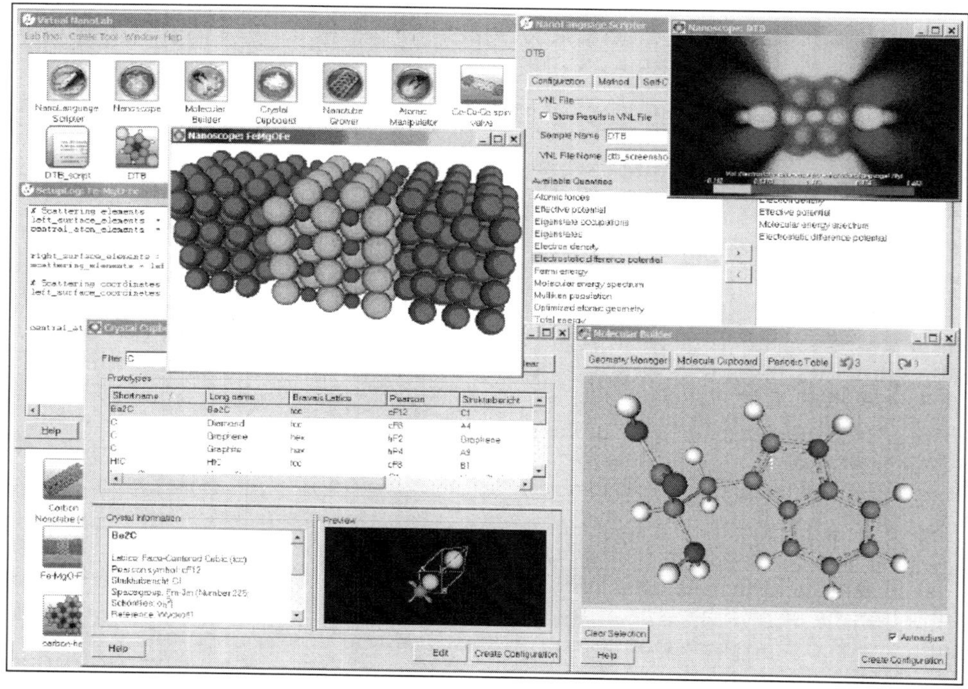

Figure 8.2. Home Page of Virtual Nano Lab.
Courtesy: QuantumWise - ATK Virtual Nano Lab Academic.

Nano Language is a scripting language interface to ATK that is fully integrated in VNL. Virtual Nano Lab interactive builders include the tools for Modeling of Molecules, Crystals, Nanotubes and two-probe systems (general and specialized, like MTJs). 3D visualization of structures shows results such as electron density, potentials as isosurfaces or contour plots and superimposing plots on geometry. The provision for export of plots as images in common formats is provided.

Schrodinger's Materials Science Suite

Schrödinger is the scientific leader in developing state-of-the-art chemical simulation software for use in pharmaceutical, biotechnology, nanotechnology and materials science research. Schrodinger's Materials Science Suite (Figure 8.3) was founded in 1990 by Richar A. Friesner and William A. Goddard. Before beginning the synthesis and testing of nanomaterials, the atomic-scale simulation can accelerate the development of new nanomaterials by identifying the most promising structures and compositions. This platform expands exploration of chemical space and suggests unexpected solutions. It saves time on analysis and synthesis by prioritizing the most promising solutions, so that reducing the cost. They provide deep insights into the mechanisms and structure-property relationships that determine performance in nanomaterials.

Figure 8.3. Home Page of Schrodinger's Materials Science Suite.
Courtesy: https://www.schrodinger.com/materials-science

Maestro

Maestro is the interface portal (Figure 8.4) to all of Schrödinger's computational technology applications. It is not only an user interface, but also it helps researchers organize and analyze data in an intuitive fashion. This platform has a building

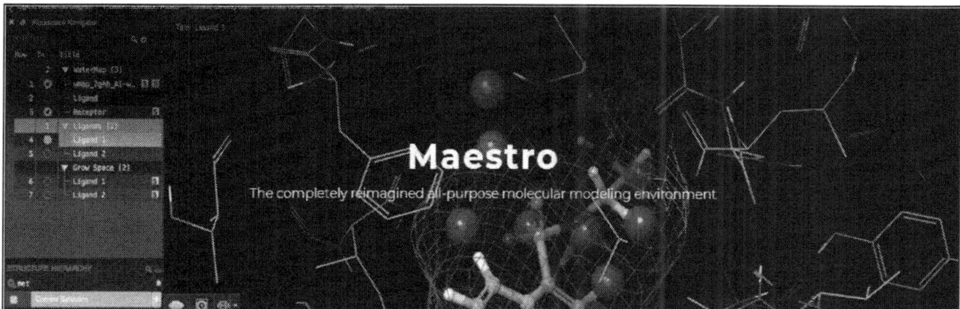

Figure 8.4: Schrodinger's Masetro Interface.
Courtesy: https://www.schrodinger.com/maestro

tool for constructing molecular models, as it supports many common file. They have a wide range of viewing option. The other advantages include **3D realism, Quantitative structural analysis, Customization scripts, Molecular properties, Data management and analysis.** Maestro runs natively on Linux, Windows, and Mac and brings clarity to a wide range of modeled nanosystem as it includes versatile measurement tools.

LAMMPS (Large-scale Atomic/Molecular Massively Parallel Simulator)

The LAMMPS (Figure 8.5) was developed from Sandia National Laboratories Temple University, 1995. It is written in C++ and it works in Linux, MacOS and windows. It is a molecular dynamics program which is free and open-source software for simulation of nanomaterials. It focuses on nanomaterials modelling at atomic, meso or continuum scale besides nano-scale metals, semiconductors, biomolecules and polymer simulation. This runs on single processors or in parallel using message-passing techniques. The code is designed to be easy to modify or extend with new functionality.

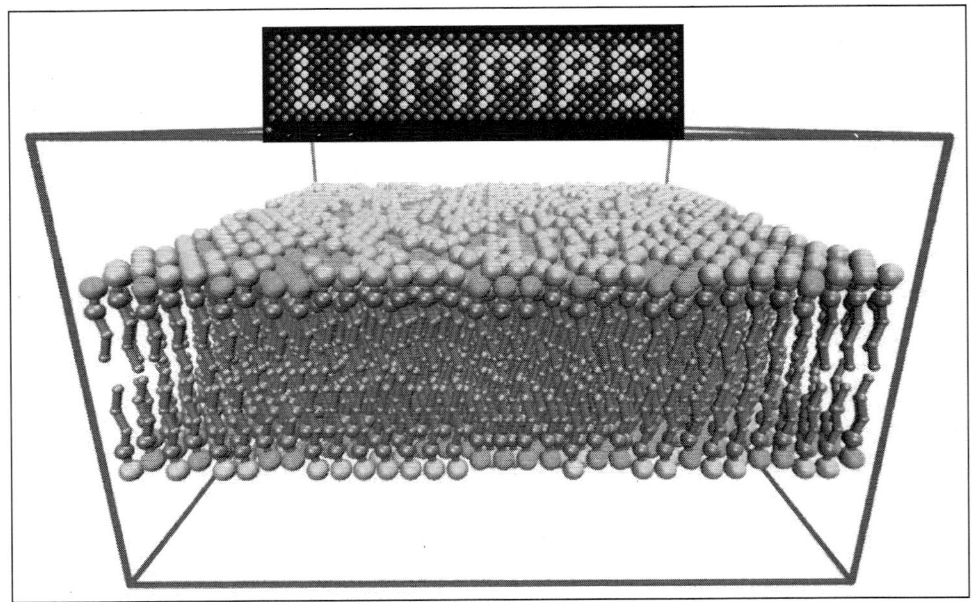

Figure 8.5. Large-scale Atomic/Molecular Massively Parallel Simulator of Nanomaterials.
https://en.wikipedia.org/wiki/LAMMPS#/media/File:Lammps-logo.png.

General Features of LAMMPS

- Runs on a single processor or in parallel
- Distributed-memory message-passing parallelism (MPI)

- Spatial-decomposition of simulation domain for parallelism
- Open-source distribution
- Highly portable C++
- Optional libraries used: MPI and single-processor FFT
- GPU (CUDA and OpenCL), Intel Xeon Phi, and OpenMP support for many code features
- Easy to extend with new features and functionality
- Runs from an input script
- Syntax for defining and using variables and formulas
- Syntax for looping over runs and breaking out of loops
- Run one or multiple simulations simultaneously (in parallel) from one script
- Build as library, invoke LAMMPS through library interface or provided Python wrapper
- Couple with other codes: LAMMPS calls other code, other code calls LAMMPS, umbrella code calls both.

Force Fields and Interatomic Potentials Used in LAMMPS for Nanoscale Models

- Pairwise potentials: Lennard-Jones, Buckingham, Morse, Born-Mayer-Huggins, Yukawa, soft, class 2 (COMPASS), hydrogen bond, tabulated
- Charged pairwise potentials: Coulombic, point-dipole
- Many-body potentials: EAM, Finnis/Sinclair EAM, modified EAM (MEAM), embedded ion method (EIM), EDIP, ADP, Stillinger-Weber, Tersoff, REBO, AIREBO, ReaxFF, COMB, SNAP, Streitz-Mintmire, 3-body polymorphic
- Long-range interactions for charge, point-dipoles, and LJ dispersion: Ewald, Wolf, PPPM (similar to particle-mesh Ewald)
- Polarization models: QEq, core/shell model, Drude dipole model
- Charge equilibration (QEq via dynamic, point, shielded, Slater methods)
- Coarse-grained potentials: DPD, GayBerne, REsquared, colloidal, DLVO
- Mesoscopic potentials: granular, Peridynamics, SPH
- Electron force field (EFF, AWPMD)
- Bond potentials: harmonic, FENE, Morse, nonlinear, class 2, quartic (breakable)
- Angle potentials: harmonic, CHARMM, cosine, cosine/squared, cosine/periodic, class 2 (COMPASS)
- Dihedral potentials: harmonic, CHARMM, multi-harmonic, helix, class 2 (COMPASS), OPLS
- Improper potentials: harmonic, cvff, umbrella, class 2 (COMPASS)

- ✰ Polymer potentials: all-atom, united-atom, bead-spring, breakable
- ✰ Water potentials: TIP3P, TIP4P, SPC
- ✰ Implicit solvent potentials: hydrodynamic lubrication, Debye
- ✰ Force-field compatibility with common CHARMM, AMBER, DREIDING, OPLS, GROMACS, COMPASS options
- ✰ Access to the OpenKIM Repository of potentials via kim_init, kim_interactions, and kim_query commands
- ✰ Hybrid potentials: multiple pair, bond, angle, dihedral, improper potentials can be used in one simulation
- ✰ Overlaid potentials: superposition of multiple pair potentials

Modeling of Silver Nanoparticles using LAMMPS

Edison da Silva and collaborators (Institute of Physics, UNICAMP, Campinas, SP, Brazil) studied the sintering processes of Ag nanoparticles (Figure 8.6) using LAMMPS. Silver tungstate is irradiated by an electron beam generates metallic silver filaments, with concomitant formation of silver nanoparticles. The effect of the electron beam produces surface plasma resonances (SPR) to the nanoparticles. These silver nanoparticles therefore, become interacting electric nano-dipoles that undergo sintering processes. Connecting theory, experiment, and long-time atomistic molecular dynamics simulations provides a deep insight into the observed experimental phenomena.

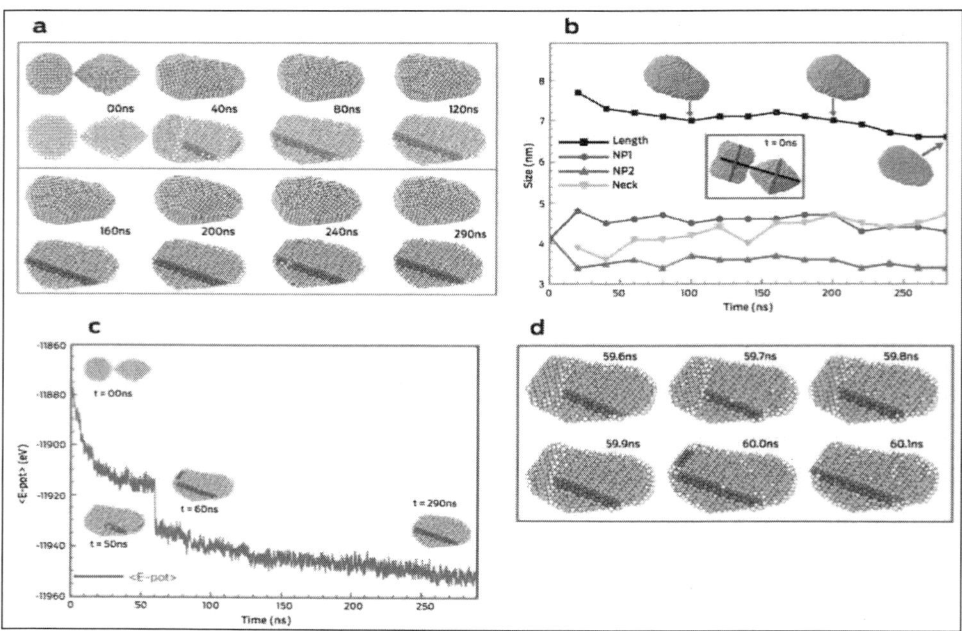

Figure 8.6. Modeling of Silver Nanoparticles
https://lammps.sandia.gov/pictures.html#silvernanoparticles
Courtesy: Da Silva *et al.*, 2019.

Avogadro

Avogadro (Figure 8.7) is an advanced molecule editor and visualizer designed for cross-platform use in molecular modeling, materials science, bioinformatics and computational nano chemistry in 2008 and updated regularly. It offers flexible high quality rendering and a powerful plug-in architecture for nanocluster modeling. It works on cross-platform like Windows, Linux, and Mac OS. It is free and open source. It is built for both advanced researchers and students as it supports multithreaded computation and rendering.

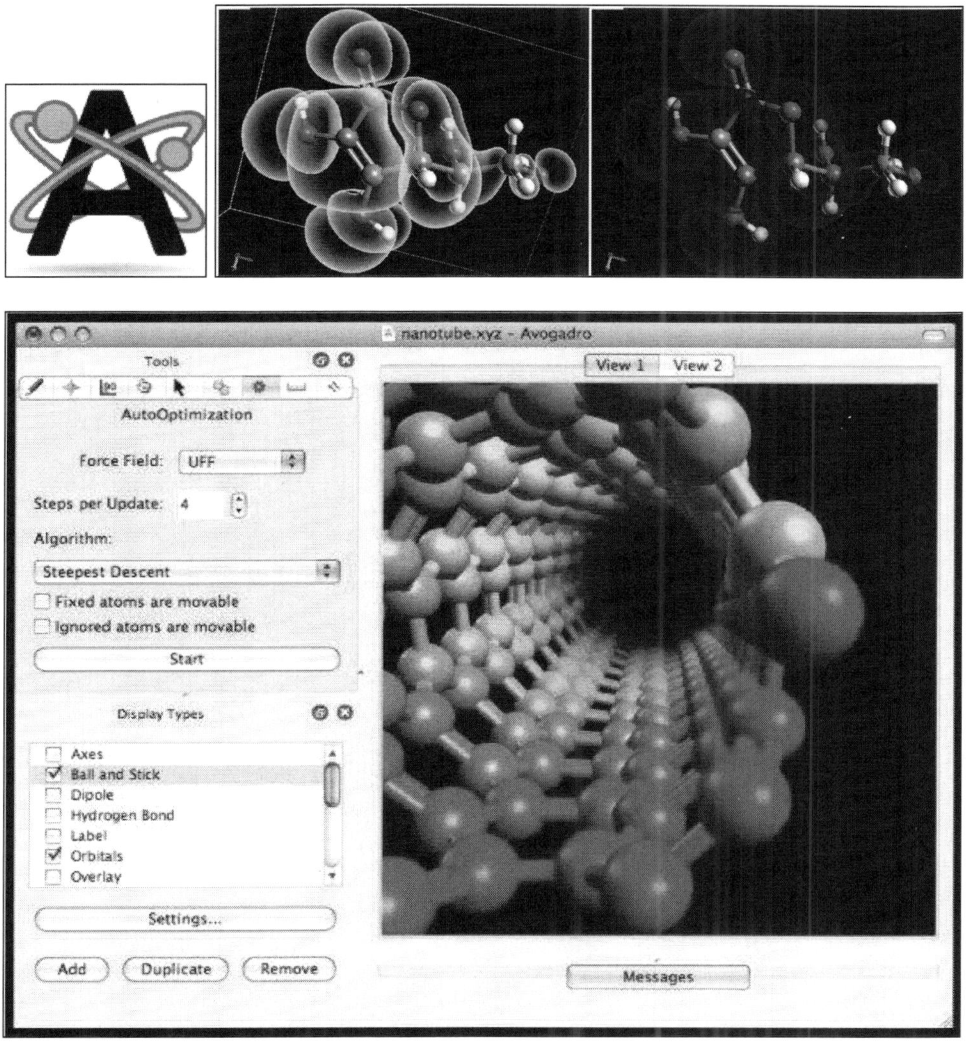

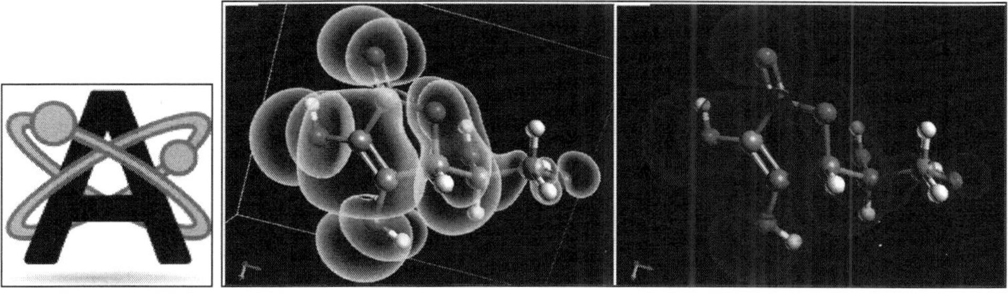

Figure 8.7. Avogadro for Nanotube Modeling.
Courtesy: Avogadro-Hanwell *et al.*, 2012.

Desmond

Desmond is used to execute high-speed molecular dynamics simulations (Figure 8.8) of biological nanosystems, developed at D. E. Shaw research. The software code helps in high performance and accuracy by having numerical techniques and new parallel algorithms. The software provides options to calculate energies and forces in condensed matter and biomolecular research. It works on Linux. Also it performs solvent simulations and can model membrane systems in various conditions. The analysis tools make it possible to visualize and examine computed results within the Schrodinger's Maestro modeling environment.

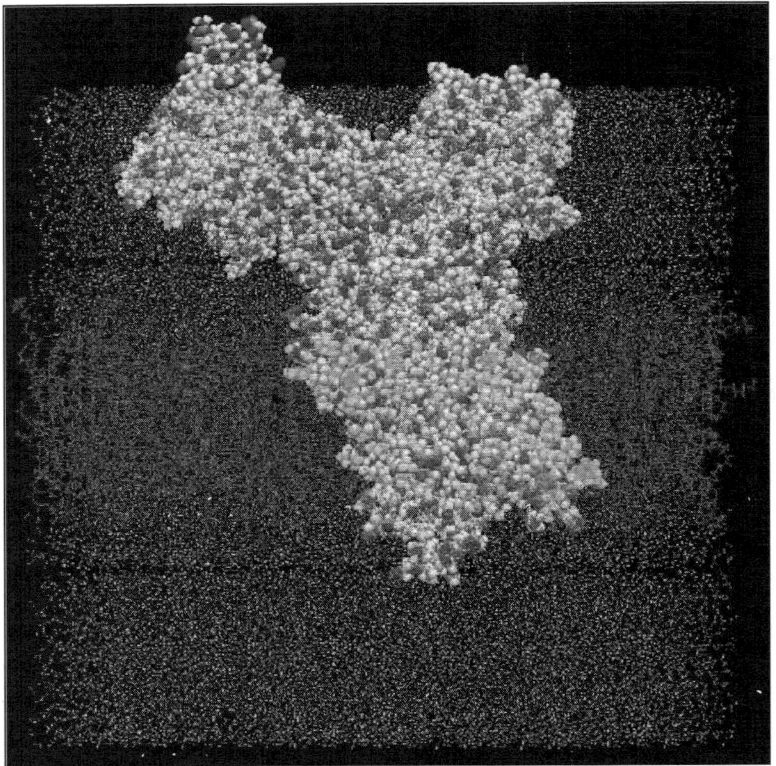

Figure 8.8. Desmond-Transmembrane Modeling.
Courtesy: http://geminiweb.slu.edu/wp-content/uploads/desmond_tutorial.pdf.

Nano-Biocomposite Simulator

This is used to develop multiscale models for bio-inspired nanomaterials. Deformation and failure mechanisms of the hierarchical nanostructure of hard biological materials were identified through different length scale. Biomimetics can be applied to the intelligent designing of nanomaterials. The system uses bio-inspired damage and mitigation strategy also. Nacre is found in abalone shells and has high strength and toughness. In biocomposite simulation of nacre

(Figure 8.9), the researchers developed a computational model of composites by incorporating key features of nacre's microstructure. After the analysis, they compared the simulations to experiments performed on similar material. By combining simulations and experiments it gives a great reach between material parameters and microstructure/nanostructure, thus developing a novel synthetic nanomaterials with superior performance.

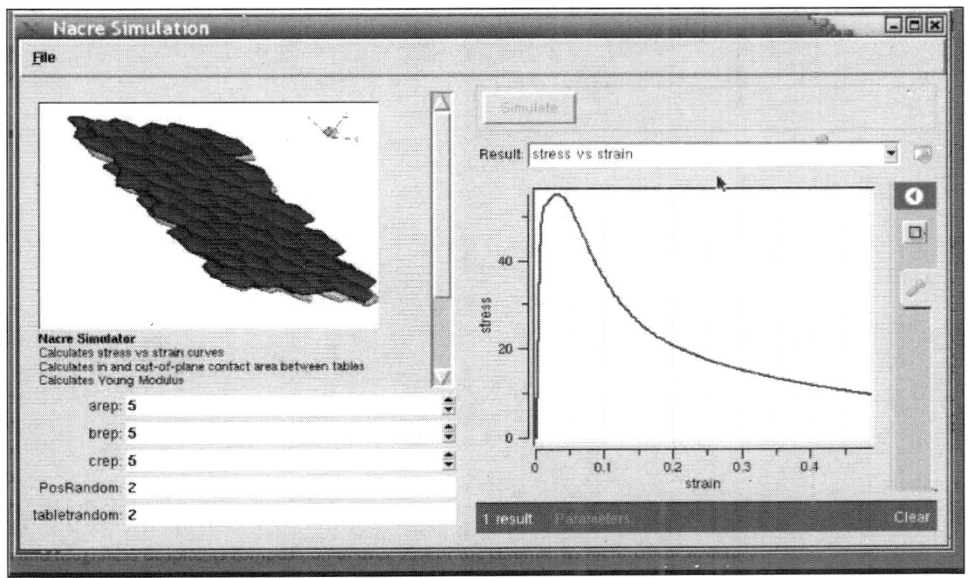

Figure 8.9. Biocomposite Simulator.
Courtesy: https://nanohub.org/resources/12273/about#citethis (Lee *et al.*, 2011).

BioSensor Modeling Tool

"BioSensorLab" is a tool (Figure 8.10) to evaluate and predict the performance parameters of biosensors and to study the time dynamics of biosensors. The performance metrics for various types of label-free, electronic biosensors can be predicted by this numerical simulator. Currently, the tool focuses on sensors that can detect the presence of charged biomolecules near the sensor surface by electrostatic interaction. Bio-molecules like DNA carry negative charge under normal physiological conditions, while the net charge of a protein molecule depends on the pH of the solution. The excess charge of the receptor-bound target biomolecule modulates the conductivity of FET channel electrons via coulomb interaction. And this change in conductivity signals the presence of complementary target molecules in the solution. Besides to electrostatic nanobiosensing, other types of biosensors based on optical detection, magnetic focusing, surface plasmon response will be also become a part of this toolset in the near future. The response of a sensor is characterized in terms of its Settling time, Sensitivity and Selectivity. The time taken by the sensor to produce a stable signal change defines the settling time. It

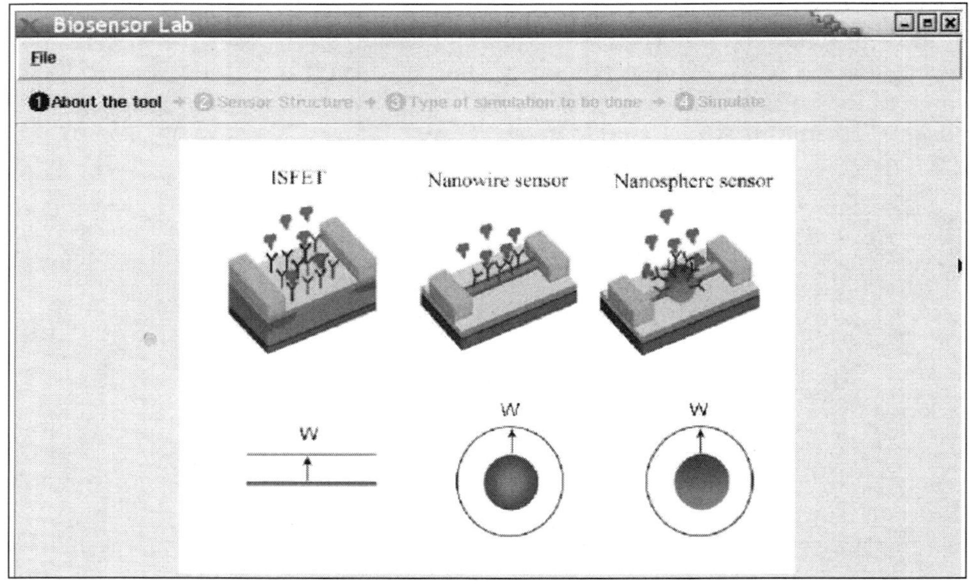

Figure 8.10. BioSensor Modeling Tool.
Courtesy: https://nanohub.org/resources/2929/about#citethis (Nair *et al.*, 2008).

is determined by bio-molecule concentration, their diffusion coefficients, and their conjugation affinity to the receptor molecules.

The performance parameters of nanobiosensors (Settling time, Sensitivity and Selectivity) can be estimated using this tool. Also we can analyze the performance of a wide variety of sensors like: Planar ISFETS, cylindrical NWs, Nanosphere, magnetic particle based schemes and Double gate FETs.

The sensor consists of a field effect device. Its surface is functionalized with a capture probe receptor molecules. The target molecules introduced to the system diffuse through the solution and reach the field effect device. They later get captured by the receptors thus by binding them close to the surface. The coulomb interaction between the charge of the target bio- molecule and the filed effect device can result in a change in conductivity of the device.

Polymer Modeler

The Polymer Modeler tool (Figure 8.11) helps to build thermoplastic polymer chain nanomaterials and can run LAMMPS to relax the chains to study their mechanical properties. It has options to specify monomers, their arrangements, torsion angles, tacticity, other parameters such as density and temperature. The resulting structures can be used as input to LAMMPS to run molecular dynamics simulations and much more options that are available.

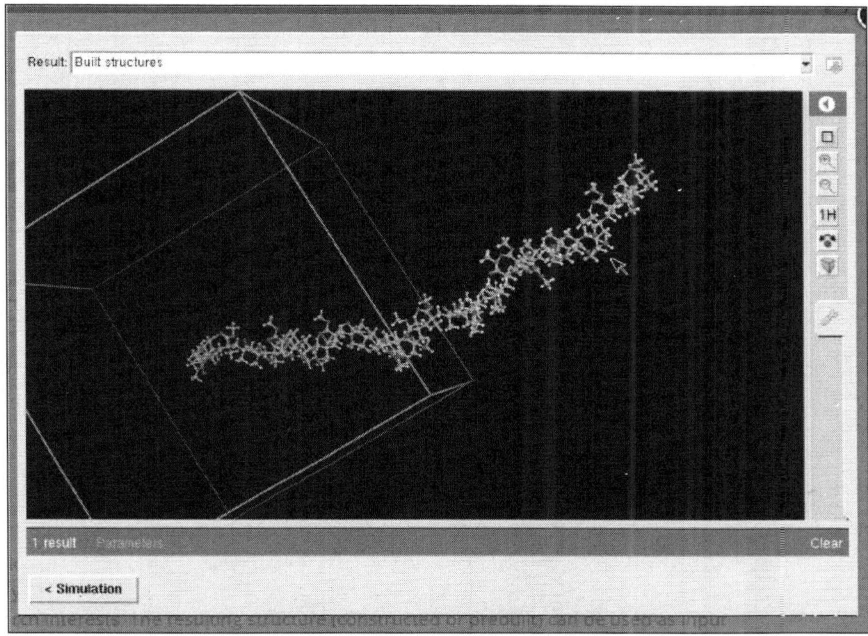

Figure 8.11. Polymer Modeler.
Courtesy: https://nanohub.org/resources/9230/about#citethis (Haley *et al.*, 2010).

PhysiCell Nano-Biorobots Simulation

PhysiCell Biorobots simulation (Figure 8.12) is a tool which helps in 2D simulation of biorobots delivering drug to cancer cells. A model was built and cloud-hosted demo are part of a computational multicellular systems biology course created and taught by Dr. Paul Macklin in the Department of Intelligent Systems Engineering at Indiana University. The models are built using Physicell, a C++ framework for multicellular systems biology. It consists of an app which demonstrates how simple rules could be used for active delivery of compounds for chemical and contact based communication, while overcoming traditional biotransport limits and using hypoxia to target delivery.

MIT Atomic-Scale Modeling Toolkit

'MIT Atomic- Scale Modeling Toolkit' is a tool (Figure 8.13) for Atomic-Scale modelling of nanomaterials. The development team of this tool are; David Strubbe, Daniel Richards, Elif Ertekin, Jeff Grossman and Justin Riley. They developed this tool for study in the basic computational problem-solving techniques to understand the nanoscale systems and predict their properties. It is suitable for both experimental and theoretical researchers as it presents a broad overview of computational nanosciencce. The tool has the following simulations: Lennard-Jones Molecular Dynamics, Monte carlo, Quantum chemistry, Density Functional Theory and Quantum Monte Carlo simulations of nanomaterials.

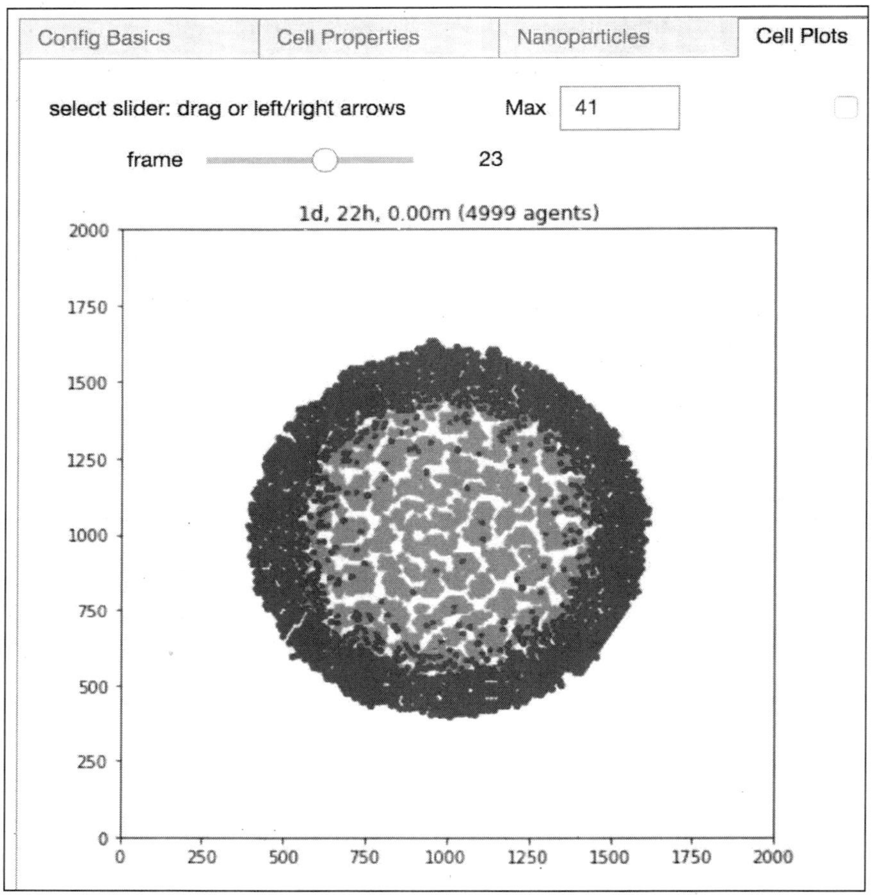

Figure 8.12: PhysiCell Biorobots Simulation.
Courtesy: https://rheiland.github.io/PhysiCell; https://nanohub.org/resources/29512/about#citethis (Heiland and Macklin, 2019).

VASP

The Vienna Ab initio Simulation Package (VASP) (Figure 8.14) is a computer program for atomic scale nanomaterials modelling, *e.g.* electronic structure calculations and quantum-mechanical molecular dynamics, from first principles. Initially VASP was based on code written by Mike Payne and then brought to University of Vienna in 1989. The main program was written by Jurgen Furthmuller and Georg Kresse in 1993.It can perform electronic structure calculations and quantum- mechanical molecular dynamics, ab initio quantum mechanical calculations using Vanderbilt pseudopotentials or the projector augmented wave method, and a plane wave basis set. The tool computes solution to many Schrödinger equations within density functional theory (DFT) or within the Hartree-Fock (HF) approximation. Green's functions methods and many-body perturbation are

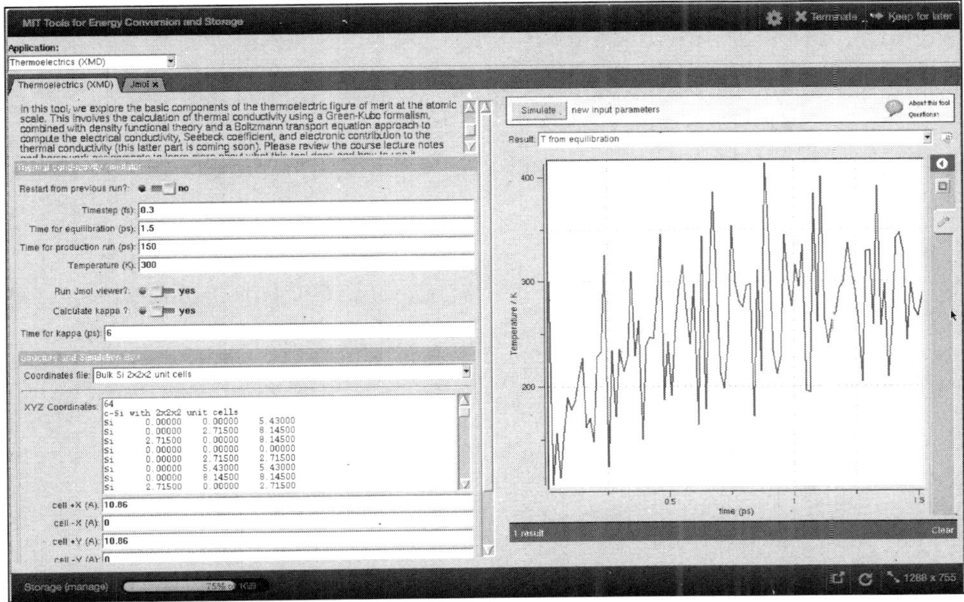

Figure 8.13. MIT Atomic-Scale Modeling Toolkit.
Courtesy: https://nanohub.org/resources/3842/about#citethis (Richards *et al.*, 2008).

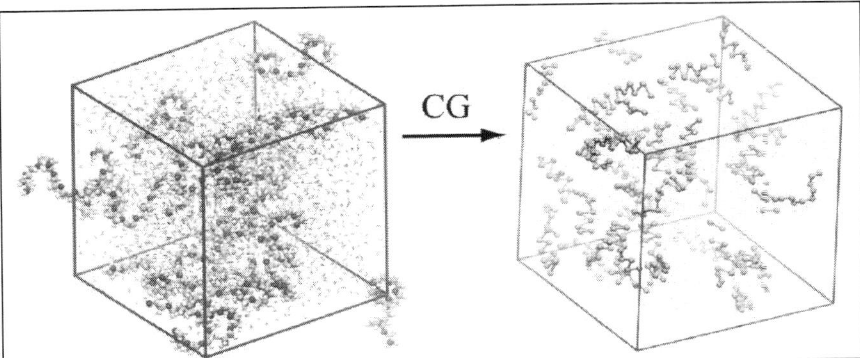

Figure 8.14. Coarse Grain Simulation using VASP Tool.
Courtesy: Gao et al., 2018.

available in VASP. This tool is currently used by more than 1400 research groups in academia and industry worldwide on the basis of software licence agreements with the University of Vienna. Norm-conserving or ultrasoft pseudopotentials, or the projector-augmented-wave method are used to explain the interactions between the electrons and ions in nanomaterials. The tool can compute an efficient iterative matrix diagonalisation techniques to determine the electronic groundstate of nanomaterials.

Quantum Espresso (QE)

QE is an integrated suite (Figure 8.15) of computer codes for electronic-structure calculations and materials modeling, based on density-functional theory, plane waves, and pseudopotentials at the nanoscale. It is written in Fortran and C language and released in 2019, under the GNU General Public License. ESPRESSO is an acronym for **open-Source Package for Research in Electronic Structure, Simulation, and Optimization**. The tool builds upon newly-restructured electronic-structure codes that have been developed and tested by some of the original authors of novel electronic-structure algorithms and applied in the last twenty years by some of the leading materials modeling groups worldwide. Innovation and efficiency are still its main focus, with special attention paid to massively parallel architectures, and a great effort being devoted to user friendliness. Quantum Espresso is evolving towards a distribution of independent and interoperable codes in the spirit of an open-source project, where researchers active in the field of electronic-structure calculations are encouraged to participate in the project by contributing their own codes or by implementing their own ideas into existing codes.

Figure 8.15. QE Model of an Atomistic Surface of Nanomaterials.
Courtesy: https://www.mri.psu.edu/materials-computation-center/methods/quantum-espresso
https://www.quantum-espresso.org/

Nano Hub.org

NanoHUB.org is the premier repository for computational nanotechnology research (Figure 8.16), education, and collaboration tools. This site hosts a rapidly growing collection of Simulation Programs for nanoscale phenomena that run in the cloud and are accessible through a web browser. NanoHUB also provides a vast array of resources that help users learn about the modelling and simulation programs for nanotechnology in general. **NanoHUB.org** is a science and engineering gateway comprising community-contributed resources and geared toward education, professional networking, and interactive simulation tools for nanotechnology. Funded by the United States National Science Foundation (NSF), it is a product of the Network for Computational Nanotechnology (NCN). NCN supports research efforts in nanomaterials. The Web portal of NCN is nanoHUB.org and it offers simulation tools, course materials, lectures, seminars, tutorials, user groups, and online meetings. Interactive simulation tools are accessible from web browsers and run via a distributed computing network at Purdue University, as well as the TeraGrid and Open Science Grid. These resources are provided by hundreds of member contributors in the nanoscience community. The NanoHUB provides in-browser simulation tools geared toward nanotechnology, materials science, chemistry and semiconductor education. NanoHUB simulations are available to users as both stand-alone tools and part of structured teaching and learning curricula comprising numerous tools. Users can develop and contribute their own tools for live deployment in the field of computational nanotechnology.

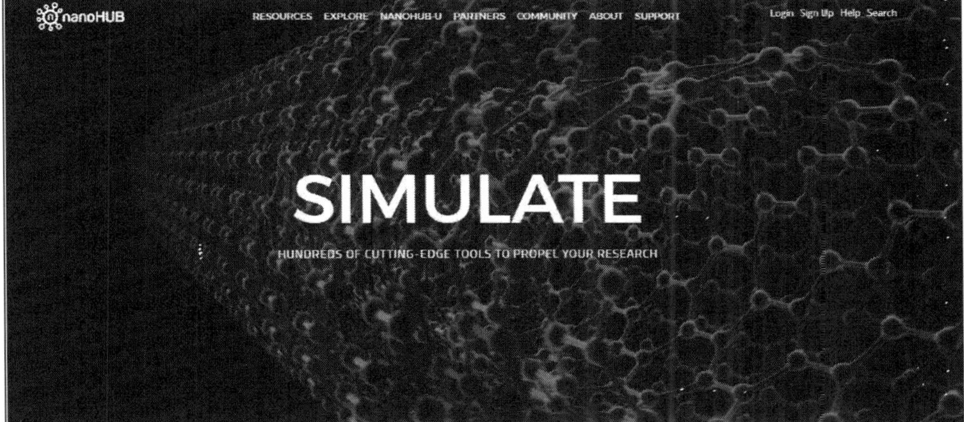

Figure 8.16: NanoHub.org.
Courtesy: https://nanohub.org/

Consolidated list of software available for Molecular / Atomistic Modeling and Dynamics simulations of Nanostructures in various operating system are given in Table 8.1.

Table 8.1. Molecular/Atomistic Modeling and Dynamics Simulation Software

Software Name	*Operating System*
AMBER	Unix- like systems
NAMD	Windows/GNU Linux/Mac OS X/Solaris
GROMACS	Solaris, Linux, OS X, Windows
CHARMM	Unix-like
DESMOND	Linux/Windows
LAMMPS	Windows/GNU Linux/Mac OS X
TINKER	Windows/GNU Linux/Mac OS X
CP2K	GNU Linux
QUANTUM ESPRESSO	GNU Linux
VASP	GNU Linux

Courtesy: Panneerselvam and Choi, 2014.

REFERENCES

1. Da Silva, E. Z., Faccin, G. M., Machado, T. R., Macedo, N. G., De Assis, M., Maya-Johnson, S., San-Miguel, M. A. (2019). Connecting Theory with Experiment to Understand the Sintering Processes of Ag Nanoparticles. *The Journal of Physical Chemistry C, 123*(17), 11310-11318.
2. Gao, H.-M., Liu, H., Qian, H.-J., Jiao, G.-S., and Lu, Z.-Y. (2018). Multiscale simulations of ligand adsorption and exchange on gold nanoparticles. Physical Chemistry Chemical Physics, 20(3), 1381–1394.
3. Haley, B. P., Wilson, N., Li, C., Arguelles, A., Jaramillo, E., and Strachan, A. (2010). Polymer Modeler.
4. Hanwell, M.D., Curtis, D.E., Lonie, D.C., Vandermeersch, T., Zurek, E., Hutchison, G.R. (2012) Avogadro: an advanced semantic chemical editor, visualization, and analysis platform. Journal of Cheminformatics 4: 17.
5. Heiland, R., and Macklin, P. (2019). PhysiCell biorobots simulation.
6. Lee, S., Zavattieri, P. D., and Strachan, A. (2011). Bio Composite Simulator.
7. Nair, P. R., Go, J., Landells, G. J., Pandit, T. R., Alam, M., Jin, X. Khan, M. R. (2008). BioSensorLab.
8. Panneerselvam, S., and Choi, S. (2014). Nanoinformatics: emerging databases and available tools. *International journal of molecular sciences, 15*(5), 7158-7182.
9. Richards, D., Ertekin, E., Grossman, J. C., Strubbe, D., and Riley, J. (2008). MIT Atomic-Scale Modeling Toolkit.
10. Zeng, H., Long, M.Q., Ori, O., Putz, M. V., and de Sousa, J.S. (2015) Computational Modeling of Physical and Chemical Properties of Nanomaterials. Journal of Nanomaterials, Article ID 915820.

Chapter 9

Synthesis of Nanomaterials using Physical Methods

S. Marimuthu[1], N. Swathika[1] and M.S. Kumuthan[1]*

[1]Department of Nano Science and Technology, Tamil Nadu Agricultural University Coimbatore – 641 003, Tamil Nadu

**e-mail: sm20@tnau.ac.in*

Nanotechnology is the study of engineering of materials with novel properties through the controlled manipulation, synthesis and assembly of the material at the nanoscale level. Nanotechnology aims primarily on the synthesis of nanomaterials with desired size and stability. There are two main approaches to synthesize nanomaterial or nanoparticle. One is a top-down approach and another is the bottom-up approach. A top-down approach where nanoparticles are synthesized by *etc.*hing out crystal planes of bulk materials. Breaking down of the bulk material into nano-sized structures or particles is the top-down approach (Etching and Ball milling) which are then subsequently assembled themselves to form nanomaterials. The Bottom-up approach involves self-assembly and coalescence of atoms, which gives rise to crystal planes, resulting in the formation of nanoparticles. The physical forces, which operate at the nanoscale, combine atoms into larger stable structures (Figure 9.1). Chemical and biological synthesis of nanomaterials follows the bottom-up approach.

The bottom-up approach is more advantageous in producing nanoparticles with fewer defects and more homogenous in nature while the imperfection of surface structure and significant crystallographic damage to the processed nanoparticles are common in the top-down approach. The bulk production of nanomaterials is possible in the top-down approach, but it involves higher machine costs and cross-contamination while synthesis is possible. Production rates of nanoparticles through the bottom-up approach are very less rate but it is a lengthy process. Synthesis of nanomaterials are classified into physical, chemical and biological methods based

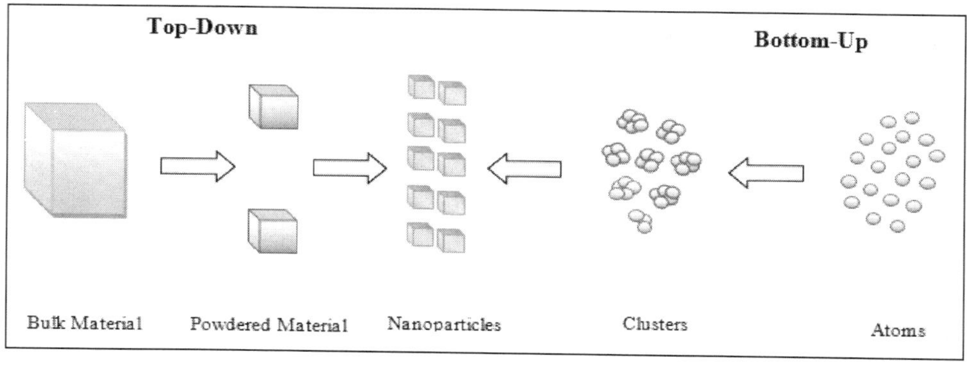

Figure 9.1. Top-down and Bottom-up Approach of Synthesizing Nanomaterials.
Courtesy: Pandey *et al.*, 2016.

on the source of energy used during the synthesis, nature of chemical reaction and source of raw materials used and this chapter discusses the physical methods of synthesizing nanomaterials.

Physical Methods

Physical methods involve the use of mechanical energy, high-energy radiations, thermal energy or electrical energy to effect material abrasion, melting, evaporation of target materials and condense as nanofilm over the wafer. The physical methods are advantageous as the methods are free of solvent contamination. The physical methods follow both top down and bottom up approaches to produce uniform monodispersed nanoparticles. High-energy ball milling is top down approach. The cross-contamination and production of abundant waste materials are limitations of the physical synthesis of nanoparticles. Laser ablation, electro spraying, inert gas condensation, physical vapor deposition, laser pyrolysis, flash spray pyrolysis, molecular beam epitaxy, high energy ball milling, melt mixing are some of the commonly used physical methods for synthesizing nanoparticles.

Physical Vapor Deposition

In general, deposition of gaseous materials over a substrate is referred to as Vapor Deposition, which is further classified into physical and chemical vapor deposition. Physical Vapor Deposition is a coating process, involving the conversion of solid material into a gaseous phase by physical processes, which is then deposited on a substrate with some modification. The deposition of material over a wafer (substrate) is typical in the range of few nanometers to several micrometers. Physical vapor deposition involves physical removal of molecules or atoms from a source by evaporation or sputtering, which are transported through a vacuum or partial vacuum by the energy of the vapor particles, and condensed as a thin film on the surfaces of substrates.

The basic steps involved in physical vapor deposition are converting metal from solid to vapor phase (by either melting and evaporation or direct sublimation);

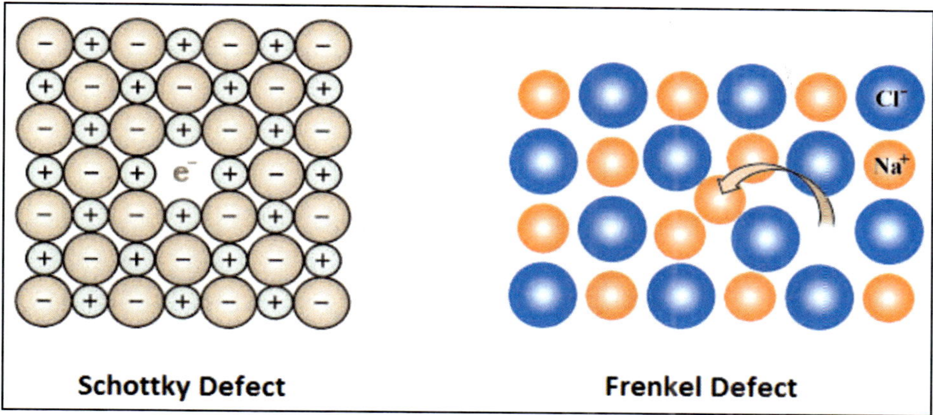

Figure 1.13. Schematic Representation of (a) Schottky and (b) Frenkel Defect.
Courtesy: https://en.wikipedia.org/wiki/Schottky_defect; https://en.wikipedia.org/wiki/Frenkel_defect **P.14**

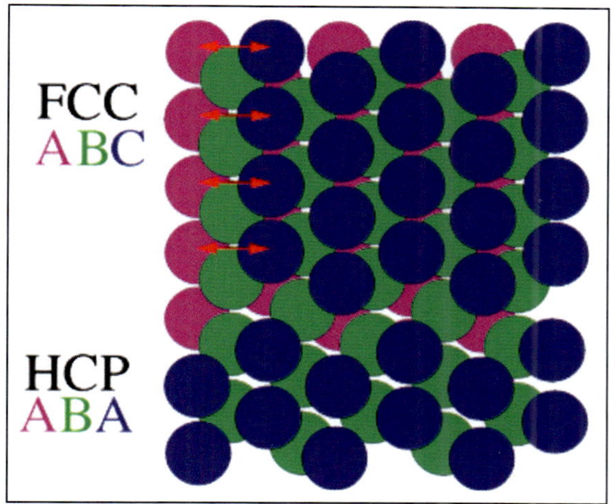

Figure 1.16. Comparison of FCC and HCP Lattices, Explaining the Formation of Stacking Faults in Close-Packed Crystals.
Courtesy: https://en.wikipedia.org/wiki/Stacking_fault **P.17**

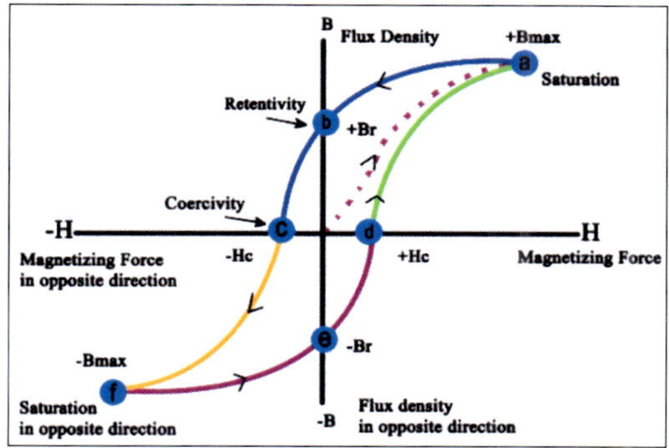

Figure 2.3. Schematic Illustration of Magnetic Hysteresis Loop.
Courtesy: Electrical Academic 2010. **P. 26**

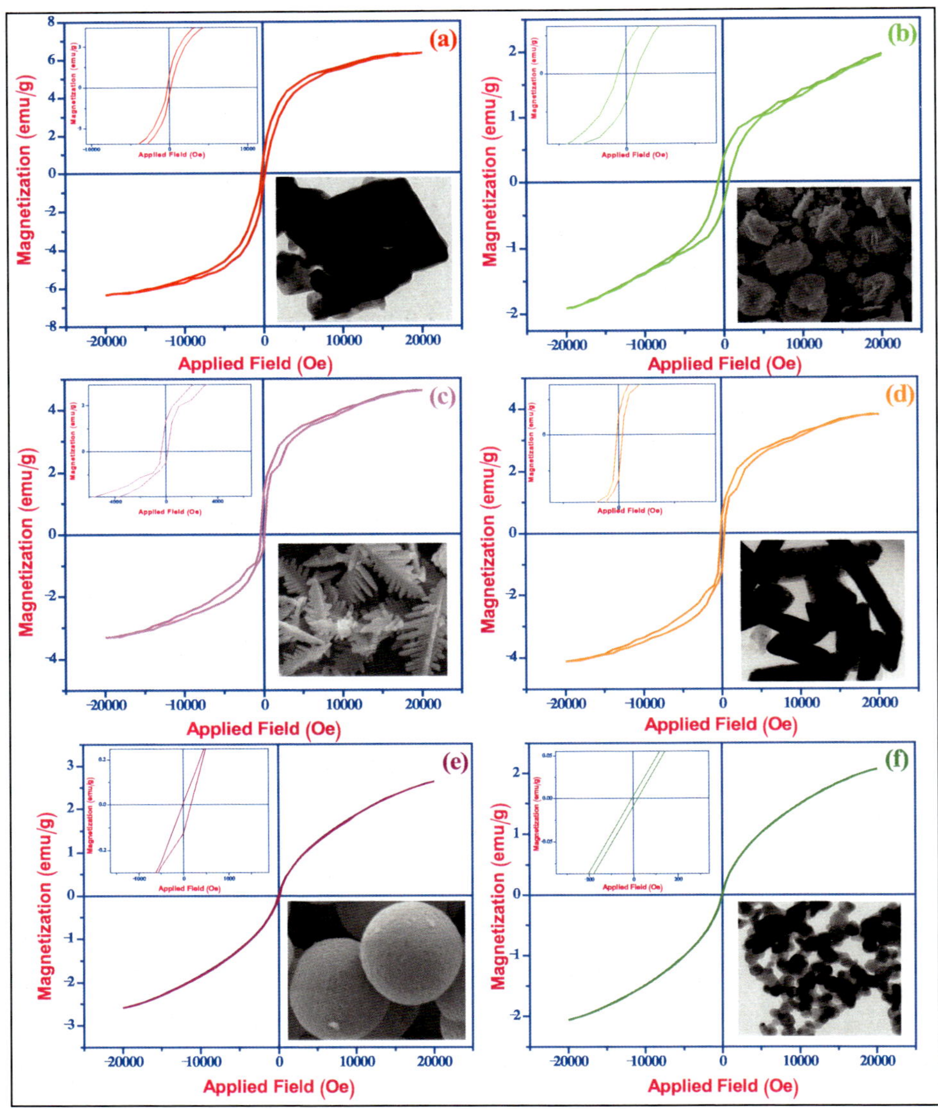

Figure 2.5. Room Temperature M-H loop of the different $LaFeO_3$ Nanostructures and Inset (left) shows the Magnified View of the M-H Loop at Low Applied Magnetic Field (Right) the Corresponding Nanostructures, (S. Thirumalairajan *et al.*, 2016). P. 32

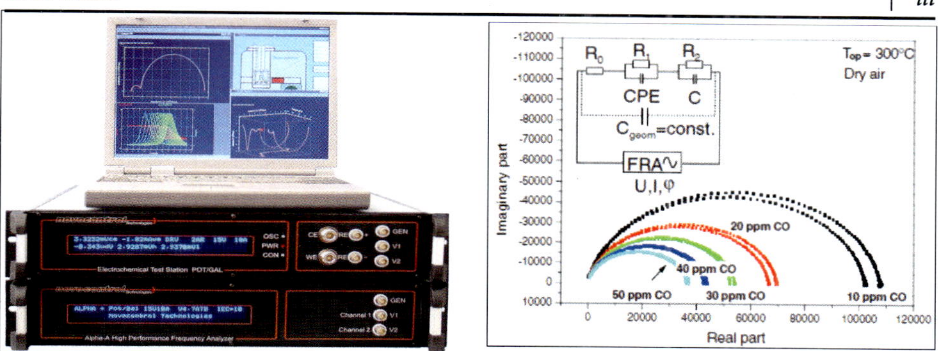

Figure 2.10. (a) Photograph of Impendence Spectroscopy, (b) Cole – Cole Plot of Impedance Spectrum.

Courtesy: Novo control Technology, Germany, 2019. **P.42**

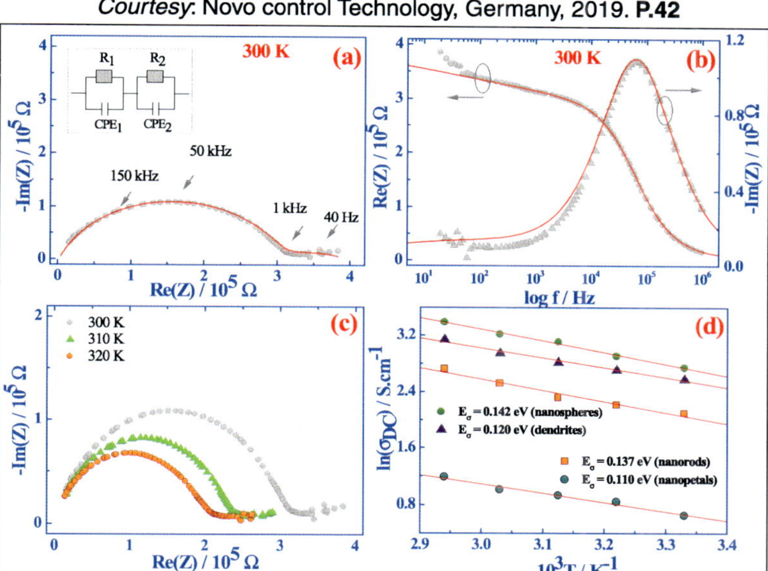

Figure 2.11. Impedance Spectra for $LaFeO_3$ Nanopetals (a) Room Temperature, (b) Real Part Re(Z*) and the Opposite of the Imaginary Part–Im(Z*) as a Function of Frequency at RT (c) Impedance Spectra at different Temperatures and (d) Arrhenius Plot of the Conductivity for different $LaFeO_3$ Nanostructures (S. Thirumalairajan, *et al.*, 2016). P. 44

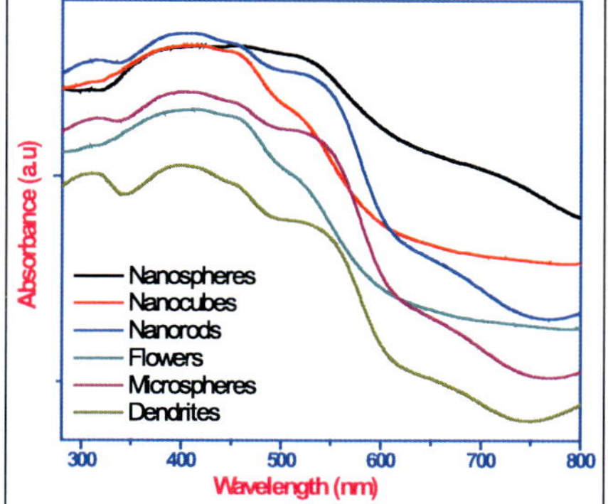

Nanostructures	*Cut-off Wavelength (nm)*	*Bandgap Energy (eV)*
Nanocubes	582	2.14
Flowers	591	2.10
Dendrites	601	2.07
Nanorods	609	2.05
Microspheres	614	2.03
Nanospheres	619	2.00

Figure 2.15. UV-Vis DRS of different $LaFeO_3$ Nanostructures and Table 2.3 Representative Corresponding band gap energy (eV). P.51

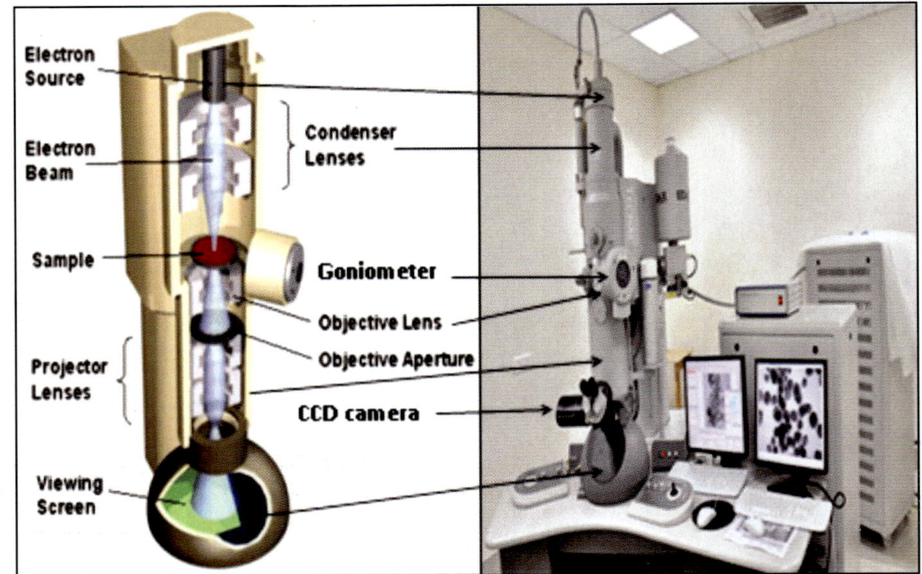

Figure 6.26. Structure of the Transmission Electron Microscope (TEM).
Courtesy: Kannan, 2018. **P. 149**

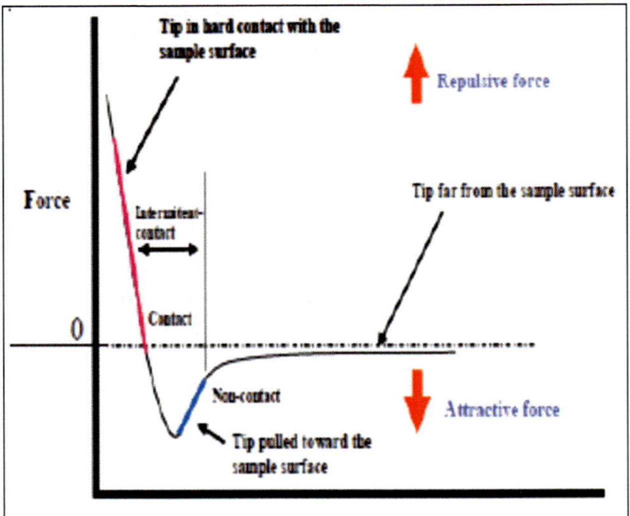

Figure 6.30. Graphical Representation Force Distance Relationship.
Courtesy: Kannan, 2013. P. 160

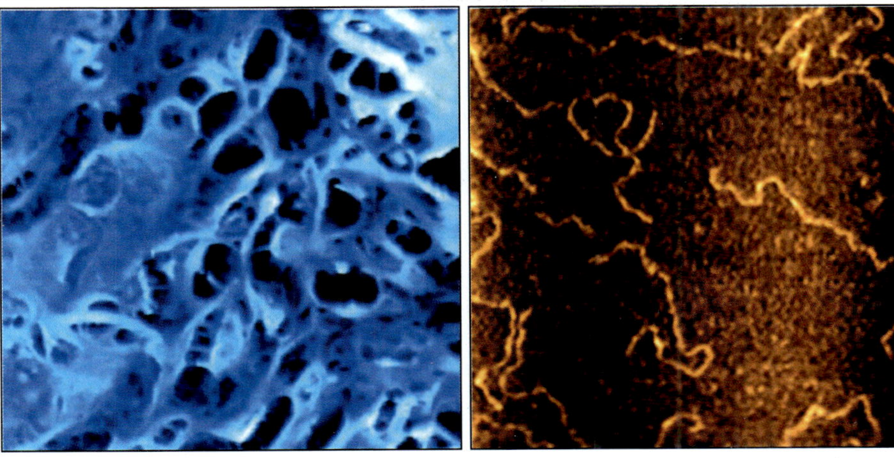

Figure 6.33. Breast Cancer Cells 78 x 78 microns.

Figure 6.34. DNA Image 1.2u x 1.2u.

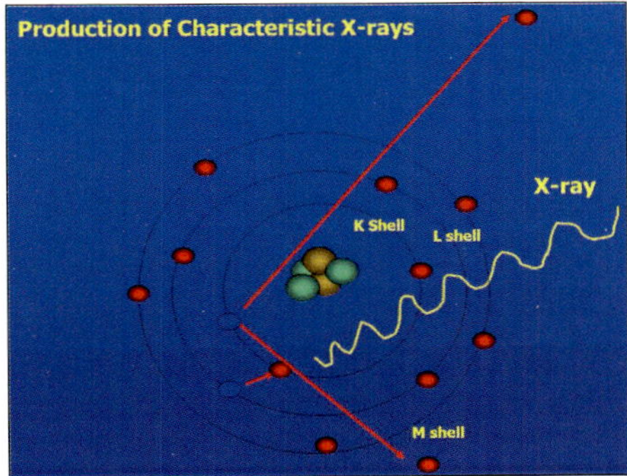

Figure 7.4. Production of Characteristic X-rays. p. 171

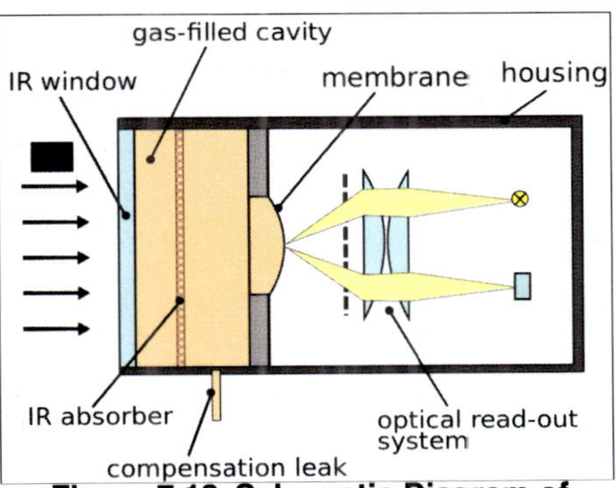

Figure 7.13. Schematic Diagram of Pneumatic Detector. P. 180

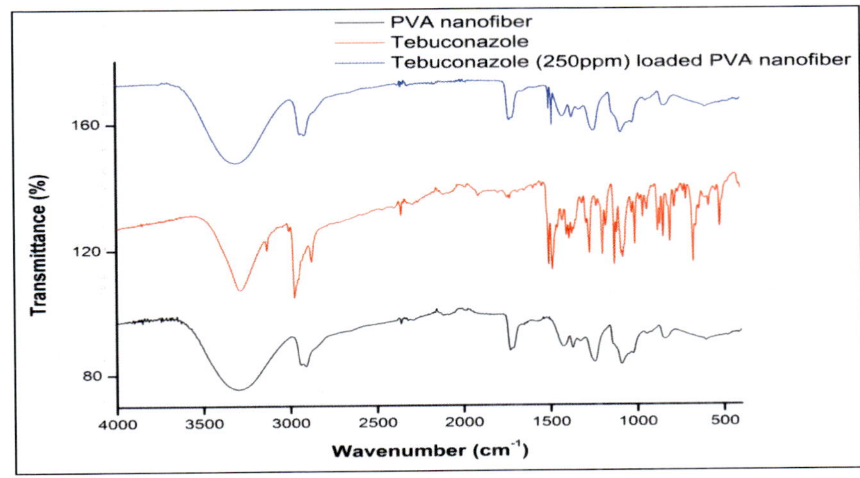

Figure 7.16. FTIR Spectra of Tebuconazole Loaded Nanofibres. P. 188

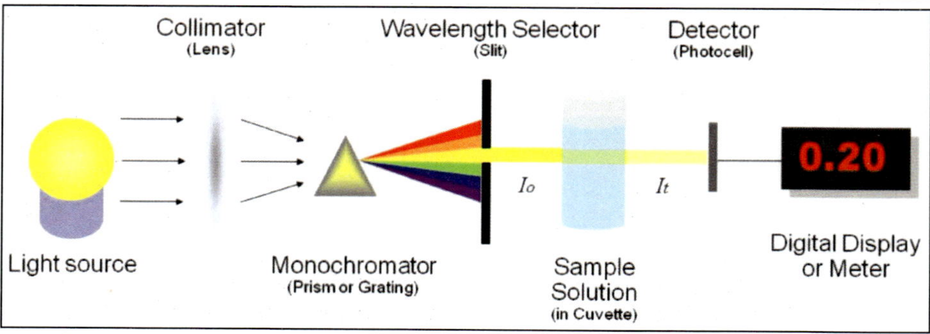

Figure 7.17. Schematic Diagram of UV-VIS Spectrometer. P.191

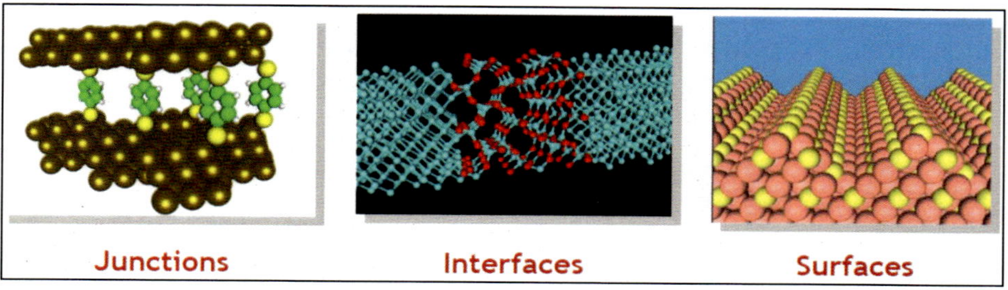

Figure 8.1. Modeling of Junctions, Interfaces and Surfaces of Nanomaterials.
Courtesy: QuantumWise - ATK Virtual Nano Lab Academic. **P. 202**

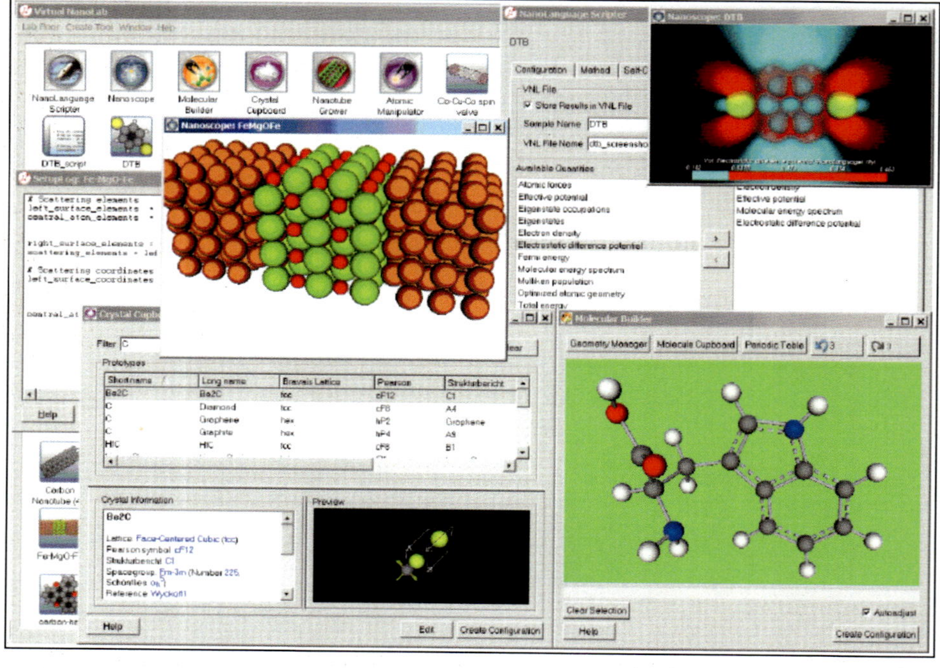

Figure 8.2. Home Page of Virtual Nano Lab.
Courtesy: QuantumWise - ATK Virtual Nano Lab Academic. **P. 202**

Figure 8.3. Home Page of Schrodinger's Materials Science Suite.
Courtesy: https://www.schrodinger.com/materials-science **P. 203**

Figure 8.4: Schrodinger's Masetro Interface.
Courtesy: https://www.schrodinger.com/maestro **P 204**

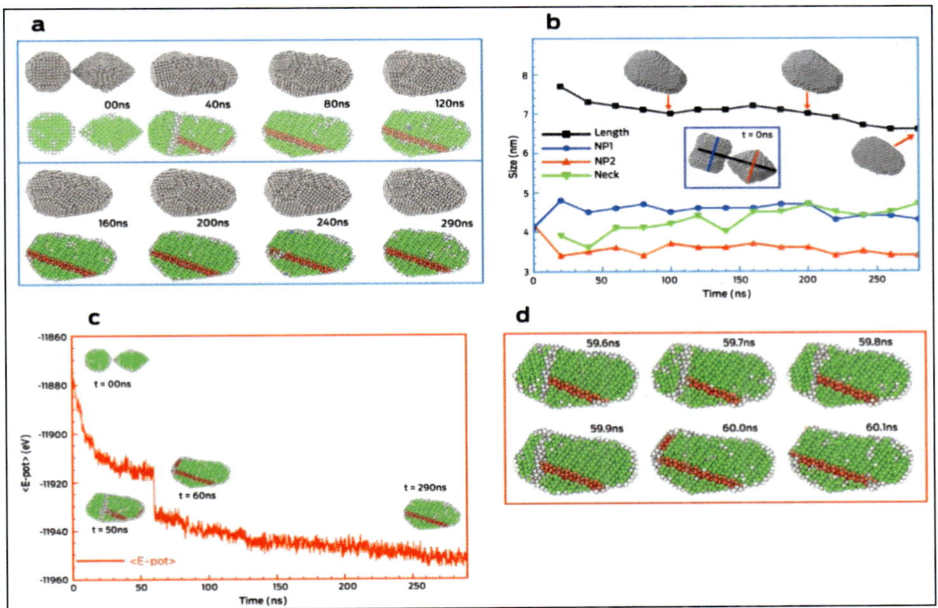

Figure 8.6. Modeling of Silver Nanoparticles
https://lammps.sandia.gov/pictures.html#silvernanoparticles *Courtesy*: Da Silva *et al.*, 2019. **P. 206**

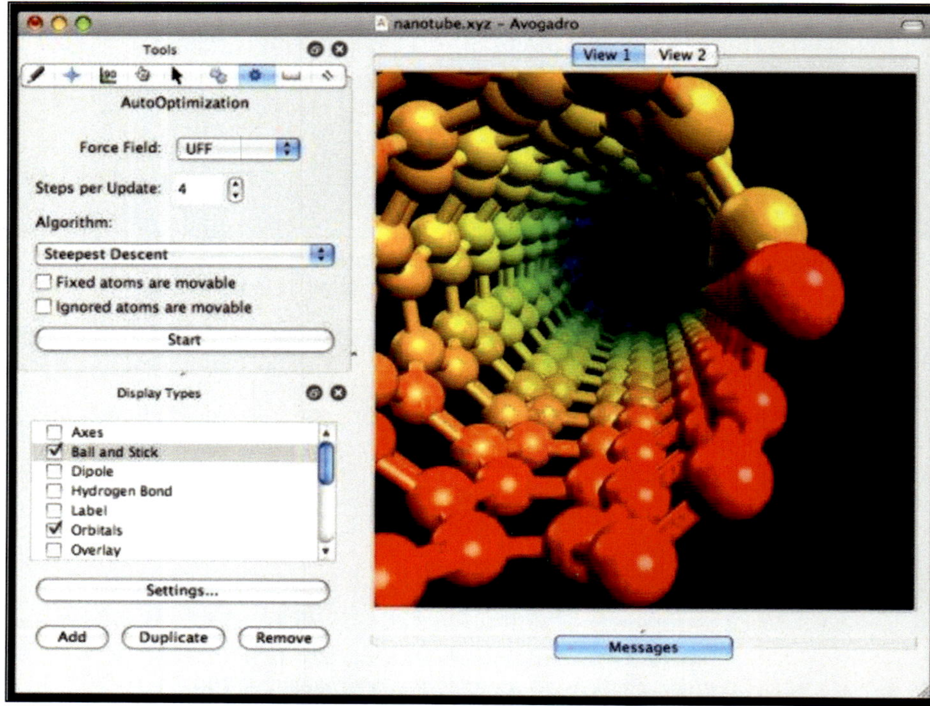

Figure 8.7. Avogadro for Nanotube Modeling.
Courtesy: Avogadro-Hanwell *et al.*, 2012. **P. 207**

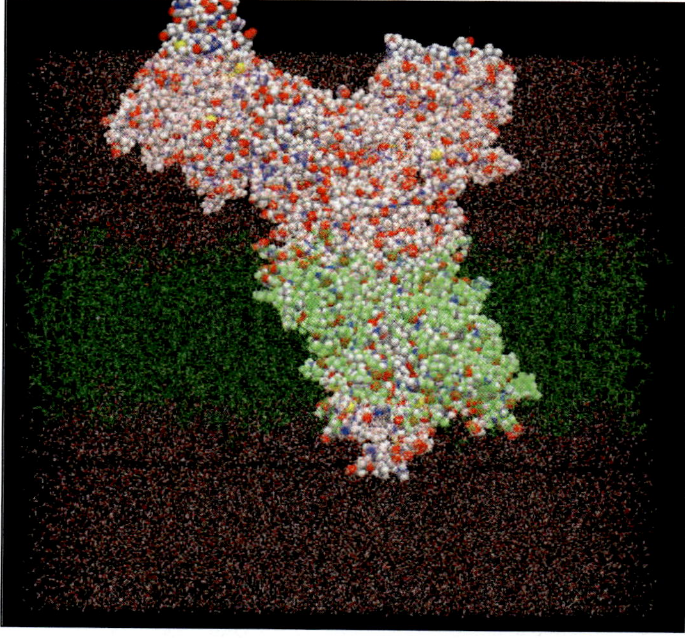

Figure 8.8. Desmond-Transmembrane Modeling.
Courtesy: http://geminiweb.slu.edu/wp-content/uploads/desmond_tutorial.pdf. **P. 208**

transporting vapor phase material to the substrate and condensing gaseous material on the substrate through nucleation and growth. Physical vapor deposition methods transfer kinetic energy to atoms in a solid or liquid, which is sufficient to overcome their binding energy causing the removal of atoms. Evaporation is also used to heat the source material directly (resistive) or indirectly (electron beam and laser) until atoms to vaporize from the target materials. Sputtering is a process of physical impact transferring kinetic energy to atoms in a target to liberate from source materials. Most commonly used techniques in physical vapor deposition are Sputtering, Electron beam evaporation, Pulsed Laser Deposition, Molecular Beam Epitaxy and Inert Gas Condensation. Physical vapor deposition techniques are carried out in a vacuum system to avoid uncontrolled oxidation of source materials (target material that is to be evaporated) and final product of deposition as well as that of components of the synthesis system. Atoms in the vapor phase from the source material do not collide with each other under vacuum prior to arrival at substrate for deposition, since the mean free path of the particles increases in the vacuum system. The transport of atoms or molecules from the source to the growth substrate is straightforward along the line of sight under vacuum conditions, and therefore the conformal coverage and a uniform film over a large area are possible.

Electron Beam Evaporation

Electron beam evaporation is a vacuum-based physical vapor deposition technique. The Electron beam is used to evaporate the atoms from the source materials. Electron beam evaporation is a process of evaporating source material, which is to be deposited on the wafer (substrate) by using the beam of electrons to evaporate the target materials. The electrons are created from cathode materials through a heating process.

The Electron beam evaporation consists of ingot, which is the source to be deposited, filament (source of electrons), magnet (to bend an electron beam and direct towards the source) and wafer over which target material is deposited as a thin film (Figure 9.2). The ingot is positioned at a positive potential ensuring that there are no chemical interactions between the filament and the ingot material, where the filament is kept out of sight. A magnetic field is used to bend and direct the electron beam from its source to the ingot location, where an additional electric field is used to steer the beam over the ingot surface allowing uniform heating. The chamber is maintained at vacuum @ 10^{-8} to 10^{-4}Torr, which is required for steering electrons properly over the source. The power requirement for this technique is high (D.C. power @ 10 to 30 kV).

The first stage of the technique is heating filament (mostly tungsten), which will produce a beam of electrons on excitation due to heating. This process is achieved through thermionic emission, which is a process of discharging electrons from heated material. The energy is applied to the filament, enough to overcome the attractive force that holds electrons together in the filament. Once the electrons are excited which move about randomly in the chamber.

The second stage is heating ingots that are to be deposited, which is generally referred to as an evaporant kept in the water-cooled grate. The source of the heating

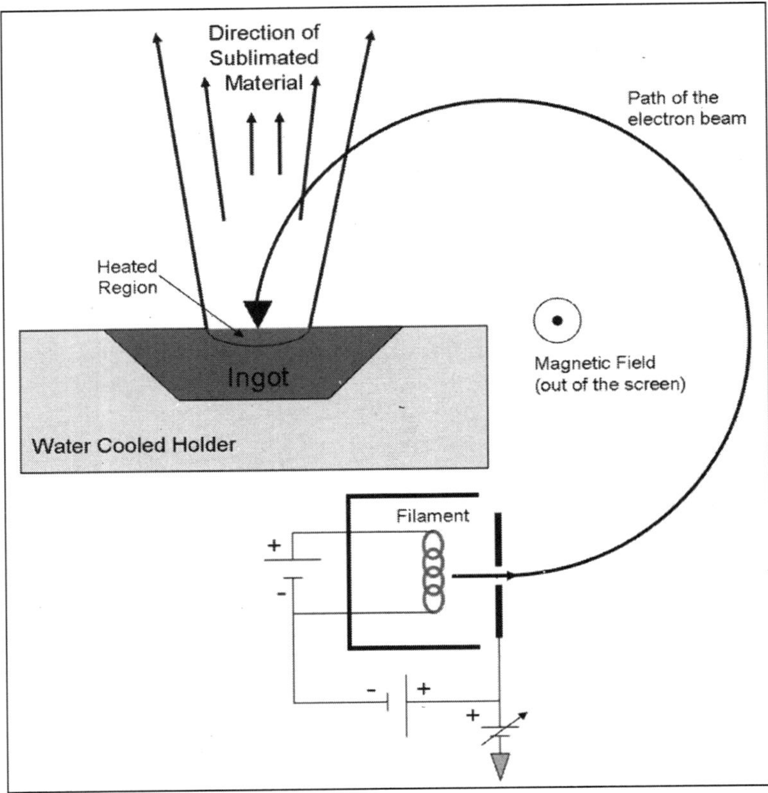

Figure 9.2. Electron Beam Evaporation Unit where Ingot (Evaporant) is Positioned over Hearth with Water-Cooling Mechanism and Filament is used to Generate Electrons Directed through a Magnetic Field to the Ingot Location.
Courtsey: http/en.wikipedia.org/wiki/Electroeam_physical_vapor_deposition#/media/File:Electron_Beam_Deposition_001).

is a beam of electrons produced from the filament. The beam of electrons must have a specific path to achieve a direction so that to concentrate at a point over the ingots for effective heating. Excited electrons move randomly in the high vacuum chambers because there is no force of attraction. Hence, a pair of electromagnets are placed, to attract the electrons at the desired direction and deflect the route of the beam of electrons towards the material to be evaporated. The material will be heated on exposure to the electron beam to reach the boiling point. Then the evaporant will be converted to the vapor phase, which started moving towards the wafer, where it is condensed on the surface of the wafer (substrate).

To summarize the electron beam evaporation technique, the electron beam hits the target and heat the target material, which evaporates when the temperature reaches above its boiling point. The evaporated material then transported and condensed onto the substrate.

The advantage of the technique is the high rate of film deposition with less surface damage from the impinging atom. The technique achieves high purity deposition (due to high vacuum) and avoids unintentional heating of wafer. Alloys are also evaporated by this technique from an alloyed evaporant to produce composite thin films of materials. The disadvantage of the technique is the production of X rays during the electron beam evaporation.

Laser Ablation

The laser ablation method uses energy from a high power laser beam to evaporate particles from a solid source. A laser is used either continuously or pulsed mode for evaporating materials. A pulsed laser beam of 30 ns pulses with energy in the range of 0.1-1 J and a frequency of 1-20 Hz is focused onto the target during the synthesis of nanomaterials. Generally, the pulsed laser is employed to remove the material from the target. The laser pulses hit the surfaces of the target (evaporant) leading to melting, followed by evaporation and ionization of the material. The evaporated materials (plasma plume) deposited onto the suitably positioned substrate (Figure 9.3). If the power density of a laser beam is greater than ablation thresh hold material (minimum energy density for removal of material), the materials will be evaporated. Usually shorter wavelength lasers including ArF (Argon fluoride), KrF (Krypton fluoride) excimer lasers, and Nd: YAG (Neodymium-doped Yttrium Aluminium Garnet) lasers used for pulsed laser deposition. The focus of laser over material causes the extremely high heating rate of the target surface (10^8 K/s) due to irradiation. Pulsed Laser Deposition is used for the preparation of the variety of materials including polymers, oxides, metallic systems, fullerenes, carbides, nitrides, *etc.* The laser energy is transferred to the charge carriers of free electrons in the case of metals while valance electrons in semiconductors, which get excited to the conduction band. After a short time of carrier thermalization, the carriers start transferring their energy to the lattice via the electron-phonon coupling. The deformation or displacement of crystal lattices of atoms from the equilibrium position is described as phonons. When electrons interact with displacements is referred to as electron-phonon coupling. The laser continues to pass the energy into the material even after finishing charge carries transferred energy to the lattice during electron-phonon coupling. This process results in laser energy being dissipated as heat, where the laser beam incident over the material's surface, resulting in the melting of the material. The melting of materials has caused the area of pronounced heat effect and mechanical cracks and other defects. The ablated materials form plasma plume reaching substrate for deposition. To summarize, pulsed laser deposition involves laser absorption on the target surface followed by ablation of the target material and creation of a plasma, sustainability of plasma, deposition of ablated materials along with plasma on the substrate and nucleation and growth of thin film over the substrate (wafer). Pulsed laser deposition is a simple technique requires the adjustment of laser power density and pulse repetition rate for the deposition. The process achieves multi-layered films of different materials by sequential ablation. The laser ablation technique produces the film thickness of atomic monolayer over substrate. The most important feature

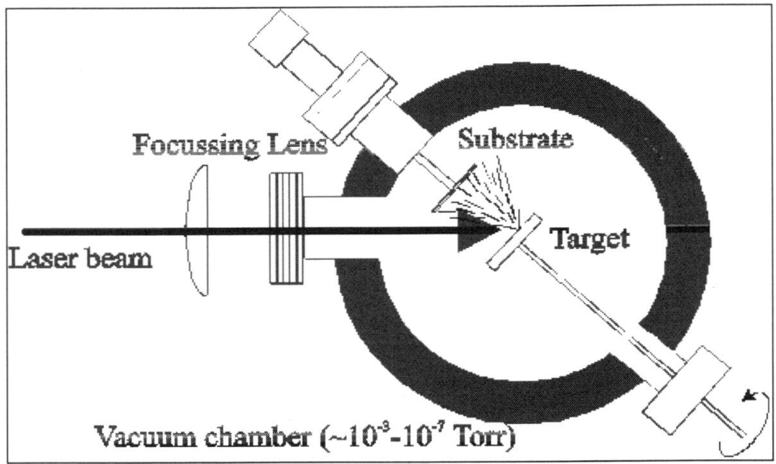

Figure 9.3. Pulsed Laser Ablation Unit is Carried Out in a Vacuum Chamber (10^{-3} to 10^{-2} torr) where the Laser is Focused to Target Materials through Focusing Lenses.
Courtsey: Adopted from www1.phys.vt.edu accessed on 27th November 2019.

of the technique is the retention of the stoichiometry of the target in the deposited films (Table 9.1)

Inert Gas Condensation

Inert gas evaporation–condensation technique synthesizes nanoparticles through the evaporation of a metallic source in inert gas. This technique involves heating of target materials either through resistive heating or indirect heat (electrons and laser) until evaporation, an inert gas or reactive gas for collisions with material vapors, a cold finger on which clusters or nanoparticles are condensed, a scraper to scrape the nanoparticles and piston-anvil for compacting nanoparticles (Figure 9.4). The Insert gas condensation technique is the only technique, which produces nanoparticles as product. Initially, the chamber is evacuated to a pressure of about 2 x 10^{-6} Torr by operating an oil diffusion pump. Metals or metal oxides are evaporated or sublimated by placing in the boat of refractory metals like tungsten (W), tantalum (Ta) and Molybdenum (Mo). The temperature is set at a predetermined value for the initial evaporation of materials. After the evacuation, an inert gas (Helium, Xenon, or Argon) is backfilled into the chamber at low pressure, usually, 0.5– 4.0 Torr and the boar is heated rapidly (at constant temperature and inert gas pressure). On evaporation of source materials, the particles of small size (<5nm) are formed near the vicinity of source materials which in turn interact with other develop into bigger sized particles. The formed particles are removed from the source by forcing the inert gas. The rate of evaporation and gas pressure in the chamber determines the particle size and distribution. The evaporated cluster of particles colloid with inserts gas molecules forming bigger sized molecules over gas molecules and condensed over cold fingers. On reaching the cold finger, insert gas molecules escape and reach the gaseous phase of the chamber. If reactive gases

Table 9.1. Fabrication of Nanofilm through Laser Ablation, Electron Beam Evaporation and Sputtering

Sl. No	*Physical method*	*Precursor and parameters*	*Nature of nanoparticles*	*Significance of nanoparticles*	*Authors*
1.	Ultra-Shot (fs) Laser Ablation Yb:KGW laser (Amplitude Systems with wavelength of 1025 nm and power of 1–100 kHz)	Hot pressed TiN target in a glass vessel containing 7 mL of ultrapure water or acetone of thickness 10 mm as synthesizing media. A laser beam of 2.3 mm diameter was focused onto the target by using 75 mm lens, where the target was moved continuously at a speed of 2 mm s^{-1} To avoid ablation being focused on the same point, Laser fragmentation technique was employed additionally to control the size of TiN NPs process during the synthesis, Oxygen dissolved in the solution to reduce the oxidation	Titanium Nitride Nanoparticles (TiN NPs)	The solution turned into blue color indicating the formation of TiN nanoparticles The stability of TiN colloids is very stable during its storage in ambient conditions. Surface potential of TiN NPs was –30 mV, higher than the stability threshold (20 mV) for colloidal solutions. Shape of TiN nanoparticles were spherical and polycrystalline in nature. The size of single crystal grain was 1-3 nm, where nanoparticle had hollow structure with cavities around 5-10 per cent and 60 per cent for acetone and water respectively as synthesizing medium.	Popov *et al.* (2019)
2.	Pulsed Laser Ablation (PLA) method (1070 nm of fibre pulsed laser)	10 g of Nickel powder in 30 mL of acetone mixed uniformly for laser irradiation. Laser power of 0.4 W with 4.8 J cm^{-2} Spot size of the beam diameter was 72 μm for one hour.	Nickel nanoparticles	The extinction spectra for Ni nanospheres were confirmed at 350 nm with the size radius of 10 nm with spherical shape	Ganjali *et al.* (2015)
3.	Picosecond laser ablation (532 nm of laser operating at 100 KHz repetition rate with pulse width of 6-8 ps.)	Laser focused on Au and Ag solid targets of purity (99.9 per cent) in deionized water. Laser power of 1.5 W was focused on Ag target for 10 min, while it was 4.5 W and 5 min for Au targets Spot size of the laser on the target was 80 μm in diameter with scanning speed of 800 mm s^{-1} for an area of 10 x 10 mm^2	Ligand free Au and Ag nanoparticles	Highly efficient green synthesis of Ag and Au nano colloids were optimized using pulsed laser ablation method. Surface Enhanced Raman Scattering (SERS) was at excitation wavelength 785 nm attributed to aggregation phenomena of nanoparticles sprayed on the support.	Fazio *et al.* (2017)

Sl. No	*Physical method*	*Precursor and parameters*	*Nature of nanoparticles*	*Significance of nanoparticles*	*Authors*
4.	Laser ablation Source of Nd: YAG (Quantel YG981E) with wavelength of 1064 nm (9 ns) and pulsed at 10 J cm^{-2} at a 10 Hz repetition rate for 10 min. Second harmonic laser: Nd:YAG laser at 532 nm and fluences of 1-5 J cm^{-2} at 1 Hz for 5-10 min.	A beam of laser is focused using 10 cm focus lens on to the gold plate placed at the bottom of the cell containing solvents such as dimethyl sulfoxide, acetonitrile, and tetrahydrofuran Size reduction and reshaping was achieved using second harmonic laser.	Gold Nanoparticles (Au NPs)	Stable Au NPs were synthesized using different organic solvents through laser ablation. High stability of nanoparticles were achieved without any capping agents.	Amendola *et al.* (2006)
5	Pulsed laser ablation Nd: YAG laser source (wavelength 532 nm) with	Yttrium barium copper oxide/isopropanol dispersion in a holder frozen with liquid nitrogen Laser frequency of 10 Hz with spot size was 0.56 cm^2. The distance from substrate to target was 4.5 cm and the substrate and target were controlled to rotate at 25 rpm and 10 rpm, respectively. The experiment was conducted under 1 x 10^{-6} Torr	Yttrium barium copper oxide nanoparticles over graphite substrate	The average Yttrium barium copper oxide particle size was 40 ± 15 nm. The average particle size of Yttrium barium copper oxide nanoparticles decreases when the duration of laser irradiation time increases. More smaller particles (diameter < 30 nm) deposited on graphene when the duration increases from 0.5 h to 2 h.	Yang and Zhang, (2019)
6	Electron beam evaporation	Target of stainless steel ball in graphite crucible (8cm^3) and target-to-sample distance was 26 cm. Deposition was carried at pressure of pressure of 6 x 10^{-6} Torr. 3 to 10 per cent of the electron beam power was maintained to achieve uniform coating	Stainless Steel (SS 316 L) coating over copper substrate	150 nm stainless steel deposited film at a fixed rate of 0.05 Ås^{-1}. The film is ultrathin, uniform, conformal, and controllable	Ali *et al.* (2019)

Sl. No	*Physical method*	*Precursor and parameters*	*Nature of nanoparticles*	*Significance of nanoparticles*	*Authors*
7	Electron beam evaporation	An iron ingot and sodium chloride were placed separately into a water-cooled crucible of 50 mm diameter. Vacuum of (1.3–2.6) x 10^{-2} Pa was created, where electron beam guns heated ingot and crucible. Mixtures of iron and sodium chloride vapors were formed and condensed on a water-cooled substrate at 40–50 °C.	Deposition of Magnetite in sodium chloride crystals	The average particle of Magnetite (Fe_3O_4) nanoparticles size increases from 3–4 nm up to 15–20 nm as the substrate temperature rises from 20 to 200 °C	Kurapov *et al.* (2018)
8	Physical Vapor Deposition through direct heating	Nickel Oxide nanoparticles deposited on aluminum oxide was kept a substrate for the deposition of zinc oxide nano wires Metal zinc powders (100 mesh, 99.999 per cent, Sigma-Aldrich) was kept in quartz tube furnace. The vertical distance between the source and the aluminum oxide substrate was about 3-5 mm. The Zn source was physically vaporized to release ZnO nanowires in atmospheric pressure under an Ar (99.99 per cent) flow rate of 500 sccm for 60 min at the temperature range of 450-600°C.	Zinc oxide nanowires	Radial direction of zinc oxide nano wires on the substrate with a high density at temperature of 450 to 600°C The average diameters of ZnO nanowires grown at 450, 500, 550, and 600°C were 55, 65, 84, and 100 nm, respectively The growth rate of ZnO nanowires increases as a function of growth temperature. The growth rates of ZnO nanowire at temperatures of 450, 500, 550, and 600°C were 2.6, 4.8, 5.4, and 7.2 μm respectively	Lyu *et al.* (2003)
9.	Direct Current high vacuum dual magnetron sputtering system (Kurt J. Lesker)	Metal targets used for coating are of high purity (99.99 per cent). Target power during the process of deposition was 250W for copper, silver and aluminum and 175W for titanium and zirconium. Vacuum @ 1.066 x 10^{-3} Pa (8.0 x 10^{-6} Torr) with argon flow rate of 40 sccm to reach the pressure of 2.66 x 10^{-1} Pa (2.0 x 10^{-3} Torr).	Metallic nanoparticles deposited on water Filter paper	Antibacterial performance were better with filter paper coated with copper and silver of 5nm thickness	Mihut *et al.* (2017)

Sl. No	*Physical method*	*Precursor and parameters*	*Nature of nanoparticles*	*Significance of nanoparticles*	*Authors*
10.	Polygonal Barrel-Sputtering Method (Radio Frequency sputtering)	Platinum (High purity of 99.99 per cent) was the target which was sputter off to deposit over the support material of TiO_2 particles (Size-21 nm). The support material was heated to 180°C introducing 2g of preheated TiO_2. Once the pressure reaches less than 8x 10^{-4} Pa, high purity argon gas was introduced slowly and the sputtering was performed at an Argon gas pressure of 0.8. RF power generator (13.56 MHz) supplied with AC of 25 W. Platinum deposition was controlled by altering the duration between 5-15 min.	Nano film of Platinum over Titanium oxide	Platinum particles deposited over TiO_2 powder support are highly dispersed. The particle size of the deposited platinum are smaller (1.8nm), thus increasing its surface area, which attributes high water splitting activities of TiO_2.	Matsubara *et al.*,(2019)
11.	Modified Reactive Magnetron Sputtering Technique through Inert gas Condensation	Vacuum chamber of 12-inch diameter chamber was used for the synthesis of WO_3 nanoparticles. Copper cold finger was maintained @ -192°C where Tungsten oxide nanoparticles are deposited over cold finger. Specific ratio of Oxygen and Argon were maintained. The nanoparticles are collected and treated thermally in furnace 400-700°C for 2 h with the heating rate of 10°C min^{-1}	Tungsten oxide nanoparticles	Spherical tungsten oxide nanoparticles with monoclinic structure Particle size ranged between 1±0.92 to 48 ± 1.72 nm.	Verma *et al.* (2019)

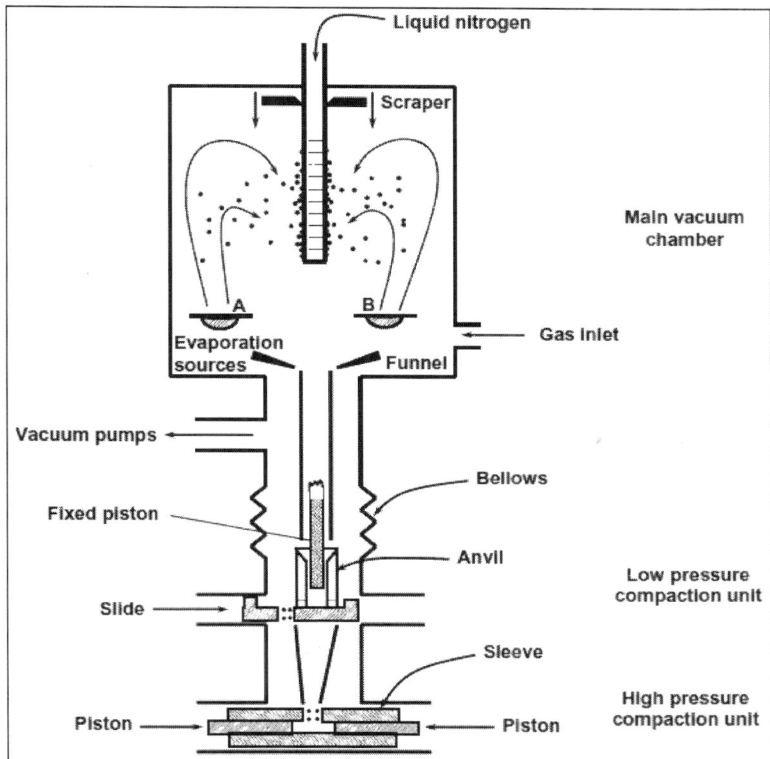

Figure 9.4. Inert Gas Evaporation and Condensation Unit for the Synthesis of Nanoparticles.
Courtsey: Adopted from https://shellzero.wordpress.com/2012/05/14/inert-gas-condensation-method).

like oxygen, hydrogen, and ammonia are used instead of inert gases in the chamber, it will result in oxide, hydride and nitride particles. Clusters of nanoparticles are condensed on the cold finger (water or liquid nitrogen cooled) are scraped off inside the vacuum system, which falls through a funnel in the collection chamber where piston anvil mechanism helps to pelletize the collected particles.

A wide range of nanoparticles including metals, alloys, intermetallic compounds, ceramics, semiconductors, and composites are synthesized through the inert gas condensation technique, which is a flexible technique for controlling the size and size distribution of the clusters/nanoparticles. The insert gas condensation produces nanoparticles with clean surfaces free from contamination. The possibilities for reacting, mixing, and coating various types, sizes, and morphologies source materials helps to synthesize new versatile materials. The technique is a continuous process, where scraping and compaction are carried out under ultra-high vacuum conditions, and hence highest purity is maintained. The inert gas condensation technique allows *in-situ* diagnostics in the system Agglomeration is a problem in the inert gas condensation technique, which is also a costly process because it requires ultra-high vacuum throughout the synthesis. Ensuring repeatability of parameters for the synthesis of nanoparticles and scaling up the process is difficult in the technique.

Sputtering

Sputtering is a thin film deposition technique, especially for retaining the composition of the original target material in the tin film (stoichiometric thin films). Sputtering is a process whereby particles are ejected from a solid target material due to bombardment of the target by energetic particles. The process is carried out by bombarding the surface of the target with gaseous ions under high voltage acceleration. In sputter deposition, usually, Ar^+ ions are incident on a target at high energy. The ionized inert gas molecules on interaction with target material may become neutral at the surface but due to their energy, incident ions may get implanted, get bounced back, create collision cascades in target atoms thus sputtering out target atoms/molecules.

Sputtering is carried out using Direct Current (DC) sputtering, Radio Frequency (RF) sputtering or magnetron sputtering. The mechanism of sputtering is using discharge or plasma of inert gas atoms or reactive gases to remove atoms or molecules physically from the target materials through the momentum transfer from ions into the target materials. The sputtering is carried out in a high vacuum or ultra-high vacuum system equipped with electrodes of which one of them is a sputter target and the other is a substrate, gas introduction (Figure 9.5).

Direct sputtering, in which the sputter target is held at a high negative terminal while the substrate is maintained at a positive potential and grounded. The required base pressure is <10 P in the chamber where argon gas is introduced. A visible glow is observed when electrodes are biased, is called plasma (a mixture of electrons, ions, neutral atoms and photons released in various collisions). When a sufficiently large

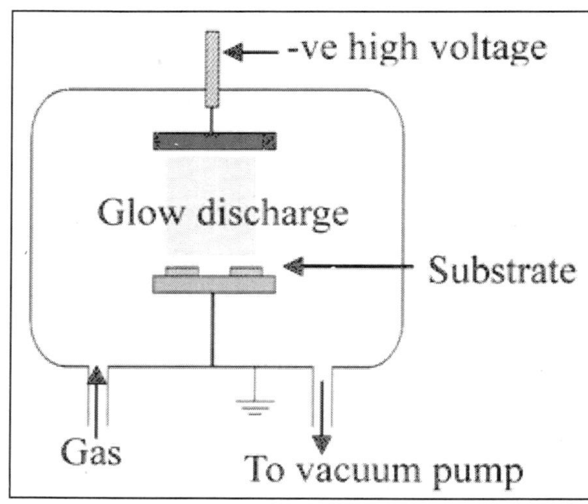

Figure 9.5. DC Sputtering for Thin Film Deposition.

number of ions is generated, this will be used to sputter off the target. Sputtering is effective in producing non-porous compact films and multilayer films for mirrors or magnetic films for spintronics applications. Virtually any materials are deposited by sputtering since the coating material is passed into the vapor phase by a mechanically rather than a chemical or thermal process. Argon is usually used as a sputtering gas. The nanoparticles synthesized through various physical methods are illustrated in Table 9.1.

High Energy Ball Milling

The reduction of particle size by high-energy ball milling is termed as mechanical milling which is a top-down approach. John Benjamin developed mechanical milling in 1970, who used the technique to synthesize oxide dispersion strengthened (ODS) Nickel based alloys to withstand high temperature and pressure. High-energy ball milling is one of the simplest ways of making nanoparticles of metals and alloys in the form of powder. The objective of milling is to reduce the particle size and blending of particles in new phases. The ball milling uses balls to impact upon the powder where balls fall freely and impact the powder.

Nanomaterials formed through a mechanical device, are referred to as a mill, where the energy is imparted to a coarse-grained material to effect a reduction in particle size. The process of mechanical attrition of bulk materials through mechanical devices is called milling. Coarse-grained materials (usually metals but also ceramics and polymers) in the form of powders, are crushed mechanically in rotating drums by hard steel or tungsten carbide balls, under controlled atmospheric condition Mechanical deformation under shear conditions and high strain rates leads to the formation of nanostructures, wherein energy is being continuously pumped into crystalline structures to create lattice defects.

The mechanism in mechanical milling is the ball-powder-ball collision, where powder particles are trapped between the colliding balls during milling and undergo deformation and/or fracture processes that define the ultimate structure of the powder. The series of steps during high energy ball milling are powder mixture placed in the ball mill is subjected to high-energy collision from the balls (milling media) and high energy milling causes the structural disintegration of the coarse-grained structure due to plastic deformation. High-energy milling consists of repeated deformation (welding, fracturing, and re-welding) of powder particles under a protective atmosphere. High-energy milling induces structural changes and chemical reactions at room temperature due to mechanical energy rather than thermal energy.

The first stage of ball milling starts with compaction followed by the rearrangement and restacking of particles, producing some fine and irregularly shaped particles. The second stage of compaction involves elastic and plastic deformation of particles. The third stage involves particle fracture, and resulting in deformation and/or fragmentation of the particles. The processes that define the milling are flattening of the elemental grains by plastic deformation followed by the welding of elemental grains. Then the formation of equilibrium size grains with parallel lamellae because of competition between welding and fracture processes occur followed by randomization of lamellae orientation because of repeated fracture and welding. Further refinement into grain microstructure takes place during the final stage of milling.

There are many types of mills such as planetary, vibratory, rod, tumbler mills, *etc.* Two types of ball mills, the attritor, and the horizontal ball mill are commonly used for the preparation of large quantities of material. The attritor consists of a stationary vertical cylindrical chamber, where the powder to be milled is placed along with the grinding medium (typically hardened steel balls). The powder and balls are agitated by rotating impellers. The attritors produce moderate quantities of material in moderate milling times. The horizontal ball mill consists of a large (typically 1 m diameter) chamber that is rotated on a horizontal axis at a frequency slightly less than the critical frequency at which the balls are pinned to the inner wall of the chamber. The horizontal ball mill produces large quantities but the milling times are quite long.

The planetary ball mill and the vibratory ball mill are used extensively for ball milling in the research environment. The planetary ball mill utilizes two (or more) small counter-rotating vials located near the perimeter of a turntable. The planetary ball mill owes its name to the planet-like movement of vials. The vials are arranged on a rotating support disk, and a special drive mechanism causes them to rotate around their own axes (rotary motion around the mill axis and a planetary motion around the vial axis). The centrifugal force produced by the vials rotating around their own axes and the rotating support disk both act on the vial contents, consisting of material to be ground and the grinding balls. The vials and the supporting disk rotate in opposite directions; the centrifugal forces alternately act in like and opposite directions. This causes the grinding balls to run down the inside of the vial causes the friction effect and followed by grinding of materials (Figure 9.6). The vibratory ball mill where the vibrations are experienced in all three directions for milling The most commonly used vibratory ball mill is the SPEX Model 8000.

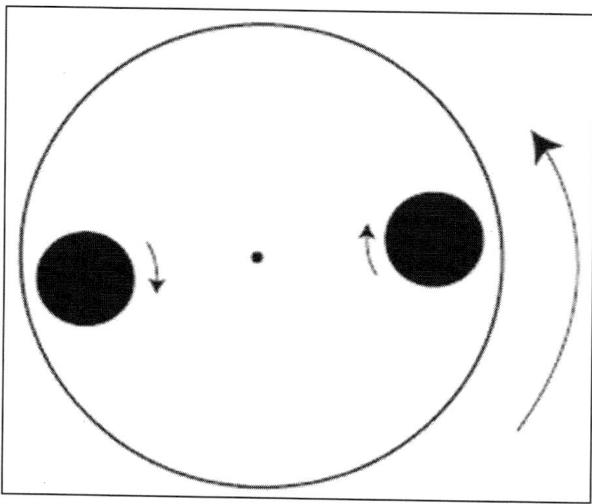

Figure 9.6. Illustration on Planetary Motion in Pulverisette Mill.

Table 9.2. Synthesis of Nanoparticles through High-energy Ball Milling

Sl. No	*Nature of Ball Mill*	*Precursor and Parameters for Ball Milling*	*Nature of Nanoparticles*	*Significance of Nanoparticles*	*Authors*
1.	High-Energy Ball Milling	Milling for 5 h, 15 h, 20 h, 25 h, and 30 h	Magnetite (Fe_3O_4) nanoparticles	22.4 nm, 25.4 nm, 19.9 nm, 23.5 nm, 20.5 nm respectively. As the duration of ball milling increases, the magnetic purity of nanoparticles is also improved	Ramli *et al.* (2019)
2.	High Energy Ball Milling (HEBM) (8000D mixer SPEX)	Ball powder ratio is 20:1 in argon atmosphere for 200 min (Tin and selenium in phase followed by $SnSe_2$ and graphite 100 min each)	Tin Di Selenide ($SnSe_2$) nanoparticles	5 nm $SnSe_2$/graphite nanosheet nanocomposites	Liu *et al.* (2019)
3	High energy planetary ball mill (Model DECO-PBM-H-0.4L)	Ruthenium chloride, Sodium hypophosphite and commercial carbon black milled with 20 Nos. of PTFE balls (6 mm in diameter), 600 rpm for 2h Pyrolyzed at 600ºC (for 2 h under an inert N_2 atmosphere.	Ruthenium Phosphide nanoparticles embedded over carbon	Exhibits impressive hydrogen evolution reaction activity with a small over potential of 36 mV at the current density of 10 mA cm^{-2} in 1 M KOH	Liu *et al.* (2019)
4	Wet ball milling	Indium oxide and tin oxide in 9:1 ration in wet milling with ammonium polyacrylate as dispersal agent and polyvinyl alcohol as binder in 200 rpm for 48h	Indium Tin Oxide (ITO) ceramic particles	High density of 7.135 g/cm^3 and low resistivity of 0.137mDINU123- cm at 1600 ºC in an oxygen atmosphere at the heating rate of 5 ºC/min for 8 h.	Liu *et al.* (2019)
5	Cryogenic ball milling (Union Process 01-HD Attritor)	Silver nitrate and sodium di hydrogen phosphate, urea and boric acid in liquid nitrogen (77K) 1000 rpm, 2 h Ball to powder ratio-20:1	Silver phosphide and boric composite nanocomposites	Average particle size was 87.5 nm	Liu *et al.*, 2019
6	Planetary Ball Mill (Pulverisette 6)	Titanium (IV) ethoxide, tetraethyl orthosilicate, aluminum isopropoxide as precursors, milling with 10 balls, for 240 minute at the milling velocity of 100 rpm	Homogeneous coating of titania, silica, alumina on the surface of Multiple walled carbon nano tubes	Uniform homogeneous coating over carbon nanotubes	Major *et al.* (2019)

Sl. No	*Nature of Ball Mill*	*Precursor and Parameters for Ball Milling*	*Nature of Nanoparticles*	*Significance of Nanoparticles*	*Authors*
7	Planetary ball mill instrument (Eloquent technologies, Bangalore, India).	2.5 g of graphene oxide with 2.5 g of iron powder in the chromium steel vial Ball-to-powder ratio at 20:1. Toluene was used as the medium for milling	Average crystallite size is 20 nm	At 10 h of milling appearance of Fe_3O_4 where at 25h, dominance of Fe_3O_4 (311) over Fe (110) phase, while at 40 h of milling complete disappearance of Fe phase	Narayana-swamy *et al.*,(2019)
8	High Energy Planetary ball mill (PBM Retsch PM 400)	Enriched wolframite ore of below 15μm with calcium carbonate (100:24) were homogenized Tungsten carbide of 700 g of 3.0mm using 2 propanol, the duration of milling for 8 h at 82 rpm	Scheelite (Calcium Tungsten Oxide) nanoparticles	Crystallite size of nanocomposite was 40nm	Polini *et al.* (2019)
9	High energy ball milling (Retsch 400 planetary ball mill)	Pure Aluminum powder with average particle size of 75 μm milled with MWCNTs of 10–12 nm under Argon atmosphere. Ball-to-powder weight ratio of 5:1 @ 400 rpm using ethyl alcohol as process control agent.	Aluminum-carbon nanotube nanocomposites	Irregular morphologies and rougher surfaces in nano composite. Compressive strength of 242.3 MPa and hardness values ranged from 0.42 GPa at the core reaching 1.0 GPa at the outer layer of the nanocomposite	Salama *et al.* (2019)
10	Ball Mill	MnO_2 to Fe @ 1:2 were milled with milling balls of 1.5 mm and balls to powder ratio of 100:1, Milling velocity @ 265 rpm. Using microwave power and frequency of 1 KW and 2450 MHz, respectively.	Ferrite manganese iron oxide nano composites	Diffraction peak of ferrite manganese iron oxide nano composites appeared at 35h of milling	Chen *et al.* (2013)
11	Planetary mill	Ball to powder mass ratio of 10:1 for 30 min	Titanium oxide nanoparticles	With sintering temperature of 400ºC, crystallite and grain size of nanoparticles were 33 and 40 respectively	Farbod and Khademal-rasool (2011)
12	Planetary ball mill (Fritsch P7)	Spray pyrolysis @500ºC followed by wet milling $LiFePO_4$ for 3 h at a ball-to-powder ratio of 40/0.5 followed by heat treatment at 600ºC for 4 h in a N_2 +3 per cent H_2 atmosphere	Carbon-coated Lithium Iron Phosphate nano composites	146 nm nanoparticle with amorphous carbon coating of 5–20nm	Konarova and Taniguchi (2010)
13	Conventional planetary ball milling	Graphite and Silica powders @1:1 in steel vials were milled with ball to powder ratio-10:1 for 240 and 168 h in argon atmosphere followed by annealing temperature of 150ºC	Silicon nanoparticle	Si core of 5nm covered with one nm of amorphous silicon dioxide layer	Lam *et al.* (2000)

The high specific gravity and larger balls give better milling because of the high impact forces on the powders. However, the balls should be denser than the material to be milled. The highest collision energy is obtained if balls of different diameters are used for milling. The use of same sized milling balls produces tracks and consequently, the balls roll along a well-defined trajectory instead of hitting the end surfaces randomly. The ball to powder ratio (BPR) or charge ratio ranges from values as low 1:1 to as high 220:1 and usually, it is 10:1 for planetary mills. The filling in the container is about 50 percent of the volume of the chamber. The milling atmosphere contaminates the powder and hence milling is carried out in argon, or helium charged milling chambers. Contamination is avoided by milling the powders with a milling media made up of the same material as that of the powders being milled.

When a mixture of powders (of different metals or alloys/compounds) are milled together, during which material transfer is involved to obtain a homogeneous alloy is mechanical alloying. Mechanical Milling refers to grinding of uniform composition powders, such as pure metals, intermetallics, or pre-alloyed powders, where the material transfer is not required for homogenization. Mechanical milling requires half the time required for mechanic alloying to achieve the same effect. Some of the metallic nanoparticles synthesized through ball milling are illustrated in Table 9.2.

Conclusion

Synthesis of nanomaterials is the important component of nanotechnology. Significant progress has been made in fabrication techniques of nanomaterials with required size and shape for diverse applications. The chapter has provided the overview of commonly employed physical strategies for the synthesis of nanomaterials. The material of interest is converted into gaseous phase in physical methods, which deposit as thin film over the substrate. The techniques help to achieve very thin film of monolayer of atoms or multilayers of similar or different materials staked over each other. The absence of solvent contamination and uniform distribution of nanoparticles are unique features of physical synthesis methods, while the mechanical synthesis of nanomaterials isscalable for bulk production but cross contamination is the limitation.

REFERENCES

1. Ali, N., Teixeira, J. A., Addali, A., Saeed, M., Al-Zubi, F., Sedaghat, A., and Bahzad, H. (2019). Deposition of Stainless Steel Thin Films: An Electron Beam Physical Vapour Deposition Approach. *Materials, 12*(4), 571.

2. Amendola, V., Polizzi, S., and Meneghetti, M. (2006). Laser ablation synthesis of gold nanoparticles in organic solvents. *The Journal of Physical Chemistry B, 110*(14), 7232-7237.

3. Chen, D., Zhang, Y., and Kang, Z. (2013). A low temperature synthesis of MnFe2O4 nanocrystals by microwave-assisted ball-milling. *Chemical Engineering Journal, 215*, 235-239.

4. Darvina, Y., Yulfriska, N., Rifai, H., Dwiridal, L., and Ramli, R. (2019, April). Synthesis of magnetite nanoparticles from iron sand by ball-milling. In *Journal of Physics: Conference Series* (Vol. 1185, No. 1, p. 012017). IOP Publishing.

5. Farbod, M., and Khademalrasool, M. (2011). Synthesis of TiO^2 nanoparticles by a combined sol–gel ball milling method and investigation of nanoparticle size effect on their photocatalytic activities. *Powder technology, 214*(3), 344-348.

6. Fazio, E., Spadaro, S., Santoro, M., Trusso, S., Lucotti, A., Tommasini, M. and Ossi, P. M. (2018). Synthesis by picosecond laser ablation of ligand-free Ag and Au nanoparticles for SERS applications. In *EPJ Web of Conferences* (Vol. 167, p. 05002). EDP Sciences.

7. Ganjali, M., Vahdatkhah, P., and Marashi, S. M. B. (2015). Synthesis of Ni nanoparticles by pulsed laser ablation method in liquid phase. *Procedia Materials Science, 11*, 359-363.

8. Konarova, M., and Taniguchi, I. (2010). Synthesis of carbon-coated LiFePO4 nanoparticles with high rate performance in lithium secondary batteries. *Journal of Power Sources, 195*(11), 3661-3667.

9. Kurapov, Y. A., Vazhnichaya, E. M., Litvin, S. E., Romanenko, S. M., Didikin, G. G., Devyatkina, T. A. and Oranskaya, E. I. (2019). Physical synthesis of iron oxide nanoparticles and their biological activity *in vivo*. *SN Applied Sciences, 1*(1), 102.

10. Lam, C., Zhang, Y. F., Tang, Y. H., Lee, C. S., Bello, I., and Lee, S. T. (2000). Large-scale synthesis of ultrafine Si nanoparticles by ball milling. *Journal of Crystal Growth, 220*(4), 466-470.

11. Liu, H., Wang, X., and Lan, Z. (2019). Cryogenic ball milling synthesis of Ag3PO4/h-BN nanoparticles with increased performance for photocatalytic oxygen evolution reaction. *Ceramics International.*

12. Liu, X., Guo, Y., Zhan, W., and Jin, T. L. (2019). Ball Milling-Assisted Synthesis of Ultrasmall Ruthenium Phosphide for Efficient Hydrogen Evolution Reaction. *Catalysts, 9*(3), 240.

13. Liu, Y., Shu, Y., Sun, B., Zeng, X., Zhu, J., Yi, J., and He, J. (2019). Electrochemical synthesis of In2O3 nanoparticles for fabricating ITO ceramics. *Ceramics International.*

14. Liu, Y., Xu, Y., Han, Y., Zhang, Z., Xu, J., Du, Y. and Zhou, X. (2019). Facile synthesis of SnSe2 nanoparticles supported on graphite nanosheets for improved sodium storage and hydrogen evolution. *Journal of Power Sources, 436*, 226860.

15. Lyu, S. C, Zhang, Y. and Lee C.J. (2003). Low-Temperature Growth of ZnO Nanowire Array by a Simple Physical Vapor-Deposition Method. *Chemistry of Materials,15*, 3294-3299

16. Major, J., Seo, J. W., Horváth, E., Kukovecz, Á., Forró, L., and Hernadi, K. (2019).

Electron Microscopy Investigation of Coated Multiwall Carbon Nanotubes Prepared by Reactive Ball Milling. *Journal of nanoscience and nanotechnology, 19*(1), 502-508.

17. Matsubara, K., Inoue, M., Hagiwara, H., and Abe, T. (2019). Photocatalytic water splitting over Pt-loaded TiO^2 (Pt/TiO^2) catalysts prepared by the polygonal barrel-sputtering method. *Applied Catalysis B: Environmental, 254*, 7-14.

18. Mihut, D. M., Afshar, A., Lackey, L. W., and Le, K. N. (2019). Antibacterial effectiveness of metallic nanoparticles deposited on water filter paper by magnetron sputtering. *Surface and Coatings Technology, 368*, 59-66.

19. Narayanaswamy, V., Obaidat, I. M., Kamzin, A. S., Latiyan, S., Jain, S., Kumar, H. and Issa, B. (2019). Synthesis of Graphene Oxide-Fe_3O^4 Based Nanocomposites Using the Mechanochemical Method and in Vitro Magnetic Hyperthermia. *International journal of molecular sciences, 20*(13), 3368.

20. Pandey, G., Rawtani, D., and Agrawal, Y. K. (2016). Aspects of nanoelectronics in materials development. In *Nanoelectronics and Materials Development*. IntechOpen.

21. Polini, R., Paci, B., Generosi, A., and Marcheselli, G. (2019). Synthesis of scheelite nanoparticles by mechanically assisted solid-state reaction of wolframite and calcium carbonate. *Minerals Engineering, 138*, 133-138.

22. Popov, A. A., Tselikov, G., Dumas, N., Berard, C., Metwally, K., Jones, N. and Da Silva, A. (2019). Laser-synthesized TiN nanoparticles as promising plasmonic alternative for biomedical applications. *Scientific reports, 9*(1), 1194.

23. Salama, E. I., Morad, S. S., and Esawi, A. M. (2019). Fabrication and mechanical properties of aluminum-carbon nanotube functionally graded cylinders. *Materialia, 7*, 100351.

24. Verma, M., Singh, K. P., and Kumar, A. (2019). Reactive magnetron sputtering based synthesis of WO^3 nanoparticles and their use for the photocatalytic degradation of dyes. *Solid State Sciences.*

25. Yang, S., and Zhang, J. (2019). Deposition of YBCO nanoparticles on graphene nanosheets by using matrix-assisted pulsed laser evaporation. *Optics and Laser Technology, 109*, 465-469.

Chapter 10

Chemical Reduction Approaches for the Synthesis of Nanomaterials

Pon. Sathya Moorthy

Department of Nano Science and Technology, Tamil Nadu Agricultural University, Coimbatore – 641 003, Tamil Nadu

Overview of Synthesis of Nanomaterials

Synthesis of nano material can be achieved using two different approaches namely (i) Top down approach and (ii) Bottom up approach as shown in the Figure 10.1.Top down approach means breaking down process *i.e.* bringing down the

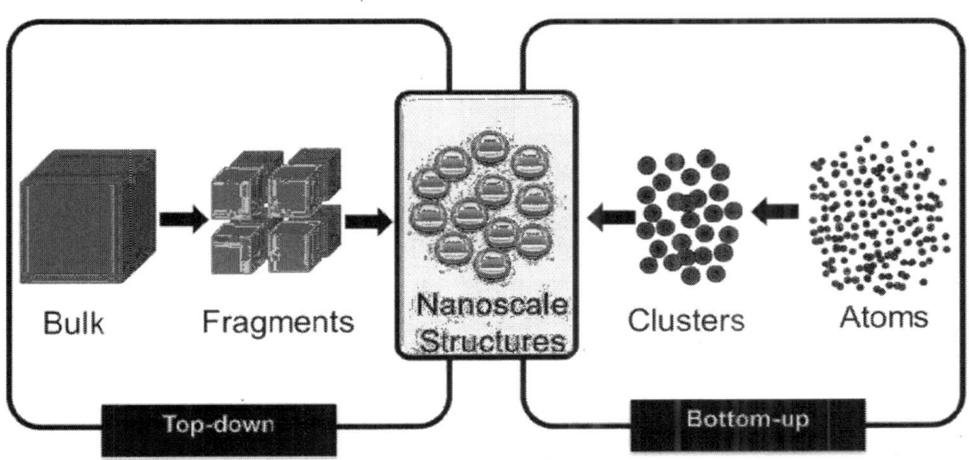

Figure 10.1. Top Down and Bottom Up Approach of Synthesis of Nano Particles.
Courtesy: R S Rawat 2015 J. Phys.: Conf. Ser. 591, 012021.

bulk materials to nanoscale either using the physical/mechanical route. Whereas, the bottom up approach means the building up process (*i.e.*) growth of sub nano particles to form a nano material either by chemical or biological route.

Various methods are available under physical, chemical and biological routes of synthesis of nanomaterials as shown in the Figure 10.2. Synthesis of nanomaterials is achieved (i) by employing external mechanical force and causing the material to scrape or wrap away and (ii) by applying the thermal and electrical energy to the bulk material and causing them to melt, evaporate and condense down to nanoscale is named as physical route of synthesis of nanomaterials. This method is a time consuming process. In the chemical route of synthesis of nanomaterials, the precursor is treated with the reducing agent to get reduced to nano scale. Stabilizing agents may be used in this method to prevent aggregation of synthesized nanomaterials and to stabilize its structure to extend its shelf life of the nanomaterials. Biological/

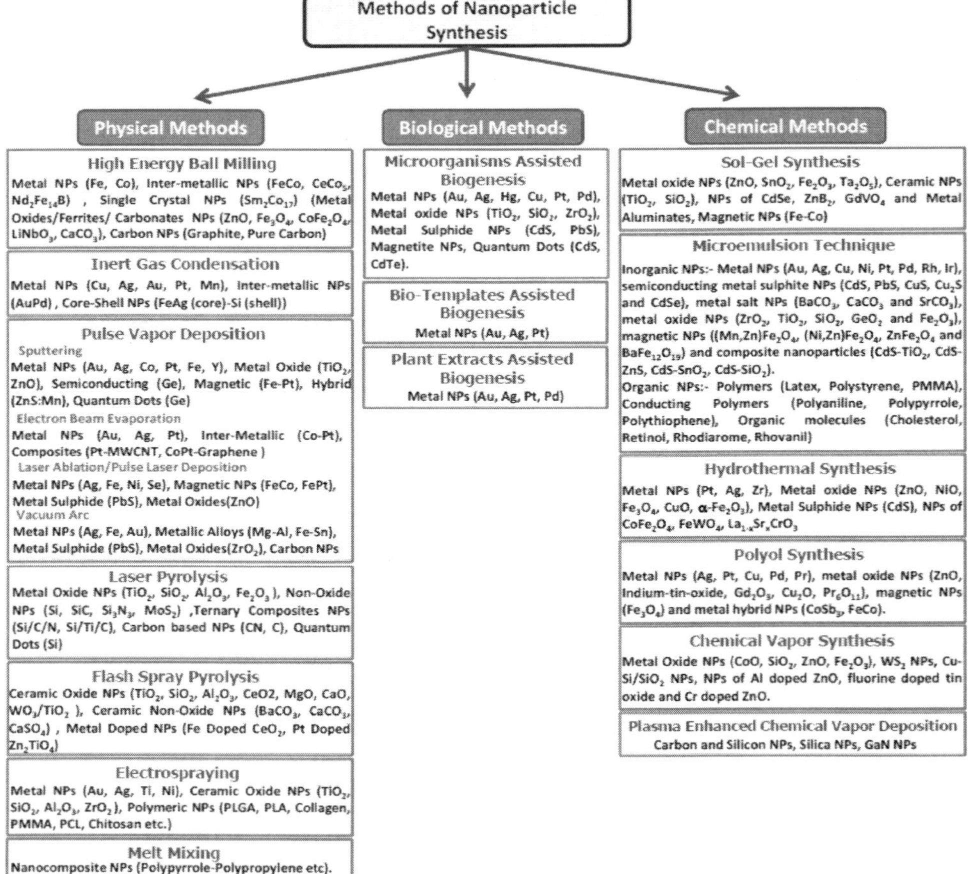

Figure 10.2. Overview on different Type of Synthesis Methods to Produce Variety of Nano Particles

Courtesy: Rsc Advances, 5(127), 105003-105037.

green/biogenic synthesis of nanomaterials is achieved by using microorganisms, plant extracts and biological molecules. The biological extracts are the cocktail of reducing, oxidizing and stabilizing agents. They act on the precursor materials in certain proportions to attain nanomaterials. Biological synthesis is an eco friendly route to synthesize nano material.

Chemical synthesis is a general route of synthesis of nanomaterials. In general, the reaction takes place in the solution form and the product formed are colloidal in nature. In chemical route of synthesis of nanomaterials, the precursor is allowed to react with reducing agent to get reduced to nano scale as shown in the Figure 10.3. Reduction, nucleation, particle growth and/or agglomeration of the particle are the critical steps involved in chemical synthesis.

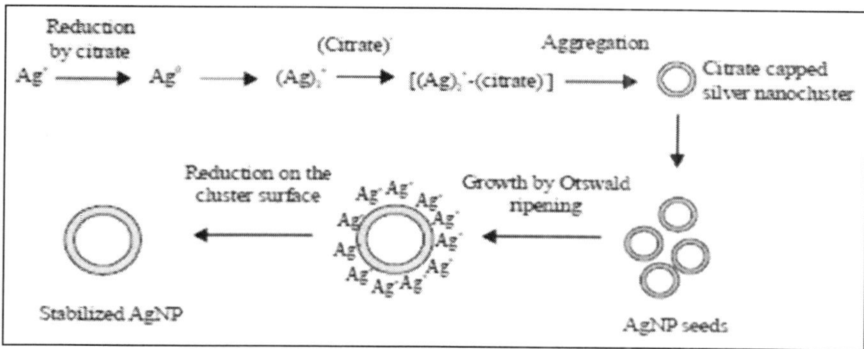

Figure 10.3. Mechanism of Silver Nitrate Reduction using Sodium Citrate as Reducing Agent.
Courtesy: Asi. J. Sc. Res.Vol 12 (1): 1-17, 2019.

Chemical route of synthesis of nanomaterials can be classified into two main categories namely (i) reduction methods and (ii) sedimentation methods. Reduction method can be further classified into two types namely (i) chemical reduction and (ii) Indirect reduction methods as shown in the Figure 10.4.

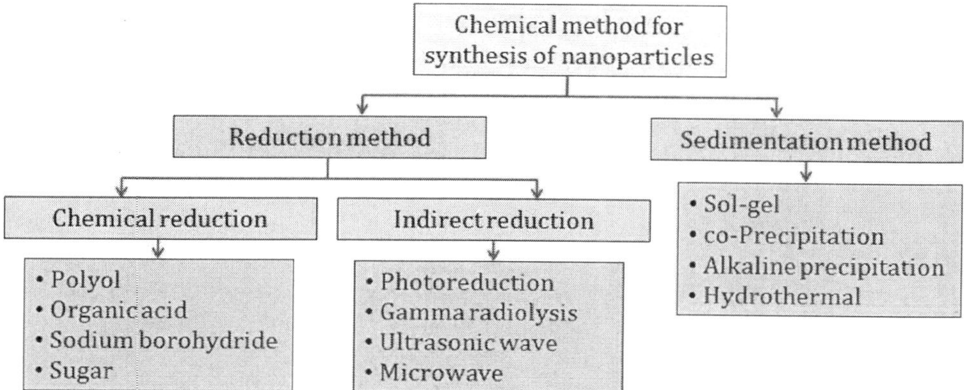

Figure 10.4. Various Chemical Methods of Synthesis of Nanomaterials.
Courtesy: https://www.researchgate.net/publication/317600016

SOL-GEL

Sol-Gel is a wet chemical technique. This technique also called as chemical solution deposition. Following critical steps are involved during the process namely, hydrolysis, polycondensation, gelation, aging, drying, densification and crystallization as seen in the Figure 10.5.

Hydrolysis :

$$\geqslant Si-OR + H_2O \xrightarrow{H+ \text{ or } OH-} \geqslant Si-OH + ROH,\ R = \text{alkyl group}$$

polycondensation :

$$\geqslant Si-OH + OH-Si \leqslant \xrightarrow{H+ \text{ or } OH-} \geqslant Si-O-Si \leqslant + H_2O$$

and/or

$$\geqslant Si-OH + RO-Si \leqslant \xrightarrow{H+ \text{ or } OH-} \geqslant Si-O-Si \leqslant + ROH$$

Figure 10.5. Mechanism Involved in the Synthesis of Nano Particles
Courtesy: Nanoscale *Res Lett* 10, 87 (2015).

SOL and GEL are the two different components involved in this process. SOL is termed as a colloidal suspension of solid particles in a solution whereas, GEL is a polymer containing liquid. This method is well suitable for the synthesis of metal oxide, metal aluminates, ceramic and magnetic nano particles. Metal alkoxide on treatment with water undergoes hydrolysis followed by condensation leading to the formation of gel phase. Excess of water molecules is removed from the gel phase to form the final nano structured materials as shown in the Figure 10.6.

Hydrothermal Synthesis

Hydrothermal synthesis method is used to fabricate metal oxide nano particles keeping control over the characteristic of particles by varying the properties of near or super critical water by using different temperature and pressure conditions. This can be achieved using the either batch hydrothermal of continuous hydro thermal process. In a chemical solution, nano particles are produced from a colloidal system that consist of two or more phases (Solid, liquid or gas states) of matter mixed together under controlled temperature and pressure is shown in the Figure 10.7.

Hydrothermal synthesis of nano particles is an affordable, less time consuming process that enables the synthesis of huge amount of nano particles with optimized size, morphology, composition and surface chemistry that is rationally inexpensive. This method is a facile and fast process synthesizing various other nano particles like Zr, ZnO, Ag, CdSect.

Polyol Synthesis

Poly ethylene glycol (PEG) is used as an reaction medium for the synthesis

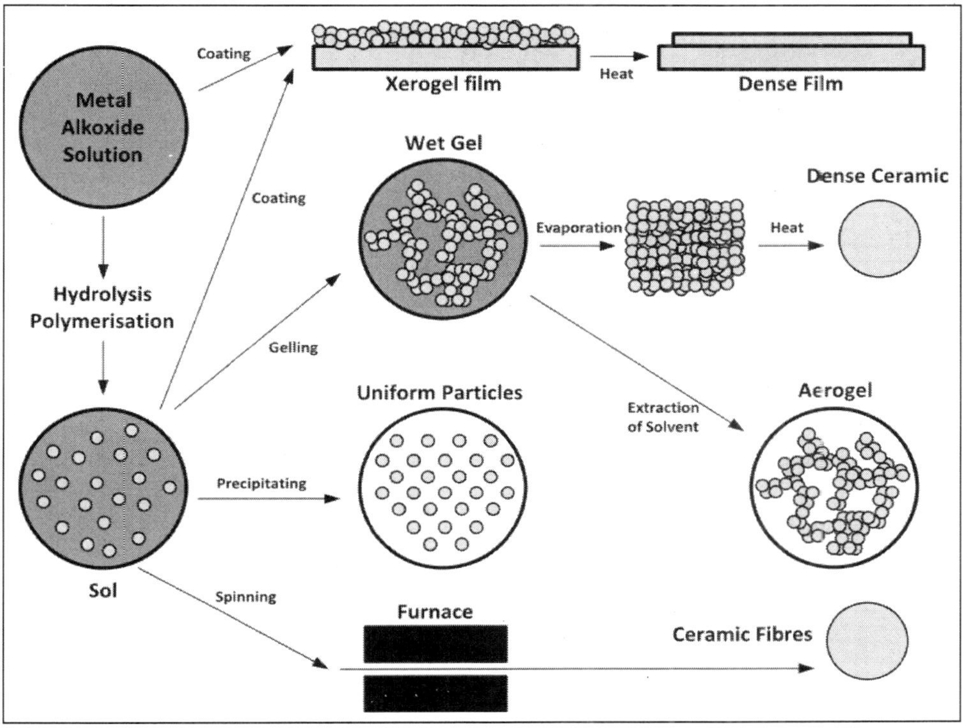

Figure 10.6. Schematic Representation of the different Stages and Routes of the Sol-gel Technology.
Courtesy: https://en.wikipedia.org/wiki/Sol%E2%80%93gel_process.

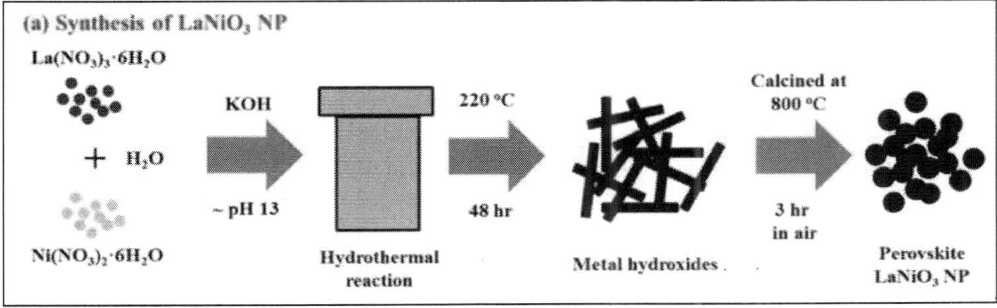

Figure 10.7. Schematic Representation of Hydrothermal Synthesis of $LaNiO_3$ Nano Particle.
Courtesy: J. Mater. Chem. A, 2015, 3, 10693-10697.

of metal nano particles. PEG serves as a solvent, complexing and reducing agent simultaneously a stabilizing agent. Generally this method is preferable for the synthesis of metal, metal oxides, magnetic as well as hybrid nano particles.Using this method tailor made synthesis of nano particles with definite size, shape, morphology and crystallinity of nano particles can be achieved (Moon *et al.*). Highly monodispersive spherical Ag nanoparticle with controlled size is established using the polyol method. Morphological evolution of iron oxide particles using polyol method using two different polyols method is shown in Figure 10.8. Different polyols like ethylene glycol as well as propylene glycol has various degrees of reducing capabilities to alter the morphology and size of the nano particles. Propylene has higher degree of reductive ability since faster nuclei formation and growth rate of nano particle like Fe_3O_4 is observed with generation of more misaligned mesocrystals due to ostwald ripening. Nanoparticle formation is quite slow while using ethylene glycol results in the formation of self assembled structures. One pot facile chemical synthesis can also be achieved using the method for synthesis of $CoSb_3$, FeCo, *etc.* (226-227)

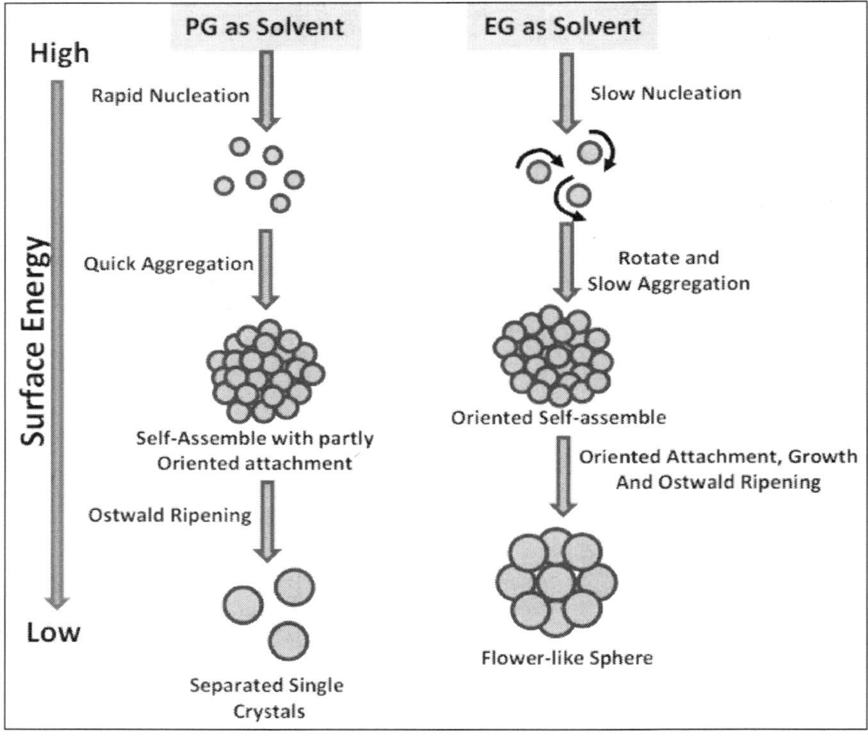

Figure 10.8. Morphology Evolution of Iron Oxide Nano Particles in Polyol Process using Two different Polyols.
Courtesy: RSC Advances, 5(127), 105003-105037.

Microemulsion Technique

The term microemulsion was first assigned by schulman in 1959. Microemulsion technique is defined as the thermally stable, macroscopically homogenous, optically transparent and isotropic dispersions constituting minimum of three components namely polar phase *i.e.* water, non-polar phase *i.e.*oil, and surfactant. Surfactant molecules creates an interfacial layer between water and oil phase reducing interfacial tension between the microemulsion and the excess phase act as a steric barrier preventing the coalescence of the droplets. Microemulsion system consists of monodispersed spherical droplet of size ranging from 600nm -8um of oil in water or water in oil depending upon the nature of surfactant used. Polar head group of the surfactant are generally hydrophilic in nature whereas the tails are hydrophobic. The formation of water in oil micro emulsion is called as reverse micelle where the polar head group creating the aqueous core and resides towards inside whereas the organic tails of the surfactants molecules directed towards outside as shown in Figure 10.9. Microemulsion technique is most commonly used for the synthesis of the inorganic metal nanoparticles (Au, Pt, Pd *etc*), semiconducting metal sulphite nano particles like (Cds, Pbs, Cus, CdSe *etc.,*), metal salt nano particles ($BaCO_3$, $CaCO_3$, *etc.,*), metal oxide nano particles (ZrO_2, TiO_2, Sio_2, *etc.,*), magnetic nano particles (($(Mn,Zn)Fe_2O_4$, (Ni,Zn) Fe_2O_4, $ZnFe_2O_4$ *etc.,*). Metal nano particles are easily prepared using micro-emulsion techniques using the reduction strategy. The growth of the nano particles with different shapes and size are found to be linked with surfactant to metal ion molar ratio.

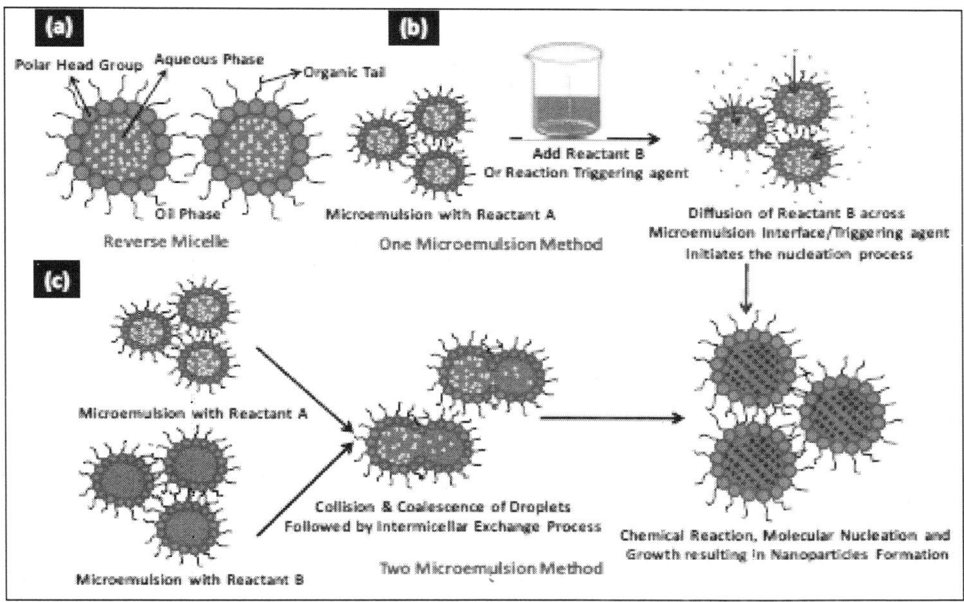

Figure 10.9. (a) Typical Reverse Micelle System, (b) Various steps involved in one microemulsion process and (c) Reaction sequence involved in the two micro emulsion nanoparticle synthesis.

Courtesy: RSC Advances, 5(127), 105003-105037.

Sono Chemical Synthesis

Ultra sound irradiations are used in the synthesis of nanoparticles. Sonochemical synthesis is advantageous method due to its non-hazardous nature, rapid reaction rates and production of very small nanoparticles. Ultrasonic wave irradiation inside the solution causes acoustic cavitations resulting in formation, growth and implosive collapse of bubbles inside the solution is noted as primary sonochemistry. Implosive collapse of the bubbles generates localized hotspots with extreme high temperature and pressure causing increase in rapid reaction rates which results in formation of nanoparticles known as secondary sonochemistry. Schematic representation of acoustic cavitation is shown in Figure 10.10. Synthesis of TiO_2 nanoparticles by sonochemical synthesis is schematically shown in the Figure 10.11.

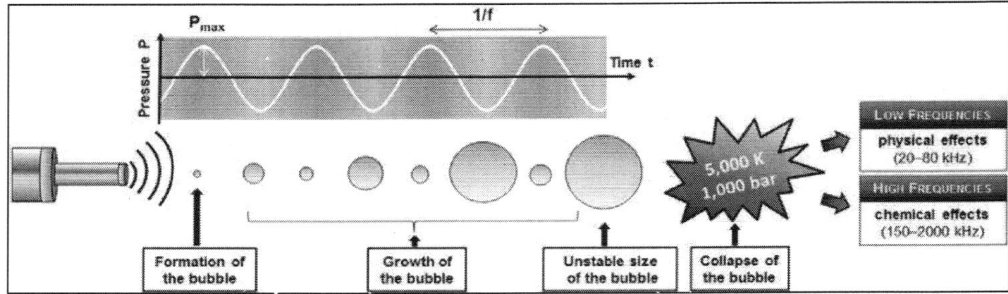

Figure 10.10 Acoustic Cavitation Showing the Formation, Growth and Implosion of Nano Particles Inside the Solution Due to Ultrasonic Irradiation.
Courtesy: Applied Clay Science 119 (2016) 193–201

TIP: Titanium isopropoxide ● Titanium Oxide Sol Particle **UIA**: Ultrasound Induced Agglomeratic

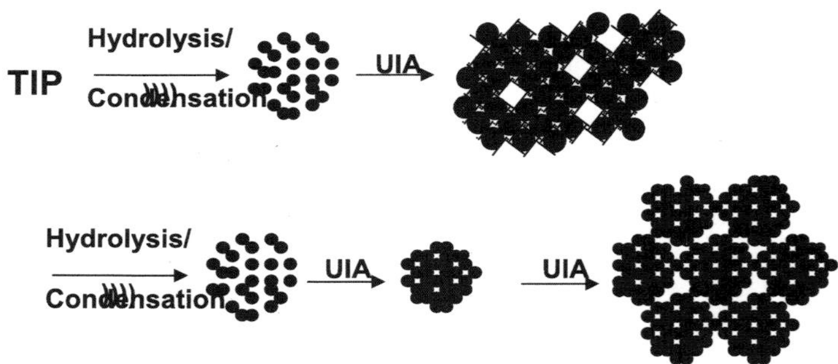

Figure 10.11. Synthesis of TiO_2 Nanoparticles by Sonochemical Synthesis.
Courtesy: Yu J. C. *et al.*, Chem. Commun. 2003, 2078.

Microwave Assisted Synthesis

Microwaves are a form of electromagnetic energy with frequencies in the range of 300 MHz to 300 GHz. The commonly used frequency is 2.45 GHz. Interaction between the precursor materials and the oscillating electric field of the microwave is used in the synthesis of nanoparticles. Solvents are heated much higher than their normal boiling point without vaporizing. Interactions between materials and microwaves are based on two specific mechanisms: dipole interactions and ionic conduction. Both mechanisms require effective coupling between components of the target material and the rapidly oscillating electrical field of the microwaves. Dipole interactions occur with polar molecules. The polar ends of a molecule tend to re-orientate themselves and oscillate in step with the oscillating electrical field of the microwaves. Heat is generated by molecular collision and friction. Generally, the more polar a molecule, the more effectively it will couple with the microwave field. Microwave assisted synthesis causes rapid rise in temperature of the solution to the reaction temperature causing no cross reaction, whereas in the conventional method temperature of the solution happens gradually and the possibilities of cross reaction ant lower temperature may be possible causing non homogeneity in size and shape of nanoparticles. Synthesis of zinc nanocluster using microwave assisted method is schematically represented in the Figure 10.12.

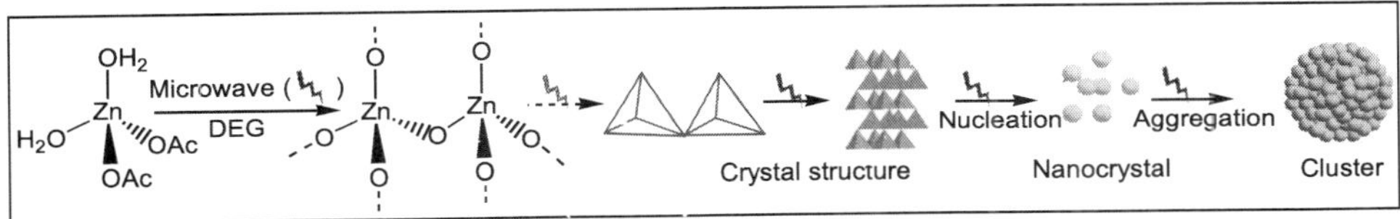

Figure 10.12. Synthesis of Zinc Nano Clusters using Microwave Assisted Synthesis.
Courtesy: Yu, J. C. *et al.*, Adv. Mater. 2008.

Co-precipitation Technique

Coprecipitation reactions involve the simultaneous occurrence of nucleation, growth, coarsening, and/or agglomeration processes. Coprecipitation reactions exhibit the following characteristics:

- ✰ The products are generally insoluble species formed under conditions of high supersaturation.
- ✰ Nucleation is a key step, and a large number of small particles will be formed.
- ✰ Secondary processes, such as Ostwald ripening and aggregation, dramatically affect the size, morphology, and properties of the products.
- ✰ The supersaturation conditions necessary to induce precipitation are usually the result of a chemical reaction.

Coprecipitation of the nano particles are achieved using (i) metals formed from aqueous solutions, by reduction from nonaqueous solutions, electrochemical reduction, and decomposition of metal organic precursors; (ii) oxides formed from aqueous and nonaqueous solutions; (iii) metal chalconides formed by reactions of molecular precursors; (iv) microwave/sonication-assisted co precipitation. Diagrammatic representation of monodisperse nano crystals is shown in Figure 10.13.

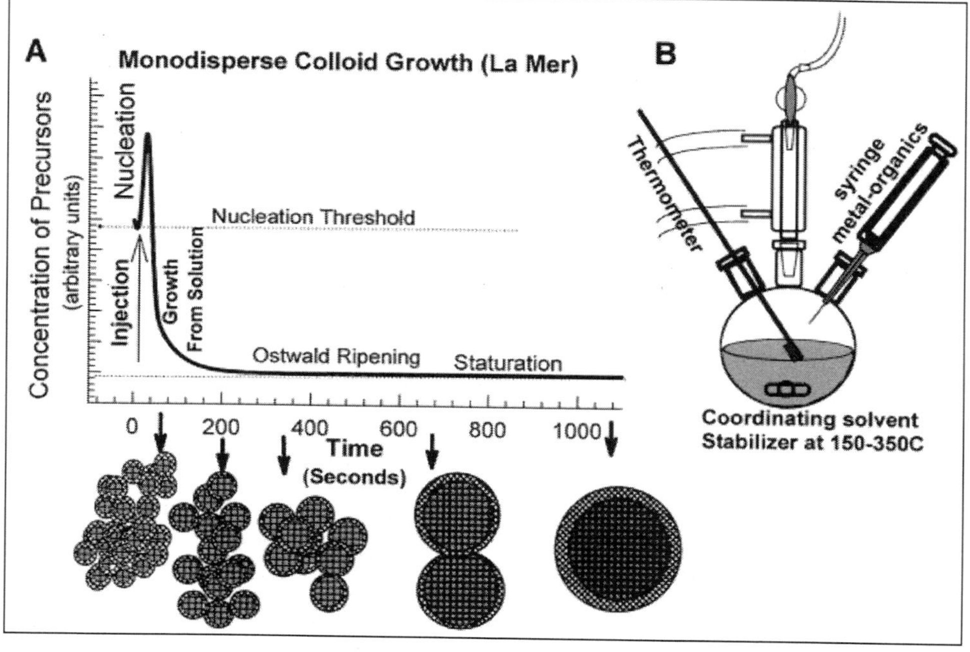

Figure 10.13. A. Stage of nucleation and growth for the preparation of monodispersednano crystals. B. simple synthetic apparatus employed in preparation of mono diapersenanocrystals samples.
Courtesy: Murray C.B. *et al.*, Annu. Rev. Mater. Sci. 2000, 30, 545.

Chemical Vapor Deposition

Chemical vapor deposition (CVD) is a widely used method for depositing thin films of a large variety of materials. Growing thin layer of advanced materials on the surface of the substrate. The process is often used in the semiconductor industry to produce thin films. Chemical vapour deposition (CVD) is a process where one or more volatile precursors are transported via the vapour phase to the reaction chamber, where they decompose on a heated substrate". Many materials may be deposited using CVD and related techniques. Metals, oxides, sulfides, nitrides, phosphides, arsenides, carbides, borides, silicides. Critical steps involved in the chemical vapour deposition are:

1. Transport of the reagents (*e.g.* $TiCl_4$, BCl_3, H_2) in the gas phase (often with carrier gas) to the reaction zone
2. Diffusion (or convection) through the boundary layer
3. Adsorption of precursors on the substrate
4. Surface diffusion of the precursors to growth sites. Reaction without diffusion is not wanted, as this may lead to rough growth surface.
5. Surface chemical reaction, formation of a solid film and formation of byproducts.

6. Desorption of by-products
7. Transport of gaseous by-products out of the reactor. Reactions in the gas-phase is most often unwanted.

In a typical CVD process, reactant gases (precursor) at room temperature enter the reaction chamber. The gas mixture is heated as it approaches the deposition surface The reactant gases undergo homogeneous chemical reactions in the vapor phase before striking the surface forming the deposited material. Frequently, volatile by-products are also produced, which are removed by gas flow through the reaction chamber. Schematic representation of typical CVD reactor is shown in the Figure 10.14. CVD was first used for hard coatings (cutting tools *etc.*) Microelectronics, 3D-structures Glass (SnO_2, TiN, SiO_2, TiO_2) Solar cells, catalysis, membranes, waveguides, mirrors, synthetic gold *etc.*

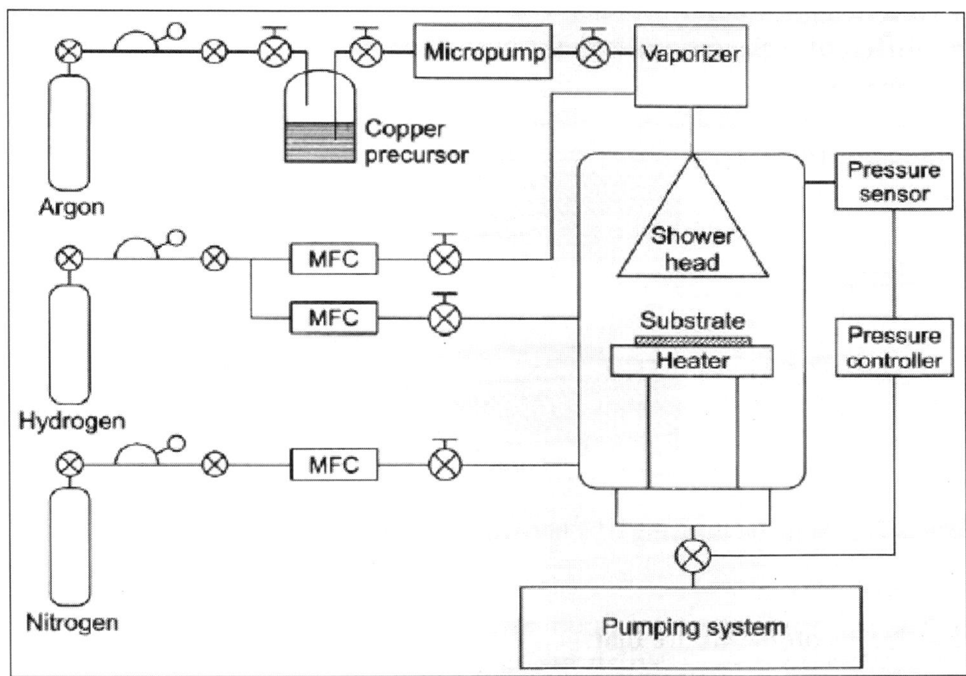

Figure 10.14. Schematic Representation of CVD Reactor.

CVD can be further classified into different categories based on the energy sources Figure 10.15.

Advantages of Chemical Route of Synthesis of Nanomaterials

1. Production of large quantities within short period is possible.
2. Control over the size, shape, crystallinity and dimension of the nanomaterials are possible by adjusting the reactant concentration, reaction time, reaction temperature, pH. pressure ect.

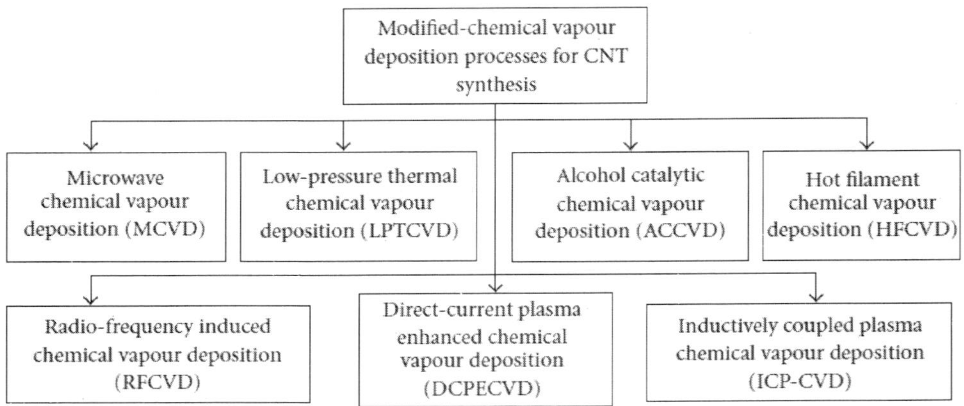

Figure 10.15. Classification of CVD.

3. The synthesized material can be stabilized by adding appropriate stabilizing agent during the chemical reaction to prevent agglomeration.

REFERENCES

1. Budnyak, T. M., Pylypchuk, I. V., Tertykh, V. A., Yanovska, E. S., and Kolodynska, D. (2015). Synthesis and adsorption properties of chitosan-silica nanocomposite prepared by sol-gel method. *Nanoscale research letters*,10(1), 87.
2. Chatel, G., Novikova, L., and Petit, S. (2016). How efficiently combine sonochemistry and clay science?.*Applied Clay Science*, 119, 193-201.
3. Cheng, C., Xu, F., and Gu, H. (2011). Facile synthesis and morphology evolution of magnetic iron oxide nanoparticles in different polyol processes. *New Journal of Chemistry*, 35(5), 1072-1079
4. Dhand, C., Dwivedi, N., Loh, X. J., Ying, A. N. J., Verma, N. K., Beuerman, R. W. and Ramakrishna, S. (2015). Methods and strategies for the synthesis of diverse nanoparticles and their applications: a comprehensive overview. *Rsc Advances*, 5(127), 105003-105037.
5. Handoko, C. T., Huda, A., and Gulo, F. (2019). Synthesis pathway and powerful antimicrobial properties of silver nanoparticle: A critical review. *Asian J. Sci. Res*, *12*, 1-17.
6. Hu, X., Gong, J., Zhang, L., and Yu, J. C. (2008). Continuous Size Tuning of Monodisperse ZnO Colloidal Nanocrystal Clusters by a Microwave Polyol Process and Their Application for Humidity Sensing. *Advanced Materials*, 20(24), 4845-4850.
7. Murray, C. B., Kagan, A. C., and Bawendi, M. G. (2000). Synthesis and characterization of monodispersenanocrystals and close-packed nanocrystal assemblies. *Annual review of materials science*, 30(1), 545-610.

8. Rawat, R. S. (2015). Dense plasma focus-from alternative fusion source to versatile high energy density plasma source for plasma nanotechnology. *In Journal of Physics: Conference Series* (Vol. 591, No. 1, p. 012021). IOP Publishing.
9. Ren, X., Pang, L., Zhang, Y., Ren, X., Fan, H., and Liu, S. F. (2015). One-step hydrothermal synthesis of monolayer MoS 2 quantum dots for highly efficient electrocatalytic hydrogen evolution. *Journal of Materials Chemistry A*, 3(20), 10693-10697.
10. Zhang, L., and Jimmy, C. Y. (2003). A sonochemical approach to hierarchical porous titania spheres with enhanced photocatalytic activity. *Chemical Communications*, (16), 2078-2079.
11. https: //www.researchgate.net/publication/317600016
12. https: //en.wikipedia.org/wiki/Sol per cent E2 per cent 80 per cent 93gel_ process.

Chapter 11

Green Synthesis of Nanoparticlares Using Biological Reductants

Badal Kumar Biswal, Paladugu Krishnadev and A. Lakshmanan

Department of Nano Science and Technology, Tamil Nadu Agricultural University, Coimbatore – 641 003, Tamil Nadu

Introduction

Nanotechnology, especially nanoparticles have various applications in different fields of science such as engineering, medicine, pharmaceutical and agriculture. According to EU standards, the particle size should be less than 100 nm to be said as nanoparticles. Due to high surface area to volume ratio, novel magnetic, optoelectronic and physiochemical properties nanoparticles are of great interest in number of applications ranging from sensors, electronics and medical field. For example, gold nanoparticles have been extensively used in pharmaceuticals and disease diagnostics. Similarly, silver nanoparticles have both anti-microbial and anti-inflammatory properties which can helps to promote faster wound healing. Some other nanoparticles such as platinum, palladium are used as biomedical applications and sensors applications, respectively. In addition, non-noble metallic nanoparticles such as iron, copper, zinc oxide, and selenium have also been used in medical treatments, cosmetic formulations, and anti-bacterial applications.

Traditionally, nanoparticles have been synthesized by either physical or chemical approaches. The physical approach includes laser ablation, lithography, high energy irradiation while chemical approach includes chemical reduction, electrochemistry and photochemical reduction. The size shape, stability, and physicochemical properties of the nanoparticles are strongly affected by various factors such as temperature, concentration of the precursor material and process kinetics involving the interplay between the precursors and reductant during the synthesis process. Furthermore, the traditional synthesis methods often use corrosive toxic chemicals with potential hazards which might induce carcinogenicity, cytotoxicity and environmental toxicity. Hence, there is a growing interest towards developing a clean, reliable, biologically compatible and eco-friendly green processes to synthesize nanoparticles with potentially controlled size and shape. Synthesis of different nanoparticles by using various biological entity (reductant) and the chemistry behind precision bio-reduction are discussed in this chapter.

Biological Synthesis of Nanoparticles

Biosynthesis involves using an environment-friendly green chemistry-based approach that employs unicellular and multicellular biological entities such as actinomycetes, bacteria, fungus, plants, viruses, and yeast. Synthesising nanoparticles via biological entities acting as biological factories offers a clean, nontoxic and environment-friendly method of synthesizing nanoparticles with a wide range of sizes, shapes, compositions, and physico-chemical properties (Mohanpuria *et al.*, 2008). According to the study biological organism or parts of them have the ability to act as a template for nanoparticle synthesis. The biological way of nanoparticle synthesis is carried out by plants, animals and microbes which include bacteria, fungi, viruses and yeasts *etc.* As microorganism-based nanoparticle synthesis needs a multistep process such as isolation, culture preparation and culture maintenance *etc.*, plant-based approach is more advantageous. Also, synthesis by plant tends to be faster, cost-effective and relatively easy to scale up the production (Iravani, 2011).

Both microorganisms and plants have showed the ability to absorb and accumulate metallic ions from environmental and industrial wastes transforming them to metal nanoparticles through various biochemical reactions through oxido-reduction (Baker *et al.*, 2013). These property of biological entities gives us scope to apply on various biotechnological applications such as bioremediation and bioleaching (Stephen and Macnaughtont, 1999).

Mechanism of Biological Synthesis

The mechanism of synthesis of nanoparticles via biological entities is not yet clearly understood. Microbial bodies as well as plants contain many microbial exudates and phytochemicals which are used in reducing metal ions to metal nanoparticles. Biological synthesis of nanoparticles occurs in 3 phases: (1) activation phase, (2) growth phase and (3) termination phase. During activation phase, the metal ions reduced to zero oxidation state and self-organization of metal atoms started through nucleation process. The second phase involves the growth of the newly formed structure and further reduction of the metal ions with an increase in the thermodynamic stability of the formed nanoparticles. Finally, the proteins and enzymes in the extracts act as a capping agent and stabilize the growth of nanoparticles to give proper shape and size (Figure 11.1)

Microbial Biosynthesis of Nanoparticles

Microorganisms of both unicellular and multicellular have the ability to

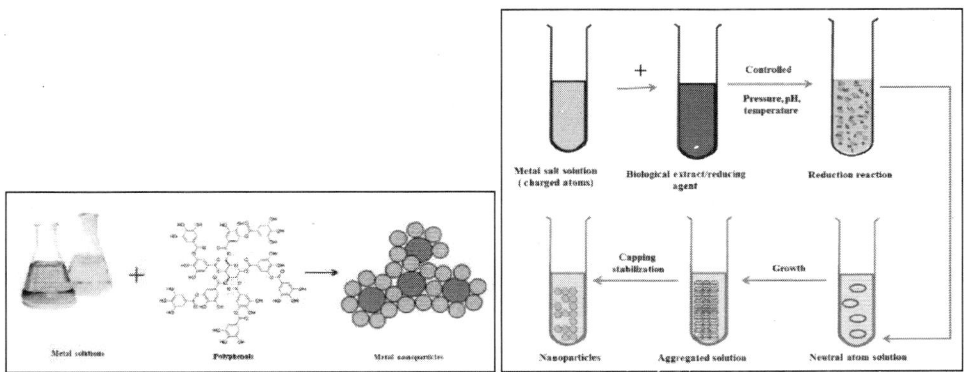

Figure 11.1. Process of Synthesis of Metal Nanoparticles from Plant Extract
Courtesy: Devatha and Thala, 2018.

synthesize nanoparticles via bottom-up approach by oxidation and reduction of metallic ions using various biomolecules such as enzymes, sugars and proteins secreted by them (Prabhu and Poulose, 2012). However, the detailed mechanism regarding nanoparticle production is yet to be studied. And also, the production mechanism of specific nanoparticle varies depending upon the interaction between the metallic ions and the microorganisms. Further, the environmental factors such as pH and temperature also affect the nanoparticles size and shape during biological synthesis (Makarov *et al.*, 2014). Microbial synthesis of nanoparticles can be of two types such as extracellular and intracellular depending upon the routes of synthesis (Arun *et al.*, 2013). The nanoparticles synthesis by microorganisms are discussed in detail in the following sections.

Actinomycetes

Actinomycetes are a group of microorganisms in between bacteria and fungi, carry out both intracellular and extracellular routes of nanoparticle synthesis. *Rhodococcus* sp. absorb Au ions intracellularly to reduce it to Au nanoparticles. The Au nanoparticles were predominantly observed at cell membrane and cell wall but not in cytosol. The results suggest that enzymes released from cell membrane and cell wall were responsible for the formation of nanoparticle while capping proteins stabilize the formed nanoparticles. The formed Au nanoparticle size ranges from 5 to 15 nm and spherical in shape which is found to be non-toxic to cells (Abdeen *et al.*, 2014). Similarly, reports confirmed that Au and Ag ions successfully reduced by intracellular enzymes secreted by cell wall to Ag/Au seeds (monomers) which consequently grown to nanoparticles (Chauhan *et al.*, 2013). Silver nitrate ($AgNO_3$) was reduced extracellularly using Streptomyces sp. LK-3 to Ag nanoparticles as reported by Karthik *et al.* (2014). As we know, nitrate reductase (NiR) is generally involved in nitrogen cycle in cells which reduce nitrate to nitrite (Korbekandi *et al.*, 2009). The study indicates that Nicotinamide adenine dinucleotide (NADH) dependent nitrate reductase enzyme is responsible for the conversion of Ag ions to Ag nanoparticle via electron transfer mechanism (Shah *et al.*, 2012). In another study, an aqueous solution containing gold chloride ($AuCl_4$) was reduced to Au_0 and subsequently to metallic Au due to electron transfer mechanism by NADH (He *et al.*, 2007). Microbially synthesized nanoparticles showed better antimicrobial activity as compared to chemically synthesized nanoparticles. This might be due to synergistic effect of stabilising proteins attached to nanoparticles during the synthesis process (Kumar *et al.*, 2013).

Algae

Algae are aquatic microorganisms that have ability for photosynthesis. Some studies suggest that algae not only help in formation of nanoparticles from the precursor metal ions, but also it accumulates and produces metal nanoparticles intracellularly. *Chlorella vulgaris*, a dried unicellular algae, synthesize tetra-chloroaurate ions and subsequently reduce it to gold nanoparticles which were accumulated near cell surface with tetrahedral, decahedral and icosahedral shape (Luangpipat *et al.*, 2011). *Sargassum wightii*, a marine algae, was capable of synthesizing Au, Ag and Au/Ag bimetallic nanoparticles (Govindaraju *et al.*, 2009). Similarly, *Kappaphycus alvarezii* and *Fucus vesiculosus* were also able to produce gold nanoparticles (Mata *et al.*, 2009; Rajasulochana *et al.*, 2010) whereas red *Chondrus crispus* and *Spyrogyra insignis* are able to synthesize both Au and Ag nanoparticles (Senapati *et al.*, 2012). In a recent study, Ramamoorthy *et al.* (2019) reported that silver nanoparticles of 9-80 nm with spherical shape synthesized from red seaweed. It showed antifungal activity against *B. bassiana* and *M. anisopliae* for 100µl of silver nanoparticles synthesized from red seaweed.

Bacteria

Bacteria have the better survivability than any microorganisms under extreme conditions. The natural defence system in bacteria deals with the toxicity due to high metallic ions include changes in metal ion concentration by redox state changes, efflux system, intracellular precipitation, accumulation of metals and extracellular formation of complexes (Dhillon *et al.*, 2012). The major bacterial species used for the synthesis of metallic nanoparticles include *Acitinobactor* sp., *Escherichia coli, Klebsiella pneumonia, Lactobacillus* spp., *Bacillus cereus, Corynebacterium* sp. and *pseudomonas* sp. Bacteria use extracellular or intracellular mode of routes for nanoparticle synthesis (Nvkv Prasad *et al.*, 2011). For example, silver (Ag) nanoparticles have been formed using *Pseudomonas stutzeri* AG259 as a biological reductant. The NADH which is essential for the reduction of Ag ions get oxidised to form NAD^+. In this process it donates an electron to cytosol which can be readily transferred to Ag ions to form metallic silver and subsequently it grows into silver nanoparticles (Ahmad *et al.*, 2003). In a similar mechanism, *Pseudomonas aeruginosa* able to produce gold nanoparticles (Husseiny *et al.*, 2007). Moreover, without involvement of bacterial enzymes, the nanoparticles can be synthesized using dead bacterial cells. Dried cells of *Bacillus megaterium* and *Corynebacterium* sp. are used in the production of Au nanoparticles (Sneha *et al.*, 2010). The reduction of metallic precursors might have happened due to interaction with the functional groups present on the cell wall at right pH and temperature (Jin-Kun *et al.*, 2001). Metal nanoparticles such as silver NP can be synthesized with different shape and sizes by altering the bacterial species, pH, temperature and culture media (Hulkoti and Taranath, 2014). For an example, dilute concentrations of gold chloride produce spherical Au nanoparticles ranging 10 to 20 nm whereas increasing the salt concentration gives nanowires at pH 6.0 using *Rhodopseudomonas capsulate*. Further, at pH 4.0 it was observed to produce both spherical and triangular plates (He *et al.*, 2007). This studies clearly indicated that altering concentration and medium pH changes the morphology of the nanoparticle during formation.

Fungi

Fungi such as *Aspergillus* sp., *Fusarium* sp. and *Penicillium* sp. have been reported to synthesize monodispersed and wide size ranges of nanoparticles. Fungi can uptake more volume of metallic ions and can secrete a large amount of enzymes and proteins per unit biomass which will lead to heavy output of biologically synthesized nanoparticles (Narayanan and Sakthivel, 2010).

The extracellular biological reduction is carried out by *Trichothecium* sp. to produce gold nanoparticles, however agitating the culture induces the formation of nanoparticles intracellularly. While in non-agitation state the fungi might have promoted the release of enzymes and proteins. Further the nanoparticles are synthesized and stabilized by bioactive reducing agents, enzymes and proteins secreted from the cell wall during the process. However, the intracellular yield of nanoparticles are lesser as compared to extracellular yield which may be due to complex extraction processes associated with intracellular production (Ahmad *et al.*, 2005). Apart from gold and silver nanoparticles, there have been reports for production of luminescent carbon quantum dots (Kumar *et al.*, 2007), magnetite (Bharde *et al.*, 2006), zirconia (Bansal *et al.*, 2004) and oxide nanoparticles (Bansal *et al.*, 2006) through various fungal species.

Viruses

Viruses, often considered as living nanoparticle made up of a nucleic acid (DNA/RNA) encapsulated within a protein coat. Viruses were generally used to synthesize semiconductor quantum dots. The specific feature of viruses that the capsid protein having high surface area leads to higher number of reactive sites for interacting with metal ions (Makarov *et al.*, 2014). In a case study it was observed that when plant extracts of *Nicotiana benthamina* or *Hordeum vulgare* was added with tobacco mosaic virus (TMV) to the gold or silver salts resulted in significantly smaller nanoparticles as compared to plant extract based nanoparticle synthesis (Love *et al.*, 2014). Further, higher concentration of TMV act as a template and induces metallization of the nanoparticles to synthesize respective nanowires.

Yeasts

Like other microorganisms, yeasts also have abilities to absorb and accumulate toxic metals due to their detoxification mechanisms such as chelation, bio precipitation and bio-sorption (Hulkoti and Taranath, 2014). CdS and PbS quantum dots were synthesized by *Schizosaccharomyces* sp. and *Torulosis* sp. effectively (Kowshik *et al.*, 2002b; Kowshik *et al.*, 2002c). *Pichia jadinii* was used in synthesis of gold nanoparticle up to 100 nm in size. The shape and size of gold nanoparticles can be regulated by controlling both the growth and cellular activities of the yeast (Gericke and Pinches, 2006b). Similarly, *Yarrowia lypolytica* was used in the synthesis of gold nanospheres by regulating bio-mass and Au salt concentration. Further increase in gold salt concentration leads to production of gold plates (Iravani, 2014). The silver tolerant yeast strain MKY3 was also helpful in accumulating high concentration silver and yielding silver nanoparticles (Kowshik *et al.*, 2002a). Some examples regarding microbial synthesis of nanoparticles are presented in Table 11.1 and 11.2.

Table 11.1. Nanoparticles Synthesized from different Fungal Sources

Nanoparticles	*Fungus Species*	*Inferences*	*References*
Silver	*Rhizopus nigricans*	35-40 nm, Round Bactericidal, catalytic	(Ravindra and Rajasab, 2014)
	Verticillium	21–25 nm, Spherical Catalysis	(Mukherjee *et al.*, 2001)
	Cariolus versicolor	25-75 nm, Spherical Water-soluble metallic catalysts, labels for living cells and tissues	(Sanghi and Verma, 2009)
	Penicillium brecompactum	23-105 nm, Crystalline spherical Antimicrobial agent	(Shaligram *et al.*, 2009)
Gold	*Verticillium luteoalbum*	< 10 nm, Triangular, hexagonal Optics, sensor, coatings	(Gericke and Pinches, 2006b)
	Trichothecium sp.	10–25 nm, Spherical, rod-like and triangular	(Ahmad *et al.*, 2005)
Zinc oxide	*Aspergillus terreus*	8 nm, Spherical Catalysis, biosensing, drug delivery, molecular diagnostics, solar cell, optoelectronics, cell labeling, and imaging	(Raliya and Tarafdar, 2014)
Titanium dioxide	*Aspergillus flavus* TFR7	12–15 Spherical Plant nutrient fertilizer	(Raliya *et al.*, 2015)

Table 11.2. Nanoparticles Synthesized from different Bacterial Sources

Nanoparticles	*Bacteria Species*	*Inferences*	*References*
Silver	*Bacillus cereus*	20–40 nm, Spherical, Antibacterial activity against *Escherichia coli, Pseudomonas aeruginosa, Staphylococcus aureus, Salmonella typhi*, and *Klebsiella pneumonia* bacteria	(Sunkar and Nachiyar, 2012)
	Lactobacillus casei	20–50 nm, Spherical, Drug delivery, cancer treatments, bio-labeling	(Korbekandi *et al.*, 2012)
	Klebsiella pneumonia, Escherichia coli, Enterobacter cloacae	28–122 nm, Spherical Optical receptors, electrical batteries, antimicrobial	(Ahmad *et al.*, 2010)
	E. coli DH 5dinna	8–25 nm, Spherical Direct electrochemistry of hemoglobin	(Du *et al.*, 2007)
	Desulfovvibrio desulfuricans	20–50 nm, Spherical Catalysis	(Deplanche and Macaskie, 2008)

Nanoparticles	Bacteria Species	Inferences	References
Iron Oxide	*Magnetospirillum magnetotacticum*	47 nm, Handle shaped cluster	(Philipse and Maas, 2002)
	Aquaspirillum magnetotacticum	40–50 nm, Octahedral prism	(Mann, 1985)

Biological Synthesis of Metal Nanoparticles via Plants

Plants being placed at higher tier in the hierarchical division of all living being, they show much advanced metabolic activities as compared to microbes. Because of the reason, plants can accumulate a huge amount of toxic metal ions and reduce them to nanoparticles in faster, cost-economic and friendly manner. The produced nanoparticles shape and size depends upon the bioactive compound and their concentration in plant extracts. Prior to nanoparticle synthesis by plant sources, the bioactive compounds from plants were extracted by using various solvents, mostly aqueous. To dilute salt solution few millilitres of plant extract was added and nanoparticle production was confirmed by the change in colour of solution. The biochemical reduction of metal ions was initiated by bioactive compounds such as phenols, alkaloids and enzymes and transforms them to zero valent states. Subsequently, the nucleation and growth take place to form nanoparticles which are further stabilized by proteins and bioactive compounds present in the plant extracts. In fact, there are many factors which drives the reaction from metal ions to get precise size and shape of nanoparticles.

Factors Affecting Biological Synthesis of Metal Nanoparticles

pH

As discussed earlier, pH plays a significant role in the synthesis of nanoparticles. Generally, pH is the amount H^+ ions present in the solution. As H^+ ions can receive and transfer electrons with high affinity, their concentrations decide the shape and size of the nanoparticles. It was observed that larger sized particles were produced in low pH values. For example, rod shaped gold nanoparticles synthesized by *Avena sativa* were larger at pH 2 (25-85 nm) as compared to pH 3 and 4 (5-20 nm). As per reports, the maximum availability of functional groups of bioactive compounds observed between pH 3 to 4 whereas fewer functional groups made available at pH 2 for the nucleation and stabilization of gold nanorods (Armendariz *et al.*, 2004).

Reactant Concentration

The concentration of reactant as well as reductant have a significant role in deciding the size and shape of nanoparticles. As reported by Huang *et al.* (2007), varying the amount of *Cinnamomum camphora* leaf extract influence the shape of gold and silver nanoparticles. The shape changed from triangular to spherical when the concentration of leaf extracts was increased in the solution. Again, Prathap Chandran *et al.* (2009) observed triangular plates of gold nanoparticles changed to spherical when increased amount of *Aloe vera* extract was added in the solution.

The presence of carbonyl groups in the *Aloe vera* extract might have played a role in regulating the shape and size of gold nanoparticles.

Reaction Time

The reaction time always depends upon the pH of the medium, temperature, reactant and catalysis rate. Plant extracts contain various bio-compounds and enzymes which are able decrease the reaction time significantly. Silver nanoparticles produced by pineapple extract takes only 2 minutes (Ahmad and Sharma, 2012) whereas it takes 15 minutes to 2 hours in case of *Chenopodium album (Dwivedi and Gopal, 2010)*. Furthermore, increasing the reaction time tends to give more bigger particle as in case of silver nanoparticles produced via *Azadirachta indica* (Prathna *et al.*, 2011) (Figure 11.2).

Figure 11.2. Synthesis of Nanoparticles from Neem Leaf Extract.
Courtesy: Devatha and Thala, 2018.

Reaction Temperature

Reaction temperature is an essential factor in determining the size, shape and yield of nanoparticles. Silver nanoparticle was synthesized via sweet orange peel extract at 25°C and 60°C, respectively. The particles synthesized at 25°C yielded average diameter of 35 nm whereas in 60°C it produced 10 nm spherical Ag nanoparticles (Kaviya *et al.*, 2011). As the temperature increases the particle formation rate increases. However, the particles get smaller and deformed to rod and plate like structure instead of spheres (Gericke and Pinches, 2006a).

Plant Derived Nanoparticles

There is a high demand for the plant derived nanoparticles as their production is environmentally safer, cost-economic, reuse of biological waste, reducing the overuse of hazardous chemicals and faster and easier to scale up the production. Nanoparticles such as gold, silver, copper and copper oxide, palladium, platinum, titanium dioxide, zinc oxide, iron oxide, indium, lead and selenium have been synthesized via various plant-based extracts (Table 11.3).

Characterisation Techniques Adopted for Nanoparticles

Nanoparticles are prepared by top-down approach or bottom-up approach. Mostly biological green synthesis of nanoparticles categorised as bottom-up approach where the nanoparticles are made from atoms and molecules. The bottom-up approach includes nucleation and growth which depends upon the biological

Table 11.3. Nanoparticles Synthesized from different Plant Sources

Nanoparticles	*Plant species*	*Inferences*	*References*
Gold	*Eucalyptus macrocarpa*	20-80 nm	(Poinern *et al.*, 2013)
	Coleus amboinicus	4.6 – 55 nm	(Narayanan and Sakthivel, 2010)
	Allium cepa (onion)	100 nm	(Parida *et al.*, 2011)
silver	*Cymbopogan citratus*	15 – 65 nm Cube and rectangular Antimicrobial effect against *E. coli*, *P. aeruginosa*, *P. mirabilis*, S. flexaneri, *S. sonnei* and *K. pneumonia*	(Sotiriou and Pratsinis, 2010)
	Nelumbo nucifera	25-80 nm, spherical, triangular, truncated triangular, decahedral Larvicidal activity against malaria and filariasis	(Santhosnkumar *et al.*, 2011)
	Moringa oleifera	57 nm, antimicrobial	(Prasad and Elumalai, 2011)
Copper	Magnolia leaf extract	40 – 100 nm, antibacterial activity against *E. coli*	(Lee *et al.*, 2011)
	Syzygium aromaticum (Clove)	40 nm and a spherical to granular morphology	(Subhankari and Nayak, 2013)
	Sterculia urens (Karaya gum) extract	Stable spherical, mean particle size of 4.8 nm, antimicrobial activity against E. coli and Staphylococcus aureus	(Padil and erník, 2013)
Palladium and Platinium Nanoparticles	*C. zeylanicum* (cinnamon) bark	Particle size (15 to 20 nm)	(Sathishkumar *et al.*, 2009)
	Annona squamosa (Custard apple) peel extract	Size from 75 to 85 nm	(Roopan *et al.*, 2012)
	Coffea arabica (Coffee) and *Camellia sinensis* (Tea) extracts	20 to 60 nm Faced centred cubic crystal symmetry	(Nadagouda and Varma, 2008)
Titanium Dioxide	*Nyctanthes arbor-tristis* leaf extracts	Have been found to produce spherical particles ranging in size from 100 to 150 nm	(Sundrarajan and Gowri, 2011)
	Catharanthus roseus	Irregular in shape and ranged in size from 25 up to 110 nm.	(Velayutham *et al.*, 2012)
Zinc Oxide	*Physalis alkekengi*	72.5 nm, pseudo-spherical in shape	(Qu *et al.*, 2011b)
	Sedum alfredii,	53.7 nm, pseudo-spherical in shape	(Qu *et al.*, 2011a)
Iron Oxide,	*Euphorbia milii*	Spherical nanoparticles size range (13 to 21 nm)	(Shah *et al.*, 2014)
Lead	*Jatropha curcas*	Spherical shaped particles ranging in size from 10 to 12.5 nm	(Joglekar *et al.*, 2011)
Selenium	*Citrus reticulata*	To produce spherically shaped particles with a mean particle size of 70 nm	(Sasidharan *et al.*, 2014)

entity, concentration of precursor material, pH, temperature and stabilising compound. The nanoparticle formation in the reaction can be confirmed by the gradual changes in colour of solution. After the reaction, nanoparticles are filtered or centrifuged at about 6000 – 10000 rpm and characterized to study their chemical composition, size, shape and other parameters.

The property of the nanoparticle produced by biological synthesis varies with their size. The size of nanoparticles can be estimated by particle size analyser. Initially, the nanoparticles are dispersed into a colloidal solution through which a laser beam is passed. Nanoparticles in a colloidal solution in a Brownian motion show Tyndall effect and scatter light with some angle towards the detector at 90° angle. Further, the colloidal stability of the nanoparticle is measured by zeta analyser which uses capillary electrode phenomena. The higher magnitude zeta potential confirms more stability and uniform distribution.

The morphology of the synthesized nanoparticles can be observed by Scanning electron microscope (SEM), Transmission electron microscope (TEM), Atomic force microscope (AFM). The SEM analyses the surface morphology whereas TEM analyses the anatomy of the nanoparticle. Atomic force microscope distinguishes the surface roughness with high precision. The elemental composition of the nanoparticle can be measured by Energy dispersive spectroscopy (EDS) which uses X-ray emitted during the analysis of SEM and TEM.

Spectroscopy based techniques such as X-ray diffraction (XRD), UV-Vis spectroscopy, Fourier transform infrared spectroscopy (FTIR), Raman spectroscopy and Nuclear magnetic resonance (NMR) are used to analyse crystal phase, chemical composition and chemical structure of the synthesized nanoparticle.

Conclusion

Overall, in twenty-first century nanoparticles have a huge role to play in different sectors of science and technology including electronics, photonics, medicine and agriculture. As discussed, among various techniques in producing nanoparticles, the chemical methods give well distributed uniform shape and size. But, the amount of hazardous chemicals used in the process slowly pollutes air, water and soil and have a negative impact on agriculture, terrestrial and marine diversity. Over the years of research, it is found that microbes and plants contain many bioactive compounds which can act as reductant to metal ions in converting them to respective nanoparticles. Plant based green synthesis is inexpensive, environment friendly and much easier to scale up as compared to microbial green synthesis due to multistep processes like isolation, culture maintenance *etc.* Plant based nanoparticles have been used in fluorescent labelling in immunoassays, targeted delivery of therapeutic drugs, tumour destruction via heating (hyperthermia), and as antibacterial agents in bandages. In addition, these nanoparticles are also used in delivering biochemical compounds in plants as pesticides and insecticides. As synthesis of nanoparticles via plants and microbes varies depending on species to species, there is a need for more elaborate studies to understand and evaluate their synthesis mechanism

REFERENCES

1. Abdeen, S., Geo, S., Sukanya, S., Praseetha, P., and Dhanya, R. (2014). Biosynthesis of silver nanoparticles from Actinomycetes for therapeutic applications. *International Journal of Nano Dimension, 5*(2), 155-162.

2. Ahmad, A., Senapati, S., Khan, M. I., Kumar, R., and Sastry, M. (2005). Extra-/ intracellular biosynthesis of gold nanoparticles by an alkalotolerant fungus, Trichothecium sp. *Journal of Biomedical Nanotechnology, 1*(1), 47-53.

3. Ahmad, A., Senapati, S., Khan, M. I., Kumar, R., Ramani, R., Srinivas, V., and Sastry, M. (2003). Intracellular synthesis of gold nanoparticles by a novel alkalotolerant actinomycete, *Rhodococcus* species. *Nanotechnology, 14*(7), 824.

4. Ahmad, N., and Sharma, S. (2012). Green synthesis of silver nanoparticles using extracts of Ananas comosus. *Green and Sustainable Chemistry, 2*(04), 141.

5. Ahmad, N., Sharma, S., Alam, M. K., Singh, V., Shamsi, S., Mehta, B., and Fatma, A. (2010). Rapid synthesis of silver nanoparticles using dried medicinal plant of basil. *Colloids and Surfaces B: Biointerfaces, 81*(1), 81-86.

6. Armendariz, V., Herrera, I., Jose-Yacaman, M., Troiani, H., Santiago, P., and Gardea-Torresdey, J. L. (2004). Size controlled gold nanoparticle formation by Avena sativa biomass: use of plants in nanobiotechnology. *Journal of nanoparticle research, 6*(4), 377-382.

7. Arun, P., Shanmugaraju, V., Ramanujam, J. R., Prabhu, S. S., and Kumaran, E. (2013). Original Research Article Biosynthesis of Silver Nanoparticles from Corynebacterium sp. and its antimicrobial activity. *Int. J. Curr. Microbiol. App. Sci, 2*(3), 57-64.

8. Baker, S., Harini, B., Rakshith, D., and Satish, S. (2013). Marine microbes: invisible nanofactories. *Journal of Pharmacy Research, 6*(3), 383-388.

9. Bansal, V., Poddar, P., Ahmad, A., and Sastry, M. (2006). Room-temperature biosynthesis of ferroelectric barium titanate nanoparticles. *Journal of the American Chemical Society, 128*(36), 11958-11963.

10. Bansal, V., Rautaray, D., Ahmad, A., and Sastry, M. (2004). Biosynthesis of zirconia nanoparticles using the fungus *Fusarium oxysporum*. *Journal of Materials Chemistry, 14*(22), 3303-3305.

11. Bharde, A., Rautaray, D., Bansal, V., Ahmad, A., Sarkar, I., Yusuf, S. M., Sanyal, M., and Sastry, M. (2006). Extracellular biosynthesis of magnetite using fungi. *Small, 2*(1), 135-141.

12. Chauhan, R., Kumar, A., and Abraham, J. (2013). A biological approach to the synthesis of silver nanoparticles with *Streptomyces* sp. JAR1 and its antimicrobial activity. *Scientia pharmaceutica, 81*(2), 607-624.

13. Deplanche, K., and Macaskie, L. (2008). Biorecovery of gold by *Escherichia coli* and *Desulfovibrio desulfuricans*. *Biotechnology and bioengineering, 99*(5), 1055-1064.

14. Dhillon, G. S., Brar, S. K., Kaur, S., and Verma, M. (2012). Green approach for nanoparticle biosynthesis by fungi: current trends and applications. *Critical reviews in biotechnology, 32*(1), 49-73.

15. Du, L., Jiang, H., Liu, X., and Wang, E. (2007). Biosynthesis of gold nanoparticles assisted by Escherichia coli DH5α and its application on direct electrochemistry of hemoglobin. *Electrochemistry Communications, 9*(5), 1165-1170.

16. Dwivedi, A. D., and Gopal, K. (2010). Biosynthesis of silver and gold nanoparticles using Chenopodium album leaf extract. *Colloids and Surfaces A: Physicochemical and Engineering Aspects, 369*(1-3), 27-33.

17. Gericke, M., and Pinches, A. (2006a). Biological synthesis of metal nanoparticles. *Hydrometallurgy, 83*(1-4), 132-140.

18. Gericke, M., and Pinches, A. (2006b). Microbial production of gold nanoparticles. *Gold Bulletin, 39*(1), 22-28. doi: 10.1007/bf03215529

19. Govindaraju, K., Kiruthiga, V., Kumar, V. G., and Singaravelu, G. (2009). Extracellular synthesis of silver nanoparticles by a marine alga, *Sargassum wightii* Grevilli and their antibacterial effects. *Journal of Nanoscience and Nanotechnology, 9*(9), 5497-5501.

20. He, S., Guo, Z., Zhang, Y., Zhang, S., Wang, J., and Gu, N. (2007). Biosynthesis of gold nanoparticles using the bacteria *Rhodopseudomonas capsulata. Materials Letters, 61*(18), 3984-3987.

21. Huang, J., Li, Q., Sun, D., Lu, Y., Su, Y., Yang, X., Wang, H., Wang, Y., Shao, W., and He, N. (2007). Biosynthesis of silver and gold nanoparticles by novel sundried Cinnamomum camphora leaf. *Nanotechnology, 18*(10), 105104.

22. Hulkoti, N. I., and Taranath, T. (2014). Biosynthesis of nanoparticles using microbes–a review. *Colloids and Surfaces B: Biointerfaces, 121*, 474-483.

23. Husseiny, M., El-Aziz, M. A., Badr, Y., and Mahmoud, M. (2007). Biosynthesis of gold nanoparticles using Pseudomonas aeruginosa. *Spectrochimica Acta Part A: Molecular and Biomolecular Spectroscopy, 67*(3-4), 1003-1006.

24. Iravani, S. (2011). Green synthesis of metal nanoparticles using plants. *Green Chemistry, 13*(10), 2638-2650.

25. Iravani, S. (2014). Bacteria in nanoparticle synthesis: current status and future prospects. *International scholarly research notices, 2014.*

26. Jin-Kun, L. Z.-Y. F., Yue-Ying, W. J.-M. L., and Hu, C. (2001). Preliminary Study on the Mechanism of Non-enzymatic Bioreduction of Precious Metal Ions [J]. *Acta Physico-chimica Sinica, 5.*

27. Joglekar, S., Kodam, K., Dhaygude, M., and Hudlikar, M. (2011). Novel route for rapid biosynthesis of lead nanoparticles using aqueous extract of *Jatropha curcas* L. latex. *Materials Letters, 65*(19-20), 3170-3172.

28. Karthik, L., Kumar, G., Kirthi, A. V., Rahuman, A., and Rao, K. B. (2014). *Streptomyces* sp. LK3 mediated synthesis of silver nanoparticles and its biomedical application. *Bioprocess and biosystems engineering, 37*(2), 261-267.

29. Kaviya, S., Santhanalakshmi, J., Viswanathan, B., Muthumary, J., and Srinivasan, K. (2011). Biosynthesis of silver nanoparticles using *Citrus sinensis* peel

extract and its antibacterial activity. *Spectrochimica Acta Part A: Molecular and Biomolecular Spectroscopy, 79*(3), 594-598.

30. Korbekandi, H., Iravani, S., and Abbasi, S. (2009). Production of nanoparticles using organisms. *Critical reviews in biotechnology, 29*(4), 279-306.

31. Korbekandi, H., Iravani, S., and Abbasi, S. (2012). Optimization of biological synthesis of silver nanoparticles using *Lactobacillus casei* subsp. casei. *Journal of Chemical Technology and Biotechnology, 87*(7), 932-937.

32. Kowshik, M., Ashtaputre, S., Kharrazi, S., Vogel, W., Urban, J., Kulkarni, S. K., and Paknikar, K. (2002a). Extracellular synthesis of silver nanoparticles by a silver-tolerant yeast strain MKY3. *Nanotechnology, 14*(1), 95.

33. Kowshik, M., Deshmukh, N., Vogel, W., Urban, J., Kulkarni, S. K., and Paknikar, K. (2002b). Microbial synthesis of semiconductor CdS nanoparticles, their characterization, and their use in the fabrication of an ideal diode. *Biotechnology and bioengineering, 78*(5), 583-588.

34. Kowshik, M., Vogel, W., Urban, J., Kulkarni, S. K., and Paknikar, K. M. (2002c). Microbial synthesis of semiconductor PbS nanocrystallites. *Advanced Materials, 14*(11), 815-818.

35. Kumar, A., Kaur, K., and Sharma, S. (2013). Synthesis, characterization and antibacterial potential of silver nanoparticles by Morus nigra leaf extract. *Indian J Pharm Biol Res, 1*(4), 16-24.

36. Kumar, S. A., Ansary, A. A., Ahmad, A., and Khan, M. (2007). Extracellular biosynthesis of CdSe quantum dots by the fungus, *Fusarium oxysporum*. *Journal of Biomedical Nanotechnology, 3*(2), 190-194.

37. Lee, H.-J., Lee, G., Jang, N. R., Yun, J. H., Song, J. Y., and Kim, B. S. (2011). Biological synthesis of copper nanoparticles using plant extract. *Nanotechnology, 1*(1), 371-374.

38. Love, A. J., Makarov, V., Yaminsky, I., Kalinina, N. O., and Taliansky, M. E. (2014). The use of tobacco mosaic virus and cowpea mosaic virus for the production of novel metal nanomaterials. *Virology, 449*, 133-139.

39. Luangpipat, T., Beattie, I. R., Chisti, Y., and Haverkamp, R. G. (2011). Gold nanoparticles produced in a microalga. *Journal of Nanoparticle Research, 13*(12), 6439-6445.

40. Makarov, V., Love, A., Sinitsyna, O., Makarova, S., Yaminsky, I., Taliansky, M., and Kalinina, N. (2014). "Green" nanotechnologies: synthesis of metal nanoparticles using plants. *Acta Naturae, 6*(1 (20)).

41. Mann, S. (1985). Structure, morphology, and crystal growth of bacterial magnetite *Magnetite biomineralization and magnetoreception in organisms* (pp. 311-332): Springer.

42. Mata, Y., Torres, E., Blazquez, M., Ballester, A., González, F., and Munoz, J. (2009). Gold (III) biosorption and bioreduction with the brown alga Fucus *vesiculosus*. *Journal of hazardous materials, 166*(2-3), 612-618.

43. Mohanpuria, P., Rana, N. K., and Yadav, S. K. (2008). Biosynthesis of nanoparticles: technological concepts and future applications. *Journal of nanoparticle research, 10*(3), 507-517.

44. Mukherjee, P., Ahmad, A., Mandal, D., Senapati, S., Sainkar, S. R., Khan, M. I., Parishcha, R., Ajaykumar, P., Alam, M., and Kumar, R. (2001). Fungus-mediated synthesis of silver nanoparticles and their immobilization in the mycelial matrix: a novel biological approach to nanoparticle synthesis. *Nano Letters, 1*(10), 515-519.

45. Nadagouda, M. N., and Varma, R. S. (2008). Green synthesis of silver and palladium nanoparticles at room temperature using coffee and tea extract. *Green Chemistry, 10*(8), 859-862.

46. Narayanan, K. B., and Sakthivel, N. (2010). Biological synthesis of metal nanoparticles by microbes. *Advances in colloid and interface science, 156*(1-2), 1-13.

47. Nvkv Prasad, T., Subba Rao Kambala, V., and Naidu, R. (2011). A critical review on biogenic silver nanoparticles and their antimicrobial activity. *Current Nanoscience, 7*(4), 531-544.

48. Padil, V. V. T., and Èerník, M. (2013). Green synthesis of copper oxide nanoparticles using gum karaya as a biotemplate and their antibacterial application. *International journal of nanomedicine, 8*, 889.

49. Parida, U. K., Bindhani, B. K., and Nayak, P. (2011). Green synthesis and characterization of gold nanoparticles using onion (*Allium cepa*) extract. *World Journal of Nano Science and Engineering, 1*(04), 93.

50. Philipse, A. P., and Maas, D. (2002). Magnetic colloids from magnetotactic bacteria: chain formation and colloidal stability. *Langmuir, 18*(25), 9977-9984.

51. Poinern, G. E. J., Chapman, P., Le, X., and Fawcett, D. (2013). Green biosynthesis of gold nanometre scale plates using the leaf extracts from an indigenous Australian plant *Eucalyptus macrocarpa. Gold Bulletin, 46*(3), 165-173.

52. Prabhu, S., and Poulose, E. K. (2012). Silver nanoparticles: mechanism of antimicrobial action, synthesis, medical applications, and toxicity effects. *International nano letters, 2*(1), 32.

53. Prasad, T., and Elumalai, E. (2011). Biofabrication of Ag nanoparticles using Moringa oleifera leaf extract and their antimicrobial activity. *Asian Pacific Journal of Tropical Biomedicine, 1*(6), 439-442.

54. Prathap Chandran, S., Chaudhary, M., Pasricha, R., Ahmad, A., and Sastry, M. (2009). Synthesis of gold nanotriangles and silver nanoparticles using Aloe vera leaf extract. *NEW METHODS TOWARDS SYNTHESIS AND CONTROLLED FUNCTIONALISATION OF INORGANIC NANOPARTICLES*, 91.

55. Prathna, T., Chandrasekaran, N., Raichur, A. M., and Mukherjee, A. (2011). Kinetic evolution studies of silver nanoparticles in a bio-based green synthesis process. *Colloids and Surfaces A: Physico-chemical and Engineering Aspects, 377*(1-3), 212-216.

56. Qu, J., Luo, C., and Hou, J. (2011a). Synthesis of ZnO nanoparticles from Zn-hyperaccumulator (Sedum alfredii Hance) plants. *Micro and Nano Letters, 6*(3), 174-176.

57. Qu, J., Yuan, X., Wang, X., and Shao, P. (2011b). Zinc accumulation and synthesis of ZnO nanoparticles using *Physalis alkekengi L. Environmental pollution, 159*(7), 1783-1788.

58. Rajasulochana, P., Dhamotharan, R., Murugakoothan, P., Murugesan, S., and Krishnamoorthy, P. (2010). Biosynthesis and characterization of gold nanoparticles using the alga *Kappaphycus alvarezii. International Journal of Nanoscience, 9*(05), 511-516.

59. Raliya, R., and Tarafdar, J. (2014). Biosynthesis and characterization of zinc, magnesium and titanium nanoparticles: an eco-friendly approach. *International Nano Letters, 4*(1), 93.

60. Raliya, R., Biswas, P., and Tarafdar, J. (2015). TiO^2 nanoparticle biosynthesis and its physiological effect on mung bean (*Vigna radiata L.*). *Biotechnology Reports, 5*, 22-26.

61. Ramamoorthy, R., Vanitha, S., and Krishnadev, P. (2019). Green synthesis of silver nanoparticles using red seaweed *Portieria hornemannii* (Lyngbye) PC silva and its antifungal activity against silkworm (*Bombyx mori L.*) Muscardine pathogens. *Journal of Pharmacognosy and Phytochemistry, 8*(3), 3394-3398.

62. Ravindra, B., and Rajasab, A. (2014). A comparative study on biosynthesis of silver nanoparticles using four different fungal species.

63. Roopan, S. M., Bharathi, A., Kumar, R., Khanna, V. G., and Prabhakarn, A. (2012). Acaricidal, insecticidal, and larvicidal efficacy of aqueous extract of *Annona squamosa* L peel as biomaterial for the reduction of palladium salts into nanoparticles. *Colloids and Surfaces B: Biointerfaces, 92*, 209-212.

64. Sanghi, R., and Verma, P. (2009). Biomimetic synthesis and characterisation of protein capped silver nanoparticles. *Bioresource technology, 100*(1), 501-504.

65. Santhoshkumar, T., Rahuman, A. A., Rajakumar, G., Marimuthu, S., Bagavan, A., Jayaseelan, C., Zahir, A. A., Elango, G., and Kamaraj, C. (2011). Synthesis of silver nanoparticles using *Nelumbo nucifera* leaf extract and its larvicidal activity against malaria and filariasis vectors. *Parasitology research, 108*(3), 693-702.

66. Sasidharan, S., Sowmiya, R., and Balakrishnaraja, R. (2014). Biosynthesis of selenium nanoparticles using *Citrus reticulata* peel extract. *World J. Pharm. Res, 4*, 1322-1330.

67. Sathishkumar, M., Sneha, K., Kwak, I. S., Mao, J., Tripathy, S., and Yun, Y.-S. (2009). Phyto-crystallization of palladium through reduction process using *Cinnamom zeylanicum* bark extract. *Journal of Hazardous materials, 171*(1-3), 400-404.

68. Senapati, S., Syed, A., Moeez, S., Kumar, A., and Ahmad, A. (2012). Intracellular synthesis of gold nanoparticles using alga *Tetraselmis kochinensis. Materials Letters, 79*, 116-118.

69. Shah, R., Oza, G., Pandey, S., and Sharon, M. (2012). Biogenic fabrication of gold nanoparticles using *Halomonas salina*. *J Microbiol Biotechnol Res, 2*(4), 485-492.

70. Shah, S., Dasgupta, S., Chakraborty, M., Vadakkekara, R., and Hajoori, M. (2014). Green synthesis of iron nanoparticles using plant extracts. *Int J Biol Pharm Res, 5*, 549-552.

71. Shaligram, N. S., Bule, M., Bhambure, R., Singhal, R. S., Singh, S. K., Szakacs, G., and Pandey, A. (2009). Biosynthesis of silver nanoparticles using aqueous extract from the compactin producing fungal strain. *Process biochemistry, 44*(8), 939-943.

72. Sneha, K., Sathishkumar, M., Mao, J., Kwak, I., and Yun, Y.-S. (2010). *Corynebacterium glutamicum*-mediated crystallization of silver ions through sorption and reduction processes. *Chemical Engineering Journal, 162*(3), 989-996.

73. Sotiriou, G. A., and Pratsinis, S. E. (2010). Antibacterial activity of nanosilver ions and particles. *Environmental science and technology, 44*(14), 5649-5654.

74. Stephen, J. R., and Macnaughtont, S. J. (1999). Developments in terrestrial bacterial remediation of metals. *Current opinion in biotechnology, 10*(3), 230-233.

75. Subhankari, I., and Nayak, P. (2013). Synthesis of copper nanoparticles using *Syzygium aromaticum* (Cloves) aqueous extract by using green chemistry. *World J Nano Sci Technol, 2*(1), 14-17.

76. Sundrarajan, M., and Gowri, S. (2011). Green synthesis of titanium dioxide nanoparticles by Nyctanthes arbor-tristis leaves extract. *Chalcogenide Lett, 8*(8), 447-451.

77. Sunkar, S., and Nachiyar, C. V. (2012). Biogenesis of antibacterial silver nanoparticles using the endophytic bacterium *Bacillus cereus* isolated from Garcinia xanthochymus. *Asian Pacific Journal of Tropical Biomedicine, 2*(12), 953-959.

78. Velayutham, K., Rahuman, A. A., Rajakumar, G., Santhoshkumar, T., Marimuthu, S., Jayaseelan, C., Bagavan, A., Kirthi, A. V., Kamaraj, C., and Zahir, A. A. (2012). Evaluation of *Catharanthus roseus* leaf extract-mediated biosynthesis of titanium dioxide nanoparticles against Hippobosca maculata and *Bovicola ovis*. *Parasitology research, 111*(6), 2329-2337.

Chapter 12

Nanofertilizers: Concept, Synthesis and Applications

K.S. Subramanain[1], C. Sharmila Rahale[2] and N. Viji[2]*

[1]Director of Research, [2]Department of Nano Science and Technology, Tamil Nadu Agricultural University, Coimbatore – 641 003, Tamil Nadu

**e-mail: kssubra2001@rediffmail.com*

Introduction

The outburst of world population in the last 10 -15 years has imposed the necessity for higher agriculture productivity to satisfy the food needs of billions of people. The increasing nutrient deficiency in soils causes significant economic losses for farmers on the one hand and considerable decreases in nutritional quality of grain for food and feed. The crop productivity can be enhanced through application of fertilizers, although they have an additional role in enhancing the food production especially after the introduction of high yielding and fertilizer responsive crop varieties. Conventional fertilizers are generally applied on the crops by either spraying or broadcasting. An important factor, on which the mode of application depends, is the real final concentration of the fertilizers in the plants. Conventional fertilizers offer nutrients in chemical forms that are not fully accessible to plants. Additionally, the inversion of these chemicals to insoluble form in soil is the reason for the very low utilization of most of the macronutrients. A concentration much below the minimal desired one reaches to the targeted site due to leaching of chemicals, drift, runoff, evaporation, hydrolysis by soil moisture, and photolytic and microbial degradation. It has been calculated that in applied fertilizers, around 40 - 70 per cent of nitrogen, 80 - 90 per cent of phosphorus, and 50 - 90 per cent of potassium content are lost in the environment and never reach the plant. These problems superimpose repeated use of fertilizers. The repeated use, in turn, adversely affects the inherent nutrient balance of the soil and results in environmental pollution affecting normal flora and fauna. The repeated use of fertilizers also increases pathogen and pest resistance, reduces soil microflora, diminishes nitrogen fixation, contributes to bioaccumulation of pesticides, and destroys habitats for birds. This vicious circle causes sustainable and economic losses.

It is well known that the yields of many crops have begun to drop as a result of imbalanced fertilization and decrease in soil organic matter. Moreover, excessive applications of nitrogen and phosphorus fertilizers affect the groundwater and also lead to eutrophication in aquatic ecosystems. The remaining minerals may either leach down and/or leak and become fixed in the soil, or contribute to air pollution. Considering these facts, the large-scale application of chemical fertilizers to increase the crop productivity is not an acceptable option for sustainability. Especially in a long-term perspective, although the conventional fertilizers increase the crop production, they disturb the soil mineral balance and decrease the soil fertility. In addition to the irreparable damage that the excess use of chemical fertilizers causes to the soil structure and mineral cycles, it spoils the soil microflora, plants, and consequently, the food chains across ecosystems, leading to heritable mutations in future generations of consumers. Thus, there is an urgent need to optimize the use of chemical fertilization to fulfil the crop nutrient requirements and to minimize the risk of environmental pollution. Accordingly, it is very important to develop smart materials that can systematically release chemicals to specific targeted sites in plants which could be beneficial in controlling the nutrition deficiency in agriculture, while keeping the natural soil structure and contributing to clean environment. Nano-fertilizers are promising alternative in this context.

Concept of Nanofertilizers

A nano-fertilizer refers to a product in nanometer scale that delivers nutrients to crops. Nano-fertilizer technology is a recent innovation. Nanofertilizers are one potential output that could be a major innovation for agriculture; the large surface area and small size of the nanomaterials could allow for enhanced interaction and efficient uptake of nutrients for crop fertilization. The integration of nanotechnology in fertilizer products may improve release profiles and increase uptake efficiency, leading to significant economic and environmental benefits. Substituting traditional methods of fertilizer application by nano-fertilizers is an approach to release nutrients into the soil both gradually and in a controlled way. Nano-fertilizers show controlled release of agrochemicals through site-targeted delivery, reduction in toxicity, and enhanced nutrient utilization of delivered fertilizers. They possess unique features that enhance plants' performance in terms of ultrahigh absorption, increase in production, rise in photosynthesis, and significant expansion in the leaves' surface area. Besides, the controlled release of nutrients contributes to preventing eutrophication and pollution of water resources.

Nanofertilizers are classified into three types.

1. Nanoscale fertilizer - Contains nanoparticles which has nutrients
2. Nanoscale additives - Contains conventional fertilizers with nanoscale additives
3. Nanoscale coating - Conventional fertilizers loaded with nanoparticles

Synthesis of Nanofertilizer

The nano-fertilizers should be formulated in a way that they retain important properties such as high solubility, stability, effectiveness, time-controlled release, enhanced targeted activity with effective concentration, and less eco-toxicity due to the safe, easy mode of delivery and disposal.

Nanoparticles possess a great potential in targeted delivery of nutrients to living systems. Nanoparticles can be loaded by nutrients most commonly through one of the following ways:

- ☆ Absorption on the nanoparticles
- ☆ Attachment on the nanoparticles mediated by ligands
- ☆ Encapsulation in nanoparticles in polymeric shell
- ☆ Entrapment in nanoparticles

Nanoscale materials can be fabricated using two approaches *i.e.* "bottom-up" or "top-down" fabrication. In bottom-up methods (Chemical method), atoms or molecules are being buildup to form nanomaterials through self-assembly under a controlled environment (Ferrari, 2005). But in case of top-down approach (Physical method), the nanostructure is being formed by the removal of building blocks from the substrate. This included photolithography, dip pen lithography and nanomolding (Peppas, 2004).

The nano-fertilizers are synthesized by either direct loading or encapsulation technique. In direct loading, single or multiple nutrients are loaded into an adsorbent which has nano-dimension. Based on the surface charge of the adsorbent, surface modification will be done for leading both the cationic (NH_4^+, K^+, Ca^{2+}) and anionic nutrients (NO^{3-}, PO_4^{2-}, SO_4^{2-}). In another way, the fertilizers are encapsulated within the nanoparticles by three ways i) Nutrients are loaded into a nanomaterial ii) Nutrients are coated with a thin film iii) emulsion formulation (Rai *et al.*, 2012).

Potassium based nano-fertilizer is developed by encapsulating potassium in nano-dimension alginate-chitosan carrier. Encapsulation of potassium is done by ionotropic pre-gelation technique to regulate the release. In this method, the ratio of potassium to alginate and calcium chloride as well the time of the reaction was taken into account. After pre-gelation, the product is stabilized and equilibrated. After synthesis, the characterization studies proved the encapsulation of potassium into the beads and can serve as a slow release nanofertilizer (Princess *et al.*, 2019).

Marivic *et al.* (2019) is developed potassium based nano-fertilizer by encapsulating potassic fertilizer into a carrier called kappa-carrageenan through crosslinking technique. Carrageenan is a food grade carrier which has wider availability and high cross-linking capacity. The SO_3^- group of carrageenan was developed a linkage between potassium ion to form the encapsulated final gel product. The major factor to be considered in the process is the potassium-carrageenan ratio. When compared to conventional potassic fertilizers, carrageenan encapsulated fertilizer releases potassium in a slow way.

Among the micronutrients, zinc plays a vital role in plant growth, chlorophyll synthesis and carbohydrate formation. Tarafdar *et al.* (2014) were synthesized zinc nanoparticles by biological synthesis and used as a nano-fertilizer. Zinc nanoparticles were synthesized from the fungi *Rhizoctonia bataticola* by agitation, incubation, filtration followed by shaking technique. The synthesized cell free filtrate was used to prepare zinc oxide salt solution and tested in pearl millet crop. When compared to conventional fertilizer treated plants, zinc nanoparticles treated plants have shown significant improvement in growth *i.e.* 15.1 per cent of more shoot length, 4.2 per cent root length and 24.2 per cent root area.

Zeolites are the naturally available clay mineral, which has a honeycomb network like structure. The inner networks of zeolite can be used for loading all the major, minor and trace nutrients. The loaded nutrients will be released from the inner network in a slow manner based on crop demand (Millan *et al.*, 2008). It also increases the retention of nutrients. Clinoptilolite type of zeolites has unique crystalline structure with high pores and channels. It has immense surface area and can hold wide range of cationic and anionic nutrient ions (Agger, 2003). Zeolite based nanofertilizer can be synthesized by further surface modification using hexadecyltrimethyl ammonium to improve the release efficiency of loaded cations. Phosphorus solubility was improved by using ammonium loaded zeolite. It will improve the uptake of phosphorus and yield of plants by improving the solubility of phosphate minerals. Application of zeolite loaded rock phosphate showed a slow release of phosphorus compared to conventional fertilizer application. This was achieved due to the improved dissolution and iron exchange reactions (Li, 2003).

Chitosan is a naturally available biodegradable polymer. Chitosan based nanomaterial is highly soluble in water that its bulk form. It has a high affinity towards the biological systems due to its cationic surface area. Besides, chitosan contains nearly 9-10 per cent of nitrogen. Chitosan based nanomaterials are synthesized by various methods including cross-linking, gelation and precipitation technique (Saharan *et al.*, 2015).

Nanofertilizers can also be synthesized from the biological waste materials. Banana peel is one of the farm waste which is rich in growth promoters and elements. Lee *et al.* (2010) have synthesized a nanofertilizer from the Egyptian banana peels var. William Ripe peel. The banana peel was shredded and treated with alkaline solution to form slurry. The slurry was filtered and pH was adjusted to 5 by adding urea and citric acid. The resultant sludge was dried and ground to fine powder to obtain nanofertilizer. The size of the nanofertilizer is ranging from 19 to 55 nm.

Nanofertilizers Uptake, Translocation and Fate in Plants

The uptake, fate and interactions between the plants and applied nanomaterials is one of the important area of research. Because the accumulation of nanoparticles in plant system is used to determine the phytotoxicity of the particular nanomaterials in ecosystem. But still the researches on uptake and translocation mechanism of nanomaterials in plants remains very limited. The uptake, translocation and accumulation of nanoparticles depend on plant nature, stage and environment. It also depends on physico-chemical properties, stability and mode of delivery of the nanoparticles.

Metal nanomaterials such as TiO_2 and SiO_2 particles are the most stable compounds during uptake, but nanomaterials such as ZnO and CuO are liable to transformation. Rico *et al.* (2018) proved that the root system uptakes and translocate ZnO^{2+}, Cu^{2+}, Al^{3+}, Ag^{2+} and Fe_3O_4 nanoparticles to the foliar part of a plant, regardless of its species. He also mentioned species dependent uptake of nanomaterials in plants such as Cu, ZnO, Al and Ag nanomaterials in leaves, $Ni(OH)_2$ nanomaterial in the stem, and CeO_2 nanomaterials in both the stem and leaves. Besides, the magnetic nanomaterials behave in a different way in plants system. The uptake of magnetic nanoparticles is depending upon the type of growth medium. A higher uptake is noticed in hydroponic medium than sand, it may be due to the adherence of magnetic nanomaterials to soil.

The entry of nanomaterials into the plants is depend on the cell wall pore diameter. The translocation efficiency is depending upon the size of the applied nanomaterials. The nanoparticle having smaller size than the cell wall pores, may easily enter into the plant system and reach the plasma membrane. The particles having size larger than the cell wall pores may agglomerate over the cell wall that hinders the uptake of plants. In such cases, the nanomaterials can be functionalized which may facilitate the enlargement of cell wall pore size also it may induce new cell wall pores. This will assist the larger particles to get entered in the plant system. The uptake of nanoparticles can be mediated by binding to carrier protein through aquaporin, iron channels or endocytosis. Also in some cases the nanomaterials may form complex with the membrane transporter proteins. The nanomaterials are also translocated into the plants by cellular internalization or intracellular biotransformation. The solutes are entered into the plant system either via diffusion or through aqueous pores (Figure 12.1). After entering into the plant cell, the nanoparticles are targeted to different cytoplasmic organelles and interfere with the plant metabolic process.

Comparing the application methods, foliar application of nanomaterials is more efficient in delivery and transport than soil drenching. In soil application, the synthetic root exudate influenced the oxidation and reduction of particles that may significantly decrease the uptake and transformation. In foliar application, nanomaterials are entering into the plants through stomata or trichome of leaves. But the entry of nanomaterials may have hindered by the cuticle of leaves, which may be due to the waxy coating. The approximate size of stomatal aperture is about a length of 25 μm and a width of 3–10 μm. But each plant stomata have unique construction, the actual size exclusion limit of stomata for nanomaterial penetration is still unknown (Eichert and Goldbach, 2008).

The mechanism of uptake of nanomaterial in root tissues was explained by Yin *et al.* (2011). The have tested silver nanomaterial uptake in *Lolium multiflorum*

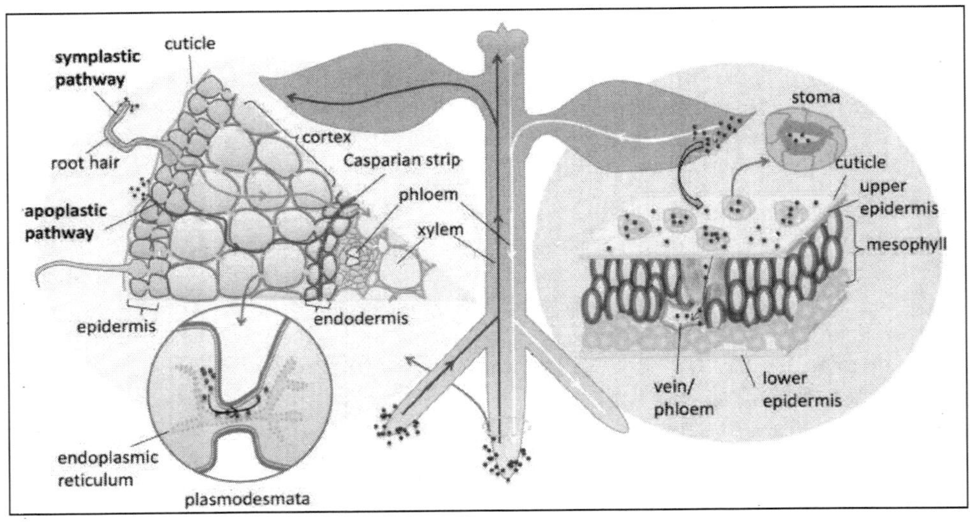

Figure 12.1. Schematic Diagram of the Uptake and Translocation Pathways of Nanomaterials in Plants.
Courtesy: Eichert *et al.*, 2008.

and given two hypotheses in uptake such as i) direct uptake of silver particles by roots through oxidative transformation ii) dissolution of silver nanomaterials in the root surface. In recent studies, the nanoparticles are being tagged with fluorescent dyes to track their movement in plants parts. Mesoporous silica nanomaterials are labeled with fluorescent dyes and the translocation pattern is tracked. The particles are penetrating the roots via symplastic and apoplastic pathways and getting translocated via xylem tissues to the aerial parts of plants (Sun *et al.*, 2014). Copper oxide nanomaterial is used in rice plants and the study proves that the particle is transported from root to shoot also it made a complex with cysteine, citrate, and phosphate ligands (Peng *et al.*, 2015). Studies on the uptake and translocation of TiO_2–alizarin red S complex in *Arabidopsis thaliana* seedlings have revealed that the mucilage released by the roots develops a pectin hydrogel complex around the root, which is most probably responsible for the entry of the nanoparticle–dye complex. In wheat plants, the uptake of TiO_2 nanoparticles includes localization in parenchyma and vascular tissues of the root.

Some analytical methods are available to obtain knowledge on uptake and translocation of nanomaterials in plants. Inductively coupled plasma mass spectroscopy (ICP-MS) and X-ray analysis are the important analytical method used to determine metal distribution in biological samples. Koelmel *et al.* (2013) investigated the uptake and distribution of gold nanomaterials in rice plants. The have identified the tissue level distribution of elemental gold in rice. Using ICP-MS. Quantitative determination of uptake of nanomaterials is also possible through enzyme digestion method.

Percolation Studies to Assess Nutrient Release Pattern

To study the release pattern of nano nutrients, percolation reactor (Hemandez *et al.*, 1992) is used. This reactor (Figure 12.2) consists of a Teflon cylinder (internal diameter= 2.5 cm, height = 15 cm) through the top of which deionized water is continuously pumped at a flow rate of 66 ml per day. Inside the reactor, 5 g of soil overlaid with the fertilizer is placed and solution is collected periodically to determine the released nutrients. The mean temperature during the experiment is 25 ± 0.2°C.

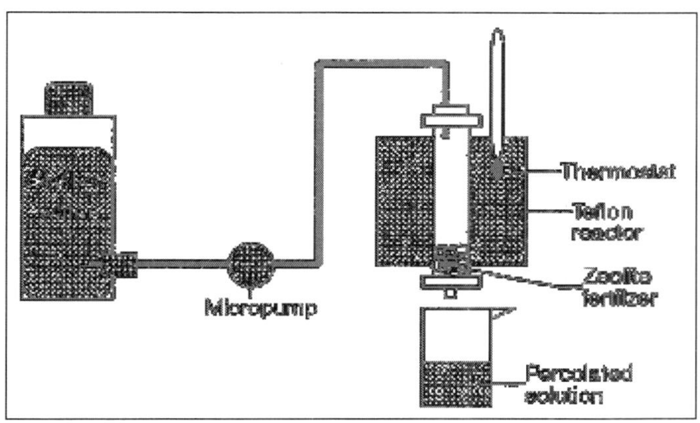

Figure 12.2. Percolation Reactor
Courtesy: Hernadez *et al.* 1992.

Application

Allen *et al.* (1993) has developed urea fortified zeolite chips which is used as a slow release nitrogen fertilizer. The adsorbed form of nitrogen was released much slower than that of ionic form. This is the reason why the zeolite encapsulated nitrogen was released in a slow rate. Urea-modified hydroxyapatite encapsulated nanocomposite was developed by Kottegoda *et al.*, 2011 and tested over gliricidia crop. The nanocomposite has shown initial rapid release of nitrogen and subsequent slow release over 60 days. The nanocomposite has recorded temporal release of nitrogen by the way it enhances nitrogen uptake and use efficiency. Abdel-Aziz *et al.* (2016) developed a NPK-nanochitosan composite by polymerization of methyl acrylic acid and chitosan. Foliar exposure of this nanocomposite in wheat plants significantly shortened the plant life span by 45 days, and increased the grain yield by 56 per cent over conventional fertilizer application.

Hydroxyapatite nanomaterials based P fertilizer is developed and tested, shown 20 per cent more seed productivity and 33 per cent more growth on soybean plants compared to conventional fertilizers. The column test of this nanofertilizer shown improved nutrient delivery and controlled mobility compared to water soluble conventional P fertilizers (Liu and Lal, 2014). Clay based P fertilizer is developed and tested for their release kinetics. Zeolite and montmorillonite clay is chosen, surface modified and used for nanoformuation. The synthesized nanoformulation have recoded an extended P release up to 40-50 days compared to conventional fertilizers where the release is only up to 10-12 days.

Lee *et al.* (2010) synthesized a banana peel based nanofertilizers and tested its nutrient content. The synthesized nanofertilizer contained most of the major elements such as nitrogen, chelated potassium, chelated iron, amino acids and proteins. The effect banana peel based nanofertilizer was checked over tomato and fenugreek crop, showed increased germination rate from 17 per cent to 97 per cent and 25 per cent to 93.17 per cent respectively. A potassium based nanofertilizers are developed by encapsulation technique and release efficiency is being studied. The release pattern studies showed that potassium was released from the encapsulated complex in a controlled way as only 14.6 per cent K was released in Britton-Robinson buffer solution after 7 days of loading (Monet 2019).

Chitosan nanomaterial based nanofertilizers enhance seed germination, nutrient uptake, photosynthetic rate and crop yield. Van *et al.* (2013) tested chitosan nanofertilizers in coffee crop under greenhouse. The treated plants have recorded an increment of 30-50 per cent, 30-60 per cent, 9.8-27.4 per cent, 17.3-30.4 per cent and 30-45 per cent in chlorophyll content, photosynthetic rate and N, P and K uptake, respectively. Chitosan based copper nanofertilizers are tested in tomato crop, recorded increases germination percentage and growth parameters. Choudhary *et al.* (2017) tested chitosan based nanofertilizers in maize plants, have shown increased growth, chlorophyll a, b and yield parameters. Due to enhance nutrient uptake, the nutrient availability includes copper, potassium, calcium, iron, magnesium and zinc for maize plants are in higher level. This nanofertilizer has also used as growth promotor in plants. Though the chitosan based nanofertilizer show better potential in crop growth and yield parameters, the mechanism involved the process of uptake is still not clearly understood.

Calcium based nanofertilizer is developed and tested over peanut crop. It recorded increased shoot biomass by 15 per cent and 0.43 per cent, 0.04 per cent, 0.014 per cent and 0.32 per cent enhanced nutrient content in roots compared to application of conventional source of calcium. The study also demonstrated the improved translocation of calcium through phloem of groundnut plants (Liu *et al.*, 2005). Sulfur nanofertilizer is developed from chinaberry and tree of heaven with sodium thiosulfate precursor and tested over vegetables. It recorded 127, 78 per cent and 133, 220 per cent of more root and shoot growth in tomato and pumpkin plants, respectively. The effect is highly concentration dependent (Salem *et al.*, 2015).

Iron based nanofertilizer have tested in black-eyed pea crop have recorded increased chlorophyll content compared to control plants. The mung bean plants are foliar-sprayed with manganese based nano ferilizers, have recorded 52 per cent increase in root length, 38 per cent in shoot length and 38 per cent in biomass production (Delfani *et al.*, 2014). Zinc based nanofertilizers are also available to enhance the nutrient availability. Maize plants have recorded 177 per cent increment of biomass production and 83 per cent in plant height under the treatment of zinc based nanofertilizers (Subbaiah *et al.*, 2016). Nanoscale micronutrient fertilizers may increase the plant metabolic rate by improving the uptake and availability of micronutrients.

The role of nanofertilizers under drought condition have been studied by Dimkpa *et al.* (2017). The study revealed that a composite micronutrient nanofertilizer includes zinc, boron and copper have recorded 33 per cent growth and 36 per cent yield enhancement in soybean crop under drought stress. They have also tested foliar application of zinc based composite nanofertilizer in sorghum plants. The study proved that it not only increases grain yield but also enhances nitrogen and potassium accumulation. This study also indicates that the fortification of N and K macronutrient fertilizers with nano-scale micronutrient fertilizers may increase overall nutrient use efficiency in plants. Dimkpa *et al.* (2019) reported the effect of manganese based nanofertilizer in wheat plants, not only having impact on grain yield but also have subtle effects in nutrient acquisition rate and uptake pattern.

Specific Properties of Nanofertilizers

The nano-fertilizers possess certain specific properties, which facilitate higher nutrient use efficiency. The important properties are:

1. The nano-fertilizers have a higher surface area, which is mainly due to the very smaller size of the nanoparticles that provide more sites to facilitate the different metabolic process in the plant system. This results in the production of more photosynthesis with less consumption of nutrient elements.
2. They have high solubility in different solvents such as water.
3. The particle size of nano-fertilizers is very small (less than 100 nm), which facilitates more penetration of nanoparticles into the plant system.
4. Nanofertilizer elements have the large surface area and particle size smaller than the pores of root and leaves of the plants. This increases penetration into the plant system from applied surfaces and thus improves uptake and nutrient use efficiency of the nano-fertilizers.
5. The reduced particle size of Nanofertilizer results in increased specific surface area and a number of particles per unit area of fertilizer, which provide more opportunity for contact of nano-fertilizers and it leads to more penetration and uptake.
6. The fertilizer elements encapsulated in nano-particles increases the availability and hence uptake of the plant nutrients to the crops.

Zeolite-based nano-fertilizers are capable of releasing the nutrients slowly to the crop plant which increase the availability of nutrient elements to the crops though out the growth period. This prevents loss of nutrients through denitrification, volatilization, leaching and fixation in the soil especially NO_3 and NH_4 forms of Nitrogen.

Biosafety Issues of Nanofertilizer

Undoubtedly nanotechnology has incredible potential to revolutionize many aspects of human life. However, the advancement of this multidisciplinary branch of science, especially the benefits from the practical application of nanoparticles have to be considered with some precautions. The major concern at world scale is whether the unknown risks of nanoparticles involving their environmental and health impact prevail over their potential benefits. Thus, the risks associated with the application of nanoparticles are yet to be evaluated before the application of nanoparticles is fully accepted and implemented. Hence, "nanotoxicology" has been developed, which is responsible for assessment of the toxicological potential and promoting safe design and use of nanoparticles. Literally all substances can be toxic to plants, animals or humans at some exposure level. However, this does not limit their use in various applications which are formulated minding the critical exposure concentration. As mentioned above, the promoting effect of nanoparticles on plant growth and physiology is expressed at very low concentrations; hence, it is hard to believe that these concentrations will pose significant health and environmental damage. The studies conducted at Tamil Nadu Agricultural University on Nano zeolites based N fertilizer indicated no inhibitory effects on soil beneficial microorganisms and earth worms. The study on the effect of nanozeolite on microorganisms and earthworm showed that the population counts of *Rhizobium sp.* and *Pseudomonas* sp. were higher in nanozeolite treatments than control. Genotoxicity effect of nano-zeolite was analyzed using *Rhizobium* as test organism and the micro slides image depicted that there was no quantifiable DNA damage in nano zeolite treated cells in comparison to the control cultures. Zeolites regardless of dose or size did not cause any mortality of earthworms

Nanofertilizer Studies at Tamil Nadu Agricultural University

Clay based nanofertilizer was formulated by Sharmila Rahale and Subramanian (2011) using Zeolite, montmorillonite, halloysite and bentonite as carriers. The zeolites (clinoptilolites) are of vulcanoclastic origin with a high cationic interchange capacity (270 C mol(p^+) kg^{-1}). Initially the zeolites were in the dimension of 1-2 µm in size. The clinoptilolite was subjected to ball milling of one hour to achieve particles of 30-40nm in diameter. Commonly occurring -clays such as montmorrilionite, halloysite and bentonite in the dimension of 30-40 nm sized particles were used for this study. All the nanoparticles were characterized using SEM, FTIR, AFM and EDAX before and after nutrient loading of NO_3–N, PO_4^{3-}, K and Zn ions. All the nanoparticles were charatacterised using SEM, FTIR, AFM and EDAX before and after nutrient loading of NO_3–N, PO_4^{3-}, K and Zn ions (Figure 12.3). These studies were intended to develop a composite nano-formulations to fertilize maize crop. The nitrate release pattern of nonfertilizer showed that at the start of the experiment, the NO_3-N was the highest concentration of 500 mmolL^{-1} N in the leachate from nano-zeolite followed by nano-montmorillonite (350 mmolL^{-1}), nano-halloysite (200 mmolL^{-1}), nano-bentonite (120 mmolL^{-1}) and pure urea (70 mmolL^{-1}). Furthermore, it can be seen that at the initial stage, nitrate release, from all the nano-clays occurred rapidly and after 240 h slow release was detected. The data revealed that the entire NO_3^- - N from urea is exhausted after 200 h beyond which the concentration of NO_3-N becomes undetectable. However, the release of nitrate from nano-zeolite continued even after 1176 h, with concentrations ranging from 110 to 114 mmol L^{-1}. The results clearly demonstrated slow and steady release of N from nano-zeolite for more than

45 days while conventional fertilizer does it for only 8 days. It has been found that the slow release of fertilizers normally follows first-order kinetics. The results showed that the release takes place in different stages in nanoparticles. However, only one stage is observed for pure urea, characterized by a steep slope with a decay time of 200 h after which the nitrate concentration reaches below detectable limits. It can be assumed that the available N has been completely released.

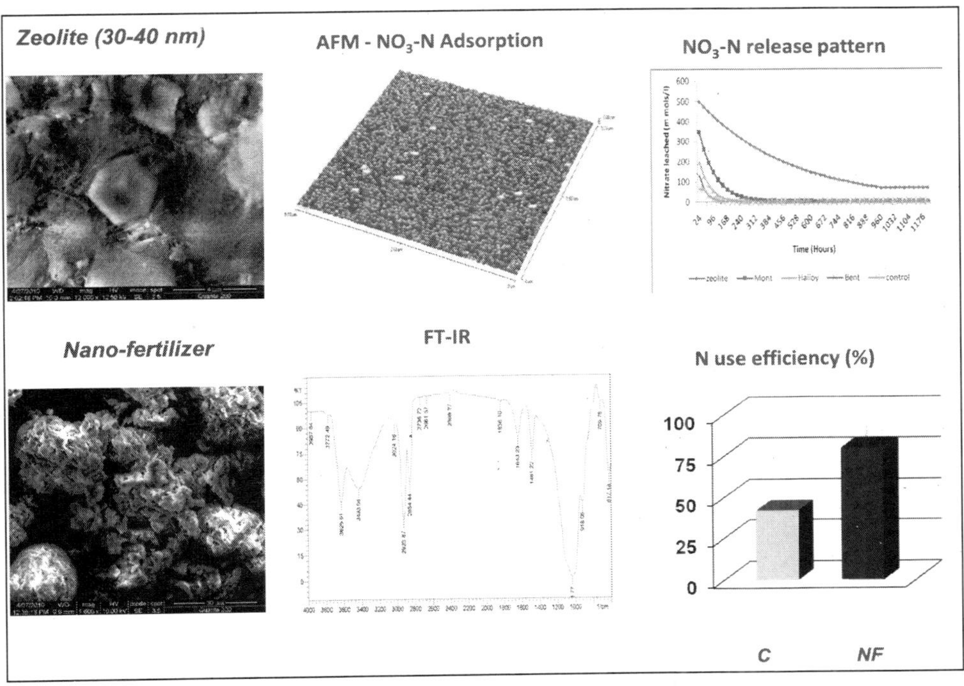

Figure 12.3. Nitrate Loaded Nano Zeolite - Characterization, Nutrient Release, N Use Efficiency.

The comparative NUE of conventional and nanofertilizer was determined using ^{15}N tracer. The results showed that nanofertilizer recorded highest NUE (74.03 per cent) when compared to conventional fertilizer. The nanofertilizer increased the NUE up to 45 per cent over control.

The sorption experiment showed that the sorption of phosphate ions on zeolitic surfaces can be well-described by the Langmuir equation. The result showed that the highest sorption of 923.7 mg kg^{-1} was observed for nano-zeolite. The release pattern of phosphate ions showed that at the start of the experiment, a maximum concentration of 650 mmolL^{-1} P is observed in the leachate from nano-zeolite followed by nano- montmorillonite (60 mmolL^{-1}), nano-halloysite (59 mmolL^{-1}), nano-bentonite (55 mmolL^{-1}) and pure SSP (49 mmolL^{-1}). In the initial stage, phosphate release, from all the nano-clays, occurs rapidly and attains a static at about 216 h, after which slow release was observed. But in SSP, the entire P was exhausted after 384 hours beyond which the concentration of phosphate ions reached below detectable limits. However, the release of phosphate ions from nano-zeolite is continued even after 1176 h, with a concentration of 20 mmol L^{-1}.

Nano-zeolite showed leaching of five stages. Phosphate levels decrease sharply in the first eighteen days (k_1 = 0.43), and stabilize thereafter to give several fractions with different rate constants. After that it follows linear release pattern with time The result showed that from 400 to 800 hours it follows second stage (k_2 = $0.8x10^{-3}$). In this stage it supplies 2.5 mmol L^{-1} of phosphate ions. The next stage falls between 800 and 1000 hours (k_3 = $0.7x10^{-3}$) and it supplies 2.0 mmol L^{-1} of phosphate ions. From 1000 hours onwards, it releases 2.0 mmol L^{-1} of phosphate ions and it continues.

The release pattern of potassium showed that the highest sorption of 1200.5 mg kg^{-1} was observed for nano-zeolite. The It can be observed from the data that at the start of the experiment, the maximum concentration of 650 mmolL^{-1} K is observed in the leachate from nano-zeolite followed by nano-montmorillonite (390 mmolL^{-1}), nano-halloysite (310 mmolL^{-1}), nano-bentonite (300 mmolL^{-1}) and pure potassium chloride (150 mmolL^{-1}). In the initial stage, potassium release from all the nano-clays, occurs rapidly and attained a static at about 216 h, after which slow release was observed. But the in potassium chloride, the entire K was exhausted after 216 hours beyond which the concentration of potassium reached below detectable limits. However, the release of potassium from nano-zeolite is continued even after 1176 h, with a concentration of 110 mmol L^{-1}. The potassium release take place in several stages and follows a first order kinetic law, whose rate constants were determined. The results indicated that from 400 to 700 hours it followed second stage (k_2 = $9.44x10^{-3}$). In this stage it supplied 3.9 mmol L^{-1} of potassium. The Next stage falls between 700 and 1000 hours (k_3 = $1.7x10^{-3}$) and it supplies 3.0 mmol L^{-1} of potassium. From 1000 hours onwards, it released 2.7 mmol L^{-1} of potassium and it continued.

The results of desorption experiment of Zn showed that at the start of the experiment, a maximum concentration of 22 ppm Zn was observed in the leachate from nano-zeolite followed by nano-montmorillonite (12 ppm), nano-halloysite (5 ppm), nano-bentonite (3ppm) and zinc sulphate (2.5 ppm). Furthermore, it can be seen that in the initial stage, zinc release, from all the nano-clays, occurs rapidly and attains a static at about 312 h, after which slow release is observed. The data revealed that the entire available Zn from zinc sulphate was exhausted after 120 h beyond which the concentration of zinc reached below detectable limits. However, the release of zinc from nano-zeolite was continued even after 1176 h, with a concentration of 5 ppm.

Thirunavukkarasu and Subramanian, (2015) developed a sulfur nanofertilizer using surface modified zeolite. Surface modification is done by hexadecyl trimethyl ammonium bromide and it is enriched with ammonium sulfate as a sulfur source. The surface modified nano-zeolite has absorbed 23 per cent more sulphate ions through binuclear bridging complex. This may be due to the preferential adsorption at the positive site of zeolite. It has recorded a slow and sustained release of sulfur from the complex. The desorption of the developed nanofertilizer has recorded a linear release of sulfur from the clay mineral. The slow release is may be due to the extensive surface area and porous nature of zeolite.

In conventional method of nitrogenous fertilizer application, nitrogen is lost by many losses such as volatilization and leaching. In order to reduce the losses and to get slow and controlled release, nitrogen based nanofertilizers are prepared. Nano-sized adsorbents such as zeolite and biochar were impregnated with urea at 1:1 ratio and tested in Alfisols and Inceptisols. The prepared nanofertilizers have recorded reduced volatilization ammonia loss (VAL) in both the soils. The VAL of different treatments are, Zeo urea (13.5 days, 15.1 days), Nano-zecurea (15.5 days, 16 days), biourea (7.5 days, 7.1 days), Nano-biourea (9 days, 9.7 days) and conventional Urea (5.5 days, 4.6 days) in Alfisols and Inceptisols respectively. The study concluded that the surface amendment of physically mixed fertilizers not influenced any change on both soils. Urea impregnation influenced the days and cumulative VAL percentage. It proves that the nano porous adsorbents are potential substrate to reduce VAL rate of urea in both Alfisols and inceptisols (Manikandan and Subramanian, 2017).

In conventional application, the use efficiency of zinc fertilizers is only about 2-3 per cent due to fixation of zinc in the soil. In order to improve the Zn use efficiency of crops, Yuvaraj and Subramanian, (2017) has developed a nano-zeolite based zinc fertilizer. They found that the zinc sorption on nano-zeolite and other nano-carrier such as manganese core shell, carbon sphere and ZnO showed that the amount of zinc sorbed increases with the increase in equilibrium Zn concentration. Among them, nano-zeolite adsorbed more zinc (126.7 ppm) followed by manganese core shell (120.5 ppm), carbon sphere (106.1) and ZnO (98.5) in concentration of 200 ppm of $ZnSO_4$ added. The zinc desorption of nano zeolite (61.2 ppm), manganese core shell (55.7 ppm), carbon sphere (50.6 ppm) and ZnO (41.9 ppm) were detected when added DTPA extractable zinc.

Zinc release pattern revealed that Zn release from nano-zeolite continued up to 888 h, with a concentration of 1.3 ppm while the entire available Zn from zinc sulphate exhausted after 504 h beyond which the concentration of zinc reached below detectable limits.

The results of the greenhouse experiments using nano-Zn fertilizer showed that the zeo-Zn and core shell Zn had increased the Zn availability, growth and yield parameters besides grain Zn content. In this study, Zn concentrations were assessed in whole grain (25.1 mg kg^{-1}), dehusked grain (21.0 mg kg^{-1}) besides polished grain (18.2 mg kg^{-1}). Same set of rice in control treatment had registered 20.5, 16.2 and 13.5 mg kg^{-1}, respectively. Interestingly, the data have shown clearly that nano-Zn forms are capable of enhancing the Zn content of rice grains at all levels of processing. This should be studied in depth in order to gain insights into the mechanisms involved in Zn fortification in rice grain.

Overall, the data clearly demonstrated that the Zn loaded nano- zeolites can be employed as an effective carrier to fortify $ZnSO_4$ to develop nano-fertilizer formulations. These forms of fertilizers are capable of releasing nutrients slowly and steadily for a period of 37 days that eventually improved the Zn use efficiency of loaded fertilizer formulations. These nano-fertilizers appear to be efficient in improving Zn use efficiencies besides Zn fortification in grains without associated environmental hazard. More pot cultures and field experiments are required prior to commercialization.

Conclusion

Nanofertilizers have a significant impact in the agriculture sector for achieving enhanced productivity and the potential benefits of nanofertilizers have stimulated a great interest to increase the production potential of agricultural crops under the current climate change scenario. The basic economic benefits of the use of nanofertilizers are reduced leaching and volatilization associated with the use of conventional fertilizers. Simultaneously, the well-known positive impact on yield and product quality has a tremendous potential to increase growers' profit margin through the utilization of this technology. However, despite the exciting outcomes of nanofertilizers in the field of agriculture, so far, their relevance has not yet been focused towards marketability. The interaction of nanomaterials with the environment and beneficial flora and fauna must be explored in detail before promoting nanofertilizers at a commercial scale. Carrying out the life cycle analysis of Nanofertilizers would provide more insight to the biosafety concerns. Furthermore, an in-depth evaluation of the effect of nanofertilizers in the soils with different physio-chemical properties is necessary in order to recommend a specific nanofertilizer for a specific crop and soil type. The mechanism behind the release and uptake of nutrients from the nano formulation needs more elaborated studies to prove the concept more lucidly.

REFERENCES

1. Abdel-Aziz, H. M., Hasaneen, M. N. and Omer, A. M. (2016). Nano chitosan-NPK fertilizer enhances the growth and productivity of wheat plants grown in sandy soil, *Spanish Journal of Agricultural Research, 14*(1), 0902.

2. Allen E. R, Hossner L.R, Ming D.W, Henninger D.L. (1993). Solubility and cation exchange in phosphate rock and saturated clinoptilolite mixtures. *Soil Science Society ofe America Journal. 57*(5): 1368-74.

3. Choudhary, R. C., Kumaraswamy, R. V., Kumari, S., Sharma, S. S., Pal, A., Raliya, R. and Saharan, V. (2017). Cu-chitosan nanoparticle boost defense responses and plant growth in maize (*Zea mays* L.), *Scientific Reports, 7*(1), 9754.

4. Delfani, M., B aradarn., Firouzabadi, M., Farrokhi, N. and Makarian, H. (2014). Some physiological responses of black-eyed pea to iron and magnesium nanofertilizers, *Communication in Soil Science and Plant Analysis, 45*(4), 530 - 540.

5. Dimkpa, C. O., Bindraban, P. S., Fugice, J., Agyin- Birikorang, S., Singh, U. and Hellums, D. (2017a). Composite micronutrient nanoparticles and salts decrease drought stress in soybean, *Agronomy for Sustainable Development, 37*(1), 5.

6. Dimkpa, C. O., White, J.C., Elmer, W. H. and Gardea- Torresdey, J. (2017b). Nanoparticle and ionic Zn promote nutrient loading of sorghum grain under low NPK fertilization, *Journal of Agricultural and Food Chemistry, 65*(39), 8552–8559.

7. Dimkpa, C. O., Singh, U., Bindraban P. S, Adisa, I. O., Elmer, W. H., Gardea-Torresdey, J. L. and White, J. C. (2019). Addition-eomission of zinc, copper,

and boron nano and bulk oxide particles demonstrate element and size-specific response of soybean to micronutrients exposure, *Science of the Total Environment, 665*, 606.

8. Eichert, T. and Goldbach, H. E. (2008). Equivalent pore radii of hydrophilic foliar uptake routes instomatous and astomatous leaf surfaces-further evidence for a stomatal pathway, *Plant physiology, 132*, 491 - 502.

9. Eichert, T., Kurtz, A., Steiner, U. and Goldbach, H. E. (2008). Size exclusion limits and lateral heterogeneity of the stomatal foliar uptake pathway for aqueous solutes and water suspended nanoparticles, *Plant physiology, 134,* 151 - 160.

10. Ferrari, M. (2005). Cancer nanotechnology: opportunities and challenges. *Nature Reviews Cancer, 5(3),* 161 - 171.

11. Koelmel, J., Leland, T., Wang, H. H., Amarasiriwardena, D. and Xing, B. S. (2013). Investigation of gold nanoparticles uptake and their tissue level distribution in rice plants by laser ablation-inductively coupled-mass spectrometry, *Environmental Pollution, 174,* 222 - 228.

12. Kottegoda, N., Munaweera, I., Madusanka, N. and Karunaratne, V. (2011). A green slow-release fertilizer composition based on urea-modified hydroxyapatite nanoparticles encapsulated wood, *Current Science, 101*(1), 73 - 78.

13. Lee, E. H., Yeom, H. J., Ha, M. S., Bae, D. H. (2010). Development of banana peel jelly and its antioxidant, and textural properties. *Food Science and Biotechnology, 19*, 449 - 455.

14. Li, Z. (2003). Use of surfactant-modified zeolite as fertilizer carriers to control nitrate release. *Microporous and Mesoporous Materials, 61*,181– 188.

15. Liu, R. and Lal, R. (2014). Synthetic apatite nanoparticles as a phosphorus fertilizer for soybean (*Glycine max*), *Scientific Reports, 4*, 5686.

16. Liu, X., Zhang, F., Zhang, S., He, X., Wang, Rp., Fei, Z. and Wang, Y. (2005). Responses of peanut to nano-calcium carbonate, *Zhiwu Yingyang Yu Feiliao Xuebao, 11*(3), 385 - 389.

17. Manikandan, A. and Subramanian, K. S. (2017). Study on mitigation of ammonia volatilization loss in urea through adsorbents. *Journal of Applied and Natural Science, 9*(2), 688 - 692.

18. Marivic, T., Veronica, M., Catalino, A. (2019). Monet Concepcion MaguyonDetras1, and Clint Charles Brutas: MATEC Web of Conferences 268, 05002.

19. Millan, G., Agosto, F., Vazquez, M. (2008). Use of clinoptilolite as a carrier for nitrogen fertilizers in soils of the Pampean regions of Argentina. Cien. Inv. Princess Jenine Nido, Veronica Migo, Concepcion Maguyon-Detras, and Catalino Alfafara: MATEC Web of Conferences 268, 05001 Allen ER, Hossner.

20. Peng, C., Duan, D., Xu, C., Chen, Y., Sun L., Zhang, H., Yuan, Zheng, L., Yang, Y., Yang, J., Zhen, X., Chen, Y. and Shi. J. (2015). Translocation and

biotransformation of CuO nanoparticles in rice (*Oryza sativa* L.) plants, *Environmental Pollution, 197,* 99 - 107.

21. Peppas, N. A. (2004). Intelligent therapeutics: Biomimetic systems and nanotechnology in drug delivery. *Advanced Drug Delivery Reviews, 56*(11): 1529 - 153.

22. Rai V, Acharya S, Dey N. 2012. Implications of Nanobiosensors in Agriculture. *Journal of Biomaterials and Nanobiotechnology, 3*(2A), 315-324.

23. Rico, C. M., Johnson, M. G. and Marcus, M. A. (2018). Cerium oxide nanoparticles transformation at the root-soil interface of barley (*Hordeum vulgare* L.). *Environmental Science-Nano, 5,* 1807 - 1812.

24. Saharan, V., Sharma, G., Yadav, M., Choudhary, M. K., Sharma, S. S., Pal, A. and Biswas, P. (2015). Synthesis and in vitro antifungal efficacy of Cu–chitosan nanoparticles against pathogenic fungi of tomato, *International Journal of Biological Macromolecules, 75,* 346 - 353.

25. Salem, N. M., Albanna, L.S., Awwad, A. M., Ibrahim, Q. M. and Abdeen, A. O. (2015). Green synthesis of nano-sized sulfur and its effect on plant growth. *Journal of Agricultural Science, 8*(1), 188.

26. Sharmila Rahale. C. (2011). Nutrient release pattern of nanofertilizer formulation. Ph.D (Ag.) Thesis. Tamil Nadu Agricultural University, Coimbatore.

27. Subbaiah L. V., Prasad, T. N. V. K. V., Krishna, T., Sudhakar, G. P., Reddy, B. R. and Pradeep, T. (2016). Novel effects of nanoparticulate delivery of zinc on growth, productivity, and zinc biofortification in maize (*Zea mays* L.), *Journal of Agricultural and Food Chemistry, 64*(19), 3778 - 3788.

28. Sun, D., Hussain, H., Yi, Z. and Siegele, R. (2014). Uptake and cellular distribution, in four plant species, of fluorescently labeled mesoporous silica nanoparticles. *Plant Cell Reports, 33*(8), 1389 - 1402.

29. Tarafdar, J. C., Raliya, R., Mahawaz, H. and Rathore, I. (2014). Development of Zinc Nanofertilizer to Enhance Crop Production in Pearl Millet (*Pennisetum americanum*) Journal of *Agricultural Research, 3*(3), 357-362.

30. Thirunavukkarasu, M. and Subramanian, K. S. (2015). Synthesis and characterization of surface modified nano-zeolite fortified with sulphate and its sulfate sorption and desorption pattern. *Journal of Scientific and Industrial Research, 74*(12), 671 - 675.

31. Van, S. N., Minh, H. D. and Anh, D. N. (2013). Study on chitosan nanoparticles on biophysical characteristics and growth of Robusta coffee in green house, *Biocatalysis and Agricultural Biotechnology, 2*(4): 289 - 294.

32. Yin, L, Y., Cheng, Y. W., Espinasse, B., Colman, B. P., Auffan, M., Wiesner, M., Rose, J., Liu, J. and Bernhardt, E. S. (2011). More than the ions: The effects of silver nanoparticles on *Lolium multiflorum*. *Environmental Science and Technology, 45,* 2360 - 2367.

33. Yuvaraj, M. and Subramanian, K. S. (2017). Development of Slow Release Zn Fertilizer Using Nano Zeolite as Carrier. *Journal of Plant Nutrition, 1,* 1- 8.

Chapter 13

Nanotechnology for Weed Management

*N. Viji[1], C.R. Chinnamuthu[2] and S. Marimuthu**

[1]Department of Nano Science and Technology, [2]Professor and Head, Dept. of Agronomy*
Tamil Nadu Agricultural University Coimbatore – 641 003, Tamil Nadu
**e-mail: sm20@tnau.ac.in*

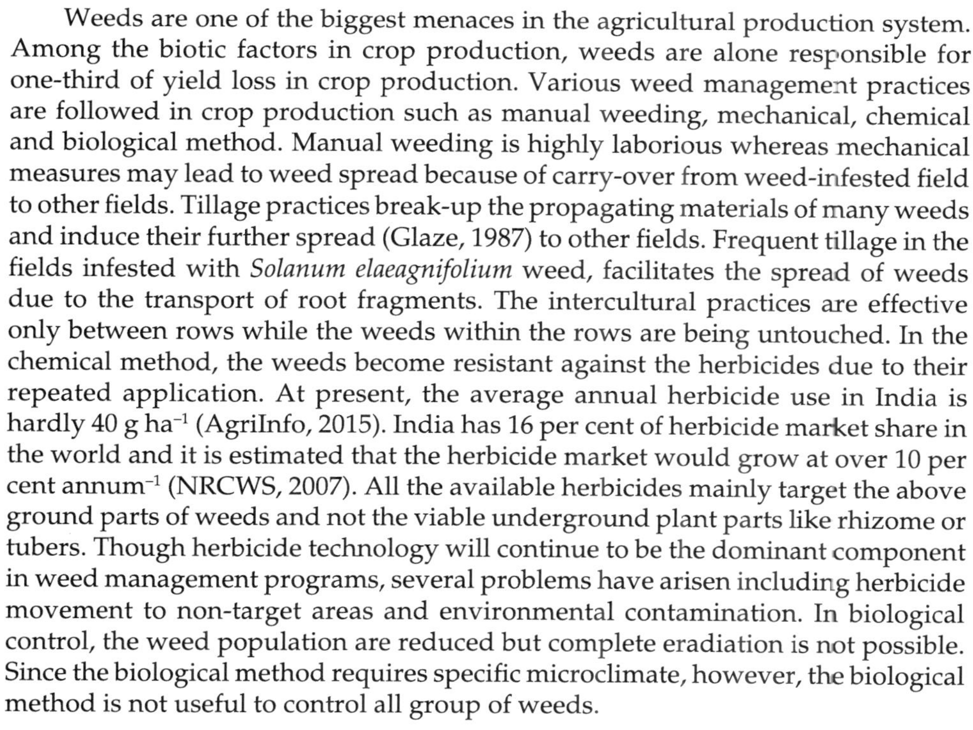

Weeds are one of the biggest menaces in the agricultural production system. Among the biotic factors in crop production, weeds are alone responsible for one-third of yield loss in crop production. Various weed management practices are followed in crop production such as manual weeding, mechanical, chemical and biological method. Manual weeding is highly laborious whereas mechanical measures may lead to weed spread because of carry-over from weed-infested field to other fields. Tillage practices break-up the propagating materials of many weeds and induce their further spread (Glaze, 1987) to other fields. Frequent tillage in the fields infested with *Solanum elaeagnifolium* weed, facilitates the spread of weeds due to the transport of root fragments. The intercultural practices are effective only between rows while the weeds within the rows are being untouched. In the chemical method, the weeds become resistant against the herbicides due to their repeated application. At present, the average annual herbicide use in India is hardly 40 g ha^{-1} (AgriInfo, 2015). India has 16 per cent of herbicide market share in the world and it is estimated that the herbicide market would grow at over 10 per cent $annum^{-1}$ (NRCWS, 2007). All the available herbicides mainly target the above ground parts of weeds and not the viable underground plant parts like rhizome or tubers. Though herbicide technology will continue to be the dominant component in weed management programs, several problems have arisen including herbicide movement to non-target areas and environmental contamination. In biological control, the weed population are reduced but complete eradiation is not possible. Since the biological method requires specific microclimate, however, the biological method is not useful to control all group of weeds.

To overcome the issues related to existing weed control methods, the latest method should be adapted to get better weed control efficiency. Nanotechnology facilitates for efficient delivery of chemicals through nanomaterials based formulations. Nanotechnology approach reduces the number of toxic herbicides used in the field, due to its slow and sustained release behaviour. A small quantity of nano-herbicide is easily blended with soil, which kill target weed seeds buried below the reach of conventional herbicides (Food quality news, 2006). Herbicides are encapsulated with various substances to get target specific and efficient release to the plants (Agri Nano biotech, 2016). Nanoparticles are designed to reach and kills target weeds. The nanoherbicides are having high penetration efficiency than conventional herbicides. Obviously, nanostructured herbicides reduce the consumption rate of conventional herbicides. Additionally, nano herbicides target only the weeds, which minimize the risk of developing herbicide resistance (Abigail and Chidambaram, 2016).

Encapsulation of Herbicides

Nanoencapsulated herbicides are known to control weeds while reducing the phytotoxicity of herbicides on crops, demonstrating the benefits of smart delivery systems in agriculture. Properly functionalized nanocapsules provide better penetration through the cuticle and allow slow and controlled release of active ingredients reaching the target weed. Nano encapsulation of chemicals with biodegradable materials makes safer handling by the growers. Efforts are undertaken to kill the notorious weeds like *Cyprus rotundus* through a smart delivery system, where root tubers are rich in starch that has to be exhausted through a suitable smart delivery system (Kanimozhi and Chinnamuthu, 2012) to reduce the further germination. Nanoencapsulated agrochemicals are designed to hold the necessary properties (effective concentration, stability and solubility) in the formulation, for time-controlled release in response to certain stimuli, enhanced targeted activity and less ecotoxicity, with safer and easy mode of delivery, avoiding repeated applications.

Fabrication of a nano-encapsulated slow-release herbicide formulation help to increase the productivity of crops by arresting the growth of new flushes of weeds appearing at all stages of crop growth. The slow-release herbicides are highly useful to attain seasonal long weed control (Agnihotri *et al.*, 2012). The herbicide active ingredients are encapsulated by several techniques including direct, indirect, solvent evaporation and spray drying methods. Encapsulation of herbicides is carried out with various substances including surfactants, polymers, nanoparticles and natural materials. The shell material is chosen based on the absorption behaviuor. In core-shell encapsulation method, the core is insoluble in nature, while the shell material is dissolved at desired conditions. The release rate generally depends on the thickness of the encapsulating shell used in the coating. Thicker shells lead to longer duration of release (Arida and Al-Tabakha, 2007), while the slow-release herbicidal formulation is synthesized by selecting a suitable carrier with the higher absorption capacity of active ingredients. The losses with such herbicide formulations are minimized in the field level.

Nano Encapsulation of Herbicides in Polymeric Capsules

Polymers are widely used as either nanospheres or nanocapsules for the delivery of herbicides. The polymeric nanoparticles having good physico-chemical properties, enhances the bioavailability of active ingredient with good biocompatibility. The polymeric nanoparticle, poly (epsilon caprolactone) was used to encapsulate atrazine herbicide. The product shows higher stability index and good polydispersity index for about 270 days. The nano herbicide formulation reduced the mobility of the herbicide in the soil while it enhanced the herbicidal activity compared to the free form of atrazine. The formulation was tested against *Brassica* sp., showed better weed control efficiency than conventional herbicide. The encapsulated triazine group of herbicide was released from the polymeric nanoparticle complex in a controlled manner through the relaxation of polymeric chains. The formulation also exhibited less genotoxicity than the conventional formulation (Pereira *et al.,* 2014). Schnoor *et al.* (2018) attempted to encapsulate atrazine herbicide using poly(lactic-co-glycolic-acid) polymer nanoparticles, which is a biocompatible and biodegradable polymer. The herbicide formulation exhibited the degradation based release profile *i.e.* the encapsulated herbicide is released from the core with the degradation of nanoparticles. This preventing the rapid release of active ingredient from the carrier after reaching the plant surface. The simulation model of poly(lactic-co-glycolic-acid) encapsulated herbicide is given below in Figure 13.1, where the polymer forms the core and the shell is made of 1,2-Distearoyl-snglycero3-phosphoethanolamine-N- [amino(polyethylene glycol). The herbicides atrazine was embedded in the polymer matrix.

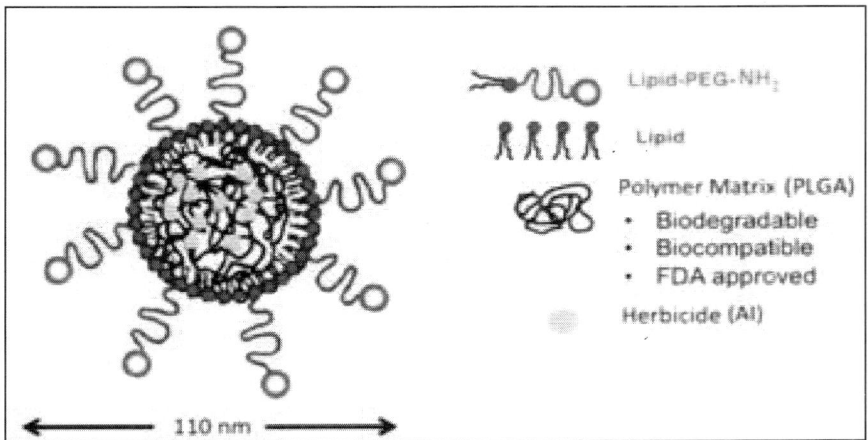

Figure 13.1. Encapsulation of Atrazine in Poly (lactic-co-glycolic-acid) for the Controlled Release of Herbicide.

Courtesy: Schnoor *et al.*, 2018.

The herbicide alachlor is susceptible to volatilization losses. Due to the loss, weed control efficiency of the herbicide was reduced. Dailey (2004) encapsulated alachlor formulations using ethyl cellulose through microencapsulation method.

It shows reduced volatilization and recorded higher weed control efficiency. The contact herbicide paraquat was encapsulated with nanoparticles to alter the release efficiency of the herbicide (Silva *et al.*, 2011). The herbicide exhibited good compatibility with the nanoparticles, which recorded a gentle release profile. They conducted soil sorption profile test, which revealed that the herbicide sorption profile was reduced when paraquat was encapsulated with the nanoparticles. Hence the herbicidal activity of nanoparticle encapsulated paraquat herbicide was increased. In order to reduce the toxicity in the soil ecosystem and to increase the herbicidal activity, paraquat herbicide was encapsulated with the alginate bead or chitosan, thereby reducing the negative impacts of herbicide while achieving the target specific weed control (Silva *et al.*, 2011).

β-Cyclodextrin is a suitable agent for improving pesticide water solubility. Due to enhanced stabilization, bioavailability and controlled release properties, a variety of new herbicidal formulation are reported using β-cyclodextrin. An experiment was conducted with 2, Methyl 4 Choloro Phenoxy Acetic acid (MCPA), which is used to control broadleaved weeds in cereals and pulses. In order to improve its solubility, MCPA was encapsulated in β-cyclodextrin and an inclusion complex is formed (Garrio *et al.*, 2012). The cyclodextrin encapsulated MCPA exhibited increased herbicide solubility and better weed control efficiency. Terbuthylazine (TBA) is a herbicide used to control grasses and broadleaved weeds, while TBA has poor aqueous solubility profile resulting in reduced bioavailability. Garrido *et al.* (2017) synthesized an inclusion complex of TBA with 5- cyclodextrin, exploiting the presence of hydrophobic cavity in cyclodextrin. The inclusion complex was noticed for its higher solubility and weed control efficiency. The herbicides such as imazapic and imazapyr are the chemicals used to control weeds in maize, soybean and groundnut. Maruyama *et al.* (2016) attempted to encapsulate imazapic and imazapyr with various materials such as alginate and chitosan tripolyphosphate nanoparticles. Compared to free form, the nanoparticle encapsulated herbicides recorded higher weed control efficiency of 60 per cent exhibiting slow and sustainable releasing pattern. The study proved that the encapsulation of the imazapic and imazapyr herbicides with the nanoparticles improved the efficiency of herbicidal action and reduced toxicity.

Pradeesh Kumar and Chinnamuthu (2014) fabricated pendimethalin herbicide nano-formulation using layer-by-layer technique. In direct method, manganese carbonate ($MnCO_3$) nano core template was synthesized by precipitation method. The mono-dispersed pendimathalin herbicide was encapsulated into the synthesized $MnCO_3$ core via layer by layer method. The characterization with electron microscopy and infrared spectroscopy confirmed the encapsulation of pendimethalin in the $MnCO_3$ template. Besides, they encapsulated pendimathalin active ingredient through solvent evaporation method. The release efficiency of pendimathalin from the encapsulated complex revealed that the herbicide was released slowly for 40 days. The encapsulated herbicide matrix was tested in legume crop, recorded seasonal long weed control and higher yield.

Nanoherbicides for the Control of Purple Nudsedge

Cyperus rotundus (Purple nutsedge) is one of the most troublesome weeds and ranked as one of the world worst weed. The weed is propagated by vegetative propagation mainly by tubers. The tubers contain many buds where only a few buds are sprouted, while the remaining buds will remain dormant and waiting for a favourable condition for germination. The dormancy of the tubers is mainly due to the presence of phenolic compounds in the tubers. Ravisankar and Chinnamuthu (2012) have attempted to eradicate the dormancy factor present in the nutsedge tuber using a growth regulator. The tubers were treated with cytokinin at a concentration of 0.01 per cent and recorded the highest phenol degradation efficiency with higher germination. Moreover, 0.01 per cent cytokinin and metolachlor herbicide at 2 kg ha^{-1} was applied, which showed the reduced viability of tubers. The application of cytokinin with the post-emergence herbicide glyphosate at 2.5 kg ha^{-1} have induced bud sprouting and the sprouts are killed through the application of herbicide.

Nanoparticles play a vital role in exhausting the weed seed bank of many weeds. The weed seed in the soil bank may be dormant for many years indicating their viability of seeds. Nanoparticles act as phenol degraders and used to exhaust the food reserves of the plant material. Brindha and Chinnamuthu (2012) synthesized zinc oxide (ZnO) nanoparticles in the laboratory and used for dormancy breaking studies of nutsedge tubers. The tubers treated with ZnO nanoparticles at 1500 mg kg^{-1} under dry dressing with tubers and 2250 mg kg^{-1} under wet treatment respectively, increased the tuber germination percentage inducing phenol degradation.

A laboratory experiment was carried out to break the dormancy of the nutsedge weed using the magnetic iron oxide nanoparticles, which act as a nano-adsorber (Viji and Chinnamuthu, 2015). Iron oxide nanoparticles at 3.0 g kg $tubers^{-1}$ recorded a higher percentage of phenol degradation (89 per cent over control). The phenolic compounds are degraded via advanced oxidation processes which are used for the removal of organic constituents. Hydroxyl radicals are generated from nanoparticles are involved in phenol degradation. Hence, the germination percentage of the iron oxide nanoparticles treated purple nutsedge tubers was increased. Besides, the zinc oxide nanoparticles at 3.0 g kg^{-1} of tubers recorded 74.5 per cent higher rate of germination over untreated control due to the degradation of phenol. Treatment with nanoparticles may lead to the production of more hydroxyl radicals results in the oxidation of phenol, thereby increasing the germination of tubers (Viji and Chinnamuthu, 2015).

Viji and Chinnamuthu (2018) studied the effect of titanium dioxide nanoparticles on nutsedge control. Titanium dioxide nanoparticles at 2.5 g kg^{-1} of nutsedge tubers recorded a higher percentage of phenol degradation *i.e.*, 69.7 per cent compared to control, which was on par with titanium dioxide nanoparticles at 2.0 and 1.5 g kg^{-1} of tubers recorded 68.8 and 67.2 per cent phenol degradation, respectively compared to control. Titanium dioxide nanoparticles at 2.5 g kg^{-1} of tubers recorded 41.7 percentage of germination over control. Then, the germinated weeds are managed by different control measures. Hence weed seed bank exhaustion of purple nutsedge is achieved using nanoparticles. Silver nanoparticles were treated at the concentration of 2.5 g kg^{-1} of nutsedge tubers recorded higher starch degradation (7.3 per cent) over control. The silver nanoparticles assist the enzymatic degradation of starch present in the tuber. Silver nanoparticle combined with alpha-amylase enzyme at the concentration of 500 ppm + 2.5 g kg^{-1} of tubers, respectively recorded a higher rate of starch degradation (25.3 per cent) over control. The enzyme and silver nanoparticle complex facilitates the degradation of starch into reducing sugars (Viji *et al.*, 2016). Vimalrajiv *et al.* (2018) conducted an experiment to study the effect of combined application of herbicide with nanoparticles in black gram crop *var.* MDU 1. Application of hydrogen peroxide at 300 ml m^{-2} followed by pendimethalin at 0.75 kg a.i. ha^{-1} + ZnO nanoparticles at 500 ppm m^{-2} registered highest weed control efficiency and yield parameters.

Encapsulation of Herbicides using Natural Carriers

Natural carriers such as clays and organic waste materials are used as a carrier for herbicide encapsulation. The pre-emergence herbicide oxadiargyl was encapsulated in natural carriers and its effect on rice *var.* TKM 13 was studied (Bommayasamy *et al.*, 2018a). The herbicide was encapsulated in various templates such as zeolite, biochar and starch. Zeolite loaded oxadiargyl showed better weed control efficiency, significantly reducing the weed dry weight of 9.6, 19.6 g m^{-2} at 20 and 40 days after transplanting respectively. The weed control efficiency of zeolite-encapsulated herbicide was on par with the application of butachlor + hand weeding in rice. The encapsulated herbicide and conventional herbicides recorded 77.1 and 86.7 per cent grain yield, respectively over the control. Abigail *et al.* (2016) developed a nanosized rice husk and used as a carrier for the encapsulation 2,4dichlorophenoxyacetic acid herbicide (2,4D). The rice husk was milled and brought down to nanosized, which was used for surface adsorption of 2,4D herbicide. It showed enhanced and reversible sorption of 2,4D. The improved herbicidal activity of this product was noticed in *Brassica* sp.

Nanoherbicides for Rainfed Agriculture

In rainfed agriculture, weed control is achieved through the broadcasting of herbicide along with seed at the time of sowing. However, chemical weed management in rainfed areas depends on moisture availability during the application of herbicides. The prior applied herbicide will become available whenever rain occurs, during which herbicides are released in a controlled way depends on the soil moisture. Hence, herbicide become diffused into the soil due to the presence of soil moisture and killed the weeds germinating immediately after the receipt of rainfall. Therefore, the nanoherbicides are ideal for rainfed farming, where pendimathalin was encapsulated using a suitable template for rainfed agriculture. Manganese carbonate core material was coated with suitable polymers such as sodium Poly Styrene Sulfonate (PSS) and Poly Allylamine Hydrochloride (PAH) by Layer-by-Layer method (LBL) to obtain water-soluble core-shell particles. Hollow-shell particles were formed from the core-shell particles by an acid *etc.*hing, where the hollow-shell particles were loaded with pendimethalin herbicide using the passive method to get controlled release of herbicide active ingredient. (Kanimozhi and Chinnamuthu, 2012) triggered through the presence of soil moisture.

Table 13.1. Nano Carriers Used for Herbicide Encapsulation

Sl.No.	Herbicide	Method of encapsulation	References
1	Atrazine	PLGA	Chen and Wang, 2019
		PCL	Grillo *et al.*, 2012
		Starch	Marvesh *et al.*, 1995
		Nanoclay	Cea *et al.*, 2010
		Alginate	Fernandez-Perez *et al.*, 2001
		Solid lipid nanoparticles	Oliveira *et al.*, 2015
		Poly(hydroxybutyrate-co-hydroxyvalerate) microspheres	Grillo *et al.*, 2010
2	Simazine	Solid lipid nanoparticles	Oliveira *et al.*, 2015
3	Pendimethalin	Polyurea	Zhang *et al.*, 2018
4	Alachlor	EC microspheres	Sopena *et al.*, 2007
		Polymeric microcapsules	Dowler *et al.*, 1999
		Starch	Wienhold *et al*, 1993
5	2,4 D	Organoclay	Hermosin *et al.*, 2006
		β-Cyclodextrin	Pérez-Martínez *et al.*, 2000
6	Isoproturon	Alginate	Fernandez-Perez *et al.*, 2000
		Alginate-Clay- Activated carbon complexA	Garrido *et al.*, 2006
7	Terbuthylazine	β-Cyclodextrin	Garrido *et al.*, 2017
8	MCPA	β-Cyclodextrin	Garrido *et al.*, 2014
9	Thiobencarb	Alginate	Gan *et al.*, 1994
10	Picloram	Poly electrolyte biopolymers	Wang and Zhao, 2013
11	Chlorimuron-ethyl	β-Cyclodextrin	Zhang *et al.*, 2015
12	Metolachlor	Starch	Douglas *et al.*, 1994
13	Paraquat	Chitosan/tripolyphosphate nanoparticle	Grillo *et al.*, 2014

Target Specificity and Translocation of Nanoherbicides in Weeds

Target specific nanoherbicides are developed using nanoparticles. Nanoparticles are used for smart delivery due to its high surface area, sorption capacity, and controlled-release kinetics to the targeted sites. The fabrication of encapsulated herbicide active ingredient with nanoparticle aims at specific receptor present in the roots of the specific targeted weed. The nanoparticle has less diameter than the targeted cell wall could simply pass through the wall and reaches the plasma membrane. Since the encapsulated herbicide forms the core of the nanoparticle, the herbicide easily enters into the target plant system. This action creates starvation; thus it will kill the targeted weed species. Due to the controlled release of herbicide from the core, the weeds are destroyed (Chinnamuthu and Kokiladevei, 2007). The image of entry and translocation pattern of nanoherbicides in weeds is given Figure 13.2.

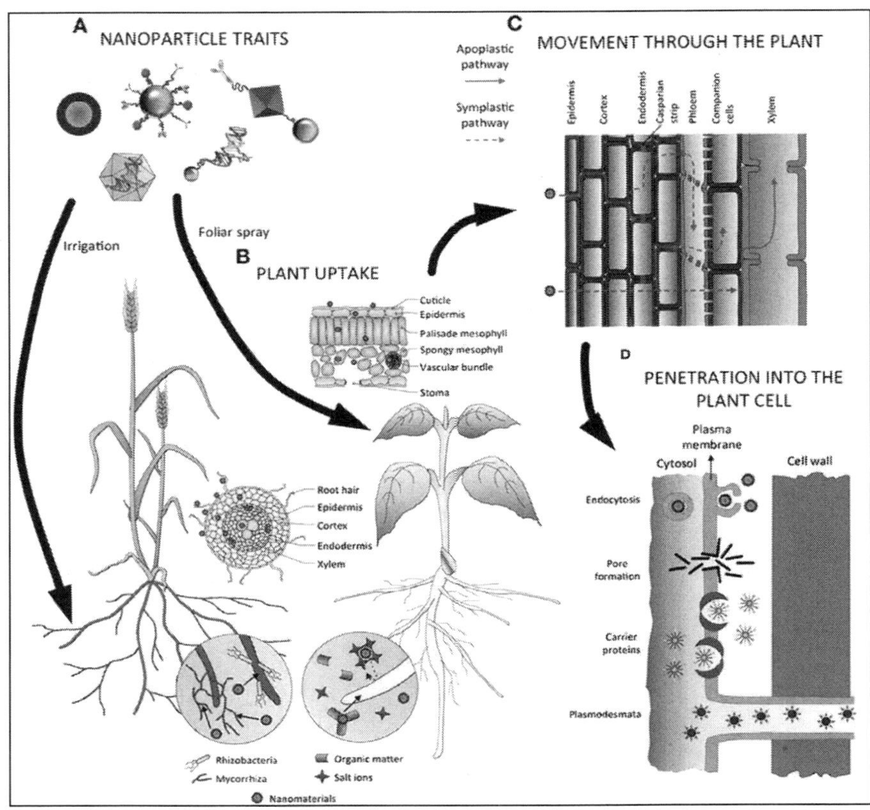

Figure 13.2. Entry of Encapsulated Herbicide into Plant System.
Courtesy: Anna *et al.*, 2019.

Conclusion

Nanotechnology has the potential scope in weed management. Generally, nanoherbicides are synthesized using a conventional herbicide, which is encapsulated within a nanoparticle's core, which facilitates the target specific control. The Nano herbicide formulation enhances the absorption and translocation efficiency of herbicides. Furthermore, nanoherbicide formulations reduce the amount of herbicide reaching the non-targets, while at the same time minimizing their negative environmental impacts. However, more research works are to be carried out to upscale the technology for field scale adoption. There is no detailed evidence regarding the mode of action of nano-herbicides in weeds. So the future researchers should focus on the mode of action and translocation pattern of nano-herbicides in weeds.

REFERENCES

1. Abigail, M, E, A, Melvin Samuel, S. and Chidambaram, R. (2016). Application of rice husk nanosorbents containing 2,4-dichlorophenoxyacetic acid herbicide

to control weeds and reduce leaching from soil. *Journal of the Taiwan Institute of Chemical Engineers, 63*, 318326.

2. Agrinanobiotech, 2016: http: / /agrinanobiotech.blogspot.tw.
3. Amns, Alharby, H. F, Hakeem, K. R. and Qureshi, M. I. (2019). Weed Control Through Herbicide-Loaded Nanoparticles. Nanomaterials and Plant Potential, 507-527.
4. Arida, A. I. and Tabakha. M. M. A. (2007). Encapsulation of ketoprofen for controlled drug release. *European Journal of Pharmaceutics and Biopharmaceutics, 66*, 48-54.
5. Bommayasamy, N, Chinnamuthu, C. R, Venkataraman, N. S, Balakrishnan, K, Rathinasamy, A and Gangaiah, B. (2018a). Effect of entrapped slow release pre-emergence herbicide oxadiargyl with zeolite, biochar, starch and water soluble polymer formulations on weed control duration and yield of transplanted rice. *International Journal of Chemical Studies, 6*(3), 1519-1523.
6. Bommayasamy, N, Chinnamuthu, C. R, Venkataraman, N. S, Balakrishnan, K, Rathinasamy, A. and Gangaiah, B. (2018b). Effect of Entrapped Preceding Rice Crop Herbicide Oxadiargyl on Growth and Yield of Succeeding Bhendi. *International Journal of Current Microbiology and Applied Sciences, 7*(6), 1915-1921.
7. Cea, M, Cartes, P, Palma, G. and Mora, M. L. (2010). Atrazine efficiency in an Andisol as affected by clays and nanoclays in ethylcellulose controlled release formulations. *Journal of Soil Science and Plant nutrition, 10*(1), 62-77.
8. Chen, X, T and Wang, T. (2019). Preparation and characterization of atrazine-loaded biodegradable PLGA nanospheres. *Journal of Integrative Agriculture, 18*(5), 1035-1041.
9. Dailey, O. 2004. Volatilization of alachlor from polymeric formulations. *Journal of Agricultural and Food Chemistry, 54*, 6742-6746.
10. Douglas, D, Buhler, Marvin, M, Schreiber and Koskinen, W. C. (1994). Weed Control with Starch-Encapsulated Alachlor, Metolachlor, and Atrazine. *Weed Technology, 8(2), 277-284.*
11. Dowler, C, Dailey, O. and Mullinix, B. (1999). Polymeric microcapsules of alachlor and metholachlor: preparation and evaluation of controlled-release properties. *Journal of Agricultural and Food Chemistry, 47*, 2908-2913.
12. Fernandez-Perez, M, Gonzalez-Pradas, E, Villafranca-Sanchez, M, Flores-Cespedes, F. (2001). Mobility of atrazine from alginate-bentonite controlled release formulations in layered soils. *Journal of Agricultural and Food Chemistry, 43*, 347-353.
13. Fernandez-Perez, M, Gonzalez-Pradas, E, Villafranca-Sanchez, M. and Flores-Cespedes, F. (2000). Mobility of isoproturon from an alginate-bentonite controlled release formulation in layered soil. *Chemosphere, 41*, 1495-1501.
14. Food quality news, (2006), http: / /www.foodqualitynews.com/R-D/Nano-herbicide-in-the-works.

15. Gan, J, Hussain, M. and Rathor, M. N. (1994). Behaviour of an alginate-kaolin based controlled-released formulation of the herbicide thiobencarb in simulated ecosystems. *Pesticide Science, 42*,265- 272.

16. Garrido E. M, Rodrigues, D, Milhazes, N, Borges, F. and Garrido. J. (2017). Molecular Encapsulation of Herbicide Terbuthylazine in Native and Modified 5-Cyclodextrin E. *Journal of Chemistry,* Article ID 8120367.

17. Garrido, E. M, Santosi, M, Silva, P, Cagide, F, Cagide, J, Garrido, J. and Borges, F. (2012). Host-guest complexes of phenoxy alkyl acid herbicides and cyclodextrins. MCPA and β-cyclodextrin. *Journal of Environmental Science and Health, Part B, 47*, 869-875.

18. Garrido, J, Cagide, F, Melle-Franco, M, Borges, F. and Garrido, E. M. (2014). Microencapsulation of herbicide MCPA with native β-cyclodextrin and its methyl and hydroxypropyl derivatives: an experimental and theoretical investigation. *Journal of Molecular Structure,1061*(1), 76-81.

19. Garrido-Herrera, F.J, González-Pradas, E. and Fernández-Pérez, M. (2006). Controlled release of isoproturon, imidacloprid, and cyromazine from alginate-bentonite-activated carbon formulations. *Journal of Agricultural and Food Chemistry. 54*,10053-10060.

20. Glaze, N.C. (1987). Cultural and mechanical manipulations of *Cyperus* spp. *Weed Technology, 1*, 82-83.

21. Grillo, R, De Melo, N. F. S, De Lima, R, Lourenco, R, W, Rosa, A. H, Fraceto, L. F. (2010) Characterization of atrazine-loaded biodegradable poly(hydroxybutyrate-co-hydroxyvalerate) microspheres. *Journal of Polymers and the Environment, 18*: 26–32

22. Grillo, R, Dos Santos, N. Z, Muruyama, C. R, Rosa, A. H, De Lima, R. and Fraceto, L. F. (2012) Poly(ε-caprolactone) nanocapsules as carrier systems for herbicides: physico-chemical characterization and genotoxicity evaluation, *Journal of Hazardous Materials, 231*, 1-9.

23. Grillo, R, Pereira, A. E. S, Nishisaka, C. S, Lima, R, Oehlke, K. and Greiner R. (2014). Chitosan/tripolyphosphate nanoparticles loaded with paraquat herbicide: an environmentally safer alternative for weed control. *Journal of Hazardous Materials, 278*, 163-171.

24. Hermosin, M. C, Celis, R, Facenda, G, Carrizosa, M. J, Ortega-Calvo, J. J. and Cornejo, J. (2006). Bioavailability of herbicide 2,4-D formulated with organoclays. *Soil Biology and Biochemistry, 38*, 2117-2124.

25. Kanimozhi, V. and Chinnamuthu, C. R. (2012). Engineering core/hollow shell nanomaterials to load herbicide active ingredients for controlled release. *Research Journal of Nanoscience and Nanotechnology*, 2(2), 58-69

26. Maruyama, C. R, Guilger, M, Pascoli, M, Jose, N. B, Abhilash, P. C, Fraceto, L. F. and Lima, R. (2016). Nanoparticles Based on Chitosan as Carriers for the Combined Herbicides Imazapic and Imazapyr. *Scientific Reports*, 6(1).

27. Mervosh, T. L, Stoller, E. W. Simmons, F. W and Ellsworth, T.R. (1995). Effects of starch encapsulation on clomazone and atrazine movement in soil and clomazone volatilization, *AGRIS, 43*(3), 445-453.

28. NRCWS. (2007). www.nrcws.org/Scenario.pdf accessed on 29.11.2019.

29. Oliveira, J. L, Campos, E. V. R. Silva, C. M. G, Pasquoto, T, Lima, R. and Fraceto, L, F (2015). Solid Lipid Nanoparticles Co-loaded with Simazine and Atrazine: Preparation, Characterization, and Evaluation of Herbicidal Activity. *Journal of Agricultural and Food Chemistry, 63*(2), 422-432.

30. Pereira, A. E. S, Grillo, R, Nathalie, F. S, Mello, Andre, H, Rosa, Leonardo, F, Fraceto. (2014). Application of poly(epsiloncaprolactone) nanoparticles containing atrazine herbicide as an alternative technique to control weeds and reduce damage to the environment. *Journal of Hazardous Materials,268*, 207215.

31. Perez-Martínez, J. I, Gines, J. M, Morillo, E, Arias, M. J. and Moyano. J. R. (2000) Improvement of the desorption of the pesticide 2,4-D via complexation with HP-β-Cyclodextrin. *Pesticide Science, 56*, 425-430.

32. Pradeesh Kumar, T. and Chinnamuthu, C. R. (2014). An Attempt to Synthesis a New Nanoformulation of Pendimethalin Herbicide for Slow Release using Direct Encapsulation Technique. *Trends in Bioscience, 7* (14), 1987-1692.

33. Ravisanakar, D, Chinnamuthu, C. R. and Srimathi, P. (2012). Influence of Growth Promoting Substances on Dormancy Breaking and Sprouting of Purple Nutsedge (*Cyperus rotundus* L.) Tuber. *Research Journal of Agricultural Sciences, 3*(6), 1213-121.

34. Schnoor, B, Elhendawy, A, Suzanna, J, Putman, M, Cerdas, R. C, Mora, D. F, Moraga, F. B. Nilo, F. G and Morales, C. S. (2018). Engineering Atrazine Loaded Poly (lactic-co-glycolic Acid) Nanoparticles to Ameliorate Environmental Challenges. *Journal of Agricultural and Food Chemistry, 66*(30), 7889-7898.

35. Silva, M, Cocenza, D. S, Grillo, R, Melo, N. F, Tonello, P. S, Oliveira, L. C. and Fraceto, L. F. (2011). Paraquat-loaded alginate/chitosan nanoparticles: Preparation, characterization and soil sorption studies. *Journal of Hazardous Materials, 190*(1-3), 366–374.

36. Sopena, F, Cabrera, A, Maqueda, C, Morillo, E. 2007. Ethylcellulose formulations for controlled release of the herbicide alachlor in a sandy soil. *Journal of Agricultural and Food Chemistry, 55*(20), 8200-8205.

37. Viji, N, Chinnamuthu, C. R. and Chinnusamy, C. (2016). Depriving the purple nutsedge by degrading the starch present in the tubers using immobilized amylase on the surface of organically synthesized silver nanoparticles. *Green Farming, 7*(1), 107-110.

38. Viji, N. and Chinnamuth, C. R. (2015a). Iron oxide nanoparticles to break the tubers dormancy of the world's worst weed the *Cyperus rotundus. International Journal of Agricultural Science and Research, 5*(3), 259-266.

39. Viji, N. and Chinnamuthu, C. R. (2015b). Breaking dormancy and inducing germination of the world worst weed the *Cyperus rotundus* using nanoparticles. *Annals of Plant and Soil Research, 17*(Special Issue), 361-363.
40. Viji, N. and Chinnamuthu, C. R. (2018). Titanium dioxide nanoparticles to break the tubers dormancy of the world's worst weed the *Cyperus rotundus* L. *Multilogic in science, 7*(special issue), 322-325.
41. Vimalrajiv, B, Chinnamuthu, C. R, Subramanian, E. and Senthil, K. (2018). Effect of nanoparticles in combination with pendimethalin and hydrogen peroxide on growth parameters and nodulation of blackgram (*Vigna mungo* L.). *International Journal of Chemical Studies, 6*(3), 2816-2819.
42. Wang, X. and Zhao, J. (2013). Encapsulation of the herbicide picloram by using polyelectrolyte biopolymers as layer-by-layer materials. *Journal of Agricultural Food Chemistry, 61*(16), 3789-3796.
43. Wienhold, B. J, Sadeghi, A. M. and Gish, T. J. (1993). Effect of Starch Encapsulation and Temperature on Volatilization of Atrazine and Alachlor. *Journal of Environmental Quality, 22*(1), 162-166.
44. Zhang, D, Zhang, X, Luo, J, Li, B and Wei, Y. (2018). Causation Analysis and Improvement Strategy for Reduced Pendimethalin Herbicidal Activity in the Field after Encapsulation in Polyurea. *ACS Omega*, 3(1), 706-716.
45. Zhang, J, Li, B, Bi, H. and Zhang, P. (2015). Clathration of chlorimuron-ethyl by 5-cyclodextrin for improvement of water solubility, *Journal of Pesticide Science, 40*(3), 33-37.

Chapter 14

Nanotechnology for More Crop per Drop

S. Marimuthu and Golla Gowtham*

Department of Nano Science and Technology, Tamil Nadu Agricultural University Coimbatore – 641 003, Tamil Nadu

**e-mail: sm20@tnau.ac.in*

Water management in agriculture is the crucial for achieving Country's food security. Half of the cultivated land in India are under rainfed farming even after reaching fullest irrigation potential. Rainfed agriculture contributes 40 per cent of the country's food production in India. Rainfed agriculture are practiced in semi-arid and arid regions, where the rainfall does not exceed 500 mm a^{-1}. The potential annual evaporation exceeds easily the annual precipitation. The people in the arid and semi-arid regions are facing food insecurity due to frequent crop failure. Food relief measures create a huge burden on governments and relief organizations. Crop failure in arid and semi-arid regions, is closely related to a stronger imbalance between effective rainfall and high potential evaporation (Hatibu *et al.*, 2006). The rainfed agriculture is referred as one tonne agriculture due to sub-optimal yields obtained below the annual food requirements. De (de Fraiture 2007) reported the significance of rainfed agriculture where 90 per cent of the food production originates from the rainfed areas. The investment on rainfed agriculture improves the agricultural productivity, which has direct impact in bringing down the level of poverty compared to that of irrigated agriculture, which has small impacts on the general food supply due to investment. Falkenmark 1986 demonstrated that there is a strong nexus between poverty, hunger and water stress. The UN Millennium Development Project has identified the 'hot spot' of malnourishment, which are located in the semi-arid and dry sub humid regions of the world where rainfed agriculture is the source of food.

Drought and heat stress are the most important limiting factors to crop productivity and ultimately the food security in rainfed agriculture. In changing climate scenarios, declined annual rainfall and changed rainfall distribution patterns are causing for the onset of droughts around the world (Lobell *et al.*, 2011). Droughts cause substantial decline in crop yields depending on the stage and frequency of occurrence during the crop growth (Yordanov *et al.*, 2000, Barnabás *et al.*, 2008). The yield reduction of 21 and 40 per cent in wheat (*Triticum aestivum* L.) and maize (*Zea mays* L.) respectively due to drought, were documented by (Daryanto *et al.*, 2016). Dry spells of 15 days during flowering and grain filling of maize grown in sandy soils caused greater yield losses (Barron *et al.*, 2003). Crop germination decides the plant population in unit area which contribute significantly for higher crop productivity (Fathi and Tari 2016). Water imbibition of seeds is the first stage of germination. The soil moisture, which is just below the field capacity in soil, favors the seed germination. The access of seeds to moisture is limited, when the soil moisture is nearer to wilting point and hence rate of germination is lower (Radhouane 2007).

The cost benefit ratio of rainfall management strategies in rainfed agriculture fetch 1: 2.14 in India which is considered as relatively high return for the investment (Joshi 2005). A number of options exist to reduce the impact of water scarcity on agricultural production. Rainwater harvesting is one viable option to increase the water productivity, which cushion crops against the impacts of drought and dry spells (Mupangwa *et al.*, 2006, Kahinda *et al.*, 2007). Integrated soil and water conservation measures (*in situ* and *ex situ*) ensure the availability of soil moisture in the root zone Incremental benefits of soil and water conservations are more noticeable where effective rainfall is generally low, *i.e.* in the 200 - 600 mm a^{-1} (Pacey and Cullis 1986). Short term measures including the application of super absorbent polymers as soil conditioner improve the soil water and crop productivity (Arbona *et al.*, 2005) and the foliar application of antitranspirant to reduce the leaf transpiration.

Nano Strategies to Overcome the Impacts of Drought

The changing economic scenario has a remarkable impact on the food security and water demand. India has significant spatial mismatches of the population intensity and water resources. People live and most of the crops is grown in regions where less water is available for cultivation. (Amarasinghe *et al.*, 2007) has predicted that nine river basins, habituating 75 per cent of the total Indian population, will be physically water-scarce by 2050. Vagaries of monsoon and frequent occurrence of drought affect the crop yields to the tune of 30-40 per cent depending on the intensity and stage of occurrence. In India, 60 per cent of agriculture is rainfall dependent and it is crucial to their livelihoods through drought resilient farming practices. Several drought management strategies have been tested and introduced in the past several decades but its adoption hardly happens due to the non-availability or uncertainty.

Nanotechnology deals with atom – by -atom manipulations and the products evolved helps to deliver the inputs precisely in agro-ecosystems. Use of nanotechnology applications in drought management is very limited. However, there are brighter spots of achieving success using nano options to tide over the impact of drought in agriculture. Seeds are prime input in agriculture and ensured population is a key for success in rainfed agriculture. Seeds are to be coated with thermo-sensitive polymers, which can protect the seeds from desiccation under drought (Marimuthu, 2017). Coating of hydrophilic polymers improves the moisture in the immediate microenvironment of seeds facilitating higher imbibition of moisture and early germination. Coated polymer enables seeds in a constant hydraulic conductance resulting in faster germination (Figure 14.1).

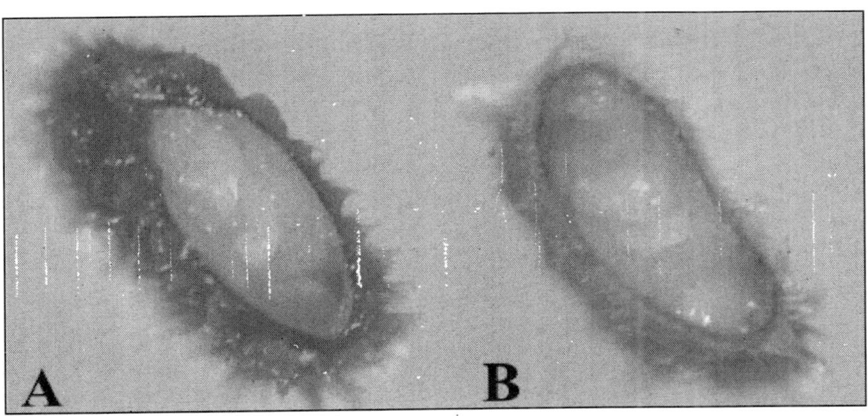

Figure 14.1. Cross Section of Tomato Seed. A. Seed coating with hydrophilic polymer maintains moisture around the surface of the seeds resulting faster germination while non-coated seeds retains less moisture (B).
Courtesy: Jacob *et al.*, 2016.

It is hypothesized that polymer solution in nanoscale emulsion ensures the complete wettability around the seed surface with less quantity of polymers compared to that of bulk conventional formulation. Micro fluidization, probe sonication and high-pressure homogenization are the few methods to fabricate nano emulsion, where the solution is subjected to higher energy thereby reducing the droplet size in the formulation

Soil moisture conservation is a key component in drought mitigation. There are several super absorbent polymers have been explored as soil conditioner for improving soil water holding capacity. Transpiration is an inevitable loss of moisture but careful mitigation strategies will minimize the water loss at critical stage of the crop growth. Chitosan nano-emulsion is explored as an anti - transpirant, which was shown to facilitate stomatal closure and conserve leaf moisture in maize (Marimuthu, 2017). Further, nano-silica spray has a key role in induced resistance against abiotic and biotic stresses (Prasad *et al.*, 2014). (Ashkavand *et al.*, 2015) demonstrated that pre-treatment of Hawthorns (Crataegus aronia L.) with silica nano particle before 45 days of planting induced drought tolerance evidenced from parameters on photosynthesis, plant water relations, malondialdehyde and membrane electrolyte leakage. (Taran *et al.*, 2017) studied the effect of a colloidal solution of Cu and Zn-nanoparticles on drought resistance of wheat seedlings. The decline in the accumulation of thiobarbituric acid reactive substances (TBARS) and simultaneous increase in the antioxidant enzymes (Superoxide Dismutase ad Catalase) was reported in wheat seedlings due to the application of Cu and Zn nanoparticles. Similarly (Pei *et al.*, 2010) used 1.0 mM of Si (sodium silicate) to ameliorate the effects of drought in wheat.

The seed treatment of nano zinc oxide in soybean improved seed germination and seedling vigour under water stress conditions (Sedghi *et al.*, 2013) due to the associated manipulation of auxin and gibberellins ratio under water stress conditions. Foliar application of maghemite nanoparticles improves the yield and oil contents in safflower cultivars, maintaining lower levels of hydrogen peroxide in the leaves (Zareii *et al.*, 2014). Similarly, (Jaberzadeh *et al.*, 2013) showed that the foliar application of titanium dioxide nanoparticles @ 0.02 per cent in wheat resulted in increased plant height, grain yield and quality parameters including gluten and starch content of under water-deficit conditions. Seed treatment with silver nanoparticles in Lentil (*Lens culinaris* Medic) decreased the negative impacts of early drought by improving germination rate, root length, root fresh and dry weights (Hojjat and Ganjali 2016).

Super absorbent polymers are in use over the last forty years in agriculture for improving the soil water holding capacity and availability of water for crops. Superabsorbent Polymers are cross-linked macromolecules, which absorb huge quantities of water, retain and release during the dry circumstances (Figure 14.2). Superabsorbent polymers are used extensively for improving the duration of water availability in the soil and thereby satisfies water requirements of crops during water deficit conditions. (Abedi-Koupai *et al.*, 2008) demonstrated the use of hydrogel for improving water availability and retention of nutrients in light textured soil. The use of hydrogel in sandy soil reduced the frequency of irrigation, soil compaction and run off. The presence of inorganic salts interferes with the water absorption of hydrogels, however, (Teodorescu *et al.*, 2009, Rudzinski *et al.*, 2002) reported that hydrogels were used as vehicle for the slow release of nutrients and pesticides respectively. (Bowman *et al.*, 1990) showed the improvement in soil physical properties amended with hydrogels. The soils amended with hydrogel have been noticed to improve aeration and drainage of the medium (Bearce and McCollum 1977).

The water and nutrient holding capacity of sandy soils are poor, where soils are characterized with excessive drainage causing leaching of below the root zone (Al-Omran *et al.*, 1998). Hydrogel absorb and retain huge quantities of water (Sannino *et al.*, 2009) thereby reducing the deep percolation of water in the soil and leaching of nutrients (Kazanskii and Dubrovskii 1992). Water and water-soluble nutrients are released uniformly to plants in soils amended with hydrogel. The higher water holding capacity in soils with hydrogel makes water available to the plants during longer period of water deficit conditions. Free pore volume in the soil is created, when water released from hydrogel during soil drying periods. This offers additional space for root growth which makes plants to establish well in water deficit conditions (Buchholz and Graham 1998). In spite of various beneficial effects of hydrogel, (Conover and Poole 1976, James and Richards 1986), (Wang and Boogher 1987) had shown little or no response with the application of hydrogel. The variation in the benefits of hydrogel application depends on the type of hydrogel. (Wallace and

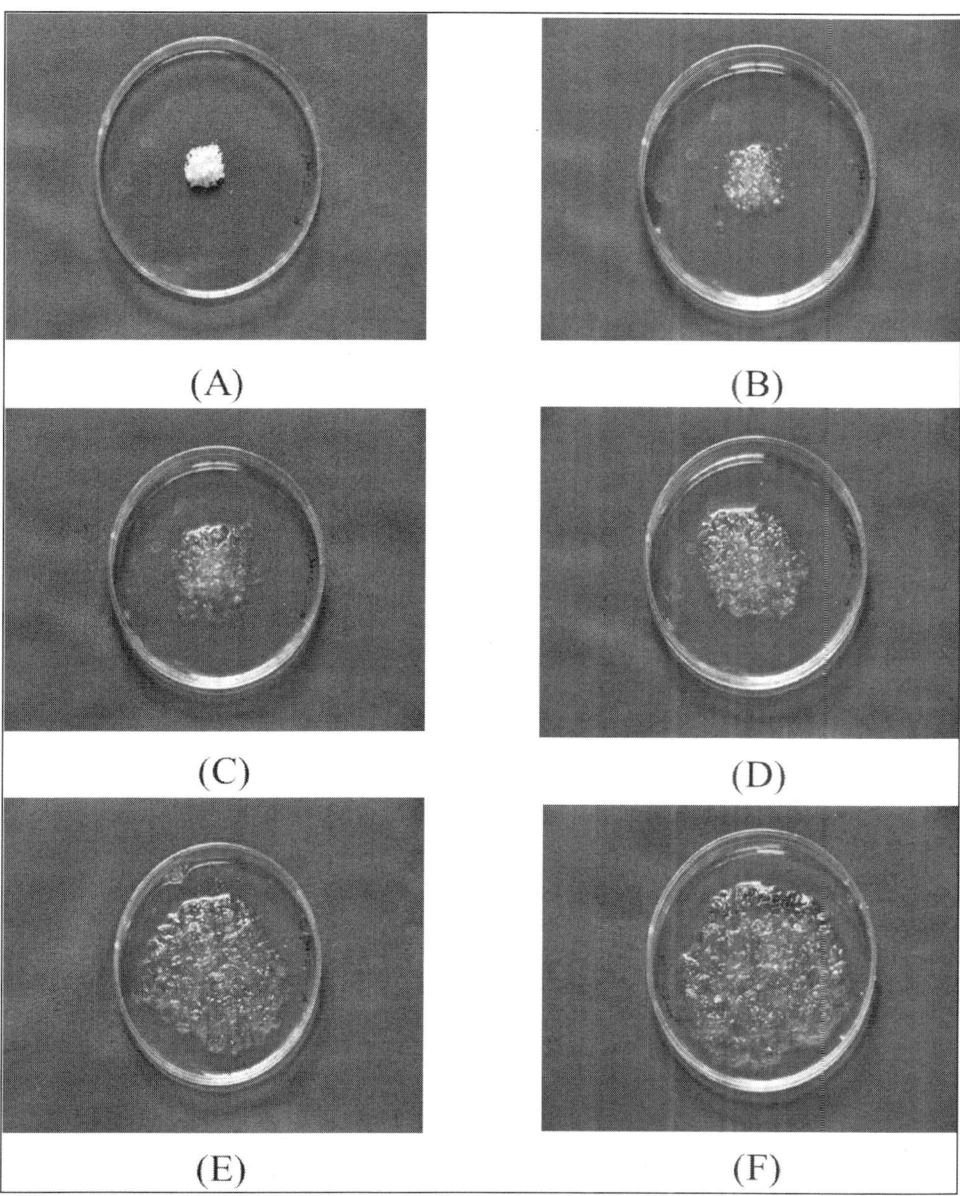

Figure 14.2. Moisture Absorbed by Cross-Linked Biopolymer at different Time in Interval. A: Suberabsorbent polymer without moisture; B: Superabsorbent polymer in water after 5 minutes; C: Superabsorbent polymer in water after 10 minutres; D: Superabsorbent polymer in water after 15 minutres; E: Superabsorbent polymer in water after 20 minutres; F: Superabsorbent polymer in water after 25 minutres.

Wallace 1986)noticed that better seed emergence and water infiltration came from soils amended with an anionic polymer.

The application rate of hydrogel ranged from 100 kg or less ha^{-1} to achieve desired effects in soils. (Reddy *et al.*, 2015) identified the range of 2.5 to 60 kg $ha^{-1}$of hydrogel application depending on the gel type and method of application. (Meena 2009) reported that the agronomic efficiency of polymers application in terms of tomato yield was 58 kg kg^{-1} of polymer application. The results indicated the postponement of irrigation in every three weeks without comprising the yield in tomato, when polymers were applied at the root zone of 10 cm soil depth. (Johnson 1984) tested different cross-linked polyacrylamides on soil moisture retention in sandy soils who showed that the evaporation losses from untreated sand was rapid reaching residual moisture of < 1 kg hg^{-1}after four days of incubation whereas, residual moisture levels were 3.9 kg hg^{-1} and 13.0 kg hg^{-1} in soils amended with polymer @ 1g kg^{-1} and 2 g kg^{-1} respectively, demonstrating the higher moisture retention in soils applied with polymers. Permanent wilting point, which is upper limit of available moisture for plants in control sand, was reached between 2–3 d, in comparison with between 6–7 d for the 1 g kg^{-1} of polymer treated sand and between 9–10 d for the 2 g kg^{-1} polymer treated sand. Koupai *et al.* (2008) estimated the residual soil moisture in sandy soil amended with hydrogel @ 8g kg^{-1} as 19.67 cm3 cm^{-3}, which is five-fold increase over the control. The higher residual moisture in sandy soils are due to their less cation holding capacity compared with the other soil textures. The expansion of polymers and associated water adsorption is increased. Dawlatzai *et al.* (2017) reported higher growth of *Coleus blumei* @ application of 40 g of Pusa Hydrogel in 5 kg of potting media with water saving of 39 per cent. Sowing of aerobic rice on beds amended with hydrogel improved the establishment of aerobic rice and growth and yield of aerobic rice (Rehman *et al.*, 2011).

The chapter discuss exclusively on the synthesis of super absorbent polymers, mechanism of water release from the hydrogel and its characterization using high-end instruments.

Synthesis of Super Absorbent Polymers (Hydrogels)

Hydrogels are long polymeric chains, which swell in free water but will not dissolve in water. The word "hydrogel", was first noticed in an article published in 1894 by Lee, Kwon and Park. Hydroxy ethyl methacrylate hydrogels are the first cross-linked polymer having wider applications. Hydrogels are given emphasis by material researchers and biomedical analysts recently because of their extensive applications. (Buwalda *et al.*, 2014) described the progress on the synthesis of hydrogels into three stages

First-generation hydrogels contains a wide scope for using cross-linking strategies to bridge monomer or polymer with an initiator. The general point is to create material with high swelling, greater mechanical properties with moderate method of preparation. The second category of the first generation hydrogels are hydrophilic polymers, which are covalently cross-linked between functional groups or by free radicals. Secondly, hydrogel developed in beginning of seventies are stimuli specific vary with temperature, pH or concentration of bio molecules in the solution. This stimuli or environmental factors are used to evoke specific events, are referred second-generation hydrogels. A third generation of hydrogels are mainly stereo complex materials such as polyethylene glycol and poly lactic acid. Physical interactions are exploited in cross-linking for hydrogel synthesis, which improves the mechanical, thermal and degradation properties. Many of the interactions provided opportunity for in situ hydrogel formation.

Super absorbent polymers are synthesized by various polymerization techniques utilizing both chemical and physical cross-linking methods. Natural and synthetic polymers with hydrophilic functional moieties are cross-linked to synthesize water reservoirs. The modification of skeletons in polymer chains with UV, γ and electron beam irradiation, change the bulk properties of polymers thereby allowing bridge between the polymer chains through cross-linking agents. Chemically cross linked hydrogels are covalently cross-linked by replacing hydrogen bond through covalent bonding, while the networks in physically cross linked hydrogels are held together in network entanglements, and secondary forces including ionic, hydrophobic interactions or hydrogen bonding (Hennink and Nostrum 2002).

Chemically cross-linked hydrogels were synthesized extensively through chain growth polymerization, which includes free radical, anionic and cationic polymerization. The synthesis of hydrogels comprises of three steps *viz.*, initiation due to the release of free radicals, propagation of free radicals through polymer chains, and termination through scavenging of free radicals at the end of the reaction. Initiators provide a pair of free radicals on dissociation, which in turn attacks the neighbouring monomer units to produce chain-initiating species that propagates along the polymer branch until the availability of active initiators. The final step is termination of polymer chain growth using high reactive electron scavengers (Minhas *et al.*, 2013)

Physical cross-linked hydrogels, which are mediated through ionic interaction, crystallization, hydrogen bond, hydrophobized polysaccharides and stereo complex formation. Ionic interaction hydrogels were prepared by optimizing the temperature, pressure and pH of the solution during the synthesis (Hoffman *et al.*, 2002). Using metallic ligands during ionic interactions produce stronger hydrogels (Hennink and Nostrum 2002). Synthetic polymers have higher rate of reproducible physical and chemical properties than that of natural polymers, which make synthetic polymers for commercial applications. Strength of the hydrogels is improved by balancing the concentration of cross linking agents and co-monomers (Nguyen and Lee 2010).

Mechanism of Water Retention and Release from Super Absorbent Polymer

Super absorbent polymers hold 99 per cent of exceptional water, which is the significance of hydrogels. The polymer chains and water are perfectly joined together, albeit rather wet. The performance of hydrogels in terms of water uptake is largely determined by the quantity, design, and attributes of water in contact and ability to interact with the solid polymer network. The equilibrium state of water influences the rate of adsorption and physical properties of the hydrogel.

Generally, a hydrogel contains three states of water that are bound with polymeric chains. Water in hydrogel exists as free water, semi-bound or interstitial water and bound water (Jhon and Andrade 1973). Free water in a hydrogel refers to water, which is not intimately linked to the polymer chain acting as bulk water, and freezes at ordinary freezing point (0°C). Interstitial water is poorly connected with polymer chains and freezes at temperatures below standard freezing point, while bound water in hydrogel strongly binds with the surface of the polymer chains with long-range order of cohesive force, which does not freeze at the usual freezing point. Bound water forms H bonds with polar groups of polymer chains, which generally interact strongly with ionic residues in the polymer chains. The structural similarities between free and bulk water causes weaker interaction of free water with polymer chains, which helps free water to move readily from hydrogel network. Intermediate water or semi bound water, which lies between the bound and free water, interacting weakly with the polymer chains. Higher the proportion of hydrophilic functional groups in the polymer chain, greater will be the bound water in hydrogel resulting in the higher swelling ratio (Katime *et al.*, 2000). Higher the hydrophobic units in hydrogel results in lower water uptake (Kim and Peppas 2003). The quantity of free and intermediate water that hold in the hydrogel will decrease with increase in the cross linking density (Hatakeyema *et al.*, 1984). (Terada *et al.*, 1993) explained the relationship between free water in hydrogel and density of cross-linking using Raman spectroscopy. The defects in hydrogen bonds network in polymer chains increases with density of cross-linking which results in poor uptake of water. Hence, the structure of the polymer network is crucial for holding water molecules. (Takushi *et al.*, 1990)explained two different stages in the dehydration process of super absorbent polymers, where free water evaporates in the first stage, followed by intermediate and bound water that evaporate in the second stage. (Koshoubu *et al.*, 1993) presented three states of hydrogels during the desorption of hydrogels, where in the second stage exhibits an intermediate state between gel and a glassy state.

The quality of hydrogels in terms of desorption depends on the quantity of semi-bound or interstitial water that hold in the hydrogel and types of interactions that occur between semi bound water and the polymer chain networks. The water held in macro - porous of hydrogels is free water, whereas water present in micro and nano pores are interstitial water. The self-assembly of nanopores presents macro - porous nature of hydrogel which plays crucial role for maintaining residual moisture in soils under water deficit conditions

The de-swelling process of hydrogels is divided into three stages based on the rate of evaporation of free water, interstitial and bound water as explained in Figure 14.3, corresponding to 5-7, 95-225, and 275-515 minutes and is shown in. At stage I of 5-7 minutes of drying most of the free water held by the hydrogel (bulk water) gets evaporated @ 12-10 g g^{-1} which is referred useful event for plants under moisture stress conditions. The stage I releases huge amount of bulk water and very minute quantities of semi- bound and bound water. The stage II corresponding to the time period of 95 – 225 minutes of drying where the hydrogel has low quantity of bulk water and moderate levels of semi-bound and bound water. The release rate of 9-5g g^{-1}, which occurs during stage II. The water released during the stage II will satisfy the water requirement of plants at permanent wilting point. The stage III occurs at 275-515 minutes of drying period where 4-0 g g^{-1} of water is released from the polymer matrix. Most of water available at this stage is bound water, which are strongly held in the active sites of hydrophilic functional groups of polymer matrix. The bound water requires high pressure to delocalize from the polymer matrix. The plant will not utilize the bound water for the normal physiological activities. To conclude, the quantity of hydrogel required for improving soil water holding capacity depends on the soil type. The release or desorption pattern of water from hydrogel are to be studied for matching the water requirement of crops, which will help in designing the quantity of hydrogel applied to the crop.

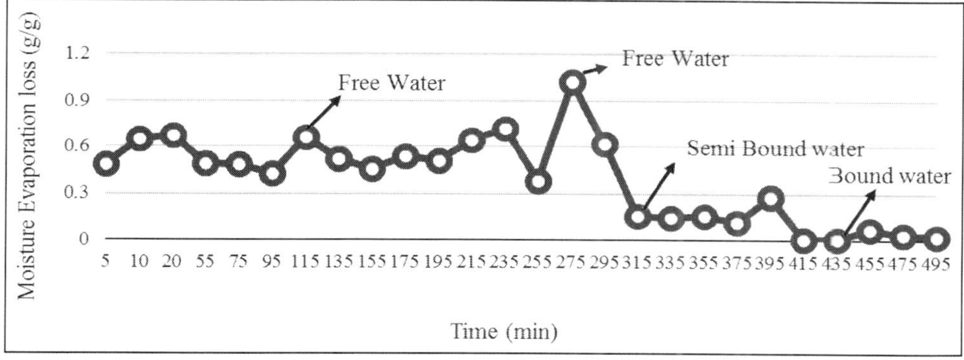

Figure 14.3. Typical Desorption of Water from the Hydrogel vs Time.

Sekine and Ikeda-Fukazawa (2009) explained the structural change during the dehydration hydrogels through infrared spectroscopy. The vibrational energy of the OH stretching in a compound depends on the bond length of the OH bond for bulk water, amorphous ice, and crystal ice. Higher shift of OH stretching mode indicates the decrease in length of OH bond. The reason for the energy shift of OH molecules is due to the coupling with the surrounding vibrational modes. The decrease in relative intensity of OH stretching mode indicates the magnitude of dehydration, which indicates the stage I dehydration in hydrogels. The vibrational energy of the C=O stretching mode in the polymer is associated with the interaction between the polymer chain and water molecules during dehydration which represents stage II and III of desorption of hydrogels.

Impact of Super Absorbent Polymer on Soil Holding Capacity

Bio super absorbent polymer is safe, biodegradable, and nontoxic and remains dormant in soil effectively. The swelling of hydrogels in the rhizosphere improves air circulation in the root zone resulting in better germination, root advancement and microbial activities. The nearness of water in soil is a fundamental factor for the survival of vegetation, which brings along essential nutrient for the growth and development. Hence, the water near the microenvironment of crops and plants helps to achieve better growth and yield (Table 14.1).

Table 14.1. Effect of Super Absorbent Polymers in Soil Water Holding Capacity and Crop Growth

Sl.No.	*Nature of Super Absorbent Polymer*	*Impact of Application on Soil Water Holding Capacity*	*Authors*
1	Potassium-based cross-linked polyacrylamide, (WaterSorb-227 S Church Ave, Fayetteville, AR. USA)	The application of hydrogel @ 0.3 per cent improved the soil water holding capacity (238.6±1.5 to 244.3±3.05 g g^{-1} of soil after 80 minutes of incubation	Abdallah, (2019)
2	Acrylamide copolymerized with convex/concave rod grafted polymers	The application of hydrogel @ 0.5 and 1.0 per cent increased soil water storage capacity 30.93 and 33.65 per cent respectively, compared to that control (26.07 per cent) after 30 days of incubation at 2cm soil depth	Zhoa *et al.*, (2019)
3	Polyacrylamide and acrylic acid by Dongying Huaye New Material Co., Ltd., Shandong.	Soil amended with hydrogel increased soil water holding capacity ranged between 2.7 to 26.5 per cent under the suction pressure between 0 to 3000 cm, where the marked increase in soil retention was noticed at suction pressure between 100 to 800 cm representing available water	Liao *et al.*, (2016)
4	Kaolinite clay based nano-composite polymer with acrylic acid, acrylamide, ammonium per sulphate as initiator and N,N'- methylene bis acrylamide as cross linker	Application of phosphorus coated kaolinite nano-composite hydrogel increased 14.4±2.5 g of soil water g^{-1} of hydrogel at 10 days of incubation	Verma *et al.*, (2017)
5	Aqua serve superabsorbent polymer	The application of Agro serve hydrogel increased the water soil holding capacity ranged from 20 to 4.5 per cent to with water having total soluble salts of 154 and 2647 ppm	Dabhi *et al.* (2014)
6	Super-AB-A-300 polymer	The hydrogel @ 0.2 and 0.4 per cent amended desert soil retained 225 and 380 g respectively while the control was completely dried	Dehkordi (2018)
7	Super AB, A-200	The application of hydrogel increased plant available water with pF 2.5 to pF 4.2, 18 per cent $percent^{-1}$ of hydrogel amended, which is approximately sixty times higher than for the pure sand	Banedjschafie and Durner (2015)
8	Super-AB-A-200	Super AB A 200 increased the saturated water content by 63.33 and 145.5 per cent compared with the controls in flat soil and 65 per cent slope soil respectively.	Dehkordi (2018)
9	Palygorskite clay based hydrogel with maize straw (MS), acrylic acid (AA), N,N-methylenebisacrylamide (MBA), potassium persulfate	Superabsorbent polymer reaches equilibrium at 3 h and retains water for four days at 35°C	Tao *et al.*, (2018)

Sl.No.	*Nature of Super Absorbent Polymer*	*Impact of Application on Soil Water Holding Capacity*	*Authors*
10	Hydrogel @ 200g of water g^{-1} of hydrogel (70 per cent acrylic-acrylamide polymers and 30 per cent Attapulgite clay) Chang'an Group Co. Ltd., Dongying, China	Application of hydrogel improved net photosynthetic rate of water deficit maize to 13.2 μ mol (CO_2) m^{-2} s^{-1} compared 11.6 μ mol (CO_2) m^{-2} s^{-1} under control	Yang *et al.*, (2019)
11	Acrylic acid monomers and polyacrylamide co-polymer	Hydrogel holds 300 to 500 times of water of their initial mass	Tripathi (2018)
12	Stockosorb	Deficit irrigation in *Atriplex lentfiformis* amended with 0.1 and 0.3 hydrogels improved available water 15.5 and 17.6 per cent respectively compared to that of 10.3 per cent in control	Yousefian *et al.*, (2018)
13	Polymer of acrylamide (N,N-methylbis-acrylamide) and mixed Na and K salts of acrylic acid	Application of hydrogel @ 0.1, 0.2 and 0.3 per cent hydrogel increased field capacity by 17, 26 and 46 per cent, respectively, compared with untreated soil. In loamy soil, an increase of 23, 36 and 50 per cent in soil field capacity was observed with addition of 0.1, 0.2 and 0.3 per cent hydrogel compared to control	Akhter *et al.*, (2004)
14	Salicyl-imine-chitosan hydrogels with urea	Hydrogel amended soil increased water retention by 154 per cent. Control soil dried completed after 7 days while those with hydrogel amended maintained wetness for 12 days	Iftime *et al.*, (2019)

The addition of soil conditioners like biochar and hydrogel polymers become crucial in arid and semiarid regions for enhancing soil water holding and nutrient use efficiency. Hydrogel have capacity to hold huge quantities of water to its original weight, from which water escapes during water deficit conditions and becomes available to the plants. The addition of hydrophilic polymers in soils influence soil water relations and physical properties such as infiltration rates, water holding capacity, bulk density, soil structure, compaction, soil texture, aggregate stability, surface crustation (Helalia and Letey 1989, 1988) and soil evaporation (Teyel and El-Hady 1981). The application of super absorbent polymers improves soil water holding capacity there by improving the access of plant to soil moisture during water deficit conditions (Agaba *et al.*, 2010). The application of hydrogels improves the water holding capacity of sandy soils, which normally has lower water holding capacity because of higher percolation rates (Bai *et al.*, 2010). The irrigation scheduling is crucial for improving crop and water productivity in agriculture. (Hüttermann *et al.*, 1999) demonstrated that application of super absorbent polymers increase soil available water in the upper soil profile, thus reducing the number of irrigations to crops.

Akhter *et al.* (2004) studied the impact of amending hydrogels in sandy loam (sand 78 per cent, silt 9 per cent, clay 13 per cent) and loam (sand 45 per cent, silt 33 per cent, clay 22 per cent) soils. The application of hydrogel increased the soil water storage in both sandy loam and loam soils compared with control (without hydrogel). The soil amended with hydrogel lowered the rate of soil moisture loss causing a delay in wilting of seedlings grown in both soils (Ekebafe *et al.*, 2016). Hydrogel as soil conditioner increases the cation exchange capacity, specific surface area and available nutrient status in soils. (El-Hady and Abo-Sedera 2006) showed that cation exchange capacity of soil treated with hydrogel were, 1.8 to 2.1 times higher than normal soils. Soil pH determines the nutrient accessibility to plants. (Mulyani *et al.*, 2019) revealed the soil acidification due to the application of hydrogels. Negative charge emerges from dissociation of H+ from hydrogel, other cations including functional group to the surface, thereby neutralizing cation exchange active sites and causing an increase in soil pH.

Callaghan *et al.* (1989) recorded higher seedlings survival and development under arid-dry conditions in sandy soils amending hydrogel, while (Viero *et al.*, 2002) observed increased seedling development when hydrogel blended with irrigation. The varied response for hydrogel application have been recorded with the soil texture. The coarse textured soil require more quantities of hydrogels to improve soil water holding capacity, while clay soil require less quantities of hydrogels to improve soil water holding capacity. The hydrogel swells on water uptake and occupy the void spaces in coarse textured soils, which in turn it releases water quickly in coarse textured soils. Hence, the quantity of hydrogels required for light textured soils is more to maintain residual moisture in soils.

Hüttermann *et al.* (1999) studied the application of polyacrylamide based hydrogels on the survival of *Pinus halepensis* (Aleppo pine) seedlings in sandy soils, which revealed that tree seedlings survived 19 days enduring dry spell in soils amended with hydrogels. (Moslemi *et al.*, 2011) revealed that co application of super absorbent polymer along with inoculation of *Azospirillum lipoferum* + *Pseudomonas putida* improved the yields in maize under moisture deficit conditions. (Singh *et al.*, 2018) studied the impact of application of pusa hydrogel and chitosan with different irrigation regimes in wheat, which revealed that irrigation with 60 per cent saturation along with pusa hydrogel and chitosan as antitranspirant recorded better results with conventional irrigation. (Roy *et al.*, 2019) studied the application of pusa hydrogel @ 5 kg ha^{-1} in wheat in North Western India, which showed that plant population, productive tillers and grain yield of wheat were more after hydrogel amendment.

Conclusion

To conclude, application of biodegradable super absorbent polymers improve soil water holding capacity thereby improving the crop yield and water productivity. The synthesis of super absorbent polymers following green chemistry should attain higher swelling ratio with free water. However, the improvement of available moisture in hydrogel-amended soils during water deficit period is crucial for crop production. The quantity of free and intermediate water state in hydrogel decides quantity of water available to the plants during water deficit conditions. Hence, the performance of super absorbent polymer technology depends on the water desorption from the hydrogel matching the water deficit conditions. The quantity of hydrogel, soil texture and cation exchange capacity of soil further decide the successful hydrogel technology

REFERENCES

1. Abdallah, A. M. (2019). The effect of hydrogel particle size on water retention properties and availability under water stress. *International Soil and Water Conservation Research.*
2. Abedi-Koupai, J., Sohrab, F., and Swarbrick, G. (2008). Evaluation of hydrogel application on soil water retention characteristics. *Journal of Plant Nutrition, 31*(2), 317-331.
3. Abedi-Koupai, Jahangir, Farahnaz Sohrab, and Gareth Swarbrick. 2008. "Evaluation of hydrogel application on soil water retention characteristics." *Journal of Plant Nutrition* 31 (2): 317-331.
4. Agaba, Hillary, Lawrence Justus Baguma Orikiriza, John Francis Osoto Esegu, Joseph Obua, John David Kabasa, and Aloys Hüttermann. 2010. "Effects of hydrogel amendment to different soils on plant available water and survival of trees under drought conditions." *Clean–Soil, Air, Water* 38 (4): 328-335.
5. Akhter, J., K. Mahmood, K. A. Malik, A. Mardan, M. Ahmad, and M. M. Iqbal. 2004. "Effects of hydrogel amendment on water storage of sandy loam and loam soils and seedling growth of barley, wheat and chickpea." *Plant Soil and Environment* 50 (10): 463-469.
6. Akhter, J., Mahmood, K., Malik, K. A., Mardan, A., Ahmad, M., and Iqbal, M. M. (2004). Effects of hydrogel amendment on water storage of sandy loam and

loam soils and seedling growth of barley, wheat and chickpea. *Plant Soil and Environment, 50*(10), 463-469.

7. Al-Omran, Abdulrasol M., Abdulaziz R. Al-Harbi, A. Wallace, and R. E. Terry. 1998. "Improvement of sandy soils with soil conditioners." *Handbook of soil conditioners: Substances that enhance the physical properties of soil. Marcel Dekker, New York*: 363-384.
8. Amarasinghe, Upali, Tushaar Shah, Hugh Turral, and B. K. Anand. 2007. *India's water future to 2025-2050: Business-as-usual scenario and deviations*. Vol. 123: IWMI.
9. Anuj Tripathi. "Superabsorbent Polymers: Artificial mini water reservoirs in soil for agriculture applications".*Acta Scientific Agriculture* 2.1(2018): 01
10. Arbona, Vicent, Domingo J. Iglesias, Josep Jacas, Eduardo Primo-Millo, Manuel Talon, and Aurelio Gómez-Cadenas. 2005. "Hydrogel substrate amendment alleviates drought effects on young citrus plants." *Plant and Soil* 270 (1): 73-82.
11. Ashkavand, Peyman, Masoud Tabari, Mehrdad Zarafshar, Ivana Tomášková, and Daniel Struve. 2015. "Effect of SiO2 nanoparticles on drought resistance in hawthorn seedlings." *Forest Research Papers* 76 (4): 350-359.
12. Bai, H., Li, C., Wang, X., and Shi, G. (2010). A pH-sensitive graphene oxide composite hydrogel. *Chemical Communications, 46*(14), 2376-2378.
13. Banedjschafie, S., and Durner, W. (2015). Water retention properties of a sandy soil with superabsorbent polymers as affected by aging and water quality. *Journal of Plant Nutrition and Soil Science, 178*(5), 798-806.
14. Barnabás, Beáta, Katalin Jäger, and Attila Fehér. 2008. "The effect of drought and heat stress on reproductive processes in cereals." *Plant, cell and environment* 31 (1): 11-38.
15. Barron, Jennie, Johan Rockström, Francis Gichuki, and Nuhu Hatibu. 2003. "Dry spell analysis and maize yields for two semi-arid locations in east Africa." *Agricultural and forest meteorology* 117 (1-2): 23-37.
16. Bearce, Bradford C., and Rebecca W. McCollum. 1977. "A comparison of peat-lite and noncomposted hardwood-bark mixes for use in pot and bedding-plant production and the effects of a new hydrogel soil amendment on their performance." *Florists' Review.*
17. Bowman, Daniel C., Richard Y. Evans, and J. L. Paul. 1990. "Fertilizer salts reduce hydration of polyacrylamide gels and affect physical properties of gel-amended container media." *Journal of the American Society for Horticultural Science* 115 (3): 382-386.
18. Bowman, Daniel C., Richard Y. Evans, and J. L. Paul. 1990. "Fertilizer salts reduce hydration of polyacrylamide gels and affect physical properties of gel-amended container media." *Journal of the American Society for Horticultural Science* 115 (3): 382-386.
19. Buchholz, Fredric L., and Andrew T. Graham. 1998. "Modern superabsorbent polymer technology." *John Wiley and Sons, Inc, 605 Third Ave, New York, NY 10016, USA, 1998. 279.*

20. Buwalda, Sytze J., Kristel W. M. Boere, Pieter J. Dijkstra, Jan Feijen, Tina Vermonden, and Wim E. Hennink. 2014. "Hydrogels in a historical perspective: From simple networks to smart materials." *Journal of controlled release* 190: 254-273.

21. Callaghan, T. V., D. K. Lindley, O. M. Ali, H. Abd El Nour, and P. J. Bacon. 1989. "The effect of water-absorbing synthetic polymers on the stomatal conductance, growth and survival of transplanted Eucalyptus microtheca seedlings in the Sudan." *Journal of Applied Ecology*: 663-672.

22. Conover, C. A., and W. Poole. 1976. "Influence of Viterra hydrogel on growth and control of wilting of 3 foliage plant species." *Fla. Foliage Grower* 13: 332-333.

23. Dabhi, R., Bhatt, N., and Pandit, B. (2014). Effect on the absorption rate of agricultural super absorbent polymers under the mixer of soil and different quality of irrigation water. *Int. J. Emerging Technol. Adv. Eng.–IJETAE, 3*(1), 1402-1406.

24. Daryanto, Stefani, Lixin Wang, and Pierre-André Jacinthe. 2016. "Global synthesis of drought effects on maize and wheat production." *PloS one* 11 (5): e0156362.

25. Dawlatzai, A. S., Jayanthi, R., and Abdiani, S. A. Efficacy of Graded Doses of Pusa Hydrogel on Growth and Quality of Coleus (Coleus Blumeil.) under Polyhouse Condition.

26. Dawlatzai, Abdul Saboor, R. Jayanthi, and Saidajan Atiq Abdiani. "Efficacy of Graded Doses of Pusa Hydrogel on Growth and Quality of Coleus (Coleus Blumeil.) under Polyhouse Condition."

27. de Fraiture, Charlotte. 2007. "Integrated water and food analysis at the global and basin level. An application of WATERSIM." *Water Resources Management* 21 (1): 185-198.

28. Dehkordi, D. K. (2018). Effect of hydrophilic polymers on seed germination and plant survival for sloping area. *Journal of Soil and Water Conservation, 73*(2), 173-178.

29. Ekebafe, Lawrence O., David E. Ogbeifun, and Felix E. Okieimen. 2016. "Effect of Reaction Conditions on the Grafting Indices of Hydrolyzed Starch Copolymers." *Nigerian Journal of Polymer Science and Technology* 11: 75-84.

30. El-Hady, O. A., and S. A. Abo-Sedera. 2006. "Conditioning effect of composts and acrylamide hydrogels on a sandy calcareous soil. II-Physico-bio-chemical properties of the soil." *International Journal of Agriculture and Biology (Pakistan).*

31. Fathi, Amin, and Davood Barari Tari. 2016. "Effect of drought stress and its mechanism in plants." *International Journal of Life Sciences* 10 (1): 1-6.

32. Hatakeyema, T., A. Yamauchi, and H. Hatakeyema. 1984. "Studies on bound water in poly (vinyl alcohol). Hydrogel by DSC and FT-NMR." *European polymer journal* 20 (1): 61-64.

33. Hatibu, N., K. Mutabazi, E. M. Senkondo, and A. S. K. Msangi. 2006. "Economics of rainwater harvesting for crop enterprises in semi-arid areas of East Africa." *Agricultural Water Management* 80 (1-3): 74-86.

34. Helalia, Awad M., and J. Letey. 1988. "Cationic polymer effects on infiltration rates with a rainfall simulator." *Soil Science Society of America Journal* 52 (1): 247-250.

35. Helalia, Awad M., and J. Letey. 1989. "Effects of different polymers on seedling emergency, aggregate stability, and crust hardness." *Soil Science* 148 (3): 199-203.

36. Hennink, W. E., and C. F. Nostrum. 2002. "The effect of network charge on the immobilization and release of proteins from chemically crosslinked dextran hydrogels." *Adv. DrugDeliv. Rev* 54: 13-36.

37. Hennink, W. E., and C. F. Nostrum. 2002. "The effect of network charge on the immobilization and release of proteins from chemically crosslinked dextran hydrogels." *Adv. DrugDeliv. Rev* 54: 13-36.

38. Hoffman, A. S. Hydrogels for biomedical applications (2002) Advanced Drug Delivery Reviews, 54 (1).

39. Hojjat, Seyed Saeid, and A. Ganjali. 2016. "The effect of silver nanoparticle on lentil seed germination under drought stress." *Int J Farm Allied Sci* 5 (3): 208-212.

40. Hüttermann, Aloys, Moitaba Zommorodi, and Kim Reise. 1999. "Addition of hydrogels to soil for prolonging the survival of Pinus halepensis seedlings subjected to drought." *Soil and Tillage Research* 50 (3-4): 295-304.

41. Hüttermann, Aloys, Moitaba Zommorodi, and Kim Reise. 1999. "Addition of hydrogels to soil for prolonging the survival of *Pinus halepensis* seedlings subjected to drought." *Soil and Tillage Research* 50 (3-4): 295-304.

42. Iftime, M. M., Ailiesei, G. L., Ungureanu, E., and Marin, L. (2019). Designing chitosan based eco-friendly multifunctional soil conditioner systems with urea controlled release and water retention. *Carbohydrate polymers, 223,* 115040.

43. Jaberzadeh, Amir, Payam Moaveni, Hamid Reza Tohidi Moghadam, and Hossein Zahedi. 2013. "Influence of bulk and nanoparticles titanium foliar application on some agronomic traits, seed gluten and starch contents of wheat subjected to water deficit stress." *Notulae Botanicae Horti Agrobotanici Cluj-Napoca* 41 (1): 201-207.

44. Jacob, S. R., Kumar, M. A., Varghese, E., and Sinha, S. N. (2016). Hydrophilic polymer film coat as a micro-container of individual seed facilitates safe storage of tomato seeds. *Scientia horticulturae, 204,* 116-122.

45. James, E. A., and D. Richards. 1986. "The influence of iron source on the water-holding properties of potting media amended with water-absorbing polymers." *Scientia Horticulturae* 28 (3): 201-208.

46. Jhon, Mu Shik, and J. D. Andrade. 1973. "Water and hydrogels." *Journal of biomedical materials research* 7 (6): 509-522.

47. Johnson, Michael S. 1984. "The effects of gel forming polyacrylamides on moisture storage in sandy soils." *Journal of the Science of Food and Agriculture* 35 (11): 1196-1200.

48. Joshi, P. K. 2005. *Meta-analysis to assess impact of watershed program and people's participation*. Vol. 8: IWMI.

49. Kahinda, Jean-marc Mwenge, Johan Rockström, Akpofure E. Taigbenu, and John Dimes. 2007. "Rainwater harvesting to enhance water productivity of rainfed agriculture in the semi-arid Zimbabwe." *Physics and Chemistry of the Earth, Parts A/B/C* 32 (15-18): 1068-1073.

50. Katime, Issa, Elena Díaz de Apodaca, Eduardo Mendizábal, and Jorge E. Puig. 2000. "Acrylic acid/methyl methacrylate hydrogels. I. Effect of composition on mechanical and thermodynamic properties."

51. Kazanskii, K. S., and S. A. Dubrovskii. 1992. "Chemistry and physics of "agricultural" hydrogels." In *Polyelectrolytes hydrogels chromatographic materials*, 97-133. Springer.

52. Khodadadi Dehkordi, D. (2018). Effect of superabsorbent polymer on soil and plants on steep surfaces. *Water and environment journal, 32*(2), 158-163.

53. Kim, Bumsang, and Nicholas A. Peppas. 2003. "In vitro release behavior and stability of insulin in complexation hydrogels as oral drug delivery carriers." *International journal of pharmaceutics* 266 (1-2): 29-37.

54. Koshoubu, Nobutatsu, Haruichi Kanaya, Kazuhiro Hara, Seiji Taki, Eisei Takushi, and Kazumi Matsushige. 1993. "Variations of mechanical properties in egg white during gel-to-glasslike transition." *Japanese journal of applied physics* 32 (9R): 4038.

55. Liao, R., Wu, W., Ren, S., and Yang, P. (2016). Effects of superabsorbent polymers on the hydraulic parameters and water retention properties of soil. *Journal of Nanomaterials, 2016*, 37.

56. Lobell, David B., Wolfram Schlenker, and Justin Costa-Roberts. 2011. "Climate trends and global crop production since 1980." *Science* 333 (6042): 616-620.

57. Marimuthu,2017. Nanotechnologies for drought management; National seminar on Nanotechnology for Evergreen revolution held during October 5 and 6,2017

58. Meena, Mukesh Kumar. 2009. "Influence of hydrophilic polymer (Luquasorb) on plant growth and physiology in tomato (Lycopersicon esculentum L. Mill).

59. Minhas, Muhammad Usman, Mahmood Ahmad, Liaqat Ali, and Muhammad Sohail. 2013. "Synthesis of chemically cross-linked polyvinyl alcohol-co-poly (methacrylic acid) hydrogels by copolymerization; a potential graft-polymeric carrier for oral delivery of 5-fluorouracil." *DARU Journal of Pharmaceutical Sciences* 21 (1): 44.

60. Moslemi, Zahra, Davood Habibi, Ahmad Asgharzadeh, Mohammad Reza Ardakani, Abdollah Mohammadi, and Azam Sakari. 2011. "Effects of super absorbent polymer and plant growth promoting rhizobacteria on yield and yield components of maize under drought stress and normal conditions." *African Journal of Agricultural Research* 6 (19): 4471-4476.

61. Mulyani, Sri, Sirojul Umah, Bella Seba, and Dimas Gilang Ramadhani. 2019. Synthesis of super absorbent polymer hydrogel based on oil palm empty fruit bunch grafted with acrylic acid and its application as a npk fertiliser carrier. *Journal of Chemical Technology and Metallurgy* 54 (4).

62. Mupangwa, Walter, David Love, and Steve Twomlow. 2006. Soil–water conservation and rainwater harvesting strategies in the semi-arid Mzingwane Catchment, Limpopo Basin, Zimbabwe. *Physics and chemistry of the Earth, Parts A/B/C* 31 (15-16): 893-900.

63. Nguyen, Thi-Hiep, and Byong-Taek Lee. 2010. Fabrication and characterization of cross-linked gelatin electro-spun nano-fibers. *Journal of Biomedical Science and Engineering* 3 (12): 1117.

64. Pacey, Arnold, and Adrian Cullis. 1986. *Rainwater harvesting: the collection of rainfall and runoff in rural areas*: Intermediate technology publications.

65. Pei, Z. F., D. F. Ming, D. Liu, G. L. Wan, X. X. Geng, H. J. Gong, and W. J. Zhou. 2010. Silicon improves the tolerance to water-deficit stress induced by polyethylene glycol in wheat (*Triticum aestivum* L.) seedlings. *Journal of Plant Growth Regulation* 29 (1): 106-115.

66. Prasad, Ram, Vivek Kumar, and Kumar Suranjit Prasad. 2014. Nanotechnology in sustainable agriculture: present concerns and future aspects. *African journal of biotechnology* 13 (6): 705-713.

67. Radhouane, Leila. 2007. "Response of Tunisian autochthonous pearl millet (Pennisetum glaucum (L.) R. Br.) to drought stress induced by polyethylene glycol (PEG) 6000." *African journal of biotechnology* 6 (9).

68. Reddy, K. Sammi, K. Srinivas, A. G. K. Reddy, K. L. Sharma, A. K. Indoria, K. Srinivas Reddy, Minakshi T. Grover, B. Srinivas, and B. Venkateswarlu. 2015. "Water Absorption and Release Characteristics of a Polymer and its Effect on Available Water Content, Tomato (*Lycopersicon esculentum*) Productivity and Water Use Efficiency in a Semi-arid Sandy Loam Soil." *Indian Society of Soil Science* 63 (4): 384-393.

69. Rehman, A., R. Ahmad, and M. Safdar. 2011. "Effect of hydrogel on the performance of aerobic rice sown under different techniques." *Plant, Soil and Environment* 57 (7): 321-325.

70. Roy, Trisha, Suresh Kumar, Lekh Chand, D. M. Kadam, Bankey Bihari, S. S. Shrimali, Rajesh Bishnoi, U. K. Maurya, Madan Singh, and M. Muruganandam. 2019. "Impact of Pusa hydrogel application on yield and productivity of rainfed wheat in North West Himalayan region." *Current Science (00113891)* 116 (7).

71. Rudzinski, W. E¬£, A. M¬£ Dave, U. H¬£ Vaishnav, S. G¬£ Kumbar, A. R¬£ Kulkarni, and T. M. Aminabhavi. 2002. "Hydrogels as controlled release devices in agriculture." *Designed monomers and polymers* 5 (1): 39-65.

72. Rudzinski, W. E¬£, A. M¬£ Dave, U. H¬£ Vaishnav, S. G¬£ Kumbar, A. R¬£ Kulkarni, and T. M. Aminabhavi. 2002. "Hydrogels as controlled release devices in agriculture." *Designed monomers and polymers* 5 (1): 39-65.

73. Sannino, Alessandro, Christian Demitri, and Marta Madaghiele. 2009. "Biodegradable cellulose-based hydrogels: design and applications." *Materials* 2 (2): 353-373.

74. Sedghi, Mohammad, Mitra Hadi, and Sahar Gholi Toluie. 2013. "Effect of nano zinc oxide on the germination parameters of soybean seeds under drought stress." *Annales of West University of Timisoara. Series of Biology* 16 (2): 73.

75. Sekine, Yurina, and Tomoko Ikeda-Fukazawa. 2009. "Structural changes of water in a hydrogel during dehydration." *The Journal of chemical physics* 130 (3): 034501.

76. Singh, Anshu Sharan, Richa Sharma, Shivam Dubey, and Vikram Kumar. 2018. "Efficacy of pusa hydrogel and chitosan on wheat (Triticum aestivum L.) physiological and biochemical parameters under water deficit condition." *Journal of Pharmacognosy and Phytochemistry* 7 (5): 1589-1591.

77. Takushi, Eisei, L. I. U. Asato, and Tadashi Nakada. 1990. "Edible eyeballs from fish." *Nature* 345 (6273): 298.

78. Tao, J., Zhang, W., Liang, L., and Lei, Z. (2018). Effects of eco-friendly carbohydrate-based superabsorbent polymers on seed germination and seedling growth of maize. *Royal Society Open Science, 5*(2), 171184.

79. Taran, Nataliya, Volodymyr Storozhenko, Nataliia Svietlova, Ludmila Batsmanova, Viktor Shvartau, and Mariia Kovalenko. 2017. "Effect of zinc and copper nanoparticles on drought resistance of wheat seedlings." *Nanoscale research letters* 12 (1): 60.

80. Teodorescu, Mircea, Anamaria Lungu, and Paul O. Stanescu. 2009. "Preparation and properties of novel slow-release NPK agrochemical formulations based on poly (acrylic acid) hydrogels and liquid fertilizers." *Industrial and Engineering Chemistry Research* 48 (14): 6527-6534.

81. Terada, Takahiko, Yasushi Maeda, and Hiromi Kitano. 1993. "Raman spectroscopic study on water in polymer gels." *The Journal of Physical Chemistry* 97 (14): 3619-3622.

82. Teyel, M. Y., and O. A. El-Hady. 1981. "Super gel as a soil conditioner." *Acta Hort* 119: 247-256.

83. Verma, M. K., Pandey, P., and De, N. (2017). Characterization of water retention and release capacity of innovative nano clay polymer composite superabsorbent. *J Pharmacogn Phytochem, 6*(6S), 42-48.

84. Viero, P. W. M., K. E. A. Chiswell, and J. M. Theron. 2002. "The effect of a soil-amended hydrogel on the establishment of a Eucalyptus grandis clone on a sandy clay loam soil in Zululand during winter." *The Southern African Forestry Journal* 193 (1): 65-75.

85. Wallace, Arthur, and Garn A. Wallace. 1986. "Effect of polymeric soil conditioners on emergence of tomato seedlings." *Soil Science* 141 (5): 321-323.

86. Wang, Yin-Tung, and Carol A. Boogher. 1987. "Effect of a medium-incorporated hydrogel on plant growth and water use of two foliage species." *Journal of Environmental Horticulture* 5 (3): 127-130.

87. Yang, W., Li, P., Guo, S., Song, R., and Yu, J. (2019). Co-application of soil superabsorbent polymer and foliar fulvic acid to increase tolerance to water deficit maize: photosynthesis, water parameters, and proline. *Chilean journal of agricultural research, 79*(3), 435-446.

88. Yordanov, I., V. Velikova, and T. Tsonev. 2000. "Plant responses to drought, acclimation, and stress tolerance." *Photosynthetica* 38 (2): 171-186.

89. Yousefian, M., Jafari, M., Tavili, A., Arzani, H., and Jafarian, Z. (2018). The effects of superabsorbent polymer on *Atriplex lentiformis* growth and soil characteristics under drought stress (Case Study: Desert Research Station, Semnan, Iran). *Journal of Rangeland Science, 8*(1), 65-76.

90. Zareii, Farinaz Davar, Arash Roozbahani, and Amir Hosnamidi. 2014. "Evaluation the effect of water stress and foliar application of Fe nanoparticles on yield, yield components and oil percentage of safflower (Carthamus tinctorious L.)." *Int J Adv Biol Biom Res* 2 (4): 1.

91. Zhao, W., Cao, T., Dou, P., Sheng, J., and Luo, M. (2019). Effect of various concentrations of superabsorbent polymers on soil particle-size distribution and evaporation with sand mulching. *Scientific reports, 9*(1), 3511.

Chapter 15

Nanotechnology for Enhancing Seed Quality and Health

K. Raja, M.S. Kumuthan and M. Latha

Department of Nano Science and Technology, Tamil Nadu Agricultural University Coimbatore – 641 003, Tamil Nadu

Application of nanomaterials or nanoproducts in agriculture is relatively new as compared to their use in biomedical and industrial sectors. Nanotechnology has the potential to revolutionize the agriculture and play an important role in food and crop production. During the past decade, a number of patents and products incorporating engineered nanoparticles (NPs) into agricultural practices, *e.g.* nanopesticides, nanofertilizers, and nano sensors, have been developed with the collective goal to promote the efficiency and sustainability of agricultural practices requiring less input and generating less waste than conventional products and approaches. In commercial agriculture, rapid and uniform seed germination and seedling emergence are important determinants of successful stand establishment and productivity of crop plants. There are several seed invigouration techniques like seed priming, coating, pelleting and designer seed technology to improve germination and seedling vigour and also for maintaining quality at storage. These interventions have different properties and to be optimized for each crop species besides practical difficulties in application. Therefore, there is a growing need to develop new seed invigouration techniques to enhance seed germination and maintain viability at senescence of various crop plants particularly.

Nanotechnological application in the field of seed science is recent and some of the metal-based nanoparticles (*e.g.*, AgNPs, AuNPs, CuNPs, FeNPs, FeS_2NPs, TiO_2NPs, ZnNPs, ZnONPs) and carbon-based nanoparticles (*e.g.*, fullerene and carbon nanotubes) have been applied as seed pre-treatment agents for promoting seed germination, seedling growth, and stress tolerance in some crop plants besides viability maintenance at ageing. Though nanomaterials have positive impact on seed quality, comprehensive studies on physiological and molecular mechanism of nano seed invigouration effects on seed germination have not been explained, and thus there are many questions remained to be addressed, especially mechanism behind NPs-induced seed germination and quality preservation. This chapter elucidates the nano interventions in improving germination and keeping the seed quality under ageing of crop plants with possible physiological, biochemical and molecular mechanisms.

Seed Invigouration of Nanoparticles for Improved Germination and Seedling Growth

The impact of nanoparticles on higher plants depend on the species and age of plants, the size and concentration of the particles, the experimental conditions such as temperature, and the duration and method of exposure. However, the interaction of nanoparticles in plant system has not yet fully understood. Nanoparticles improve the seed quality by increasing germination and seedling vigour under optimal and sub-optimal conditions. In contrast to positive effective of nanoparticles, there are reports indicating the negative effects of nanomaterials on higher plants. In recent years, several metal-based NPs and carbon-based nanoparticles seed invigoration has found to improve the quality associated with changes in morphological, physiological and molecular seed attributes. There are research findings suggested both positive and negative effects of NPs on plant growth and development, which depends on the composition, concentration, size and physical and chemical properties of NPs as well as plant species (Ma *et al.*, 2008; Khot *et al.*, 2012).

Cysteine capped ZnONPs seed priming at 250 ppm of greengram had advanced the radicle growth (Figure 15.1), improved the germination, seedling growth and vigour (Table 15.1), dry matter production, hydrolytic enzyme such as α- amylase and protease (Figure 15.2) and antioxidant enzymes *viz.*, peroxidase and superoxide dismutase enzymes activity both under normal and moisture stress conditions. The SEM images of radicle supplement the results by showing the increased cell division and cell elongation (Figure 15.3) in cysteine capped ZnONPs primed seeds as compared to non-primed and bulk source of zinc (Kumuthan, 2017).

Table 15.1. Nanoparticles Seed Invigouration on Germination and Seedling Vigour of Greengram

Particulars	*Germination (per cent)*	*Vigour Index*	*Dry Matter (mg/seedlings)*
Control	72	1681	3.98
Cysteine capped ZnO NPs	90	3064	5.80
CD (P=0.05)	1.847	54.11	1.71

Raja *et al.* (2019) further sustains the role of inorganic nanoparticles on improving seed germination and seedling growth. They observed that blackgram seeds primed in ZnO NPs dispersed water @ 600 mgl^{-1} for 3 hrs recorded maximum germination (74 per cent), root length (16.9 cm), shoot length (20.4 cm) and seedling

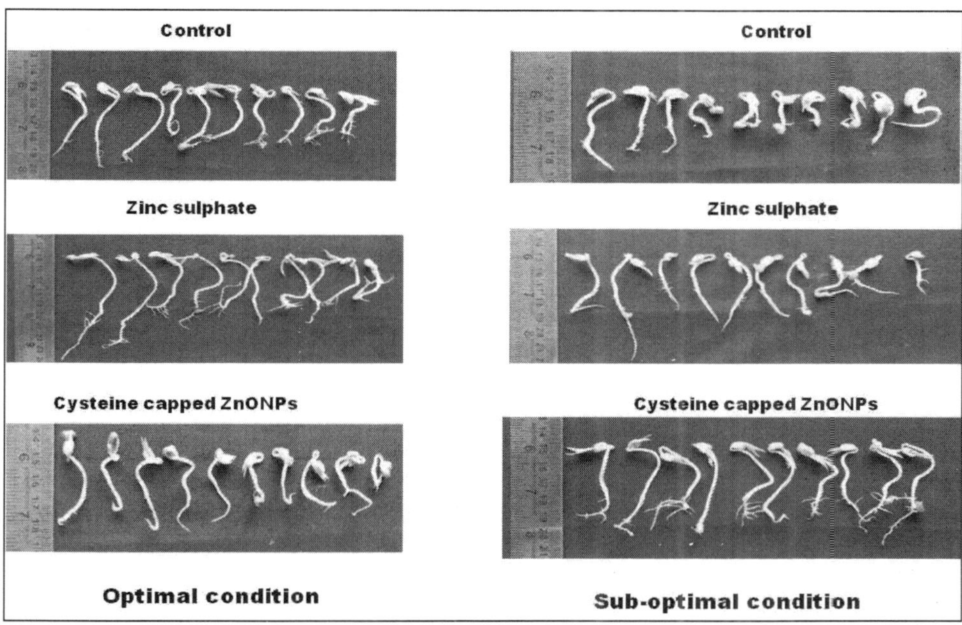

Figure 15.1. Radicle Growth of Nanoprimed Greengram at Optimal and Sub-optimal Condition.

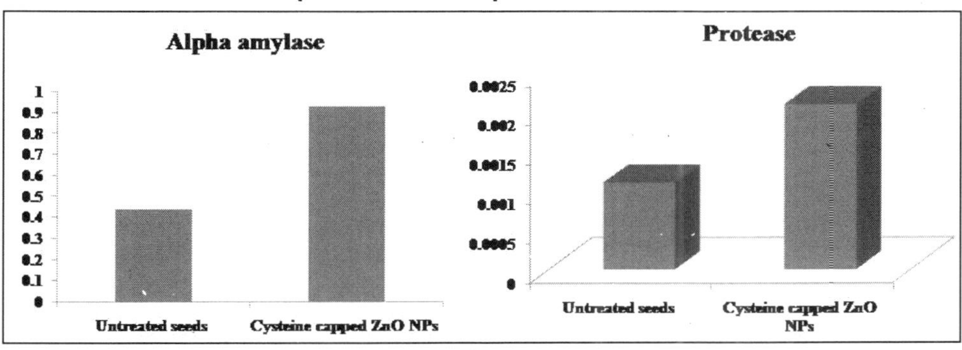

Figure 15.2. Nanopriming on Hydrolytic Enzymes Activity in Greengram.

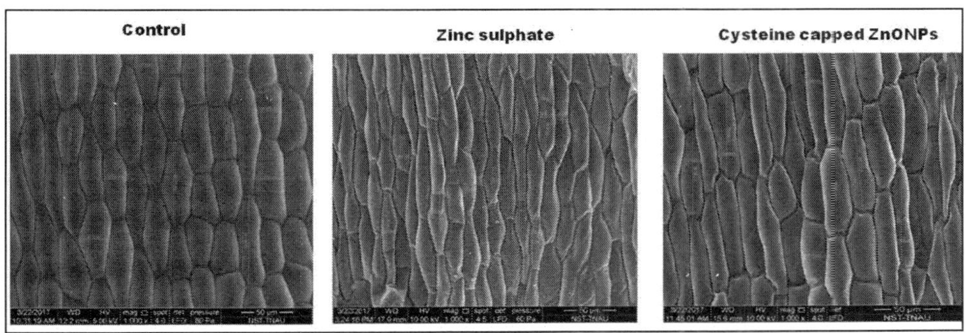

Figure 15.3. SEM Images of Radicle Cell Growth Morphology of Nanoprimed Greengram Seeds.

vigor (2760) than unprimed seeds which recorded the minimum (Table 15.2). Similarly seeds soaked in Cu NPs dispersed water for 3hr found to be superior over control in recording significantly higher germination, root and shoot length and seedling vigour (Figure 15.4). The improved effect of NPs is because of increased level IAA in the roots and thereby an increase in the growth of radicle and shoot was observed.

Table 15.2. Effect of ZnO NPs Dry Seed Invigouration on Germination and Vigour of Blackgram

Treatments (mg L^{-1})	*Shoot Length (cm)*	*Root Length (cm)*	*Germination (per cent)*	*Vigour Index*
Control	16.0	14.0	67	1995
400 mg	16.2	14.2	67	2037
500 mg	16.7	14.4	68	2099
600 mg	17.5	15.2	68	2224
700 mg	17.9	15.7	69	2302
800 mg	17.4	16.0	70	2321
900 mg	17.2	16.1	70	2324
1000 mg	17.6	15.8	70	2321
1100 mg	18.7	16.3	72	2503
1200 mg	17.6	16.2	68	2298
1300 mg	17.3	15.9	66	2191
Mean	17.3	15.4	68	2238
CD (P=0.05)	1.22	1.61	4.42	65.80

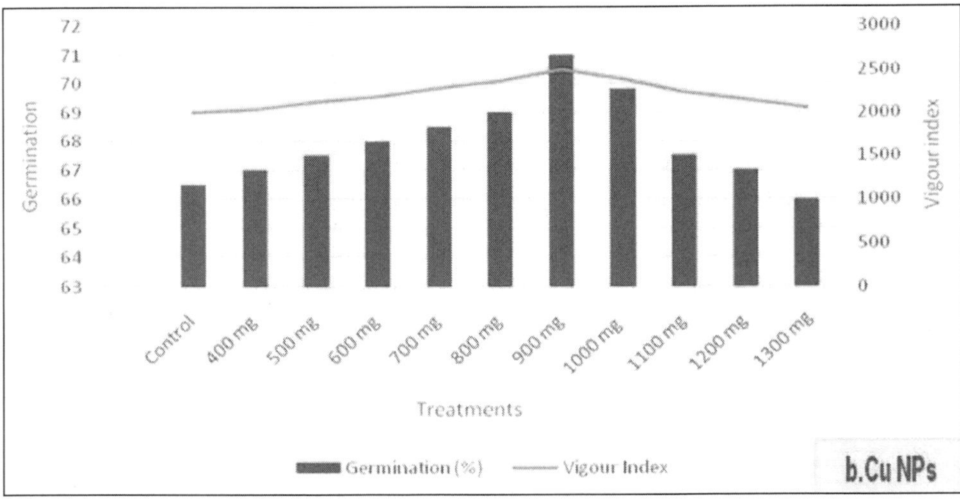

Figure 15.4. Influence of Cu NPs Priming on Germination and Seedling Growth of Blackgram.

Nano-SiO_2 and nano-titanium dioxide (nano-TiO_2) seed priming improved germination by increasing nitrate reductase (Lu *et al.*, 2002) and also by enhancing seeds ability to absorb and utilize water and nutrients (Zheng *et al.*, 2005). Zinc oxide NPs seed invigouration at 1000 ppm promoted seed germination, seedling vigor and plant growth and also proved to be effective in increasing stem and root growth in peanuts (Prasad *et al.*, 2012).

Carbon nano-tubes (CNTs) fetch much importance in biological and material sciences as they have more surface area and potential to regulate the moisture and other kind of abiotic stresses. CNTs seed invigouration improves germination and seedling growth by serving as new pores for water permeation by penetration of seed coat. Indeed CNTs have shown to enhance germination and seedling growth of several crops (Table 15.3).

The metal oxide NPs like ZnO is known to improve the germination and seedling vigor of a wide spectrum of crops such as tomato, onion, chilli, groundnut and blackgram (Table 15.3). In all the cases, the improved germination is resulted by promoting growth hormones in the germinating seeds. Silver (Ag) nano-particles are well known for their anti-microbial properties. As a result of anti-microbial and anti-pathogenic effects, seed borne pathogens are effectively controlled, that eventually has resulted in improved germination of many crop species (Table 15.3). Almutairi and Alharb (2015) observed the significant enhancement of the germination percentage in watermelon and zucchini plants with Ag NPs as compared to untreated seeds. Despite the fact that Ag NPs are beneficial, excessive use is reported to have deleterious effects on crops. Ag NPs showed a toxic effect on corn root elongation. This study showed that exposure to Ag NPs caused both positive and negative effects on plant growth and germination.

TiO_2 NPs are found to have favourable effects on seed invigoration. On entry to the seeds, TiO_2 NPs quench the reactive oxygen species and break the dormancy caused by the phenolics in the seeds. Several reports have clearly indicated the beneficial role of TiO_2 in enhancing germination (Table 15.3). Feizi *et al.* (2013) reported that TiO_2 NPs treatment @ 60 ppm improved the germination in *Foeniculum vulgare*. TiO_2 NPs enhanced the seed germination and promoted radicle and plumule growth of canola seedlings (Mahmoodzadeh *et al.*, 2013). Zero valent Iron (ZVI) NPs also exhibited similar effects in several crops. Other NPs such as copper, silica, alumina and cerium had reported to improve germination of a wide array of crops (Table 15.3).

Nanoparticles for Extended Shelf Life of Seeds during Ageing

Seed storage is the integral part of the seed production programme. Loss of seed vigour and viability are said to be associated with the ageing phenomenon and results in poor stand and performance in the field. The storage potential of any seed depends on several factors like type of seed, period of storage, chemical composition, seed treatment, seed moisture and temperature and storage containers. The seed deterioration is inevitable, irreversible and inexorable but it can be delayed to some extent by proper management practices like preservation under controlled conditions and by some seed invigouration techniques. Preservation of seeds under

Table 15.3. Effect of Nanoparticles on Germination and Seedling Growth of Crop Plants

Nanoparticle(s)	*Plant*	*Concentration(s)*	*Impact Observed on*	*Reference(s)*
1. Carbon nano-tubes (CNTs)				
Graphene oxide	*Vicia faba* L.	400 and 800 mg/L	Improved germination	Anjum *et al.*, (2014)
Carbon Nanotubes (CNTs)	*Lycopersicon esculentum* Mill	40 µg/mL	Enhanced germination and seedling growth	Morla *et al.*, (2011)
	Medicago sativa L. *Triticum aestivum*	75 wt per cent CNTs	Improved root elongation and growth	Miralles *et al.*, (2012)
	Medicago sativa L. *Triticum aestivum*	75 wt per cent CNTs Impurities	Improved root elongation and growth	
Single walled Carbon Nanotubes (SWCNTs)	*Allium cepa* L. *Cucumis sativus* L.	315 and 1,750 mg/L	Improved root elongation and growth	Canas *et al.*, (2008)
Multi-walled Carbon Nanotubes (MWCNTs)	*Hordeum vulgare* L. *Glycine max* L. *Zea mays* L.	25–100 µg/mL	Improved germination	Lahiani *et al.*, (2013)
	Lycopersicon esculentum Mill	50 and 200 µg/ml	Increased plant height and No. of flowers	Khodakovskaya *et al.*, (2013)
	Nicotiana tabacum L.	5 up to 500 µg/ml	Improved growth of plants	Khodakovskaya *et al.*, (2012)
o-MWCNTs	*Triticum aestivum* L.	10–160 µg/mL	Enhanced root growth and vegetative biomass	Wang *et al.*, (2012)
wsCNTs	*Cicer arietinum* L.	6.0 µg/mL	Improved growth rate	Tripathi *et al.*, (2011)
MWCNTs, dMWCNT	Lycopersicon esculentum Mill	40 µg/mL	Increased the nutrient uptake efficiency (K, Ca, Fe, Mn and Zn)	Tiwari *et al.*, (2013)
Pristine MWCNTs	*Zea Mays* L.	20 mg/L	Improved the nutrient transport and biomass	Tiwari *et al.*, (2014)

Nanoparticle(s)	*Plant*	*Concentration(s)*	*Impact Observed on*	*Reference(s)*
2. Zinc Oxide NPs				
ZnO NPs	*Cucumis sativus* L.	400 mg/kg	Increased the micronutrients (Cu, Mn and Zn) content and growth	Zhao *et al.*, (2014)
	Arachis hypogea L.	1,000 ppm	Improved germination, stem, root growth and yield	Prasad *et al.*, (2012)
	Vigna radiata Tomato) Onion Groundnut Chilli	1000mg kg^{-1}	Improved germination, seedling growth and vigour index in naturally aged seeds. Increased the lipid peroxidase activity	Sridhar (2012) Shyla *et al.*, (2014) Senthilkumar (2011)
	Vigna radiata L. Wilczek	1000ppm	Increased dry weight of plant	Patra *et al.*, (2013)
3. Silver NPs				
Ag NPs	*Boswellia ovaliofoliolata*	10–30 μg/mL	Improved germination and seedling growth	Savithramma *et al.*, (2012)
	Phaseolus vulgaris L. *Zea mays* L.	60 ppm	Increased root and shoot length, and dry weight of seedling	Salama (2012)
	Vigna radiata L.	100 μM	Antagonistic inhibition by 2,4-dichlorophenoxyacetic acid (2,4-D) at 500 μM of plant growth	Karuppanapandian *et al.*, (2011)
SiO_2 NPs	*Zea mays L.*	15 kg/ha	Improved all the growth parameters of plant	Suriyaprabha *et al.*, (2012)

Nanoparticle(s)	*Plant*	*Concentration(s)*	*Impact Observed on*	*Reference(s)*
4. Titanium oxide NPs				
TiO_2 NPs	*Arabidopsis thaliana*,	400 mg/L	Enhanced root length	Lee *et al.*, (2010)
	Foenicutum vulgare	60 ppm	Improved germination and seedling growth	Feizi *et al.*, (2013)
	Lemna minor L.	lower than 200 mg/L	Improved plant growth	Song *et al.*, (2012)
	Triticum aestivum L.	1,000 mg/L	Increased chlorophyll content	Mahmoodzadeh *et al.*, (2013)
	Spinacia oleracea L.	0.25 per cent	Protected chloroplasts from aging	Hong *et al.*, (2005a and b)
	Lycopersicon esculentum Mill	50-200 mg/L	Net photosynthetic rate, conductance of H_2O_2 and transpiration rate, Regulation of photosystem II (PSII)	Qi *et al.*, (2013)
5. Zero Valent Iron NPs				
ZVI	*Arabidopsis thaliana*	500 mg/L	Improved root elongation	Kim *et al.*, (2014)
	Vigna radiata	1000 mg/kg of seed	Improved germination, seedling growth and vigour index in naturally aged seeds	Senthilkumar (2011)
6. Other NPs				
CeO_2 NPs	*Arabidopsis thaliana*	250 ppm	Increased the biomass content	Ma *et al.*, (2013)
	Raphanus sativus L.	5000 mg/L	Improved root elongation and growth	Wu *et al.*, (2012)
CuO NPs	*Triticum aestivum* L.	500 mg/kg (sand culture)	Increased the biomass content	Dimkpa *et al.*, (2012)
Hydroxyapatite suspension	*Lactuca sativa* L.	100–2,000 mg/L	Improved root length and growth	Wang *et al.*, (2012)
Sulfur NPs	*Vigna radiata* L.	2000 and 4000 ppm	Increased dry weight of seedling and plant	Patra *et al.*, (2013)
Au NPs	*Arabidopsis thaliana*	10 and 80 µg/mL	Improved germination, root and shoot length, early flowering and yield	Kumar *et al.*, (2013)
Aluminum oxide NPs	*Arabidopsis thaliana*	400–4000 mg/L	Improved root length	Lee *et al.*, (2010)
	Lemna minor L.	10 mg/L	Enhanced root elongation and length	Juhel *et al.*, (2011)

Courtesy: Subramanian *et al.*, 2018.

controlled conditions is costlier and not feasible in the developing countries like India, under such situation seed invigouration techniques are best alternatives, and applications of nanotechnology in maintaining seed quality is one of the most promising interventions during storage.

Nanoparticles can be one of the ways to retain the vigour and viability during storage by preventing the losses due to biotic and abiotic stress. The chemical synthesized nanoparticles such as Zinc oxide (ZnO), Silver (Ag), Copper oxide (CuO) and Titanium oxide (TiO_2), dry seed treatment shown to preserve the viability of wide array of crops under ageing. Blackgram seeds invigourated with ZnO nanorods @ 1000 and 1250 mg kg^{-1} and stored under ambient conditions found to be preserved with germination of 88 per cent and vigour index (1973) as against untreated control where the germination was 75 per cent and vigour index was 1673 (Senthilkumar, 2011). Similarly the ZnO and silver NPs treated groundnut seeds significantly outperformed over control in terms of recording germination (Figure 15.5) and seedling vigour index (Figure 15.6) even after 12 months of natural ageing. This is also supported by registering higher dehydrogenase enzyme activity (0.699 OD) and minimum electrical conductivity of 0.347 dSm^{-1} (Krishna Shyla, 2014). Dileep Kumar (2015) further established the use of metal oxide nanoparticles for preserving the storability of chilli seeds even after 12 months of storage under ambient conditions by recording 75 per cent germination which is above the minimum required germination standard (Table 15.4). In all these cases, such phenomenal response is attributed to the quenching of reactive oxygen species by donating electron by the metal oxide nanoparticles which slow down the lipid peroxidation process (Table 15.5). The increased activity of antioxidant enzymes due to nanoparticles seed invigoration during natural ageing further proved that the NPs could extend the shelf life of seeds. Herein the groundnut seeds treated

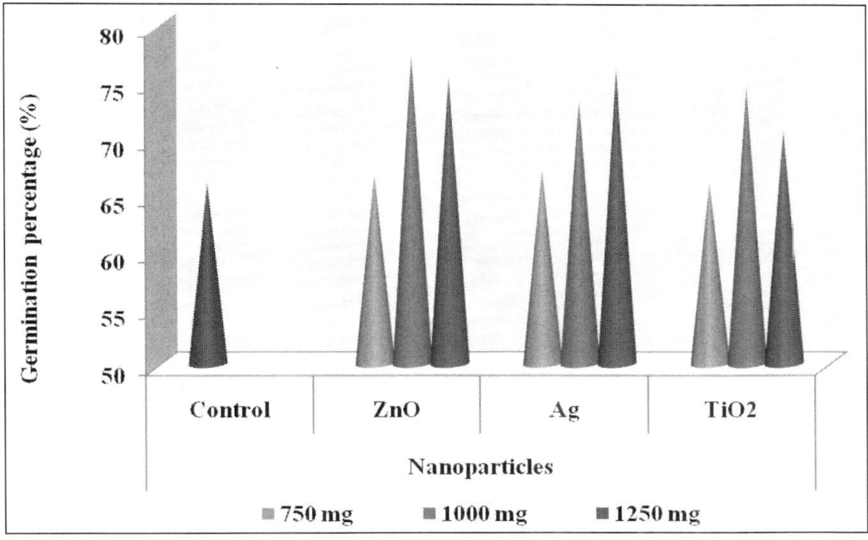

Figure 15.5. Effect of Inorganic Nanoparticles on Germination of Groundnut during Ageing.

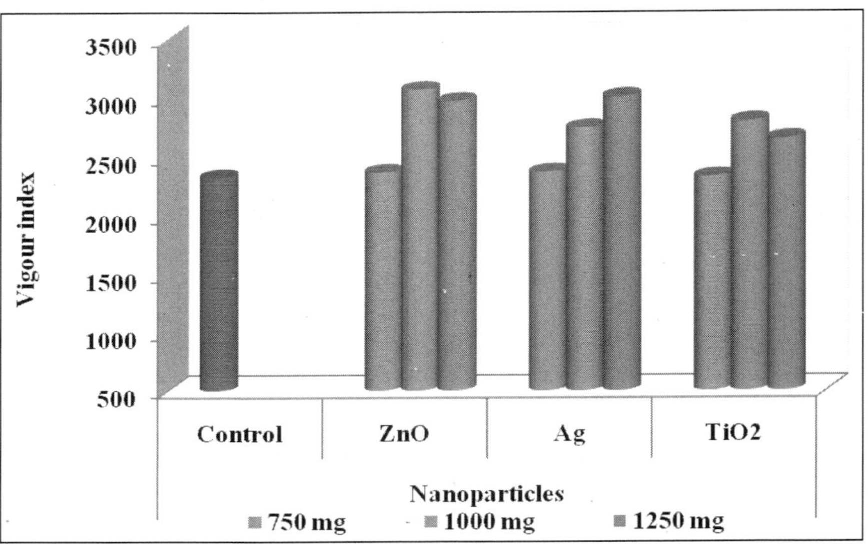

Figure 15.6. Inorganic Nanoparticles on Seedling Vigor of Groundnut during Ageing.

with ZnO NPs @ 1000 mg kg^{-1} registered higher activity of catalase (0.421 µg H_2O_2 mg^{-1} min^{-1}) and peroxidase enzymes (0.312 OD 10 min^{-1}) at 12th month of storage (Figure 15.7). These antioxidant enzymes scavenge the free radicals generation and hence reduce the lipid autooxidation processes thus maintain the cell membrane integrity and improve the storability of seeds.

Table 15.4. Nanoparticles on Germination and Vigour Index of Chilli Seeds under Ageing

Treatments (mg/kg^{-1})	*Germination (per cent)*			*Vigour Index*		
	TiO_2	*ZnO*	*Ag*	*TiO_2*	*ZnO*	*Ag*
750	66 (54.33)	70 (56.79)	67 (54.79)	897	1044	1044
1000	68 (55.55)	75 (60.00)	72 (58.05)	970	1285	1137
1250	67 (54.79)	73 (58.70)	68 (55.55)	1111	1202	1045
Mean	67 (54.79)	73 (58.70)	69 (56.17)	993	1177	1075
Control	66 (54.33)			861		
	T	D	T x D	T	D	T x D
SED	0.65	0.56	1.12	23.25	20.14	40.85
CD (P=0.05)	1.34	1.16	2.32	47.17	40.85	81.71

Effect of Inorganic Nanoparticles on Seed Volatile Emission

A diverse array of volatile compounds is produced by seeds during storage suggesting that several reactions occur in seeds during deterioration process. The identity of volatile compounds produced by seeds can indicate the kinds of chemical

Table 15.5. Effect of Inorganic Nanoparticles on Lipid Peroxidation (OD value) of Groundnut

Treatments	*Period of storage (p)*					
	P 0	*P 3*	*P 6*	*P 9*	*P 12*	*Mean*
Control	0.048	0.068	0.106	0.152	0.195	0.112
ZnO 750 mg	0.050	0.061	0.097	0.141	0.189	0.105
ZnO 1000 mg	0.048	0.062	0.081	0.113	0.155	0.089
ZnO 1250 mg	0.049	0.060	0.087	0.121	0.163	0.094
Ag 750 mg	0.052	0.069	0.099	0.145	0.186	0.108
Ag 1000 mg	0.047	0.066	0.095	0.130	0.177	0.101
Ag 1250 mg	0.048	0.061	0.084	0.115	0.159	0.091
TiO_2 750 mg	0.054	0.072	0.101	0.147	0.190	0.110
TiO_2 1000 mg	0.046	0.065	0.089	0.131	0.168	0.097
TiO_2 1250 mg	0.051	0.065	0.093	0.136	0.180	0.102
Mean	0.049	0.065	0.093	0.133	0.176	
	Treatments		*Period*		*Treatments x Period*	
SE d	0.0017		0.0019		0.006	
CD (P=0.5)	0.0033**		0.0038**		0.012**	

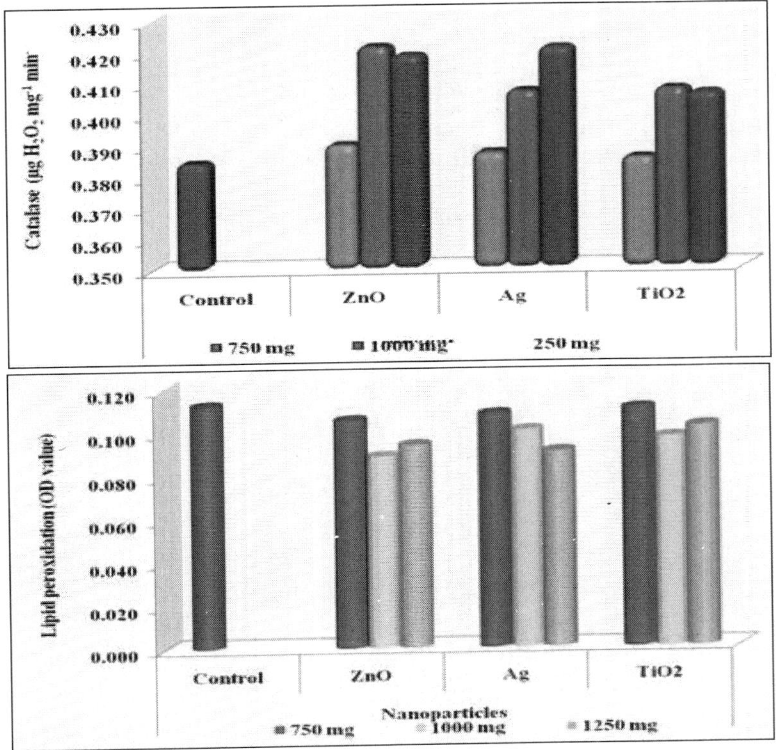

Figure 15.7. Effect of Inorganic Nanoparticles on Catalase and Peroxidase Activity of Groundnut under Ageing.

reactions that occur during storage and those reactions may lead to loss of viability. Metal and metal oxide nanoparticles seed invigouration found effective in reducing the Volatile Organic Compound emission during storage. Krishna Shyla (2014) established the role of inorganic NPs on volatile compounds release under natural ageing where the groundnut seeds treated with nanoparticles emitted lesser amount of VOCs during ageing. The number of volatiles emitted by the seeds treated with ZnO and Ag was relatively lower with 16.3 and 16.0 respectively than the control which had as much as 36 compounds. In general seed treated with NPs resulted in the emission of reduced number of volatile compounds than the control. Upon seed treatment with NPs, the 12 aldehydes recovered in the control got reduced to 4.3, each 5 of ester and carboxyl to 3.5 and 3.1 respectively, and 7 unknown to 1. A significant of reduction in the number of compounds emitted due to NPs treatment was evident which had 55 per cent reduction in case of both the ZnO and Ag while 39 per cent in the case of TiO_2. Area of the peak in the graphical output of GCMS analysis is indicative of the strength of the compound. The analysis of the area of the peak enclosed that the inorganic and organic NPs treated seeds had minimum area and the least was with ZnO (10.5 per cent) compared to control (21.9 per cent) accounting for 52 per cent reduction (Figure 15.8). Germination of seeds treated with NPs when correlated with volatile compounds revealed a negative relationship with aldehydes, carboxyl and unknown compounds while other was non-significant. Dileep Kumar (2015) further demonstrated the nanoparticles on VOCs production in chilli seeds during storage wherein ZnO and Ag NPs produced about 11.7 VOCs than the control (29 compounds). A significant reduction in the number of compounds emitted due to NPs treatment was evident which had 59.6 per cent reduction in case of ZnO and Ag while 58.6 per cent in the case of TiO_2 as compared to control which clearly revealed that nanoparticles invigorated seeds emitted low amount of

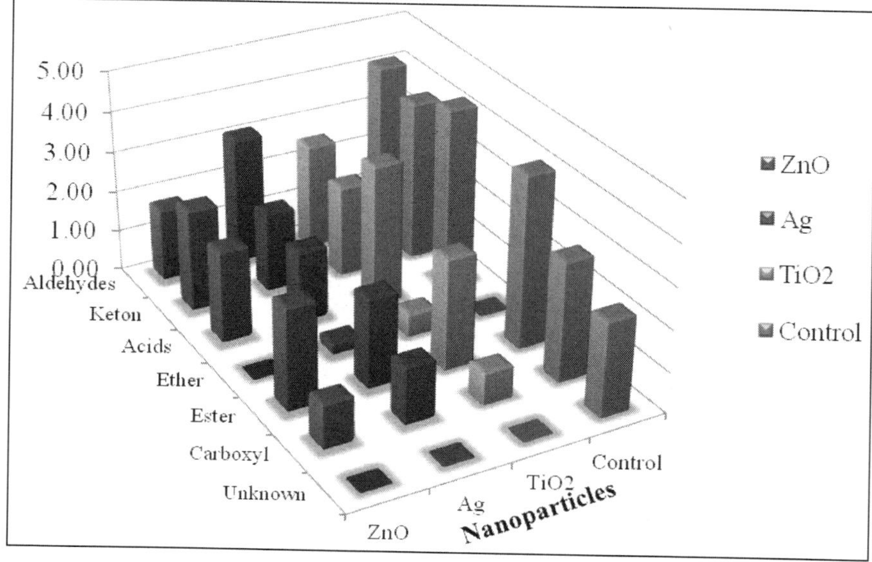

Figure 15.8. NPs on Production of Volatile Aldehydes in Aged Groundnut Seeds.

volatiles indicating the integrity of cell membrane by quenching the ROS damage on lipid bilayer in the cell membrane under ageing.

Nanoparticles on Seed Health

Maintenance of seed quality during storage of seed is very much essential. Though the initial seed quality and storage environment are important to prolong the shelf life of seeds, invasion of fungal pathogen and infestation of storage insects also play a major role in decreasing the viability of seed lot that shows storage pathogens and insects play a major role in determining the shelf life of seed with their shorter life span. The antimicrobial properties of nanoparticles have found effective in controlling the growth of pathogens like *Penicillium expansum* (He *et al.*, 2011) and pathogenic yeast *Candida albicans* (Lipovsky *et al.*, 2011) and storage pest *viz.*, rice weevil (Debnath *et al.*, 2010). In an experiment, NPs seed invigoration before storage brought down the seed pathogen infection to 32.3per cent compared to control (44.0 per cent) in groundnut under natural ageing (Table 15.6). In this case, Ag NPs @ 1250 mg kg^{-1} protected the seeds with minimum seed pathogen infection of 18.0 per cent (Krishna Shyla, 2014). This antifungal activity has further confirmed with invitro antifungal activity test of NPs on mycelial growth of *Macrophomina phaseolina* where the Ag and ZnO NPs significantly reduced the mycelia growth at 750 and 1750 ppm respectively. But the complete growth of mycelia observed at 1250 ppm Ag NPs (Figure 15.9). The microscopic image supplements the results by observing hyphal damage due to NPs which destroy the cell membrane structure and sclerotia formation of fungus (Figure 15.10). Moreover, Dileep Kumar (2015) also demonstrated the NPs on storage pathogens in chilli seeds stored under ambient environment. The results indicated that Ag NPs reduced the fungal infection to 6.0 per cent from 18.2 per cent observed in the control. Herein the bioassay test confirms

Table 15.6. Effect of Inorganic Nanoparticles on Pathogen Infection (per cent) in Groundnut under Natural Ageing

Treatments	*Period of storage (p)*				
	P 0	*P 3*	*P 6*	*P 12*	*Mean*
Control	0.0	10.0	23.0	44.0	21.23
ZnO 750 mg	0.0	0.0	15.0	37.0	15.15
ZnO 1000 mg	0.0	0.0	10.0	32.0	12.23
ZnO 1250 mg	0.0	0.0	10.0	30.0	11.31
Ag 750 mg	0.0	0.0	10.0	34.0	12.31
Ag 1000 mg	0.0	0.0	5.0	28.0	10.08
Ag 1250 mg	0.0	0.0	5.0	27.0	9.54
TiO_2 750 mg	0.0	5.0	15.0	38.0	16.38
$TiO_2$1000 mg	0.0	5.0	15.0	33.0	14.38
$TiO_2$1250 mg	0.0	0.0	10.0	32.0	12.15
Mean	**0.0**	**2.0**	**11.8**	**33.5**	
	Treatments	*Period*		*Treatments x Period*	
CD (P=0.5)	0.23**	0.27**		0.85**	

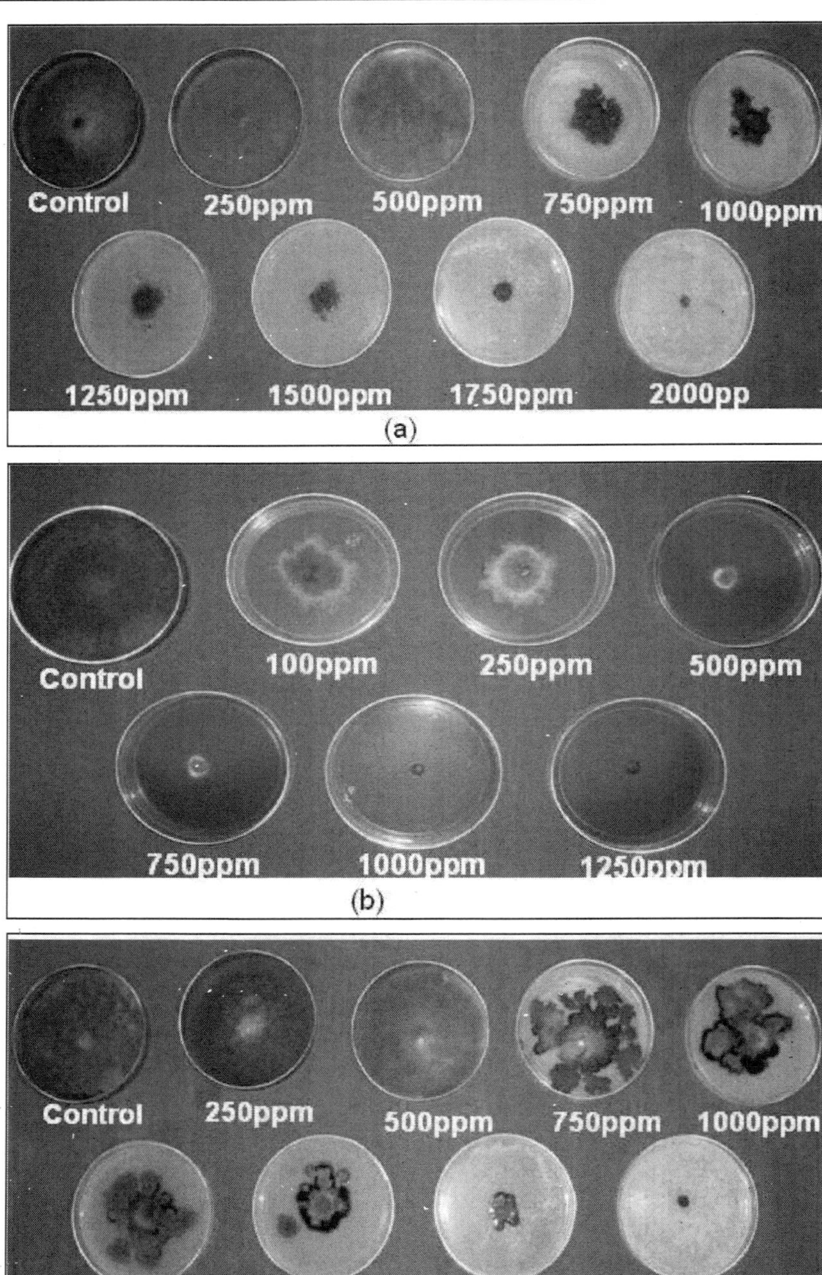

Figure 15.9. Inhibition Zones Caused by different Concentrations of Nanoparticles of (a) ZnO, (b) Ag and (c) TiO_2 against *Macrophomina phaseolina.*

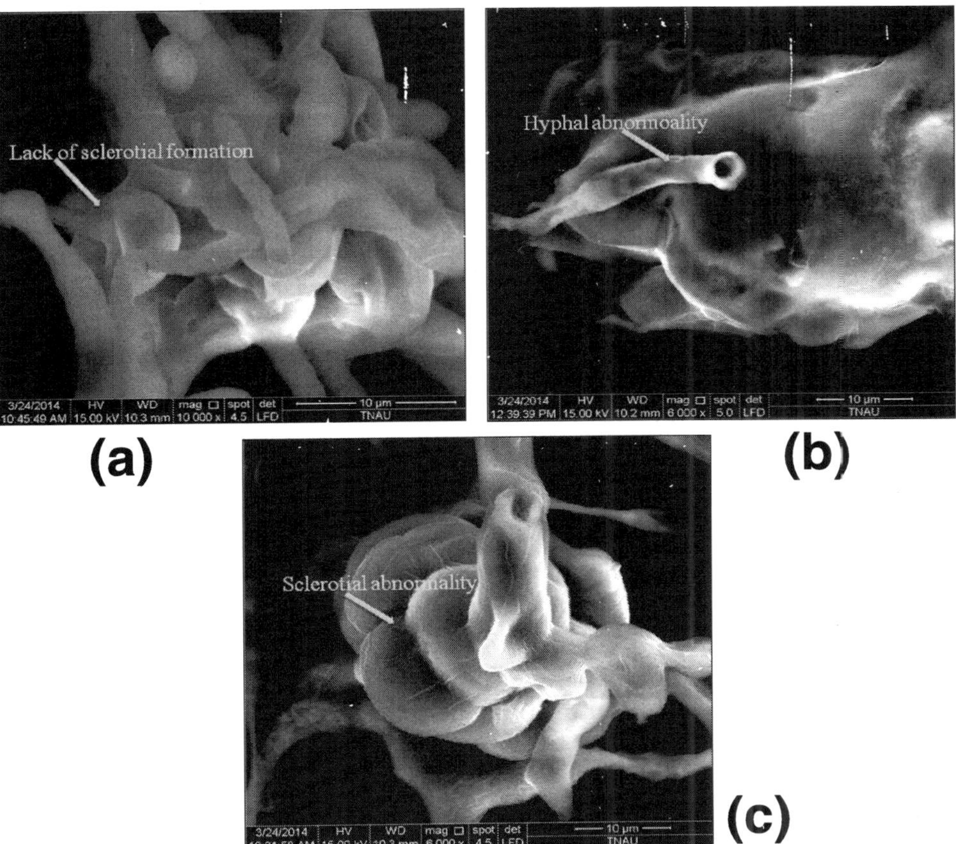

Figure 15.10. Scanning Electron Images of *Macrophomina phaseolina* Surface Ultrastucture in NPs Treated (a) Lack of sclerotia formation, (b) Hyphal abnormality and (c) Sclerotia abnormality.

the antifungal activity of NPs on *S. rolfsii* where the complete inhibition of mycelial growth was observed in Ag NPs at 750ppm which is coincided with the induced mycelial malformation and inhibited sclerotia production.

NPs have proved on controlling the seed storage pests in wide spectrum of crops. Zinc Oxide NPs revealed high efficacy against *S. oryzae* (Hamza and Amany, 2012; Sabbour, 2013). Goswami *et al.* (2010) studied the applications of different kind of nanoparticles *viz.* Ag, aluminium oxide, zinc oxide and titanium dioxide in the control of rice weevil (*Sitophilus oryzae*). Effect of NPs in the management of *L. serricorne* in stored chilli seeds revealed that Ag and ZnO NPs were highly effective (100 per cent) at 1250 mg kg^{-1} on 14DAT The SEM image of *L. serricorne* neck, leg and antennal base and last abdominal segment indicated the presence of Ag and ZnO NPs (Dileep Kumar, 2015). studies by Rouhani *et al.* (2012) reported that the silica and silver nanoparticles with a particle size range 20–60 nm was much effective on larvae than adults of *Calosobruchus maculatus*. Compared to commercially available insecticides, inorganic nanostructured materials may provide an inexpensive and dependable alternative for stored product insect pest control. Silver nanoparticles effectively control the seed storage pest of *Callosobruchus maculatus* in pulses (Arumugam *et al.*, 2016), and *Corcyra cephalonica* in paddy (Vani and Brindhaa, 2013).

Mechanism of Improving Germination by Nanoparticles/Nanomaterials Seed Invigouration

The mechanism involved in improved germination, seedling growth and vigour due to NPs seed treatment has not yet been thoroughly studied. The possible mechanism for enhanced germination is increased water uptake or imbibition rate due to penetration of nanoparticles through the natural pores over the seed coat or creation of small pores by nanoparticles themselves and making the seed coat as soften. The increased water imbibition enhances the synthesis of aquaporin water loving protein molecules by up-regulating the expression of aquaporin genes (Khodakovsakaya *et al.*2016). The silver nanoparticles primed seed enhances the expression level of aquaporin genes *PIP1-1* and *PIP2-1* in rice (Figure 15.11) which resulted higher water uptake and improved germination in rice (Wuttipong Mahakham *et al.*, 2017). There is higher accumulation of Reactive Oxygen Species (ROS) observed in germinating seeds of NPs fortification (Wuttipong Mahakham *et al.*, 2017) which indicates that ROS stimulates the seed germination. There are some scientific evidences explain that the synthesis of aquaporin protein not only facilitate the seed to imbibe water but also regulate the ROS movement across the biological membrane, thus, the root bud elongates and grow out through seed coat. The elevated level of ROS is controlled and brought down in the optimum range of oxidative window by the increased activity of antioxidant enzymes in nanoprimed seeds.

The increased water uptake due to nanoparticle invigouration triggers metabolic events connected with seed germination. Increase in imbibition rate, augment the hydrolytic enzymes activity such as alpha amylase, protease and lipase *etc.*, which degrade the stored food materials in seeds in to simple sugars

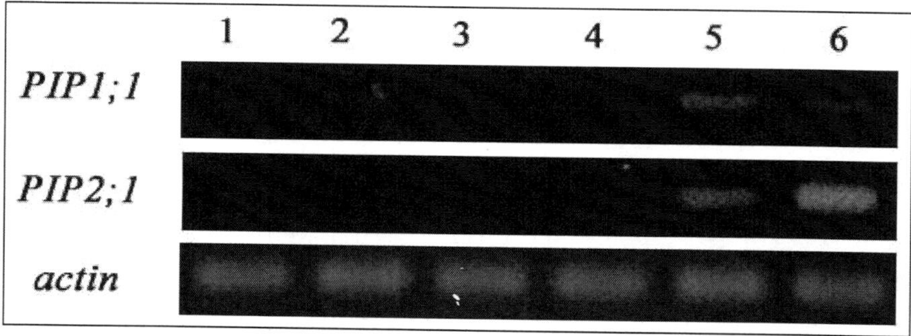

Figure 15.11. AgNPs Nanopriming Enhanced Aquaporin Gene Expression in Aged Rice Seeds. 1. Unprimed seeds, 2. Hydroprimed seeds, 3. $AgNO_3$@10ppm primed seeds, 4. $AgNO_3$@20ppm primed seeds, 5. AgNPs@10ppm primed seeds and 6. AgNPs@20ppm primed seeds.

Courtesy: Wuttipong Mahakham *et al.*, 2017.

and energy in the form ATP, and this energy is transferred to germinating and growing embryo. Moreover, increasing simple sugar concentration in the seed cells due increased hydrolytic enzymatic degrading activity of food materials leads to reduces the osmotic potential and hence the water potential, as a consequence the gradient difference between the inside and outside seed tissues water potential facilitates the water movement in to the seeds by osmosis. This physiological event accelerates the germination and early emergence of seedlings. Seeds primed with cysteine capped ZnO nanoparticles induced the hydrolytic enzymes such as α- amylase and protease activity in greengram which coincided with improved radicle growth, germination, seedling growth and vigour dry matter production (Kumuthan, 2017). The amplified activity of alpha amylase (Figure 15.12) boost up the starch hydrolysis in germinating silver nanoparticles primed rice seeds, which provided more of soluble sugars necessary for energy generation required for seedling emergence and growth, and this effect is linked with enhanced degradation of starch during germination wherein the nanoprimed seeds had no compound starch granules (CSGs) and contained smaller and well decomposed (Figure 15.13) starch granules. According to Wuttipong Mahakham *et al.* (2017) there are three possible mechanisms namely seed coat penetration of NPs by interacting with cell walls of seed coat to create small pores which facilitate quick water uptake that can induce metabolic events and starch hydrolysis of seed during initial phase of germination. Secondly, the nanoparticles found to enhance the α-amylase activity to fasten the starch hydrolysis and thus seed can generate higher available sugars to support embryo growth. Thirdly, nanoparticles could generate ROS being in oxidative window range for acting as signalling molecules and participating in cell wall loosening and endosperm weakening. The proposed mechanisms are depicted in the Figure 15.14.

Emerging Nanotechnologies in Seed Quality Enhancement

Seed is only input has a direct significance in sustaining the farm productivity and profitability. In developing country like India, traditionally, seeds of the

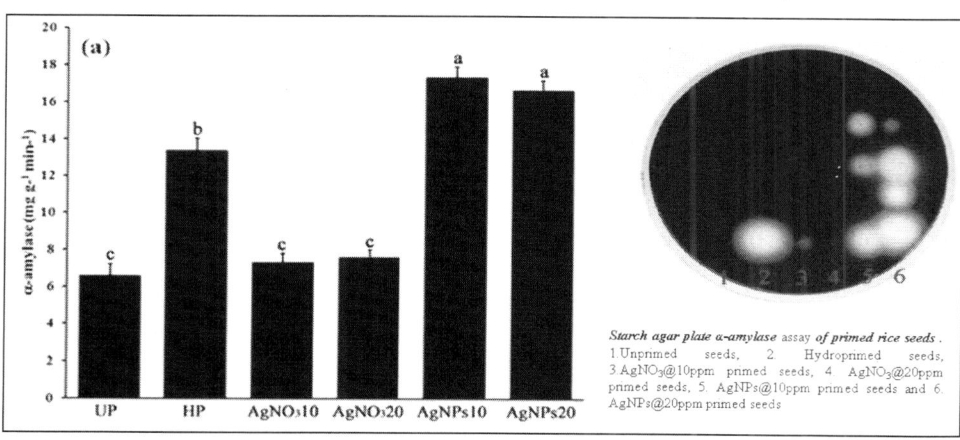

Figure 15.12. Nanopriming on Alpha Amylase Activity in Rice Seeds.
Courtesy: Wuttipong Mahakham *et al.*, 2017.

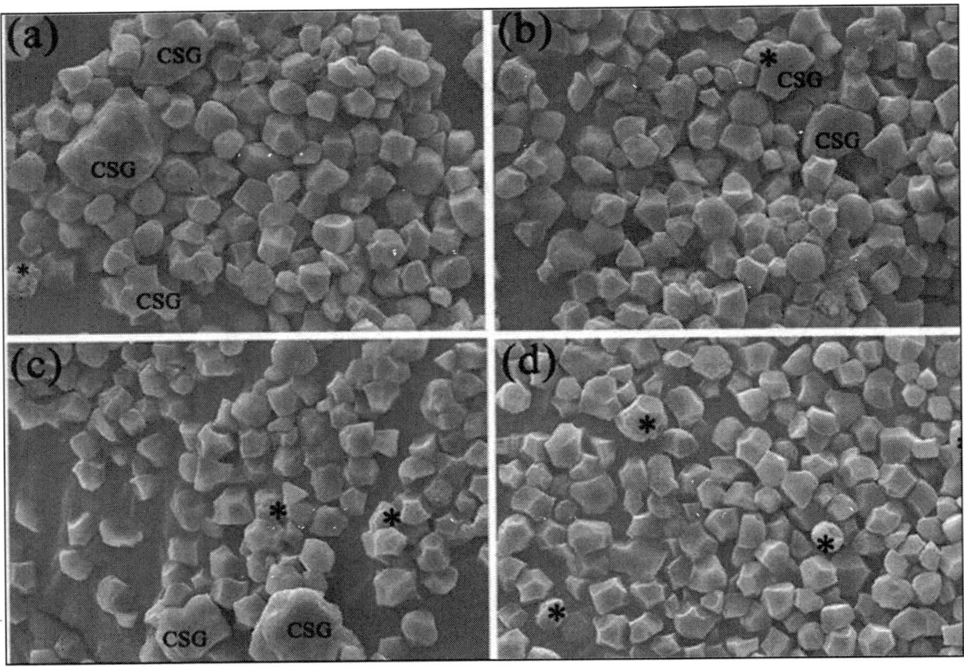

Figure 15.13. Nanopriming on Starch Metabolism in Rice Seeds.

SEM micrographs of starch granules of unprimed seeds (a), $AgNO_3$-primed seeds (b), hydroprimed seeds (c), and AgNPs-primed seeds (d). Starch granules were isolated from seeds after 24 h of imbibition. Asterisk indicates starch granule with small pits. CSGs, compound starch granules. (Scale bar = 20 μm).

Courtesy: Wuttipong Mahakham *et al.*, 2017.

preceding season will be used for sowing for succeeding cropping season. Presently, most of the farmers are shifted from using of own saved seed source to procurement from the market. The time between the cropping decision and procurement of seeds from authentic source is very narrow and hardly possible for the Indian farmers to check the seed quality prior to sowing. This situation necessitates infusing innovative technologies for early detection and improving the seed quality prior to sowing in the field. These innovative technologies are often referred as the third generation seed treatments for quality enhancement and assurance that encompass seed quality detection using e-nose for quick detection of seed viability and quality enhancement through polymer nanoemulsion coating and smart delivery of agri-inputs through seeds. Since, seed is a "nano-input" and a miniaturized laboratory that can serve as a single solution to address complex and multi-dimensional field problems.

Polymer Nanoemulsion Seed Coating

Water soluble polymers are widely used for coating the seeds. These polymers at a threshold of moisture gets dissolved allowing the coated seed to absorb the

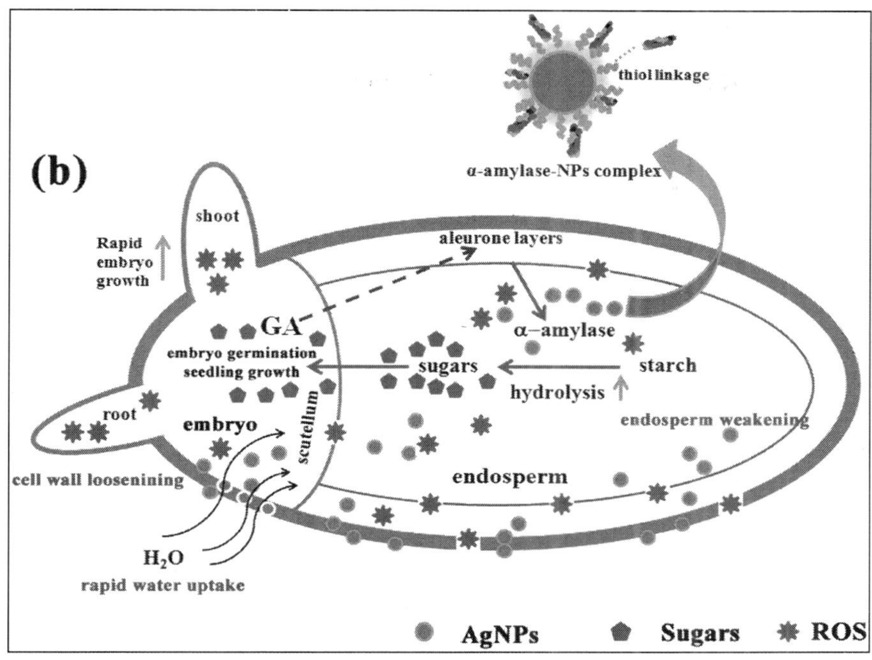

Figure 15.14. Proposed Hypothesis on Mechanisms of Nanoparticles Seed Invigouration during Germination.
Courtesy: Wuttipong Mahakham *et al.*, 2017.

moisture and promote germination besides allowing the plants to withstand drought. This technology if employed in the pre-monsoon condition will help to ensure successful establishment of the treated seeds. When the seeds are coated with specific water soluble nano-polymers, they will break open only when the moisture content of the soil is above 45-50 per cent. This will ensure continued growth of the germinating seedling for the success establishment under dry land conditions. Another type of polymer utilized for enhancing/delaying the seed germination is temperature responsive. These polymers exhibit temperature induced phase transitions that change the permeability properties of the polymer. Seeds treated with polymer were found to take less water at 10°C compared to 25°C. These facilities the sowing of seeds coated with polymer during cooler months which will germinate once the time reaps up.

Most of the polymer technologies were developed and utilized in the temperate countries especially in corn seed production and hybrid development. However, tropical countries like India, these technologies are to be tailored to suit our requirements wherein temperature dependent, water loving polymers to be developed and employed in coating with the seeds to make them germinate under favourable conditions. For example, thermosensitive nano-polymer coated seeds if sown in high temperature regimes will not germinate and desiccate until the favourable conditions for the germination set in owing to the lowered temperature.

Similarly, hydrophilic nano-polymers can help in getting a better crop stand through uniform germination of the coated seed whenever the soil moisture attains desirable level.

Nano-polymers is one of the emerging areas of interest in seed science to promote germination ability through customization. TNAU is the early bird in nanotechnological research for agriculture applications and there is an elegant work done for developing nano polymer matrix to encapsulate bio-active molecules for seed coating application. Plant growth promoting hormones (GA_3 and IAA) impregnated polymer (carboxymethyl cellulose) emulsion were prepared with specified particle size (29.7 and 47.7nm) and stability (-63.0 and -69.0 mV) at high energy homogenization. The hormones loaded nanoemulsions coated at the rate of 15ml per kg of seeds improved the germination and seedling vigour in cotton (Figure 15.15) (Sakthivel, 2016), groundnut (Table 15.7 and Figure 15.16) (Tamilarasan, 2017) and blackgram (Table 15.8 and Figure 15.16) (Raja, 2019) while tested under in vitro environment. The enhanced effect of nanoemulsions on germination and seedling growth is due to increased imbibition rate that activates the hydrolytic enzymes (alpha amylase, lipase and protease) which act upon the stored food reserves in seeds and disintegrate them in to simple sugars and eventually the energy to the embryo. In all the seeds, these hydrolytic enzymes activity were increased in hormones infused nanoemulsion coated seeds. It is also additive effect of hydrophilic polymer carboxylmethyl cellulose and hormones where CMC coating had improved

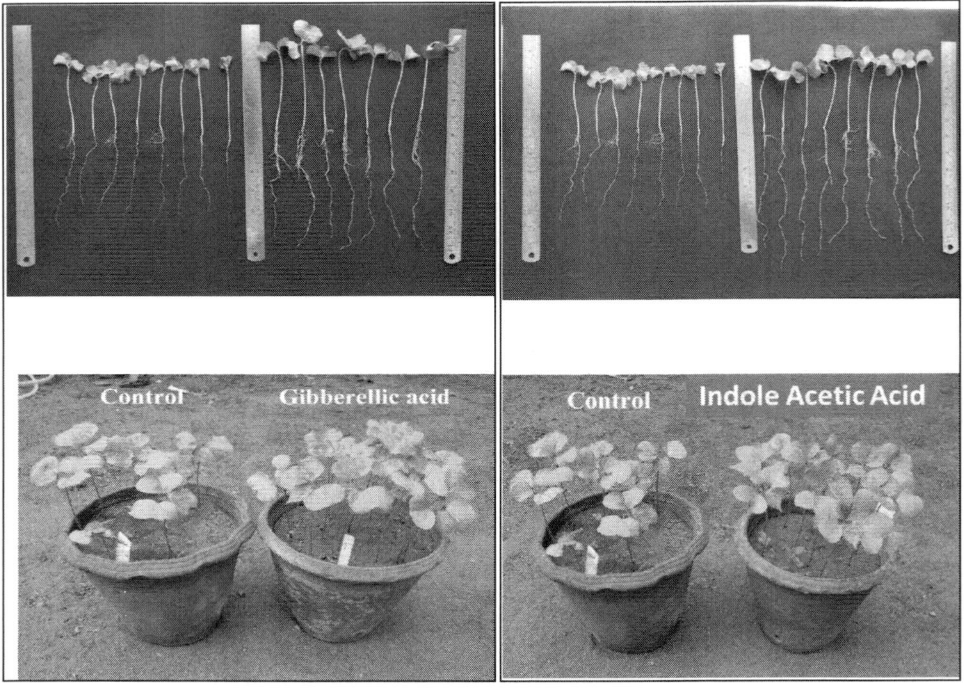

Figure 15.15. Hormones Loaded Nanoemulsion Seed Coating on Germination and Seedling Growth in Cotton.

germination and seedling vigor in cotton by retaining water for a longer period of time. Enhanced cell division and cell elongation as a result of GA_3 and IAA infusion advanced the radicle emergence and growth (Clayton, 1988).

Table 15.7. Effect GA_3 Loaded Nanoformulation Coating on Seed Quality of Groundnut (Tamilarasan, 2017)

Treatments (NF ml/kg)	*Speed of Germination*	*Germination (per cent)*	*Root Length (cm)*	*Shoot Length (cm)*	*Vigour Index*	*DMP (g/10 seedlings)*
Control	3.2	76 (60.66)	7.9	11.8	1497	2.748
5 ml/kg	3.7	80 (63.43)	8.4	14.2	1808	2.768
10 ml/kg	4.6	85 (67.21)	8.9	14.7	2006	3.177
15 ml/kg	5.5	87 (68.57)	9.4	15.3	2142	3.451
20 ml/kg	4.3	82 (64.89)	8.5	13.6	1812	3.098
Mean	4.3	82 (64.89)	8.6	13.9	1853	3.048
CD (P=0.05)	0.139	2.125	0.295	0.251	54.100	0.071

Table 15.8. IAA Loaded Nanoformulation Seed Coating on Quality of Blackgram

Treatments (NF ml/kg)	*Germination (per cent)*	*Vigour Index*	*α-amylase Activity (mg maltose min^{-1})*
Control	70	1972	2.50
5 ml/kg	74	2201	3.01
10 ml/kg	82	2418	3.91
15 ml/kg	84	2578	4.22
20 ml/kg	78	2414	3.76
Mean	78	2270	3.450
CD (P=0.05)	3.57	88.87	2.65

The bio-efficacy of hormones loaded nanoemulsion seed coating was tested under PEG induced moisture stress. The results showed that seeds coated with IAA loaded nanoformulation at 15 ml per kg germinated at higher rate and produced the longest radicle over the untreated control in groundnut (Figure 15.17). The germination (84 per cent) and radicle length (1.8 cm) were higher in IAA loaded formulation coated seeds @ 15 ml kg^{-1} at -5 bar whereas it was low in uncoated seeds. Even under high moisture stress of -15 bar, there was a significant improvement in germination and radicle length due to nanoformulation coating was observed. IAA loaded formulation coated seeds @ 15 ml kg^{-1} had higher germination rate (72 per cent) and radicle length (1.6 cm) at -15 bar than untreated control. The positive results of hormones loaded nanoformulation coated seeds were probably due to increased activity of antioxidant enzymes (Figure 15.18) which reduced the ROS during germination under stress. This is supported with present findings wherein the SEM analysis shows (Figure 15.18) that the radicle produced in nanoformuation

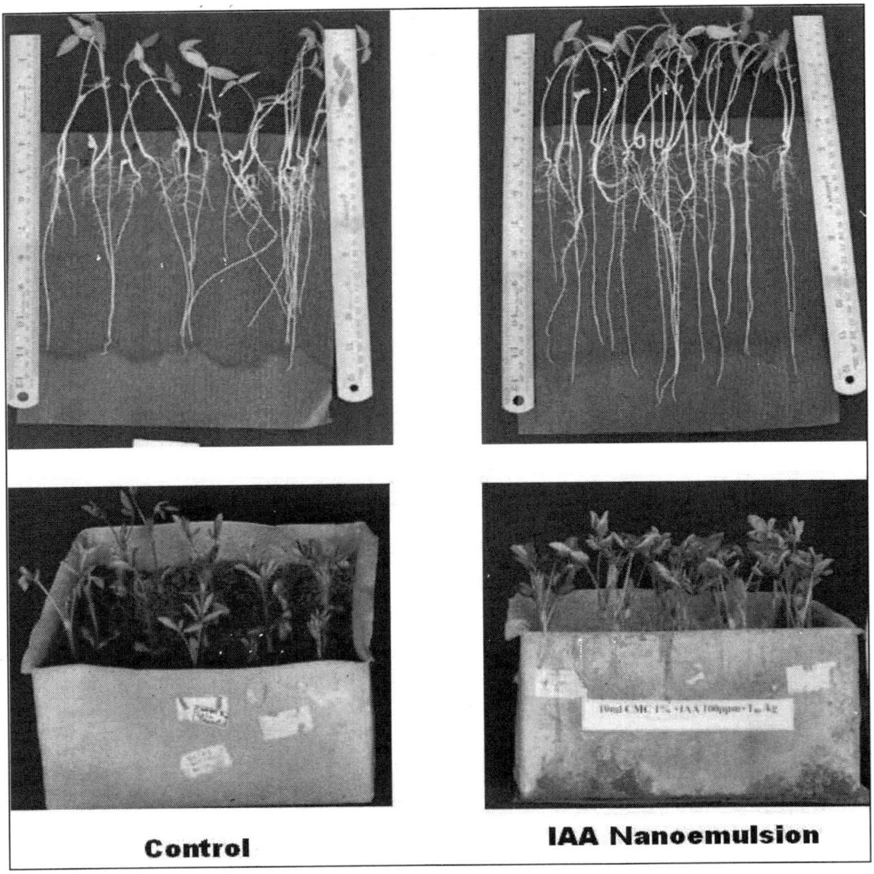

Control
IAA Nanoemulsion

Figure 15.16. IAA Loaded Nanoformulation Coating on Germination of Blackgram and Groundnut.

coated seeds sown at -15 bar had distinct and higher cell elongation at middle of the radicle as compared to uncoated seeds where the length of cell elongation was minimum. Besides, the cell proliferation due to enhanced cell division at the tip of radicle was more in nanoformulation coated seeds than uncoated seeds (Figure 15.18). The same kind of impact was noticed in blackgram (Table 15.9 and Figure 15.19).

Control
CMC + IAA coated
Control
CMC + IAA coated

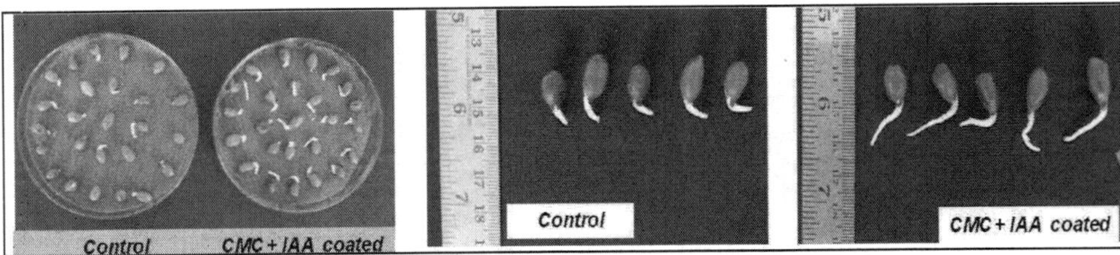

Figure 15.17. IAA Nanoformulation on Germination and Radicle Growth of Groundnut.

Table 15.9. Effect of Hormones Loaded Nanoformulations on Germination of Blackgram under PEG Induced Moisture Stress

Treatment	*Germination (per cent)*			
	-5 bar	*-10 bar*	*-15 bar*	*Mean*
Control	68 (54.81)	60 (48.36)	52 (42.91)	**59** (47.55)
GA_3NF@10ml/kg	74 (62.05)	64 (51.58)	60 (48.36)	**66** (53.34)
GA_3NF@15ml/kg	74 (62.05)	68 (55.67)	64 (51.58)	**68** (55.67)
IAANF@10ml/kg	76 (61.25)	68 (54.81)	64 (51.58)	**71** (57.22)
IAANF@15ml/kg	80 (63.43)	80 (63.43)	72 (58.05)	**77** (61.77)
Mean	**74** (62.06)	**68** (55.67)	**62** (50.48)	68 (55.66)
	T	B		T x B
CD (p=0.05)	0.923	0.642		1.316

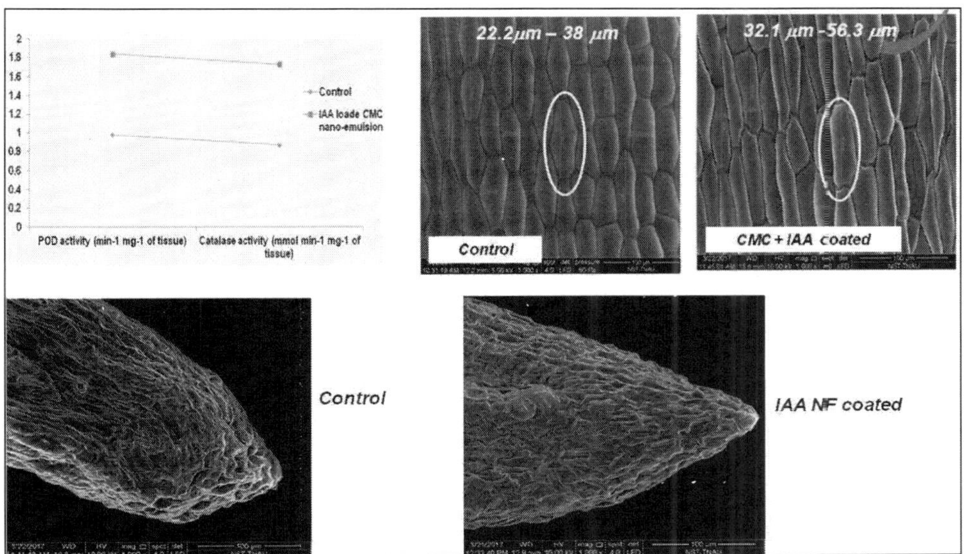

Figure 15.18. IAA Nanoformulation on Antioxidant Enzyme Activity and Radicle Cell Elongation and Division of Groundnut.

Nanofibre Seed Coating: An Innovative Seed Invigouration Technique

Nano Science is a promising field wherein an atom by atom manipulation can be made for synthesizing well structured products/materials which could act as a seed invigourant or as smart delivery vehicles that carry the active molecules and release at the targeted site. Electrospinning is a versatile and effective technology for obtaining continuous polymer fibres with a diameter ranging from a few nanometers to micron (Yang *et al.*, 2004). Electrospun nanofibre is an advanced technology

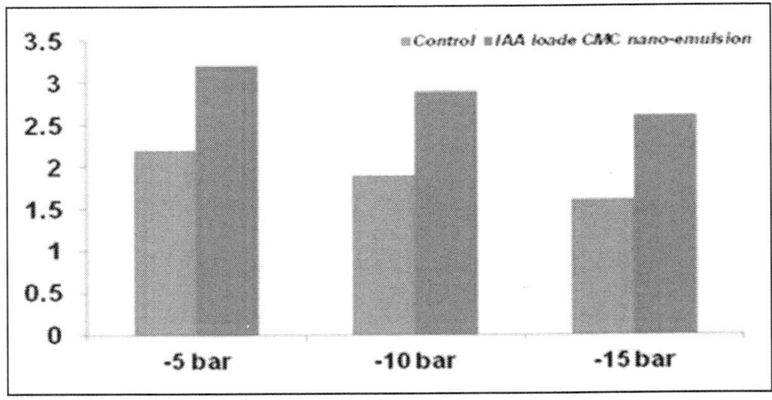

Figure 15.19. IAA Nanoformulation on Radicle Growth of Blackgram under Moisture Stress.

wherein the inputs can be encapsulated and delivered to the specific target site as a smart delivery system, and electrospun fibre has been exploited as a smart and sustained delivery of agricultural inputs through seed to improve germination and seedling growth in cotton (Ramesh *et al.*, 2014), rice (Casteneda *et al.*, 2014).

Seed coating with rhizobia loaded electrospun polyvinyl alcohol (PVA) nanofibres in improving germination, seedling growth, number of nodules and plant biomass of soybean (Damasceno *et al.*, 2013). Seed coating with PVA nanofiber-immobilized rhizobacteria *P. agglomerans* ISIB55 and *B. caribensis* ISIB40 survived on seeds for 30 days and contributed to the successful colonization of both bacteria on the plant root. Moreover, it increased germination, seedling length and dry weight of root, leaf number in soybean (Gregorio *et al.*, 2017).

In cotton, seeds coated with PVP electrospun polymer fibre matrix loaded with GA_3 and IAA registered higher germination (98 and 96 per cent) and seedling vigour (2734 and 2670) under laboratory test (Sakthivel, 2016). Raja (2017) further established the nanofibre seed invigouration technique for smart delivery of hormones through seed. The PVA nanofibre was successfully infused with growth promoting hormones *viz.*, GA_3 and IAA, and the bioefficacy was evaluated through seed coating in groundnut and blackgram. The data showed that hormones encapsulated nanofibre coated seeds excelled the uncoated seeds in recording improved seed quality both in blackgram and groundnut. Hormones loaded nanofibre coated seeds registered higher germination (78 and 88 per cent) and seedling vigour (2987 and 3458) in groundnut tested under *in vitro* condition (Table 15.10). Similar trend of result was observed in blackgram (Table 15.11).

Subsequently microbial cells immobilized PVA nanofibre matrix was developed for seed coating to improve the germination and seedling vigour in groundnut (Mukri Chinna, 2018). The immobilization of *Methylorubrum aminovorans* cells in PVA nanofibre demonstrate that the viability of microbial cells could be protected more than 30 days under ambient environment, and this is because of polymeric matrix which acts as protective shell against environmental stress and dehydration

Table 10. Effect of Hormones Loaded E-spun Fibre Coating on Seed Quality in Groundnut

Treatment	*Germination (per cent)*	*Seedling Length (cm)*	*Vigour Index*
Uncoated seeds	68	31.9	2169
PVA Coating	74	33.8	2501
GA_3 infused PVA nanofibre	78	38.3	2987
IAA infused PVA nanofibre	88	39.3	3458
Mean	77	35.8	2779
CD (P=0.05)	1.642	0.524	69.17

Table 15.11. Effect of Hormones Loaded E-spun Fibre Coating on Seed Quality in Blackgram

Treatment	*Germination (per cent)*	*Seedling Length (cm)*	*Vigour Index*
Uncoated seeds	70	31.9	2233
PVA Coating	74	32.1	2375
GA_3 infused PVA nanofibre	78	38.3	2987
IAA infused PVA nanofibre	86	39.9	3431
Mean	77	35.5	**2756**
Sed	0.854	0.283	49.0
CD (P=0.05)	1.831	0.609	98.12

of microbial cells, and the efficacy test on seed quality showed that seeds inoculated with *Methylorubrum aminovorans* immobilized PVA electrospun nanofibres recorded higher germination (84 per cent) and seedling vigor (Figure 15.20) while tested under *invitro* conditions. It is also supplemented with the results of pot culture

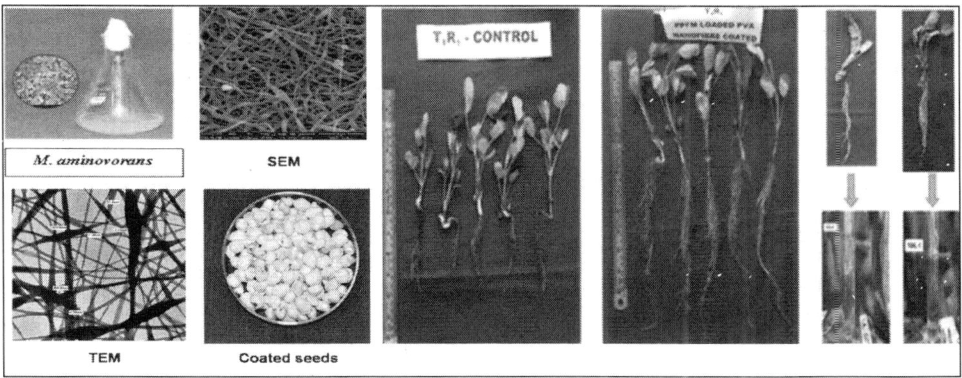

Figure 15.20. *Methylolorubrum aminovorans* Immobilized Nanofibre Seeds Coating on Germination, Seedling Growth and Root Volume per Plant in Groundnut.

study wherein the results expressed that the seeds inoculated with microbial cells immobilized nanofibre registered higher seedling emergence (83 per cent), higher root volume of 181.8 and 186.1 cc plant^{-1}, more number of root nodules per plant 7.5 and 20.5 (Figure 15.21) and maximum biomass of 615.22 mg plant^{-1} and 642.02 mg plant^{-1} at 25 and 45 days after sowing, respectively.

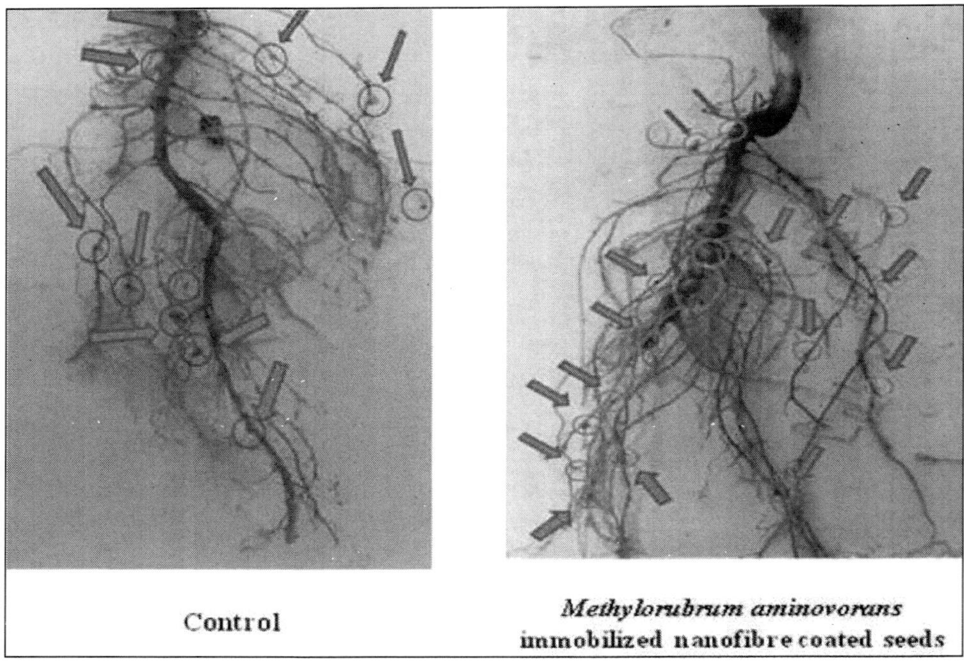

Figure 15.21. *Methylolorubrum aminovorans* Immobilized Nanofibre Seeds Coating on Root Nodules per Plant in Groundnut.

The efficiency of polymer nanofibre matrix for entrapping of agri-inputs to control and targeted release was further tested with tebuconazole fungicide molecule loading. Latha (2019) has encapsulated the tebuconazole molecule in PVA nanofibre for smart delivery of the fungicide molecules to control soil born pathogen *Macrophomina phaseolina* which causes dry root rot disease in pulses. The bioefficacy test results of disc diffusion assay exhibited that the mycelia growth of fungi *Macrophomina phaseolina* was found to be almost arrested in 750ppm tebuconazole infused nanofibres added plates with 96.7 per cent inhibition zone whereas full growth of *Macrophomina phaseolina* observed in plates without fungicide molecules showing zero percentage inhibition (Figure 15.22). The output of poisoned food technique illustrated that the media mixed with 750ppm tebuconazole infused nanofibres had no mycelia growth of fungus associated with 100 per cent inhibition zone. However, the media without fungicide molecules had shown zero percentage inhibition with complete growth of *Macrophomina phaseolina* (Figure 15.22), and the data of artificial inoculation demonstrated that there was zero per cent disease

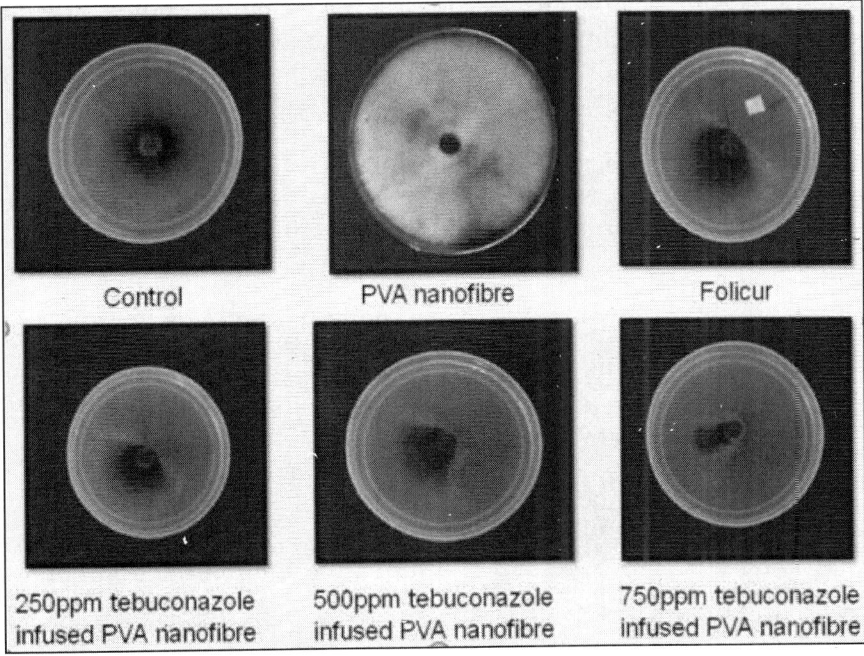

Poisoned food technique.

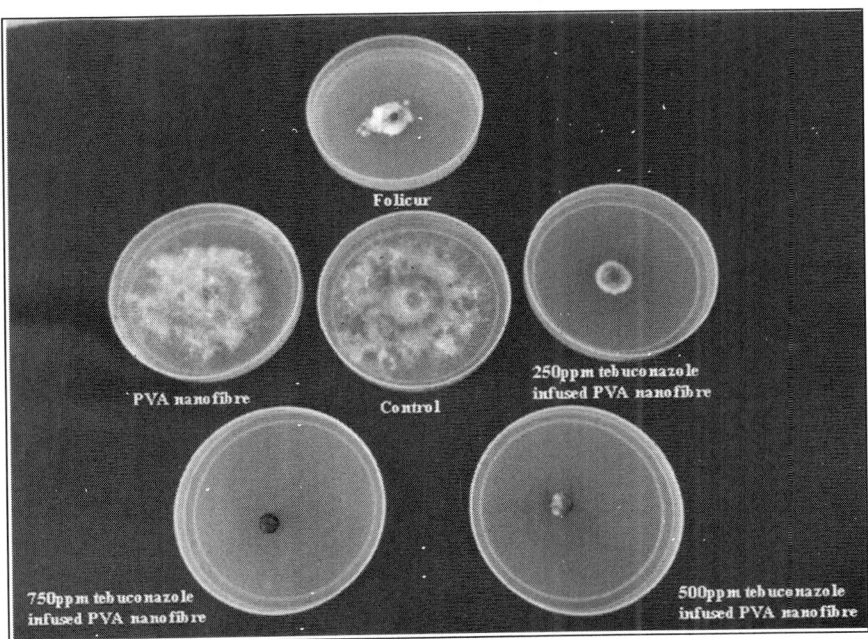

Disc diffusion assay.

Figure 15.22. Bio-efficacy of Tebuconazole Infused PVA Nanofibres on *Macrophomina phaseolina.*

incidence noticed in plants raised from pots without the pathogen inoculation whereas the plants inoculated with *Macrophomina phaseolina* alone and folicur treated seeds have shown 96.0 and 12.0 per cent dry root rot incidence, respectively. With respect to tebuconazole encapsulated nanofibres seed invigouration, the plants raised from seeds coated with 750ppm fungicide loaded nanofibres have accounted 7.0 per cent incidence (Table 15.12). The results of seed quality exhibited that seeds coated with 750ppm tebuconazole loaded nanofibres found to record significantly higher germination, root and shoot length, dry matter production and seedling vigour (Figure 15.23)

Table 15.12. Bio-efficacy of Tebuconazole Infused Nanofibres Seed Invigouration on *Macrophomina phaseolina* Incidence

Treatments	*Disease Incidence (per cent)*
Control	0 (0.00)
Macrophomina phaseolina inoculation	96 (77.28)
Folicur (Commercial formulation) @ 2g/kg of seeds	12 (9.66)
PVA nanofibre + Tebuconazole @ 250ppm	10 (8.05)
PVA nanofibre + Tebuconazole @ 500ppm	8 (6.44)
PVA nanofibre + Tebuconazole @ 750ppm	7 (5.64)
Mean	22.17 (17.85)
CD (P=0.05)	0.843

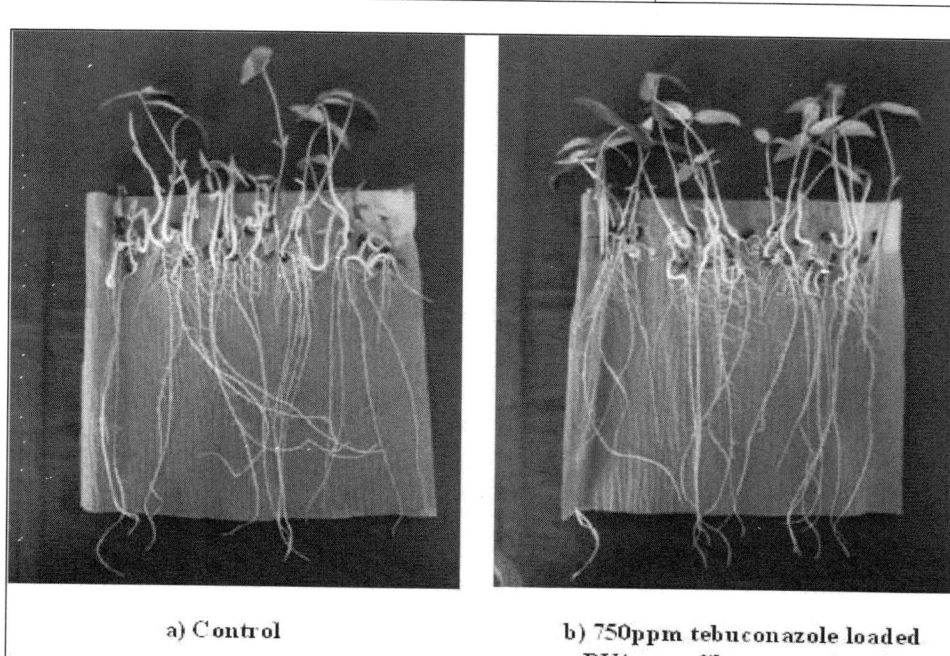

a) Control

b) 750ppm tebuconazole loaded PVA nanofibre coated seeds

Figure 15.23. Bio-efficacy of Tebuconazole Infused PVA Nanofibres on Germination and Vigour.

Conclusion

Nanotechnology is a fascinating and widely exploited field of science and the nano particle seed invigouration technique is a potential nano intervention in agriculture to enhance germinability and vigour and also to extend the shelf life of seeds without affecting the prevailing environment. The existing and conventional seed invigouration technologies have not produced expected results and warrants alternate concepts through nanotechnology to make significant impact in seed sector to enhance germination and vigour. The nanotechnology applications in seed science have just started gaining importance and many research results have vividly indicated that there is a vast scope of inclusion of nanotechnology for seed quality enhancement. There is an abundance of scope to exploit nanotechnological approaches for improved seed quality attributes, and also encapsulating the bioactive molecules for smart delivery of agricultural inputs through seeds that facilitates enhanced use efficiency and ensure environmental sustainability.

REFERENCES

1. Arumugam, G., Velayutham, V., Shanmugavel, S., and Sundaram, J. (2016). Efficacy of nanostructured silica as a stored pulse protector against the infestation of bruchid beetle, *Callosobruchus maculatus* (Coleoptera: Bruchidae). *Applied Nanoscience, 6*(3), 445-450.
2. Cañas, J. E., Long, M., Nations, S., Vadan, R., Dai, L., Luo, M. and Olszyk, D. (2008). Effects of functionalized and nonfunctionalized single walled carbon nanotubes on root elongation of select crop species. *Environmental Toxicology and Chemistry: An International Journal, 27*(9), 1922-1931.
3. Castañeda, L. M., Genro, C., Roggia, I., Bender, S. S., Bender, R. J., and Pereira, C. N. (2014). Innovative rice seed coating (*Oryza sativa*) with polymer nanofibres and microparticles using the electrospinning method. *Journal of Research Updates in Polymer Science, 3*(1), 33-39.
4. Damasceno, R., Roggia, I., Pereira, C., and de Sá, E. (2013). Rhizobia survival in seeds coated with polyvinyl alcohol (PVA) electrospun nanofibres. *Canadian journal of microbiology, 59*(11), 716-719.
5. De Gregorio, P. R., Michavila, G., Muller, L. R., de Souza Borges, C., Pomares, M. F., de Sa, E. L. S and Vincent, P. A. (2017). Beneficial rhizobacteria immobilized in nanofibers for potential application as soybean seed bioinoculants. *PloS one, 12*(5), e0176930.
6. Debnath, N., Das, S., Seth, D., Chandra, R., Bhattacharya, S. C., and Goswami, A. (2011). Entomotoxic effect of silica nanoparticles against *Sitophilus oryzae* (L.). *Journal of Pest Science, 84*(1), 99-105.
7. Dimkpa, C. O., McLean, J. E., Latta, D. E., Manangón, E., Britt, D. W., Johnson, W. P. and Anderson, A. J. (2012). CuO and ZnO nanoparticles: phytotoxicity, metal speciation, and induction of oxidative stress in sand-grown wheat. *Journal of Nanoparticle Research, 14*(9), 1125.

8. Dimkpa, C. O., McLean, J. E., Latta, D. E., Manangón, E., Britt, D. W., Johnson, W. P. and Anderson, A. J. (2012). CuO and ZnO nanoparticles: phytotoxicity, metal speciation, and induction of oxidative stress in sand-grown wheat. *Journal of Nanoparticle Research*, 14(9), 1125.

9. Feizi, H., Kamali, M., Jafari, L., and Moghaddam, P. R. (2013). Phytotoxicity and stimulatory impacts of nanosized and bulk titanium dioxide on fennel (*Foeniculum vulgare Mill*). *Chemosphere*, 91(4), 506-511.

10. Goswami, A., Roy, I., Sengupta, S., and Debnath, N. (2010). Novel applications of solid and liquid formulations of nanoparticles against insect pests and pathogens. *Thin solid films*, 519(3), 1252-1257.

11. Hamza, A. M. (2012). Efficacy and Safety of Non-traditional Methods as Alternatives for Control of *Sitophilus oryzae* (L.)(Coleoptera: Curculionidae) in Rice Grains. *Egyptian Journal of Biological Pest Control*, 22(2).

12. Hong, F., Yang, F., Liu, C., Gao, Q., Wan, Z., Gu, F. and Yang, P. (2005). Influences of nano-TiO 2 on the chloroplast aging of spinach under light. Biological trace element research, 104(3), 249-260.

13. Juhel, G., Batisse, E., Hugues, Q., Daly, D., van Pelt, F. N., O'Halloran, J., and Jansen, M. A. (2011). Alumina nanoparticles enhance growth of *Lemna minor*. *Aquatic toxicology*, 105(3-4), 328-336.

14. K.Raja. 2017. "Teri International conference on Research to Innovations."

15. K.Subramanian, K.Raja and S. Marimuthu. Multivarious Applications of Nanotechnology for Enhanced Productivity in Agriculture. Emerging trend in Agriclture: Fundamental and applied aspects. Edited by Harikesh B. Sing *et al.* (2018). CABI international publication (UK and USA). pp. 56 - 77. (ISBN 13: 978 1786 391445)

16. Karuppanapandian, T., Wang, H. W., Prabakaran, N., Jeyalakshmi, K., Kwon, M., Manoharan, K., and Kim, W. (2011). 2, 4-dichlorophenoxyacetic acid-induced leaf senescence in mung bean (*Vigna radiata L.* Wilczek) and senescence inhibition by co-treatment with silver nanoparticles. *Plant Physiology and Biochemistry*, *49*(2), 168-177.

17. Khodakovskaya, M. V., De Silva, K., Biris, A. S., Dervishi, E., and Villagarcia, H. (2012). Carbon nanotubes induce growth enhancement of tobacco cells. *ACS nano*, 6(3), 2128-2135.

18. Khodakovskaya, M. V., Kim, B. S., Kim, J. N., Alimohammadi, M., Dervishi, E., Mustafa, T., and Cernigla, C. E. (2013). Carbon nanotubes as plant growth regulators: effects on tomato growth, reproductive system, and soil microbial community. Small, 9(1), 115-123.

19. Khodakovskaya, M., Dervishi, E., Mahmood, M., Xu, Y., Li, Z., Watanabe, F., and Biris, A. S. (2009). Carbon nanotubes are able to penetrate plant seed coat and dramatically affect seed germination and plant growth. *ACS nano, 3(10)*, 3221-3227.

20. Kim, J. H., Lee, Y., Kim, E. J., Gu, S., Sohn, E. J., Seo, Y. S. and Chang, Y. S. (2014). Exposure of iron nanoparticles to *Arabidopsis thaliana* enhances root elongation by triggering cell wall loosening. *Environmental Science and Technology*, 48(6), 3477-3485.

21. Kumar, V., Guleria, P., Kumar, V., and Yadav, S. K. (2013). Gold nanoparticle exposure induces growth and yield enhancement in Arabidopsis thaliana. Science of the Total Environment, 461, 462-468.

22. Kumuthan. 2017. "Morphological and biochemical responses of green gram (*Vigna radiata*) exposed to ZnO nanoparticles." M. Sc, Tamil Nadu Agricultural University.

23. Lahiani, M. H., Dervishi, E., Chen, J., Nima, Z., Gaume, A., Biris, A. S., and Khodakovskaya, M. V. (2013). Impact of carbon nanotube exposure to seeds of valuable crops. *ACS Applied Materials and Interfaces*, 5(16), 7965-7973.

24. Lee, C. W., Mahendra, S., Zodrow, K., Li, D., Tsai, Y. C., Braam, J., and Alvarez, P. J. (2010). Developmental phytotoxicity of metal oxide nanoparticles to *Arabidopsis thaliana. Environmental Toxicology and Chemistry: An International Journal, 29(3)*, 669-675.

25. Ma, C., Chhikara, S., Xing, B., Musante, C., White, J. C., and Dhankher, O. P. (2013). Physiological and molecular response of Arabidopsis thaliana (L.) to nanoparticle cerium and indium oxide exposure. ACS Sustainable Chemistry and Engineering, 1(7), 768-778.

26. Mahakham, W., Sarmah, A. K., Maensiri, S., and Theerakulpisut, P. (2017). Nanopriming technology for enhancing germination and starch metabolism of aged rice seeds using phytosynthesized silver nanoparticles. *Scientific reports*, 7(1), 8263.

27. Mahmoodzadeh, H., Nabavi, M., and Kashefi, H. (2013). Effect of nanoscale titanium dioxide particles on the germination and growth of canola (*Brassica napus*). *Journal of Ornamental and Horticultural Plants* 3(1), 25-32.

28. Miralles, P., Johnson, E., Church, T. L., and Harris, A. T. (2012). Multiwalled carbon nanotubes in alfalfa and wheat: toxicology and uptake. *Journal of the Royal Society Interface*, 9(77), 3514-3527.

29. Patra, P., Choudhury, S. R., Mandal, S., Basu, A., Goswami, A., Gogoi, R. and Gopal, M. (2013). Effect Sulfur and ZnO Nanoparticles on Stress Physiology and Plant (Vignaradiata) Nutrition. *In Advanced Nanomaterials and Nanotechnology* (pp. 301-309). Springer, Berlin, Heidelberg.

30. Prasad, T. N. V. K. V., Sudhakar, P., Sreenivasulu, Y., Latha, P., Munaswamy, V., Reddy, K. R., and Pradeep, T. (2012). Effect of nanoscale zinc oxide particles on the germination, growth and yield of peanut. *Journal of plant nutrition*, 35(6), 905-927.

31. Qi, M., Liu, Y., and Li, T. (2013). Nano-TiO 2 improve the photosynthesis of tomato leaves under mild heat stress. *Biological trace element research*, 156(1-3), 323-328.

32. Raja, K., Sowmya, R., Sudhagar, R., Moorthy, P. S., Govindaraju, K., and Subramanian, K. S. (2019). Biogenic ZnO and Cu nanoparticles to improve seed germination quality in blackgram (*Vigna mungo*). Materials Letters, 235, 164-167.

33. Ramaesh. 2014. "Biodegradable nanoparticles as carrier in smart delivery of pesticides." 3rd International Conference on Agriculture and Horticulture.

34. Rouhani, M., Samih, M. A., and Kalantari, S. (2012). Insecticide effect of silver and zinc nanoparticles against Aphis nerii Boyer De Fonscolombe (Hemiptera: Aphididae). *Chilean journal of agricultural research*, 72(4), 590.

35. Sakthivel. (2016). Development of hormone loaded nanoformulations (nanoemulsions) for improving the germination and vigour of cotton (M. Sc), Tamil Nadu Agricultural University.

36. Salama, H. M. (2012). Effects of silver nanoparticles in some crop plants, common bean (*Phaseolus vulgaris* L.) and corn (*Zea mays* L.). *Int Res J Biotechnol*, 3(10), 190-197.

37. Savithramma, N., Ankanna, S., and Bhumi, G. (2012). Effect of nanoparticles on seed germination and seedling growth of *Boswellia ovalifoliolata* an endemic and endangered medicinal tree taxon. *Nano Vision*, 2(1), 2.

38. Senthilkumar, S. (2011). Customizing nanoparticles for the maintanence of seed vigour and viability in Blackgram (*Vigna mungo*) cv. VBN 4. M. Sc.(Agri.) Thesis.

39. Shyla, K. K., Natarajan, N., and Nakkeeran, S. (2014). Antifungal activity of zinc oxide, silver and titanium dioxide nanoparticles against Macrophomina phaseolina. *J. Mycol. Plant Pathol*, 44(3), 269-274.

40. Song, G., Gao, Y., Wu, H., Hou, W., Zhang, C., and Ma, H. (2012). Physiological effect of anatase TiO^2 nanoparticles on Lemna minor. *Environmental toxicology and chemistry*, 31(9), 2147-2152.

41. Tamilarasan. 2017. "Studies on the efficacy of hormones loaded nanoformulation on seed quality of groundnut (*Arachis hypogaea* L.)." M. Sc, Tamil NAdu Agricultural University.

42. Tiwari, D. K., Dasgupta-Schubert, N., Cendejas, L. V., Villegas, J., Montoya, L. C., and García, S. B. (2014). Interfacing carbon nanotubes (CNT) with plants: enhancement of growth, water and ionic nutrient uptake in maize (*Zea mays*) and implications for nanoagriculture. *Applied Nanoscience*, 4(5), 577-591.

43. Tiwari, D. K., Dasgupta-Schubert, N., Villasenor, L. M., Tripathi, D., and Villegas, J. (2013). Interaction of carbon nanotubes with mineral nutrients for the promotion of growth of tomato seedlings. *Nano Studies*, 7, 87-96.

44. Tripathi, S., Sonkar, S. K., and Sarkar, S. (2011). Growth stimulation of gram (Cicer arietinum) plant by water soluble carbon nanotubes. *Nanoscale*, 3(3), 1176-1181.

45. Vani, C., and Brindhaa, U. (2013). Silica nanoparticles as nanocides against *Corcyra cephalonica* (S.), the stored grain pest. *International Journal of Pharma and Bio Sciences*, 4(3).

46. Wang, X., Han, H., Liu, X., Gu, X., Chen, K., and Lu, D. (2012). Multi-walled carbon nanotubes can enhance root elongation of wheat (*Triticum aestivum*) plants. *Journal of Nanoparticle Research*, 14(6), 841.

47. Wu, S. G., Huang, L., Head, J., Chen, D. R., Kong, I. C., and Tang, Y. J. (2012). Phytotoxicity of metal oxide nanoparticles is related to both dissolved metals ions and adsorption of particles on seed surfaces. *J Pet Environ Biotechnol*, 3(4), 126.

48. Zhao, L., Peralta-Videa, J. R., Rico, C. M., Hernandez-Viezcas, J. A., Sun, Y., Niu, G. and Gardea-Torresdey, J. L. (2014). CeO^2 and ZnO nanoparticles change the nutritional qualities of cucumber (*Cucumis sativus*). *Journal of agricultural and food chemistry*, 62(13), 2752-2759.

Chapter 16

Nanotechnological Approaches in Plant Protection

Malaichamy Kannan[1], *Kolanthasamy Elango*[2], *and Kasivelu Govindaraju*[3]

[1]*Department of Plant Protection, HC&RI, Periyakulam, Tamil Nadu*
[2]*Department of Entomology, Tamil Nadu Agricultural University, Coimbatore – 641 003, Tamil Nadu*
[3]*Centre for Ocean Research, Sathyabama Institute of Science and Technology, Chennai – 600 119, Tamil Nadu*
**e-mail: kanento@gmail.com*

Introduction

Nanotechnology is a promising field of multi-disciplinary research and opens up a wide array of opportunities in various fields like agriculture, food, medicine, electronics *etc.* In agriculture, development of nanotechnology creates an excellent opportunity to address complex technical and economic issues in pest management. Nanotechnology has developed immensely in the last decade and was able to create many innovative materials with an enormous range of forthcoming applications (Bhattacharyya *et al.*, 2010). Nanoscale structures like nanomaterials, encapsulation and entrapment of agrochemicals such as fertilizers, pesticides, herbicides, plant growth regulators and other active substances by using surfactants, polymers, dendrimers, surface ionic attachments and other related mechanisms can be utilized in smart delivery of plant protection inputs, which allow the slow uptake of active ingredients and in turn reduces the amount of agrochemical application by minimizing the input and waste. The prominence of nanoscale delivery system in agriculture is because of its improved solubility and stability to degradation in the environmental factors (Johnston, 2010; Chen and Yada, 2011; Sastry *et al.*, 2010 and Jha *et al.*, 2011).

Globally insect pests cause a huge crop loss of 14 per cent and plant pathogens cause an estimated loss up to 13 per cent with a value of US $2,000 billion per year (Pimentel, 2009). Similarly, in agriculture, fungi are responsible for more than 70 per cent of all major crop diseases (Agrios, 2005), annual crop losses due to pre and post-harvest fungal diseases exceed 200 billion euros and in the United Stated alone, over $600 million are annually spent on fungicides (Fernandez *et al.*, 2010) for management of fungal diseases. Nearly, one quarter of food crops worldwide are affected by mycotoxins such as aflatoxins, ergot toxins, Fusarium toxins, patulin and tenuazonic acid (Schneider and Ullrich, 1994). Nanomaterials are used efficiently for safe administration of pesticides, herbicides, and fertilizers at lower doses. Pesticides cause adverse effects on human health, aquatic animal, and on pollinating insects. So, nanomaterials play an important role in decreasing toxicity and in turn help in increasing the efficacy of pesticides (Mousavi and Rezaei, 2011). Nano-pesticide formulations increase the solubility of poorly soluble active ingredient and helps in releasing the active ingredient slowly. The bioavailability of poorly water soluble agrochemicals can be increased through the use of additives or by nanoparticulate formation of agrochemicals (Kah *et al.*, 2012). Nanoparticles are loaded with pesticides and released slowly based on environmental trigger. Because of their high reactivity at nanoscale compared to their bulk counterparts, lesser quantity of nanocides show enhanced effect in crop protection (Debnath *et al.*, 2011).

The potential uses and benefits of nanotechnology are enormous and include management of pest and diseases through novel nanomaterial-based insecticides formulations. Traditional strategies like integrated pest management used in agriculture are insufficient, and application of chemical pesticides have adverse effects on animals and human beings apart from the decline in soil fertility (Goswami *et al.*, 2010). Therefore, nanotechnology would provide green and efficient alternatives for crop protection in agriculture without harming the nature.

Nanomaterials as a Novel Insecticides against Insect Pests of Crops

Nanomaterials such as carbon, titanium dioxide, zeolites, silver, silica, copper alumina have been suggested as potential candidates for controlling a range of agricultural pests. Nanomaterials may come in many diverse forms (surprisingly often >100 nm), from solid doped particles to (often non persistent) polymer and oil–water based structures, besides as additives (mostly for controlled release) and active constituents. Plant protection products containing nanomaterials alter the functionality or risk profile of active ingredients (nano-enabled pesticides) and assure many benefits over conventional pesticide products. These benefits may include improved formulation characteristics, easier application, better targeting of pest species, increased efficacy, lower application rates, and enhanced environmental safety (Walker *et al.*, 2017).

Non-metal Nanoparticles

The alumina-silicate filled nanotube can stick to the surface hair of insect pests and ultimately enters the body, interfere certain physiological functions. Stadler *et al.* (2010) demonstrated that the insecticidal activity of nanostructured alumina against two stored grain insect pests *viz. Sitophilus oryzae* L. and *Rhyzopertha dominica* (F.), and suggested the potential of inorganic nanostructured alumina as a cheaper and reliable alternative for control of such insect pests. Nanoscale silica, a type of unique material, prepared from bio-resources and used as nano-pesticide applications. Application of silica from different sources to sugarcane as nutrient can be absorbed translocate and deposited in the tissues, especially in the leaf epidermis, which can enhance the extent of resistance to early shoot borer (Prabakaran *et al.*, 2017). Kannan *et al.* (2017) reported that biologically derived nano-silica from wild sugarcane, *Erianthus arundinaceus* proved insecticidal activity against larvae of diamond backmoth, *Plutella xylostella* with LC_{50} was (30.03 µg/mL). Debnath *et al.* (2011) reported higher insect mortality from treatment with smaller silica nanoparticles (15–30 nm) than with bulk silica (100–400 nm) which indicated that smaller nanoparticles have more efficiency (90 per cent mortality for 2 g/kg of seeds) compared to bulk silica against *Sitophilus oryzae*. Borie *et al.* (2015) findings stated that silica nanoparticles were highly effective against *Sitophilus oryzae, Rhyzopertha dominica* and *Callosobruchus maculatus* with an effective dosage of 2.5 g/kg, 1.5 g/kg and 1 g/kg of seeds. Ravikumar *et al.* (2017) stated that nanosilica applied at 0.5 and 0.25 g/kg doses were found to be effective against cigarette beetle *Lasioderma serricorne* on turmeric rhizomes.

Metal and Metal Oxide Nanomaterials

Several metal nanoparticles like silver (Ag), gold (Au), aluminum (Al), silica (Si), zinc (Zn), metal oxides, zinc oxide (ZnO) and titanium dioxide (TiO_2) are being developed and deployed for crop pest management (Sahayaraj, 2017) The applications of different metal-based nanoparticles *viz.*, silver nanoparticles, aluminium oxide, zinc oxide and titanium dioxide were evaluated by Goswami *et al.* (2010) for control of rice weevil *Sitophilus oryzae,* using solid and liquid formulations of the nanoparticles and reported that hydrophilic nanoparticles was most effective then hydrophobic nano particles. Rouhani *et al.* (2012) confirmed that silver and zinc nanoparticles showed good insecticidal efficacy against the oleander aphid, *Aphis nerii* a common pest of several ornamental plants. Sahayaraj *et al.*(2016) reported that Ag^+ aqueous solution as well as pungam oil-based gold nanoparticles caused more mortality, while aqueous solutions of Ag^+-AuNPs treatment drastically enhanced the food consumption, assimilation but reduced the conversion and affected the growth in *Pericallia ricini.* Similarly, TiO_2 nanoparticles also expressed its insecticidal property in *Plutella xylostella* (LC_{50} of 37.57 µg/mL) in cauliflower (Preetha *et al.*, 2018). The use of PEG coated nanoparticles loaded with garlic essential oil improved the efficacy against adults of *Tribolium castaneum* in which the efficacy remained over 80 per cent even after five months due to slow and persistent release of active compounds from nanoparticles (Yang *et al.*, 2009).

Nanomaterials for Controlled Release of Semiochemicals

Nanomaterial based formulations of semiochemicals facilitate controlled release of volatile molecules for insect attraction and repellent action. Dharanivasan *et al.* (2017) explored the scope of using metal oxide nanoparticles (SiO_2, TiO_2, and

ZnO) to regulate discharge of methyl eugenol (ME) from lure dispensers. Among the samples, ME with TiO_2 nanoparticles showed a higher number of fruit fly catches than the others for up to 12 weeks. The scope for combining both cue lure and methyl eugenol parapheromones in enhancing the attractiveness to *Bactrocera cucurbitae* has been well documented in India (Amsa *et al.*, 2014). Prakash *et al.* (2015) suggested that slow release of pheromone was observed in nano zeolite formulation (7 weeks) followed by rubber septa (4 weeks) in Rice yellow stem borer, *Scirpophaga incertulas*. Similarly, Kannan and Soundararajan (2017) loaded the male attracting female sex pheromone of rice yellow stem borer, *Scirpophaga incertulas* in a polymer based electrospun fiber matrix for extended release and enhanced moth caches. The Synthetic pheromone delivery for the red palm weevil at the Central Plantation Crops Research Institute (CPCRI), Kasaragod, Kerala revealed that setting up of nanomatrix pheromone trap @ one/ha significantly attracted more number of red palm weevils of both the sexes with highest longevity than other commercially available Chemtica® and PCI® pheromone lures. Similarly, Bhagat *et al.* (2013) has developed a pheromone-based nano-gel in order to regulate the release of methyl eugenol against fruit fly (*Bactrocera dorsalis*) in guava. The nano-gel has been developed using supramolecular principles and nanotechnological approaches that are chemically, mechanically and thermally stable and offered extended release of eugenol under field conditions then eugenol alone.

Nanoencapsulated Pesticides Formulations for Pest Control

Pesticide formulation is a mixture of various materials such as active ingredients, solvents, surface active ingredients, stabilizers, *etc.* Among these materials, the active ingredients alone responsible for controlling target pests, and other materials *viz.*, organic solvents, surface active ingredients (stickers or spreaders) and stabilizers help maintain their solubility, stability and improve their pesticidal activity. Nano-encapsulation implies the use of a different kind of nanoparticle with an insecticide wherein the insecticide is slowly but efficiently released to a particular host plant for insect pest control. Nano-encapsulation with nanoparticles in form of pesticide allows for proper absorption of the chemical into the plants unlike the case of larger particles (Scrinis and Lyons 2007). Release mechanisms of nano encapsulation include diffusion, dissolution, biodegradation and osmotic pressure with specific pH (Vidhyalakshmi *et al.*, 2009). A range of nanoparticles could be used in pesticides controlled release formulations. Ragaei and Sabry (2014) classified the more popular nanomaterials that are being used in controlled release formulations for pesticides.

Importance of Nano-encapsulation of Pesticides in Agriculture

The paramount in plant protection would be the ability of chemicals to remain active or survive in the spray environment (sun, heat, rain) of the target organism (*e.g.* insects and fungus) effectively, preferably possess a new mode of action and provide economic returns. Currently, available pesticides and biocides are unable to maintain these properties, resulting in repeated application, increased production costs and ultimately the environment being more polluted. Nanoscale carriers or nanoencapsulated materials can enhance the bioavailability of active substances and reduce toxicity, improving the timed release of active substances, reduces the amount being applied and enabling the precise targeting of active substances by encapsulating the bioactive compounds (Smith *et al.*, 2008). The nanoencapsulation process should enhance the following characters to get success. They are 1.Increase the controlled release properties of the pesticide, 2. Increase the solubility of active ingredients, 3. Protect against premature degradation and 4. Increase the stability of the pesticide

Nano-encapsulated Pesticides against Crop Pests

Nanoencapsulated pesticides have exhibited better pest control efficacy than commercially pesticides. Nanoencapsulation materials allow the release of active ingredients in a controlled way, resulting in the retention of pest control efficacy over a longer period. Various controlled release formulations (CRFs) have already been prepared using different nanoencapsulation materials and their release behaviour has been well described. Nanoencapsulated CRFs pesticides or biocides are found to be an effective tool to control stored grain pests. Loha *et al.* (2012) evaluated efficacy of nanoencapsulated CRFs of β-cyfluthrin with PEG showed higher mortality of *Callosobruchus maculatus* than commercial formulation. The EC50 of commercial and nano CRFs with PEG β-cyfluthrin on the 7th were 43.24 mg L^{-1} and 1.89 mg L^{-1}, respectively, Similar results were also observed in the CRFs of carbofuran against the root-knot nematode (*Meloidogyne incognita*) infecting tomato plant (Pankaj *et al.*, 2012) and also CRFs of pyrifluquinazon with polymeric chitosan on *Myzus persicae* (green peach aphid) (Kang *et al.*, 2012).

Adak *et al.* (2012) synthesized nanoencapsulated CRFs of imidacloprid using polyethylene glycol, l aliphatic diacids and tested against major pests of soybean, namely *Melanagromyza sojae* (stem fly) and *Bemisia tabaci* (white fly). The results exhibited that the CRFs managed the insects better than commercial formulations at lower dose and do not leave any terminal residues in crops and soil. The spinosad insecticide excites the insect nervous system and its efficiency depended on particle size. When particle size of the spinosad nanosuspension applied to spider mites was 404 nm, the LC50 was 15 mg L^{-1}, which further decreased to 11 mg L^{-1} for 372 nm, 7.6 mg L^{-1} for 332 nm and 4 mg L^{-1} for 163 nm and showed better efficiency at lower dose. Similarly, the nanosuspension of etopfenprox (as a liposome) showed higher efficacy against *Spodoptera litura* than as a commercial pesticide (Hwang *et al.*, 2011).

Nano Formulations of Botanical Biopesticides

An outcry is exhibited against the use of pesticides due to their hazardous effects on human as well as environment (Sparks *et al.*, 2012). Biopesticides and plant based botanicals are considered as alternative to synthetic chemical insecticides. Yang *et al.* (2009) reported on insecticidal activity of polyethylene glycol-coated nanoparticles loaded with garlic essential oil against adult *Tribolium castaneum* that exhibited 80 per cent mortality. Nanoparticles containing essential oils of geranium or bergamot showed contact toxicity against *T. castaneum* and *Rhizopertha dominica* (Gonzalez *et al.*, 2014). Forim *et al.* (2013) reported cent per cent mortality of *Plutella xyllostella* for the nanoparticles loaded with neem extract. Nanoemulsions of the essential oils from three local plants *Ageratum conyzoides, Achillea fragrantissima* and *Tagetes minuta* prepared using the High Pressure Homogenization showed ovicidal, adulticidal and residual activities against the cowpea beetle, *C. maculatus* (Nenaah *et al.*, 2014). Hemat *et al.* (2015) prepared nanoemulsion of Eucalyptus oil by ultrasonication method, reported insecticidal and biological activity against newly-hatched larvae of *Pectinophora gossypiella* and *Earias insulana*. Nanoformulation of the botanical biopesticide based on sweet flag (*Acorus calamus*) has been tested for comparative toxicity to the red flour beetle, *Tribolium castaneum* and indicated that the lower doses (below the 1.0 per cent) of nano formulation was sufficient to kill 100 per cent mortality (Prabakaran *et al.*, 2017). Similar studies by Kalyanakumar *et al.* (2017) with the same two products have shown comparable results on ovicidal efficacy on eggs of the rice moth, *Corcyra cephalonica*.

Nanotechnological Approaches for Plant Diseases Management

Productivity of food crops grown for human consumption is at risk due to the incidence of pathogens, insect pests and weeds. Crop management requires a multidisciplinary approach because pathogens (fungi, bacteria, viruses, nematodes) not only interact with each other, but with other biotic and abiotic factors to create heavy yield loss. Nanotechnology offers an important role in improving the existing crop management techniques especially in plant disease management by controlled and targeted release of agrochemicals to enhance their efficiency. Potential applications of nanotechnology in crop diseases management include development of nano devices for early detection of plant disease and nanomaterials as a novel fungicides, bactericides, anti-viral agents and nematicides for the management of fungal, bacterial, viral and nematode associated diseases in crop plants (Ghormade *et al.*, 2011; Mahendra *et al.*, 2012).

Nanoparticles in Plant Disease Management

Due to emerging plant diseases, agricultural production is reduced worldwide. Every year millions of dollars have been spent to control plant diseases. Various natural and artificial control measures have been used for plant protection. Use of pesticides is the most prevalent method for disease control at present. Scientists are searching for alternative measures against diseases control application. As an alternative to chemical pesticides, nanoparticles alone have the potential to be directly applied to plant seeds, foliage, or roots for protection against disease causing pathogens, such as fungi, bacteria, viruses and nematodes. Various metal nanoparticles such as silver, copper, zinc oxide and titanium dioxide have been intensively researched for their anti-fungal, anti-bacterial, anti-viral and nematicidal properties (Gogos *et al.*, 2012; Kah and Hofmann, 2014).

Biopolymer Based Nanoparticles

Development of nanoformualtions for field application of agrochemicals requires the use of readily biodegradable, non-toxic, environment friendly, safe and low-cost materials. Natural biopolymers by natural sources with good physical and chemical properties is a fascinating approach to prevent the use of petrochemical and toxic chemical substances in production of nanomaterial.

Chitosan

Chitosan based nanoparticles have easy biodegradability, bio-compatibility, non-allergenicity, and anti-microbial activity, with non-toxicity to animals and humans. Chitosan nanoparticles are protecting against infections caused by the mosaic virus of alfalfa, snuff, peanut, potato, and cucumber (Kochkina *et al.*, 1994; Pospieszny *et al.*, 1991). In another study, chitosan nanoparticles have shown anti-microbial properties against *Fusarium* crown, root rot in tomato, *Botrytis* bunch rot in grapes, and *Phyricularia grisea* in rice, Fusarium root rot of peas (Chirkov, 2002), but are less effective against bacteria. Malerba and Cerana (2016) summarized potential mechanisms that lead to the anti-microbial effects of chitosan, *via* agglutination, disruption of the cell membrane, inhibition of H^+-ATPase activity, inhibition of toxin production, inhibition of the synthesis of messenger RNA and proteins, and blockage of nutrient flow. Anti-viral effects have been observed in beans against bean mild mosaic virus and, in tobacco, against tobacco mosaic virus and Tobacco necrosis virus. Chitosan has also shown effectiveness against oleander aphid (*Aphis nerii*), cotton leaf worm (*Spodoptera littoralis*), root-knot nematode (*Meloidogyne javanica*), and nymphs of the spear psylla (*Cacopsylla pyricola*). Chitosan holds promise as both nanoparticles alone and has shown to have an enormous potential as a nano-carrier to entrap, encapsulate, absorb or attach active molecule to develop effective agricultural formulations (Malerba and Cerana, 2016). Chitosan could penetrate fungal cell wall and bind to its DNA and inhibit the synthesis of mRNA and, in turn, affect the production of essential proteins and enzymes (Kong *et al.*, 2010).

Silver Nanoparticles

Silver has been used as an anti-microbial agent since ancient civilizations. Silver has been demonstrated to possess effective antimicrobial effect and has been vastly used in medicine (Wei *et al.*, 2009). Moreover, silver can be manufactured into silver nanoparticles using nanotechnology approach to have improved physical, chemical and biological properties. Kim *et al.* (2008) evaluated the antifungal efficacy of colloidal silver nanoparticles solution, against rose powdery mildew. Silver nanoparticles are more adhesive on bacterial and fungal cell surface due to charge interactions. Lamsal *et al.* (2011) showed the effective usage of silver nanoparticles instead of commercial fungicides against six *Colletotrichum* species associated with pepper anthracnose and found that 100 ppm of silver nanoparticles inhibited the growth of fungal hyphae as well as conidial germination *in vitro* when compared to the control. The effect of silver nanoparticles on growth inhibition of sclerotium-forming species *Rhizoctonia solani*, *Sclerotinia sclerotiorum* and *S. minor* showed severe damage of hyphal wall and collapse of fungal hyphae (Min *et al.*, 2009). Anti-fungal activities of silver nanoparticles were studied against *Alternaria alternata*, *Sclerotinia sclerotiorum*, *Macrophomina phaseolina*, *Rhizoctonia solani*, *Botrytis cinerea*, and *Curvularia lunata* (Krishnaraj *et al.*, 2012). Jain and Kothari, (2014) demonstrated, the silver nanoparticles were sprayed onto bean leaves, complete suppression of sun-hemp rosette virus was observed. Elbeshehy *et al.* (2015) showed that faba bean plants challenged with bean yellow mosaic virus, and sprayed with silver nanoparticles, produced remarkably better results on the plant.

Silica Nanoparticles

Silica nanoparticles can be easily synthesised with a controlled size, shape, and structure, making them highly advantageous delivery vehicles. They are commonly produced in a spherical shape with pore-like holes; for example, porous hollow silica nanoparticles (PHSNs) or mesoporous silica nanoparticles (MSNs). PHSN and MSN commonly load the pesticide into the inner core to protect the active molecules and, therefore, provide a sustained release. The shell structure of PHSNs protects the active molecules inside the nanoparticles against degradation by UV light. The literature suggests that silicon has already been used to enhance plant tolerance against various abiotic and biotic stresses and, therefore, silica nanoparticles seem to be the natural choice for the development of agri-products for pests (Barik *et al.*, 2008). Torney *et al.* (2007) used honeycomb MSN system to deliver DNA and chemicals into plant cells and intact leaves. They loaded the gene, its chemical inducer into the MSN system and capped the ends with gold nanoparticles and studied the release pattern of chemicals and induction of gene expression in the plants under controlled-release conditions. This evident that silica nanoparticles used in target-specific delivery of proteins, nucleotides and chemicals in plant biotechnology.

Copper Nanoparticles

Copper-based fungicides produce highly reactive hydroxyl radicals which can damage lipids, proteins, DNA, and other biomolecules. It plays an important role in disease prevention and treatment of large variety of plants (Borkow and Gabbay, 2005). Complexation of copper with chitosan nanogels was shown to have strong synergistic effect between chitosan and copper in inhibiting the growth of phytopathogenic fungus *Fusarium graminearum.* Because of its bio-compatibility, these nano hydrogels are included as a new generation of copper-based bio-pesticides and it could also be developed into an efficient delivery system for copper based fungicides for plant protection (Brunel *et al.*, 2013). Low melting point soda-lime glass powder containing copper nanoparticles showed efficient antimicrobial activity against gram-positive, gram-negative bacteria, yeast and fungi, key reason for the increased antimicrobial activity is because of inhibitory synergistic effect of the Ca^{2+} lixiviated from the glass (Esteban- Tejeda *et al.*, 2009).

Nanocomposites

Nanocomposites are materials that incorporate nano sized particles into a matrix of standard material. The result of the addition of nanoparticles is a drastic improvement in properties that can include mechanical strength, toughness and electrical or thermal conductivity. Pinto *et al.* (2013) described preparation and antifungal activity of composite films of pullulan and silver nanoparticles against *Aspergillus niger* as a model system. They found that these composite films show strong inhibitory action on fungal sporulation, which was confirmed by disruption of the spores cells when observed under SEM. Park *et al.* (2006) developed a new nano-sized silica-silver composite material showed good antifungal activity where pathogens disappeared from the infected leaves within three days of spraying and the plants remained healthy thereafter. Nanocomposites showed 100 per cent growth inhibition of *Pythium ultimum, Magnaporthe grisea, Colletotrichum gloeosporioides, Botrytis cinerea* and, *Rhyzoctonia solani* at 10 ppm concentration. Whereas, *Bacillus subtilis, Azotobacter chrococuum, Rhizobium tropici, Pseudomonas syringae* and *Xanthomonas compestris* pv. *vesicatoria* showed 100 per cent growth inhibition at 100 ppm concentration.

Nanotechnology for Plant Disease Diagnostics

The use of nanoparticles promises to help promote in vitro diagnostics to the next level of performance. Nanotechnology has been integrated into the next generation diagnostic techniques: (i) to improve assay sensitivity, specificity, and limits of detection (ii) to increase sample throughput; and (iii) to reduce assay complexity and cost. Nanoparticles often require less analyte to register a response, which is largely due to the small area of the sensing surface. Smaller sensing areas generally allow higher-density arrays to be fabricated, and this feature maximizes the number of analytes (*e.g.*, pathogens or biomarkers) that can be interrogated in a single test without increasing sample requirements. The broad ranges of nanotechnologies applied to pathogen detection are nano-arrays, nano-fluidics and nano-transducers. They improve sensitivity, specificity, and throughput while decreasing analysis time and sample volume. Quantum dots, gold nanoparticles and superparamagnetic nanoparticles are the most promising nanostructures for *in vitro* diagnostic applications. These nanoparticles can be conjugated to recognition moieties such as antibodies, oligonucleotides for detection of target pathogens.

Plant diseases are caused by microorganism such as fungi, bacteria, viral, viroids and phytoplasma. Most of the plant pathogens are transmitted through seed or planting material, which can spread secondarily through vectors and lead to severe loss to the crop. Among plant diseases, viral diseases cause serious crop losses and affect the quality of products. Diagnostics of plant pathogens is vital for the production of quality planting material especially vegetatively propagated crops such as banana, citrus, potato, cassava, sugarcane, taro, amorphophalus, black pepper and cardamom. Nucleic acid based diagnostics techniques such as PCR, real time PCR and nucleic acid spot hybridization methods are available for most of these plant pathogens in India. ELISA is known popular detection technique used for the efficient detection of plant viruses and this robust, high- throughput technique can be handled for the detection of large number of samples at a time. Further, it cannot be used on-site or in the field and again this technique is less sensitive and require special skill and require at least a day in the laboratory.

Recently, lateral flow immune assay (LFIA) are used for qualitative or semi-quantitative detection and monitoring of pathogens in non-laboratory environments. This LFIA works with chromatographic principle coupled with immunological recognition system. LFIA is based on the interaction between the target virus and immune- reagents (antibodies and their conjugates with colloidal nanoparticles) applied on the membrane carriers (test strips). When the test strip is dipped into the sample being analyzed, the sample liquid flows through membranes and triggers immunochemical interactions resulting in visible coloration in test and reference lines (Geertruda *et al.*, 2007) (Figure 16.1).

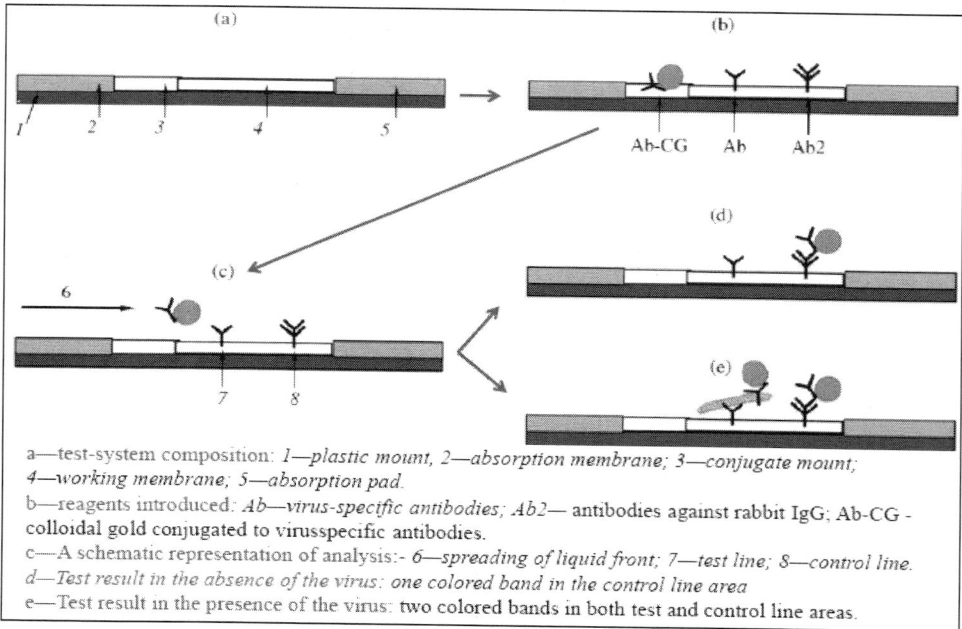

Figure 16.1. Lateral Flow Immunological Assay (LFIA) using Gold Nano-Particles.

In agriculture, this technology has not yet been exploited and very limited information is available. The LFIA strip technique has been reported for the detection of few plant viruses (Boonham *et al.*, 2007). The use of metallic nanoparticles composed of gold/silver have many optical and electronic properties, derived from their size and composition (Nath *et al.*, 2008). Other nanoparticles including fluorescent quantum dots and carbon nanotubes have been used in various applications including DNA detection, and the development of immunoassays for the detection of pathogens (Baptista *et al.*, 2006; Alivisatos *et al.*, 2005). Express immune-chromatographic test strip assays were developed for detection of five plant viruses varying in shape and size of virions: spherical carnation mottle virus, bean mild mosaic virus, rod shaped tobacco mosaic virus, and filamentous potato viruses X and Y. Multi-membrane composites (test strips) with immobilized polyclonal antibodies against viruses and colloidal gold-conjugated antibodies were used for the analysis.

Conclusion

Nanomaterials in different forms can be used for efficient management of insect pests and plant diseases through novel formulations of potential pesticides, thereby providing green and eco-friendly alternatives for crop protection. With the advancement of nanotechnology, application of green chemistry in synthesis of nanomaterials by using various biological resources has reduced the use of toxic chemicals/solvents. The controlled release properties of nanoencapsulation materials to release the active ingredients to the target area using autosensing power needs further investigation. Although complete features (*e.g.*, synthesis, efficacy and their fate) related to these nanomaterials are rarely found and promising nanoencapsulated pesticides are at a very early stage of development, it is expected that this technology will reduce firstly, the dosage of pesticides needed for crop protection, and secondly, human exposure to pesticides. A major contribution that is expected to emerge from the auspicious results of green pesticides is the application of nanoparticles to encapsulate and stabilize bio-products, which will reduce environmental hazards. Even though the toxicity of nanomaterials has not yet clearly understood, it plays a significant role in crop protection because of its unique physical and chemical properties. The application of nanomaterials is relatively new in the field of agriculture and it needs further research investigations.

REFERENCES

1. Adak, T., Kumar, J., Dey, D., Shakil, N. A and Walia, S. (2012). Residue and bio-efficacy evaluation of controlled release formulations of imidacloprid against pests in soybean (G lycine max). *Journal of Environmental Science and Health, Part B, 47*(3), 226-231.
2. Agrios, G.N. 2005. Plant Pathology, Elsevier Academic Press, San Diego, California, USA.
3. Alivisatos, A.P., W. Gu and C. Larabell. (2005). Quantum dots as cellular probes. *Biomed. Eng.* 7: 55– 76.
4. Amsa.T, A.Venkatachalam, S.Sithanantham and M.S. Nalina Sundari. 2014. Exploratory study of dispenser types and blending of parapheromones in field attraction of cucurbit fruit fly, *Bactrocera cucurbitae. Hexapoda,* 21(2): 38-41.
5. Baptista, P.V., M. Koziol-Montewka, J. Paluch - Oles,G. Doria and R. Franco.2006. Goldnanoparticle- probe based assay for rapid and direct detection of Mycobacterium tuberculosis DNA in clinical samples, *Clin. Chem.* 52: 1433–1434.
6. Barik, T.K., Sahu, B and Swain, V. (2008). Nano-silica-from medicine to pest control. *Parasitol Res.,* 103: 253–258.
7. Bhagat, D., Samanta, S.K. and S. Bhattacharya.2013. Efficient Management of Fruit Pests by Pheromone Nanogels. *Sci. Rep.* 3, 1294; DOI: 10.1038/srep01294.
8. Bhattacharyya, A., Bhaumik, A., Usha Rani, P., Mandal, S and Epidi, T.E. (2010). Nano-particles - A recent approach to insect pest control. *African Journal of Biotechnology.* 9(24): 3489-3493.
9. Boonham, N., Tomlinson, J., and Mumford, R. 2007. Micro arrays for rapid detection of plant viruses. *Annu. Rev. Phytopathol.* 45, 307–328
10. Borie, H.A. M.F. El-Samahy and O.A. Galal and A.F. Thabet. 2014. The efficiency of silica nanoparticles in control cotton leaf worm, *S.littorals* (Lepidoptera: Noctuidae) in soybean plants under laboratory conditions.*Glob.J. Agric.Food safety.Sci.* 1 : (2) 161-168.

11. Borkow, G. and Gabbay, J. (2005). Copper as a biocidal tool. *Current medicinal chemistry*, 12: 2163-2175.

12. Brunel, F.,ElGueddari, N.E. and Moerschbacher, B.M. (2013). Complexat ion of copper (II) with chitosan nanogels: Toward control of microbial growth. *Carbohydrate Polymers*, 92: 1348-1356.

13. Chen, H. and Yada, R. (2011).Nanotechnologies in agriculture: New tools for sustainable development. *Trends in Food Science and Technology*, 22: 585-594.

14. Chirkov, S. The antiviral activity of chitosan (review). Appl. Biochem. Microbiol. 2002, 38, 1–8.

15. Debnath, N., Das, S., Seth, D., Bhattacharya, R.C and Goswami, A. (2011). Entomotoxic effect of silica nanoparticles against *Sitophilus oryzae* (L.) *Journal of Pest Science*, 84(1): 99-105.

16. Dharanivasan, G., S. Sithanantham, M. Kannan, S. Chitra, K. Kathiravan and S. Janarthanan. (2017). Metal Oxide Nanoparticles Assisted Controlled Release of Synthetic Insect Attractant for Effective and Sustainable Trapping of Fruit Flies, 28(4): 2167-2183.

17. Elbeshehy, E.K.F.; Elazzazy, A.M., Aggelis, G.2015. Silver nanoparticles synthesis mediated by new isolates of Bacillus spp., nanoparticle characterization and their activity against Bean Yellow Mosaic Virus and human pathogens. *Front. Microbiol*, 6, 453.

18. Es teban-Tejeda, L., Malpart ida, F., Es teban-Cubillo, A., Pecharromán, C. and Moya, J. S. (2009).Antibacterial and antifungal activity of a soda-lime glass containing copper nanoparticles. *Nanotechnology*, 20: 505-701.

19. Fernandez, R.G., Prats, E., Jorrin-Novo, J.V. (2010). Proteomics of Plant Pathogenic Fungi, *J. Biomed. Biotechnol.*, Vol. 2010, Pp. 36.

20. Forim, M. R., E. S. Costa, M. F. Grac, F. Silva '/ J. Fernandes, J. M. Mondego and A.L.l Boica.(2013). Development of a new method to prepare nano-/ microparticles loaded with extracts of *Azadirachta indica*, their characterization and use in controlling *Plutella xylostella*, *J. Agric. Food Chem.*, 61: 9131–9139.

21. Geertruda.A, Posthuma-Trumpie, Jakob Korf and Aart van Amerongen. 2009. Lateral flow (immuno) assay: its strengths, weaknesses, opportunities and threats. A literature survey. *Anal Bioanal Chem.*, 393: 569–582

22. Ghormade, V., Deshpande, M. V. and Paknikar, K. M. 2011. Pers pect ives for nano-biotechnology enabled protection and nutrition of plants. *Biotechnology Advances*, 29: 792-803.

23. Gogos, A., Knauer, K and Bucheli, T. D. (2012). Nanomaterials in plant protection and fertilization: current state, foreseen applications, and research priorities. *Journal of agricultural and food chemistry*, *60*(39), 9781-9792

24. Gonzalez, J.O.W. M.M. Gutierrez, A. A.Ferrero and B.F.Band. (2014). Essential oils nanoformulations for stored-product pest control – Characterization and biological properties, *Chemosphere*, 100: 130–138.

25. Goswami, A., Roy, I., Sengupta, S and Debnath, N. (2010). Novel applications of solid and liquid formulations of nanoparticles against insect pests and pathogens. *Thin Solid Films.* 519: 1252–1257.

26. Hemat Z.M., G.M. M.Tahany, and H. Torkey. (2015). Effect of formulated nanoemulsion of eucalyptus oil on the cotton bollworms, *J. Biol. Chem. Research,* 32 (2): 478-484

27. Hwang, I. C.; Kim, T. H.; Bang, S. H.; Kim, K. S.; Kwon, H. R.; Seo, M. J.; Youn, Y. N.; Park, H. J.; Yasunaga-Aoki, C.; Yu, Y. M. (2011). Insecticidal Effect of Controlled Release Formulations of Etofenprox Based on Nano-bio Technique. J. Fac. Agric. Kyushu Univ., 56, 33-40.

28. Jain, D and Kothari, S. L. (2014). Green synthesis of silver nanoparticles and their application in plant virus inhibition. *J Mycol Plant Pathol*, *44*(1), 21-24.

29. Jha, Z., Behar, N., Sharma, S. N., Chandel, G., Sharma, D. and Pandey, M. (2011). Nanotechnology: prospects of agricultural advancement. *Nano Vision,* 1: 88-100.

30. Johnston, C. T. (2010). Probing the nanoscale architecture of clay minerals. *Clay Minerals*, 45: 245-279.

31. Kah, M and Hofmann, T. (2014). Nanopesticide research: current trends and future priorities. *Environment international, 63,* 224-235.Kalyanakumar, R., Sithananthan, S., Jeyarajan Nelson S and M.Prabakar. 2017. *Proceedings of National Seminar on Nanotechnology for Evergreen Revolution,* October, 5-6. Tamil Nadu Agricultural University, Coimbatore.

32. Kah, M., Beulke, S., Tiede, K. and Hofmann, T. (2012). Nanopesticides: State of knowledge, environmental fate, and exposure modeling. *Critical Reviews in Environmental Science and Technology,* 43: 1823-1867.

33. Kang, M. A., Seo, M. J., Hwang, I. C., Jang, C., Park, H. J., Yu, Y. M., and Youn, Y. N. (2012). Insecticidal activity and feeding behavior of the green peach aphid, *Myzus persicae,* after treatment with nano types of pyrifluquinazon. *Journal of Asia-Pacific Entomology, 15*(4), 533-541.

34. Kannan, M. and R. P. Soundarajan. (2017). Development of Electrospun Nanofiber Sex Pheromone for Yellow Stem Borer, *Scirpophaga incertulas* in Rice, ICAR-IIRR *News Letter,* 15 (1): 43-44.

35. Kannan, M., Bhaskaran, A., Prithiva.J.N. Preetha, S. and Lokesh, S. (2017). Biogenic silica nanoparticles from *Erianthus arundinaceus* as novel insecticide against diamond back moth, *Plutella xylostella* L. 224-227. *Proceedings of the International Symposium on Sugarcane.18-20* September,2017. *Hotel Le Meridian,Coimbatore.*: Sugarcane Breeding Institute and Tamil Nadu Agricultural University,Coimbatore.

36. Kim, H. S., Kang, H. S., Chu, G. J. and Byun, H. S. (2008). Antifungal effectiveness of nanosilver colloid against rose powdery mildew in greenhouses. *Solid State Phenomena,* 135: 15-18.

37. Kochkina, Z. M., Pospeshny, G., and Chirkov, S. N. (1995). Inhibition by chitosan of productive infection of T-series bacteriophages in the Escherichia coli culture. *Mikrobiologiia, 64*(2), 211-215.

38. Kong, M., Chen, X. G., Xing, K. and Park, H. J. (2010). Antimicrobial properties of chitosan and modeof action: a state of the art review. *International Journal of Food Microbiology*, 144: 51-63.

39. Krishnaraj, C., Ramachandran, R., Mohan, K and Kalaichelvan, P. T. (2012). Optimization for rapid synthesis of silver nanoparticles and its effect on phytopathogenic fungi. *Spectrochimica Acta Part A: Molecular and Biomolecular Spectroscopy, 93*, 95-99.

40. Lamsal, K., Kim, S.W., Jung, J. H., Kim, Y. S., Kim, K. S. and Lee, Y. S. (2011). Application of silver nanoparticles for the control of Colletot richum species *in vitro* and pepper anthracnose disease in field. *Mycobiology*, 39: 194-199.

41. Loha, K. M.; Shakil, N. A.; Kumar, J.; Singh, M. K.; Srivastava, C. (2012). Bio-efficacy evaluation of nanoformulations of beta-cyfluthrin against *Callosobruchus maculatus* (Coleoptera: Bruchidae). *J. Environ. Sci. Heath.*, 47, 687-91.

42. Mahendra R, Shivaji D, Aniket G, Kamel-Abd-Elsalam.(2012). Strategic nanoparticle-mediated gene transfer in plants and animals - a novel approach. *Current Nanoscience*, 8: 170-179.

43. Malerba, M and Cerana, R. (2016). Chitosan effects on plant systems. *International journal of molecular sciences, 17*(7), 996.

44. Min, J. S., Kim, K. S., Kim, S. W., Jung, J. H., Lamsal, K., Kim, S. B., Jung, M. and Lee, Y. S. 2009. Effects of colloidal silver nanoparticles on sclerotium-forming phytopathogenic fungi. *Journal of Plant Pathology*, 25: 376-380

45. Mousavi, S. R. and Rezaei, M. (2011). Nanotechnology in agriculture and food production. *Journal of Applied Environmental and Biological Sciences*, 1: 414-419.

46. Nenaah G.E., S.I.A. Ibrahim and B.A. Al-Assiuty. (2014). Chemical composition, insecticidal activity and persistence of three Asteraceae essential oils and their nanoemulsions against *Callosobruchus maculatus* (F.), *Journal of Stored Products Research*, 61: 9 – 16.

47. Pankaj; Shakil, N. A.; Kumar, J.; Singh, M. K.; Singh, K. (2012). Bioefficacy evaluation of controlled release formulations based on amphiphilic nano-polymer of carbofuran against *Meloidogyne incognita* infecting tomato. *J. Environ. Sci. Health*, B, 47, 1971 520-528.

48. Park, H.-J., Kim, S. H., Kim, H. J. and Choi, S.-H. (2006). A New Composit ion of nanosized silica-silver for control of various plant diseases. *Journal of Plant Pathology*, 22: 295-302.

49. Pimentel, D. (2009). Pesticides and Pest Control. In: PESHIN, R. and DHAWAN, A. (eds.) Integrated Pest Management: Innovation-Development Process. Springer Netherlands.

50. Pinto, R. J., Almeida, A., Fernandes, S. C., Freire, C. S., Silves tre, A. J., Neto, C. P. and Trindade, T. (2013). Antifungal activity of transparent nanocomposite thin films of pullulan and silver against *Aspergillus niger*. *Colloids and Surfaces B: Biointerfaces*, 103: 143-148.

51. Pospieszny, H., Chirkov, S and Atabekov, J. (1991). Induction of antiviral resistance in plants by chitosan. *Plant Science*, *79*(1), 63-68.

52. Prabakaran, M., Sithanantham, S., Subramanian, K.S., Vijayaprasad, P., Punna rao B.V and Babu. B. (2017). Silica application effect on sugarcane early shoot borer. p393-396. *Proceedings of the International Symposium on Sugarcane.18-20* September, *2017.Hotel Le Meridian,Coimbatore.*: Sugarcane Breeding Institute and Tamil Nadu Agricultural University,Coimbatore.

53. Prakash,V. G., M. Kannan, R.Philipsridhar, G.Ravi. (2015). Preliminary Studies on developing Nano-Matrices for regulated release of pheromone to monitor yellow stem borer, Scirpophagaincertulas (Walker) in rice. In the proceedings of the International Conference on "Innovative Insect Management Approaches for Sustainable Agro-Ecosystem" to be held at Agricultural College and Research Institute, Madurai between Jan 27 and 30, (2015).

54. Preetha, S., M. Kannan, S. Lokesh, N. Viji, J. N. Prithiva and V. Gowtham. 2018. Titanium Dioxide (TiO^2) Nanoparticles as a Novel Insecticide against Diamondback Moth, *Plutella xylostella* L. in Cauliflower. *Trends in Biosciences* 11(21): 2999-3003.

55. Ragaei, M and Sabry, A.H. (2014). Nanotechnology for insect pest control. *International Journal of Science, Environment and Technology*, 3(2): 528–54.

56. Rouhani, M., Samih, M.A. and Kalantari, S. (2012). Insecticide effect of silver and zinc 2nanoparticles against *Aphis nerii* Boyer de fonscolombe (Hemiptera: Aphididae). *Journal of Nanomaterials*, 590-594.

57. Sahayaraj, K., Madasamy, M and Anbu Radhika, S. (2016). Insecticidal activity of bio-silver and gold nanoparticles against *Pericallia ricini* Fab. (Lepidaptera: Arctiidae). *J.Biopest.*, 9(1): 63-72.

58. Sastry, R. K., Ras hmi, H., Rao, N. and Ilyas, S. (2010). Integrat ing nanotechnology into agri-food systems research in India: a conceptual framework. *Technological Forecasting and Social Change*, 77: 639-648.

59. Schneider, S., Ullrich, W.R. (1994). Differential induction of resistance and enhanced enzyme activities in cucumber and tobacco caused by treatment with various abiotic and biotic inducers. *Physiol. Mol. Plant Pathol.*, 45: 291 304.

60. Scrinis, G and Lyons, K. (2007). The emerging nano-corporateparadigm nanotechnology and the transformation of nature, food and agri-food systems. *Int. J. Sociol. Food Agric.*, 15(2): 22-44.

61. Smith, K., Evans, D. A and El-Hiti, G. A. (2007). Role of modern chemistry in sustainable arable crop protection. *Philosophical Transactions of the Royal Society B: Biological Sciences*, *363*(1491), 623-637.Sparks TC, Dripps JE, Watson GB and Paroonagian D. (2012). Resistance and cross resistance to the spinosyns-A review and analysis. *Pestic. Biochem. Physiol.*, 102: 1-10.

62. Stadler, T., Buteler, M and Weaver, D.K. (2010). Novel use of nanostructured alumina as an insecticide. *Pest Manage Sci.*, 66(6): 577–579.

63. Thomas, S. and McCubbin, P. (2003). A comparison of the antimicrobial effects of four silver-containing dressings on three organisms. *Journal of Wound Care*, 12: 101-107.

64. Torney, F., Trewyn, B. G., Lin, V. S.Y. and Wang, K. (2007).Mesoporous silica nanoparticles deliver DNA and chemicals into plants. *Nature Nanotechnology*, 2: 295-300

65. Vidhyalakshmi, R., Bhakyaraj, R and Subhasree, R.S. (2009). Encapsulation: The Future of Probiotics-A Review. *Adv. Biol. Res.*, 3(3-4): 96-103.

66. Walker, G. W., Rai Ş. Kookana, Natalie E. Smith Melanie Kah, Casey L. Doolette, Philip T. Reeves, Wess Lovell, Darren J. Anderson, Terence W. Turney and. Divina A. Navarro. (2017). Ecological Risk Assessment of Nano-enabled Pesticides: A Perspective on Problem Formulation. *J.Agric.FoodChem.*, Article.

67. Wang, L., Li, Z., Zhang, G., Dong, J and Eastoe, J. (2007). Oil-in-water nanoemulsions for pesticide formulations. *J Colloid Interface Sci.*, 314: 230–235.

68. Wei, D., Sun, W., Qian, W., Ye, Y. and Ma, X. (2009). The synthesis of chitosan-based silver nanoparticles and their antibacterial activity. *Carbohydrate Research*, 344: 2375-2382.

69. Yang, F.L, Li, X.G, Zhu, F, Lei C.L. (2009). Structural characterization of nanoparticles loaded with garlic essential oil and their insecticidal activity against *Tribolium castaneum* (Herbst) (Coleoptera: Tenebrionidae). *J Agric Food Chem.*, 57(21): 10156–10162.

Chapter 17

Nano-Scale Materials as Probes, Sensors and Delivery Vehicles

*Badal Kumar Biswal, Paladugu Krishnadev, D. Jeya Sundara Sharmila and A. Lakshmanan**

Department of Nano Science and Technology, Tamil Nadu Agricultural University, Coimbatore – 641 003, Tamil Nadu
**e-mail: microlaxman@gmail.com*

Introduction

Materials can be engineered to nanoscale with improved functionalization to show different properties as compared to its bulk counterpart which can be used in various applications. With the advent in nanobiotechnology many nanoscale materials such as nanotubes, nanochannels, nanocapacitors, nanopores, nanofibers *etc.* can be functionalised which is used in drug delivery, emulsions, imaging, biomaterials, food, optical and electronics, pathogens, biosensors, and in vitro diagnostics. Nanoscale materials are miniature in size and high surface to volume ratio gives it the opportunity to enter to the cell without any barriers as well as interact with the target materials very efficiently. In this chapter, we will discuss about the use of nanoscale materials as probes and sensing material to image or diagnose the genetic and physiological defects. Also, different kinds of delivery vehicles made up of nanoscale materials for targeted delivery of therapeutics to get rid of disease and disorders are being discussed. Prior to that, we need to know how nanoscale materials are functionalized and used in different nanobiotechnological approaches.

Biomolecule Functionalization of Nanoparticles as Nano Probes/ Sensors/Delivery Vehicles

Nanoparticle – Biomolecule Conjugate

A nanoparticle–biomolecule conjugate is a nanoparticle with biomolecules attached to its surface. Nanoparticles are minuscule particles, typically measured in nanometers (nm), that are used in nanobiotechnology to explore the functions of biomolecules (Figure 17.1). Properties of the ultrafine particles are characterized by the components on their surfaces more so than larger structures, such as cells, due to large surface area-to-volume ratios. Large surface area-to-volume-ratios of nanoparticles optimize the potential for interactions with biomolecules.

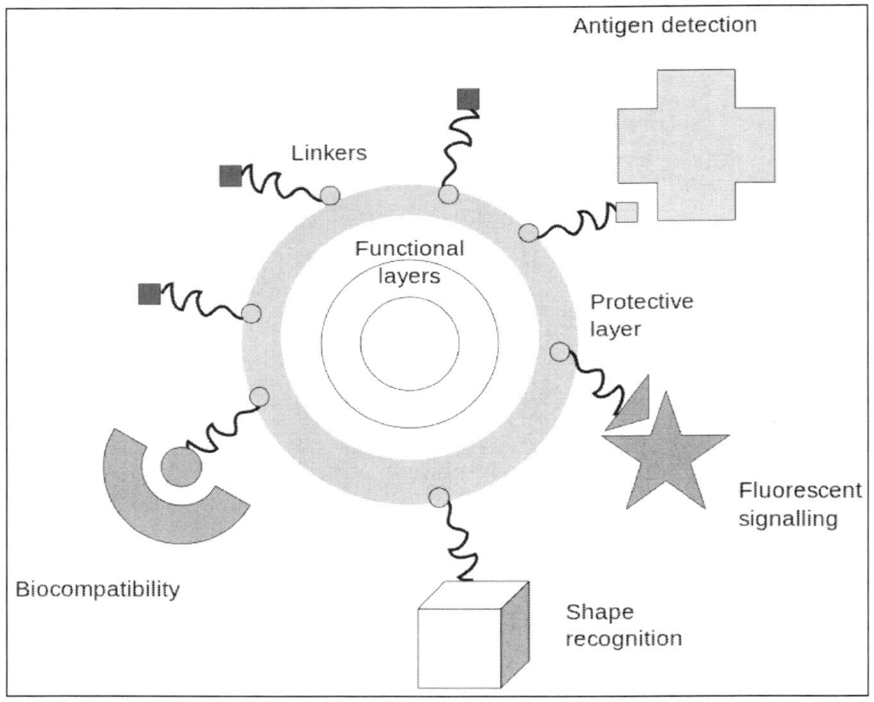

Figure 17.1. Nanoparticle – Biomolecule Conjugate.
Courtesy: https://en.wikipedia.org/wiki/Nanoparticle%E2%80%93biomolecule_conjugate.

Functionalization of Biomolecules

Nanoparticles exhibit unique physical properties that are coupled with size range commensurate with biomolecular and cellular systems. These features make them attractive materials for therapeutic delivery and diagnostic applications. Using different strategies, nanoparticles have been functionalized with a variety of ligands such as small molecules, surfactants, dendrimers, polymers, and biomolecules. Different multivalent surface structures fabricated with small molecules and polymeric ligands offer the capability to incorporate multiple therapeutic drugs or bio-macromolecules by covalent or non-covalent conjugation. Similarly, biomolecule-conjugated nanoparticles can impart desired properties such as specific recognition or biocompatibility. The ease of such functionalization allows chemists to create the desired functionalities for their application in clinics.

Bi-functional Linkers (BFL)

BFLs require strongest covalent linkage necessary to attach the biomolecule with the substrate of delivery system or sensor. In (Figure 17.2a), BFL is covalently bond with substrate at one end and protein at another end. Example of a BFL is glutaraldehyde (Figure 17.2b).

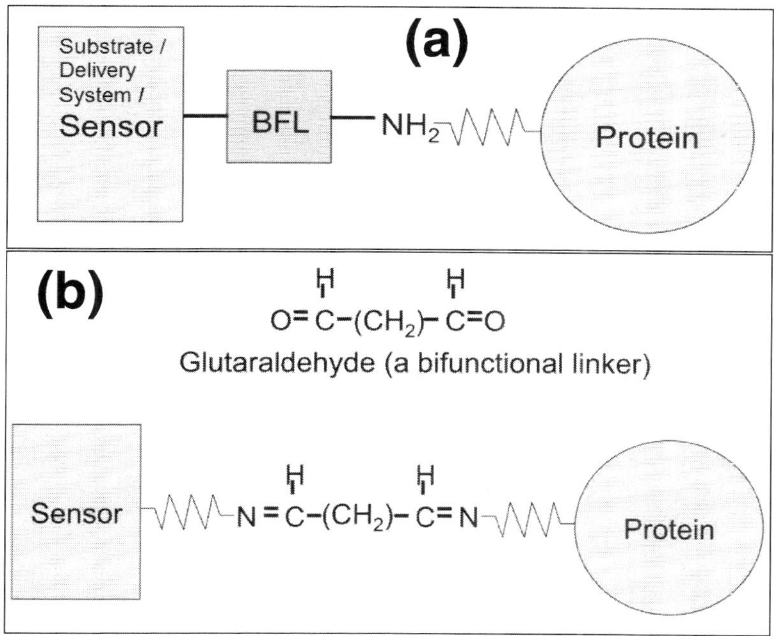

Figure 17.2. (a) BFL in Functionalization of Biomolecule; (b) Glutaraldehyde is a BFL through Amine-Aldehyde bond.
Courtesy: https://www.slideshare.net/

Types of Functionalization of Biomolecules

- ☆ Functionalization by Electrostatic Adsorption
- ☆ Functionalization by Chemisorption of Thiol-Derivatized Biomaterials
- ☆ Functionalization by Specific Interactions
- ☆ Receptor-Induced Aggregation of Guest-Functionalized Nanoparticles

Functionalization by Electrostatic Adsorption

Functionalization by electrostatic adsorption (Figure 17.3a) is carried out for biomolecules ranging from low-molecular-weight organic substances (*e.g.*, vitamin

C) to large protein/enzyme molecule besides the simplest case such as nanoparticles that are stabilized by anionic ligands such as carboxylic acids (citrate, tartarate, lipoic acid), the adsorption of proteins originates from electrostatic interactions.

Functionalization by Chemisorption of Thiol-Derivatized Biomaterials

Functionalization by Chemisorption of Thiol-Derivatized Biomaterials (Figure 17.3b) is established by strong chemisorption of proteins on nobel metal nanoparticles that can originate from the binding of thiol groups (from cysteine residues) existing in the proteins (*e.g.*, immunoglobulins, serum albumin) to the nanoaprticles surface. If no such residues are available in the native proteins, thiol groups can be incorporated by chemical means or by genetic engineering.

Functionalization by Specific Interactions

Anchor groups such as, disulfides, phosphine ligands, or thiols are often used for the binding of the bifunctional linkers to Au, Ag, CdS, and CdSe nanoparticles in the functionalization by specific interactions (Figure 17.3c). These anchor groups readily substitute weakly adsorbed molecules stabilizing the nanoparticles, or may be incorporated in the nanoparticle synthesis, resulting in a nanoparticle

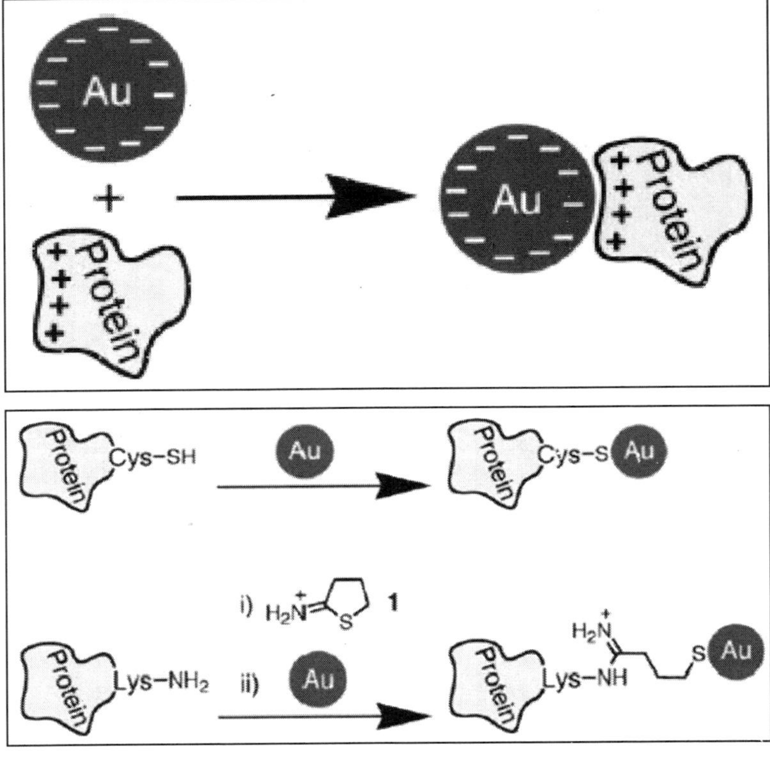

Figure 17.3. (a) Electrostatic Adsorption; (b) Chemisorption of Thiol-Derivatized Biomaterials.
Courtesy: https://www.slideshare.net/

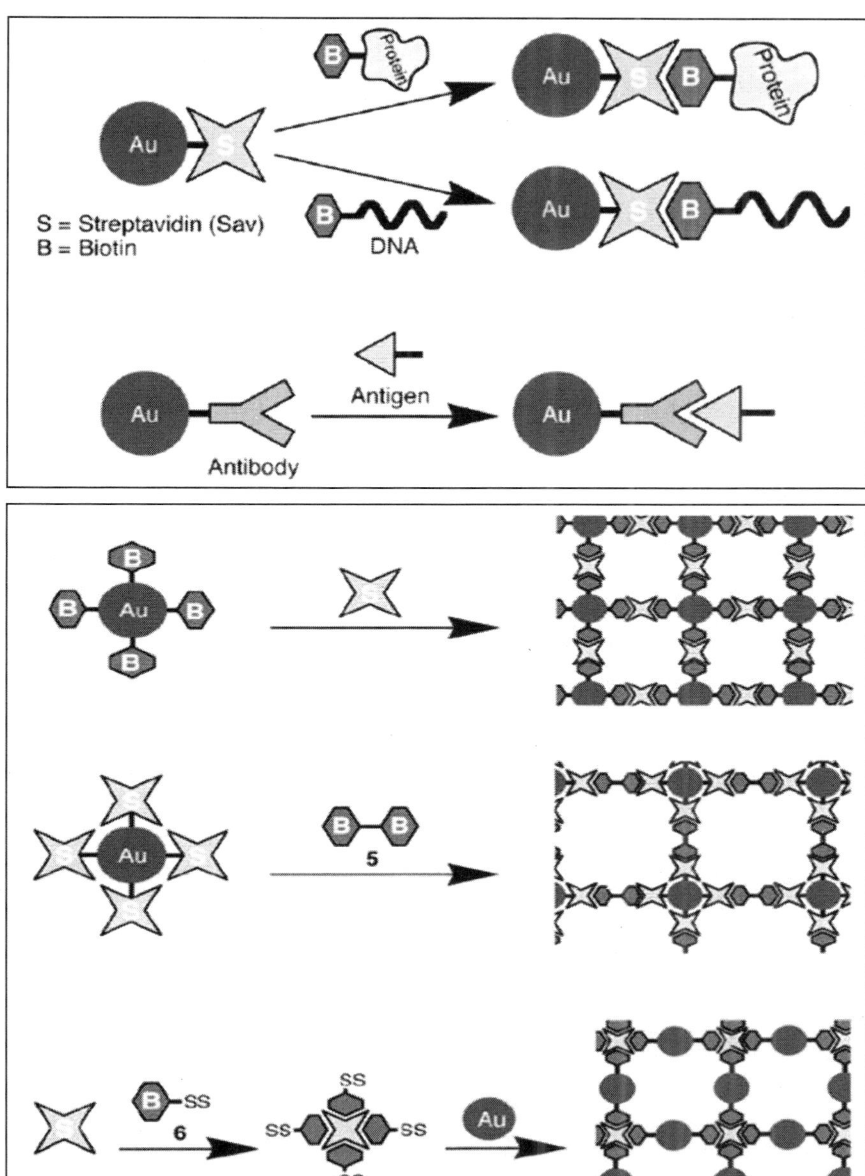

Figure 17.3. (c) Specific Interactions; (d) Receptor-Induced Aggregation of Guest Functionalized Nanoparticles.
Courtesy: https://www.slideshare.net/

surface providing functional groups for further reactions. These structures provide unique synthetic routes for the covalent binding of a single target biomolecule per nanoparticle. By covalently attaching proteins to nanoparticle surfaces, problems of instability, reversibility and inactivation can be overcome.

Receptor-Induced Aggregation of Guest-Functionalized Nanoparticles

Protein-based recognition systems can be used to organize inorganic nanoparticles into network-aggregates, for instance with the interaction between D-biotin and the biotin-binding protein streptavidin. The recognition between water-soluble biotin and the homotetrameric protein Sav is characterized by an extraordinary affinity constant of $Ka > 10^{14}\ M^{-1}$, which makes it the strongest ligand-receptor binding interaction presently known in guest-functionalized nanoparticles (Figure 17.3d).

Nanoprobes

In twenty first century, due to advancement of biotechnology and nanotechnology approaches it is possible to detect a particular disease or disorder both in-vitro and in-vivo. The nanoscale materials having the properties of bioluminescence and fluorescence can be used as nanoprobes for labelling and imaging of living entity. To be used as nanoprobes the material should be non-toxic, biocompatible and stable. Also, nanoprobes should have narrow excitation band which ultimately gives high-resolution imaging. Photobleaching is an important factor in picking a nanomaterial as nanoprobe as it degrades the quality of the image which gives unclear information about the sample. In this section, we will see about the classification of different types of nanoprobes with some examples.

Nanoprobes for Sensing and Imaging

Nanoprobes have great potential in ultrasensitive biodetection when nanomaterials functionalised with suitable biorecognition element. For example, gold and quantum dots are frequently used for their high optical and electronic properties. Apart from that various proteins, enzymes and nucleic acid have been frequently used in combination with nanoparticles for the detection of biomolecules.

Nanoamplifier Probes

Due to biocompatibility properties of some nanoparticles they can be attached to various biomolecules such as antibodies, proteins and nucleic acid. These biomolecules in fact amplify the optoelectronic properties of nanoparticles to get better signal detection. For example, gold nanoparticle was functionalised with repeated array of nucleotides terminated with a fluorophore molecule which is also known as nanoflare system used to detect targeted mRNA and miRNA sequences. By scanometric method the detection limit was strengthen to 1fM. Like gold nanoparticle, graphene and graphene oxide can be used as a suitable platform for functionalization of several biomolecules. An aptasensor made up of nanoprobe includes graphene oxide (GO) functionalised with aptamers labelled with flourescein amidite (FAM) used to detect thrombin by fluorescence resonance energy transfer (FRET) technique (Figure 17.4). Due to high fluorescence quenching ability of GO, the detection limit reached as low as 31.3 pM. Also, GO based probes can be modified using various biomolecules such as DNA probe for label free DNA detection and streptavidin as well as antibody complexes for the identification of biotinylated proteins and specific antigens, respectively.

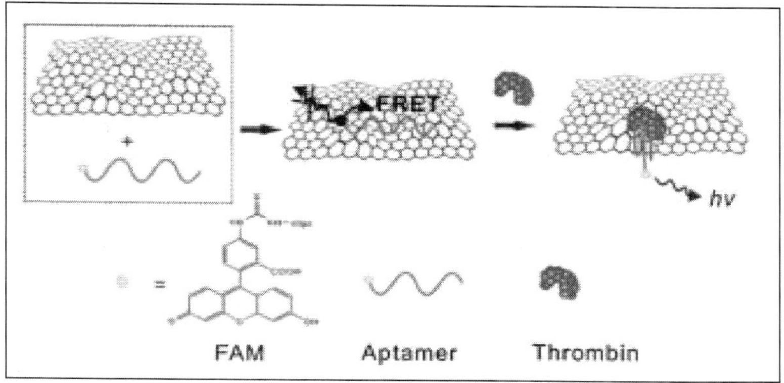

Figure 17.4. Graphene Based Probe.
Courtesy: Chang *et al.*, 2010.

Quantum Dots Probes

Quantum dots shows broad absorption spectra, narrow emission spectra, long fluorescence lifetime and enhanced photostability due to smaller size (> 10 nm) and unique electrical and optical properties. The self-fluorescence ability of quantum dots gives opportunity to monitor through label free fluorescent assay. Further, G-quadraplex nanostructure in combination with CdSe-ZnS quantum dot showed chemiluminescence resonance energy transfer (CRET) to detect targeted DNA sequences with a detection limit of 50 fM (Figure 17.5). Quantum dots also possess electro-chemiluminoscent (ECL) property which gives excellent temporal and spatial controllability and does not involve the problems of scattered light and luminescent impurities. A nanoprobe for ultrasensitive detection of DNA was developed using quantum dots and gold nanoparticle by isothermal circular amplification process, which can detect DNA as low as 5 aM.

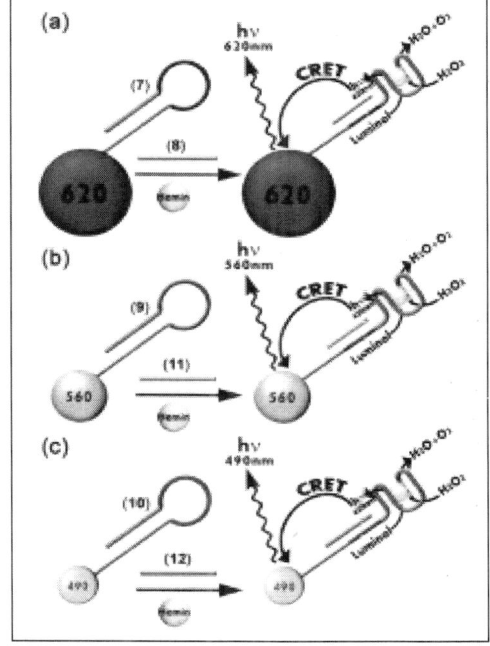

Figure 17.5. Quantum Dots Based Target DNA Detection.
Courtesy: Freeman *et al.*, 2011.

Superquenching Nanoprobes

The sensitivity of fluorescent probes depends on low intensity emission and high background contrast. Quenching technique usually consists of a fluorophore and a quencher in which fluorophore act as a donor while quencher as an acceptor. The quenching process depends on the distance between quencher and fluorophore which occurs via either energy transfer or electron transfer. Reports confirmed that gold nanoparticle showed ultrahigh quenching ability than the conventional dyes. In a study, they used Mn:ZnSe-ZnS core-shell nanocrystals tagged with goat antihuman IgG act as fluorophore and gold nanoparticle tagged with rat antihuman IgG which acts as acceptor which confirmed superquenching ability of gold nanoparticle. Gold nanoparticle can also used as a probe for Hepatitis B virus (HBV) detection. Here, gold nanoparticle was fused with the HBV surface antigen epitope as an acceptor and quantum dots are labelled with Fab antibody as donors. As the antibody interact with the antigen, the gold nanoparticle and QDs come closer to each other which causes energy transfer reactions and quenching takes place (Figure 17.6).

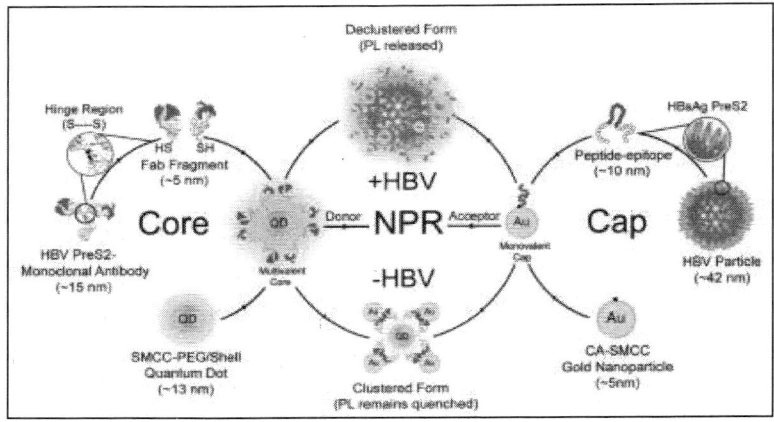

Figure 17.6. Detection Hepatitis B Virus by Quenching Nanoprobes.
Courtesy: Draz *et al.*, 2012.

Raman Nanoprobes

In recent days, Surface-Enhanced Raman Scattering (SERS) attracted great attention in the field of sensors for its ultrasensitive biodetection. Novel SERS-active nanostructures can be used as nanoprobes for the detection of various biomolecules. As single nanoparticle of gold and silver are inefficient to produce required level of SERS effect, the SERS capability was enhanced by functionalising with Raman active biomaterials for the biorecognition event.

Nanoprobes for Live Cell Imaging

Nanostructured fluorescent materials are used as probes for imaging live cell efficiently than conventional dyes. There are 3 types of fluorescent nanoprobes such as in first type nanomaterial possesses intrinsic fluorescent abilities. The examples of type 1 nanoprobes are semiconductor quantum dots, silicon dots, carbon dots, nanodiamonds, graphene quantum dots, near infrared single walled carbon nanotubes, metal nanoclusters, lanthanide doped nanocrystals dye encapsulated silica and polymer nanoparticles. Type 2 materials have the property of quenching of fluorescence which can enhance the background contrast suitable for imaging. So far, gold nanoparticles, carbon nanomaterials, MoS_2 nanosheets, copper oxide nanobelts and polyaniline fibres are regarded as the best quencher. Type 3 nanoprobe materials acts as a filler or encapsulating material for the type 1, type 2 or both type 1 and type 2 material. Common type 3 nanomaterials include polymer nanoparticles, nanofibers, magnetic nanoparticles, silica nanoparticles and porous nanomaterials.

Sensors

Probes or dye are essential for detecting a disease or abnormality in the organism but to estimate in quantitative terms it has to be integrated to a device, which is also known as sensors. A sensor. According to the principle sensors are of 2 types *i.e.* a sensor which uses chemical or physical materials for sensing whereas in another type recognition is done either by biomolecule or part of the sensory organ or whole organism to sense the analyte with precision. The working principle of biosensors occurs at 5 steps. Initially bioreceptors bind to the analyte. Later an electrochemical interface created to give signal. After that a transducer converts biological sensing to electrical signal. The signal processes by a signal processor and displayed to the operator. Nanoparticles and nanostructured materials integrated with biorecognition molecule enhances the precision, sensitivity and multiplicity.

Nanoparticles Based Biosensors

Nanoparticles such as gold, silver, carbon dots are used in combination with biorecognition element to enhance sensitivity. According to the properties of nanoparticles, they are used in optical, magnetic, electrochemical and microgravimetry sensors. For example, gold nanoparticle has been used in the microgravimetry based biosensors. The fabrication of this type of sensor was done by incorporating gold nanoparticles in quartz crystal microbalance (QCM) for the detection of *E. coli*. In this study, AuNPs were immobilized on the surface of the QCM to absorb more single-stranded DNA probes, which then captured the biotinylated E. coli DNA. After that, a second amplification step was performed using outer avidin-coated AuNPs (Figure 17.7).

Nanostructured Thin films

Nanostructured thin films are used as a very good platform to fabricate electrochemical sensors and biosensors with high sensitivity due to high surface area. The use of natural polymers gives compatibility as well as more functional groups for the attachment of probes. For example, chitosan is a better choice for the use in biosensors due to its excellent biocompatibility, complete biodegradability and non-toxic in nature. Chitosan is the second most abundant natural polymer after cellulose which is mostly found in crustaceans, insect and some fungi. Chitosan film is transparent and porous in nature. Due to that it is highly suitable for optical and electrochemical biosensors which allows light and movement of ions through the film. The presence of amine groups in chitosan gives opportunity for selective functionalization which enhances sensitivity for the specific analyte.

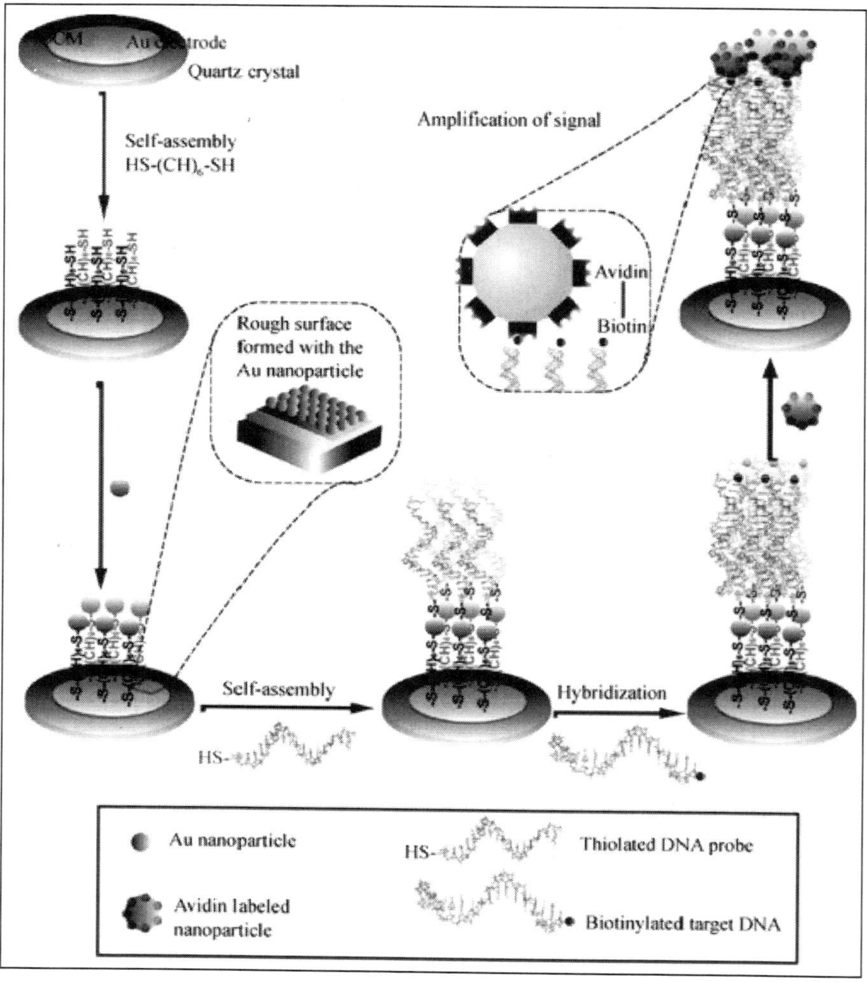

Figure 17.7. QCM Sensor with Gold Nanoparticle to Detect *E. coli*.
Courtesy: Draz *et al.*, 2012.

Dendrimers

Dendrimers are hyperbranched, monodispersed, star-shaped, and nanometer-scale three dimensional macromolecules with a very high density of surface functional groups. They are composed of three distinct components, *i.e.*, the core, the interior dendron, and the exterior surface with terminal functional groups. They have been used extensively in various biosensors and diagnostics, such as those based on electrochemistry, fluorescence, surface enhanced Raman scattering, impedimetric, and surface plasmon resonance, mainly as they increase the analytical sensitivity, stability, and reproducibility.

Lipids

Lipid based nanoparticles such as vesicles, thin films and liposomes are used to encapsulate the biorecognition elements such as proteins, enzymes, receptors and nucleic acid which can help in bypassing the plasma membrane without hassle and prevent denaturation. As lipid nanomaterials are amphiphilic in nature, they often used as an amplifier of signal. For example, a hybrid nanoparticle composed of lipid and polydiacetylene (PDA) have been employed for the development of colorimetric biosensor as when it interacted with the analyte, confirmation changes in PDA was induced which causes blue to red chromatic transition.

Miniaturized Devices

With the advancement in nanotechnology and electronics, now-a-days scientists were putting up miniatured biosensors which can work and detect very low current up to picoamperes and femtoamperes. Several types of electrodes such as single-walled carbon nanotubes (SWNTs), boron-doped silicon nanowires (SiNWs) and Sn doped In_2O_3 nanowires (ITO-NWs) have been shown to be interesting for building nanodevices.

Advantages of Nanomaterials in Biosensors

- ✰ Early detection
- ✰ Rapid and high throughput detection
- ✰ Detection processes are simple, user friendly, fast and cost effective
- ✰ Reduced material requirement to fabricate and easier recycling
- ✰ Novel properties and new capabilities
- ✰ Repetitive, portable and stability
- ✰ Ultra-sensitive devices for mycotoxin detection
- ✰ Detection of pathogen in environment.
- ✰ Used in toxicity analysis

Nanoscale Delivery Vehicles

During ancient days, humans have suffered with various diseases, and are widely recovered through plant derived medicines. Coming to present or modern days of medicinal practices, it is usually followed from the traditional basis. Majorly 25 per cent of all the available compounds are from pharmaceutical derivatives as on today are obtained from natural resources. Various naturally derived compounds with their wide molecular backgrounds always give us a challenging task to develop a novel drug delivery medicine. These natural based products give us a characteristic properties *viz.*, good chemical, biological, and less toxicity to specificity at macro-molecular level.

Some of the drug delivery challenges are if the drug material size is larger means it exhibits instability, less solubility, poor bioavailability, and absorption in the human body systems, which issues a poor target-specific delivery, tonic effectiveness and adverse effects of drugs. Therefore, by implementing and following the various drug delivery systems could solve the major and critical issues related to illness and diseases. Hence, by mimicking the principles of nanotechnology tools, one can design the suitable, controlled drug delivery with significant success.

Nanotechnology can always bridge the conventional barrier of physical and biological sciences by choosing the nano based structures and phases for various field of sciences. Nowadays attention is paved in case of nano based drug delivery system and nanomedicine, where nano scale particles with unique properties are showing a tremendous interest. Nano-biomaterials are derived from biological sources which are typically measured in the nano range between the sizes of 1 and 100 nm, which can be exploited much and influence more in the frontiers areas of nanomedicine like biosensors, microfluidics, microarray tests, drug delivery, and tissue engineering. Basically, nanomedicine can be developed by employing nano- scale (curative agents) from nanotechnology tools. The biomedicine field is the complex of one or more disciplines such as nanobiotechnology, biosensors, drug delivery, and tissue engineering besides been powered by nano-scale particles.

Nanoscale Nutrient Delivery Systems Types

For sustaining, a balanced nutrition is more essential for humans to have various health benefits with nutritional function. For example, these nano scale nutrients will inhibit some of the common chronic diseases (heart, diabetes, and cancer) besides improving overall health status such as stamina increase, energy levels, and attention. Differentnanoscale delivery systems with their potential types are shown in Figure 17.8.

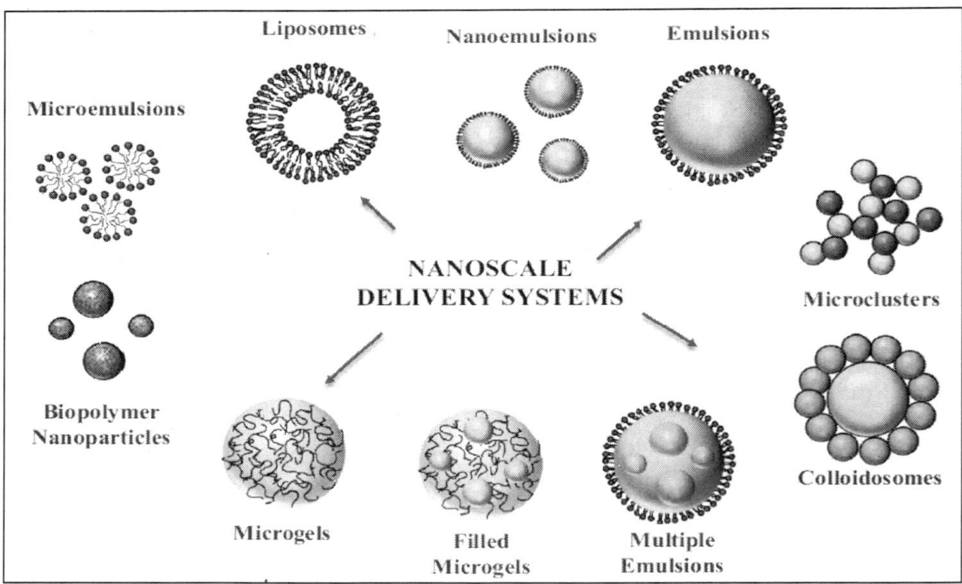

Figure 17.8. Different Types of Nanoscale Delivery Systems.
Courtesy: McClements, 2015.

Lipid Based Nanoscale Delivery System

The lipid based nanoscale delivery systems offers a unique platforms such as liposomes, solid lipid nanoparticles, emulsions, and, micelles *etc.* These platforms exhibit various properties like biodegradability, biocompatibility, scaling-up, incorporating both lipophilic and hydrophilic drugs for better dispersions of water soluble drugs with poorly and, low cost as compared to polymer based system which will pave as an alternate drug delivery system; moreover, the applications of these lipid based drug delivery systems have been favoured by getting the GRAS (Generally Recognized as Safe) status and their conventional usage of products such as food and pharmaceutical.

Solid Lipid Nanoparticles

Solid lipid nanoparticles (SLN) came into effect in the year of 1990s as an alternative to the existing available carrier systems such as liposomes, emulsions, and, polymeric nanoparticles.

These solid lipid core based drug carriers will possess with a numerous advantages over the conventional drug carriers by exhibiting the better targeting, advantageous, higher physical stability, lower toxicity, biocompatibility, and scaling up. Curcumin is the important drug plant with unique benefits. Difruloyl methane a constituent of turmeric possesses a strong anti-inflammatory, antioxidant, and anti-cancerous properties. The applicability of this drug for curing many forms of diseases is limited and this is due to their poor systemic bioavailability.

Wang *et al.,* fabricated a complex of curcumin-SLN nanoformulation by employing the principle of sol-gel method with particle sizes ranging from 20 to 80 nm for targeting the lung tumour inhibition, besides giving a opening platform for new anti-cancerousdrug development.

The solid lipid nanoparticles are having desirable features, as well as they also have some limitations including purity of SLN, and poor drug loading. Furthermore, the drug may undergo transition storage which may increases the size and releasing the encapsulated entity. Though, the conventional methods of drug delivery platforms owe a merit for future exploration.

Food Grade Delivery Systems

The food-grade based nanoparticles delivery systems are in the size range of 100-300 nm and attracted much in recent times. These drug delivery systems are formulated based on nature of the bioactive, release rate and compatibility with the carrier.

Protein-Based Nanodelivery Systems

Food hydrocolloids such as proteins are widely used to formulate the drug containing bioactives. These protein-based delivery systems exhibit with unique properties *viz.*, biodegradability, excellent water binding capacity, more amount of nutritional content, cost efficient, and GRAS status.

The delivery of bioactives with their both physico-chemical attributes of multiple functional groups present are generally in the polypeptides sequence. Furthermore, studies have confirmed that the bioactive peptides generation via enzymatic degradation *in vivo* are subsequently increase the no of physiological effects, leading to an advantage with acceptability of protein-based carrier drug delivery. The various preparation methods generally employed for producing protein-based carriers showing much attention as a targeted delivery vehicle. Some of the most common techniques are, emulsification, gelation, spray drying, extrusion, and coacervation areused as a deliverysystem such as hydrocolloids, hydrogels, micelles, nano-assemblies, micro-and nanoparticles, and films.

Some of the protein carriers which are generally investigated from the animal sources such as collagen, casein, elastin, gelatin, whey proteins, and egg albumin and some of the protein from plant sources are wheat gliadin, glycinin, and zein *etc.* Further, from the vegetal origin source, nanoparticulate based protein carriers are showing potential interest in recent times with huge advantages. The nanoparticles can be hardened by exploiting with the low solubility and high hydrophobicity of particular proteins by negating the further chemical and physical treatments.

Carbohydrate-Based Nanodelivery Systems

Most commonly both nutraceuticals and pharmaceuticals industries seek a novel delivery system through carbohydrate for an efficient targeted and controlled release. Again, these carbohydrates are classified into mono, oligo, and polysaccharides. Among these mono and oligosaccharides are used generally for encapsulating matrix, most often polysaccharides are being used as one of the carrier components because of their characteristics of molecular structure and showing ability of bioactive entrapment efficiency.

Some of the most common examples employed for the different delivery systems are cellulose, starch, alginate and xanthan gum, and dextran-spermine and chitosan with their magnificent electrical and charge dependent on the pH relative to the p*K*a. Once the nanocarrier system is formulated with the complex of alginate which are cross-linked with the magnesium or calcium chloride (divalent ions), which will employ successfully as alginate-based nanoparticles. These alginate-based nanoparticles are applied mostly for delivering the insulin and drugs and vitamin D_3.

Chitosan-Based Delivery Systems

Chitosan is the emerging polysaccharide which are widely used for different application purposes because of it non-toxic, biodegradable, film forming, antimicrobial, biocompatible in nature. Recently much attention is given in chitosan based oral drug delivery systems and as well as also applied in biomedical field. Chitosan is Generally Recognized as Safe Status (GRAS), because it can be applied in different fields of applications and besides delivering of nutraceuticals. It is used in medicine, for example, bioimaging, diagnosis, tissue engineering, would healing and finally drug delivery formulations (hydrogel system, films, and particulate system). A functionalised chitosan-based hydrogel (Liu *et al.*, 2018) was reported as a wound dressing and drug delivery system in the treatment of wound healing (Figure 17.9).

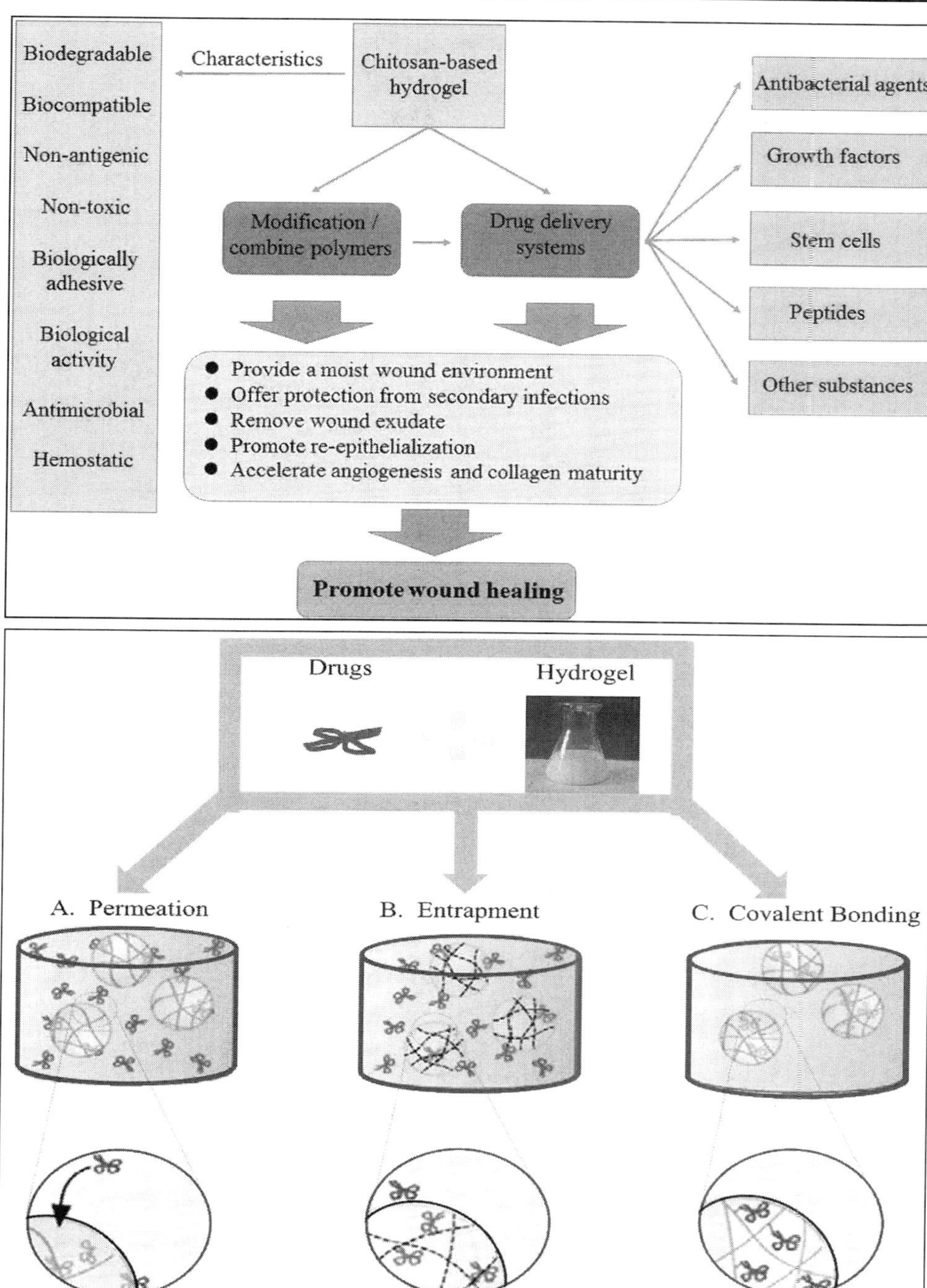

Figure 17.9. Mechanism of Preparation and Application of Chitosan-based Hydrogel for Wound Dressing.
Courtesy: Liu *et al.*, 2018.

Common Oral Delivery Systems

Delivery Systems	*References*
Phospholipid-based delivery systems Amphipathic lipids can be self-assembled into lipid bilayers, *e.g.* liposomes, phytosomes. Increases loading efficiency of both hydrophilic and lipophilic compounds.	Chen *et al.* (2010)
Emulsion-based delivery systems Mixture of two or more immiscible liquids stabilized by amphipathic compounds, *e.g.* microemulsions, nanoemulsions and solid lipid nanoparticles. Increases the dose efficiency and potency of lipophilic compounds.	Davidov-Pardo and McClements (2014)
Chemical modification Involves the association of nonbioactive promoieties (such as functional groups and polymers) with bioactive compounds that can subsequently be dissociated within the body by the action of enzymes or other chemicals. Increases the potency and dose efficiency of the parent compound from the molecular level.	Zhou *et al.* (2014)
Chitosan-based delivery systems Chitosan is formulated in both micron sized and nanosized particles to deliver bioactives. It exhibit mucoadhesive properties, facilitates tight junction widening by increasing the BA of target nutraceutical.	Mukhopadhyay *et al.* (2013)
Nanodispersions Homogeneously Bioactives >100 nm Allow minimal processing with higher dosing of bioactive Increases the potency and dose efficiency of bioactive compounds.	Adjonu *et al.* (2014)
Mesoporous silica particles Bioactives >10 nm distributed within the nanopores of a microscale silica particle; allow for the further use of pH or redox potential sensitive gates to maximize dosing of bioactive in a specific area of the GIT.	Pérez-Esteve *et al.* (2015)

Nanomaterials for Plant Genetic Engineering

Nanoparticles are very much useful materials for delivering various biomolecule, because of they are having ability to cross the barrier membranes of biological materials, protect and release the cargoes, and finally it is targeted multiple times via physical or chemical tunability. These properties enabled nanoparticles to revolutionize for a better targeted delivery and sustained release especially in mammalian systems. However, nano-based carrier systems in plants are underexplored because of the cell wall, which are aided chemically or mechanically (Figure 17.10).

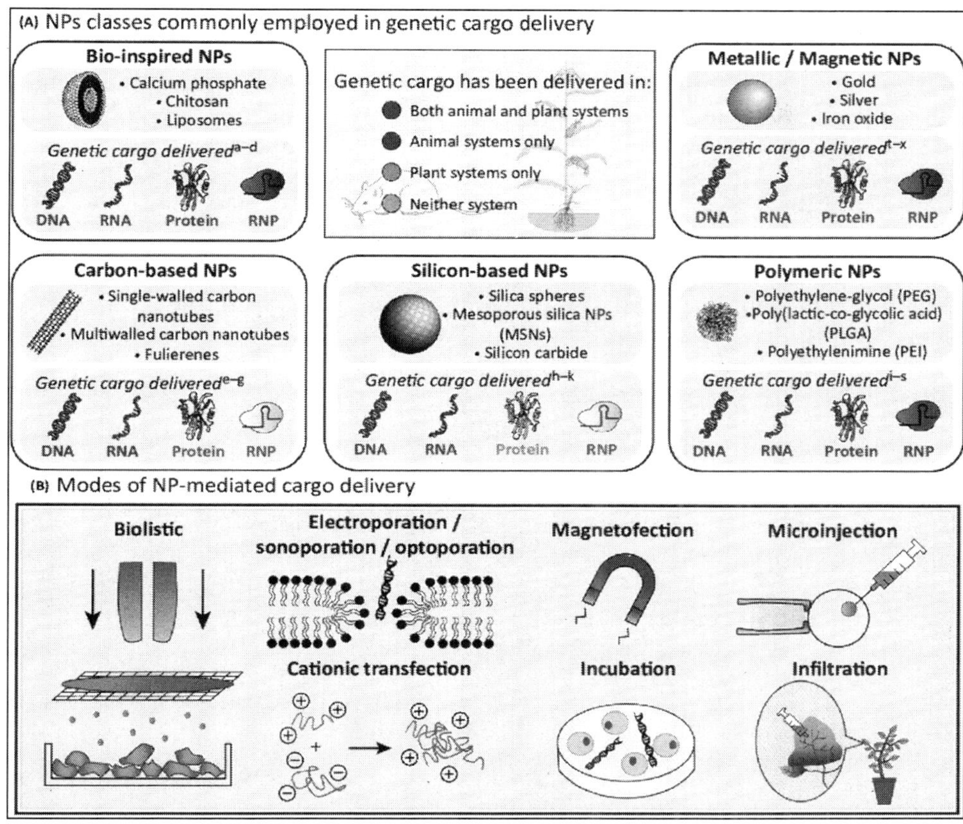

Figure 17.10. Nanoparticle (NP)-Mediated Genetic Cargo Delivery to Animals and Plants.
Courtesy: Cunningham *et al.*, 2018.

The nanoparticles-based plant delivery methods are having some limitations of such as in vitro plant studies, genotypes, regeneration capacity across the species, and developmental age of source tissue. Thus, parallel optimization of delivery and regeneration is necessary to improve efficiency and expand stable transformation capabilities to all plant species,

Characteristics of Nanoscale Delivery Systems

Before designing any nanoscale nutrient delivery systems, there are some factors should be taken into consideration for various application purposes such as pharmaceutical and, food industry:

Safety

The safety of nanoscale delivery system plays a vital role before the food or drug products are commercialized. These nanoscale sized particles are incorporated into either food matrix, drugs, and beverages are consumed bydifferent age groupsand to improve overall health benefits of the class of persons. The compositions and quantities of nanoscale delivery systems are safe and environment benign. Typical example, the rate of absorption, distribution, digestion excretion, and, metabolism is considered and altered based on the size. It is important to ensure the safety of the nanoscale delivery systems by testing potential toxicity level.

Commercial Viability

Once, after developing the food or drug-based delivery systems should possess all kinds of commercial status. Generally, all the delivery systems are fabricated and designed within the food-grade ingredients should be feasible and economically viable.

Food Matrix Compatibility

Basically, various food products which are available abundantly are exhibiting physic-chemical properties and sensory attributes, *viz.*, shelf-life, colour, flavour, taste, textures, and overall appearance. For example, some of the food products are in low viscosity state (fortified waters), while some are optically opaque (yogurts and spreads) are either in solids or semi solids. It is so much important for us to encapsulate and protect the bioactive agent besides improving the delivery system as well as quality attributes. Thus, attention is paved to nanoscale delivery system, where one of the particle diameter is <50 nm and protects from scattering of light for food and, drug products.

Robustness

Once, after the incorporation of food grade compounds, which will experience environmental stresses during the tune of production, processing, storage, transportation, and, consumer utilization. Any kind of delivery system should be physically and chemically stable when they are exposed to various environmental stresses, and retaining the functional properties.

Performance

The nanoscale delivery system should possess all kinds of functional attributes for the specific requirements of the drug and, food application. For example, it may be tested based on the particular ingredient which can be protection, handling, storing, and, delivering. So, the drug or bioactive compounds are protected from degradation during course of storage, besides improving the bioavailability, and, controlled or sustained release of bioactive ingredient can be successful through oral route of delivery. The overall performance of the delivery systems should be fabricated through precision application, where it meets the requirements of nanoscale delivery systems.

Labelling Requirements

Labelling of the products is important to meet the specific ingredient requirements such as "all-natural," "vegetarian," "vegan," "Kosher," "dairy free," *etc.* So, this type of application is so important for delivering the food or drug products to consumers with detailed ingredient requirements, for example, some of the vegetarian products do not use gelatin or milk protein in case of vegan products.

Advantages of Nanoscale Materials-Based Drug Delivery Systems

- ☆ Increase the aqueous solubility of the drug
- ☆ Produces prolonged release of the drug
- ☆ Offers systems for combine therapy
- ☆ Offers appropriate forms for all routes of administration
- ☆ Improve the bioavailability of the drug
- ☆ Targeted delivery of the drug
- ☆ Decreased toxic side effects of the drug
- ☆ Allow rapid formulation development
- ☆ Improves pharmacokinetic and pharmacodynamics of drugs
- ☆ Protect the drug from degradation

REFERENCES

1. Adjonu, R., Doran, G., Torley, P., and Agboola, S. (2014). Whey protein peptides as components of nanoemulsions: A review of emulsifying and biological functionalities. *Journal of Food Engineering, 122,* 15-27.
2. Chen, Z.-P., Sun, J., Chen, H.-X., Xiao, Y.-Y., Liu, D., Chen, J., Cai, H., and Cai, B.-C. (2010). Comparative pharmacokinetics and bioavailability studies of quercetin, kaempferol and isorhamnetin after oral administration of Ginkgo biloba extracts, Ginkgo biloba extract phospholipid complexes and Ginkgo biloba extract solid dispersions in rats. *Fitoterapia, 81*(8), 1045-1052.
3. Cunningham, F. J., Goh, N. S., Demirer, G. S., Matos, J. L., and Landry, M. P. (2018). Nanoparticle-mediated delivery towards advancing plant genetic engineering. *Trends in biotechnology, 36*(9), 882-897.
4. Davidov-Pardo, G., and Mcclements, D. J. (2014). Resveratrol encapsulation: designing delivery systems to overcome solubility, stability and bioavailability issues. *Trends in food science and technology, 38*(2), 88-103.
5. Knauth, P., and Schoonman, J. (2004). *Nanostructured Materials Selected Synthesis Methods, Properties and Applications*: Kluwer Academic Publishers.
6. Liu, H., Wang, C., Li, C., Qin, Y., Wang, Z., Yang, F.,. and Wang, J. (2018). A functional chitosan-based hydrogel as a wound dressing and drug delivery system in the treatment of wound healing. *RSC advances, 8*(14), 7533-7549.
7. Mazumder, S., and Pavurala, N. (2016). A review on nanoprobes for sensing, imaging and disease detection. *J. Mater. Sci. Nanotechnol, 4*(1).
8. Mcclements, D. J. (2015). Nanoscale nutrient delivery systems for food applications: improving bioactive dispersibility, stability, and bioavailability. *Journal of food science, 80*(7), N1602-N1611.

9. Mihai, G. (2016). *Nutrient Delivery* (Vol. 5): Nikky Levy.
10. Morais, M. G. D., Martins, V. G., Steffens, D., Pranke, P., and Da Costa, J. a. V. (2014). Biological applications of nanobiotechnology. *Journal of nanoscience and nanotechnology, 14*(1), 1007-1017.
11. Mout, R., Moyano, D. F., Rana, S., and Rotello, V. M. (2012). Surface functionalization of nanoparticles for nanomedicine. *Chemical Society Reviews, 41*(7), 2539-2544.
12. Mukhopadhyay, P., Sarkar, K., Chakraborty, M., Bhattacharya, S., Mishra, R., and Kundu, P. (2013). Oral insulin delivery by self-assembled chitosan nanoparticles: *in vitro* and *in vivo* studies in diabetic animal model. *Materials Science and Engineering: C, 33*(1), 376-382.
13. Patra, J. K., Das, G., Fraceto, L. F., Campos, E. V. R., Del Pilar Rodriguez-Torres, M., Acosta-Torres, L. S., Diaz-Torres, L. A., Grillo, R., Swamy, M. K., and Sharma, S. (2018). Nano based drug delivery systems: recent developments and future prospects. *Journal of nanobiotechnology, 16*(1), 71.
14. PérezEsteve, É., RuizRico, M., MartínezMáñez, R., and Barat, J. M. (2015). Mesoporous silicabased supports for the controlled and targeted release of bioactive molecules in the gastrointestinal tract. *Journal of food science, 80*(11), E2504-E2516.
15. Sanvicens, N., Pastells, C., Pascual, N., and Marco, M.-P. (2009). Nanoparticle-based biosensors for detection of pathogenic bacteria. *TrAC Trends in Analytical Chemistry, 28*(11), 1243-1252.
16. Schmid, G. (2004). *Nanoparticles: From Theory to Application*: WILEY-VCH Verlag gmbh and Co. Kgaa.
17. Zhou, H.-F., Yan, H., Hu, Y., Springer, L. E., Yang, X., Wickline, S. A., Pan, D., Lanza, G. M., and Pham, C. T. (2014). Fumagillin prodrug nanotherapy suppresses macrophage inflammatory response via endothelial nitric oxide. *ACS nano, 8*(7), 7305-7317.

Chapter 18

Nanotechnology in Plant Genome Engineering

*M. Praghadeesh[1], V. Karthika[1], K.S. Subramanian[2] and A. Lakshmanan[1]**

[1]Department of Nano Science and Technology, [2]Director of Research, Tamil Nadu Agricultural University, Coimbatore – 641 003, Tamil Nadu
**e-mail: microlaxman@gmail.com*

Introduction

Over 10000 years ago, human beings embarked on an ever-ending quest to attain self-sufficiency of foods and survival by domesticating wild plant species and natural flora. In general, every single attempt of domestication was largely focused on traits like increasing yield and productivity, resistance to biotic and abiotic stress, delayed flowering, senescence, short duration, early maturing, reduced dormancy, climate-resilient, and consumer preference (color, shape, and size) (Lenser and Theiben, 2013). In the course of the domestication process, the tools and approaches *viz.*, selective breeding, hybridization, crossbreeding, mutagenesis, marker-assisted selection, gene pyramiding, protoplast fusion, plant tissue culture, transgenesis vis-à-vis transgenic crops, genomics, and genome editing had also evolved in pace with inventions and breakthrough technologies in plant biology and biotechnology. Each breakthrough technology has furthered our understanding and comprehension of complex and intricate biological mechanisms that lead to obtaining the desired phenotype in plants (Saveland, 1992). The level of resolution and manipulation in biological systems had progressively decreased from macro, micro, nano and more recently to the atomic scale. Gene and genome in total is the basic and crucial unit that was manipulated or engineered in all the domestication and breeding approaches. This revelation of 'gene' as the fundamental unit governing a specific phenotype and its targeted manifestation results in desirable traits constituted the baseline for all modern-day gene modification and genome editing techniques. Genome engineering is a type of modification in genetic material (functional gene) of an organism in which DNA is inserted, modified, deleted or replaced in a highly precise and targeted manner resulting in achieving molecular level precision breeding. Nanotechnology with its unique properties of reduced size, increased surface area and ability to reengineer carrier materials at atomic level had open up a plethora of potential applications in genome engineering. Size of the payload (DNA molecule) (Raj and Khan, 2016); mode of delivery or delivery module of the payload (Cunningham *et al.*, 2018); and tracking of its movement in biological system (Lahiani *et al.*, 2015) are the three major domains in which nanotechnology can synergistically complement genome engineering. This chapter aims to elucidate the mechanisms involved in conventional genome engineering approaches/technologies, its limitations, and the myriad of new and innovative targeted genome engineering tools emerging at the nexus of nanotechnology, nanobiotechnology and genome engineering to overcome those limitations.

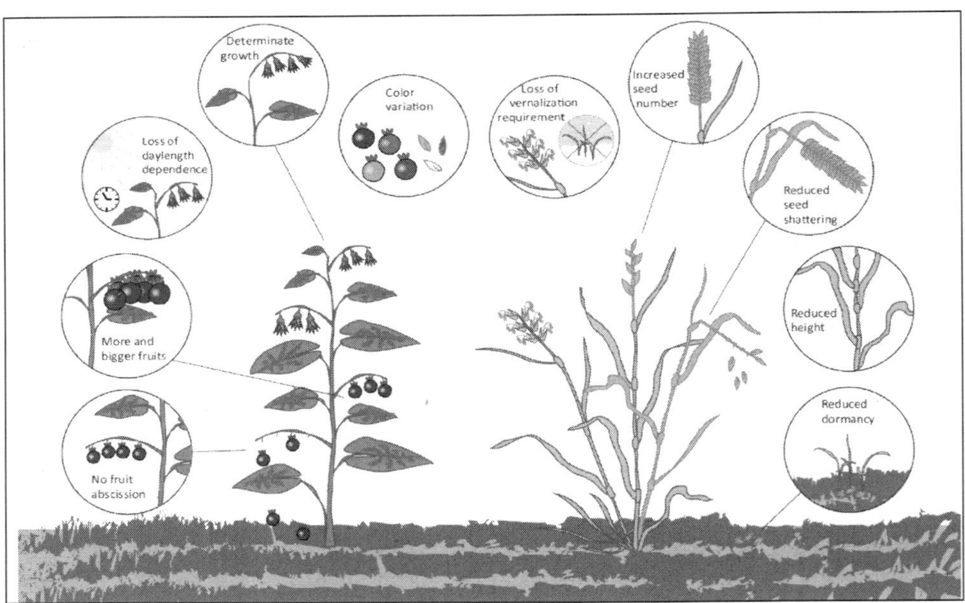

Figure 18.1. Major Crop Traits that have been the Target of Domestication Strategies.
Courtesy: Lenser and Theiben, 2013.

Genome Engineering Methods: Past and Present

Agrobacterium tumefaciens (*Agrobacterium*), a humble rod-shaped gram-negative bacterium that in nature infects plant roots and causes crown gall symptom. The bacterium houses *Ti* (Tumour inducing) plasmid which upon infection gets transferred and followed by expression into host plant system to excrete opines from plant system on which the bacterium continues to thrive. Researches had exploited the phenomenon by inserting a gene of choice in the *Ti* plasmid thereby effectively converting agrobacterium as a transporter system to deliver foreign genes into host plants. Transgenic plants developed using incorporating genes from same or related plant species are termed as cisgenesis (*e.g.* Blight resistance in potato) and the genes sourced from distantly related species were termed as transgenesis (*e.g.* Resistance to insect pest using Bt gene) (Schubert and Williams, 2006). This approach catapulted the genome engineering research towards the production of many transgenic crop species with numerous traits and potentially increasing the area under cultivation of transgenic plants/genetically modified crops to 191.7 million hectares encompassing 70 countries worldwide. India ranks 5th in the area under production (11.6 million hectares) and adoption rate next to USA, Brazil, Argentina, and Canada (ISAAA, 2018). The major transgenic crops developed and adopted worldwide include maize, soybean, canola, cotton, sugarbeets, potato, canola, brinjal and alfalfa and the most common traits of interest were herbicide tolerance, insect resistance, virus resistance, delayed fruit ripening and stacked multiple traits (ISAAA, 2017). Genetically modified *Bt* Cotton (Monsanto, Bollgard) exhibiting insect resistance is the only crop approved for commercial cultivation in India. Genetically modified food crops, *Bt* Brinjal (Resistance to fruit shoot borer) (Kumar *et al.*, 2011) and Dhara Mustard (DMH-11) genetically modified for enhanced breeding process are yet to be deregulated and cleared for cultivation (Shukla *et al.*, 2018).

In spite of the advancement in *Agrobacterium* mediated gene transfer, this method suffers from sharp drawbacks like limited host range (dicots); difficulty in transforming cells with the capacity to regenerate and transformation that requires target DNA to be delivered into chloroplasts and mitochondria. Particle bombardment or gene gun method of transformation enables us to overcome the above drawbacks which involve delivery of desired DNA or gene directly into plant cells using gold or tungsten particles. The particles are coated with DNA followed by bombardment of coated DNA in high velocity for lodging and integrating into plant cells leading to stable and transient expression. The biolistic method has several merits like wide host range (including monocots), varied cell types amenable for transformation, multiple transformation events (cotransformation), delivery of large DNA fragments and relatively simpler transformation protocol (Baltes *et al.*, 2017). Both approaches result in random integration of target DNA into the plant genome (Gao and Nelson, 2013).

In parallel to the extensive research in the area of agrobacterium and gene gun mediated gene transfer, genetic transformation methods based on harnessing the capacity of cells to undergo targeted double-strand breaks (DSBs) and innate repair machinery (Non homologous end joining (NHEJ) and homology-directed repair (HDR) were also pursued (Chen and Gao, 2014). The widely used approaches of such genome engineering platform are Zinc Finger Nuclease (ZFNs) (Kim *et al.*, 1996), transcription activator-like effector nuclease (TALENs) (Christian *et al.*, 2010) and Clustered regularly interspaced palindromic repeats (CRISPR/Cas) systems (Jinek, *et al.*, 2012, Doudna and Charpentier, 2014). The common and unique principle in all three approaches involves two components. First, proteins programmed to target specific DNA binding domain of one's interest (Recognition) and second fusion of cleavage domain/protein (*FokI*, Cas9) with that of the recognition domain which acts as molecular scissors. Irrespective of the choice of delivery system or genome engineering method, the ability to precisely and effectively transfer the foreign DNA surpassing the rigid and multi-layered cell wall and lipid bilayer of organelles (chloroplast and mitochondria) remains a major bottleneck in executing targeted genome engineering.

Nanoparticle – A Potential Delivery Vehicle of DNA

The documented size exclusion limit of the cell wall and plant organelle is in the range of ~10 – 20 nm and 300 to 500 nm respectively (Zhang *et al.*, 2019) that needs to be circumvented for the engineering of nuclear and plastid genomes. Wang *et al.* (2019) hypothesize that similarity in the size of plant plastids and microparticles (gold and tungsten) used in gene gun method often results in the destruction of organelle and pose difficulty to deliver DNA by bombardment. Nanoparticles (NP) are largely defined as materials of size less than or equal to 100 nm in any one of its three-dimension. These nanoparticles are highly amenable for manipulation in physical and chemical properties when functionalized with target DNA molecule can seamlessly deliver biological agents/cargo (DNA, RNA, and proteins) in a force independent and targeted manner.

Extensive research attempts were made during the past decade to explore and utilize nanomaterial-DNA conjugate for gene transfer. It has proven beyond doubt that, passive or force independent mode of transport of NP-DNA conjugate is highly successful in the mammalian system. Translation of the similar success in the plant system is still at its nascent stage due to the complex cellular structure and cell barrier. Nevertheless, insights gained in conjugating techniques of NP-DNA in mammalian systems serve as a blueprint and guideline for plant transformation systems. Although initial attempts involved biolistic mode of introducing functionalized silica-based nanomaterial with DNA (Torney *et al.*, 2007) and proteins (Martin-Ortigosa *et al.*, 2012) in plant system, force independent nanomaterial delivery vehicle proved to be relatively more effective and had high success rate in retrieving transformants. Chang *et al.* (2013) developed a novel genome engineering strategy using functionalized mesoporous silica nanoparticles (MSNs). Target DNA was loaded on to functionalized MSNs and was delivered into roots of *Arabidopsis thaliana* in a passive manner. Transient gene expressions were confirmed in the epidermal layer, inner cortical layer and endodermal root tissue using antibody labeling and fluorescence tagging. Mitter and co-workers demonstrated that topical application of conjugated dsRNA with layered double hydroxide clay nanosheet had increased the stability of dsRNA for over a period 30 days unlike the naked spray of dsRNA had exhibited stability for 5-7 days only. The authors had also demonstrated the efficacy of BioClay (dsRNA-clay conjugate) in rendering extended RNAi mediated protection against cauliflower mosaic virus (CMV) in cowpea (*Vigna unguiculata*) and *Nicotiana tabacum* cv. xanthi even after 20 days of conjugate spray (Mitter *et al.*, 2017).

Zhang *et al.* (2010) had employed chitosan nanoparticles as carrier material to deliver SiRNA targeting precise control of insect pests. Chitosan, owing to its unique binding potential with RNA and ability to penetrate into the cell membrane

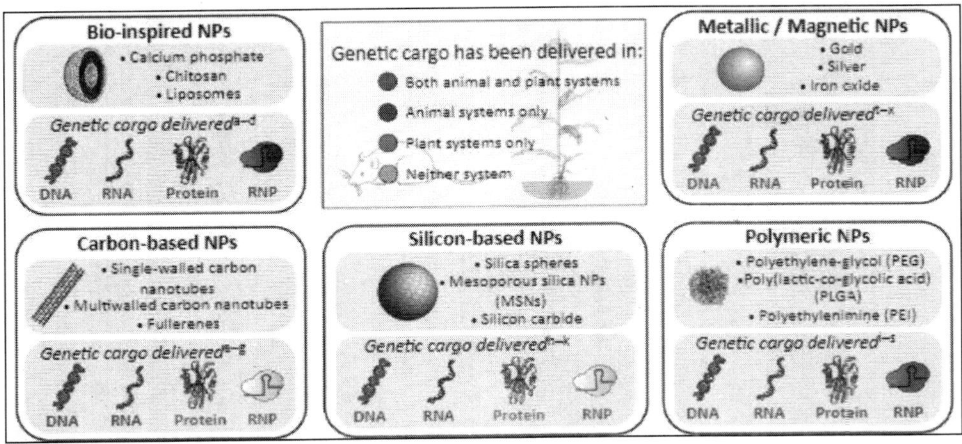

Figure 18.2. Nanoparticle (NP) Mediated Genetic Cargo Delivery to Animals and Plants.
Courtesy: Cunningham *et al.*, 2018.

of the insect host system renders increased host plant resistance and biotic stress on plants. Similarly, Mout *et al.* (2017) augmented the efficiency of CRISPR-Cas9 by conjugating cationic arginine gold nanoparticle (ArgNPs) resulted in a 30 percent increased efficiency in delivery of sgRNA into cytoplasmic and nuclear genome of cell lines. In addition, the use of ArgNPs conjugated sgRNA reduced the chances of unintended off-target modification and thereby increasing the specificity of the CRISPR recognition module (Mout *et al.*, 2017).

Advantages of using Nanoparticles as DNA Carrier Molecule

Nanoparticles possess numerous advantages over macromolecules in terms of agility and ease of use. i. Nanoparticles can be synthesized and functionalized with uniform shape and size for the effective delivery of the target molecule. ii. Nanoparticles remain stable in different biochemical and physiological conditions of plants (root, leaf, shoot, *etc.*) and protect the DNA molecule from degradation iii. Nanoparticles can be conjugated with naked DNA or in combination with other functional components (fluorescence dye, enzymes) for co-delivery at the target site and thus becomes a multi-functional unit in genome engineering. iv. Nanoparticles can be programmed to release the DNA molecule at desired time interval or in response to specific internal or external stimuli. v. Nanoparticles synthesized from biologically derived materials like chitosan, cellulose, mesoporous silica could be safely degraded in plant system after DNA delivery (Jiang *et al.*, 2019). The conventional nanoparticles that are widely used in gene modification applications that are derived from lipids, biopolymers and inorganic materials are summarized (Table 18.1).

Table 18.1. List of Nanoparticles, their Delivery Method and Outcome in the Host Plant System

Nanoparticle Type	*Plant Species Cell/Tissue Type*	*Delivery Method*	*Target Outcome*	*Reference*
PAMAM dendrimer	*A. thaliana* roots	Passive	Developmental gene silencing led to systemic phenotypes	Jiang *et al.*, 2014
Gold-plated MSNs	*Z. mays* embryos	Biolistic DNA and protein co-delivery	Both transient and stable expression	Martin-Ortigosa *et al.*, 2014
Organically functionalized CNTs	*N. tabacum* protoplasts and leaf explants	Passive	Both transient and stable expression	Burlaka *et al.*, 2015
Dimethyl aminoethyl methacrylate (DMAEM) polymer NPs	*N. tabacum* and *Ceratodon purpureus* protoplasts	PEG Transfection	Both transient and stable expression	Finiuk *et al.*, 2017
Magnetic Fe_3O_4 NPs	*Gossypium hirsutum* pollen	Magnetic field	~1 per cent efficiency for generating stable transgenic seeds	Zhao *et al.*, 2017
Cationic polyamidoamine (PAMAM) dendrimers	*Alfalfa* cells	Transfection	Ultrasound-enhanced gene delivery and expression	Amani *et al.*, 2018
Polyethylenimie NPs	*Crocus sativus L.* cells	Transfection	Enhanced Transfection efficiency	Firoozi *et al.*, 2018
DNA nanostructures	*Nicotiana benthamiana* leaves	DNA nanostructure-mediated biomolecule delivery	Gene silencing	Zhang *et al.*, 2019
Mesoporous silica	Tomato leaves and shoots	MSN-mediated gene delivery (spray and injection)	Transient gene expression	Hajiahmadi *et al.*, 2019
Chitosan-complexed single-walled carbon nanotubes	*Eruca sativa*, *Nasturtium officinale*, *Nicotiana tabacum*, *Spinacia oleracea*, and *Arabidopsis thaliana* mesophyll protoplasts	Chloroplast-targeted transgene delivery	Transient expression	Kwak *et al.*, 2019
Single-walled nanotubes	*Nicotiana benthamiana* leaves	Passive delivery	Post-transcriptional gene silencing and Protection of siRNA against nuclease degradation	Demirer *et al.*, 2019
Nanosheets	*Arabidopsis thaliana*, cells	Passive delivery, Nonendocytic pathway	ssDNA and fluorescent dye (FITC)	Bao *et al.*, 2017

Reengineered DNA Nanostructures – Future of Genome Engineering

Genome engineering at the crossroads of nanotechnology had evolved from NP based DNA delivery system to an elegant DNA based nanostructure for precise delivery of aptamers or payload. The specific Watson and crick base pairing principle had powered the development of geometrically defined self-assembling and dynamic DNA nanostructures. The requirement of a large amount of synthetically derived DNA material with defined nucleotide sequences that remained as a bottleneck in developing DNA nanostructure was overcome by the advancement in nano biotechnological toolsets (Seeman and Sleiman, 2017). Bacteriophage proliferation is the versatile and most commonly utilized approach to replicate a defined nucleotide sequence into 'n' number of the copies in bacterial system, followed by isolation and purification for downstream applications. DNA nanostructure by virtue of its composition *i.e.* nucleotide sequence, once introduced into host system can autonomously scan for corresponding target site in nuclear and mitochondrial genome leading to introduction of precise replacement or alteration in genes. Besides, demonstrating high level of accuracy in targeted genome engineering, DNA nanostructure with defined geometry could be fabricated or conjugated with organic (enzymes, proteins) or inorganic material (quantum dots, fluorescent dyes, magnetic nanoparticles) to serve additional purposes of tracking or inciting site-specific responses. Numerous iterations of DNA nanocarriers with different structural properties and tailored *in vivo* functionalities have been developed and synthesized. Among different methods, i. Self-assembled DNA based nanostructures, ii. DNA based rolling circle amplification, iii. Multifunctional DNAsomes and iv. DNA origami-based nanostructures display potential for applications in genome engineering (Figure 18.3) (Jiang *et al.*, 2019).

Self-Assembled DNA Based Nanostructures

DNA holds the repository of genetic information and plays a key role in expression and regulation of biomolecules (proteins, enzymes, transcription factors) in all living organisms. In the backdrop of inventions like polymerase chain reaction, DNA cloning, DNA sequencing, Seeman and Co-workers proposed that DNA molecules can be selectively manipulated or bioengineered into multifunctional and programmable nanostructures (Seeman 1982, 1998; Kallenbach *et al.*, 1983). The stringent base pairing (purines and pyrimidines) and self-recognition properties enable DNA molecules to serve as building blocks of different sizes, shape, and geometries in nanostructures (Okholm and Kjems, 2016). By means of rational design of appropriate nucleotide sequences, DNA strands play role as stick and glue to spontaneously orient into desired 2D or 3D shapes which has made DNA architecture a new and promising tool in genome engineering (Smith *et al.*, 2013, Li *et al.*, 2013).

DNA Based Rolling Circle Amplification

Rolling circle amplification (RCA) is an enzymatic process of replicating multiple copies of the DNA template using a short RNA primer resulting in a single-

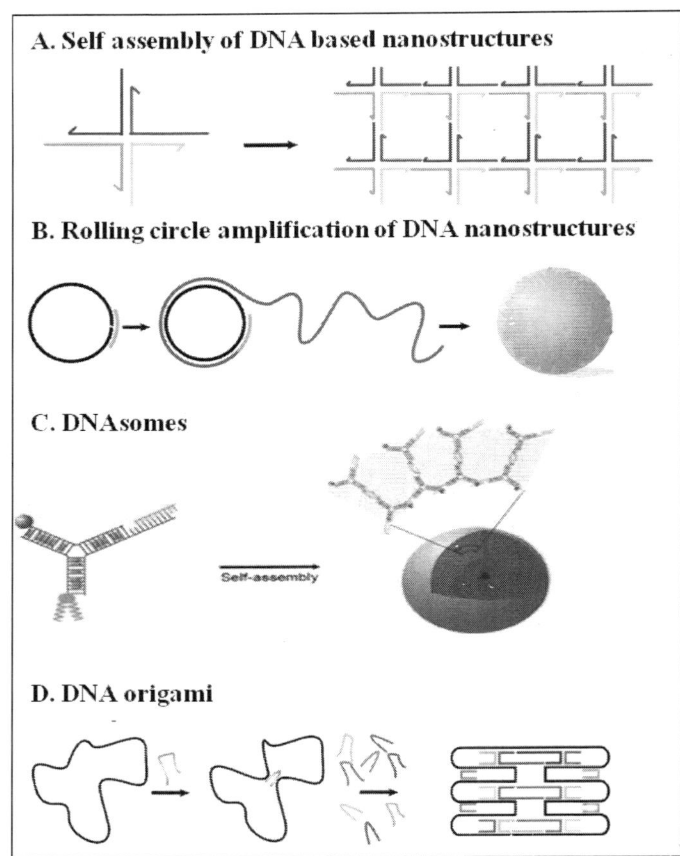

Figure 18.3. Diagrammatic Representation of Various Nanostructures and its Synthesis Methodology.
Courtesy: Jing *et al.*, 2019.

stranded DNA or RNA. The enzymatic conversion is carried out by unique type of DNA and RNA polymerases (Phi29, Bst, and Vent exo-DNA polymerase for DNA, and T7 RNA polymerase for RNA) in specific isothermal conditions (37°C). Resulting RCA product is a long stretch of DNA concatemer housing thousands of tandem repeats of desired nucleotide sequences (Figure 18.4) (Ali *et al.*, 2014). The potential and versatility of the RCA technique have made it a promising tool in generating complex DNA nanostructures *viz.*, DNA microsponges, nanoclaws, nanoribbons, nanoflowers, and nanocapsules. The DNA nanostructure developed from RCA products can be hybridized and tethered with multifunctional components like biorecognition (biotin, antibody), tracking (fluorophores) and triggers for delivery (Ouyang *et al.*, 2017). Sleiman and co-workers have demonstrated the use of RCA products as a framework to house DNA nanotubes. They synthesized DNA nanotubes using framework from RCA product scaffolds leading to a triangular, stimuli-responsive copolymer with interlinking of strands. Fluorescent labeling

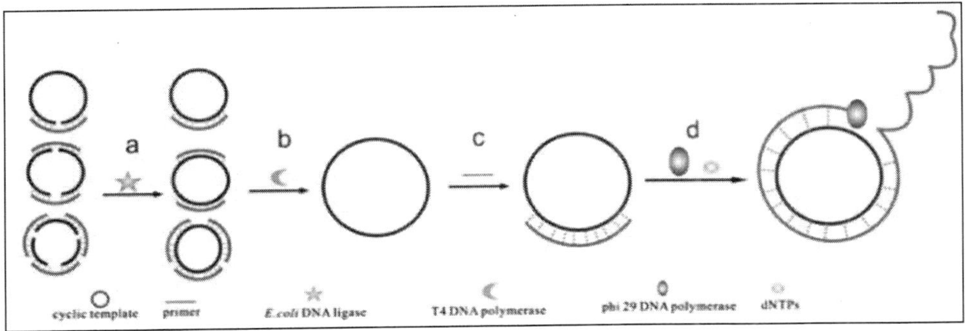

Figure 18.4. Stages Involved in Production of DNA Banostructure through RCA.
Courtesy: Ali *et al.*, 2014.

study of DNA nanotube exhibited resistance to nuclease enzyme in the host, increased penetration efficiency in cells and amenable for encapsulation and targeted and controlled release (Hamblin *et al.*, 2012).

Zhang and co-workers have reported similar findings related to the synthesis of DNA nanotubes and in addition, proved that the dimensions and configuration of DNA nanostructure can be altered by manipulating the length of the RCA circular template and concentration of the RCA products. Further, they had demonstrated the feasibility of the approach in developing complex layered 3D nanoribbons by modifying the RCA template length and molar ratio of RCA product (Ouyang *et al.*, 2013; Zhang *et al.*, 2017).

Multifunctional DNAsomes

DNAsomes are a series of Y-shaped biological entities sequentially made from three strands of DNA. A lipid molecule, N- glutaryl phosphatidylethanolamine, NGPE was functionalized and attached to the base arm of the 'Y' shaped individual unit and in addition, the fluorophore or any inorganic nanoparticles can be chemically attached to the two 'Y' arms. In *in vivo* the 'Y' shaped DNA nanostructure exhibit hydrophilic and the lipid component exhibits hydrophobic nature. The combination of both factors results in the formation of hollow spheres consisting of DNA, lipid molecule and desired inorganic or organic component in the size range of 100 to 5000 nm. Though the exact mechanism of transport and delivery of the desired target remains elusive, the evidence to prove the targeted release in cytoplasm and nuclease was confirmed using fluorophore dye studies. Roh and co-workers had developed DNA based anisotropic, branched, cross-linkable monomer units which under conducive conditions self assembles into liposome like core-shell DNAsome structure. The size of the DNAsomes can be differentially altered based on the target site and desired payload (Roh *et al.*, 2011).

DNA Origami-based Smart Nanostructures

Discovery of DNA nanostructures and DNA origami by Rothemund (2006) is an important milestone in Nanobiotechnology and genome engineering. It is the process of fabricating a linear scaffold of DNA molecule (ssDNA) to produce an arbitrary architecture and numerous support strands hold the scaffold in a definite shape. The support strands should be meticulously designed to be complementary to the specific regions in the long DNA scaffold which plays major role in sequentially initiating the self-assembly process. In addition to the physical support of the strands, they also serve as hotspots or domains for functionalization depending on the intended purpose (Yang *et al.*, 2015). To date, numerous variants of DNA origami nanostructures had been constructed. *viz.*, curved surface (Han *et al.*, 2011); DNA box with lid (Andersen *et al.*, 2009); honeycomb lattice and 2D structure (Douglas *et al.*, 2009); polyhedral mesh (Benson *et al.*, 2015); twisted and curved nano bundles (Dietz *et al.*, 2009); higher-order structures (Hong *et al.*, 2017) with predetermined geometries and intrinsic biocompatibility for targeted genome engineering.

Besides the role of forming scaffold and payload delivery, DNA nanostructures can act as miniature molecular devices (customizable bio-machinery) regulating specific functions in cellular systems. The DNA nanostructure-based devices can be programmed to execute assembly of proteins and nucleic acids governing specific enzymatic reactions in the cell (Simmel, 2012). Similarly, a series of nanomolecular devices could be interlinked to form 'nano factories' fully programmed to assemble target compounds and result in programmable cascade of enzymatic reaction in cells (Wilner *et al.*, 2009; Delebecque *et al.*, 2011) and even leading to introduction of targeted modifications in genome of living organisms.

Voigt and coworkers demonstrated the ability of DNA miniature devices to perform designated chemical reactions by hybridizing them on the rectangular origami-based platform (Figure 18.5A). The advantage of hybridizing the DNA nanostructure on the origami platform enables to program sequential execution of enzymatic reaction (Voigt *et al.*, 2010). Liu and coworkers had further advanced the nanodevices by localizing enzyme and cofactor in two arms of Tweezer like DNA origami structure (Figure 18.5B). They were able to regulate the enzymatic reaction by modulating the movement of tweezers by using DNA oligonucleotide as fuel or thrust for the tweezer's movement. The enzymatic reaction happens when both ends meet and interrupted when moved apart (Liu *et al.*, 2013, Castro *et al.*, 2015). Linko

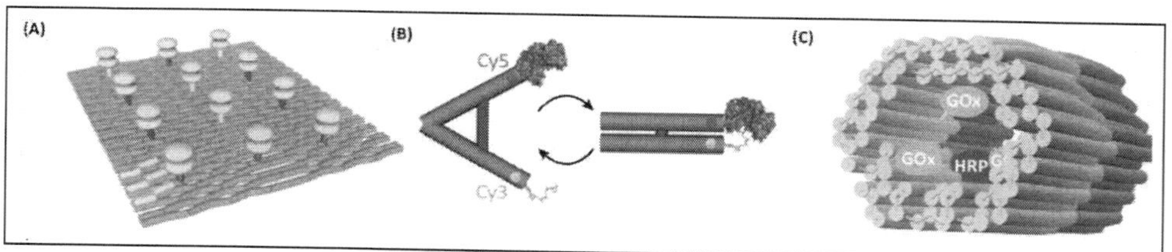

Figure 18.5. Smart DNA Origami Nanodevices and Molecular-Scale Platforms

(A) A DNA origami platform for controlling single-molecule chemical reactions. (B) A switchable tweezer-like enzymatic nanoreactor; Cy3/5, cyanine dyes 3/5. (C) A tubular DNA origami-based enzyme cascade nanoreactor for biosensing; GOx, glucose oxidase; HRP, horseradish peroxidase.

Courtesy: Linko *et al.*, 2015a.

and coworkers had combined the efficiency of many tubular DNA nanostructures building blocks into a single component to detect glucose in the molecular level. The DNA nanostructure houses horse radish protein (HRP) which would serve as visible indicator of glucose level. The DNA nanostructure in addition to serving as the backbone of the device, they are loaded with recognition sequences to remobilize to target organelle and perform its intended functions (Linko *et al.*, 2015a,b).

All of these features and potential of DNA origami *vis-a-vis* DNA nanostructure approach make it a versatile and promising candidate for loading and targeting wide range of genome engineering and live tracking in the plant and animal system.

Regulatory Uncertainty towards Genetically Engineered Crops

Genetic engineering had evolved and taken much iteration in terms of payload/gene and also in the mode of delivery, all in an effort to overcome the limitations of traditional breeding approaches like cost, time, the expertise required precise modification and unintended off-targets. Modern biotechnology and nanotechnology tools enable the rapid development of genetically modified crop variants with resistance to pests and diseases and tolerant to many abiotic stress. Since the discovery and introduction of GM crops, the global cultivation in terms of area has increased 110 fold to 191.7 million hectares. United States of America, Canada, and Brazil are the leading countries in cultivation of GM crops had formulated and put in place a stringent and foolproof regulatory mechanism for development and commercialization of new crop variants and genome engineering tools. US GMO regulatory mechanism adheres to product-based approach with specific focus on plant pests. Thereby any mechanism involving *Agrobacterium* should by default subjected to regulation whereas biolistic method based genome editing is exempted from subjecting to regulatory process (depending on the source of gene) (Kamle *et al.*, 2017; ISAAA, 2018). Whereas, European union GMO regulation is majorly focused on process-based approach wherein the modern-day DNA based nanostructures and DNA origami-based approach will be subjected to regulation on case by case. The advent of nuclease-based genome engineering (CRISPR, DNA nanostructures, *etc.*,) has set forth global regulatory body to redefine the regulatory mechanisms and key aspects that define GM and Non-GM organisms. Recently, the United States Department of Agriculture (USDA) stated that it has no course of direction to include genome-edited plants and genome engineering tools under the current US regulatory umbrella (Davison and Ammann, 2017; Raman, 2017, Cunningham *et al.*, 2018). The regulatory body of GM crops in India is becoming obsolete and has to keep abreast of the rapidly evolving technologies apart from regulating and deregulating crops in the current development pipeline.

Conclusion and Future Prospects

Genetic engineering of plants has significantly propelled the scientific breakthroughs and resulted in the development of new crop variants with improved growth characteristics, disease and pest resistance, abiotic stress tolerance, and enhanced nutritional value. In parallel, advances in highly targeted genome editing technologies have optimized the accuracy and precision of introducing new or altered genes without off-targets or unintended gene transfers. In spite of what is accomplished to date in traditional methods of plant genetic engineering and genome editing, many challenges lie ahead in developing groundbreaking technologies to meet the requirement of growing population and gaining consumer acceptance across the globe. *Agrobacterium* based versatile approach and preferred method for plant genetic transformation are becoming absolute owing to its limited host range, inability to control over integration location and time taken to develop into commercial lines. Biolistic particle delivery is effective, host-independent transformation methods but due to the cumbersome process of regenerating healthy plant tissue leading to transgenic plants hindered it from using it as high throughput genome editing technique. Exponential growth and knowledge on nano dimension of materials had given additional momentum in using NPs as a carrier for target DNA molecules. The potential for designing NP carriers with multi-functional cargo loading (DNA, RNA, proteins) and release capabilities to effectively circumvent the cell wall and membranes an exciting field of ongoing research. Further, the application of DNA is not restricted to genetics anymore. Means to develop synthetic DNA or DNA nanostructure has advanced the genome engineering and target/ payload delivery into plant systems. Further advancement in the development of DNA structures with desirable mechanical behavior will enable to fabricate custom-designed DNA nanostructures/origami with a multitude of functions like DNA recognition, site-specific DNA replacement, target environmental condition based delivery of payload, tagging of the fluorophore to monitor the movement of the target molecule in the host system, *etc.* There has been a shift in the applications of tools of genome engineering from using single DNA/gene-based modifications towards systematic and synchronized execution of multiple gene alteration/genome engineering, which has been highly attributed to the discoveries in developing DNA origami and DNA nanostructures. Further, the DNA nanostructures armed with various countermeasures can be preprogrammed and introduced into the host system and at the event of stress, the DNA vehicle/nanostructure will sense the environment, manipulate, assemble and transport molecular component to the target site and execute the appropriate gene modifications to evoke countermeasures in the plant in an independent manner. With prospects of fabricating such many DNA nanostructure entities into a large layered DNA origami system, functional scope and application of nanotechnology in genome engineering will continue to expand for years to come with exciting new innovations in the genome engineering process and product.

REFERENCES

1. Ali, M. M., Li, F., Zhang, Z., Zhang, K., Kang, D. K., Ankrum, J. A., and Zhao, W. (2014). Rolling circle amplification: a versatile tool for chemical biology, materials science and medicine. *Chemical Society Reviews*, *43*(10), 3324-3341.

2. Amani, A., Zare, N., Asadi, A., and ZAKARIA, R. A. (2018). Ultrasound-enhanced gene delivery to alfalfa cells by hPAMAM dendrimer nanoparticles. Turkish Journal of Biology, 42(1), 63-75.

3. Andersen, E. S., Dong, M., Nielsen, M. M., Jahn, K., Subramani, R., Mamdouh, W., and Pedersen, J. S. (2009). Self-assembly of a nanoscale DNA box with a controllable lid. *Nature, 459*(7243), 73.

4. Baltes, N. J., Gil-Humanes, J., and Voytas, D. F. (2017). Genome engineering and agriculture: opportunities and challenges. In *Progress in molecular biology and translational science* (Vol. 149, pp. 1-26). Academic Press.

5. Bao, W., Wan, Y., and Baluška, F. (2017). Nanosheets for delivery of biomolecules into plant cells. *Trends in plant science, 22*(6), 445-447.

6. Benson, E., Mohammed, A., Gardell, J., Masich, S., Czeizler, E., Orponen, P., and Högberg, B. (2015). DNA rendering of polyhedral meshes at the nanoscale. *Nature, 523* (7561), 441.

7. Briefs, I. S. A. A. A. (2017). Global Status of Commercialized Biotech/GM Crops in 2017: Biotech crop adoption surges as economic benefits accumulate in 22 years.

8. Burlaka, O. M., Pirko, Y. V., Yemets, A. I., and Blume, Y. B. (2015). Plant genetic transformation using carbon nanotubes for DNA delivery. Cytology and genetics, 49(6), 349-357.

9. Castro, C. E., Su, H. J., Marras, A. E., Zhou, L., and Johnson, J. (2015). Mechanical design of DNA nanostructures. *Nanoscale, 7*(14), 5913-5921.

10. Chang, F. P., Kuang, L. Y., Huang, C. A., Jane, W. N., Hung, Y., Yue-ie, C. H., and Mou, C. Y. (2013). A simple plant gene delivery system using mesoporous silica nanoparticles as carriers. *Journal of Materials Chemistry B, 1*(39), 5279-5287

11. Chen, K., and Gao, C. (2014). Targeted genome modification technologies and their applications in crop improvements. *Plant Cell Reports, 33*(4), 575-583.

12. Christian, M., Cermak, T., Doyle, E. L., Schmidt, C., Zhang, F., Hummel, A. and Voytas, D. F. (2010). Targeting DNA double-strand breaks with TAL effector nucleases. *Genetics, 186*(2), 757-761.

13. Cunningham, F. J., Goh, N. S., Demirer, G. S., Matos, J. L., and Landry, M. P. (2018). Nanoparticle-mediated delivery towards advancing plant genetic engineering. *Trends in biotechnology, 36*(9), 882-897.

14. Davison, J., and Ammann, K. (2017). New GMO regulations for old: determining a new future for EU crop biotechnology. *GM crops and food, 8*(1), 13-34.

15. Delebecque, C. J., Lindner, A. B., Silver, P. A., and Aldaye, F. A. (2011). Organization of intracellular reactions with rationally designed RNA assemblies. *Science, 333*(6041), 470-474.

16. Demirer, G., Zhang, H., Goh, N., Chang, R., and Landry, M. (2019). Nanotubes effectively deliver siRNA to intact plant cells and protect siRNA against nuclease degradation. Available at SSRN 3352632.

17. Dietz, H., Douglas, S. M., and Shih, W. M. (2009). Folding DNA into twisted and curved nanoscale shapes. *Science, 325*(5941), 725-730.

18. Doudna, J. A., and Charpentier, E. (2014). The new frontier of genome engineering with CRISPR-Cas9. *Science, 346*(6213), 1258096.

19. Douglas, S. M., Dietz, H., Liedl, T., Högberg, B., Graf, F., and Shih, W. M. (2009). Self-assembly of DNA into nanoscale three-dimensional shapes. *Nature, 459*(7245), 414.

20. Finiuk, N., Buziashvili, A., Burlaka, O., Zaichenko, A., Mitina, N., Miagkota, O. and Yemets, A. (2017). Investigation of novel oligoelectrolyte polymer carriers for their capacity of DNA delivery into plant cells. Plant Cell, Tissue and Organ Culture (PCTOC), 131(1), 27-39.

21. Firoozi, B., Nasser, Z., Sofalian, O., and Sheikhzade-Mosadegh, P. (2018). Enhancement of the transfection efficiency of DNA into Crocus sativus L. cells via PEI nanoparticles. Journal of integrative agriculture, 17(8), 1768-1778.

22. Gao, C., and Nielsen, K. K. (2013). Comparison between Agrobacterium-mediated and direct gene transfer using the gene gun. In *Biolistic DNA Delivery* (pp. 3-16). Humana Press, Totowa, NJ.

23. Hajiahmadi, Z., Shirzadian-Khorramabad, R., Kazemzad, M., and Sohani, M. M. (2019). Enhancement of tomato resistance to Tuta absoluta using a new efficient mesoporous silica nanoparticle-mediated plant transient gene expression approach. Scientia horticulturae, 243, 367-375.

24. Hamblin, G. D., Carneiro, K. M., Fakhoury, J. F., Bujold, K. E., and Sleiman, H. F. (2012). Rolling circle amplification-templated DNA nanotubes show increased stability and cell penetration ability. *Journal of the American Chemical Society, 134*(6), 2888-2891.

25. Han, D., Pal, S., Nangreave, J., Deng, Z., Liu, Y., and Yan, H. (2011). DNA origami with complex curvatures in three-dimensional space. *Science, 332*(6027), 342-346

26. Hong, F., Zhang, F., Liu, Y., and Yan, H. (2017). DNA origami: scaffolds for creating higher order structures. *Chemical reviews, 117*(20), 12584-12640.

27. Jiang, L., Ding, L., He, B., Shen, J., Xu, Z., Yin, M., and Zhang, X. (2014). Systemic gene silencing in plants triggered by fluorescent nanoparticle-delivered double-stranded RNA. Nanoscale, 6(17), 9965-9969.

28. Jiang, Q., Zhao, S., Liu, J., Song, L., Wang, Z. G., and Ding, B. (2019). Rationally designed DNA-based nanocarriers. *Advanced drug delivery reviews.*

29. Jinek, M., Chylinski, K., Fonfara, I., Hauer, M., Doudna, J. A., and Charpentier, E. (2012). A programmable dual-RNA–guided DNA endonuclease in adaptive bacterial immunity. *Science,337*(6096), 816-821.

30. Kallenbach, N. R., Ma, R. I., and Seeman, N. C. (1983). An immobile nucleic acid junction constructed from oligonucleotides. *Nature, 305*(5937), 829.

31. Kamle, M., Kumar, P., Patra, J. K., and Bajpai, V. K. (2017). Current perspectives on genetically modified crops and detection methods. *3 Biotech, 7*(3), 219.

32. Kim, Y. G., Cha, J., and Chandrasegaran, S. (1996). Hybrid restriction enzymes: zinc finger fusions to Fok I cleavage domain. *Proceedings of the National Academy of Sciences, 93*(3), 1156-1160.

33. Kumar, S., Misra, A., Verma, A. K., Roy, R., Tripathi, A., Ansari, K. M., and Dwivedi, P. D. (2011). *Bt* Brinjal in India: A long way to go. *GM crops, 2*(2), 92-98.

34. Kwak, S. Y., Lew, T. T. S., Sweeney, C. J., Koman, V. B., Wong, M. H., Bohmert-Tatarev, K., and Strano, M. S. (2019). Chloroplast-selective gene delivery and expression in planta using chitosan-complexed single-walled carbon nanotube carriers. Nature nanotechnology, 14(5), 447.

35. Lahiani, M. H., Chen, J., Irin, F., Puretzky, A. A., Green, M. J., and Khodakovskaya, M. V. (2015). Interaction of carbon nanohorns with plants: uptake and biological effects. *Carbon, 81*, 607-619.

36. Lenser, T., and Theiβen, G. (2013). Molecular mechanisms involved in convergent crop domestication. *Trends in plant science, 18*(12), 704-714.

37. Li, J., Fan, C., Pei, H., Shi, J., and Huang, Q. (2013). Smart drug delivery nanocarriers with self assembled DNA nanostructures. *Advanced materials, 25*(32), 4386-4396.

38. Linko, V., Eerikäinen, M., and Kostiainen, M. A. (2015a). A modular DNA origami-based enzyme cascade nanoreactor. *Chemical communications, 51*(25), 5351-5354.

39. Linko, V., Ora, A., and Kostiainen, M. A. (2015b). DNA nanostructures as smart drug-delivery vehicles and molecular devices. *Trends in biotechnology, 33*(10), 586-594.

40. Liu, M., Fu, J., Hejesen, C., Yang, Y., Woodbury, N. W., Gothelf, K., and Yan, H. (2013). A DNA tweezer-actuated enzyme nanoreactor. *Nature communications, 4*, 2127.

41. Martin Ortigosa, S., Valenstein, J. S., Lin, V. S. Y., Trewyn, B. G., and Wang, K. (2012). Gold functionalized mesoporous silica nanoparticle mediated protein and DNA codelivery to plant cells via the biolistic method. *Advanced Functional Materials, 22*(17), 3576-3582

42. Martin-Ortigosa, S., Peterson, D. J., Valenstein, J. S., Lin, V. S. Y., Trewyn, B. G., Lyznik, L. A., and Wang, K. (2014). Mesoporous silica nanoparticle-mediated intracellular cre protein delivery for maize genome editing via loxP site excision. Plant physiology, 164(2), 537-547.

43. Mitter, N., Worrall, E. A., Robinson, K. E., Li, P., Jain, R. G., Taochy, C. and Xu, Z. P. (2017). Clay nanosheets for topical delivery of RNAi for sustained protection against plant viruses. *Nature plants, 3*(2), 16207.

44. Mohammadinejad, R., Karimi, S., Iravani, S., and Varma, R. S. (2016). Plant-derived nanostructures: types and applications. *Green Chemistry, 18*(1), 20-52.

45. Mout, R., Ray, M., Yesilbag Tonga, G., Lee, Y. W., Tay, T., Sasaki, K., and Rotello, V. M. (2017). Direct cytosolic delivery of CRISPR/Cas9-ribonucleoprotein for efficient gene editing. *ACS nano, 11*(3), 2452-2458.

46. Okholm, A. H., and Kjems, J. (2016). DNA nanovehicles and the biological barriers. *Advanced drug delivery reviews, 106*, 183-191.

47. Ouyang, X., Chao, J., Su, S., and Fan, C. (2017). DNA Nanostructure as Smart Carriers for Drug Delivery. In *3D DNA Nanostructure* (pp. 121-132). Humana Press, New York, NY.

48. Ouyang, X., Li, J., Liu, H., Zhao, B., Yan, J., Ma, Y., and Fan, C. (2013). Rolling circle amplification based DNA origami nanostructrures for intracellular delivery of immunostimulatory drugs. *Small, 9*(18), 3082-3087.

49. Raj, D. B. T. G., and Khan, N. A. (2016). Designer nanoparticle: nanobiotechnology tool for cell biology. *Nano convergence, 3*(1), 22.

50. Raman, R. (2017). The impact of Genetically Modified (GM) crops in modern agriculture: A review. *GM crops and food, 8*(4), 195-208.

51. Roh, Y. H., Lee, J. B., Kiatwuthinon, P., Hartman, M. R., Cha, J. J., Um, S. H., and Luo, D. (2011). DNAsomes: multifunctional DNA based nanocarriers. *Small, 7*(1), 74-78.

52. Rothemund, P. W. (2006). Folding DNA to create nanoscale shapes and patterns. *Nature, 440*(7082), 297.

53. Saveland, R. N. (1992). National Geographic Society," Exploring Your World: The Adventure of Geography"(Book Review). *Journal of Geography, 91*(6), 282

54. Schubert, D., and Williams, D. (2006). 'Cisgenic'as a product designation. *Nature biotechnology, 24*(11), 1327.

55. Seeman, N. C. (1982). Nucleic acid junctions and lattices. *Journal of theoretical biology, 99*(2), 237-247.

56. Seeman, N. C. (1998). DNA nanotechnology: novel DNA constructions. *Annual review of biophysics and biomolecular structure, 27*(1), 225-248.

57. Seeman, N. C., and Sleiman, H. F. (2017). DNA nanotechnology. *Nature Reviews Materials, 3*(1), 1-23.

58. Shukla, M., Al-Busaidi, K. T., Trivedi, M., and Tiwari, R. K. (2018). Status of research, regulations and challenges for genetically modified crops in India. *GM crops and food, 9*(4), 173-188.

59. Simmel, F. C. (2012). DNA-based assembly lines and nanofactories. *Current opinion in biotechnology, 23*(4), 516-521.

60. Smith, D., Schüller, V., Engst, C., Rädler, J., and Liedl, T. (2013). Nucleic acid nanostructures for biomedical applications. *Nanomedicine, 8*(1), 105-121.

61. Torney, F., Trewyn, B. G., Lin, V. S. Y., and Wang, K. (2007). Mesoporous silica nanoparticles deliver DNA and chemicals into plants. *Nature nanotechnology, 2*(5), 295.

62. Voigt, N. V., Tørring, T., Rotaru, A., Jacobsen, M. F., Ravnsbæk, J. B., Subramani, R., and Gothelf, K. V. (2010). Single-molecule chemical reactions on DNA origami. *Nature nanotechnology, 5*(3), 200.

63. Wang, J. W., Grandio, E. G., Newkirk, G. M., Demirer, G. S., Butrus, S., Giraldo, J. P., and Landry, M. P. (2019). Nanoparticle-Mediated Genetic Engineering of Plants. *Molecular plant, 12*(8), 1037-1040.

64. Wilner, O. I., Weizmann, Y., Gill, R., Lioubashevski, O., Freeman, R., and Willner, I. (2009). Enzyme cascades activated on topologically programmed DNA scaffolds. *Nature nanotechnology, 4*(4), 249.

65. Yang, Y. R., Liu, Y., and Yan, H. (2015). DNA nanostructures as programmable biomolecular scaffolds. *Bioconjugate chemistry, 26*(8), 1381-1395.

66. Zhang, H., Demirer, G. S., Zhang, H., Ye, T., Goh, N. S., Aditham, A. J., and Landry, M. P. (2019). DNA nanostructures coordinate gene silencing in mature plants. *Proceedings of the National Academy of Sciences, 116*(15), 7543-7548.

67. Zhang, X., Zhang, J., and Zhu, K. Y. (2010). Chitosan/double stranded RNA nanoparticle mediated RNA interference to silence chitin synthase genes through larval feeding in the African malaria mosquito (*Anopheles gambiae*). *Insect molecular biology, 19*(5), 683-693.

68. Zhang, K., Deng, R., Sun, Y., Zhang, L., and Li, J. (2017). Reversible control of cell membrane receptor function using DNA nano-spring multivalent ligands. *Chemical science, 8*(10), 7098-7105.

69. Zhao, X., Meng, Z., Wang, Y., Chen, W., Sun, C., Cui, B., and Luo, D. (2017). Pollen magnetofection for genetic modification with magnetic nanoparticles as gene carriers. *Nature plants*, 3(12), 956.

Chapter 19

Nanotechnology in Chemical and Optical Sensors

S. Thirumalairajan[1] and K. Girija[2]*

[1]Ramalingaswami Re-entry Faculty Fellow, Department of Nano Science and Technology, Tamil Nadu Agricultural University, Coimbatore – 641 003, Tamil Nadu
**e-mail: sthirumalairajan@gmail.com*
[2]Assistant Professor, Department of Physics, Dr. N.G.P. Arts and Science College, Coimbatore – 641 048, Tamil Nadu

CHEMICAL SENSOR

Introduction

Chemical (gas) and optical sensors play an important role in many aspects of science, technology, industry and in daily life. Accurate, sensitive and reliable identification of various characteristic gases or specific components in low concentration is mandatory in some areas such as industrial, agricultural plants, automotive technologies, food sciences, environmental monitoring, or air quality security. Sensors convert measurement of physical phenomena into an electrical signal. A device which converts one form of energy to another, *i.e.* as when input is a physical quantity and output is an electrical signal is called Sensor, other side when input is electrical signal and output a physical quantity is called Actuator. In the last few years, chemical and optical sensors have received a close attention of the sensor community throughout the world because of some unusually superior sensing performance. However, research and development of chemical and optical devices continue to be faced with numerous challenges in terms of sensitivity, selectivity, promptness of response, robustness, and many other aspects.

Chemical sensors are essential need for industrial, health safety, environmental monitoring and medical applications. Semiconducting gas sensors act as gas sensitive resistor that is simple and cost effective for atmosphere monitoring (Franke, *et al.*, 2006). The use of new sensor materials is underway along with the efforts to enhance the performance of traditional devices, such as resistive metal oxide sensors, through nano-engineering. Metal oxide gas sensors have been used for several decades for low-cost detection of combustible and toxic gases. The sensitivity, selectivity, and stability are the important parameters in sensing applications. Nanomaterials provide the opportunity to dramatically increase the response of these materials, as their performance is directly related to exposed surface to volume ratio, different shape and controlled size (Niu, X., *et al.*, 2004).

Chemical Sensor – Definition

A gas sensor is a transducer that detects gas molecules and produces an electronic signal with a magnitude proportional to the concentration of the gas.

Classification of Chemical Sensors

Solid Electrolyte Sensors

Solid electrolytes are the materials that allow the conduction of ions but not the conduction of electrons. Similar to liquid electrolytes, they support the function of electrochemical cells, in which chemical reactions are allowed to proceed for completion, only if separate paths are provided for the flow of ions (through the electrolyte) and electrons (through an electronic conductor).

Catalytic Sensors

Potentially explosive mixtures of methane, or other flammable gases, with air can be monitored by means of a catalytically active solid state sensor. The device, often referred to as the 'pellistor', is essentially a catalytic micro calorimeter. It consists of a catalytic surface constructed around a temperature sensor, and a heater which maintains the catalyst at a sufficiently high temperature to ensure rapid combustion of any flammable gas molecules present. However, these sensors suffered from several problems, including loss of sensitivity with time due to poisoning and burning out when exposed to high gas concentrations (Na, N, Zhang *et al.*, 2006)

Semiconducting Oxide Gas Sensors

A third type of gas sensor, used widely and effectively is a gas sensitive resistor. A sensing element normally comprising a semiconducting material, having a high surface to volume ratio, is deployed on a heated insulating substrate between two metallic electrodes. The reaction of gas molecules can take place at the semiconductor surface to change the density of charge carriers available on the surface. Hence, the conductance of the device changes progressively with changing atmospheric composition

Sensor Mechanism

Gas-sensing mechanism combined with the gas sensing behaviour of the perovskite oxide can be demonstrated in detail from Figure 19.1. $CuAlO_2$ is a *p*-type

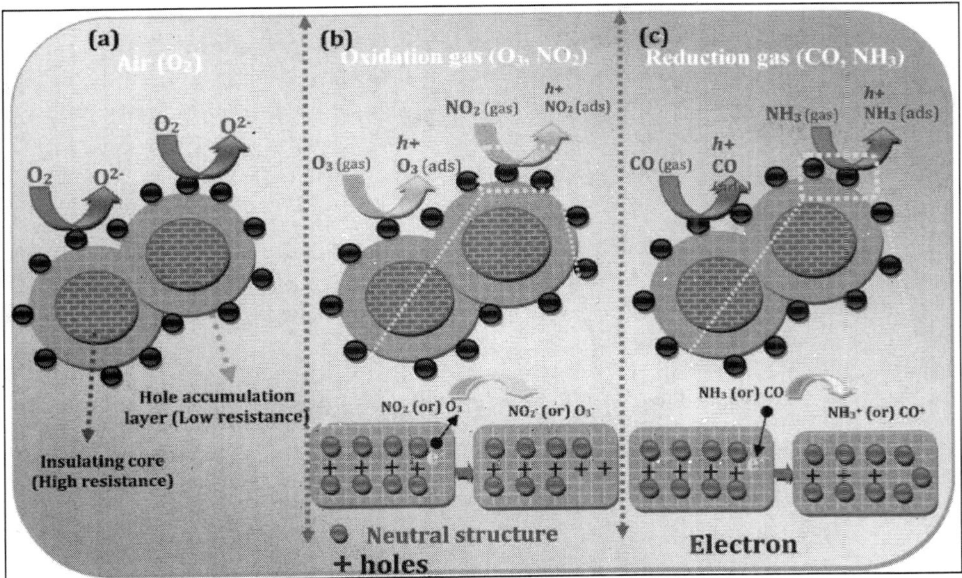

Figure 19.1. Schematic Diagrams for Oxidation (O_3 and NO_2) and Reducing (CO and NH_3) Type Gas Sensing Operating Mechanism on *p*-type $CuAlO_2$ Microspheres Constituted of Nanoparticles (S. Thirumalairajan *et al.*, 2015).

semiconductor, the acceptor level lies near the valance band and at operating temperature all the acceptor levels are ionized (filled) leaving holes in the valance band, the $CuAlO_2$ ion vacancies are the acceptors. When operated at a moderate temperature, the oxygen molecules in air can trap electrons from the $CuAlO_2$ conduction band and form chemisorbed oxygen species (O_2^-, O^-, and O^{2-}) on the surface of the sensing layer, which has been represented in the following equation. *i.e* the adsorption of negatively charged oxygen can generate the holes.

$$O_2\,(gas) \Longleftrightarrow O_2(ads) + h^+ \Longleftrightarrow O_2^-\,(ads) + h^+ \Longleftrightarrow 2O^-\,(ads) + h^+ \Longleftrightarrow O^{2-}(ads) + h^+ \tag{1}$$

When oxidation gas NO_2 are introduced to the surface of $CuAlO_2$ sample as shown in Figure 19.1b, the NO_2 molecules adsorb on the $CuAlO_2$ surface acting as acceptors. This surface trapping of lone-pair electrons causes band bending which results in the formation of free holes and an increase in hole concentration near the interface forming the hole accumulated layer at the $CuAlO_2$ surface which decrease the resistance of $CuAlO_2$ sample. The following reaction (equation 2) takes place on the surface of $CuAlO_2$ sensing film for oxidation gases.

$$CuAlO_2 + NO_2\,(gas) \Longleftrightarrow CuAlO_2^+ - NO_2^-\,(ads) \Longleftrightarrow CuAlO_2 - NO_2^-(ads) + h \tag{2}$$

Further, the accepted electrons in the reactions are annihilated with neutral structure and same reaction has been represented in the schematic diagram shown in Figure 19.1b (bottom).

Then, for *p*-type semiconductors, oxygen adsorption produces an increase in the number of holes in the valence hand, which increases their conductivity (Franke, M. E., *et al.*, 2006). When the sensor was exposed to the reducing gases such as CO at 200 °C, the interaction between the reducing gas and the adsorbed oxygen ions on the $CuAlO_2$ sensors surface can be explained as:

$$CuAlO_2 + CO\,(gas) \Longleftrightarrow CuAlO_2^{+} - CO_3^{+}\,(ads) \Longleftrightarrow CuAlO_2\text{-}CO_3^{+}\,(ads) + h \tag{3}$$

The released electrons in the reactions are annihilated with holes (equation 5) and have been represented in the schematic diagram as shown in Figure 19.1c (bottom). Electrons are injected into the conduction band and then coupled with holes in the valence band, producing a decrease in the conductivity or increased resistance. This change in the conductivity indicates that the examined $CuAlO_2$ microspheres sample posses a typical *p*-type semiconducting feature.

Working Principle of Semiconductor Metal Oxide Based Sensor

The process involved in gas sensing has the following steps

- ✰ Diffusion of gas/vapour (analyte) to the active region
- ✰ Surface interaction between analyte and sensing materials
- ✰ Desorption of analyte from active region
- ✰ Diffusion of analyte away from the active site
- ✰ Adsorption of gas/vapour onto the active site on the surface of sensing material.

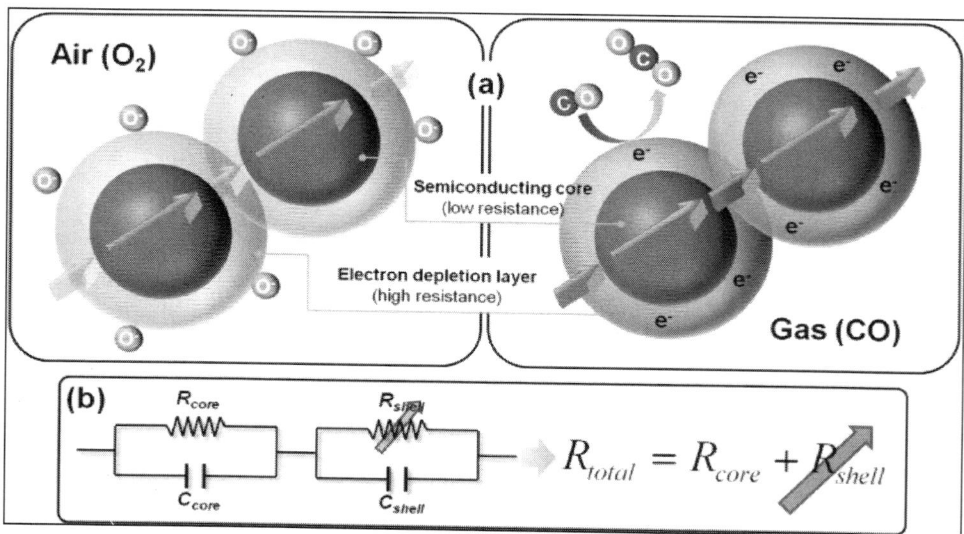

Figure 19.2. Working Principle and Equivalent Circuit of *n*-type Oxide Semiconductors.
Courtesy: H.J Kim *et al.*, 2014.

The working principle for gas sensing in metal oxide materials is based on reactions that occur at the sensor surface, resulting in a change in the concentration of adsorbed oxygen. Oxygen ions adsorbed in the material's surface, removing electrons from the bulk creates a potential barrier that limits electron movement and conductivity. The height of the barrier is reduced, when reactive gases combine with this oxygen, thereby the conductivity increase as shown in Figure 19.2. This change in conductivity is directly related to the amount of a specific gas present in the environment, resulting in a quantitative determination of the presence of gas and its concentration. The diffusion of analyte involves the physical movement of the analyte onto active sites of the metal oxide sensing materials. Diffusion depends on the ambient temperature of the measured atmosphere, which is generally assumed to be at room temperature. The second step in the process namely, the interaction between the analyte and the metal, determine the functioning of the gas sensor. There were attempts to propose the reaction mechanism on the surface. From the well-known Langmuir-Hinshelwood mechanism (Gong, H., *et al.*, 2006), the reactant species CO and oxygen would be adsorbed on the surface to facilitate the reaction. Usually, O_2 adsorbs dissociativly on the surface as atomic oxygen species and these atomic oxygen species were more reactive than molecular species. It was well known that CO is a reducing agent and reacts with adsorbed oxygen to form CO_2. The reaction rate would be controlled by the rate of CO_2 desorption from the surface. The simplified reaction scheme could be written as follows

$$S_{(s)} + ½\,O_{2\,(g)} \longrightarrow S \ldots.1/2\,O_{2\,(ad)} \quad (4)$$

$$S \ldots.\, O_{2\,(ad)} \longrightarrow S \ldots.\, O^{-}_{(ad)} \quad (5)$$

$$S_{(s')} + CO_{(g)} \longrightarrow S \ldots.CO^{+}_{(ad)} \quad (6)$$

$$S\ldots O^{-}_{(ad)} + S\ldots CO^{+}_{(ad)} \longrightarrow S \ldots.CO_{2\,(ad)} + S_{(s)} \quad (7)$$

$$S \ldots.CO_{2\,(ad)} \longrightarrow S\,(_{s'}) + CO_{2\,(g)} \quad (8)$$

Here, (s) and (s2) represent active sites on the surface for oxygen and CO adsorption, respectively; (g) and (ad) are gas and adsorbed from species. The active sites (s) and (s2) may or may not be the same, depending on the type of materials. When the sensor response measurement was carried out at higher temperatures, the adsorbed species of oxygen and CO reacts on the surface which requires low activation energy. This is because the adsorption was exothermic process and the reaction activation energy becomes lowered. The reaction was possible at lower temperatures. Another mechanism is also possible as CO being a reducing gas have the tendency to take lattice oxygen from the surface. First, CO had to adsorb on the surface and then it will react with the lattice oxygen to form CO_2. The reaction scheme could be written as follows:

$$M\text{-}O_{(s)} + CO_{(g)} \longrightarrow M\text{-}O \ldots.\, CO^{+}_{(ad)} \quad (9)$$

$$M\text{-}O\ldots CO^{+}_{(ad)} \longrightarrow M^{+}\ldots\ldots O\,CO_{(ad)} \quad (10)$$

$$M^{+}\ldots\ldots O\,CO_{(ad)} \longrightarrow M - [\,]^{+} + CO_2\,(g) \quad (11)$$

$$M - [\,]^{+} + ½\,O_2(g) \longrightarrow M - [\,]^{+}\ldots O^{-}_{(ad)} \quad (12)$$

$$M - [\,]^{+}\ldots O^{-}_{(ad)} \longrightarrow M\text{-}O_{(s)} \quad (13)$$

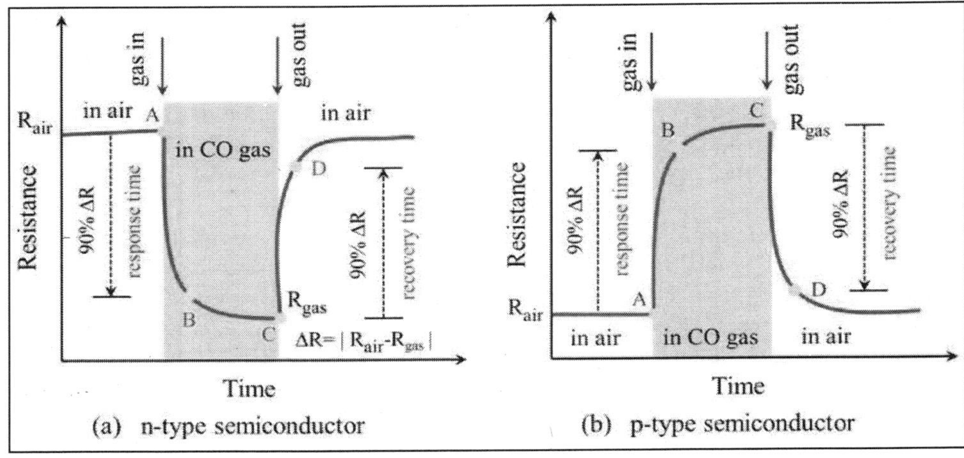

Figure 19.3. Sensing Response of n-type and p-type in Reducing Gas Environment (CO).
Courtesy: Santhaveesuk, T., *et al.*, 2015.

Here, (s) represents an active site on the metal oxide surface, M - [ ]+ is a reduced metal oxide site on the surface, (g) and (ad) are gas and adsorbed from species. Eq. (9) represents the adsorption of CO on the active site of the metal oxide and it would react with surface lattice oxygen (Eq. (10)). Adsorbed CO_2 would get desorbed as shown in Eq. (11) and reduced active site will get regenerated by taking oxygen from air as shown in Eqs. (12) and (13) to continue the adsorption and desorption cycle. The rate of the reaction was controlled by active site reduction process (Yu, J.H., *et al.*, 2002)

After the surface interaction and change in electrical resistance, the adsorbed gas can remain adsorbed on the sensor indefinitely (at low temperatures) or react with the sensing material (sensor poisoning) or desorption and diffuse back into the surrounding atmosphere. Desorption refers to disengagement of a molecule or an atom from a surface. Like adsorption, desorption is also a thermally activated process. When the gas desorbs and diffuses away from the sensor surface, the sensor is ready for the next measurement. The time between one measurement and the other, at which the sensor remains unusable is known as the dead time. Based on the response of the sensing element to the incoming gas, they can be classified as and shown in Figure 19.3.

p-Type Gas Sensors

- ☆ Surface resistance decrease on exposure to oxidizing gases
- ☆ Surface resistance increases on exposure to reducing gases

n-Type Gas Sensors

- ☆ Surface resistance increases on exposure to oxidizing gases
- ☆ Surface resistance decrease on exposure to reducing gases

Sensor Parameters

Important sensor parameters are namely, sensitivity, selectivity, stability, response time, recovery time and life time, which determine the efficiency of the gas sensors.

(a) Sensitivity

Sensitivity refers to the variations in the physical or chemical properties of the sensor on exposure to a gas. It can be defined as the minimum level of concentration that can be detected by the sensor and also the smaller increment in the gas concentration that the sensor can distinguish. In case of metal oxide based gas sensors or chemiresistors, it is the resistance which changes on exposure to gas. In such cases, the sensitivity is mathematically, defined by the following equation

$$S = R_a/R_g$$

Here, R_a is the resistance of the sensing material in the reference or air atmosphere

R_g is the resistance of the material on exposure to gas

Sensitivity is generally enhanced either by doping, which modified the carrier concentration and mobility or by microstructure changes such as reduction of the oxide particle size to the nanometre scale.

(b) Selectivity

Selectivity is defined as the ability of the sensor to discriminate various gases. In other words, selectivity is the ability of a sensor to respond only for a particular gas for which it is designed to sense. In other words, selectivity represents the specificity of the sensor towards a particular gas. But inherently, the perovskite oxide gas sensors show broad sensitivity and lack selectivity. An improvement in the selectivity of the perovskite oxide based gas sensor is achieved by the following steps,

- ☆ Different particle size and porosity
- ☆ By choosing appropriate working temperature
- ☆ By the application of doping or catalysts
- ☆ Different ways of formation of the sensing material

(c) Response Time

It is the time taken by the sensor to reach 90 per cent of the steady state conductivity reached by the sensor when exposed to the full scale concentration. Usually in perovskite/metal oxide gas sensors, though the resistance value chages immediately on exposure to gas, it takes a quiet long time to reach the steady state.

(d) Recovery Time

The response time, the recovery time of a gas sensor is also a very important parameter. It is the time taken by the sensor to regain 10 per cent of its resistance, on the removal of sensor from the gas environment to clean air/atmosphere. The steady state value of resistance of the sensor in air is sometimes referred to as baseline. But on repeated usage and due course, the baseline drift is quiet natural in perovskite/metal oxide based sensors. It is required for efficient sensor that this recovery time should also be as small as possible.

(e) Lifetime and Stability

The lifetime and stability of a gas sensor are very important parameters. This is the ability of a sensor to maintain its properties even after several times of usage. Only when a sensor has a good life time and stability, it can be used in making sensor arrays especially in real time sensing measurements.

(f) Operating Temperature

The operating temperature should be specified for any sensor. It is a function of the semiconductor oxide composition and of the gas which is to be detected. It is important to define this parameter because with respect to single gas, the sensor reaches the maximum of its sensitivity at a precisely fixed temperature.

Designing of a Gas Sensor Setup and Chemical Sensor Measurement

Perovskite oxide powders (10 mg) were dispersed in 1mL isopropyl alcohol by an ultrasonic cleaner for 30 minutes and the suspension was dropped onto a appropriate substrate containing 100 nm thick Pt electrodes separated by a distance of 50 µm. The substrates were then heated to 90 °C for 10 min to evaporate the solvent, followed by treatment at 300 °C for 60 min in an electric furnace in air, to stabilize the sample before the gas sensing measurements were performed as shown in Figure 19.4. Sensing reaction (S) defined as $S = R_{air}/R_{gas}$, is the electric resistance of the sensor exposed to the resistance in the air (R_{air}) and to the target oxidizing gas (R_{gas}). The response time of the sensor was defined as the time required for a change in the electrical resistance to reach 92 per cent of the initial value when exposed to ozone gas. Similarly, the time required for the electrical resistance of the sensor to reach 90 per cent of the initial value after the ozone gas has been turned off is the recovery time. During the measurements the humidity was under controlled condition within the range 50–60 per cent RH. The detailed report of sensitivity, selectivity stability, response and recovery time have been reported in our previous publication (S. Thirumalairajan, *et al.*, 2014).

Application of Chemical Sensor

The following is a list of classification of sensors that are commonly used in various applications and the same is represented in the image. All these sensors are used for measuring one of the physical properties like Temperature, Resistance, Capacitance, Conduction, Heat Transfer *etc.* Various types of sensor are listed below,

- ☆ Temperature Sensor
- ☆ Proximity Sensor
- ☆ Accelerometer
- ☆ Infrared Sensor

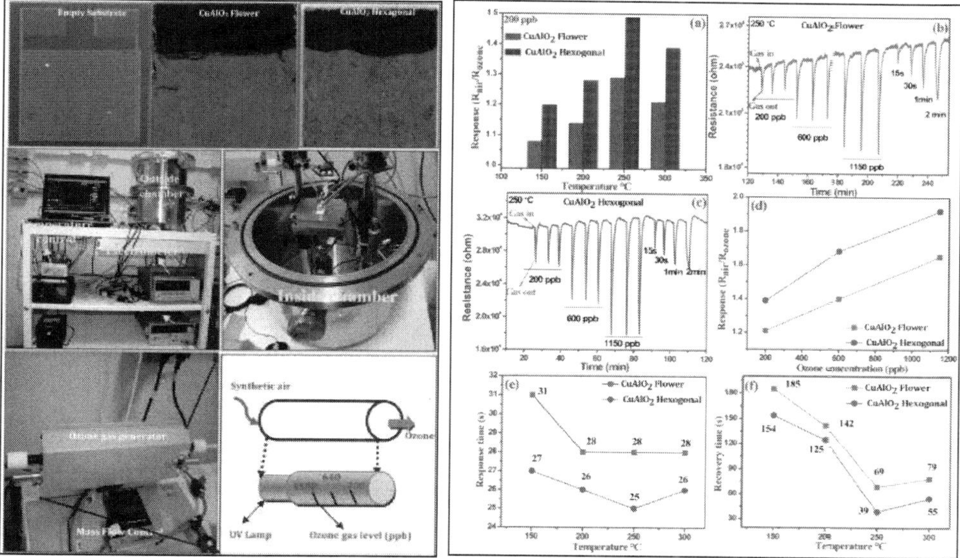

Figure 19.4. (a) Photograph of sensing films and chemical sensor measurement setup and (b) analysed chemical sensor parameter.
Courtesy: S. Thirumalairajan *et al.*, 2014.

- ✰ Pressure Sensor
- ✰ Light Sensor
- ✰ Ultrasonic Sensor
- ✰ Smoke, Gas Sensor
- ✰ Touch Sensor
- ✰ Color Sensor
- ✰ Humidity Sensor

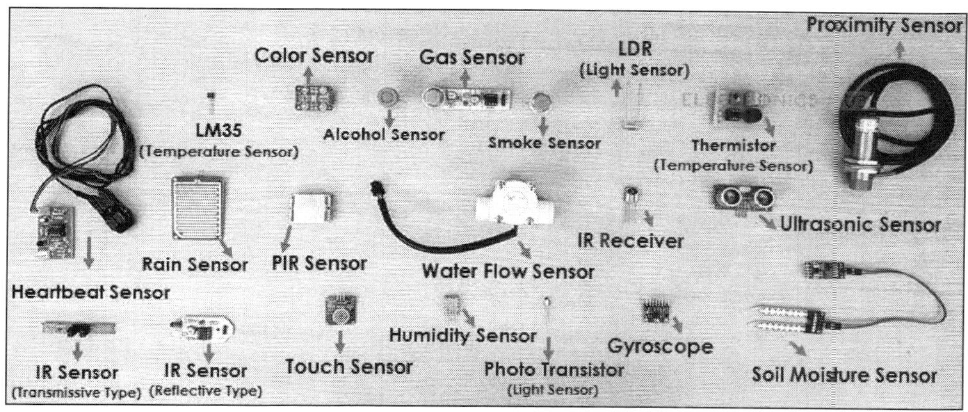

Figure 19.5

OPTICAL SENSOR

Introduction

Optical sensors, represent a group of chemical sensors in which electromagnetic (EM) radiation is used to generate the analytical signal in a transduction element. The interaction of this radiation with the sample is evaluated from the change of a particular optical parameter and is related to the concentration of the analyte (Krioukov, *et al.*, 2002). Typically, an optical chemical sensor consists of a chemical recognition phase (sensing element or receptor) coupled with a transduction element. The receptor identifies a parameter, *e.g.*, the concentration of a given compound, pH, *etc.*, and provides an optical signal proportional to the magnitude of this parameter. The function of the receptor is fulfilled in many cases by a thin layer that is able to interact with the analyte molecule, catalyse a reaction selectively, or participate in a chemical equilibrium together with the analyte. The transducer translates the optical signal produced by the receptor into a measurable signal that is suitable for processing by amplification, filtering, recording, display, *etc.*

Optical sensors are used in numerous researches and commercial applications such as for quality and process control, medical technologies, metrology, imaging, and remote sensing. There are many types of optical sensors; many based on the use of lasers, imaging systems, and/or fibers. In addition, novel sensor methods that enable more advanced sensing are continuously being developed by using novel materials, such as meta materials, micro and nanostructured materials or by employing new frequency bands as for example THz radiation (Jin, Z., *et al.*, 2001). Other features of optical sensors include the distinction of whether it is placed internally or externally in a device. The comparison of the two types is given in Table 19.1.

Table 19.1. Comparison of Intrinsic and Extrinsic Optical Sensors

Intrinsic	*Extrinsic*
Applications- rotation, acceleration	Applications- temperature, pressure
More sensitive	Less sensitive
Tougher to multiplex	Easily multiplexed
More elaborate signal demodulation	Ingress/egress connection problems
More expensive	Less expensive
Reduces connection problems	Easier to use

Optical Sensor Working Principle

In general, there are two components in optical sensing *viz.*, transmitter (*i.e.* optical source) and receiver (optical detector). The concept is depicted in the figure with the example of optical fiber. As shown in Figure 19.6, light beam changes its parameters when any object comes in between transmitter and receiver. There are five useful parameters of light which are being measured in optical sensing *viz.* intensity, phase, wavelength, polarization, spectral distribution.

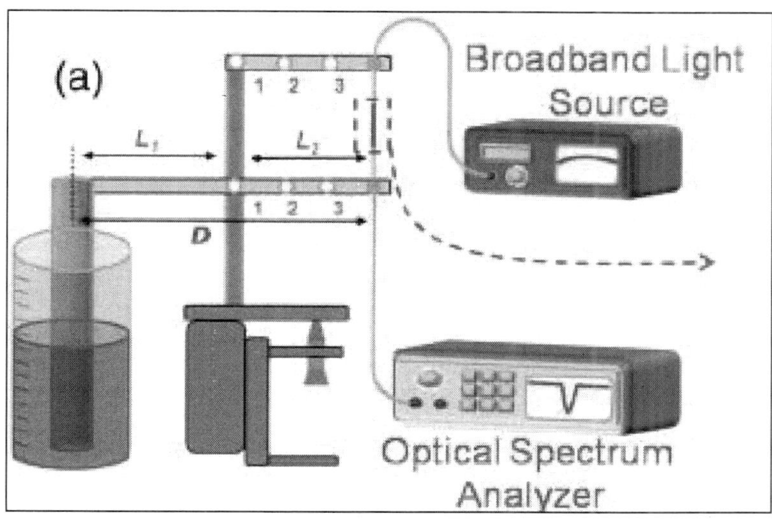

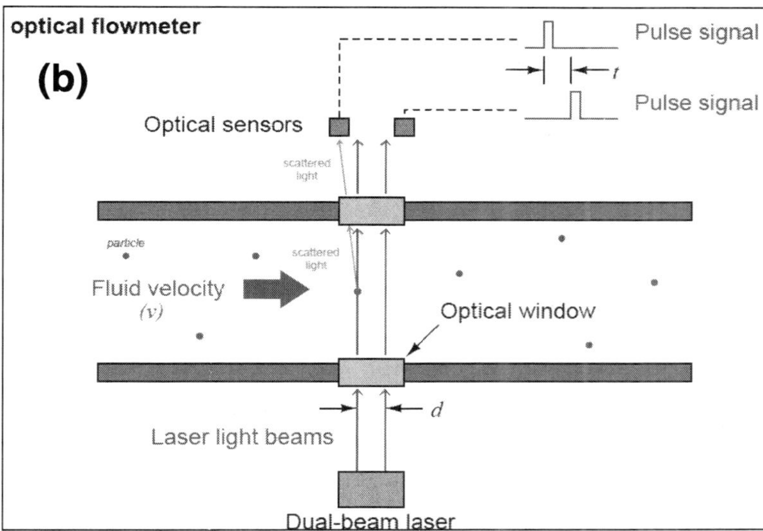

Figure 19.6. Schematics of Fiber Optic Liquid Level Sensor System Based on a Combination of Optical Interferometry and (b) Optical Sensor Working Principle.
Courtesy: Zhang, X.B., *et al.*, 2014.

Means of Sensing

In this category, sensors are generally based either on measuring an intensity change in one or more light beams or on looking at phase changes in the light beam by causing them to interact or interfere with one another. Thus, sensors in this category are termed either intensity sensors or interferometric sensors. Techniques used in the case of intensity sensors include light scattering (both Rayleigh and Raman), spectral transmission changes (*i.e.*, simple attenuation of transmitted light due to absorption), microbending or radiative losses, reflectance changes, and changes in the modal properties of the fibers.

Extent of Sensing

This category is based on whether sensors operate only at a single point or over a distribution of points. Thus, sensors in this category are termed either point sensors or distributed sensors. In the case of a point sensor, the transducer may be at the end of a fiber, the sole purpose of which is to bring a light beam to and from the transducer. Examples of this sensor type are interferometers bonded to the ends of fibers to measure temperature and pressure. In the case of a distributed sensor, as the name implies, sensing is performed all along the fiber length. Examples of this sensor type are fiber Bragg gratings distributed along a fiber length to measure strain or temperature (Lutti, J., *et al.*, 2008).

Role of Optical Fiber

Further distinction is often made in the case of fiber sensors as to whether measurands act externally or internally to the fiber. Where the transducers are external to the fiber and the fiber merely registers and transmits the sensed quantity, the sensors are termed *extrinsic sensors*. Where the sensors are embedded in or are part of the fiber and for this type there is often some modification to the fiber itself, the sensors are termed *internal* or *intrinsic sensors* (Xu, Z., Chen *et al.*, 2010). Examples of extrinsic sensors are moving gratings to sense strain, fiber-to-fiber couplers to sense displacement, and absorption cells to sense chemistry effects. Examples of intrinsic sensors are those that use microbending losses in the fiber to sense strain, modified fiber claddings to make spectroscopic measurements, and counter-propagating beams within a fiber coil to measure rotation. A wide variety of physical phenomena are used to sense the quantity to be measured.

Types of Optical Sensors

There are different kinds of optical sensors, the most common types which we have been using in our real world applications are given below.

- ✰ Photoconductive devices used to measure the resistance by converting a change of incident light into a change of resistance.
- ✰ The photovoltaic cell (solar cell) converts an amount of incident light into an output voltage.
- ✰ The Photodiodes convert an amount of incident light into an output current.

Phototransistors are a type of bipolar transistor where the base-collector junction is exposed to light. This results in the same behavior of a photodiode, but with an internal gain. The operating principle is the transmitting and receiving of light in an optical sensor, the object to be detected reflects or interrupts a **light beam sent out by an emitting diode**. Depending on the type of device, the interruption or reflection of the light beam is evaluated. This makes it possible to detect objects independently of the material they are constructed from (wood, metal, plastic or other). Special devices even allow for a detection of transparent objects or those with different colors or variations in contrast. Different types of optical sensors are explained below and is shown in Figure 19.7.

Through-Beam Sensors

The system consists of two separate components the transmitter and the receiver are placed opposite to each other. The transmitter projects a light beam onto the receiver. An interruption of the light beam is interpreted as a switch signal by the receiver. It is irrelevant where the interruption occurs.

Retro-Reflective Sensors

Transmitter and receiver are both in the same house, through a reflector the emitted light beam is directed back to the receiver. An interruption of the light beam initiates a switching operation, where the interruption occurs is of no importance.

Diffuse Reflection Sensors

Both transmitter and receiver are in one housing. The transmitted light is reflected by the object to be detected.

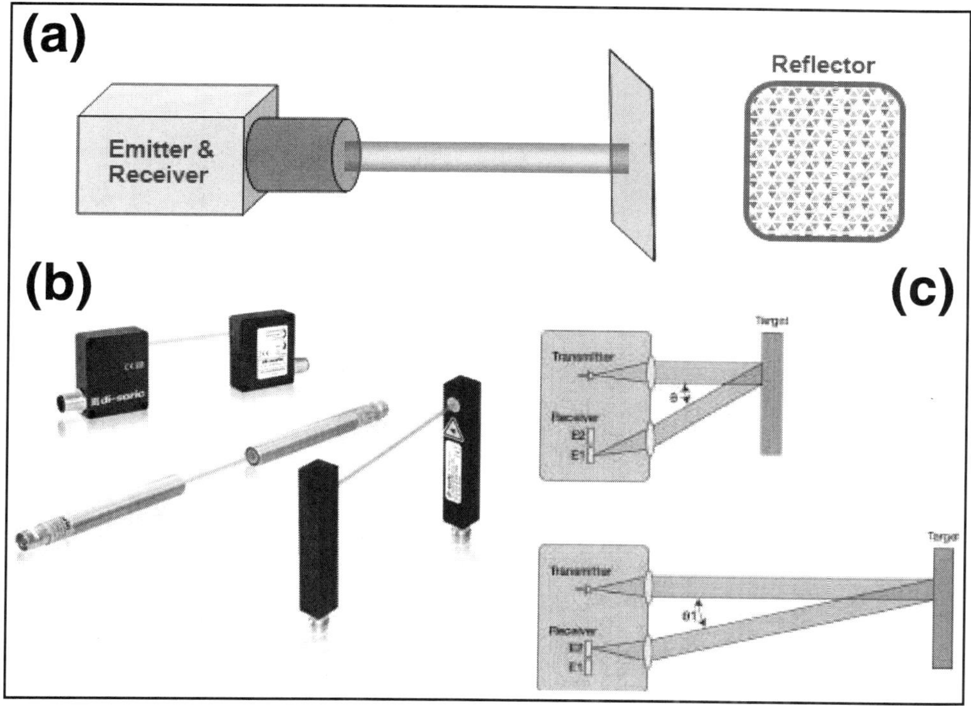

Figure 19.7. Photograph of (a) Through-Beam Sensors (b) Retro-Reflective Sensors and (c) Diffuse Reflection Sensors.
Courtesy: Senasys photosensor, 2018.

Different Light Sources for Optical Sensors

There are many types of light sources. The sun and light from burning torch flames were the first light sources used to study optics. As a matter of fact, light coming from certain (exited) matter (*e.g.*, iodine, chlorine, and mercury ions) provides the reference points in the optical spectrum (Feng, S., *et al.*, 2012). One of the key components in optical communication is the monochromatic light source. In optical communications, light sources must be monochromatic, compact, and long lasting. There are two different types of light source.

LED (Light Emitting Diode)

During the recombination process of electrons with holes at the junctions of *n*-doped and *p*-doped semiconductors, energy is released in the form of light. The excitation takes place by applying an external voltage and the recombination may be taking place, or it may be stimulated as another photon. This facilitates coupling the LED light with an optical device as shown in Figure 19.8a.

LASER (Light Amplification by Stimulated Emission Radiation)

A laser is created, when the electrons in the atoms in special glasses, crystals, or gasses absorb energy from an electrical current and gets excited. The excited electrons move from a lower-energy orbit to a higher-energy orbit around the atom's nucleus. When they return to their normal or ground state this leads to the electron emitted photons (particles of light). These photons are all at the same wavelength and coherent. The ordinary visible light comprises multiple wavelengths and is not coherent as shown in Figure 19.8b.

Advantages of Optical Sensor

- ✰ The optical signal is immune to electro-magnetic interference and hence it is ideal to be employed in microwave environment.
- ✰ It has very high sensitivity, range and resolution compare to non-optical sensors.
- ✰ It is small in size and light in weight.
- ✰ It provides complete electrical insulation from high electrostatic potential.
- ✰ Multiplexed sensors provide measurements at large number of distributed point.

Disadvantages of Optical Sensor

- ✰ The only problem with both optical sensor and non-optical sensor is interference from multiple effects. For example, sensor used for pressure/ strain measurement is very sensitive to temperature variation.
- ✰ The other concern is the different types of losses involved in optical domain. The common among them are absorption loss, scattering loss, coupler loss, insertion loss, reflection loss, impurity loss, *etc.*

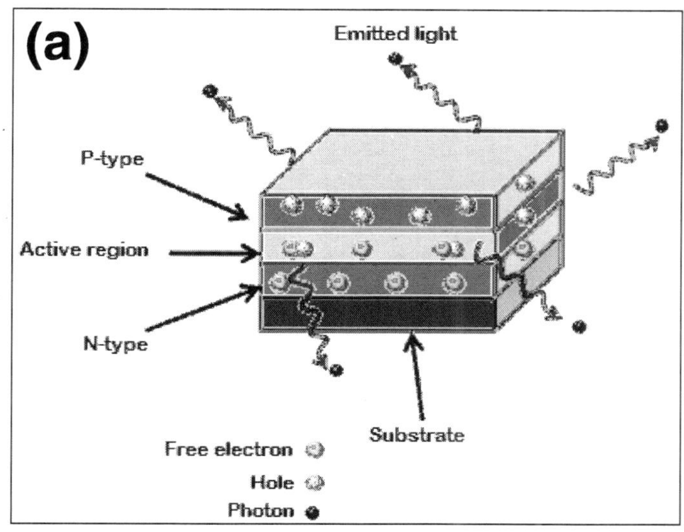

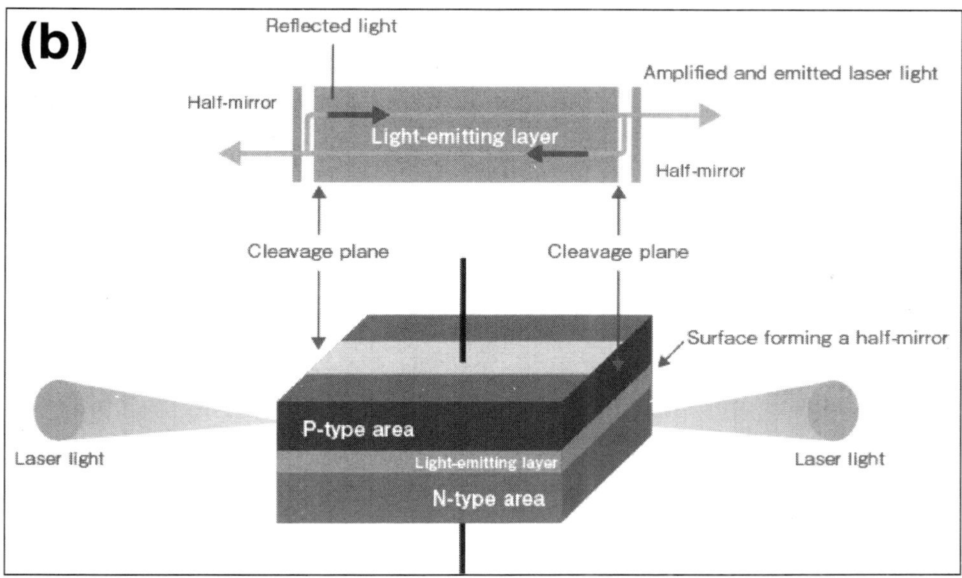

Figure 19.8. Photograph of (a) Light Emitting Diode (LED) (b) LASER (Light Amplification by Stimulated Emission Radiation).
Courtesy: Senasys photosensor, 2018.

Applications of Optical Sensor

Industrial Applications

- ☆ Temperature measurement by infrared sensors
- ☆ Surface and subsurface detection of defects and delaminations

- ✰ Surface inspection for gauging roughness and thickness
- ✰ Characterization of particle flows (temperature, velocity and diameter)
- ✰ In-process composition analysis by laser-induced plasma spectroscopy

Biometrics Applications

Optical sensor scans our finger and functions like a camera, obtains our fingerprint image indirectly. We may have similar kind of experience: when a government agency takes our fingerprint, someone will check our fingerprint to see if it is visible. Then, they will put some lotion on our finger, and the operator will press our finger on scanner to take our fingerprint. Optical sensor works in a similar way, so it needs good visible and clean finger surface.

Biomedical Applications

Flow monitoring by laser Dopplerimetry is used in several biomedical sensing applications, including dermatology for testing of skin irritants, gastroenterology via endoscopes for making blood perfusion measurements in the stomach and duodenum, *etc.*, and dentistry for contact probes to measure blood flow in the teeth and gums. Sensors are used in internal medicine for angiology and vascular surgery to monitor blood flow during vascular reconstruction and the degree of arteriosclerosis in arteries, and they are used in orthopedics for monitoring the blood perfusion in tissues during and after surgery.

REFERENCES

1. Feng, S., Darmawi, S., Henning, and Zhang, X. (2012). A miniaturized sensor consisting of concentric nanorings on the end facet of an optical fiber. *Small, 8*(12), 1937-1944.
2. Franke, M. E., Koplin, T. J., and Simon, U. (2006). Metal and metal oxide nanoparticles in chemiresistors: does the nanoscale matter. *small, 2*(1), 36-50.
3. Gong, H., Hu, J. Q., Wang, J. H., and Zhu, F. R. (2006). Nano-crystalline Cu-doped ZnO thin film gas sensor for CO. *Sensors and Actuators B: Chemical, 115*(1), 247-251.
4. Jin, Z., Su, Y., and Duan, Y. (2001). Development of a polyaniline-based optical ammonia sensor. *Sensors and Actuators B: Chemical, 72*(1), 75-79.
5. Kim, H. J., and Lee, J. H. (2014). Highly sensitive and selective gas sensors using p-type oxide semiconductors: Overview. *Sensors and Actuators B: Chemical, 192*, 607-627.
6. Krioukov, E., Klunder, D. J. W., Driessen, A., Greve, J., and Otto, C. (2002). Sensor based on an integrated optical microcavity. *Optics letters, 27*(7), 512-514.
7. Lutti, J., Langbein, W., and Borri, P. (2008). A monolithic optical sensor based on whispering modes in polystyrene microspheres. *Applied Physics Letters, 93*(15), 151103

8. Na, N., Zhang, S., and Zhang, X. (2006). A catalytic nanomaterial-based optical chemo-sensor array. *Journal of the American Chemical Society, 128*(45), 14420-14421.

9. Niu, X., Du, W., and Du, W. (2004). Preparation, characterization, and gas-sensing properties of rare earth oxides. *Sensors and Actuators B: Chemical, 99*(2-3), 399-404.

10. Santhaveesuk, T., Gardchareon, A., Wongratanaphisan, D., and Choopun, S. (2015). Ethanol sensing properties of Zn_2TiO_4. *Ceramics International, 41*, S809-S813.

11. Thirumalairajan, S., and Mastelaro, V. R. (2016). A novel organic pollutants gas sensing material p-type CuAlO2 microsphere constituted of nanoparticles for environmental remediation. *Sensors and Actuators B: Chemical, 223*, 138-148.

12. Thirumalairajan, S., Mastelaro, V. R., and Escanhoela Jr, C. A. (2014). In-depth understanding of the relation between CuAlO2 particle size and ozone gas sensor detection at a nanoscale level. *ACS applied materials and interfaces, 6*(23), 21739-21749.

13. Xu, Z., Chen, X., Kim, H. N., and Yoon, J. (2010). Sensors for the optical detection of cyanide ion. *Chemical Society Reviews, 39*(1), 127-137.

14. Yu, J. H., and Choi, G. M. (2002). Selective CO gas detection of Zn 2 SnO 4 gas sensor. *Journal of electroceramics, 8*(3), 249-255.

15. Zhang, X. B., Guo, C. C., Li, Z. Z., Shen, G. L., and Yu, R. Q. (2002). An optical fiber chemical sensor for mercury ions based on a porphyrin dimer. *Analytical chemistry, 74*(4), 821-825.

Chapter 20

Bio and Nano Sensors: Theory and Applications in Agriculture

V. Ramamoorthi[1], S. Srivignesh[2], S. Marimuthu[1] and K.S. Subramanian[3]*

[1]*Department of Nano Science and Technology, Tamil Nadu Agricultural University, Coimbatore – 641 003, Tamil Nadu*
[2]*Department of Horticulture and Floriculture, School of Life Sciences Central University of Tamil Nadu, Thiruvarur – 610 005, Tamil Nadu*
[3]*Director of Research, Tamil Nadu Agricultural University, Coimbatore – 641 003, Tamil Nadu*
**e-mail: vasudevanramamoorthi@gmail.com*

Biosensors are a special type of bioelectronic device that are suitable for detection and quantifying the wide range of chemical and biological substances present in various practical application (biosecurity, environmental, agricultural, pharmaceuticals and food and animal industries). The term biosensor was first introduced by Clark and Lyons in the year 1962 after the invention of enzyme based electrode system. Biosensors can be defined as a compact and highly specific analytical device that transforms biological signals into electrical outputs, when an analyte and bio-recognizing material combines with the presence of signaling transducers (Nigam and Shukla, 2015). The electrical outputs recorded from the biomolecular interactions were amplified, processed and displayed through the support of computer-aided readouts; in that way, the users can understand the association that occurred between ligand and targeted material (Figure 20.1). However, this strength of an electrical current produced from the interaction was directly proportional to the concentration of targeted analyte, which are capable to afford semi-quantitative and quantitative analytical information (Wang *et al.*, 2014). According to IUPAC recommendations "a biosensor is an independently integrated receptor transducer device, which is capable of providing selective

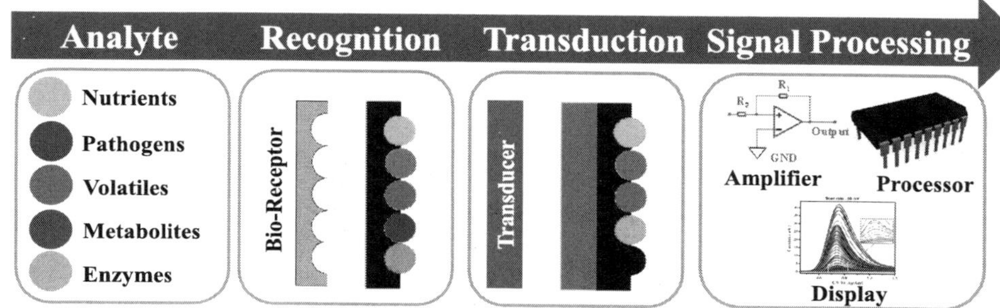

Figure 20.1. Schematic Workflow and Components of Typical Biosensor.

quantitative or semi quantitative analytical information using biological recognition element (Thevenot *et al.*, 1999)". The fundamental working principles involved in the biosensors are "bio-recognition" (biological component) and "bio-sensing" (electronic component) (Karunakaran *et al.*, 2015). The main objectives of the biosensor are to provide accurate, reliable, and quick information about the concerned analyte in real time. In this chapter, the current status of different biosensors and their mechanisms involved in the detection of various analyte with respect to agricultural applications are detailed.

Basic Principles of Biosensor

Biosensor works on the basic principle of signal transduction (Neethirajan *et al.*, 2005). The physico-chemical (pH, temperature, mass change, *etc.*,) changes produced during the binding of analyte to the biological material forms a bound analyte, which in turn produces biological signals that can be converted and amplified with the help of signal transducers (Grieshaber *et al.*, 2008).

Components of Biosensor

A biosensor consists of four main components: (a) analyte (b) receptor-biological recognizing system, (c) physico or chemical transducer – convert biological signals into electrical signals, and (d) microelectronics – amplifier, processor and display (David *et al.*, 2008). The basic components involved in the construction of typical biosensor are represented in Figure 20.1.

Analyte

A substance of interest that is exposed for measurement (or) analysis. For example, nitrate is an "analyte" for biosensor designed to measure environmental quality.

Bioreceptor

A biological molecular species that utilizes biochemical mechanism for specific interaction. Antibody/antigen, nucleic acid, DNA/RNA, enzyme and cell organelles are some of the commonly used forms of bioreceptors. A significant and distinguishing feature of biosensor that comprises biorecognition towards the target analyte.

Transducer

Transducer is an electronic circuit that converts one form of energy (biological signals) into another form (electrical signals). This practice of energy exchange is popularly called as signalization.

Electronic Systems

The electronic system consists of complex electronic circuits that perform signal amplification, processing and displaying of transduced signals from analogue into digital form. The indications on the display might be graphic, tabular, numeric or an image depending on the requirements.

Classification of Biosensor

The biosensors can be classified on the basis of biological sensing element (enzyme, microbe, proteins, antibodies *etc.,*), method of bio sensing (affinity, metabolic or catalytic) and type of transducer (electrochemical, optical or mass based) engaged in the detection procedure. The various classifications of biosensors are depicted in Figure 20.2.

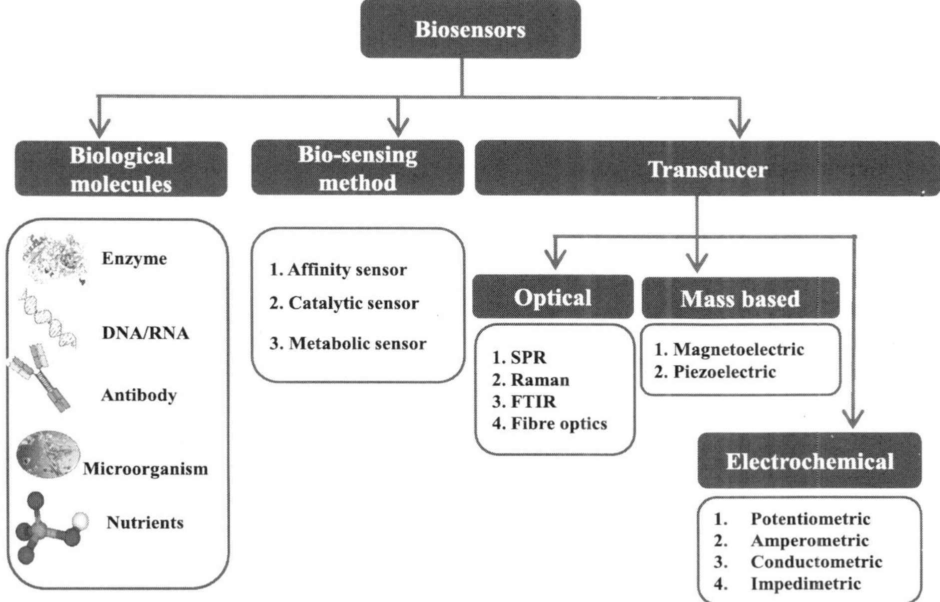

Figure 20.2. Classification of Biosensors Based on Biological Molecule, Biosensing Method and Type of Transducer.

Classification of Biosensors Based on Bioreceptor

A bioreceptor is molecular species that utilizes a biochemical mechanism for specific recognition and responsible for the binding of targeted analyte to the surface of signal transducer. Bioreceptors are generally divided into five major classes: antibody/antigen, enzyme, cellular structure/cell, nucleic acid/DNA and biomimetic or synthetic (Salgado *et al.*, 2011). The most common forms of bioreceptors used in biosensing are: a. enzymatic interactions; b. antibody-antigen interactions; c. nucleic acid interactions; d. microbial or cellular interactions; e. interactions using synthetic materials. The antibodies and enzymes are the leading groups of biological molecule/bioreceptor that are widely used in biosensor applications (Monosik *et al.*, 2012).

Enzyme Bioreceptor: Biosensor

Enzyme is a widely used protein molecule that catalyzing a particular substrate into a product without being consumed in any chemical/biological reaction (Figure 20.3). The bioreceptor are mounted on the transducer surface through covalent attachment, adsorption, gel entrapment or an electrochemically engendered polymer. Compared with chemical reactions, enzymes are fairly fast and highly selective and they can be used in combination with different transduction mechanisms. The mechanisms of these bioreceptors can comprise: (a) transformation of the bound analyte into a sensor measurable product (b) detection of enzyme activated or inhibited analyte and (3) evaluation of modified enzyme properties acted upon interaction.

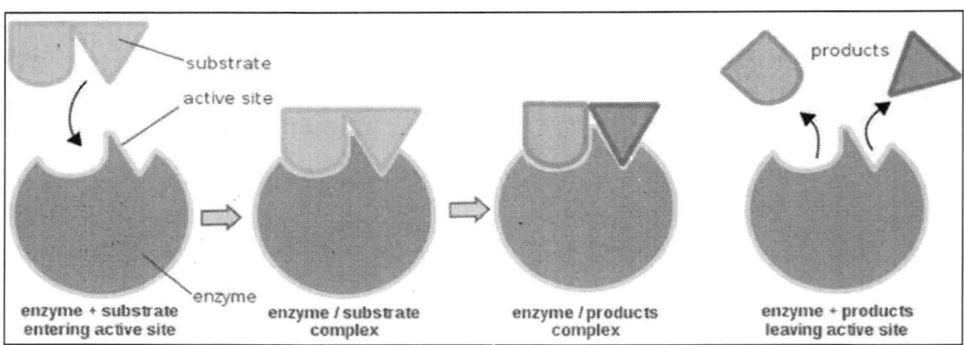

Figure 20.3. Catalytic Mechanism of Enzyme Bioreceptor and Targeted Substrate.
Courtesy: Shivangi Soni, 2016.

Antibody Bioreceptor: Immunosensor

Antibody bioreceptors works on the principle of antigen-antibody interactions that can be transduced directly into a quantifiable electrical signal. Antibodies (immunoglobulin/glycoproteins) are heavy globular plasma proteins (50 kDa) that are composed of two light chains and two heavy chains forming the well-known Y shaped structure. Schematic demonstration of Y-shaped structure of an antibody is shown in Figure 20.4. These antibodies secreted in the animal cells

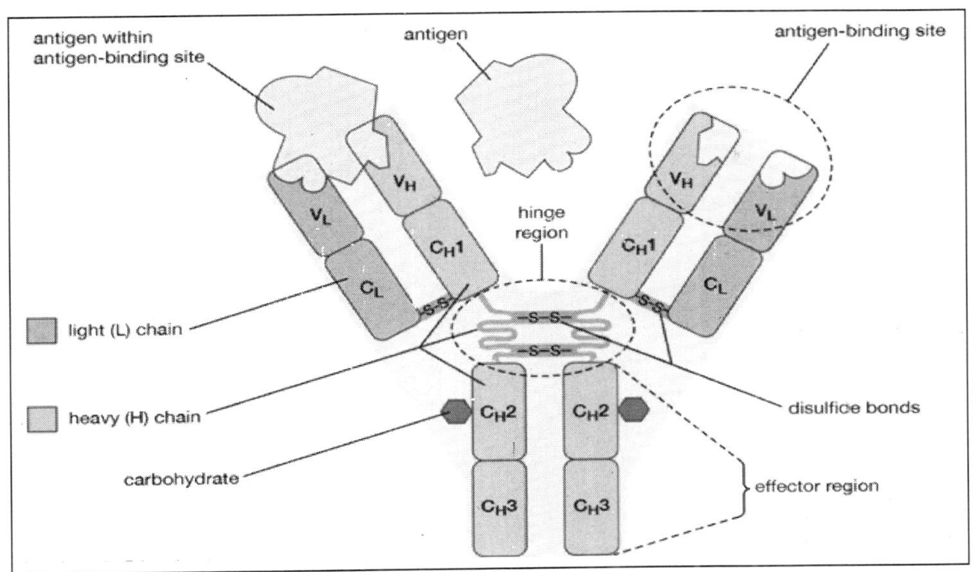

Figure 20.4. Schematic Demonstration of Y-shaped Structure of an Antibody
Courtesy: https://www.britannica.com/science/antibody.

by immunological responses to the foreign (antigen) agent (Sharma *et al.*, 2016). Antibodies are sensitive to changes in ionic strength, pH, temperature and chemical inhibitors. Monoclonal (highly specific but identical) and polyclonal (highly sensitive but less specific) antibodies are commonly employed in the construction of Immunosensors. Monoclonal antibodies produce significantly less background staining than polyclonal antibodies, because of their specificity towards specific antigens in the presence of interfering molecules (Karunakaran *et al.*, 2015).

Nucleic Acid/DNA Bioreceptor: Genosensor

Nucleic acids biosensors are used to detect the individual nucleotides based on nucleic acid sequencing processes and that are being developed for rapid and inexpensive testing of genetic and pathogenic diseases. DNA is highly suitable nucleic acid for biosensing applications due to the nature of base-pairing interactions among complementary sequences. DNA biosensor is a device that can combine with specific DNA probe and signal transducer through hybridization techniques then the target DNA is captured on the recognition layer, and the resulting hybridization signal is transduced into a usable electronic signal for display and analysis (Liu *et al.*, 2012). This recognition mechanism depends on the method of signal transduction (electrochemical, mechanical or optical). DNA techniques like recombination, hybridization and amplification are all based on the structure of double helix DNA. The general structure of DNA and DNA hybridization is displayed in Figure 20.5.

Microbial Bioreceptor: Microbial Biosensor

The enzymes or proteins present in the living organisms can act as a bioelement to detect specific molecule or state surrounding environment based

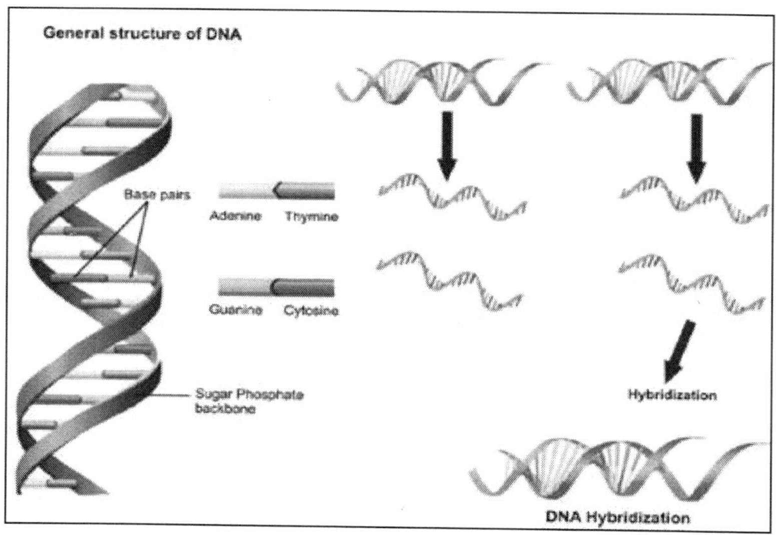

Figure 20.5. General Structure of DNA and DNA Hybridization.

on the measurement of their mechanism, in many cases accompanied by the consumption of O_2 or CO_2. Microbial biosensor is an analytical device that immobilizes microorganisms (bacterial, fungi or virus) onto a transducer surface to generate electrical signals proportionate to the concentration of specific substrate. The immobilization process determines the quality of the signal transferred from microorganisms to the transducer. Microorganisms have been integrated with a variety of transducers such as conductometric, potentiometric, amperometric, colorimetric, voltammetry, fluorimetric and luminometric to fabricate biosensor devices (Su *et al.*, 2011; Thouand, 2018).

Classification of Biosensors Based on Biosensing Method

Biosensors are broadly classified into three different types based on its biorecognition mechanism. **Affinity sensors** rely on the mechanism of reversible binding of the analyte to a specific receptor (Beyer *et al.*, 2014); **Metabolic sensors** measures the cell signals or concentration of a substrate during the interaction of analyte and biomolecule through the incorporation of metabolic signals into the signaling networks (Krejci, 2012); **Catalytic biosensors** works on the nature of secondary substrates produced during the interaction of analyte and biological elements (Li and Lu, 2000).

Classification of Biosensors Based on Transducers

A transducer is the device that converts biological signals into detectable electrical signals. The detectable electrical signals were electrochemical, calorimetric, optical magnetic or mass change in nature. The commonly accomplished biosensors based on different transduction methods are classified in one of the three main classes:

1. Optical biosensor - SPR, Raman, FTIR and Fibre optics
2. Mass based biosensor - Magnetometry and Piezometry
3. Electrochemical biosensor - Potentiometry, Conductometry, Amperometry, Impedimetry

Optical Biosensor

Optical biosensors are the most commonly used class of biosensors. The optical detection is performed by manipulating the interaction of biorecognition element with an optical field. The biorecognition procedure induces a change in the amplitude, phase, frequency or polarization of the incident light with response to the physical or chemical change (Damborsky *et al.,* 2016). The major components of optical biosensors were; light source, biological recognition element, optical transmission medium and wavelength detection system. Optical biosensors are broadly divided into methods based on the different parameter: Label free method directly measures the signals, formed during the interaction of analyte with the transducers. In contrast label based method involves the use of fluorescent tags signals to detect the target molecule (Perumal and Hashim, 2014).

Fiber optic biosensors (FOBS) are optical fiber-derived devices which use optical field to measure biological species such as proteins, cells, and RNA/DNA (Leung *et al.,* 2007). Optical sensors based on excitation of surface plasmon (electromagnetic wave), generally mentioned as **Surface plasmon resonance sensors** (SPR) and it was first demonstrated for biosensing in 1983 by Liedberg *et al.* (1983). SPR is a noninvasive label free method of optical biosensor used for observing real time interactions between injected analyte and immobilized biomolecule. SPR phenomenon occurs on the surface of conducting material and at the interface of two media when it was illuminated by polarized light at a specific wavelength. The reduction of reflected light intensity at a specific angle is known as resonance angle (Tang *et al.,* 2010). The principle of surface plasmon resonance (SPR) is characterized in Figure 20.6.

Mass Based Biosensor

Piezoelectric biosensors is a class of micro electromechanical systems (MEMS) highly suitable for label-free portable biosensing, which is depends on the principle of changes in oscillating crystal resonance frequency due to the interaction of analyte and bioreceptor (Pohanka, 2018). The biosensing material coated on the surface of piezoelectric material (transducer) produces electrical signal when applying mechanical stress. A crystal oscillates at a certain frequency depends on the mass of the crystal and coating. Quartz is the most commonly used single crystal which can withstand extreme thermal, mechanical and chemical stresses (Nicu *et al.,* 2005). There are two classes of piezoelectric sensors based on the type of travelling wave; **Surface acoustic wave (SAW)** piezoelectric sensors - propagating through the interior of the substrate; **Bulk acoustic wave (BAW)** piezoelectric sensors - propagating on the surface of the substrate.

Magnetoelastic biosensors are made from amorphous ferro-magentoelastic metal flim ribbons coated on a pH sensitive polymer. The magnetoelastic substance

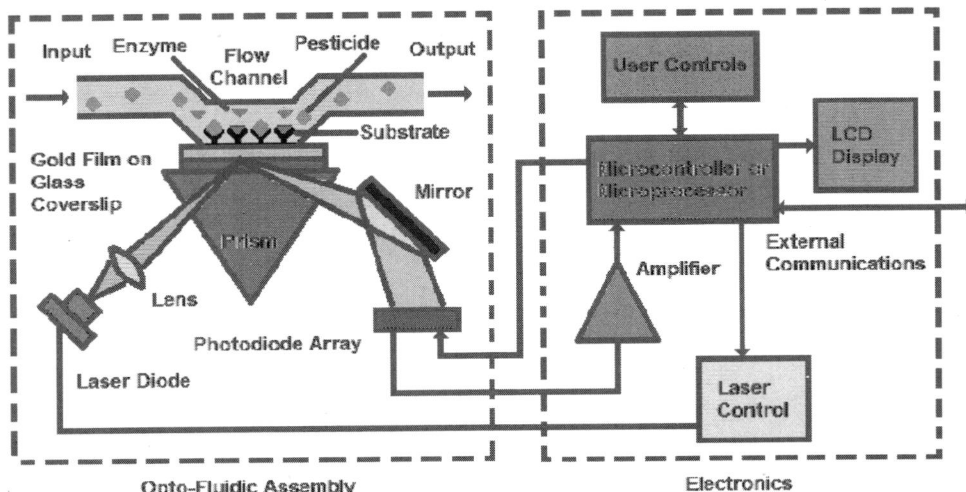

Figure 20.6. Optical Biosensor Based on Surface Plasmon Resonance (SPR).
Courtesy: Vargas-Bernal *et al.*, 2012.

changes its magnitudes (starts to oscillate emitting) when it is exposed to a magnetic field. The amplitude, frequency and damping of the oscillating magnetic field give information about the property of coated material (Zourob *et al.*, 2007). The components and schematic sketch of mass based biosensors are detailed in Figure 20.7.

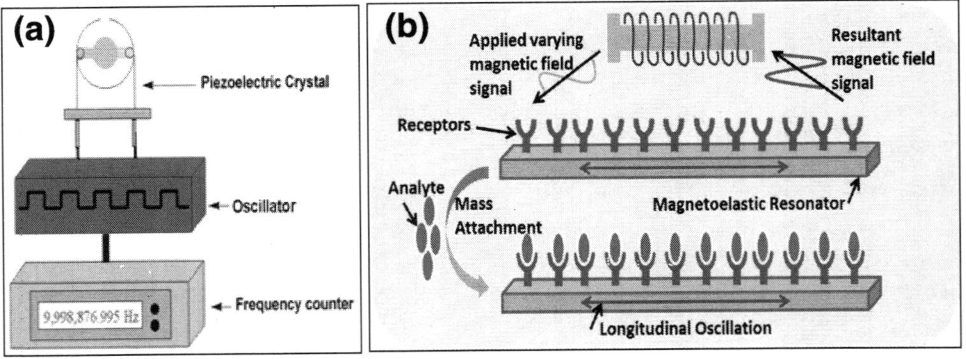

Figure 20.7. Schematic Representation of Piezoelectric (a) and Magnetoelastic Biosensors (b)
Courtesy: Shruthi *et al.*, 2014.

Calorimetric Biosensors

Calorimetric biosensors works on the principle of calorimetric transduction that are designed to detect changes in temperature between analyte and a biorecognition element during biological or chemical reactions (Xie *et al.*, 1999). This change in temperature directly related to the amount of reactant or products formed. The evolved heat change is measured using either a thermophile (ceramic semiconductor) or thermistor (MOF) and widely used for label free screening of biomolecules with more stability and sensitivity.

Electrochemical Biosensors

According to the 1999 IUPAC recommendation, "an electrochemical biosensor is a self-contained integrated device that is capable of providing specific quantitative or semi quantitative analytical information using a biological recognition element (biochemical receptor) that is retained in direct spatial contact with an electrochemical transduction element". Electrochemical transducers used to measure the current produced from oxidation and reduction reaction (Rotariu *et al.*, 2016). The electrical signal resulted by the recognition process proportional to the analyte concentration. Electrochemical biosensors can be classified into five types depending upon the nature of biorecognition event: potentiometric, voltametric, amperometric, conductometric and Impedimetric.

Electrochemical cell is a device that can generate electrical energy from chemical reactions which is comprises working electrode (WE), reference electrode (RE) and counter electrode (CE). Electrochemical analyzer is an instrument used to measure the potential or current of an analyte in an electrochemical cell. Schematics of representation of electrochemical cell and electrochemical work station are detailed in Figure 20.8.

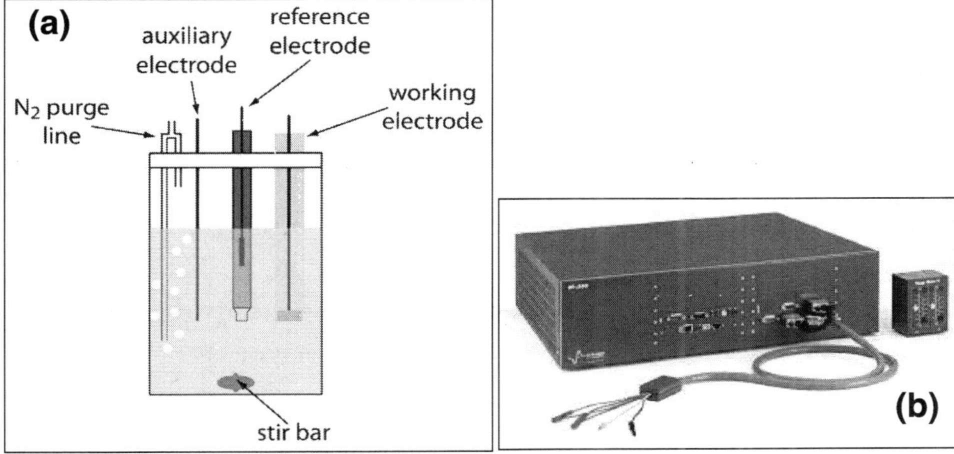

Figure 20.8. Schematic Representation of Electrochemical Cell (a) and Work Station (b).
Courtesy: mmrc.caltech.edu.

Microfluidic devices are a miniaturized version of conventional electrochemical cells that containing tunnels and chambers through which fluids are confined or flow. The behavior of microelectrode differs from conventional electrodes in that nonlinear diffusion is the predominant mode of transport. (Aryasomayajula *et al.*, 2017). In microfluidic cell three different materials were optimized as electrode:

silver-epoxy composite as pseudo-reference electrode, graphite-epoxy composite as auxiliary electrode and gold film or graphite-epoxy composite as working electrode. The performance of the microfluidic cell was categorized using cyclic voltammetry. The microfluidic electrochemical cell is represented in Figure 20.9.

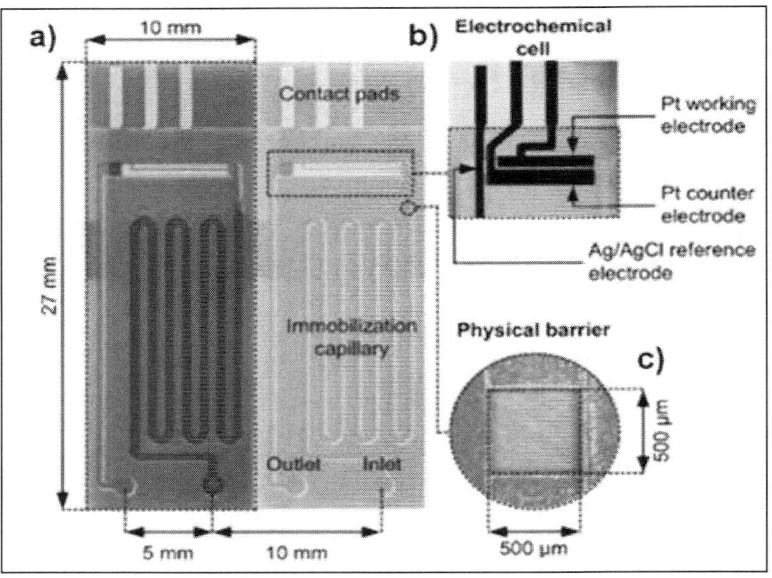

Figure 20.9. Microfluidic Electrochemical Cells.
Courtesy: Horak *et al.*, 2014.

Lab-On-Chip (LOC) or Screen Printed Electrode (SPCE) is a device that has several integrated laboratory functions on a single chip, which is normally few millimeters to several centimeters in size. The Lab-On-Chip (LOC) or Screen Printed Electrode (SPCE) device is shown on Figure 20.10.

Application of Bio and Nano Sensors in Agriculture

Agriculture includes the production of crops, which have been susceptible to damage in the form of biotic and edaphic that causes severe loss in profit. Therefore, the effective way of increasing profits would be to reduce the loss of crops from natural threats (Fletcher *et al.*, 2007). Biosensors are a powerful alternative technology to conventional analytical techniques, harnessing the sensitivity and specificity of biological systems in low cost and portable devices. The need for quick and accurate sensing opens up opportunities for biosensors in many different agricultural areas -in situ analysis of nutrient and pollutants, detection and identification of pest and diseases in crops and the summary of various bio and nano sensors technologies involved in the agricultural applications are depicted in Table 20.1.

Deoxyribonucleic acid (DNA) is basic building block for all living organisms inclusive of plants providing codes for the synthesis for genes and proteins. Nanofabrication of DNA is relatively simple and complementary base pairing

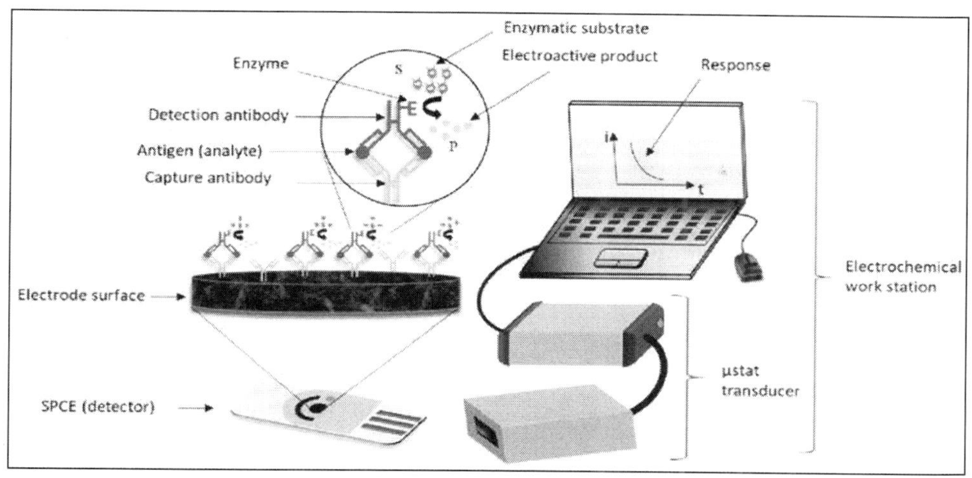

Figure 20.10. Lab-On-Chip or Screen Printed Electrode.
Courtesy: Vasilescu *et al.*, 2016.

can be effectively controlled which paved way for the DNA biosensors and DNA microarrays. DNA biosensors and DNA microarrays (DNA chips) application has successfully employed in the medical diagnostic field and potential application in agriculture is evolving fast. DNA based detections are needy in detection of forensic substances, biological contaminants and analysis gene expression patterns. The old system of DNA chip based detection systems are mainly based on binding the DNA materials to its complimentary sequence/probe which is a single stranded DNA (ssDNA) or chemically synthesized oligo nucleotides attached on the solid substrate (glass, silicon, special plastics) or in a buffer solution, allowing the target DNA to get hybridized. Generally, 10,000 to 40,000 probes are placed in each zones of 10-100 μm (Hahn *et al.*, 2005). The synthesized oligos can be directly in-situ synthesized on the surface or deposited on the chemically activated surfaces (Dufva, 2005). The Hybridized DNA is detected by means of electrochemically or by optical transduction modes which allows to study gene expression profiles of thousand genes simultaneously. Biosensor microarrays primarily used for biological detections, complete transcriptome and proteome changes of biological samples. Whereas DNA biosensors the DNA probes were directly immobilized on the transducer surface. The DNA biosensors are more robust, highly sensitive, parallel detection on multiple modes, reusable, and specifically miniature size (Lamartine, 2006). This lacks simultaneous analysis of a given phenomenon, possibility of one single application and hard to miniature these devices. DNA biosensors are capable of detecting the single base pair genetic mutations. This is highly useful in identifying the single nucleotide polymorphisms (SNP), for developing the molecular markers in plants. The genetic mutation in single gene can alter the resistance level in plants, or to identify the mutation leading to new trait (for example ovate gene in tomato). Recent technological advancements paved way to design localized surface plasmon resonance (LSPR) biosensors which have been subjected to a great scientific interest

Table 20.1. Application of Nano and Biosensors in Agriculture (Adopted from Khater *et al.*, 2017)

Target Analyte	*Sensing Strategy*	*Transduction Method*	*Nanomaterial Exploited*	*LOD*	*Reference*
Atrazine	Inhibition of tyrosinase	Amperometric	TiO_2 nanotubes	0.1 ppt	Yu *et al.*, 2010
Methyl	Inhibition of AChE	Differential pulse voltammetry	Multi-walled carbon nanotubes-chitosan nanocomposite	7.5 x 10^{-13}M	Dong *et al.*, 2013
Acetamiprid	aptamer	Colorimetric analysis	Gold nanoparticles	5 μM	Shi *et al.*, 2013
Glyphosate and glufosinate	Specific Double template imprinted polymers	Differential pulse anodic stripping voltammetry	Multi walled carbon nanotubes	0.35 μg mL^{-1} 0.19 μg mL^{-1}	Prasad *et al.*, 2014
Nitrate	Ion selective electrode	Potentiometric	Graphene oxide	10^{-5} M	Pan *et al.*, 2016
Nitrate	Nitrate reductase	Impedimetric	PEDOT nanofibers - Graphene oxide	0.68 mg/L	Ali *et al.*, 2017
Urea and Urease activity	Gold nanoparticles acting as a catalyst imitating horseradish peroxidase	Colorimetric pH indicator	Gold nanoparticles	5 μM 1.8 U/L	Deng *et al.*, 2016
Ganoderma boninense	DNA probe	Fluorescence resonance	Quantum dots	3.55 x 10^{-9}M	Bakhori *et al.*, 2013
Pantoea stewartii sbusp. *stewartii*-NCPPB 449	Immunosensor	Enzyme-linked immunosorbent assay (ELISA)	Gold nanoparticles	7.8 x 10^3 cfu/mL	Zhao *et al.*, 2014
Cymbidium mosaic virus	Fibre optic particle Plasmon resonance	Immunosensor	Gold nanorods	48 pg/mL 42 pg/mL	Lin *et al.*, 2014
Trichoderma harzianum	Single stranded DNA probe	Electrochemical analysis	ZnO nanoparticles - chitosan nanocomposite	1.0 x 10^{-19} mol/L	Siddiquee *et al.*, 2014

in the last few years. The LSPR biosensors offer quick and multiplexing options paved way to design of lab-on-a-chip (LOC) optical biosensor platforms. Recently it was shown that the hybrid plasmonic biosensor could specially detect *Bacillus thuringiensis* Cry1Ab protein in crop samples (Ming *et al.*, 2014). For instance, Lee *et al.*, 2016 developed nanoplasmic biochip coupled with smartphone for quantitative detection of imidacloprid pesticide, similarly a LSPR based biosensor was developed to detect shinga and cholre toxins Nagatsuka *et al.*, 2013.

Masrie *et al.* (2017) developed an optical transducer; LED and photodiode based biosensor for soil nitrogen, phosphorus and potassium monitoring. Fibre optical biosensing devices have been proposed to monitor plant nutrient and water stresses (Asundi *et al.*, 2006). A low-power, low-cost dual probe heat pulse techniques (DPHP) have been proposed for monitoring real-time soil moisture status. A DPHP sensor has two probes: a temperature sensor probe and a heater probe with a power AC supply of 3.3V. To detect volumetric water content the sensor was validated with red and white clay soil samples (Jorapur *et al.*, 2015).

A graphene foam titanium nitrate nanofibers (GF–TiN NFs) based microfluidic electrochemical sensors for the detection of nitrate in agricultural soils was proposed by Ali *et al.* (2016). The microfluidic device provides a high sensitivity of 683.3 µA mg–1 L cm^{-2}, a wide dynamic range from 0.01 to 442 mg L^{-1} cm^{-2} and an ultralow limit of detection of 0.01 mg L^{-1} for nitrate ions in real soil sample solutions. Soil diseases were diagnosed through quantitative measurement of differential oxygen consumption between good and bad microbes resident in the soil (Rai *et al.*, 2012). Dorji *et al.* (2017) introduced a portable electronic nose (e-nose) wireless sensing station for online soil monitoring. Volatile organic compounds (VOCs) footprints obtained from the farmer's field were correlated with physical and chemical properties of the soil, thereby ensuring precise soil nutrient management.

Niguyen and Angeles (2012) determined the changes of the nitrate concentration in pineapple using nitrate strip to establish the critical values of NO_3 in sap related toNO_3 of the sap leaf and the vegetative yield. Carlson *et al.* (2000) conducted a fluorimetric biosensor experiments to quantify aflatoxins, which are commonly found in a variety of agricultural products. A high density microarray biosensor was established to detect *E. coli* O157:H7 colony forming units. The change in impedance of the biosensor is proportional to the number of bacteria on the sensor surface (Radke and Alocilja, 2005). A simple and sensitive dual amplified electrochemical immunoassay was developed using gold nanoparticle and horseradish peroxidase (HRP) for the control of Stewarts vascular wilt of maize with a limit of detection (LOD) of 7.8 x 103 CFU/mL. HRP method of identification increased the detection sensitivity by 20-fold compared with conventional ELISA (Zhao *et al.*, 2014). An Impedimetric platform based biosensor was developed for the detection of Prunus Necrotic Ringspot Virus (PNRSV) in plant extracts using anti-PNRSV IgG polyclonal antibody, which is capable to discriminate healthy plants from infected plants (Jarocka *et al.*, 2013). Similarly Lau *et al.* (2017) presented a new recombinase polymerase amplification (RPA) based isothermal electrochemical biosensor for plant pathogen DNA detection, which was composed of colloidal gold nanoparticle and disposable screen printed electrode. The developed sensor was 10,000 times more sensitive than conventional gel electrophoresis or PCR.

Xanthomonas axonopodis pv. vesicatoria bacterial spot diseases were successfully detected using Fluorescent silica nanoparticles (FSNP) conjugated antibodies Etefagh *et al.* (2013). Highly sensitive QCM and selective (quartz crystal microbalance) biosensor were introduced for the detection of maize chlorotic mottle virus (MCMV) with a detection limit of 250 µg/mL. A mixture of 1:10 11-Mercaptoundecanoic acid and 3-Mercaptopropanoic acid was applied on gold surfaces of QCM for specific recognition of MCMV and anti-MCMV (Huang *et al.*, 2014). Mendes *et al.* (2009) introduced a biosensor that can detect soyabean rust *Phakopsora pachyrhizi* which has been reported as pathogenic fungus. The acoustic-based biosensor (QCM) could sense the plant pathogens like *Ralstonia solanacearum*, *Xanthomonas campestris pv. Vesicatoria* and *Pseudomonas syringae pv tomato* (Papadakis *et al.*, 2015).

Tang *et al.* (2015) developed a simple organophosphorus detecting biosensor in Fuji apple and cucumber by utilizing unmodified screen printed silver electrode, which are characterized by thiocholine generated by acetylcholinesterase (AChE). A detection limit of 2.5 µg/L chlorpyrifos was obtained during the study. Nanomaterials like multi walled carbon nanotubes, gold, silver and graphene nanoparticles, SnO_2-chitosan, silver-graphene nanocomposite have also been used for this purpose (Charoenkitamorn *et al.*, 2015; Chen *et al.*, 2015; Chatzipetrou *et al.*, 2016 and Govindasamy *et al.*, 2017).

The potential challenges, that would arise due to increase in global population and climate change, warrant intervention of nano technolgiocal approaches in agriculture, to harness the sustainable crop production. The use of nanomaterials has been evidenced in potential applications such as Nanofertilizers, nano pesticides to control emerging pest and diseases and to tackle the resistance evolution, nanomaterials to improve the quality of soil, nano biosensors for precision agriculture and packaging systems, nano technologies in waste management. However, the full potential of nano biosensors in agriculture and food sector is yet to be realized. In addition, any new technological interventions need a long term risk and impact assessment on non-targeted organisms (bees, soil microbes) and its associated environmental effects.

REFERENCES

1. Ali, M. A., Mondal, K., Wang, Y., Jiang, H., Mahal, N. K., Castellano, M. J.,. and Dong, L. (2017). In situ integration of graphene foam–titanium nitride based bio-scaffolds and microfluidic structures for soil nutrient sensors. *Lab on a Chip*, *17*(2), 274-285.

2. Aryasomayajula A., Bayat P., Rezai P., and Selvaganapathy P.R. (2017) Microfluidic Devices and Their Applications. In: Bhushan B. (eds) Springer Handbook of Nanotechnology. Springer Handbooks. Springer, Berlin, Heidelberg

3. Asundi, A., Liew, O. W., Li, B., and Chong, J. (2006). Optical sensors enhance plant nutrient monitoring.

4. Beyer, U., Wyss, T., Robin, F., and Heinemann, L. (2014). Glucose measurement by affinity sensor and pulsed measurements of fluidic resistances: proof of principle. *Journal of diabetes science and technology, 8*(1), 100-108.
5. Carlson, M. A., Bargeron, C. B., Benson, R. C., Fraser, A. B., Phillips, T. E., Velky, J. T.,. and Ko, H. W. (2000). An automated, handheld biosensor for aflatoxin. *Biosensors and Bioelectronics, 14*(10-11), 841-848.
6. Charoenkitamorn, K., Chailapakul, O., and Siangproh, W. (2015). Development of gold nanoparticles modified screen-printed carbon electrode for the analysis of thiram, disulfiram and their derivative in food using ultra-high performance liquid chromatography. *Talanta, 132*, 416-423.
7. Chatzipetrou, M., Milano, F., Giotta, L., Chirizzi, D., Trotta, M., Massaouti, M.,. and Zergioti, I. (2016). Functionalization of gold screen printed electrodes with bacterial photosynthetic reaction centers by laser printing technology for mediatorless herbicide biosensing. *Electrochemistry Communications, 64*, 46-50.
8. Chen, D., Jiao, Y., Jia, H., Guo, Y., Sun, X., Wang, X., and Xu, J. (2015). Acetylcholinesterase biosensor for chlorpyrifos detection based on multi-walled carbon nanotubes-SnO_2-chitosan nanocomposite modified screen-printed electrode. *Int. J. Electrochem. Sci, 10*, 10491-10501.
9. Clark Jr, L. C., and Lyons, C. (1962). Electrode systems for continuous monitoring in cardiovascular surgery. *Annals of the New York Academy of sciences, 102*(1), 29-45.
10. Damborský, P., Švitel, J., and Katrlík, J. (2016). Optical biosensors. *Essays in biochemistry, 60*(1), 91-100.
11. David, W.G., Gonsalves, K., Halberstadt, C., and Laurencin, C.T. (2008) Biomedical Nanostructures: Clinical Applications of Micro- and Nanoscale Biosensors. Wiley, LN.
12. Dorji, U., Pobkrut, T., and Kerdcharoen, T. (2017, February). Electronic nose based wireless sensor network for soil monitoring in precision farming system. In *2017 9th International Conference on Knowledge and Smart Technology (KST)* (pp. 182-186). IEEE.
13. Dufva, M. Biomol. Eng. 2005, 22, 173. (42) Castellino, A. M. Genome Res. 1997, 7, 943.
14. Etefagh, R., Azhir, E., and Shahtahmasebi, N. (2013). Synthesis of CuO nanoparticles and fabrication of nanostructural layer biosensors for detecting Aspergillus niger fungi. *Scientia Iranica, 20*(3), 1055-1058.
15. Ferreira, A. A. P., Uliana, C. V., de Souza Castilho, M., Pesquero, N. C., Foguel, M. V., dos Santos, G. P.,. and Yamanaka, H. (2013). Amperometric biosensor for diagnosis of disease. *State of the Art in Biosensors-Environmental and Medical Applications, 12*, 253-289.
16. Fletcher, M. P., Diggle, S. P., Crusz, S. A., Chhabra, S. R., Cámara, M., and Williams, P. (2007). A dual biosensor for 2alkyl4quinolone quorumsensing signal molecules. *Environmental microbiology, 9*(11), 2683-2693.

17. Govindasamy, M., Mani, V., Chen, S. M., Chen, T. W., and Sundramoorthy, A. K. (2017). Methyl parathion detection in vegetables and fruits using silver@ graphene nanoribbons nanocomposite modified screen printed electrode. *Scientific reports, 7*, 46471.

18. Grieshaber, D., MacKenzie, R., Voros, J., and Reimhult, E. (2008). Electrochemical biosensors-sensor principles and architectures. *Sensors, 8*(3), 1400-1458.

19. Hahn, S., Mergenthaler, S., Zimmermann, B., and Holzgreve, W. (2005). Nucleic acid based biosensors: the desires of the user. *Bioelectrochemistry, 67*(2), 151-154.

20. Horak, J., Dincer, C., Bakirci, H., and Urban, G. (2014). Sensitive, rapid and quantitative detection of substance P in serum samples using an integrated microfluidic immunochip. *Biosensors and Bioelectronics, 58*, 186-192.

21. https: / /www.britannica.com/science/antibody

22. https: / /www.slideshare.net/ShivangiSoni6/enzyme-biosensor-final

23. Huang, X., Xu, J., Ji, H. F., Li, G., and Chen, H. (2014). Quartz crystal microbalance based biosensor for rapid and sensitive detection of maize chlorotic mottle virus. *Analytical Methods, 6*(13), 4530-4536.

24. Jarocka, U., Radecka, H., Malinowski, T., Michalczuk, L., and Radecki, J. (2013). Detection of Prunus Necrotic Ringspot Virus in plant extracts with impedimetric immunosensor based on glassy carbon electrode. *Electroanalysis, 25*(2), 433-438.

25. Jorapur, N., Palaparthy, V. S., Sarik, S., John, J., Baghini, M. S., and Ananthasuresh, G. K. (2015). A low-power, low-cost soil-moisture sensor using dual-probe heat-pulse technique. *Sensors and Actuators A: Physical, 233*, 108-117.

26. Karunakaran, C., Bhargava, K., and Benjamin, R. (2015). *Biosensors and bioelectronics*. Elsevier.

27. Khater, M., de la Escosura-Muñiz, A., and Merkoçi, A. (2017). Biosensors for plant pathogen detection. *Biosensors and Bioelectronics, 93*, 72-86.

28. Krejci, A. (2012). Metabolic sensors and their interplay with cell signalling and transcription.

29. Lamartine, J. (2006). The benefits of DNA microarrays in fundamental and applied bio-medicine. *Materials Science and Engineering: C, 26*(2-3), 354-359.

30. Lau, H. Y., Wu, H., Wee, E. J., Trau, M., Wang, Y., and Botella, J. R. (2017). Specific and sensitive isothermal electrochemical biosensor for plant pathogen DNA detection with colloidal gold nanoparticles as probes. *Scientific reports, 7*, 38896.

31. Lee, K. L., You, M. L., Tsai, C. H., Lin, E. H., Hsieh, S. Y., Ho, M. H.,. and Wei, P. K. (2016). Nanoplasmonic biochips for rapid label-free detection of imidacloprid pesticides with a smartphone. *Biosensors and Bioelectronics, 75*, 88-95.

32. Leung, A., Shankar, P. M., and Mutharasan, R. (2007). A review of fiber-optic biosensors. *Sensors and Actuators B: Chemical, 125*(2), 688-703.

33. Li, J., and Lu, Y. (2000). A highly sensitive and selective catalytic DNA biosensor for lead ions. *Journal of the American Chemical Society, 122*(42), 10466-10467.

34. Liedberg, B., Nylander, C., and Lunström, I. (1983). Surface plasmon resonance for gas detection and biosensing. *Sensors and actuators, 4,* 299-304.

35. Liu, A., Wang, K., Weng, S., Lei, Y., Lin, L., Chen, W., Lin, X., and Chen, Y. (2012). Development of electrochemical DNA biosensors. Trends Anal. Chem. 37, 101e111.

36. Marazuela, M., and Moreno-Bondi, M. (2002). Fiber-optic biosensors–an overview. *Analytical and bioanalytical chemistry, 372*(5-6), 664-682.

37. Masrie, M., Rosman, M. S. A., Sam, R., and Janin, Z. (2017, November). Detection of nitrogen, phosphorus, and potassium (NPK) nutrients of soil using optical transducer. In *2017 IEEE 4th International Conference on Smart Instrumentation, Measurement and Application* (*ICSIMA*) (pp. 1-4). IEEE.

38. Mendes, R. K., Carvalhal, R. F., Stach-Machado, D. R., and Kubota, L. T. (2009). Surface plasmon resonance immunosensor for early diagnosis of Asian rust on soybean leaves. *Biosensors and Bioelectronics, 24*(8), 2483-2487.

39. Ming, H., Wang, M., and Yin, H. (2015). Detection of Bacillus thuringiensis Cry1Ab protein based on surface plasmon resonance immunosensor. *Analytical biochemistry, 468,* 59-65.

40. mmrc.caltech.edu

41. Monosik, R., Stredansky, M., and Sturdik, E. (2012). Biosensors-classification, characterization and new trends. *Acta Chimica Slovaca, 5*(1), 109-120.\

42. Nagatsuka, T., Uzawa, H., Sato, K., Kondo, S., Izumi, M., Yokoyama, K.,. and Nishida, Y. (2013). Localized surface plasmon resonance detection of biological toxins using cell surface oligosaccharides on glyco chips. *ACS applied materials and interfaces, 5*(10), 4173-4180.

43. Neethirajan, S., Karunakaran, C., and Jayas, D. S. (2005). Biosensors: An emerging technology for the agricultural and food industry. In Proceedings of the CSAE/SCGR Meeting, June (pp. 26-29).

44. Nigam, V. K., and Shukla, P. (2015). Enzyme based biosensors for detection of environmental pollutants-A review. *J. Microbiol. Biotechnol, 25*(11), 1773-1781.

45. Niguyen N. T. and Angeles, D. E. (2012). Foliar sap NO_3-N in pineapple [Ananas comosus (L.) Merr.] at varying N levels and its relation with dry matter yield," Asia Life Sciences, 21(2),609–623.

46. Papadakis, G., Skandalis, N., Dimopoulou, A., Glynos, P., and Gizeli, E. (2015). Bacteria murmur: application of an acoustic biosensor for plant pathogen detection. *PloS one, 10*(7), e0132773.

47. Perumal, V., and Hashim, U. (2014). Advances in biosensors: Principle, architecture and applications. *Journal of applied biomedicine, 12*(1), 1-15.

48. Pohanka, M. (2018). Overview of piezoelectric biosensors, immunosensors and DNA sensors and their applications. *Materials, 11*(3), 448.

49. Radke, S. M., and Alocilja, E. C. (2005). A high density microelectrode array biosensor for detection of E. coli O157: H7. *Biosensors and Bioelectronics, 20*(8), 1662-1667.

50. Rai, V., Acharya, S., and Dey, N. (2012). Implications of nanobiosensors in agriculture. *Journal of Biomaterials and Nanobiotechnology, 3*(2A), 315.

51. Raul, R. G., Irineo, T. P., Gerardo, G. G. R., and Miguel, C. M. L. (2016). Biosensors Used for Quantification of Nitrates in Plants. *Journal of Sensors, 2016.*

52. Rotariu, L., Lagarde, F., Jaffrezic-Renault, N., and Bala, C. (2016). Electrochemical biosensors for fast detection of food contaminants–trends and perspective. *TrAC Trends in Analytical Chemistry, 79*, 80-87.

53. Salgado, A. M., Silva, L. M., and Melo, A. F. (2011). Biosensor for environmental applications. In Environmental Biosensors. IntechOpen.

54. Sharma, S., Byrne, H., and O'Kennedy, R. J. (2016). Antibodies and antibody-derived analytical biosensors. *Essays in biochemistry, 60*(1), 9-18.

55. Shruthi, G. S., Amitha, C. V., and Mathew, B. B. (2014). Biosensors: a modern day achievement. *J Instrument Technol, 2*, 26-39.

56. Su, L., Jia, W., Hou, C., and Lei, Y. (2011). Microbial biosensors: a review. *Biosensors and bioelectronics, 26*(5), 1788-1799.

57. Tang, W., Fan, K., Zhou, J., Wu, J., and Ji, F. (2015). Unmodified screen-printed silver electrode for facile detection of organophosphorus pesticide. *Ionics, 21*(2), 587-592.

58. Tang, Y., Zeng, X., and Liang, J. (2010). Surface plasmon resonance: an introduction to a surface spectroscopy technique. *Journal of chemical education, 87*(7), 742-746.

59. Thevenot, D. R., Toth, K., Durst, R. A., and Wilson, G. S. (1999). Electrochemical biosensors: recommended definitions and classification. *Pure and applied chemistry, 71*(12), 2333-2348.

60. Thouand, G. (2018). Microbial biosensors for analytical applications.

61. Vargas-Bernal, R., Rodríguez-Miranda, E., and Herrera-Pérez, G. (2012). Evolution and expectations of enzymatic biosensors for pesticides. *Pesticides-Advances in Chemical and Botanical Pesticides.*

62. Vasilescu, A., Nunes, G., Hayat, A., Latif, U., and Marty, J. L. (2016). Electrochemical affinity biosensors based on disposable screen-printed electrodes for detection of food allergens. *Sensors, 16*(11), 1863.

63. Wang, X., Lu, X., and Chen, J. (2014). Development of biosensor technologies for analysis of environmental contaminants. *Trends in Environmental Analytical Chemistry, 2*, 25-32.

64. Xie, B., Ramanathan, K., and Danielsson, B. (1999). Principles of enzyme thermistor systems: Applications to biomedical and other measurements. In *Thermal Biosensors, Bioactivity, Bioaffinitty* (pp. 1-33). Springer, Berlin, Heidelberg.

65. Zhao, Y., Liu, L., Kong, D., Kuang, H., Wang, L., and Xu, C. (2014). Dual amplified electrochemical immunosensor for highly sensitive detection of Pantoea stewartii sbusp. stewartii. *ACS applied materials and interfaces, 6*(23), 21178-21183.

66. Zourob, M., Ong, K. G., Zeng, K., Mouffouk, F., and Grimes, C. A. (2007). A wireless magnetoelastic biosensor for the direct detection of organophosphorus pesticides. *Analyst, 132*(4), 338-343.

Chapter 21

Nano Films and Nanoformulations from Bio Sources for Extending Shelf Life of Fruits

P. Divyabharathi[1], *D. Pradeep*[1] *and K.S. Subramaian*[2]

[1]*Department of Nano Science and Technology,* [2]*Director of Research, Tamil Nadu Agricultural University, Coimbatore – 641 003, Tamil Nadu*

Introduction

Fruits are a rich source of vitamins, minerals and dietary fibers which constitute an important role in our daily diet. India's diverse climate ensures the availability of all varieties of fresh fruits and ranks second in fruit production around the world, after China (Welfare, 2016). Amongst fruits, India ranks first in the production of banana (28.92 per cent), papaya (44.03 per cent) and mango (37.57 per cent). The vast production base offers India a tremendous opportunity for export. One of the main reasons for the lower availability is the large quantity of post-harvest losses and short shelf-life of fruits at various stages of marketing (Sachdeva *et al.*, 2013).

In India, raw material production is about 450 million tons that are the rich source of various nutrients. But the major part of these raw materials is wasted as post-harvest losses. In cereals it accounts for 10 percent, in semi-perishables, it accounts for 20 percent 30-40 percent of loss is from fruits and vegetables, and 25 percent loss in other food products such as milk, meat, fish and egg. The annual loss from fruits and vegetables is esteemed to be 16 million dollar every year.

Nanotechnology has many concepts to overcome the post-harvest loses by developing advanced packaging such as nano films and food preservation formulations from biological sources. Many functional molecules from biological origin such as hexanal, chitosan *etc.* can be developed in to nano formulation in different delivery modes for enhanced preservation of agro produces such as fruits and vegetables. This chapter provides an overview of various nano systems that can be used in food packaging and preservation

Development of Nano-Films from Agro-wastes

Packaging is the act of containing, protecting and presenting the contents as fresh from processing till the consumer end. Packaging, along with better transportation, has made it possible to centralize production facilities into areas where raw materials are concentrated and, therefore, take advantage of large-scale production (Hannen, 1998). Food packaging is the final operation of fresh produce processing that allows the products to be protected and safely distributed to consumers. Together with cold chain management, packaging allows the quality of fresh products to be preserved with extension of the shelf-life (Nicola and Fontana, 2014). Fruits are living products, undergoing ripening and at the end, an ageing process, in which the plant tissue is broken down. The products undergo various biological processes, which also continue after the products have been harvested. It is essential to minimize quick degradation and decomposition of fresh produce, in order to obtain optimal shelf-life. The use of suitable packaging is vital in this respect (Thompson, 1996).

Conventional Packaging

It is essential to minimize physical damage to fresh produce in order to obtain optimal shelf-life (Thompson, 1996). The most common form of packaging in this sector is the use of the fiber board carton; however, additional internal packaging, for example tissue paper wraps, trays, cups or pads, is required to reduce damage from abrasion. For the most delicate fruits, smaller packs with relatively few layers are used to reduce compression damage. Individual fruits may also be wrapped separately in tissue or waxed paper.

Use of Plastics as Packaging Material

Plastics are being used as a packaging material due to their advances in food industries. It has better oxygen and water vapour transmission properties. The flexible nature of the commercial film (Cling) is thus easy to use. But the main disadvantage of polythene/plastic packaging films is their non-biodegradable nature and persistence in the soil (J. Li, Yang, Li, Jabeen, and Shi, 2015). Commercially available films are made up of Low Density Polyethylene (LDPE) and due to their persistence nature, they pollute the environment more easily (Usha, Sangeetha, and Palaniswamy, 2011). The plastic/polythene-based industry in India is a well-developed commercial sector industry with an expected increase of 129 per cent by 2023. India ranks the highest in the usage of polythene/plastic produce of about 4 per cent. Most of these packages are made of plastic based on fossil fuels. Since these sources are finite, and the production and waste created by these packages have a large environmental impact, an alternative is needed to secure an economically and environmentally sustainable future. The synthetic packaging materials are made from petroleum-based monomers, and constitute serious environmental concern because they are practically non-degradable.

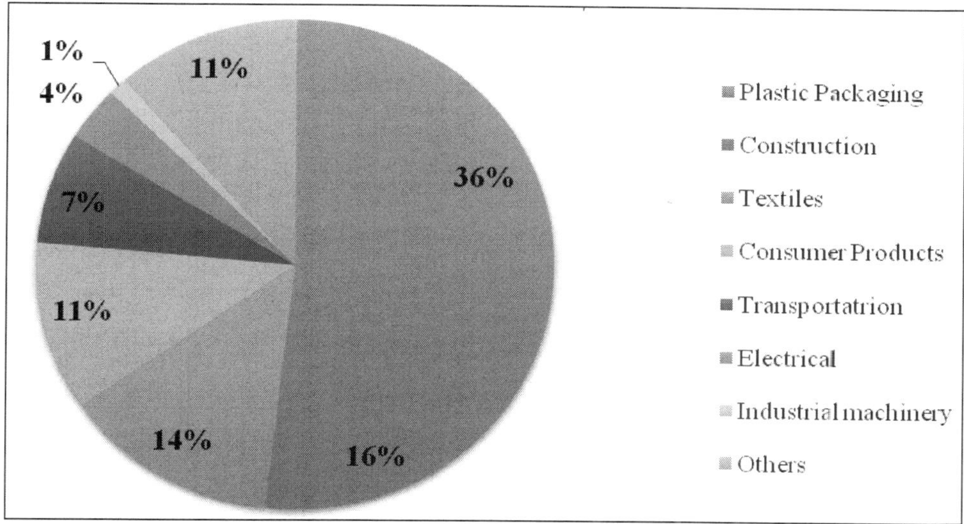

Figure 21.1. Use of Plastics in Various Fields of Applications
Courtesy: https://www.bloombergquint.com/charts/worlds-plastic-burden-weight-of-a-billion-african-elephants]

According to a report from the Central Pollution Control Board (CPCB), in September 2017 from 60 major cities in India, generates about 25,940 tonnes of plastic waste per day. Approximately 94 per cent of this contains recyclable thermoplastics, PET (polyethylene terephthalate) and PVC (polyvinyl chloride). The rest belong to thermoset and other plastics categories, such as sheet molding compound (SMC), fiber reinforced plastic (FRP) and multi-layer thermocol, which cannot be recycled.

India consumed 16.5 million tons of plastic in 2017-18 alone and according to FICCI, 43 per cent of India's plastics used in packaging are single-use plastics and the consumption clearly exceeded India's recycling capacity.

Biodegradable Packaging

Plastics that are bio-based and bio-degradable provide a promising alternative. These plastics require less or no fossil fuels and their production leads to less green-house gas (GHS) emissions than their petrol-based counterparts (Weiss *et al.*, 2012). Moreover, biodegradable plastics can fully degrade in several weeks to a few years, compared to a period lasting up to ten centuries for regular plastics. For example, polyvinyl alcohol (PVA) is found to decompose within three weeks under the right conditions (Nampoothiri, Nair, and John, 2010). Additionally, bio-plastics can be industrially composted with other biodegradable waste, such as food and garden waste, whilst petrol-based polymers are domestic waste and commonly end up in landfills or incineration ovens (Webb, Arnott, Crawford, and Ivanova, 2013)

Cellulose and cellulose derived materials show a remarkable consideration in the past decades due to the properties such as abundance, biodegradability, and renewability. Due to the properties such as high surface area, adjustable surface

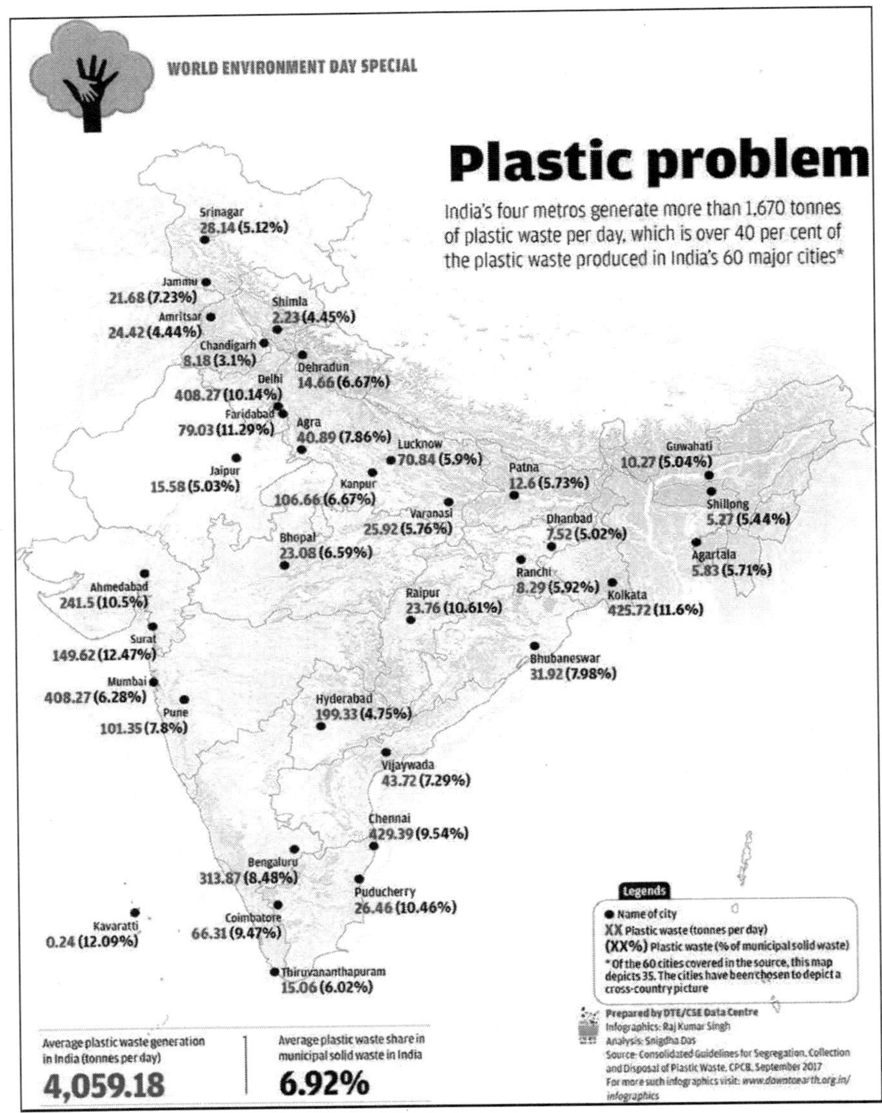

Figure 21.2. Plastic Waste Generation in Various States in India.
Courtesy: https://www.downtoearth.org.in/news/waste/an-indian-consumes-11-kg-plastic-every-year-and-an-average-american-109-kg-60745.

properties and biodegradability, nano cellulose has brought greater consideration than its macroscopic form (Brinchi, Cotana, Fortunati, and Kenny, 2013). In this concern, Nano Fibrillated Cellulose (NFC) derived from plant wastes proved to be a good alternative and can be used as packaging material for enhanced preservation of products and also against the damage caused by UV rays (Gopinathan, Subramanian, Paliyath, and Subramanian, 2017).

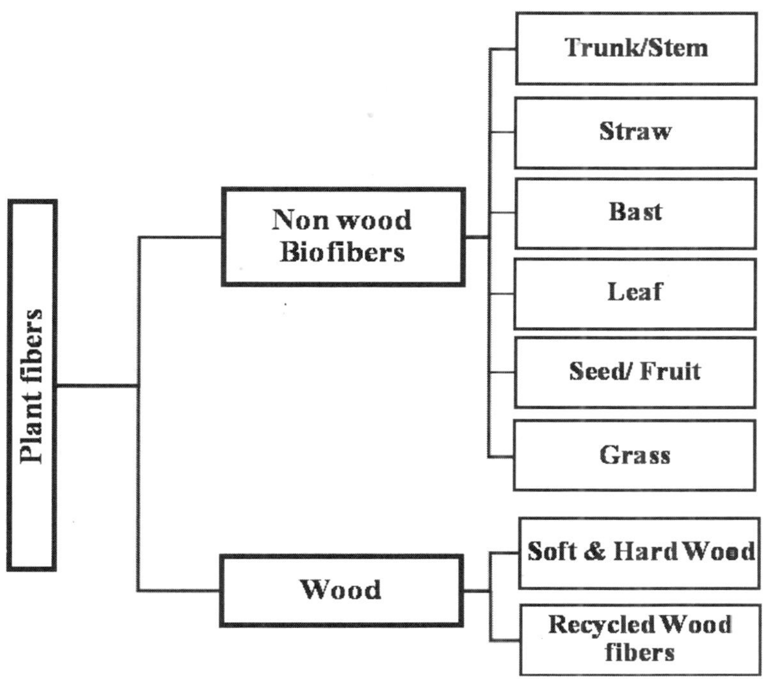

Figure 21.3. Schematic Representation of different Biofibers as Reinforcing Material.

Banana is one of the important fruit crops cultivated in tropical parts of the world. Cellulose is the major component of banana fiber and the highest cellulose content was recorded in Nendran peduncle fiber (60.27 per cent) followed by Nendran pseudostem fiber (59.23 per cent). (Mukhopadhyay *et al.*, 2008). Cellulose derived from banana pseudostem can be used in the production of micro cellulose, as well as nano cellulose.

Table 21.1. Cellulose Composition Derived from Plant Fibers

Sources	*Cellulose (per cent)*
Harwood	43-47
Softwood	40-44
Bagasse	40
Coir	32-43
Corn cobs	45
Corn stalks	35
Cotton	95
Flax	63
Hemp	70
Henequen	78

Sources	*Cellulose (per cent)*
Istle	73
Jute	71
Kenaf	36
Ramie	76
Sisal	73
Wheat straw	30
Banana fibre	70

Courtesy: Mohanty *et al.*, 2002.

Nano cellulose is referred as a nano structured cellulose. They may also be called as cellulose nanofibers (CNF) or microfibrillated cellulose (MFC) or nano fibrillated cellulose (NFC) based on the size of the fiber. Nano cellulose possesses high strength and stiffness and is abundant, biodegradable, and light weight. These advantages make nanocellulose a promising candidate for bio-nanocomposite production (Siró and Plackett, 2010). Cellulose nano fibers (CNFs) with a diameter less than 100 nm is an important type of nano cellulose. CNFs have large specific surface area and useful mechanical properties (Bhatnagar and Sain, 2005), and therefore they have been investigated for use in composites, coatings and films.

Table 21.2. Various Agro-residues for Nanocellulosic Ber Extraction

Various Sources (Nanocellulosic Fbers)	*References*
Banana pseudostem	(Abraham *et al.*, 2011)
Wood	(Chen *et al.*, 2011)
Cotton	(de Morais Teixeira *et al.*, 2010)
Sugar beet	(Dinand, Chanzy, and Vignon, 1999)
Hemp	(Wang, Sain, and Oksman, 2007)
Potato	(Dufresne, Dupeyre, and Vignon, 2000)
Soybean	(Wang and Sain, 2007)
Pineapple leaf	(Cherian *et al.*, 2010)
Banana rachis	(Zuluaga *et al.*, 2009)
Cladodes and spines	(Malainine *et al.*, 2003)
Sisal	(Morán, Alvarez, Cyras, and Vázquez, 2008)
Wheat straw and soy hulls	(Alemdar and Sain, 2008)
Coconut husk	(Rosa *et al.*, 2010)
Lemon and maize	(Rondeau-Mouro *et al.*, 2003)
Mulberry	(R. Li *et al.*, 2009)
Oil palm empty fruit bunch	(Fahma, Iwamoto, Hori, Iwata, and Takemura, 2010)

Edible films and coatings are generally based on biological materials such as proteins, lipids and polysaccharides. The main polysaccharides that can be included in edible coating formulations are starch and starch derivatives, cellulose derivatives, chitosan, pectin, alginate and other gums. Chitosan has an excellent film forming capability and is flexible and biodegradable. Polymers can be heat treated to enhance specific properties and dry heat treatment techniques to increase the water resistance for chitosan film (Viyapuri *et al.*, 2019). In the food industries, the pectin is used as gelling and stabilizing agents. These have been used in the preservation of many fruits and vegetables such as guava, apple, avocado, carrot, papaya, tomato *etc.* (Menezes and Athmaselvi, 2016). Xanthan gum is a macromolecular polysaccharide produced mainly from sugarcane, corn, or their derivatives by *Xanthomonas campestris* in aerobic conditions. The good biocompatibility of xanthan gum due to its natural origin could minimize allergic reactions against its products (Huang *et al.*, 2017).

Antimicrobial and antioxidant packaging are promising forms of active packaging. This packaging reduces the gas transmission rates and targets the microbes which contaminate the food in package. The use of edible coatings as carriers of antimicrobial compounds is another potential alternative to enhance safety of fresh-cut produce. Antimicrobial edible coatings may provide increased inhibitory effects against spoilage and pathogenic bacteria by maintaining effective concentrations of the active compounds on the food surfaces. There are several categories of antimicrobials that can be potentially incorporated into edible coatings, including organic acids (acetic, benzoic,lactic, propionic, sorbic), fatty acid esters (glyceryl monolaurate), polypeptides (lysozyme, peroxidase, lactoferrin, nisin), plant essential oils (cinnamon, oregano, lemon grass), nitrites and sulphites (Franssen and Krochta, 2003). Although several types of antimicrobials incorporated into edible coatings have been used to extend shelf-life of fresh commodities, their use in fresh-cut fruits is yet limited. Currently organic acids and plant essential oils are the main antimicrobial agents incorporated into edible coatings

Nanotechnology in Better Packaging of Food

The application of nanotechnology to the food sector offers significant potential benefits, including enhancement in production and processing. Nano-composite food packaging comprises the largest contribution of this technology in the food sector and the application is predicted to increase over the next two decades (Peelman *et al.*, 2013). Globally, the nanoparticle based industry may influence at least $3 trillion by 2020 (Duncan, 2011). According to worldwide development predictions, bio-nanotechnology will be responsible for development of innovative food packaging methods. The biggest concern of food industries is on increasing the shelf-life of the food products, which can be attained by encapsulated protective coating and advanced packaging system developed on the basis of these approaches.

Nano-composite film comprises of two phases, of which one phase acts as continuous phase/matrix system, and the second phase as the disperse phase/ filler system. Usually, polymer is a matrix system and the reinforcing material is the filler system. Commonly, nano-composites are those containing polymers with one organic or inorganic disperse phase (*i.e*) reinforcing material with a dimension lesser than 100 nm(Rhim, 2011). The assimilation of different nanoparticles such as SiO_2, TiO_2, $KMnO_4$, clay, nano cellulose, nano fibrillated cellulose and carbon nanotubes are capable of amending the packaging material in terms of mechanical and barrier properties (Mukherjee, Mondal, and Das, 2014). It has been reported that mechanical properties of packaging material such as tensile strength, shear strength, barrier property, thermal stability and optical characteristics can be enhanced by incorporating nanofillers into their matrices.

Extraction of Nano Fibrillated Cellulose

The department of Nano science and Technology, Tamil Nadu Agricultural University has standardized the extraction of NFC from banana pseudostem. The inner most portion of the banana pseudostem is chopped and used for effective fiber extraction. Prior to fiber extraction, the initial weight (IW) of the pseudostem is recorded. The fiber is extracted using a banana fiber extractor and the extracted fibers are allowed to dry under sunlight for about 24 hours. Then the fibers are chopped into small pieces to a size of 5-10 cm. Chopped fibers are treated with 2 per cent NaOH in a ratio of 1:10 and kept in an autoclave at 120°C, 20 lb pressure (Steam explosion) for about one hour (Alkali treatment). The fibers are washed with distilled water for the complete removal of alkali from the fiber. After alkali treatment, bleaching is done with the mixture of sodium hydroxide and acetic acid (2.7g and 7.5g) and sodium hypochlorite in the ratio of 1:3. The bleaching step is repeated for better removal of yellow lignin layer from banana fiber. Bleached fibers are treated with 10 per cent oxalic acid and autoclaved at a pressure of 20 lb. The Acid hydrolysis step is repeated eight times for better digestion and defibrillation. As suggested by Abraham *et al.* (2011), the fibers are taken out from autoclave and washed with distilled water to make them free from acid. The processed nano fibrils aresuspended in water and kept stirring with a mechanical stirrer at 1500 rpm for about 16 hours.

Development of Polymer Nanocomposite Films

Five types of nano-films have been developed, each with Chitosan, Pectin, Sodium alginate, Cellulose acetate and Xanthan gum + HPMC Polymer nano composite by solvent casting technique.

Research Study on Shelf Life Extension of Fruits

From the shelf-life studies of whole guava fruits, the commercial Cling film had shown only 4 days of shelf-life, whereas in nano-film developed with xanthan gum showed 3 more days enhancement, with the shelf-life of 7 days. In whole guava fruits, it was observed that, as the days increased, firmness and physiological loss in weight decreased. In case of fresh cut guava fruit, the commercial Cling film (check) had shown only three days of shelf life, whereas xanthan gum film showed three more days enhancement, with a shelf-life of five days.

Cellulose nano fibrils from the banana pseudostem fibers exhibited improved morphological, chemical and crystalline properties. These nano-fibrils can be used as a potential reinforcing component in the production of polymer nanocomposites.

Figure 21.3. Schematic Representation of Extraction of NFC.

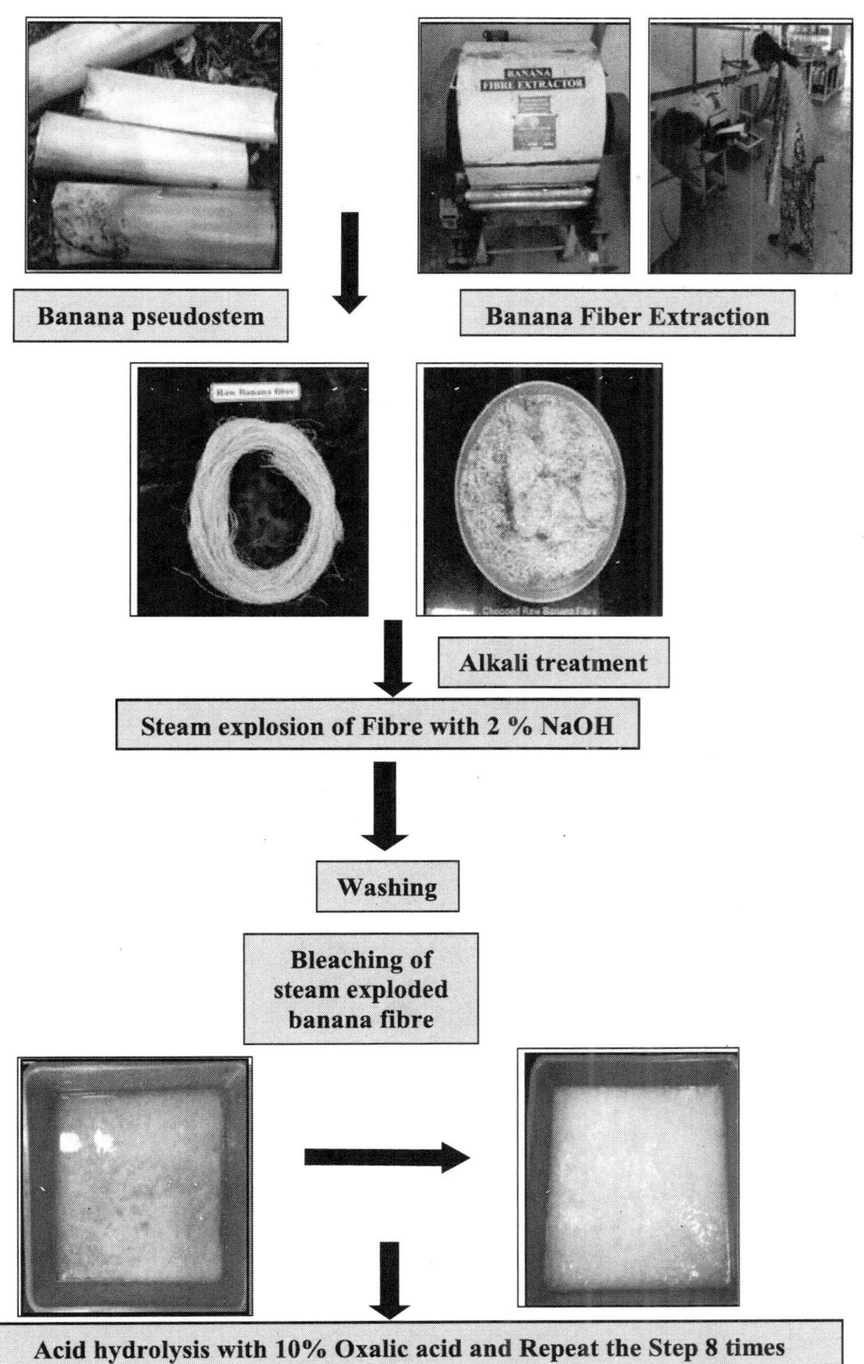

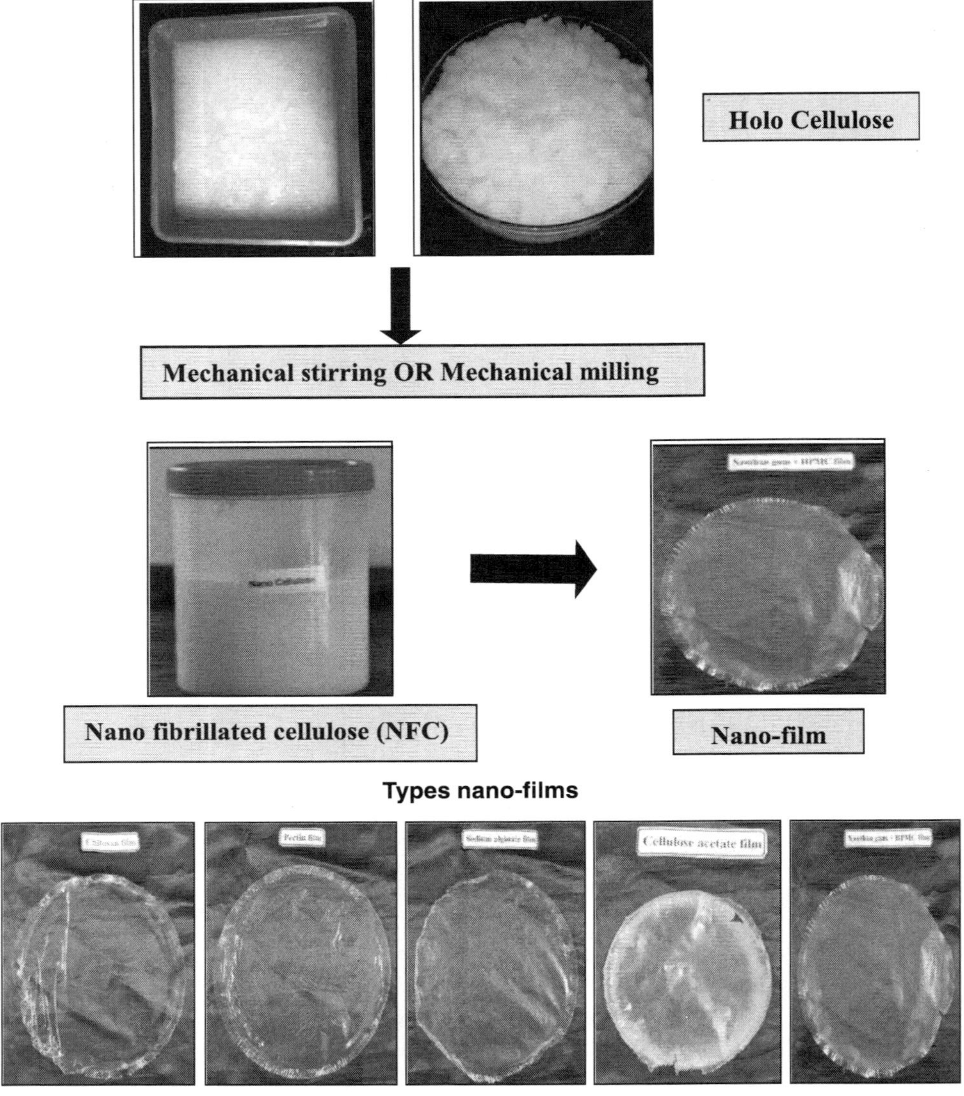

Steam coupled with alkali and acid treatment is the low cost and effective method in nano fibrils extraction. The, so produced, nano-fibrils from banana pseudostem possess desirable properties that can be used for producing bio-degradable nano packaging films. Eco-friendly nano-films from banana cellulose may serve as a better alternative for plastics, ensuring the environmental safety.

Table 21.4. Different Starch Based Edible and Biodegradable Films by Casting Technique

Starch Sources	*Other components*	*Plasticizer*	*References*
Sago starch	-	Glycerol	Abdorreza, Cheng, and Karim, 2011)
Corn starch	-	Glycerol	(Bertuzzi, Armada, and Gottifredi, 2007)
Cassava starch	Potassium sorbate	Glycerol	(Famá, Flores, Gerschenson, and Goyanes, 2006)
Tapioca starch	Potassium sorbate	Glycerol	(Flores, Conte, Campos, Gerschenson, and Del Nobile, 2007)
Pea starch	Bees wax	Glycerol	(Han, Seo, Park, Kim, and Lee, 2006)
Banana starch	Soy lecithin	Glycerol	(Osés, Niza, Ziani, and Maté, 2009)
Potato starch	Sunflower oil	Glycerol	(Pagella, Spigno, and De Faveri, 2002)
Maize starch	Wax of operculum	Glycerol	(Petersson and Stading, 2005)
Native potato starch	Grindsted Acetem 70-001	Glycerol	(Reddy and Yang, 2010)
Corn starch	Citric acid (Cross linking)	Glycerol	(Rodríguez, Oses, Ziani, and Mate, 2006)
Mango,banana, and okenia starch	Soy lecithin	Glycerol	(Romero-Bastida *et al.*, 2005)
Tapioca starch	Chitosan	Glycerol	(Vásconez, Flores, Campos, Alvarado, and Gerschenson, 2009)
Native starch from Cará roots	Mineral clay	Glycerol	(Wilhelm, Sierakowski, Souza, and Wypych, 2003)

Hexanal Based Nanoformulations in Extending the Shelf-Life of Fruits

Hexanal ($C_6H_{12}O$) is a naturally occurring volatile compound of C-6 aldehyde formed. It is usually evolved when a plant gets injured or wounded, it is also called as hexanldehyde/ caproaldehyde which is an oxidative derivative of long-chain fatty acid. Six carbon aldehyde with or without double bonds is dominant compounds released by plant material through the lipoxygenase pathway after plant damage (Vick and Zimmerman, 1987). Hexanal is an aldehyde compound with molecular weight 100.16 g mol^{-1}, the boiling point of 130-131°C and a melting point of <20°C. But having very low vapor pressure (Paliyath *et al.*, 1999).

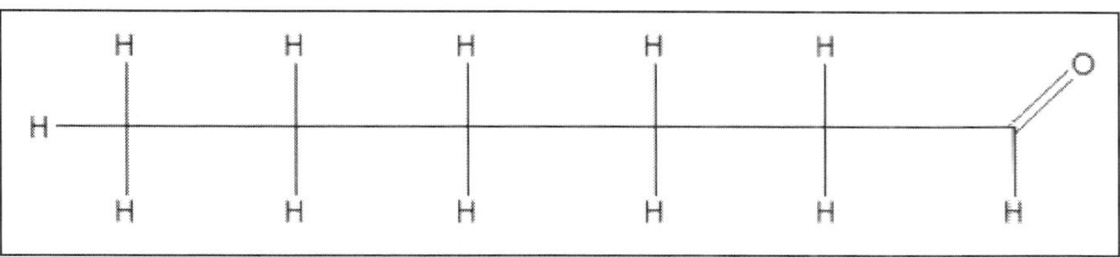

Figure 21.4: Chemical Structure of Hexanal.

Role of Phospholipase D (PLD) in the Fruit Ripening Process

The process where initial changes associated with ripening and senescence occurred at the cell membrane, affecting cellular compartmentalization and accelerating the senescence process (Paliyath *et al.*, 2008). Phospholipase D (PLD), a phospholipid degrading enzyme, catalyzes the hydrolysis of membrane phospholipids such as phosphatidylcholine, phosphatidylethanolamine, phosphatidylinositol and phosphatidylglycerol to yield phosphatidic acid, that triggers membrane deterioration. PLD is the key enzyme initiating a series of catabolic reactions that lead to the eventual deterioration of the membrane which is tightly regulated during growth and development of fruits. The onset of membrane deterioration of fruits leads to the softening of the skin and pulp of the fruits (Jincy *et al.*, 2017). However, during ripening, PLD is progressively stimulated in an autocatalytic manner resulting in massive membrane degradation and destabilization ultimately leading to senescence (Paliyath, 1992). Thus, PLD inhibition is one of the key physiological processes that convert in the preservation of fruits.

Role of Hexanal in Inhibition of PLD Enzyme

Hexanal is a compound found to be effective in delaying the ripening process in perishables like fruits and vegetables (Paliyath *et al.*, 2003). Hexanal is proved to be a more potent inhibitor due to the lack of hydroxyl group that interrupts the complete hydrolysis of substrate-enzyme intermediate. PLD inhibition technologies for the post-harvest preservation of fruits and vegetables (Paliyath and Murr, 2007) were developed based on this property (Paliyath *et al.*, 2003). Hexanal treatment showed promising results in enhancing the shelf-life of several fruits such as apple, banana, cherry, peach, strawberry as well as vegetables such as broccoli, tomato, and several fresh-cut vegetables, and flowers such as carnation and rose (Paliyath *et al.*, 2008).

Hexanal is an aldehyde compound easily volatile in nature and it has a low flash point at room temperature, so it can easily vaporize. Nanotechnology has a way to deliver hexanal via emulsion, nanofibre loaded with hexanal, cyclodextrin Metal-Organic Framework (MOF) and hexanal vapor technique. Enhanced Freshness Formulation (EFF) technically is an emulsion used in two ways as pre-harvest spray and post-harvest dip. Pre-harvest spray is one of the easy ways to deliver the hexanal on the field itself and is useful for farmers to spray hexanal in the field. The spray was done once or twice before harvest to retain the fruits on the tree for quite a long period which helps the farmer to harvest the fruit when the market value was high. Post-harvest treatment is mostly useful for the fruit pack houses, where they treat the fruit into hexanal solution which leads to retaining the fruit fresh until it reaches the consumers. Nanofiber and cyclodextrin MOF loaded with hexanal are feasible techniques for long-distance market and export markets. The sticker (nanofibre) and pellet (MOF) are placed inside the packaging boxes both these techniques release hexanal for a quite long time than other techniques. They dissipate hexanal slowly which improves the shelf life of fruits. Hexanal vapor technique may appropriate for bulk transportation of perishables. The container

Table 21.5. Effectiveness of Hexanal Based Nanoformulation in Extending Shelf Life of Temperate Fruits

Sl.No.	*Nature of Formulation*	*Fruits*	*Results*	*Reference*
1	Enhanced freshness formulation 2 (Pre-harvest spray) 0.02 per cent v/v	Wine Grape Varieties (Pinot Noir, Gamay, Cabernet Franc, and Merlot)	There was no change in total soluble solids (TSS), pH and total acidity. When came to the wine production the treated varieties cabernet and Gamay yield more than control and merlot produce less amount of wine than control because of low production of grapes	Oke *et al.*, 2018
2	Enhanced Freshness Formulation 1 (Pre-harvest spray)	Apple	The apple treated with EFF had improved in quality, free from pathogen infection and the bitter pit was reduced in hexanal treated fruit	Padmanaban *et al.*, 2018
3	Enhanced Freshness formulation 1 and 2 (Pre-harvest spray)	Blueberry	The ripening of fruit was not affected by the pre-harvest spray. The size, shape, and weight of the fruit were similar when compared to control. Fruits are kept at two different temperatures 4°C and 21°C TSS content was increased in EFF 1 and decreased in EFF 2. Decreased in total acidity was observed in fruit stored at 21°C and no difference at 4°C. coming to firmness fruit stored at 21°C loss their firmness while 4°C maintained their firmness than control	Padmanaban *et al.*, 2018
4	Enhanced Freshness Formulation 2 (Pre-harvest spray and post-harvest vapor)	Sweet cherry	The treated fruit had increased in cherry color and the intensity of the color reduced to normal cherry red which indicates the delayed ripening of fruit and there is no change in Brix and antioxidant content in the pre-harvest spray While coming to post-harvest vapor there are no significant changes among control and treated fruits	Padmanabhan, P., and Paliyath, G. (2018)

Table 21.6. Effectiveness of Hexanal Based Nanoformulation in Extending Shelf Life of Tropical Fruits

Sl.No.	*Nature of Formulation*	*Fruits*	*Results*	*Reference*
1	Pre-harvest spray Enhanced Freshness Formulation (EFF) 2 per cent	Mango variety (Alphonso and banganapalli)	The hexanal sprayed trees can retain the fruit for 2-3weeks and the fruits treated with hexanal having shelf life period of 4-6 weeks	Anusuya *et al.* (2016)
2	Pre-harvest spray followed by a post-harvest dip with Enhanced Freshness Formulation (EFF) 2 per cent	Mango varieties (Banganapalli and Alphonso)	The shelf life increased in treated fruit than control in ambient storage 3-4 days and cold storage 6-8 days than control	Preethi *et al.* (2018)
3	Pre-harvest spray hexanal at 2 per cent and 3 per cent	Banana variety grand nain	Hexanal improves the retention of fruit in the plant for 12-18 days and the shelf life of fruit is of 30 days	Hutchinson, M. (2018)

Sl.No.	*Nature of Formulation*	*Fruits*	*Results*	*Reference*
4	Post-harvest Dip	Banana cultivar red banana	The shelf life of EFF treated fruit was 12days to that of untreated is 8days.	Chitra, R. (2017)
5	Post-harvest Dip 0.02 per cent	Mango variety neelum	The shelf life of fruit increase 2 days than control	Jincy *et al.* (2017)
6	Post-harvest Dip	Banana variety grand nain at 85 per cent maturity	6 days increased shelf life for treated one over control	Venkatachalam, *et al.* (2018)
7	Electrospun Nanofibre (PVA, PVA+NFC Grand nain and PVA+NFC Poovan)	Mango	Shelf life extended up to 11-12 days	Badal Kumar biswal. (2014)
8	Electrospun Nanofibre (PVA+ β-cyclodextrin)	Mango variety Alphonso	The fruit shelf life extended up to 20 days under storage condition	Shanmugapriya, R. (2016)
9	Electrospun nanofibre (PVA+ β-cyclodextrin) overlaid with PLGA	Mango variety Alphonso	The shelf life extended up to 11 days than control	Preetha, S. (2017)
10	Electrospun nanofibre (PVA+ β-cyclodextrin) over the cellulosic substrate	Banana variety Nendran and Grand Nain	Shelf life extended up to 2 days in hexanal exposed fruit (Grand Nain) and 3 days extended shelf life than control in case of nendran variety	Pradeep, D. (2018)
11	Cyclodextrin hexanal inclusion complex	Mango variety neelum	The shelf life of fruit extended up to 14 days under ambient storage condition	Vivek Kumar, R. (2015)
12	Cyclodextrin Metal-Organic Framework (CDMOF)	Mango variety neelum	The shelf life of fruit extended up to 8 days	Jayanthi, R. (2017)
13	β-cyclodextrin Metal-Organic Framework	Banana variety Nendran	The shelf life of fruit extended up to 12 days	Prasanna Kumar, K. (2018)
14	Hexanal Vapor	Banana variety Grand Nain	Treated fruit extend shelf life up to 18 days whereas control shelf life up to 8 days	Ashwini *et al.* (2018)

or in the warehouses the fruits are fumigated towards hexanal which extends the fruit shelf life in storage houses itself.

Biosafety of Hexanal Nanoformulation

The Enhanced Freshness formulation tested in honey bee at different concentration from 0.02 per cent to 0.06 per cent showed the mortality rate of 4-7 per cent, which is lesser than the threshold level (Mohan *et al.*, 2017). Biosafety studies conducted on earthworm, beneficial microorganisms and natural enemies did not show toxicity (Karthika *et al.*, 2015). In the entire above-mentioned organism except human cell line the concentration of EFF tested was 0.02 per cent - 0.06 per cent, and there is no harmful effect recorded including human cell line even under 123 to 30000 ppm(Gunasekaran *et al.*, 2015). The results revealed that there is no adverse effect on the tested organisms. It is safe to use hexanal formulation at 0.02 per cent to enhance the shelf life of fruits.

Conclusion

Nanotechnology has the potential to minimize the post harvest losses through nano films and nano formulations. It would be advisable to use natural polymer based nano systems in food preservation than engineered nanoparticles. Using natural or biological source based nano systems offers advantages such as recycling of agro wastes in to useful products besides low or no toxicity to human beings and other flora and fauna.

REFERENCES

1. Abdorreza, M. N., Cheng, L., and Karim, A. (2011). Effects of plasticizers on thermal properties and heat sealability of sago starch films. *Food Hydrocolloids, 25*(1), 56-60.
2. Abraham, E., Deepa, B., Pothan, L., Jacob, M., Thomas, S., Cvelbar, U., and Anandjiwala, R. (2011). Extraction of nanocellulose fibrils from lignocellulosic fibres: A novel approach. *Carbohydrate Polymers, 86*(4), 1468-1475.
3. Alemdar, A., and Sain, M. (2008). Isolation and characterization of nanofibers from agricultural residues–Wheat straw and soy hulls. *Bioresource technology, 99*(6), 1664-1671.
4. Anusuya, P., Nagaraj, R., Janavi, G. J., Subramanian, K. S., Paliyath, G., and Subramanian, J. (2016). Pre-harvest sprays of hexanal formulation for extending retention and shelf-life of mango (*Mangifera indica* L.) fruits. *Scientia horticulturae, 211*, 231-240.
5. Ashwini, T., Ganapathy, S., Subramanian, K. S., Rani, C. I., and Meenakshi, G. G. (2018). Effect of Hexanal Vapour on Postharvest Quality and Shelf Life of Banana var. Grand Naine. *Int. J. Curr. Microbiol. App. Sci, 7*(2), 2441-2450
6. Badal kumar biswal. (2014). Controlled release of hexanal by using electrospun fibre matrix. (M.Sc.,), TNAU, Coimbatore.
7. Bertuzzi, M. A., Armada, M., and Gottifredi, J. (2007). Physicochemical characterization of starch based films. *Journal of Food Engineering, 82*(1), 17-25.

8. Brinchi, L., Cotana, F., Fortunati, E., and Kenny, J. (2013). Production of nanocrystalline cellulose from lignocellulosic biomass: technology and applications. *Carbohydrate polymers, 94*(1), 154-169.

9. Chen, W., Yu, H., Liu, Y., Chen, P., Zhang, M., and Hai, Y. (2011). Individualization of cellulose nanofibers from wood using high-intensity ultrasonication combined with chemical pretreatments. *Carbohydrate Polymers, 83*(4), 1804-1811.

10. Cherian, B. M., Leão, A. L., De Souza, S. F., Thomas, S., Pothan, L. A., and Kottaisamy, M. (2010). Isolation of nanocellulose from pineapple leaf fibres by steam explosion. *Carbohydrate Polymers, 81*(3), 720-725.

11. de Morais Teixeira, E., Corrêa, A. C., Manzoli, A., de Lima Leite, F., de Oliveira, C. R., and Mattoso, L. H. C. (2010). Cellulose nanofibers from white and naturally colored cotton fibers. *Cellulose, 17*(3), 595-606.

12. Dinand, E., Chanzy, H., and Vignon, R. (1999). Suspensions of cellulose microfibrils from sugar beet pulp. *Food Hydrocolloids, 13*(3), 275-283.

13. Dufresne, A., Dupeyre, D., and Vignon, M. R. (2000). Cellulose microfibrils from potato tuber cells: processing and characterization of starch–cellulose microfibril composites. *Journal of Applied Polymer Science, 76*(14), 2080-2092.

14. Duncan, T. V. (2011). Applications of nanotechnology in food packaging and food safety: barrier materials, antimicrobials and sensors. *Journal of colloid and interface science, 363*(1), 1-24.

15. Fahma, F., Iwamoto, S., Hori, N., Iwata, T., and Takemura, A. (2010). Isolation, preparation, and characterization of nanofibers from oil palm empty-fruit-bunch (OPEFB). *Cellulose, 17*(5), 977-985.

16. Famá, L., Flores, S. K., Gerschenson, L., and Goyanes, S. (2006). Physical characterization of cassava starch biofilms with special reference to dynamic mechanical properties at low temperatures. *Carbohydrate Polymers, 66*(1), 8-15.

17. Flores, S., Conte, A., Campos, C., Gerschenson, L., and Del Nobile, M. (2007). Mass transport properties of tapioca-based active edible films. *Journal of Food Engineering, 81*(3), 580-586.

18. Franssen, L., and Krochta, J. (2003). Edible coatings containing natural antimicrobials for processed foods *Natural antimicrobials for the minimal processing of foods* (pp. 250-262): Elsevier.

19. Gopinathan, P., Subramanian, K. S., Paliyath, G., and Subramanian, J. (2017). Genotypic Variations in Characteristics of Nano-Fibrillated Cellulose Derived from Banana Pseudostem. *BioResources, 12*(4), 6984-7001.

20. Gunasekaran, K., Karthika, S., Nandakumar, N. B., Subramanian, K. S., Paliyath, G., and Subramanian, J. (2015). Biosafety of Hexanal.

21. Han, J., Seo, G., Park, I., Kim, G., and Lee, D. (2006). Physical and mechanical properties of pea starch edible films containing beeswax emulsions. *Journal of Food Science, 71*(6), E290-E296.

22. Hannen, J. (1998). Method and apparatus for packaging a roll of material: Google Patents.

23. Hegazy, R. (2013). Post-harvest situation and losses in India. Accessed online at https: //www.researchgate.net/publication/301770292_Postharvest_Situation_and_Losses_in_India.

24. Hutchinson, M. (2018). Application of Hexanalcontaining Compositions and Its Effect on Shelflife and Quality of Banana Varieties in Kenya. *Postharvest Biology and Nanotechnology*, 191-198.

25. Jayanthi, R. (2017). Cyclodextrine metal organic famework (cdmof) as delivery system for hexanal to enhance the shelf life of fruits. (M.Sc.,), TNAU, Coimbatore.

26. Jincy, M., Djanaguiraman, M., Jeyakumar, P., Subramanian, K. S., Jayasankar, S., and Paliyath, G. (2017). Inhibition of phospholipase D enzyme activity through hexanal leads to delayed mango (*Mangifera indica* L.) fruit ripening through changes in oxidants and antioxidant enzymes activity. *Scientia Horticulturae, 218*, 316-325.

27. Karthika, S., Kumar, N. B. N., Gunasekaran, K., and Subramanian, K. S. (2015). Biosafety of nanoemulsion of hexanal to honey bees and natural enemies. Indian J Sci Technol, 8, 1-7.

28. Li, J., Yang, D., Li, L., Jabeen, K., and Shi, H. (2015). Microplastics in commercial bivalves from China. *Environmental Pollution, 207*, 190-195.

29. Li, R., Fei, J., Cai, Y., Li, Y., Feng, J., and Yao, J. (2009). Cellulose whiskers extracted from mulberry: A novel biomass production. *Carbohydrate Polymers, 76*(1), 94-99.

30. Malainine, M. E., Dufresne, A., Dupeyre, D., Mahrouz, M., Vuong, R., and Vignon, M. R. (2003). Structure and morphology of cladodes and spines of Opuntia ficus-indica. Cellulose extraction and characterisation. *Carbohydrate Polymers, 51*(1), 77-83.

31. Menezes, J., and Athmaselvi, K. (2016). Study on effect of pectin based edible coating on the shelf life of sapota fruits. *Biosciences Biotechnology Research Asia, 13*(2), 1195-1199.

32. Mohan, C., Sridharan, S., Subramanian, K. S., Natarajan, N., and Nakkeeran, S. (2017). Effect of nanoemulsion of hexanal on honey bees (Hymenoptera; Apidae).

33. Morán, J. I., Alvarez, V. A., Cyras, V. P., and Vázquez, A. (2008). Extraction of cellulose and preparation of nanocellulose from sisal fibers. *Cellulose, 15*(1), 149-159.

34. Mukherjee, D. K., Mondal, A., and Das, A. (2014). Alcohol Oxidation in Ionic Liquids Catalysed by Recyclable Platinum Nanoparticles: A Green Approach. *Journal of Catalyst and Catalysis, 1*(1), 20-32.

35. Nampoothiri, K. M., Nair, N. R., and John, R. P. (2010). An overview of the recent developments in polylactide (PLA) research. *Bioresource technology, 101*(22), 8493-8501.

36. Nicola, S., and Fontana, E. (2014). Fresh-cut produce quality: implications for a systems approach *Postharvest handling* (pp. 217-273): Elsevier.

37. Oke, M., Padmanabhan, P., and Paliyath, G. (2018). Effect of Hexanal Composition Treatment on Wine Grape Quality. *Postharvest Biology and Nanotechnology*, 215-223.

38. Osés, J., Niza, S., Ziani, K., and Maté, J. I. (2009). Potato starch edible films to control oxidative rancidity of polyunsaturated lipids: effects of film composition, thickness and water activity. *International journal of food science and technology, 44*(7), 1360-1366.

39. Padmanabhan, P., and Paliyath, G. (2018). Benefits of Application of Hexanal Compositions on Apples. *Postharvest Biology and Nanotechnology*, 225-230.

40. Padmanabhan, P., and Paliyath, G. (2018). Improving Shelflife and Quality of Sweet Cherry (Prunus avium L.) by Preharvest Application of Hexanal Compositions. *Postharvest Biology and Nanotechnology*, 237-242.

41. Padmanabhan, P., Krishnaraj, T., Sullivan, A., and Paliyath, G. (2018). Preharvest Spray Application of Blueberry Fruits with Hexanal Formulations Improves Fruit Shelflife and Quality. *Postharvest Biology and Nanotechnology*, 231-236.

42. Pagella, C., Spigno, G., and De Faveri, D. (2002). Characterization of starch based edible coatings. *Food and Bioproducts Processing, 80*(3), 193-198.

43. Paliyath, G., and Murr, D. P. (2007). Compositions for the preservation of fruits and vegetables: Google Patents.

44. Paliyath, G., Pinhero, R. G., Yada, R. Y., Murr, D. P. J. J. O. A., and Chemistry, F. (1999). Effect of processing conditions on phospholipase D activity of corn kernel subcellular fractions. 47(7), 2579-2588.

45. Paliyath, G., Subramanian, J. J. P. B., Technology of Fruits, V., and Flowers. (2008). Phospholipase D inhibition technology for enhancing shelf-life and quality. 240.

46. Paliyath, G., Yada, R., Muir, D. P., and Pinhero, R. G. (2003). Inhibition of phospholipase D: Google Patents.

47. Peelman, N., Ragaert, P., De Meulenaer, B., Adons, D., Peeters, R., Cardon, L.,. Devlieghere, F. (2013). Application of bioplastics for food packaging. *Trends in Food Science and Technology, 32*(2), 128-141.

48. Petersson, M., and Stading, M. (2005). Water vapour permeability and mechanical properties of mixed starch-monoglyceride films and effect of film forming conditions. *Food Hydrocolloids, 19*(1), 123-132.

49. Pradeep, D. (2018). Development of Cellulosic Nano-Matrix to Extend Shelf-life of Fruits. (M.Sc.,), TNAU, Coimbatore.

50. Prasanna kumar, K. (2018). Studies on assessing release kinetics and upscaling of â-cyclodextrin metal organic frame work- hexanal complex technology for enhancing shelf life of fruits. (M.Sc.,), TNAU, Coimbatore.

51. Preetha, S. (2017). Development of multilayer electrospun matrices for controlled release of hexanal to extend the shelf-life of fruits. (M.Sc.,), TNAU, Coimbatore.

52. Preethi, P., Soorianathasundaram, K., Sadasakthi, A., Subramanian, K. S., Paliyath, G., and Subramanian, J. (2018). Influence of Hexanal formulation on storage life and post-harvest quality of mango fruits. *Journal of Environmental Biology, 39*(6), 1006-1014.

53. Reddy, N., and Yang, Y. (2010). Citric acid cross-linking of starch films. *Food chemistry, 118*(3), 702-711.

54. Rhim, J.-W. (2011). Effect of clay contents on mechanical and water vapor barrier properties of agar-based nanocomposite films. *Carbohydrate polymers, 86*(2), 691-699.

55. Rodríguez, M., Oses, J., Ziani, K., and Mate, J. I. (2006). Combined effect of plasticizers and surfactants on the physical properties of starch based edible films. *Food Research International, 39*(8), 840-846.

56. Romero-Bastida, C. A., Bello-Pérez, L. A., García, M. A., Martino, M. N., Solorza-Feria, J., and Zaritzky, N. E. (2005). Physicochemical and microstructural characterization of films prepared by thermal and cold gelatinization from non-conventional sources of starches. *Carbohydrate Polymers, 60*(2). 235-244.

57. Rondeau-Mouro, C., Bouchet, B., Pontoire, B., Robert, P., Mazoyer, J., and Buléon, A. (2003). Structural features and potential texturising properties of lemon and maize cellulose microfibrils. *Carbohydrate Polymers, 53*(3), 241-252.

58. Rosa, M., Medeiros, E., Malmonge, J., Gregorski, K., Wood, D., Mattoso, L.,. Imam, S. (2010). Cellulose nanowhiskers from coconut husk fibers: Effect of preparation conditions on their thermal and morphological behavior. *Carbohydrate Polymers, 81*(1), 83-92.

59. Sachdeva, S., Sachdev, T. R., and Sachdeva, R. (2013). Increasing fruit and vegetable consumption: challenges and opportunities. Indian journal of community medicine: official publication of Indian Association of Preventive and Social Medicine, 38(4), 192.

60. Shanmugapriya, R. (2016). Development of electrospun fiber matrices to extend the shelf-life of fruits. (M.Sc.,), Coimbatore, TNAU.

61. Thompson, A. K. (1996). *Postharvest technology of fruit and vegetables*: Blackwell science.

62. Usha, R., Sangeetha, T., and Palaniswamy, M. (2011). Screening of polyethylene degrading microorganisms from garbage soil. *Libyan agriculture research center journal international, 2*(4), 200-204.

63. Vásconez, M. B., Flores, S. K., Campos, C. A., Alvarado, J., and Gerschenson, L. N. (2009). Antimicrobial activity and physical properties of chitosan–tapioca starch based edible films and coatings. *Food Research International, 42*(7), 762-769.

64. Venkatachalam, K., Muthuvel, I., Sundaresan, S., Subramanian, K. S., Janaki, J. G., Sullivan, J. A.,. and Subramanian, J. (2018). Post-harvest dip of enhanced freshness formulation to extend the shelf life of banana (Musa acuminata cv. Grand Naine) in India. *The Journal of the Faculty of Food and Agriculture,* 1.

65. Vick, B. A., and Zimmerman, D. C. (1987). Oxidative systems for modification of fatty acids: the lipoxygenase pathway Lipids: structure and function (pp. 53-90): Elsevier.

66. Vivek kumar, R. (2015). Hexanal and â Cyclodextrin complex for slow release of hexanal to enhance shelf life of Mango (*Mangifera indica*). (M.Sc.,), TNAU, Coimbatore

67. Wang, B., and Sain, M. (2007). Isolation of nanofibers from soybean source and their reinforcing capability on synthetic polymers. *Composites Science and Technology, 67*(11-12), 2521-2527.

68. Wang, B., Sain, M., and Oksman, K. (2007). Study of structural morphology of hemp fiber from the micro to the nanoscale. *Applied Composite Materials, 14*(2), 89.

69. Webb, H., Arnott, J., Crawford, R., and Ivanova, E. (2013). Plastic degradation and its environmental implications with special reference to poly (ethylene terephthalate). *Polymers, 5*(1), 1-18.

70. Weiss, M., Haufe, J., Carus, M., Brandão, M., Bringezu, S., Hermann, B., and Patel, M. K. (2012). A review of the environmental impacts of biobased materials. *Journal of Industrial Ecology, 16,* S169-S181.

71. Wilhelm, H.-M., Sierakowski, M.-R., Souza, G., and Wypych, F. (2003). Starch films reinforced with mineral clay. *Carbohydrate Polymers, 52*(2), 101-110.

72. Zuluaga, R., Putaux, J. L., Cruz, J., Vélez, J., Mondragon, I., and Gañán, P. (2009). Cellulose microfibrils from banana rachis: Effect of alkaline treatments on structural and morphological features. *Carbohydrate Polymers, 76*(1), 51-59.

Chapter 22

Nanotechnology for Enhanced Food Nutrition, Food Safety Assessment and Green Nanopackaging

Haripriya Shanmugam

Department of Nano Science and Technology, Tamil Nadu Agricultural University, Coimbatore – 641 003, Tamil Nadu
Email: haripriya.s@tnau.ac.in

Application of nanotechnology is penetrating different layers of food and food packaging Industries across the globe due to the multi-functional unique properties exhibited by nanomaterials. For example, researchers at Nottingham University, United Kingdom have developed nanoscale salt particles that increase the saltiness of food without increasing the amount of salt consumed. Likewise, 40 per cent less fat mayonnaise was prepared by employing nanoemulsion technique. These are considered to be a boon for people consuming less salty and fat foods. On the other hand, exploration of Nano-electromechanical systems (NEMS) embedded biosensor are bestowed with rapid detection, high sensitivity and reproducibility in sensing phenomenon against pathogenic microorganisms, pesticide residues and harmful analytes ensuring food safety. Nanotechnology is revolutionizing the food packaging Industry and an example is bottles made from nanocomposites to minimize leakage of carbon dioxide. In this chapter, the need, and recent studies on applying nanotechnology for food fortification for nutritional security, development of nanobiosensors for effective food safety assessment, environment friendly nanopackaging materials and concerns associated with food based nanomaterials are discussed.

Nanotechnology in Food Fortification

According to a report by World Health Organization, more than two billion people are suffering from vitamins and minerals deficiency across the globe, with pregnant as well as lactating women and young children, being the most vulnerable. Food fortification is recommended by the Copenhagen Consensus, the Food and Agriculture Organization and the World Health Organization, as one of the effective strategy for reducing global population suffering from micronutrient malnutrition, which is also known as "Hidden hunger". Globally, the three most common forms of micronutrient malnutrition are due to iron, vitamin A and iodine deficiency.

Food fortification is a process, wherein micronutrients are added to a particular food at required levels than what the original food provides and the concentration to be added is population specific. The common food fortifying agents are vitamins, minerals, proteins, prebiotics, probiotics, plant bioactives *etc.*, which are necessary for consumption in sufficient quantity and failure of which results in micronutrient deficiency affecting normal functioning of the human body. Generally, food fortifications are designed either through incorporation into staple foods, condiments or processed foods having preferences specific to geographic demarcation. For example, incorporation of the fortifying agents may be introduced either into staple foods (rice, wheat, corn, and oils), condiments (salt, soy sauce, sugar) and processed foods (breakfast cereals, infant complementary foods, dairy products). The major criteria to be considered while incorporating the fortifying agents into the chosen food or product is its stability without altering the color, taste and appearance of the original food. In practice, iodine was fortified with salt, folate through grains or grain products, calcium in milk and vitamin A in cooking oil (Das, Salam, Kumar, and Bhutta, 2013). Vitamin D fortification of milk reduced the risk of osteoporosis in the elderly, especially in higher latitude regions with lower ultraviolet light incident during winter (Kinyamu, Gallagher, Rafferty, and Balhorn, 1998)

Biofortification is another approach to enrich micronutrient content of the agriculture produces especially stable food crops through breeding for increased mineral and vitamin content. Iron, zinc and provitamin A are the important component of biofortification program. Indian Council of Agricultural Research, New Delhi developed bio-fortified crop varieties like rice, DRI Dhan 62 (zinc fortified) in 2013 and iron enriched pulses variety, RWR 2245 in 2010.WB02, the first zinc and iron fortified wheat lines were commercialized for test marketing in India and ICTP-8203-Fe, first iron enriched pearlmillet variety was commercialized in Maharashtra. However, biofortified crop varieties with enhanced nutritional qualities must be evaluated for the bioavailability and its effect on consumer's health.

The best way to prevent micronutrient malnutrition is to consume a balanced diet. Food fortification is likely to have a less immediate impact on short terms; however it might have a much broader and more sustained impact on long terms. Although increasing dietary diversity is generally regarded as the most desirable option, it takes the longest time needed to implement. In fact, fortification is being used in developed countries for more than 80 years as a means to restore micronutrients lost during food processing like B vitamins (thiamine, riboflavin and niacin). For example, salt iodization introduced in both Switzerland (Bürgi, Supersaxo, and Selz, 1990) and the United States of America during early 1920s has since expanded progressively all over the world to the extent that iodized salt is now used in most countries.

Multiple micronutrients fortification was thought of as an effective way to combat hidden hunger. It was reported to have a better impact on human health than use of a single fortificant. In South Africa, fortification of biscuits with iron, β-carotene and iodine improved the nutrient status of the school childrens (Van Stuijvenberg, Dhansay, Lombard, Faber, and Benade, 2001). Likewise, in Botswana, regular consumption of a twelve micronutrients enriched beverage by school childrens improved their iron, folate, riboflavin and zinc status, apart from weight gain(Latham and Jones, 2001). However, compatibility of combined essential nutrients with different release pattern and stimuli responsive nanoformulations is possible with nanofortification foods. Fortified food products and beverages prepared using nanotechnology with outcome are listed in Table 22.1.

Nutrient bioavailability refers to the proportion of a nutrient that is absorbed from the diet and used for normal body functions. Milk is considered to be a good source of lipophilic (A, D, E, and K) and hydrophilic (B Vitamins and C vitamins) vitamins as well as minerals on everyday consumption. Gaucheron (2011) reported that the presence of fat soluble Vitamins A, D, E, and K in the lipid phase and water soluble vitamins of group B and C in the aqueous phase of the milk. In this case, due to the hydrophobic nature of lipophilic vitamins, its bioavailability is affected resulting in micronutrient deficiency despite its intake. Another case of poor mineral bioavailability was recorded with calcium, wherein poor dissolution rate and poor permeability capacity into the epithelial membrane during its residence time in the gastrointestinal tract affects its bioavailability. Nanoformulation of micronutrients either as single nutrient or multinutrient composition can be designed to overcome poor bioavailability and permeability issues associated with essential vitamins and minerals.

Nanotechnology for Food Safety Assessment

Currently, consumers awareness on food safety have made the food suppliers involved in food chain distribution shoulder the responsibility to adopt practices of traceability and accountability across the globe. The need for rapid detection techniques based on reproducible methods for food quality and safety assessment is on the rise. Since Arthur C. Clark developed the first biosensor in 1962 for glucose detection, the field of biosensors have been intensively studied and designed to suit various applications, especially for food safety. A biosensor designed for assessing food quality is a device that transforms the response based on the properties of food into a detectable signal, mostly an electric signal. Among the biological elements utilized towards constructing biosensors, enzymes are the most widely used recognition element due to their unique specificity and sensitivity (Souza *et al.*, 2012). However, enzyme purification is considered a costly affair, apart from time-consuming task. Nanobiosensors are devices developed to detect pathogens and quantify low concentrations of pesticide residues and harmful chemicals. According to (Otles and Yalcin, 2010) nano sensors have the potential to exhibit high sensitivity, rapid response, and accuracy by exploitation of NEMS in the biosensing component.

Table 22.1. Reported Fortified Food Products and Beverages using Nanotechnology with the Study Outcome

Sl.No.	*Fortified Product*	*Fortificants*	*Techniques Explored*	*Outcome*	*References*
1	Vitamin E nanoparticle for beverages	Vitamin E	Microfluidizer	Improved stability of Vitamin E nanoparticle for beverage applications	Chen *et al.* (2004)
2	Ferric pyrophosphate nanoparticles based Infant cereals and yogurt	Micronized and dispersible Ferric pyrophosphate	Dispersion technique	Improved bioavailability of iron due to its very smaller size of 0.3µm	Fidler *et al.* (2004)
3	Nanoencapsulation of Beta-lactoglobulin- polysaccharide complexes for non-fat foods and drinks	Beta-lactoglobulin-polysaccharide complexes	Nanoencapsulation	Provided significant protection of vitamins against degradation and enrichment of health promoting non-fat foods with hydrophobic nutraceuticals	Ron *et al.* (2010)
4	Nano fortified Milk powder	Milk powder	High-pressure homo-genizer	Increased bioavailability and absorption of calcium from calcium citrate nano fortified milk powder compared to skimmed milk powder	Erfanian *et al.* (2015)
5	Zinc oxide nanoparticles fortified Biscuits	Zinc oxide nanoparticles	Co-precipitation	Improved intestinal absorption and bioavailability of the nano scaled zinc oxide due to its size of 25 ± 5 nm compared to bulk zinc oxide form	Raya *et al.* (2016)
6	Iron oxide, zinc oxide, and calcium phosphate nanoparticles for yogurt	Iron oxide, zinc oxide, and calcium phosphate	Core-shell method (Core- nanoparticles, shell- inulin)	Improved physicochemical and rheological property of yogurt by using these nanoparticles	Santillan *et al.* (2017)
7	Plant based vitamin D_3 fortified nanoemulsion	Vitamin D_3	Oil- in water Nano-emulsion	Enhanced chemical stability and bioavailability of vitamin D_3 with emulsion based nanodelivery systems	Schoener *et al.* (2019)

Food safety is to ensure that food will not cause any harm to the consumer during preparation and consumption (Pal, Patel, Bariya, Godishala, and Kandi, 2017) The major reason behind deterioration of food depends on a cascade of biochemical reactions that occur between various components of food and the environment surrounding the food. Different types of biosensors have been developed to detect spoilage, bacterial growth, pesticide residues, and toxins and to monitor set storage conditions. Metal nanoparticles like silver, gold, platinum and palladium nanoparticles are widely used in construction of biosensors due to their unique physico-chemical properties. Likewise, nanobiosensors like enzyme-, immuno-, microbial-, and DNA-biosensors have the potential to be developed into real-time monitoring devices to improve the food safety control system (Prakitchaiwattana and Det-udom, 2017). However, the main issue with diagnosis of *Escherichia coli* in foods is to distinguish between the pathogenic and non-pathogenic species. Due to high sensitivity and selectivity, nanobiosensors are also developed to sense chemical hazards like pesticide residues and toxins, apart from food spoilage pathogenic microorganisms. Of late, numerous nanoparticle based detection platforms have been developed to assess food quality and safety (Table 22.2).

A typical example is development of nanoporous alumina membrane based electrochemical biosensor with magnetic nanoparticles to detect histamine at limits of detection as low as 3 nM (Ye *et al.*, 2016). Histamine is widely used as one of the indicators for food quality control during production, storage and transportation (Peng *et al.*, 2008). Similarly, (Moon *et al.*, 2018) developed a portable Surface plasmon resonance based competitive immunoassay sensor with low limits of Aflatoxin B1 (AFB1) detection of 2.51 parts per billion and quantification of 16.32 parts per billion in rice, peanut, and almond samples. Among the four types of Aflatoxins, AFB1 is considered to be the most detrimental, as it is extremely heat stable, difficult to destroy (Miliæeviæ, Škrinjar, and Baltiæ, 2010) and current global climate change is favoring its spread across different countries (Medina, Rodriguez, and Magan, 2014). Many countries have even set legal limits for AFB1 ranging from 0 to 50 ppb in food commodities (Liu, Hsu, Lu, and Yu, 2013). However, the developed sensor still has its limitations in terms of reusability, and costs feasibility. This example portray that sensitivity alone is not a major criteria required for developing nanobiosensors. The developed nanobiosensors should also be techno-economically feasible for wider usability and commercialization.

Nanotechnology for Green Food Nanopackaging

Food packaging continues to evolve in response to changing market requirements like consumer's demand for safer, healthier and quality foods with extended shelflife. In general, packaging is considered significant in marketing, safety, storage and hygienic handling of produces/products on daily basis (Coles, McDowell, and Kirwan, 2003). During food processing and transportation, foods are exposed to physical, chemical, and biological risks. For food based produces/products, packaging is for protection, preservation, retaining product freshness and prevention of cross-contamination, absence of pesticide residues and harmful pathogens after harvest/processing. (Rossi *et al.*, 2017) stated that sensors can be developed by combining desirable properties like antimicrobial and antifouling

Table 22.2. Type of Nanobiosensors and the Target Microorganism, Pesticides and Chemicals

Sl.No.	*Type of Nanoparticle*	*Detection Method*	*Target Microorganism*	*References*
Detection of pathogenic bacteria				
1	Screen printed electrode coated with nano-gold membrane	Spectrofluorometry and flow cytometry	*Vibrio parahaemolyticus* with a detection limit of 7.374 X10^4 cfu/mL^{-1}	Zhao *et al.* (2004)
2	Zinc sulfite–Cadmium selenide quantum dots	Fluorescence microscopy	*Escherichia coli* with a detection limit of 10^3-10^7 cfu/mL $^{-1}$and *Salmonella typhimurium* with a detection limit of 10^5-10^6 cfu/mL^{-1}	Yang and Li (2006)
3	Single Walled Carbon nanotube	Field-effect transistor	*Salmonella infantis* with a detection limit of 100 cfu/mL	Villamizar *et al.* (2008)
4	Nanostructured zinc oxide coated with indium-tin- oxide as electrode	Electrochemical Immunosensor	Ochratoxin-A with a detection limit of 0.006 nM/dm^3, response time as 25 s and sensitivity of 189 DINU123-/nM/dm^{3}cm^{-2}	Ansari *et al.* (2010)
5	Polypyrrole nanowires	Electrochemical method: Linear sweep voltammetry	*Bacillus globigii* with a detection limit of 1-100 cfu/mL	Garcia-Aljaro *et al.* (2010)
6	Gold nanoparticles	Surface plasmon resonance	*Salmonella sp.*, with a detection limit of 10^3 to 10^6 cells mL^{-1}	Afonso *et al.* (2013)
Detection of food-contaminating pesticides and chemicals				
1	Multiwalled Carbon nanotubes	Electrochemical: cyclic voltammetry and solid-phase extraction high-performance liquid chromatography	Sudan I dyes with a detection limit of 5.0 g L^{-1}	Gan *et al.* (2008)
2	Gold nanoparticles	Colorimetry, Fluorescence spectroscopy	Organophosphorus and carbamate pesticides with a detection limit of 0.32 ppm and 2.4 ppm respectively	Liu *et al.*,(2012)
3	Silver and gold nanoparticles	Colorimetric, fluorescence and ultraviolet–visible spectroscopy	Melamine	Vasimalai and John (2013)
4	Zinc oxide and multi walled carbon nanotubes	Electrochemical: cyclic voltammetry	Bisphenol, a chemical in food samples with a detection limit of 80 nmol L^{-1}	Najafi *et al.* (2014)

properties, thermal and mechanical protection, and oxygen and moisture barrier, with barcode system to verify the status of food quality.

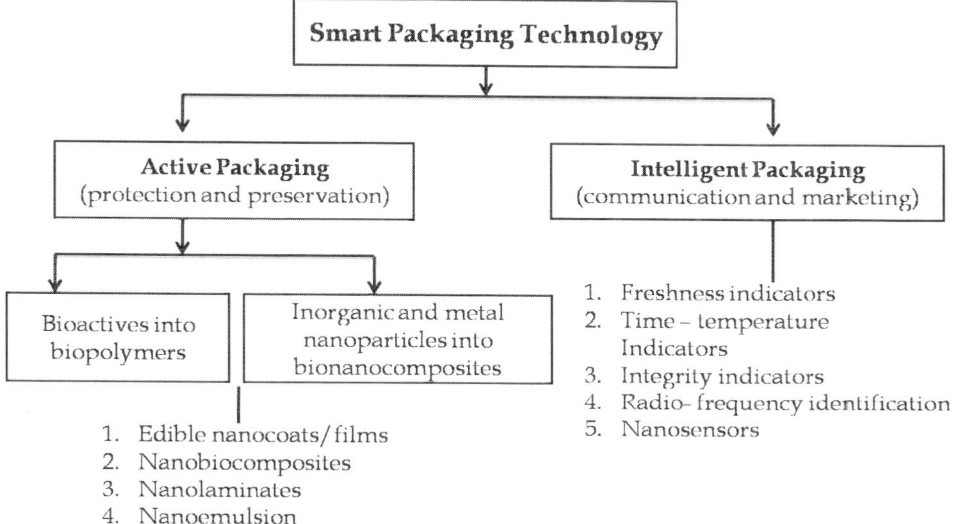

Figure 22.1. Types of Smart Packaging Technologies and the Techniques Used to Develop Nanomaterials for Food Packaging.

European regulation (EC) No 450/2009 defines active packaging as packaging systems that interact with the food in such a way as to "deliberately incorporate components that would release or absorb substances into or from the packaged food or the environment surrounding the food" (European Commission 2009). Nanomaterials are being developed in the food packaging industry due to unique functional properties they impart to a packaging material. Increasing demand for sustainability and environmental safety led to a number of investigations on development of food packaging materials that could rapidly degrade and completely mineralize in the environment. ASTM International, headquartered at West Conshohocken, Pennsylvania is an international standards organization that develops and publishes voluntary consensus technical standards for a wide range of materials. According to ASTM, the term biodegradable is defined as 'material capable of undergoing decomposition into carbon dioxide, methane, water, inorganic compounds, or biomass in which the predominant mechanism is the enzymatic action of microorganisms that can be measured by standard tests, in a specified period of time, reflecting available disposal condition (ASTM 2005). Environment friendly biodegradable packaging can reduce the negative environmental impacts generated by synthetic nanopackaging and nanocomposite materials. Use of biopolymers (cellulose, chitosan, starch, zein, guar gum, gum arabic) as food packaging materials has limitations like poor mechanical, thermal and barrier properties to that of non-biodegradable materials derived from petrochemical sources. Bionanocomposites have been thought of as a promising alternative with maximum benefits by food packaging Industries. Bionanocomposite is an active food packaging material, wherein the food packaging interacts with food in certain

Table 22.3. Different Techniques Used to Develop Bionanocomposite for Food Packaging and their Functional Characteristics

Sl.No.	*Composition of Nanobiocomposite*	*Method Employed*	*Performance Efficiency*	*References*
1	Starch with montmorillonite	Solution casting	Increased Tensile strength and decreased water vapour diffusion	Park *et al.* (2002)
2	Starch and micro crystalline cellulose	Solution casting	Increased tensile strength, thermal stability and water barrier properties	Ma, Tong, Xu, and Fang (2008)
3	Starch and cellulose nanoparticles	Solution casting	Improved thermal property, water vapour and tensile strength	Chang, Wu, Anderson, and Ma (2012)
4	Starch with xanthan gum and sodium montmorillonite	Melt chelation technique	Increased stress and shear rate; reduced water permeability and sorption capacity	Melo *et al.* (2011)
5	Protein and micro fibrillated cellulose	Solution casting	Reduced water permeation rate and absorption rate followed by enhancement in mechanical property	Fadel *et al.* (2012)
6	Chitosan and nanocrystalline cellulose	Solution casting	Reduced water permeability and enhanced tensile strength	Khan *et al.* (2012)
7	Cassava starch with sodium montmorillonite	Melt chelation technique	Increased tensile strength	Souza *et al.* (2012)
8	K-carrageenan nano fibrillated cellulose	Solution casting	Increased stretchability, reduced water vapour and oxygen intrusion	Savadekar, Karande, Vigneshwaran, Bharimalla, and Mhaske (2012)
9	Cellulose–halloysite nanoclay	Solution casting	Decreased oxygen permeability and enhanced mechanical strength	Alipoormazandarani, Ghazihoseini, and Nafchi (2015)
10	Starch and cellulose nanofiber	Solution casting	Reduced water vapour permeability and enhanced mechanical strength	Babaee, Jonoobi, Hamzeh, and Ashori (2015)
11	Chitosan and fish gelatin	Solution casting	Improved tensile strength and reduced water vapour permeability	Hosseini, Rezaei, Zandi, and Farahmandghavi (2015)
12	Protein and cellulose pulp fiber	Solution casting	Improved mechanical strength	Oliveira *et al.* (2015)
13	Gelatin/ethanolic extract of coconut husk/Cloisite sodium nanoclay	Solution casting	Increased moisture resistance with reduced lipid peroxidation	Nagarajan *et al.* (2015)
14	Whey protein isolate and montmorillonite	Solution casting	Enhanced tensile strength	Wakai and Almenar (2015)
15	Chitosan and clay platelets	Electron beam irradiation	Increased tensile strength	Shahbazi, Rajabzadeh, and Ahmadi (2017)

way by releasing bioactive ingredients like antimicrobial agent, antioxidant agent, or by reducing few undesirable elements like oxygen, ethylene and water vapor.

Another interesting type of active packaging is edible nano coats. Edible coatings are substances applied to the exterior of food so that the final product is fit for consumption. Presence of cuticular waxy layer on the surface of fruits and vegetables acts as a natural barrier against excessive transpiration and water loss (Woods, 1990). During cleaning or processing of fresh fruits and vegetables, this natural barrier gets disturbed or washed away and that need to be replaced. Coatings can also be used as carriers of functional ingredients. During storage, the edible coat should not ferment, precipitate, separate or develop off-flavors; allow restricted gaseous exchange of oxygen, carbon dioxide and water vapor; comply with legal, safety, and performance requirements. For example, fresh fruits processed in the European Union cannot be coated with morpholine containing fruit coating formulation (Hagenmaier, 2004). Edible nanopackaging technology allows incorporation of hydrophilic and lipophilic substances with antimicrobial or antioxidant properties as active components to be released during storage to increase the shelf life of the coated food produces and products. Edible nano-coats are consumed along with the food without waste generation making it an environment friendly packaging material.

Safety Concerns of Nanomaterials in Food Industry

The comprehensive risk assessment associated with any developed nanomaterial based food packaging need to be studied before up-scaling. It is also imperative to understand the potential of nanoparticle migration into food and its possible hazards, interaction between nanomaterial-based food contact materials (FCMs) upon ingestion (Silvestre, Duraccio, and Cimmino, 2011). Echegoyen *et al.* (2013) studied migration of silver in the form of ions or particles from three different types of nanocomposites into food stimulants and confirmed migration of silver nanoparticles into food stimulants. The highest level of silver nanoparticle migration was observed with acidic foods and microwave heating was also found to increase migration of silver from nanocomposite packaging material. In Europe, the legislation currently applies an overall migration limit of 10 mg constituent per dm2 surface area to all substances that can migrate from Food contact materials (FCMs) to foods (Commission Regulation (EU) No. 10/2011) and nanomaterial risk assessment has to be performed on a case-by-case basis (Silvestre *et al.*, 2011); Commission Regulation (EU) No. 10/2011). Reports from earlier studies concluded that the rate of migration of nanoparticles correlated with the percentage of nanofiller materials used in the nanocomposite packaging material.

The rapid penetration of nanotechnology based food produces and products in the Food Industries across the globe have raised public and environmental safety, ethical policy, and regulatory issues. In 2010, the House of Lords committee on science and technology conducted an in-depth investigation on nano-foods available in the market, and criticized the food industry on maintaining secrecy over use of nanotechnology. The Food and Agriculture Organization of the United Nations report on food based nanotechnology also emphasized on lack of transparency about what the food industry was doing with nanotechnology.

Table 22.4. List of Edible Coat Nanoformulations and its Improved Functionality in Studied Food Produces

Sl.No.	*Product/Produce*	*Edible coating*	*Type of Nanoformulation*	*Improved Properties*	*References*
1	Green beans	Mandarin essential oil and chitosan	Nanoemulsion	Reduced permeability to water vapor with improved anti- microbial activity	(Severino *et al.*, 2014)
2	Carrot	Microfibrillated	Nanofibers starch (matrix)	Reduced permeability to water vapor with improved functionality and compatibility	(Deng, Jung, Simonsen, Wang, and Zhao, 2017)
3	Strawberries	Curcumin and limonene with methyl cellulose	Liposomes	Increased anti-fungal activity, titratable acidity and total phenolic contents in limonene coated strawberries	Dhital *et al.*,(2017)
4	Grape berry	Lemongrass oil and chitosan	Nanoemulsion	Reduced the initial growth of *Salmonella typhimurium* and showed improved antioxidant capacity	(Oh *et al.*, 2017)
5	Low-fat cut cheese	Oregano essential oil	Nanoemulsion	Increased resistance against *Staphylococcus aureus* and preserved the outward appearance of cheese	(Artiga-Artigas, Acevedo-Fani, and Martín-Belloso, 2017)
6	Nano-film	Cinnamaldehyde with pectin (low and high methyl ester)	Nanocomposite film	Improved antimicrobial activity provided by cinnamaldehyde against food pathogens due to its increased surface area	(Otoni *et al.*, 2014)
7	Solid lipid nanoparticles	Quercetin	Glyceryl monostearate (matrix),surfactant - sorbitan monolaurate and lecithin	Increased bioavailability of quercetin in its native form	(Aditya *et al.*, 2014)
8	Solid lipid nanoparticles	Curcumin	Glyceryl behenate (matrix) surfactant - soy lecithin	Increased oral bioavailability of curcumin	(Righeschi *et al.*, 2016)

Table 22.5. Significant Legislations Framed by European Union for Use of Nanomaterials in Food

Sl.No.	*Name of Regulations/ Directives*	*Specific Feature*	*Details*
1	Regulation (EC) No 1333/2008	Food additives	States that a food additive already authorized but obtained using nanotechnology requires a re-evaluation before marketing
2	Regulation (EC) No 1332/2008	Food enzyme	States that a food enzyme already included in the Community list but prepared by different methods or using starting materials significantly different (It is specified that "Significantly different" could mean a change in particle size) from those included in the risk assessment of the Authority, should be submitted for re-evaluation
3	Regulation (EC) No 1334/2008	Food Flavouring	Flavouring must undergo a common (EU-wide) assessment and authorization prior marketing and lays down a Union list of flavouring and source materials approved for use in and on foods and their conditions of use.
4	Directive 2002/46/EC Food supplements	Food supplements	Stated that the food supplements (minerals or vitamin) can be used. The use of nanoformulation of minerals and vitamin requires a safety evaluation prior marketing which will be done under Novel Food Regulation, due to the differences in production, potential differences in nutritional value and bioavailability when compared to macro-scale counterparts
5	Regulation(EU) No 10/2011a	"Plastic materials and articles intended to come into contact with food"	States that substances in nanoformulation should be used only if listed in the Annex I of the regulation
6	Regulation(EU) No 528/2012	Biocides	As of today, nanomaterials based Biocides are not eligible for a simplified authorization procedure. For subsequent nanomaterials based product authorization and approving nanomaterials as active substances, the test methods applied to the nanomaterials shall be accompanied and standardized by an explanation addressing their exact appropriateness considering the specific characteristics of each nanomaterials

The direct and indirect effects of nanomaterials to human health and the biological fate of nanomaterials after digestion, their behavior within gastrointestinal tract, and their possible interactions with biological systems should be understood (Yu *et al.*, 2018). Interestingly in a study conducted with French and German consumers to know their acceptance among nanofortified vitamins and nanopackaging, French consumers reluctantly accepted nanopackaging, while German consumers were less inclined to accept nanofortification to nanopackaging (Bieberstein *et al.*, 2013). Consumers are more likely to accept nanotechnology in food packaging rather than for food processing. Although regulations of nanotechnology in food based applications are available in few countries, there have been significant advances in the application of nanotechnology in the food industry. Hence, it is high time that countries having nano based food materials in the market adopt suitable regulatory frameworks to enjoy the true benefits of nanotechnology.

REFERENCES

1. Aditya, N., Macedo, A. S., Doktorovova, S., Souto, E. B., Kim, S., Chang, P.-S., and Ko, S. (2014). Development and evaluation of lipid nanocarriers for quercetin delivery: a comparative study of solid lipid nanoparticles (SLN), nanostructured lipid carriers (NLC), and lipid nanoemulsions (LNE). *LWT-Food Science and Technology, 59*(1), 115-121.
2. Afonso AS, Perez-Lopez B, Faria RC, Mattoso LHC, Hernandez- Herrero M, Roig-Sagues AX, Costa MM, Merkoci A. (2013). Electrochemical detection of *Salmonella* using gold nanoparticles. *Biosens Bioelectron* 40: 121–126. doi:10.1016/j.bios.2012.06.054.
3. Alipoormazandarani, N., Ghazihoseini, S., and Nafchi, A. M. (2015). Preparation and characterization of novel bionanocomposite based on soluble soybean polysaccharide and halloysite nanoclay. *Carbohydrate polymers,* 134: 745-751.
4. Ansari, A. A., Kaushik, A., Solanki, P. R., and Malhotra, B. D. (2010). Nanostructured zinc oxide platform for mycotoxin detection. *Bioelectrochemistry,* 77(2): 75-81.
5. Artiga-Artigas, M., Acevedo-Fani, A., and Martín-Belloso, O. (2017). Improving the shelf life of low-fat cut cheese using nanoemulsion-based edible coatings containing oregano essential oil and mandarin fiber. *Food Control,* 76: 1-12.
6. Babaee, M., Jonoobi, M., Hamzeh, Y., and Ashori, A. (2015). Biodegradability and mechanical properties of reinforced starch nanocomposites using cellulose nanofibers. *Carbohydrate polymers,* 132: 1-8.
7. Bieberstein, A., Roosen, J., Marette, S., Blanchemanche, S. and Vandermoere, F. (2012). Consumer choices for nano-food and nano-packaging in France and Germany. *European Review of Agricultural Economics,* 40(1): 73-94.
8. Bürgi, H., Supersaxo, Z., and Selz, B. (1990). Iodine deficiency diseases in Switzerland one hundred years after Theodor Kocher's survey: a historical review with some new goitre prevalence data. *European Journal of Endocrinology, 123*(6), 577-590.

9. Chang, P. R., Wu, D., Anderson, D. P., and Ma, X. (2012). Nanocomposites based on plasticized starch and rectorite clay: Structure and properties. *Carbohydrate polymers, 89*(2), 687-693.

10. Chen, C. C. and Wagner, G. (2004). Vitamin E nanoparticle for beverage applications. *Chemical Engineering Research and Design*, 82(11), 1432-1437.

11. Coles, R., McDowell, D., and Kirwan, M. J. (2003). *Food packaging technology* (Vol. 5): CRC Press.

12. Das, J. K., Salam, R. A., Kumar, R., and Bhutta, Z. A. (2013). Micronutrient fortification of food and its impact on woman and child health: a systematic review. *Systematic reviews*, 2(1), 67.

13. Deng, Z., Jung, J., Simonsen, J., Wang, Y., and Zhao, Y. (2017). Cellulose nanocrystal reinforced chitosan coatings for improving the storability of postharvest pears under both ambient and cold storages. *Journal of food science, 82*(2), 453-462.

14. Echegoyen, Y. and Nerín, C. (2013). Nanoparticle release from nano-silver antimicrobial food containers. *Food and Chemical Toxicology*, 62, 16-22.

15. Erfanian, A., Mirhosseini, H., Rasti, B., Hair-Bejo, M., Mustafa, S. B. and Manap, M. Y. A. (2015). Absorption and bioavailability of nano-size reduced calcium citrate fortified milk powder in ovariectomized and ovariectomized-osteoporosis rats. *Journal of Agricultural and Food Chemistry*, 63(24), 5795-5804.

16. Fadel, S. M., Hassan, M. L. and Oksman, K. (2013). Improving tensile strength and moisture barrier properties of gelatin using microfibrillated cellulose. *Journal of Composite Materials*, 47(16), 1977-1985.

17. Fidler, M. C., Walczyk, T., Davidsson, L., Zeder, C., Sakaguchi, N., Juneja, L. R. and Hurrell, R. F. (2004). A micronised, dispersible ferric pyrophosphate with high relative bioavailability in man. *British Journal of Nutrition*, 91(1), 107-112.

18. García-Aljaro, C., Bangar, M. A., Baldrich, E., Muñoz, F. J. and Mulchandani, A. (2010). Conducting polymer nanowire-based chemiresistive biosensor for the detection of bacterial spores. *Biosensors and Bioelectronics*, 25(10), 2309-2312.

19. Gaucheron, F. (2011). Milk and dairy products: a unique micronutrient combination. *Journal of the American College of Nutrition*, 30(sup5), 400S-409S.

20. Hagenmaier, R. D. (2004). *Fruit coatings containing ammonia instead of morpholine.* Paper presented at the Proceedings of the Florida State Horticultural Society.

21. Hosseini, S. F., Rezaei, M., Zandi, M., and Farahmandghavi, F. (2015). Fabrication of bio-nanocomposite films based on fish gelatin reinforced with chitosan nanoparticles. *Food Hydrocolloids, 44*, 172-182.

22. Khan, A., Khan, R. A., Salmieri, S., Le Tien, C., Riedl, B., Bouchard, J.,. Lacroix, M. (2012). Mechanical and barrier properties of nanocrystalline cellulose reinforced chitosan based nanocomposite films. *Carbohydrate Polymers, 90*(4), 1601-1608.

23. Kinyamu, H. K., Gallagher, J. C., Rafferty, K. A., and Balhorn, K. E. (1998). Dietary calcium and vitamin D intake in elderly women: effect on serum parathyroid hormone and vitamin D metabolites. *The American Journal of Clinical Nutrition, 67*(2), 342-348.

24. Latham, L., and Jones, R. (2001). Alfalfa mosaic and pea seed-borne mosaic viruses in cool season crop, annual pasture, and forage legumes: susceptibility, sensitivity, and seed transmission. *Australian Journal of Agricultural Research, 52*(7), 771-790.

25. Liu, B.-H., Hsu, Y.-T., Lu, C.-C., and Yu, F.-Y. (2013). Detecting aflatoxin B1 in foods and feeds by using sensitive rapid enzyme-linked immunosorbent assay and gold nanoparticle immunochromatographic strip. *Food Control, 30*(1), 184-189.

26. Ma, H., Tong, L., Xu, Z., and Fang, Z. (2008). Intumescent flame retardant-montmorillonite synergism in ABS nanocomposites. *Applied Clay Science, 42*(1-2), 238-245.

27. Medina, A., Rodriguez, A., and Magan, N. (2014). Effect of climate change on Aspergillus flavus and aflatoxin B1 production. *Frontiers in microbiology, 5*, 348.

28. Melo, C. D., Garcia, P. S., Grossmann, M. V. E., Yamashita, F., Dall'Antônia, L. H. and Mali, S. (2011). Properties of extruded xanthan-starch-clay nanocomposite films. *Brazilian Archives of Biology and Technology*, 54(6), 1223-1333.

29. Miliæeviæ, D. R., Škrinjar, M., and Baltiæ, T. (2010). Real and perceived risks for mycotoxin contamination in foods and feeds: challenges for food safety control. *Toxins, 2*(4), 572-592.

30. Moon, J., Byun, J., Kim, H., Lim, E.-K., Jeong, J., Jung, J., and Kang, T. (2018). On-site detection of aflatoxin B1 in grains by a palm-sized surface plasmon resonance sensor. *Sensors, 18*(2), 598.

31. Nagarajan, M., Benjakul, S., Prodpran, T., and Songtipya, P. (2015). Properties and characteristics of nanocomposite films from tilapia skin gelatin incorporated with ethanolic extract from coconut husk. *Journal of Food Science and Technology*, 52(12), 7669-7682.

32. Najafi, M., Khalilzadeh, M. A., and Karimi-Maleh, H. (2014). A new strategy for determination of bisphenol A in the presence of Sudan I using a ZnO/CNTs/ionic liquid paste electrode in food samples. *Food Chemistry*, 158, 125-131.

33. Oh, Y. A., Oh, Y. J., Song, A. Y., Won, J. S., Song, K. B., and Min, S. C. (2017). Comparison of effectiveness of edible coatings using emulsions containing lemongrass oil of different size droplets on grape berry safety and preservation. *LWT, 75*, 742-750.

34. Oliveira, T. G. D., Makishi, G. L. D. A., Chambi, H. N. M., Bittante, A. M. Q. B., Lourenço, R. V.and Sobral, P. J. D. A. (2015). Cellulose fiber reinforced biodegradable films based on proteins extracted from castor bean (Ricinus communis L.) cake. *Industrial Crops and Products*, 67, 355-363.

35. Otles, S., and Yalcin, B. (2010). Nano-biosensors as new tool for detection of food quality and safety. *LogForum 6, 4, 7.*

36. Otoni, C. G., de Moura, M. R., Aouada, F. A., Camilloto, G. P., Cruz, R. S., Lorevice, M. V. and Mattoso, L. H. (2014). Antimicrobial and physical-mechanical properties of pectin/papaya puree/cinnamaldehyde nanoemulsion edible composite films. *Food Hydrocolloids, 41*, 188-194.

37. Pal, M., Patel, A., Bariya, A., Godishala, V., and Kandi, V. (2017). A review of biotechnological Applications in Food Processing of Animal Origin. *Am J Food Sci Tech, 5*(4), 143-148.

38. Park, H. M., Li, X., Jin, C. Z., Park, C. Y., Cho, W. J., and Ha, C. S. (2002). Preparation and properties of biodegradable thermoplastic starch/clay hybrids. *Macromolecular Materials and Engineering, 287*(8), 553-558.

39. Peng, J.-f., Fang, K.-t., Xie, D.-h., Ding, B., Yin, J.-Y., Cui, X.-m.,. Liu, J.-f. (2008). Development of an automated on-line pre-column derivatization procedure for sensitive determination of histamine in food with high-performance liquid chromatography–fluorescence detection. *Journal of Chromatography A, 1209*(1-2), 70-75.

40. Prakitchaiwattana, C., and Det-udom, R. (2017). Contaminant sensors: nano sensors, an efficient alarm for food pathogen detection *Nanobiosensors* (pp. 511-572): Elsevier.

41. Raya, S. D. H. A., Hassan, M. I., Farroh, K. Y., Hashim, S. A. and Salaheldin, T. A. (2016). Zinc oxide nanoparticles fortified biscuits as a nutritional supplement for zinc deficient rats. *J. Nanomed. Res*, 4(2).

42. Righeschi, C., Bergonzi, M. C., Isacchi, B., Bazzicalupi, C., Gratteri, P., and Bilia, A. R. (2016). Enhanced curcumin permeability by SLN formulation: The PAMPA approach. *LWT-Food Science and Technology, 66*, 475-483.

43. Ron, N., Zimet, P., Bargarum, J. and Livney, Y. D. (2010). Beta-lactoglobulin–polysaccharide complexes as nanovehicles for hydrophobic nutraceuticals in non-fat foods and clear beverages. *International Dairy Journal*, 20(10), 686-693.

44. Rossi, M., Passeri, D., Sinibaldi, A., Angjellari, M., Tamburri, E., Sorbo, A.,. Dini, L. (2017). Nanotechnology for food packaging and food quality assessment *Advances in food and nutrition research* (Vol. 82, pp. 149-204): Elsevier.

45. Santillán-Urquiza, E., Méndez-Rojas, M. Á. and Vélez-Ruiz, J. F. (2017). Fortification of yogurt with nano and micro sized calcium, iron and zinc, effect on the physicochemical and rheological properties. *LWT*, 80, 462-469.

46. Savadekar, N., Karande, V., Vigneshwaran, N., Bharimalla, A., and Mhaske, S. (2012). Preparation of nano cellulose fibers and its application in kappa-carrageenan based film. *International journal of biological macromolecules, 51*(5), 1008-1013.

47. Schoener, A. L., Zhang, R., Lv, S., Weiss, J. and McClements, D. J. (2019). Fabrication of plant-based vitamin D 3-fortified nanoemulsions: influence of carrier oil type on vitamin bioaccessibility. *Food and Function*, 10(4), 1826-1835.

48. Severino, R., Vu, K. D., Donsì, F., Salmieri, S., Ferrari, G., and Lacroix, M. (2014). Antibacterial and physical effects of modified chitosan based-coating containing nanoemulsion of mandarin essential oil and three non-thermal treatments against Listeria innocua in green beans. *International journal of food microbiology, 191*, 82-88.

49. Shahbazi, M., Rajabzadeh, G., and Ahmadi, S. J. (2017). Characterization of nanocomposite film based on chitosan intercalated in clay platelets by electron beam irradiation. *Carbohydrate polymers, 157*, 226-235.

50. Silvestre, C., Duraccio, D., and Cimmino, S. (2011). Food packaging based on polymer nanomaterials. *Progress in polymer science, 36*(12), 1766-1782.

51. Souza, A., Benze, R., Ferrão, E., Ditchfield, C., Coelho, A., and Tadini, C. C. (2012). Cassava starch biodegradable films: Influence of glycerol and clay nanoparticles content on tensile and barrier properties and glass transition temperature. *LWT-Food Science and Technology, 46*(1), 110-117.

52. Van Stuijvenberg, M., Dhansay, M., Lombard, C., Faber, M., and Benade, A. (2001). The effect of a biscuit with red palm oil as a source of b-carotene on the vitamin A status of primary school children: a comparison with b-carotene from a synthetic source in a randomised controlled trial. *European journal of clinical nutrition, 55*(8), 657.

53. Vasimalai, N. and John, S. A. (2013). Picomolar melamine enhanced the fluorescence of gold nanoparticles: Spectrofluorimetric determination of melamine in milk and infant formulas using functionalized triazole capped goldnanoparticles. *Biosensors and Bioelectronics*, 42, 267-272.

54. Villamizar, R. A., Maroto, A., Rius, F. X., Inza, I. and Figueras, M. J. (2008). Fast detection of Salmonella Infantis with carbon nanotube field effect transistors. *Biosensors and Bioelectronics*, 24(2), 279-283.

55. Wakai, M., and Almenar, E. (2015). Effect of the presence of montmorillonite on the solubility of whey protein isolate films in food model systems with different compositions and pH. *Food Hydrocolloids, 43*, 612-621.

56. Woods, J. (1990). Moisture loss from fruits and vegetables. *Postharvest news and information, 1*(3), 195-199.

57. Yang, L. and Li, Y. (2006). Simultaneous detection of Escherichia coli O1576" H7 and Salmonella Typhimurium using quantum dots as fluorescence labels. *Analyst*, 131(3), 394-401.

58. Ye, W., Xu, Y., Zheng, L., Zhang, Y., Yang, M., and Sun, P. (2016). A nanoporous alumina membrane based electrochemical biosensor for histamine determination with biofunctionalized magnetic nanoparticles concentration and signal amplification. *Sensors, 16*(10), 1767.

59. Yu, H., Park, J.-Y., Kwon, C. W., Hong, S.-C., Park, K.-M., and Chang, P.-S. (2018). An Overview of Nanotechnology in Food Science: Preparative Methods, Practical Applications, and Safety. *Journal of Chemistry, 2018*.

60. Zhao, X., Hilliard, L. R., Mechery, S. J., Wang, Y., Bagwe, R. P., Jin, S. and Tan, W. (2004). A rapid bioassay for single bacterial cell quantitation using bioconjugated nanoparticles. Proceedings of the National Academy of Sciences, 101(42), 15027-15032.

Chapter 23

Application of Nanotechnology in Solar, Wind and Hydrogen Energy

*D. Jeya Sundara Sharmila**

Department of Nano Science and Technology, Tamil Nadu Agricultural University
Coimbatore – 641 003, Tamil Nadu
e-mail: djssharmila@gmail.com

Overview of Alternate Energy Sources and Energy Challenges in the Future

The alternate energy sources include radiant solar energy, biomass energy, wind energy, hydro energy, geothermal energy and ocean energy. They are commonly referred to as "new or renewable" because they are not yet used on a wide scale, their expertise such as nanotechnology application in energy challenges are still in a rapid phase of development and they will play a much larger role in our energy use in the future. They are growing at much faster rates than conventional sources, however, since these sources began at low levels, it will take time to build them up.

Use of alternate energy sources has expanded quickly because of the energy challenges, growing concerns over diminishing fossil fuels and the environmental impacts of fossil fuel combustion. Advances in modern nanotechnology make it easier and less expensive. Benefits of the new renewables include they alleviate air pollution and greenhouse gas emissions that can cause climate change, they are inexhaustible and unlike fossil fuels they help diversify a country's energy economy besides they create jobs and become sources of income and taxes, especially in rural areas. Technological interventions such as nanotechnology in energy systems may be helpful in crossing the economic barriers and a quick switch to renewable and transitioning too slowly will cause fossil fuel depletion, and renewables will not be able to compensate.

Radiant Solar Energy

Solar power plants use the steam that is produced to turn turbine as active and passive solar heating system besides photovoltaic cells and "Solar batteries" that use special semiconductors. The advantages of solar energy are renewable, free, high energy yield, very clean source of energy with no air/water pollution during operation and low operating costs that will pay for themselves over time. However their disadvantages are due to the requirement of intermittent source for energy storage issues and low energy density that requires pretty much land.

Hydro-energy

The hydro-energy ways are cheap to operate with long life and lower operating costs than all other power plants. They are renewable and high yield with multipurpose usage of reservoirs such as flood control, drinking water, aquaculture, recreation besides accounts for less air pollution than fossil fuel combustion. However, the disadvantages of hydro-energy production are human population displacement, more significant breeding ground for disease, reduced availability of water downstream, ecosystem impacts, barriers to migrating fish, loss of biodiversity both upstream and downstream, coastal erosion, reduced nutrient flow (dissolved and particulate) and water pollution problems such as low dissolved oxygen (DO), increased H_2S toxicity *etc.* besides green house gases emission (CH_4, N_2O, CO_2). The decommissioning is a big problem and size issue of many (most) of the above problems are significantly worse for larger dams. However, small dams have shorter lifetimes, less capacity, and are more intermittent.

Wind Energy

Wind energies are having the advantages of high net energy yield, renewable and free, very clean source of energy, no pollution (air or water) during operation, long operating life, low operating/maintenance costs, can be quickly built; not too expensive, now almost competitive with hydro and fossil fuels and land can be used for other purposes that can combine wind and agricultural farms. However their limitations are due to energy storage issues as an intermittent source of energy that needs backup to store energy for low-wind days or must be connected to the electrical grid. The wind energy harvesting is only practical in areas that are windy enough. They cause visual pollution, danger to birds and affected by low energy density of wind.

Biomass Energy

Biomass energy is the use of living and recently dead biological material as an energy source that ultimately dependent on the capture of solar energy and conversion to a chemical (carbohydrate) fuel. Theoretically it is a carbon neutral and renewable source of energy. Traditionally it is based on forest management, using wood and biodegradable waste as fuel (Examples: manure, crop residue, sewage, municipal solid waste). Agricultural production of energy crops gained recent interest for showing high yield and low maintenance (examples: corn, sugarcane, switchgrass, hemp, willow, palm oil, rapeseed, and many others) for liquid or gaseous biofuel production due to the breakdown of biomass in the absence of O_2 (examples: *Bioethanol* from fermentation, Cellulosic bioethanol is usually from a grass (switchgrass) and Biodiesel from rapeseed and other sources). Even though the Biomass energy resources are versatile, renewable, with no net CO_2 emissions (ideally) and emits less SO_2 and NO_x than fossil fuels, their disadvantages are due to low energy density/yield, land conversion, biodiversity loss and usual problems associated with intensive agriculture such as nutrient pollution, soil depletion, soil erosion and other water pollution problems.

Geothermal Energy

Geothermal power plants use earth's heat to power steam turbines such as directly using of hot springs (etc) as heat source besides geothermal heat pumps. This is renewable, easy to exploit in some cases, CO_2 production less than with fossil fuels and high net energy yield. However, their disadvantages are because of non-availability everywhere, H_2S pollution and water pollution (somewhat similar to mining).

The Hydrogen Economy

The Hydrogen Economy is a hypothetical large-scale system in which elemental hydrogen (H_2) is the primary form of energy storage. Fuel cells would be the primary method of conversion of hydrogen to electrical energy that is efficient, clean and scalable in particular hydrogen (usually) plays a central role in transportation. Potential advantages are clean, renewable and probably more reliable (using distributed generation). But many roadblocks are including potential showstoppers poses great technological challenges for efficient hydrogen production, storage, and transport.

Nanotechnology Application in Solar Energy

Solar cells and photodetectors are devices that convert an optical input into current. A solar cell is an example of a photovoltaic device, *i.e.*, a device that generates voltage when exposed to light *i.e.*, solar radiation (Figure 23.1). The photovoltaic effect was discovered by Alexander-Edmond Becquerel in 1839, in a junction formed between an electrode (platinum) and an electrolyte (silver chloride).

Solar Cell Working Principle

A simple solar cell is a pn junction diode (Figure 23.2). The n region is heavily doped and thin so that the light can penetrate through it easily. The p region is lightly doped so that most of the depletion region lies in the p side. The penetration depends on the wave-length and the absorption coeffcient increases as the wavelength decreases. Electron hole pairs (EHPs) are mainly created in the depletion region and due to the built-in potential and electric field, electrons move to the n region and the holes to the p region. When an external load is applied, the excess electrons travel through the load to recombine with the excess holes. Electrons and holes are also generated with the p and n regions. The shorter wavelengths (higher absorption coeffcient) are absorbed in the n region and the longer wavelengths are absorbed

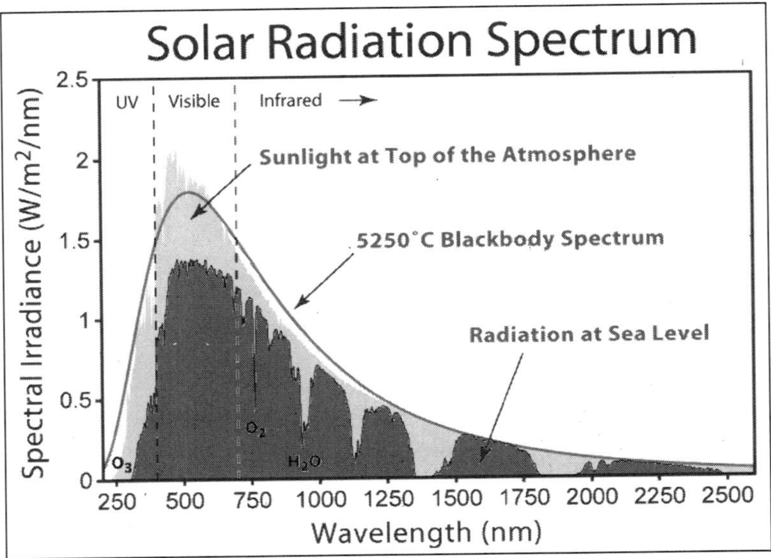

Figure 23.1: Solar Radiation Spectrum.
Courtesy: from http://en.wikipedia.org/wiki/Sunlight.

in the bulk of the p region. Some of the EHPs generated in these regions can also contribute to the current. Typically, these are EHPs that are generated within the minority.

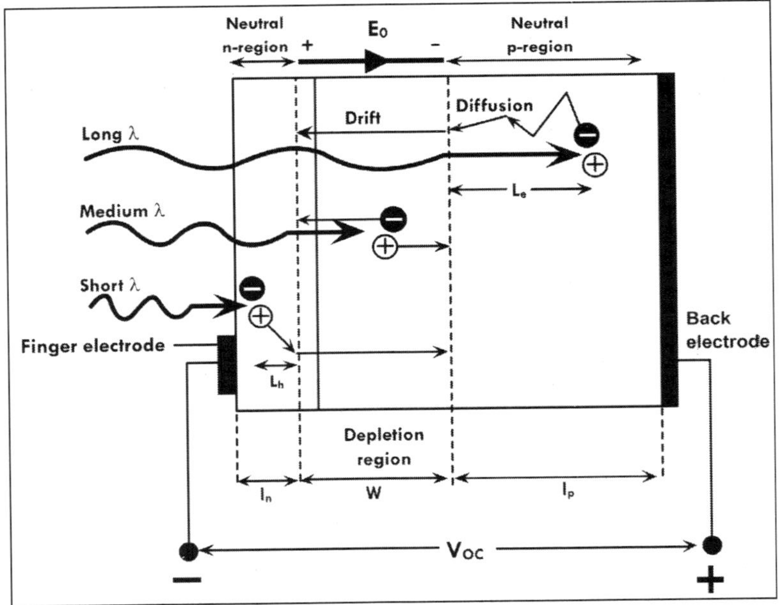

Figure 23.2: Working Principle of Solar Cell.
Courtesy: Principles of Electronic Materials - S.O. Kasap, 2006.

Nanotechnology has a significant impact on energy conversion and storage as nanomaterials exhibit distinct size dependent properties in the 1-100 nm range where quantum phenomena are involved. Nanotechnology may contribute toward efficient and low-cost systems for generating, storing, and transporting energy. Materials and structures that are designed and fabricated at a nanoscale level as well as thin films can offer a potential to produce new devices and processes that may enhance efficiencies and reduce costs in many areas, such as solar photovoltaic systems, hydrogen production, fuel cells, solar thermal systems and energy saving technologies like low e-coatings and electrochromic devices for smart windows. Nanotechnology for photovoltaic PV cells (Figure 23.3) that enable production of electricity by means of photoelectric effect. Currently, PV market is based on silicon wafer-based solar cells (thick cells of around 150–300 nm made of crystalline silicon). This technology, classified as the first generation of photovoltaic cells, accounts for more than 86 per cent of the global solar cell market. The second generation of photovoltaic materials is based on introduction of thin film layers (1–2 nm) of semiconductor materials. Inclusion of Nanotechnology components in PV cells present an ability to control the energy bandgap provides flexibility and interchangeability, while the introduction of nanostructured materials enhances the effective optical path and significantly decreases the probability of charge recombination (Pelemis and Hut, 2013).

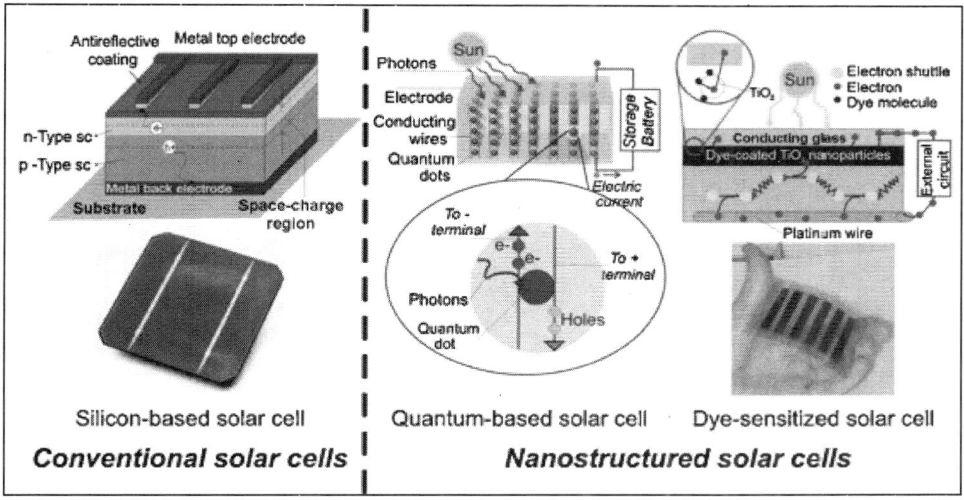

Figure 23.3: Conventional (Silicon-based) Solar Cell Vs Nanostructured Solar Cells (Quantum-based and dye-sensitized solar cells).
Courtesy: Serrano *et al.*, 2009.

Evolution of Photovoltaic Technology

Dye-Sensitized Solar Cells

Dye-sensitized solar cells (DSCs) (Figure 23.4) invented by Michael Grätzel became a very popular alternative to silicon-based solar cells because of their great

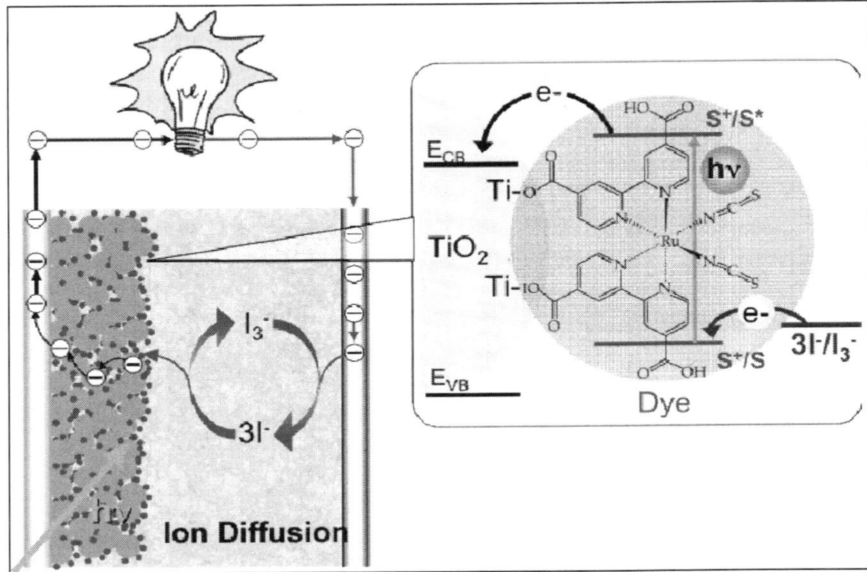

Figure 23.4: Mechanism of Dye-Sensitized Solar Cell.
Courtesy: Halim, 2013.

potential to convert solar energy into electric energy at low cost. This cell can be made from cheap materials such as inorganic and organic dyes which do not need to be highly pure as required for silicon wafer.

The operating principle of a DSC is that an inorganic dye is anchored to a wide bandgap mesoscopic semiconductor. The popular dyes used for DSC are ruthenium bipyridine and zinc porphyrin complexes. For a mesoscopic semiconductor, TiO_2 (anatase) is widely used in the solar cells; however, other alternative metal oxides such as ZnO, SnO_2 and Nb_2O_5 can be used (Halim, 2013). After excitation of dye by light, the dye releases its electron from the HOMO (highest occupied molecular orbital) to the LUMO (lowest unoccupied molecular orbital). This photoelectron then swiftly transfers from the LUMO of the dye to the conduction band of semiconductor TiO_2. The semiconductor carries the electron to photoanode which passes the electron to the platinized counter electrode (Halim, 2013).

Quantum-Dot Based Solar Cells and Shockley-Queisser Limit

A major problem with single band gap solar cells including silicon wafer, inorganic dye, and polymer is that their efficiency cannot go beyond the Shockley-Queisser efficiency limit of 33 per cent (Shockley, 1961). Theoretical calculation showed that 1.13 eV bandgap of single solar cell is lost 47 per cent of the energy as heat when the hot electrons are moving down from the higher conduction band to the conduction band edge, 2 per cent of energy is lost due the recombination when the tapped electrons from the conduction band is returned to the valence band, and 18 per cent of energy is lost when photons failed to excite electrons which reside on the upper and lower bandgap compared to the 1.13 eV bandgap.

Interestingly, a tiny material known as quantum dot can excite two or multiple electron-hole pairs (excitons) at a time while absorbing one photon. This process is known as multiple exciton generation (MEG) for quantum dots and semiconductor nanocrystals because these electron-hole pairs are not free; however, they are confined (quantum confinement) in dimension due to their tiny size. A similar process (known as impact ionization) appeared in bulk semiconductors which requires relatively higher energy compared to semiconductor nanocrystals and quantum dots for producing multiple electron-hole pairs.

Theoretical studies revealed that the multiple exciton generation (MEG) based solar cells which utilized quantum dots and semiconductor nanocrystals can exceed the S-Q limit with an efficiency of 66 per cent and even higher of 80 per cent. Moreover, changing the size of the of quantum dots one can utilize the Sun's broad spectrum of visible and infrared ranges. Another advantage with quantum dots is that they can be processed in solution and can easily be incorporated into solar cell device (Halim, 2013).

Nanotechnology Interventions in Future Generation – Printable Solar Cells

Absorption of light is inherently higher on the nanoscale materials whereas conventional materials such as Si and GaAs do a good job but are too expensive and can not be manufactured as Printable Solar Cells (Figure 23.5) like printing the newspaper.

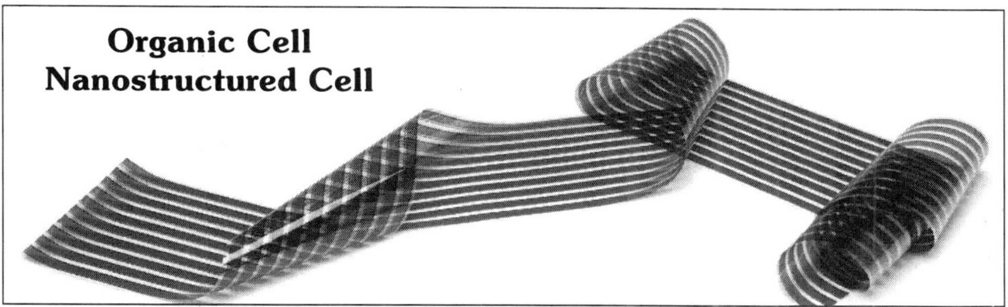

Figure 23.5: Nanostructured Printable Solar Cell.

Organic nano-structured photovoltaics convert sunlight into electrical Power. Nanomaterials offer the ability to get good light absorption in flexible formats. Interpenetrating networks on the nanoscale offer the opportunity to get high performance charge collection from low cost materials (Figure 23.6).

Nanotechnology Application in Wind Energy

Among all the renewable energy sources, wind energy is cleaner that requires less maintenance with a sustainable energy source for electricity generation. The generation of electricity from wind was started in around 1888 in Denmark. During this time, the wind turbine blade was made from stainless steel, but the blades of the turbine failed within 100 h of operation. Since then, different types of materials

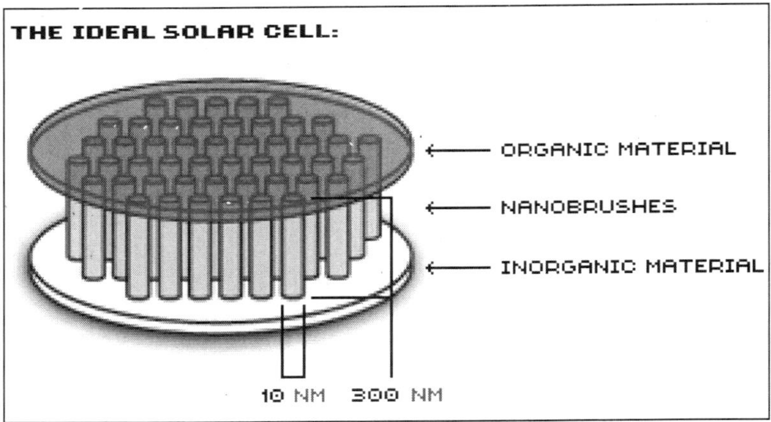

Nanotechnology Solar Cell Design

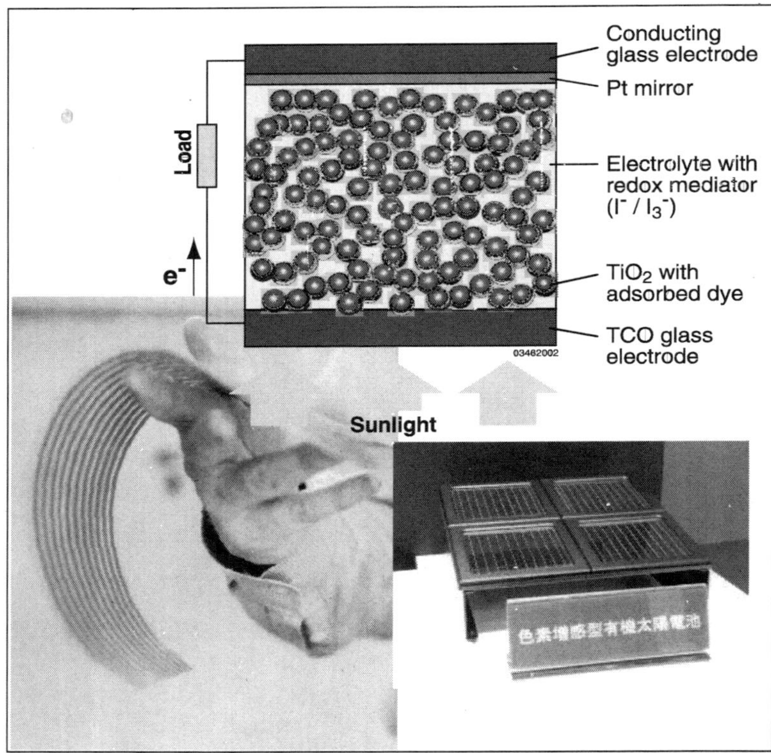

Interpenetrating Nanostructured Networks

Figure 23.6: Nanotechnology in Modern Solar Cells.
Courtesy: https://slideplayer.com/slide/4182719/

such as aluminium, plastics, and wood are used for the development of wind turbine blades, but after the 1970s, composite materials were used for the wind turbine blades (Mishnaevsky *et al.*, 2017). A wind turbine is a device that is used to convert the kinetic energy in the wind into electrical energy. Wind turbines capture the power from wind through blades, which are built on the aerodynamic principles to harness energy. The geometry and the position of the airfoil are determined to quantify the drag and lift force in the wind turbine to generate tangential force. Wind turbines can be classified in terms of the axis around which the turbine blades rotate. They are horizontal axis wind turbines (HAWT) in which the blades spin around a horizontal axis and vertical axis wind turbines (VAWT) in which the blades spin around a vertical axis (Figure 23.7). In general, the main components of the turbine are generator, tower, gearbox, and blades. Among all the components, the blade is considered as the costliest component in the wind turbine.

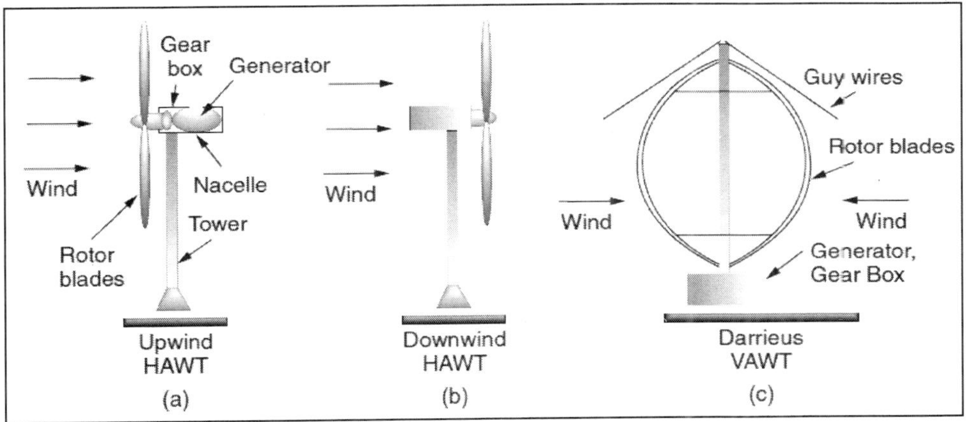

Figure 23.7: (a and b) Horizontal Axis Wind Turbines; (c) Vertical Axis Wind Turbines.
Courtesy: Masters, G. M., 2004.

Nanotechnology in Wind Turbines

"Rotor blades, in particular, need to be stronger and lighter to compensate for the ever-increasing demand for larger turbine blades. This is due to the larger swept area that can be produced by large wind turbines to produce more wind energy output. Advances in nanotechnology had been employed in developing cost-effective materials with the higher strength-to-mass ratio.

Advanced characteristics of Nano-Reinforced Materials for Rotor Blades

- High reliability and fatigue resistant nano-reinforced materials (Figure 23.8) for stable operations.
- High stiffness nano-reinforced material for better stability due to the aerodynamic loads and orientation of blade during operation.
- High strength nano-reinforced material to resist extreme wind conditions and blade gravity loads.

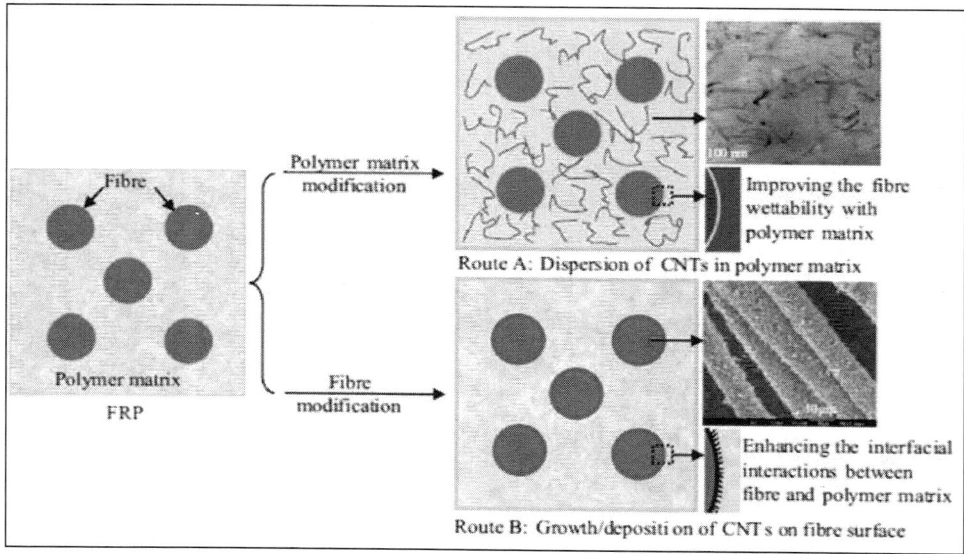

Figure 23.8: Carbon Nanotube (CNT) - Fibre Reinforced Polymer (FRP) Nanocomposite Materials for Wind Blade Materials.
Courtesy: Ma and Zhang, 2014.

- Low-density nano-reinforced material to reduce the overall load on the tower due to gravity and aerodynamic forces.
- Nano-reinforced materials with properties to resist environmental fouling effects such as icing, humidity and temperature.

The large surface area of the CNT improves the interfacial interactions between the matrix and the fibres that resulted in the enhancement of the mechanical properties of wind turbine blade materials with higher fatigue resistance, shear strength and fracture toughness compared to the conventional composites reinforced by microscale materials. Fouling of rotor blades with insects or formation of ice, as well as material erosion due to environmental effects (abrasive dust particles or diffusion of moisture/water vapour) can cause deterioration of blade health and may lead to blade failure. The application of nanocomposite coatings and nano-paints on the blade surfaces is one of the current solutions to control the fouling of rotor blades due to erosion and other adverse environmental effects" (Muzammil *et al.*, 2019).

Nanotechnology Application in Hydrogen Energy

Hydrogen is a renewable energy source and when hydrogen is used as an energy source, the only byproducts are water and heat. Once obtained, hydrogen can power virtually everything now powered by fossil fuels. Hydrogen is actually more powerful than gasoline: liquid hydrogen has a BTU (British Thermal Unit) of 60,000 per pound, where gasoline only has 18,000 per pound. NASA has used hydrogen as rocket fuel since the 1940's and as primary fuel while in space and for making drinking water (1 pound H + O = 9 pounds water) and this process generates a byproduct of usable electricity.

One advantage is that hydrogen stores approximately 2.6 times the energy per unit mass as gasoline, and the disadvantage is that hydrogen needs about 4 times the volume for a given amount of energy.

Hydrogen Production Processes

- ☆ Steam Methane Reforming
- ☆ Coal Gasification
- ☆ Partial Oxidation of Hydrocarbons
- ☆ Biomass Gasification
- ☆ Biomass Pyrolysis
- ☆ Electrolysis
- ☆ Thermochemical
- ☆ Photochemical
- ☆ Photobiological

The Steam Methane Reforming (SMR) Process

SMR is the most common method of producing commercial bulk hydrogen used in the industrial synthesis of ammonia (Figure 23.9). It is the least expensive method.

- ☆ At 700 – 1100 °C and in the presence of a nickel based catalyst (Ni), steam reacts with methane to yield carbon monoxide and hydrogen.
  - ❐ $CH_4 + H_2O \rightarrow CO + 3\,H_2$
- ☆ Additional hydrogen can be recovered by a lower-temperature gas-shift reaction with the carbon monoxide produced. The reaction is summarized by:
  - ❐ $CO + H_2O \rightarrow CO2 + H_2$

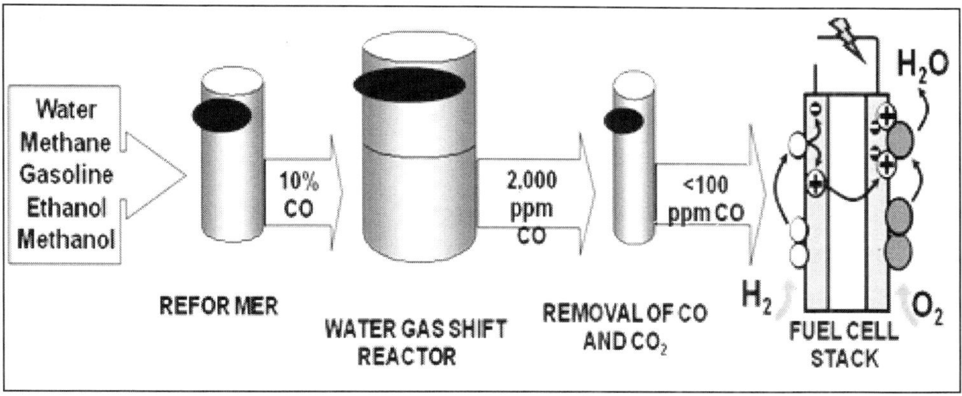

Figure 23.9. Steam Methane Reforming.

Carbon dioxide and other impurities are removed from the gas stream, leaving essentially pure hydrogen in this endothermic reaction (Heat must be added to the reactants for the reaction to occur.)

Coal Gasification

It is a well-established commercial technology competitive with SMR only where oil and/or natural gas are expensive. Here coal could replace natural gas and oil as the primary feedstock for hydrogen production, since it is so plentiful in the world.

Partial Oxidation Hydrocarbons

This process can be used to produce hydrogen from heavy hydrocarbons such as diesel fuel and residual oil. Any hydrocarbon feedstock that can be compressed or pumped may be used in this technology. methane and other hydrocarbons in natural gas are reacted with a limited amount of oxygen (typically, from air) that is not enough to completely oxidize the hydrocarbons to carbon dioxide and water.

$CH_4 + ½O_2 \rightarrow CO + 2H_2$ (+ heat) [Exothermic reaction (heat is evolved)]

Thermochemical Production of Hydrogen

When water is heated to above 2500°C, it separates into oxygen and hydrogen in a process known as thermolysis. However, at such high temperatures, it is difficult to prevent the oxygen and hydrogen from recombining to form water. Thermochemical water-splitting cycles can lower the temperature and help separate oxygen and hydrogen products to produce pure hydrogen gas. These cycles can improve the efficiency of hydrogen production from 30 per cent for conventional electrolysis to around 50 per cent efficiency. One of the most promising cycles so far is the **sulfur-iodine (S-I) cycle**. Sulfur dioxide (SO_2) and iodine (I_2) are fed into the cycle as chemical catalysts. A catalyst lowers the activation energy of a reaction without being used up by the reaction. In this cycle, sulfur dioxide (SO_2) and iodine (I_2) are feed into the cycle as a chemical catalyst. A catalyst lowers the temperature at which the reaction will occur without being used up by the reaction.

There are three steps in the S-I cycle

Step 1

$I_2 + SO_2 + 2H_2O \rightarrow 2HI + H_2SO_4$

The reaction is run at 120°C. The hydrogen iodide and sulfuric acid are separated, usually by distillation.

Step 2

Generation of oxygen and regeneration of SO_2.

$H_2SO_4 \rightarrow H_2O + SO_2 + 1/2\ O_2$

This reaction is run at 850°C.

Step 3

Generation of hydrogen and regeneration of I

$2HI \rightarrow H_2 + I_2$

This reaction is run at 450°C.

Sulfur–Iodine Cycle reactions can reduce the high temperature demands of the thermolysis of water for the production of hydrogen gas and can provide a mechanism for the separation of oxygen and hydrogen products to prevent recombination.

Biomass Production of Hydrogen

Hydrogen can be produced numerous ways from biomass. Biomass is defined as a renewable resource made from renewable materials. Examples of biomass sources include: switchgrass, plant scraps, garbage, human wastes. Gasification of biomass could be a way of extracting hydrogen from these organic sources. The biomass is first converted into a gas through high-temperature gasifying.

The hydrogen rich vapor is condensed in pyrolysis oils. These oils can be steam reformed to generate hydrogen. This process has resulted in hydrogen yields of 12 per cent - 17 per cent hydrogen by weight of the dry biomass. When biological waste material is used as a feedstock, this process becomes a completely renewable, sustainable method of hydrogen generation.

Electrolysis

Electrolysis is the technical name for using electricity to split water into its constituent elements, hydrogen and oxygen. The splitting of water is accomplished by passing a DC electric current through water. The electricity enters the water at the cathode, a negatively charged terminal, passes through the water and exists via the anode, the positively charged terminal. The hydrogen is collected at the cathode and the oxygen is collected at the anode. Electrolysis produces very pure hydrogen for use in the electronics, pharmaceutical and food industries. The hydrogen is collected at the cathode and the oxygen is collected at the anode. Electrolysis produces very pure hydrogen for use in the electronics, pharmaceutical and food industries.

Photobiological

This method involves using sunlight, a biological component, catalysts and an engineered system. Specific organisms, algae and bacteria, produce hydrogen as a byproduct of their metabolic processes. These organisms generally live in water and therefore are biologically splitting the water into its component elements. Currently, this technology is still in the research and development stage and the theoretical sunlight conversion efficiencies have been estimated up to 24 per cent.

Hydrogen Storage

- ☆ Physical storage of H_2: Compressed, Cryogenically liquified, Metal Hydride ("sponge"), Carbon nanofibers

- ✰ Chemical storage of hydrogen: Sodium borohydride, Ammonia, Methanol, Alkali metal hydrides
- ✰ New emerging methods: Amminex tablets, DADB (Diammoniate of Diborane), Solar Zinc production, Alkali metal hydride slurry.

Compressed method of hydrogen storage is volumetrically and gravimetrically inefficient, but the technology is simple, so by far the most common in small to medium sized applications *i.e.*, 3500, 5000, 10,000 psi variants. Cryogenically liquified methods are gravimetrically and volumetrically efficient, but very costly when compared to compress methods. The hydrogen is Compressed, chilled, filtered, condensed and boils at 22K (-251 °C). It is a slow "waste" evaporation that kept at 1 atm or just slightly over. Metal Hydride sponges are technically a chemical reaction, but acts like a physical storage method in which hydrogen is absorbed like in a sponge and operates at 3-30 atm, much lower than 200-700 for compressed gas tanks. Though it is comparatively very heavy, but shows with good volumetric efficiency, good for small storage, or where weight doesn't matter. Broken down by reformer, yields CO, CO_2, and H_2 gas. Methenol is another very common hydrogen transport method in which the distribution infrastructure exists same as gasoline. Ammonia exhibit slightly higher volumetric efficiency than methanol that must be catalyzed at 800-900°C for hydrogen release. It is toxic and usually transported as a liquid, at 8 atm. Some Ammonia remains in the catalyzed hydrogen stream, forming salts and many drawbacks, thus Methanol is considered to be a better solution. Alkali metal hydrides were produced by "Powerball" company, that makes small (3 mm) coated NaH spheres. These spheres are cut and exposed to water as needed for H_2 gas release that produces hydroxide solution as waste. Sodium Borohydrate is the most popular of many hydrate solutions in which the solution is passed through a catalyst to release H_2 and it is commonly a one-way process (sodium metaborate must be returned if recycling is desired.) However, some alternative hydrates are too expensive or toxic and the "Millennium Cell" company used Sodium Borohydrate technology. Amminex is essentially an ammonia storage method in which ammonia stored in a salt matrix and very stable. Ammonia separated and catalyzed for use and likely to have non-catalyzed ammonia in hydrogen stream that leads to ammonia poisoning contraindicates use with PEM fuel cells, but compatible with alkaline fuel cells.

	Gravimetric H_2 density (wt % H_2)	Volumetric H_2 density (kg H_2/l)	Energy density (MJ/l)
AMMINEX	**9.1**	**0.11**	**13.0**
US-DOE Goal (2015)	9.0	0.081	9.72

Performing quantum calculations on a supercomputer, scientists at Pacific Northwest National Laboratory (PNNL) have characterized a material that might allow on-board refueling of hydrogen powered vehicles. Researchers, led by Maciej Gutowski, looked at different crystalline structures of a compound made up of nitrogen, boron and hydrogen - NBH_6 - and found one that might be more stable compared to ammonia borane, a molecular crystal built of NH_3BH_3 molecules. Ammonia borane can hold a lot of hydrogen but isn't easily reversible - or able to be refilled with hydrogen. Ammonia borane, as a storage material, would likely have to be removed from the vehicle and be sent to some sort of processing plant and undergo a reaction to be refilled. The more stable compound, diammoniate of diborane or DADB, holds more promise for reversibility. Initial thermodynamic properties for the compound indicate that it might spontaneously uptake hydrogen fuel. This work is performed under the Grand Challenge Project "Computational studies of materials to hydrogen storage" in the Molecular Sciences Computing Facility at PNNL. Researchers plan to perform additional calculations, synthesize the diammoniate of diborane compound and test their theories on the material in the coming year.

Solar Zinc for Hydrogen Production and Storage

Since zinc is a relatively abundant metal, and is the fourth among all metals in world production – being exceeded only by iron, aluminum and copper – it can be considered a natural choice for producing and storage of hydrogen. The problem is that the current industrial production of pure zinc (Zn) from zinc oxide (ZnO) by either electrolysis or smelting furnaces is characterized by its high-energy consumption and concomitant pollution, derived mostly from the combustion of fossil fuels for heat and electricity.

To address the issue solar energy can be used as the main energetic source in the production of Zn from ZnO. In 2004 the European Union and the Swiss Federal Office of Science and Education decided to fund a joint research to explore this possibility using a 45 kW solar furnace in Villigen, Switzerland, a 75kW solar simulator in Zurich and the largest solar research facility in the world with 1MW output located at the Weitzman Institute in Israel. The need for such a large amount of power is due to the need for very high temperatures required for the production of Zn from ZnO (normally around 1750°C/3182°F). Adding small amounts of carbon in the form of coal enabled the Weitzman team to reduce the Zn production temperature to a more manageable 1200°C (2192F). For the future, the team sees the possibility of replacing the coal completely with biomass thus making the entire process completely pollution free.

The real achievement of the Weitzman team is the scale of production of Zn which reached an average of about 50kg/h during tests using the existing Solzinc solar reactor located at the center of the Institute. On a full scale industrial facility much larger amounts could be extracted using a similar process. The energy cycle developed by the researchers is very efficient and relatively self-sustaining. ZnO is mined and transported to the Solzinc solar facility where it is mixed with small amounts of coal and put inside the solar furnace located on top of a high tower. A large array of heliostats (computer guided highly reflective mirrors) follows the sun around the sky and reflects the light to a hyperbolic mirror located inside the solar tower producing highly concentrated heat inside the solar furnace. At a heat of above 1200°C (2192F) the ZnO breaks down into Zn and oxygen which in turn recombines with the carbon to create CO as a minor by-product. The Zn is then cooled down to create a fine powder which can be safely handled and transported. In order to produce hydrogen from the Zn powder a much simpler process is performed where the Zn is mixed with water at a temperature of 350°C (662F). The oxygen inside the water recombines with the Zn to produce ZnO once again and the by-product is pure hydrogen.

This process has many advantages over existing ways of producing hydrogen. First and foremost it uses a renewable form of energy - the sun. Furthermore the main material required for the process is the relatively inexpensive zinc oxide which is almost completely recycled back by the end of the process. Another important advantage is that the hydrogen could be produced where it is needed, *i.e.*, at the local fuel station instead of transporting large amounts of explosive hydrogen across the country. Trucks loaded with safe Zn powder would transport it to the fuel station where it would undergo the relatively simple treatment of extracting the hydrogen using steam. Last but not least, the entire process is relatively clean and when biomass will replace the coal as an additive to the ZnO mix, the process will be completely nonpolluting.

The superior storage density of chemical Alkaline metal hydride slurry is indicated as one unit of slurry carries the potential of generating twice as much volume of hydrogen (at about the same weight) as one unit of cryogenically cooled liquid hydrogen. Liquid hydrogen is a proven method of storing hydrogen, but it takes substantial energy to liquefy the hydrogen and there is continual "boil off" of hydrogen during storage. Slurry, on the other hand, is stored at normal temperature and normal pressure. The Safe Hydrogen is patented and proprietary chemical hydride slurry can release hydrogen as needed by the addition of water. Because hydrogen is also contributed by the water in the reaction, the Safe Hydrogen fuel carries an impressive hydrogen generation potential. The reaction, besides producing hydrogen, generates some heat and a hydroxide byproduct. It has the potential of cost and energy efficiencies with no material waste. The recycling process converts the hydroxide byproduct back to the original hydride. All other materials, constituting the slurry are also reclaimed and may be used over and over again. Recycling is achieved in very large-scale, centralized manufacturing processes. The recycling process is designed to use any energy source to generate the hydrogen carried by the slurry. A major cost advantage of this technology is based on the characteristics of the slurry. Slurry is a nonexplosive, noncorrosive, environmentally safe, pumpable "hydrogen fuel". Slurry can be stored, transported and pumped with existing tanks, pumps and pipelines and can therefore distribute hydrogen to the market utilizing the existing fossil fuel infrastructure. The only difference from existing fuel delivery systems is that the delivery devices such as trucks or rail tankers don't return empty. They are fully loaded in both directions. They return from delivery runs, loaded with depleted slurry that needs to be recycled.

Hydrogen Storage with Carbon Nanotubes (CNTs), Nano-fibers and Carbon Nanomaterials

Carbon Nanofiber complex structure presents a large surface area for hydrogen to "dissolve" into with an early claim set the standard of 65 kgH_2/m^2 and 6.5 per cent by weight as a "goal to beat". The claim turned out not to be repeatable and the research further continues. On the otherhand, CNTs are stable, lightweight, inexpensive, possess large active surface area and large internal volume that can be accessed.

The mechanism of hydrogen storage in CNT is based on the first and simplest approach: Physisorption (Figure 23.10a) and van der Waals interaction between H_2 and CNT wall (internal or external) in which there is no energy barrier to overcome, but relatively weak binding → low temperatures.

Nanotube doping is incoportaed for improved Capacity of hydrogen storage with Alkali Metals. Li-doped: peak adsorption of 15-20 wt per cent at 673 K and stable in ambient conditions. K-doped: peak adsorption of 14 wt per cent at 298 K and highly unstable in ambient conditions. Storage can be attributed to tube exterior surface also.

Simulated functionalization (Figure 23.10b) with light transition metals Sc, Ti and V on slightly larger SWNTs with sufficient interior space allows functionalization of inner surface. Each metal atom, inside or outside, can physisorb up to 4 H_2 at high coverage ~8 wt per cent storage may be possible with excellent binding energy.

Based on SWNT simulations predict chemisorption is also a more favourable approach with a forecast of fully hydrogenated SWNT could store 7.8 wt per cent hydrogen. The stability of hydrogenated SWNTs increases with diameter, however there is a large kinetic barrier to chemisorbtion *i.e.*, dissociation of H_2.

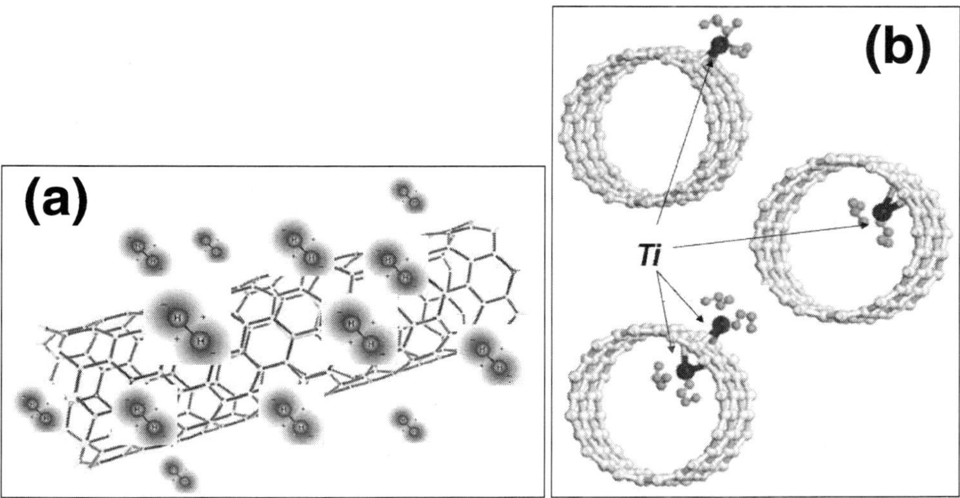

Figure 23.10. (a) CNT - Physisorption; (b) Functionalization for hydrogen storage.
Courtesy: Durgun *et al.*, Phys Rev B 2008.

Prospects for CNT and other Carbon Based Nanomaterials for Storage of Hydrogen

Synthesis of functionalized SWNT systems are investigated for the feasibility of storage through physisorption. Investigation of catalytic "spillover" mechanisms for a practical source of atomic hydrogen for chemisorption and parallel studies with other carbon nanomaterials (Figure 23.11) are going on.

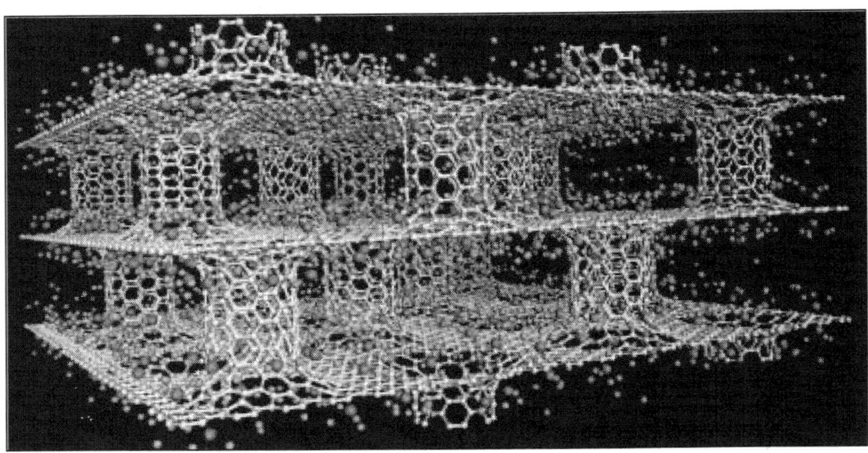

Figure 23.11: Cartoon Representation of CNT and Carbon Nanomaterials for Storage of Hydrogen.

The requirements for effective on-board hydrogen storage using nanomaterials are

- ☆ Appropriate thermodynamics for favourable enthalpies of hydrogen absorption and desorption
- ☆ Fast kinetics (quick uptake and release)
- ☆ High storage capacity (specific capacity to be determined by usage)
- ☆ Effective heat transfer
- ☆ High gravimetric and volumetric densities
- ☆ Long cycle lifetime for hydrogen absorption/desorption
- ☆ High mechanical strength and durability
- ☆ Safety under normal use and acceptable risk under abnormal conditions
- ☆ Cheap components and materials

REFERENCES

1. Durgun, E., Çýracý, S. and Yildirim, T. (2008). Functionalization of carbon-based nanostructures with light transition-metal atoms for hydrogen storage. *Physical Review B*. 77, 085405.

2. Halim, M. A. (2013). Harnessing Sun's Energy with Quantum Dots Based Next Generation Solar Cell Nanomaterials 3: 22-47.
3. Kasap, S.O. (2006). Principles of Electronic Materials 3rd Edn. McGraw-Hill
4. Ma, P. C. and Zhang, Y. (2014). Perspectives of carbon nanotubes/polymer nanocomposites for wind blade materials. Renew Sust Energ Rev 30: 651–660
5. Masters, G. M. (2004). In: Renewable and Efficient Electric Power Systems. John Wiley.
6. Mishnaevsky L *et al.* (2017). Materials for wind turbine blades: an overview. *Materials* 10(11):1285
7. Muzammil, W. K., Md. Mizanur Rahman, A. Fazlizan, M. A. Ismail, H. K. Phang, and M. A. Elias. (2019). Nanotechnology in Renewable Energy: Critical Reviews for Wind Energy. In: S.
8. Pelemiš, S. and Hut, I. (2013). Nanotechnology materials for solar energy conversion. Contemporary Materials (Renewable Energy Sources), IV–2: 145-151
9. Serrano, E., Rus, G. and Garcý´a-Martý´nez, J. (2009). Nanotechnology for sustainable energy. *Renewable and Sustainable Energy Reviews* 13: 2373–2384.
10. Shockley, W., Hans, J. and Queisser, H.J. (1961). Detailed balance limit of efficiency of *p-n* junction solar cells. *J. Appl. Phys*, *32*, 510–519.
11. Siddiquee *et al.* (eds.). Nanotechnology: Applications in Energy, Drug and Food.

Chapter 24

Nanotechnology Interventions in Batteries and Fuel Cells

*D. Jeya Sundara Sharmila**

Department of Nano Science and Technology, Tamil Nadu Agricultural University
Coimbatore – 641 003, Tamil Nadu
e-mail: djssharmila@gmail.com

Nanotechnology in Battery Applications

"Nanotechnology plays a key role for future battery applications and its market adoption. Electrolyte conductivity increases up to six times by introducing nanoparticles of alumina, silicon or zirconium to non-aqueous liquid electrolytes. Most efforts have been focused on solid state electrolytes, solid polymer electrolytes (SPE) *etc.* Poly(ethylene oxide)-based (PEO-based) SPE received most attention since PEO is safe, green and lead to flexible films. Nevertheless, polymers usually have low conductivity at room temperature and, depending on SPE compositions, their interfacial activity and mechanical stability are not high enough. In this logic, nanocomposite polymer electrolytes could help in the fabrication of highly efficient, safe and green batteries. For example, the introduction of ceramic nanomaterials as separators in polymer electrolytes increases the electrical conductivity of these materials at room temperature from 10 to 100 times compared with the corresponding undispersed SPE system. TiO_2, Al_2O_3 and SiO_2 and S-ZrO_2 (sulphate-promoted superacid zirconia) have been used for this purpose, founding that the introduction of S-ZrO_2 led to the best performance. Another example is the formation of a crystalline complex by dissolution of lithium salts (LiXF6, being X phosphorus, tin or antimony) in low molecular mass and high polydispersity PEO. The crystalline structure of PEO makes the ionic conductivity. Moreover, the short chains and polydispersity of PEO cause a high defect concentration that considerably enhances this conductivity. The nanoscaled metal alloy composites can increase the life cycle of nanocomposites through the reduction of volume change caused during the formation of the alloy. The promising interventions of nanotechnology in battery functions are:

1. The construction of an nano-architectured anode in a variety of nanoforms (*i.e.*, silicon nanopillars done by *etc.*hing of bulk materials)
2. The fabrication of intermetallic electrodes as nanoparticles through displacement reactions by deposition of nanoparticles on nanostructured electrodes (*i.e.*, Ni_3Sn_4 for $Li_4.4Sn$);
3. The use of amorphous nanostructured alloys (titanium, vanadium, chrome or cobalt in amorphous tin phase)

Nanotechnology in Li-ion Batteries

Li-ion batteries consist of a three primary functional components. The main components of a Li-ion battery device are the positive electrode, negative electrode and the electrolyte, for which a variety of materials may be used (Figure 24.1). Generally, the most useful material for the positive electrode is one of three materials: lithium cobalt oxide, lithium iron phosphate, or a spinel such as lithium MnO. On the other hand, the most common materials for the negative electrode are carbon based compounds and lithium- containing alloys. Upon charging, lithium ions are extracted from the positive electrode material and inserted into the negative electrode material. Upon discharging, the reverse process takes place. Common batteries should exhibit three characteristics: (a) high energy and power capacity, (b) high charging rate, and (c) long lifetime (cycling stability). Although Li-ion batteries are available commercially, the performance of Li-ion batteries is limited by the current electrode and electrolyte materials. For future generations of rechargeable Li-ion batteries, not only for applications in portable electronic devices but especially for clean energy storage and use in hybrid electric vehicles, further improvements of materials are essential. We need to find new, efficient and effective ways to improve the physical and chemical characteristics of the materials for use in electrochemical Li-ion batteries. One avenue that is opening up is that of NSMs for electrochemical Li-ion batteries devices. However, the selection of NSMs for use as negative electrode, positive electrode, and electrolyte, as well as their voltage, capacity, life, and safety abilities in electrochemical Li-ion battery devices can dramatically improve. Recently, great efforts have been made toward the development of improved electrolyte, negative and positive electrodes materials. It has been demonstrated that the morphology of nanostructures can be optimized to achieve higher energy and charge densities in electrochemical Li-ion battery devices. Therefore, nanoarchitecture is another consideration that can be optimized for further improvement. In this regard, research on NSMs is expanding rapidly into the assembly of nanoparticles in 0D and 1D structures. Recently, some research groups have tried to synthesize 0D nanoparticles to avoid the problem of poor Li+ ion diffusivity and electron conductivity, which would eliminate either metal doping or carbon coating. 0D nanoparticles with high-crystallinity are particularly advantageous when the aim as above mentioned is to achieve faster lithium-ion diffusion. They also display a better rate capability, which indicates its possible use in electrochemical Li-ion battery devices. However, 0D NSMs are not very stable and very difficult to make in large quantities. Various efforts have been implemented to synthesize novel 0D NSMs but have been met with very limited success in the improvement of electrochemical Li-ion battery device performance.

Due to this, researchers have fabricated new 1D NSMs that will make electrochemical Li-ion battery devices more efficient due to their unique mechanical, thermal, electrical properties and structural versatility. Various efforts have been made by the scientific community to fabricate novel 1D NSMs such as nanotubes, nanorods, and nanobelts. The figure below shows the schematic diagram of a Li+ ion pathway in 1D NSMs.

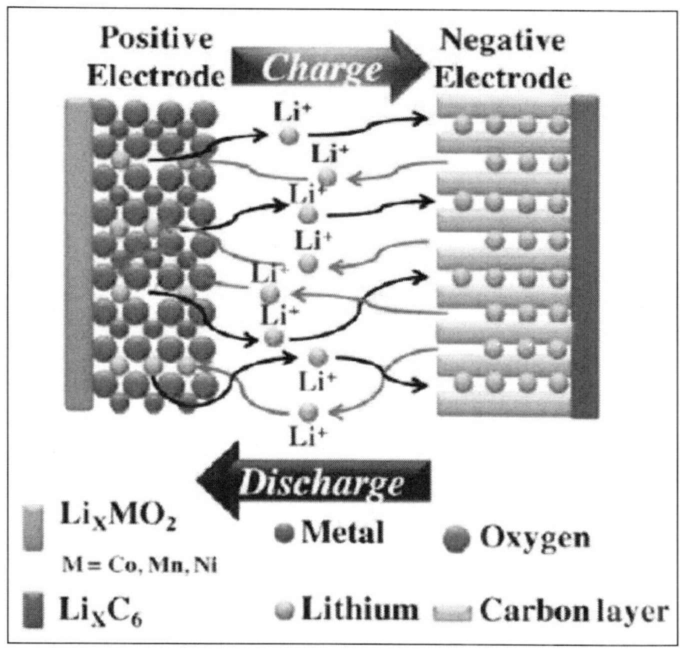

Figure 24.1: Depiction of the Li-ion Battery.
Courtesy: Tiwari *et al.*, 2012.

They also constructed a full cell composed of a $LiCoO_2$ cathode and a carbon–silicon core–shell nanowire anode. Significantly, using these core–shell nanowires they achieved high mass loading and an area capacity of 4mA/hcm^2. Importantly, the capacity of a Li-ion full cell consisting of a cathode of $LiCoO_2$ and anode of Si nanotubes shows a higher capacity than commercially available graphite even after 200 cycles. However, a new problem arose for 1D NSMs are low surface area and slow diffusion rates. Due to low surface area, slow diffusion rates and poor conductivity, 1D NSMs show the poor energy density and poor rate capacities. Consequently, improvement in surface area to volume ratio, diffusion rates and conductivity is necessary.

2D NSMs are a useful approach to improve the electrode performances in higher overall capacity, high-rate capability, and longer cycling life. 2D NSMs normally exhibit small crystalline size, high surface area, large surface-to-volume ratio, and favorable structural stability over 1D, and 0D NSMs. The development of a 2D nanostructure gives the right structure necessary for a fast mass transport, for instance. Recently, Zhou *et al.*, synthesized a $Si/TiSi_2$ heteronanostructure on Ti foil in a CVD chamber with TiCl, SiH and H as precursors. The $Si/TiSi_2$ heteronanostructure was used as an anode material for Li+ batteries. In this heteronanostructure, Si nanoparticles acts as an active component to store and release Li+ while $TiSi_2$ serves as the inactive component to support Si and to facilitate charge transport.

In order to further achieve higher energy and charge densities in electrochemical Li-ion battery devices, 3D nanostructured electrodes may be needed. As 3D NSMs electrode have a higher surface area, larges surface-to-volume ratio, and more favorable structural stability over 2D, 1D, and 0D NSMs. The 3D NSMs can be flooded with electrolyte, ensuring a high specific surface area in contact with the electrode and hence a high flux of Li^+ ion across the interface. The nanowalls between the NSMs, of equal dimensions throughout, ensure very short diffusion paths for Li ions on intercalation and deintercalation, and hence homogeneous rates of transports. The $Si/TiSi_2$ heteronanostructures show fast and superior charge/discharge capacities. An excellent reversible capacity was obtained for the following MnO/CNT nanotube array electrodes. The coaxial MnO/carbon nanotube arrays offer the following advantages:

1. Presence of carbon nanotube improves the electrical conductivity
2. Homogeneous electrochemical accessibility and better ionic conductivity is achieved by avoiding agglomerative binders and other additives
3. Well-directed coaxial alignment provides a better conductive path
4. MnO/CNT hybrid nanostructures offer a unique dual-binding mechanism of lithium storage (insertion–deinsertion in the case of carbon nanotubes and formation and decomposition of LiO in the case of MnO nanotubes).

MnO_2/CNT Nanotube Array Electrode for Li-ion Battery

The use of 3D NSMs is increasing rapidly in the field of Li-ion battery devices because of their substantial advantages in terms of mass transport (Figure 24.2). Transport in 3D NSMs typically encompasses shorter transport length for Li^+ ions and electrons, better electrode–electrolyte contact area, and great merits to permit reproducible Li^+ insertion and extraction. This verifies their promising application as electrode materials for Li-ion batteries.

The nanowalls between the NSMs, of equal dimensions throughout, ensure very short diffusion paths for Li ions on intercalation and deintercalation, and hence homogeneous rates of transports.

The coaxial MnO_2/carbon nanotube arrays offer the following advantages:

1. Presence of carbon nanotube improves the electrical conductivity

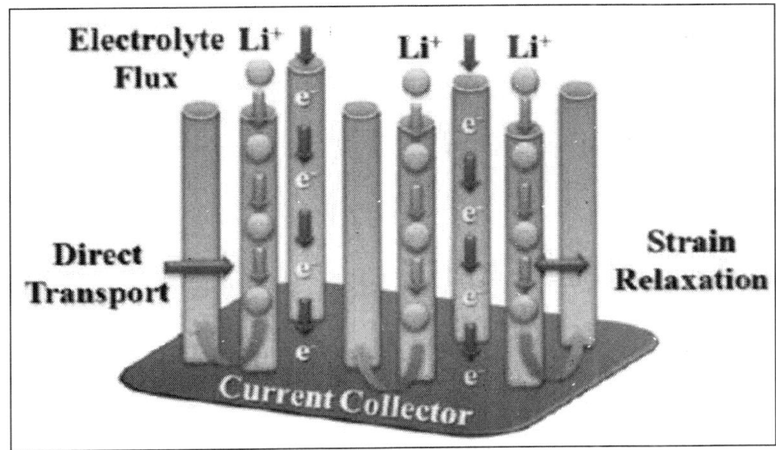

Figure 24.2: Li-ion Pathway in Ordered Nanostructures.
Courtesy: Tiwari *et al.*, 2012.

2. Homogeneous electrochemical accessibility and better ionic conductivity is achieved by avoiding agglomerativebinders and other additives
3. Well-directed coaxial alignment provides a better conductive path
4. MnO2/CNT hybrid nanostructures offer a unique dual-binding mechanism of lithium storage (insertion–deinsertion in the case of carbon nanotubes and formation and decomposition of Li2O in the case of MnO2 nanotubes).

Supercapacitors

Electrochemical capacitors (ECs), also named supercapacitors (Figure 24.3) and ultracapacitors, store electrical energy are similar to batteries, but using a different mechanism. While batteries do it chemically, ultracapacitors store electricity physically, by separating the positive and negative charges. Since the discovery of ECs, these devices have attracted considerably less attention than batteries as energy storage devices. Nevertheless, thanks to both the contribution of nanotechnology and the better understanding of charge storage mechanisms (ion behaviour in small pores) the interest on ECs has noticeably increased recently. The drawbacks of classical capacitors can be summarized in three main points: (i) the high cost of premium performance electrodes due to the miniaturization, (ii) the large requirements for long cycle life (around 105 cycles) and (iii) their low efficiency–materials that combines both high surface area with a low resistivity are required. The main determining factor for power density and maximum power output is the surface area of each electrode that makes up the capacitor. The use of nanostructured materials dramatically increases this surface area (*e.g.* up to 1000 m^2/g of carbon). Moreover, in contrast to capacitors, supercapacitors utilize a small volume of electrolyte, which interacts with the surface of each electrode to store charge. Consequently, they offer a unique combination of high power and high energy performance parameters.

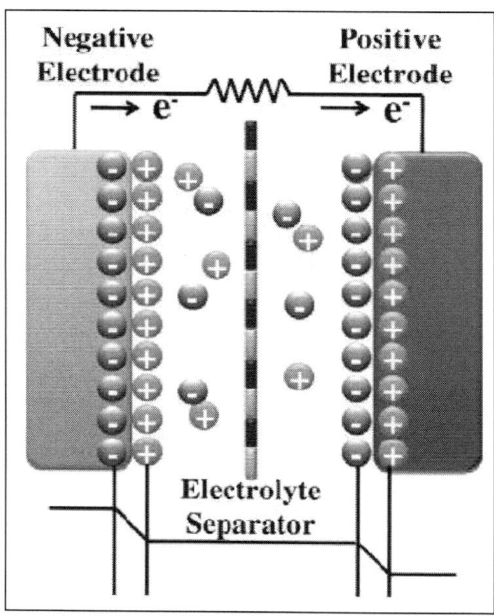

Figure 24.3: Illustration of a Supercapacitor.
Courtesy: Tiwari *et al.*, 2012.

There are three types of ECs: pseudocapacitors, also named redox supercapacitors, electrochemical double layer capacitors (EDLCs) and hybrid capacitors. The nanomaterials typically used are metal-based nanocomposites and conductive polymers, carbon-based nanostructures and hybrid inorganic/organic nanocomposites, respectively" (Tiwari *et al.*, 2012; Serrano *et al.*, 2009).

Nanotechnology in Fuel Cells

"A fuel cell is an electrochemical device that produces electricity without combustion by combining hydrogen and oxygen to produce water and heat. The concept of fuel cells were discovered German Scientist G H Shoenbein and first developed by William Grove. In 1839, Grove was experimenting on electrolysis (the process by which water is split into hydrogen and oxygen by an electric current), he observed that combining the same elements could also produce an electric current. During 1930s -1950s, Francis Thomas Bacon, a British scientist, worked on developing alkaline fuel cells. He demonstrated a working stack in 1958 and this technology was licensed to Pratt and Whitney where it was utilized for the Apollo spacecraft fuel cells.

Working Principle of a Fuel Cell

A fuel cell is a device that uses hydrogen (or hydrogen-rich fuel) and oxygen to create electricity by an electrochemical process. A single fuel cell consists of an electrolyte sandwiched between two thin electrodes (a porous anode and cathode) Hydrogen, or a hydrogen-rich fuel, is fed to the anode where a catalyst separates hydrogen's negatively charged electrons from positively charged ions (protons). At the cathode, oxygen combines with electrons and, in some cases, with species such as protons or water, resulting in water or hydroxide ions, respectively. The electrons from the anode side of the cell cannot pass through the membrane to the positively charged cathode; they must travel around it via an electrical circuit to reach the other side of the cell. This movement of electrons is an electrical current. The amount of power produced by a fuel cell depends upon several factors, such as fuel cell type, cell size, the temperature at which it operates, and the pressure at which the gases are supplied to the cell. Still, a single fuel cell produces enough electricity for only the smallest applications. Therefore, individual fuel cells are typically combined in series into a fuel cell stack. A typical fuel cell stack may consist of hundreds of fuel cells. Fuel cells are classified primarily by the kind of electrolyte they employ. This determines the kind of chemical reactions that take place in the cell, the kind of catalysts required, the temperature range in which the cell operates, the fuel required, and other factors. There are several types of fuel cells currently under development, each with its own advantages, limitations, and potential applications.

Classification of Fuel Cells

Based on the Type of Electrolyte

- ☆ Alkaline Fuel Cell (AFC)
- ☆ Phosphoric Acid Fuel Cell (PAFC)
- ☆ Polymer Electrolytic Membrane Fuel Cell (PEMFC)
- ☆ Solid Polymer Fuel Cell (SPFC) and
- ☆ Proton Exchange Membrane Fuel Cell (PEMFC)
- ☆ Molten Carbonate Fuel Cell (MCFC)
- ☆ Solid Oxide Fuel Cell (SOFC)

Fuel Cell Type	Electrolyte Used	Operating Tempearture	Electrode Reactions	
Polymer Electrolyte	Polymer Membrane	60-140°C	Anode:	$H_2 = 2H^+ + 2e^-$
			Cathode:	$1/2\,O_2 + 2H^+ + 2e^- = H_2O$
Direct Methanol	Polymer Membrane	30-80°C	Anode:	$CH_3OH + H_2O = CO_2 + 6H^+ + 6e^-$
			Cathode:	$3/2\,O_2 + 6H^+ + 6e^- = 3H_2O$
Alkaline	Potassium Hydroxide	150-200°C	Anode:	$H_2 + 2\,OH^- = H_2O - 2e^-$
			Cathode:	$1/2\,O_2 + H_2O + 2e^- = 2\,OH^-$
Phosphoric Acid	Phosphoric Acid	180-200°C	Anode:	$H_2 = 2H^+ + 2e^-$
			Cathode:	$1/2\,O_2 + 2H^+ + 2e^- = H_2O$
Molten Carbonate	Lithium/Potassium Carbonate	650°C	Anode:	$H_2 + CO_3^{2-} = H_2O + CO_2 + 2e^-$
			Cathode:	$1/2\,O_2 + CO_2 + 2e^- = CO_3^{2-}$
Solid Oxide	Yittria Stablized Zirconia	1000°C	Anode:	$H_2 + O^{2-} = H_2O + 2e^-$
			Cathode:	$1/2\,O_2 + 2e^- = O^{2-}$

Based on Types of Fuel and Oxidant

- Hydrogen (pure)-Oxygen (pure) fuel cell
- Hydrogen rich gas-air fuel cell
- Ammonia –air fuel cell
- Synthesis gas- air fuel cell
- Hydro carbon (gas)- air fuel cell

Alkaline Fuel Cells (AFC)

The alkaline fuel cell uses an alkaline electrolyte such as 40 per cent aqueous potassium hydroxide. In alkaline fuel cells, negative ions travel through the electrolyte to the anode where they combine with hydrogen to generate water and electrons.

It was originally used by NASA on space missions. NASA space shuttles use Alkaline Fuel Cells. Alkaline fuel cells (AFCs) (Figure 24.4) were one of the first fuel cell technologies developed, and they were the first type widely used in the U.S. space program to produce electrical energy and water on board spacecraft. These fuel cells use a solution of potassium hydroxide in water as the electrolyte and can use a variety of non-precious metals as a catalyst at the anode and cathode. High-temperature AFCs operate at temperatures between 100°C and 250°C (212°F and 482°F). However, more-recent AFC designs operate at lower temperatures of roughly 23°C to 70°C (74°F to 158°F).

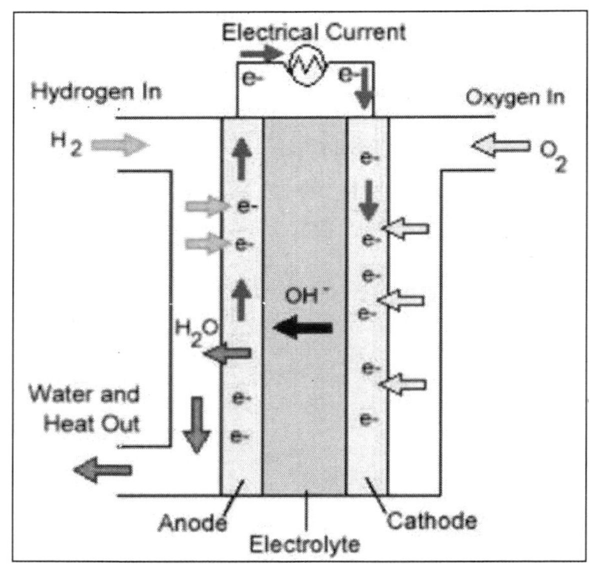

Figure 24.4: Alkaline Fuel Cell.
Courtesy: https://www.energy.gov/eere/fuelcells/types-fuel-cells.

Anode Reaction: $2H + 4OH^- \gg 4H_2O + 4e^-$

Cathode Reaction: $O_2 + 2H_2O + 4e^- \gg 4OH^-$

AFCs are high-performance fuel cells due to the rate at which chemical reactions take place in the cell. They are also very efficient, reaching efficiencies of 60 percent in space applications. The disadvantage of this fuel cell type is that it is easily poisoned by carbon dioxide (CO_2). In fact, even the small amount of CO_2 in the air can affect the cell's operation, making it necessary to purify both the hydrogen and oxygen used in the cell. CO2 can combine with KOH to form potassium carbonate which will increase the resistance. This purification process is costly. Susceptibility to poisoning also affects the cell's lifetime (the amount of time before it must be replaced), further adding to cost. Cost is less of a factor for remote locations such as space or under the sea. However, to effectively compete in most mainstream commercial markets, these fuel cells will have to become more cost effective. AFC stacks have been shown to maintain sufficiently stable operation for more than 8,000 operating hours.

Molten Carbonate Fuel Cells (MCFC)

The molten carbonate fuel cell uses a molten carbonate salt as the electrolyte. It has the potential to be fuelled with coal- derived fuel gases, methane or natural gas. These fuel cells can work at up to 60 per cent efficiency. In molten carbonate fuel cells, negative ions travel through the electrolyte to the anode where they

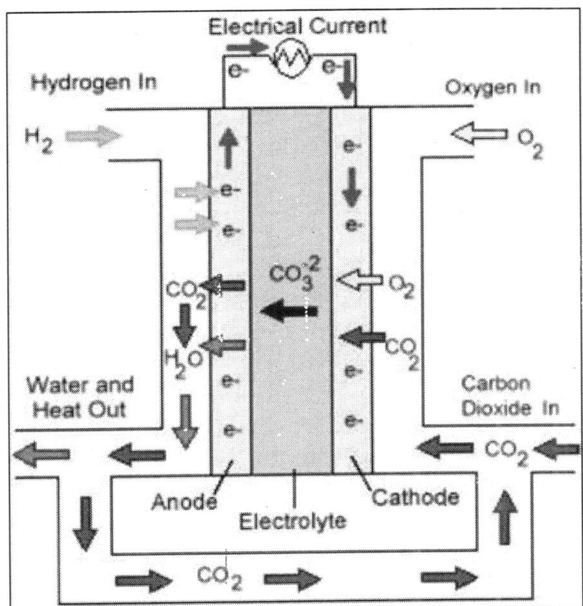

Figure 24.5: Molten Carbonate Fuel Cell.
https://www.energy.gov/eere/fuelcells/types-fuel-cells

Anode Reaction: $CO_3^{-2} + H_2 \rightarrow H_2O + CO_2 + 2e^-$

Cathode Reaction: $CO_2 + ½O_2 + 2e^- \rightarrow CO_3^{-2}$

Overall Cell Reaction: $H_2 + ½O \rightarrow H_2O$

combine with hydrogen to generate water and electrons. Molten carbonate fuel cells (MCFCs) (Figure 24.5) are currently being developed for natural gas and coal-based power plants for electrical utility, industrial, and military applications. MCFCs are high-temperature fuel cells that use an electrolyte composed of a molten carbonate salt mixture suspended in a porous, chemically inert ceramic lithium aluminum oxide (LiAlO2) matrix. Since they operate at extremely high temperatures of 650°C and above, nonprecious metals can be used as catalysts at the anode and cathode, reducing costs. Unlike alkaline, phosphoric acid, and polymer electrolyte membrane fuel cells, MCFCs don't require an external reformer to convert more energy-dense fuels to hydrogen. Due to the high temperatures at which they operate, these fuels are converted to hydrogen within the fuel cell itself by a process called internal reforming, which also reduces cost. Although they are more resistant to impurities than other fuel cell types, scientists are looking for ways to make MCFCs resistant enough to impurities from coal, such as sulfur and particulates. The primary disadvantage of current MCFC technology is durability. The high temperatures at which these cells operate and the corrosive electrolyte used accelerate component breakdown and corrosion, decreasing cell life. Scientists are currently exploring corrosion-resistant materials for components as well as fuel cell designs that increase cell life without decreasing performance.

Phosphoric Acid Fuel Cells (PAFC)

A phosphoric acid fuel cell (PAFC) consists of an anode and a cathode made of a finely dispersed platinum catalyst on carbon and a silicon carbide structure that holds the phosphoric acid electrolyte (Figure 24.6). In phosphoric acid fuel cells, protons move through the electrolyte to the cathode to combine with oxygen and electrons, producing water and heat. This is the most commercially developed type of fuel cell and is being used to power many commercial premises. Phosphoric acid fuel cells use liquid phosphoric acid as an electrolyte and the acid is contained in a Teflon-bonded silicon carbide matrix and porous carbon electrodes containing a platinum catalyst. The phosphoric acid fuel cell (PAFC) is considered the "first generation" of modern fuel cells. It is one of the most mature cell types and the first to be used commercially, with over 200 units currently in use. This type of fuel cell is typically used for stationary power generation, but some PAFCs have been used to power large vehicles such as city buses. PAFCs are more tolerant of impurities. They are 85 percent efficient when used for the co-generation of electricity and heat, but less efficient at generating electricity alone (37 to 42 percent). PAFCs are also less powerful than other fuel cells, given the same weight and volume. As a result, these fuel cells are typically large and heavy. PAFCs are also expensive. Because, PAFCs require an expensive platinum catalyst, which raises the cost of the fuel cell.

Polymer Electrolyte Membrane (PEM) Fuel Cells (PEMFC)

In polymer electrolyte membrane (PEM) fuel cells (Figure 24.6), protons move through the electrolyte to the cathode to combine with oxygen and electrons, producing water and heat. Polymer electrolyte membrane (PEM) fuel cell uses a polymeric membrane as the electrolyte, with platinum electrodes. These cells operate at relatively low temperatures. These cells are the best candidates for cars,

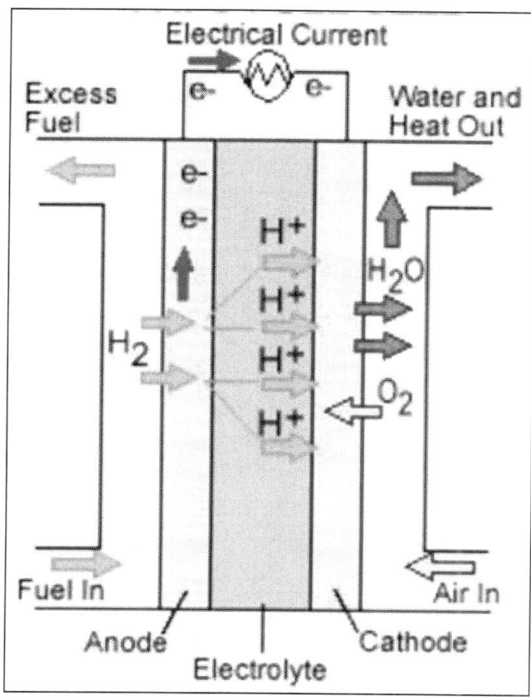

Figure 24.6: PAFC and PEM Fuel Cell.
Courtesy: https://www.energy.gov/eere/fuelcells/types-fuel-cells.

Anode Reaction: $H_2 = 2H^+ + 2e^-$

Cathode Reaction: $½O_2 + 2H^+ + 2e^- = H_2O$

for buildings and smaller applications. Polymer electrolyte membrane (PEM) fuel cells–also called proton exchange membrane fuel cells–deliver high power density and offer the advantages of low weight and volume, compared to other fuel cells. PEM fuel cells use a solid polymer as an electrolyte and porous carbon electrodes containing a platinum catalyst. They only hydrogen, oxygen from the air, and water to operate and do not require corrosive fluids like some fuel cells. They are typically fueled with pure hydrogen supplied from storage tanks or on board reformers. Polymer electrolyte membrane fuel cells operate at relatively low temperatures, around 80°C (176°F). Low temperature operation allows them to start quickly (less warm-up time) and results in less wear on system components, resulting in better durability. However, it requires that a noble metal catalyst (typically platinum) be used to separate the hydrogen's electrons and protons, adding to system cost. The platinum catalyst is also extremely sensitive to CO poisoning, making it necessary to employ an additional reactor to reduce CO in the fuel gas if the hydrogen is derived from an alcohol or hydrocarbon fuel. This also adds cost. Developers are currently exploring platinum/ruthenium catalysts that are more resistant to CO.

Solid Oxide Fuel Cells (SOFC)

SOFC Fuel Cell (Figure 24.7) works at high temperature that uses a solid ceramic electrolyte, such as zirconium oxide stabilised with yttrium oxide, instead of a liquid and operate at 800 to 1,000°C. In solid oxide fuel cells, negative ions travel through the electrolyte to the anode where they combine with hydrogen to generate water and electrons. Efficiencies of around 60 per cent and are expected to be used for generating electricity and heat in industry and potentially for providing auxiliary power in vehicles. Since the electrolyte is a solid, the cells do not have to be constructed in the plate-like configuration typical of other fuel cell types. High temperature operation removes the need for precious-metal catalyst, thereby reducing cost. They are not poisoned by carbon monoxide (CO), which can even be used as fuel and Sulphur resistant. This allows SOFCs to use gases made from coal. Scientists are currently exploring the potential for developing lower temperature SOFCs operating at or below 800°C that have fewer durability problems and cost less.

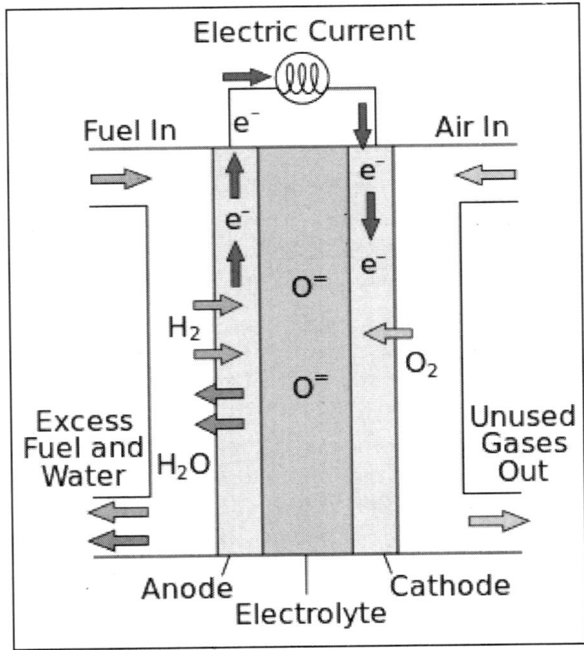

Figure 24.7: SOFC Fuel Cell.
Courtesy: https://en.wikipedia.org/wiki/Solid_oxide_fuel_cell

Regenerative Fuel Cells (RFC)

This class of fuel cells produces electricity from hydrogen and oxygen, but can be reversed and powered with electricity to produce hydrogen and oxygen; effectively storing energy or electricity.

Basic energy conversion of a fuel cell can be described as:

Chemical energy of fuel = Electrical energy + Heat energy

The input energy is that produced during reactions at the electrodes.

Performance

One measure of the energy conversion efficiency of a fuel cell is the ratio of the actual voltage at a given current density to the maximum voltage obtained under no load (open circuit) conditions.

Energy Conversion Efficiency = On load voltage/theoretical open circuit voltage

Fuel Cell Efficiency

Since fuel cells use materials that are typically burnt to release their energy, the fuel cell efficiency is described as the ratio of the electrical energy produced to the heat that is produced by burning the fuel.

From the basic definition of efficiency: $\lambda = W/Q_{in}$

The maximum efficiency of the fuel cell would be 83 per cent.

Classification of Losses in an Actual Fuel Cell

Activation Losses: These losses are caused by the slowness of the reaction taking place on the surface of the electrodes. A proportion of the voltage generated is lost in driving the chemical reaction that transfers the electrons.

Ohmic Losses: The voltage drop due to the resistance to the flow of electrons through the material of the electrodes. This loss varies linearly with current density.

Concentration Losses: Losses that result from the change in concentration of the reactants at the surface of the electrodes as the fuel is used.

Fuel Crossover Losses: Losses that result from the waste of fuel passing through the electrolyte and electron conduction through the electrolyte. This loss is typically small, but can be more important in low temperature cells" (https://www.energy.gov/eere/fuelcells/types-fuel-cells).

Nanotechnology for Catalysts and Polymer Electrolyte Membrane Fuel Cells and Applications of 0D, 1D, 2D and 3D NSMs

"The main active components are fuel electrode (anode) are oxidant electrode (cathode). The proton conducting solid membrane used as electrolyte, is sandwiched between the anode and cathode electrodes. Although the technology of such fuel cells was developed to an extent, some problems remain unclear and only their solution can result in wide-scale commercialization of fuel cells, especially as the efficient current sources for portable devices. The most important problem is the development of efficient catalysts, which would provide the long-term operation of fuel cells without sacrificing their characteristics. The CO poisoning intrinsic to Pt based cathodes pose a further barrier to fuel cell devices. Hence it is important to open up new methods for making low cost and effective electrocatalysts. It is well known that the catalytic activity attainable depends significantly on the size of the Pt catalysts and their support. However, there are two routes to overcome this problem. First, the direct synthesis of novel 0D, 1D, 2D, and 3D NSMs (Pt or Pt alloy catalysts) with high surface area for fuel cell devices. Secondly, the synthesis of novel NSMs with high surface area as a support of Pt or Pt alloys catalysts for fuel cell devices.

During the past decade, the clusters stabilizing Pt nanoparticles have attracted a considerable interest as catalysts for low temperature fuel cell devices, such as polymer electrolyte fuel cells (PEMFCs)and direct methanol fuel cells (DMFCs). The clusters stabilizing Pt nanoparticles are formed in order to maximize the active number of surface versus the inactive number of Pt bulk atoms. However, one of the major problems associated with the clusters stabilizing Pt catalysts are their stability in the acid environment of DMFCs and PEMFCs. This is because of after only 100 h of testing the reaction rate dropped to half in such fuel cell devices. During this process the CO-ads species were continuously formed, which strongly absorbs to the clusters stabilizing Pt active sites to poison the fuel cell performance devices. Due to these problems of the clusters stabilizing Pt nanoparticles, researchers decided to investigate the bimetallic clusters and colloids with nanometer-scale dimensions for fuel cell device applications. There are two main reasons for using the bimetallic clusters and colloids with nanometer-scale dimensions, firstly, because they may serve as models for studying the formation of different alloy catalysts. Secondly, it is possible to save precious metal, such as Pt, by optimizing the synthetic conditions so that only very thin surface layers occur. In most cases, Pt and Pt-based alloy nanoparticles are used in the form of small nanoparticles, since small size offers high surface area which is especially crucial for catalysis and fuel cell applications. A report by Tiwari *et al.*, showing perfect Pt nanocubes had much higher electrocatalytic activity and stability for methanol and ethanol oxidation than other nanostructures such as truncated Pt nanocubes, truncated Pt (cubes + tetrahedra) and spherical Pt nanoparticles. Wang *et al.*, reported the controlled synthesis of dendritic Au@Pt core–shell nanomaterials. Chen *et al.*, synthesized the Co–Pt catalyst with a hollow sphere structure via a very simple thermolytic reaction. The Co–Pt hollow sphere catalyst showed the high surface area with a comparison to the pure Pt nanoparticle catalyst and Co–Pt nanoparticles. In addition, the Co–Pt hollow sphere catalyst showed enhanced electrocatalytic activity for methanol oxidation with a comparison to the pure Pt nanoparticle catalyst and Co–Pt nanoparticles at the same Pt loading.

1D NSMs

In the last few years, attention has been shifted towards the synthesis of 1D NSMs because of their fundamental importance and potential applications in fuel cells. To date, 1D NSMs (such as nanowires, nanorods, nanobelts, and nanotubes) have been most extensively studied because of their simplicity in preparation and its improved catalytic properties as compared to bulk metallic materials. Ksar *et al.*, fabricated the Pd nanowires in a hexagonal mesophase (as soft templates) by electron beam irradiation. They also reported that the average lengths of the Pd nanowires are controlled by the amount of co-surfactant that assists the reduction/growth processes. In addition, the electrochemical studies indicated that the Pd nanowires had a very good electrocatalytic activity and stability for ethanol oxidation. Liu *et al.*, prepared the nanoporous Pt–Co alloy nanowires by dealloying electrodeposited by dealloying Pt-Co nanowires and examined their electrochemical properties Pt–Co nanowires (15 h) had a superior tolerance against CO poisoning. In addition, nanoporous Pt–Co alloy nanowires exhibit an enhanced electrocatalytic activity toward methanol oxidation (in comparison to state-of-the art Pt/C and PtCo/C catalysts).

2D NSMs

2D NSMs such as junctions (continuous islands), branched structures, nanoprisms, nanoplates/nanosheets, nanowalls, and nanodisks are an especially promising form of nanoscale with a rare combination of extremely high specific surface area, exceptional thermal/electrical conductivity and stability. The utilization of 2D NSMs as a support to anchor catalyst nanoparticles and facilitate electron transport opens up interesting new possibilities for designing and building of next generation catalysts for fuel cells. In addition, 2D NSMs can avoid aggregation and maintain high specific surface areas; these factors are important in increasing the accessibility of adsorbates to reactive sites. Li *et al.*, fabricated the close-packed Au nanoprism thin films on the indium tin oxide coated glass substrates. Au nanoprism are single crystalline, whose basal surfaces are atomically flat {111} planes and lateral surfaces are {110} planes. A small amount of Pt or Pd was deposited onto the Au nanoprism thin film electrodes and studies their catalytic properties. They noticed that the Pt on Au nanoprism exhibited greatly enhanced catalytic activity toward the oxidation of methanol and much better CO poison resistance than commercial Pt-based catalysts. Wong 's group have synthesized Pd nanoplates array onto Au substrate by electrochemical process. The Pd nanoplate's array electrode shows extraordinary electro-catalytic activity towards the oxidation of methanol. Choi and Woo prepared a 2D Pt–Ru nanowire network using a SBA-15 nanoreactor and showed that this nanostructured material displayed a higher performance than commercial Pt–Ru (JM) in DMFC.

The catalyst structure of 2D continuous Pt island network on a flat Si substrate by Tiwary group on the Si substrate has better electroactivity toward methanol oxidation than the blanket Pt/Si and the Ru decorated Pt film/Si electrodes. Recently many material scientists and engineers have fabricated 2D graphene nanosheets as catalyst supports for DMFCs. The electrochemical results of methanol oxidation indicate that Pt–Ru/graphene nanosheets display excellent catalytic activity over that of Pt–Ru/Vulcan. These results indicate that graphene nanosheets could significantly increase the electrochemical active surface area resulting in better catalyst utilization.

3D NSMs

Compared to 0D, 1D, and 2D NSMs, the 3D NSMs hold a unique morphology, which are expected to provide a larger surface area and to facilitate the transport of reactants and products effectively, so that electrochemical performance can be greatly improved. Several research groups have fabricated 3D Pt nanoflowers as a catalyst for fuel cell applications. Tong *et al.*, prepared 3D Pt nanosheets perpendicular to a boron-doped diamond (BDD) hybrid film surface, synthesized by double template method. Zhang *et al.*, prepared 3D Pt nanoflowers via a template-free, direct electrodeposition method. This electrochemical experiment demonstrates that Pt nanoflowers have superior performance in aspects of electrocatalytic activity and stability towards the oxidation of methanol" (Tiwari *et al.*, 2012).

REFERENCES

1. Agrawal RC, Pandey GP. Solid polymer electrolytes: materials designing and all-solid-state battery applications: an overview. *J Phys D Appl Phys* (2008); 41: 223001–19.
2. Bruce PG, Scrosati B, Tarascon JM. Nanomaterials for rechargeable lithium batteries. *Angew Chem Int Ed* (2008); 47: 2930–46.
3. Chan CK, Peng HL, Liu G, Mcilwrath K, Zhang XF, Huggins RA, *et al.*, High performance lithium battery anodes using silicon nanowires. Nat Nanotechnol (2008); 1: 31–5.
4. Chan CK, Peng HL, Twesten RD, Jarausch K, Zhang XF, Cui Y. Fast, completely reversible Li insertion in vanadium pentoxide nanoribbons. *Nano Lett.*, (2007); 7: 490–5.
5. Green M, Fielder E, Scrosati B, Wachtler M, Serra-Moreno J. Structured silicon anodes for lithium battery applications. *Electrochem Solid-State Lett.*, (2003); 6: A75–9.
6. Nam KT, Kim DW, Yoo PJ, Chiang CY, Meethong N, Hammond PT, *et al.*, Virus enabled synthesis and assembly of nanowires for lithium ion battery electrodes. *Science* (2006); 312: 885–8.
7. Park MS, Wang GX, Kang YM, Wexler D, Dou SX, Liu HK. Preparation and electrochemical properties of SnO_2 nanowires for application in lithium-ion batteries. *Angew Chem Int Ed* (2007);119: 764–7.
8. Patil A, Patil V, Shin DW, Choi JW, Paik DS, Yoon SJ. Issue and challenges facing rechargeable thin film lithium batteries. *Mater Res Bull* (2008); 43: 1913–42.
9. Taberna PL, Mitra S, Piozot P, Simon P, Tarascon JM. High rate capabilities Fe_3O_4-based Cu nano-architectured electrodes for lithium-ion battery applications. *Nat Mater* (2006); 5: 567–73.
10. Tiwari, J. N., Tiwari, R. N. and Kim, K. S. Zero-dimensional, one-dimensional, two-dimensional and three-dimensional nanostructured materials for advanced electrochemical energy devices. Progress in Materials Science 2012; 57: 724–803.
11. Serrano, E., Rus, G. and Garcý´a-Martý´nez, J. (2009) Nanotechnology for sustainable energy. *Renewable and Sustainable Energy Reviews* 13: 2373–2384.
12. Whittingham MS. Inorganic nanomaterials for batteries. *Dalton Trans.*, (2008); 40: 5424–31.

Chapter 25

Nanotechnology for Sustainable Biofuel Production

S. Marimuthu[1], N. Swathika[1] and K. Geetha Thanuja[2]*

[1]Department of Nano Science and Technology, [2]Department of Agricultural Microbiology, Tamil Nadu Agricultural University, Coimbatore – 641 003, Tamil Nadu

**e-mail: sm20@tnau.ac.in*

The global economy continues to grow driven by the pro economic policies in different countries. The growth is supported with the higher energy consumption with a challenge of less carbon emission. In India, the consumption of petroleum based productions 251.93 Ton mn in 2018, which indicates the annual incremental requirement of 6.54 per cent over the previous year (www. ceicdata.com). The World Energy Outlook (WEO) report of the International Energy Agency (IEA) forecasts that India's primary energy demand will become doubled in 2035, increasing 750 Mtoe in 2011 to 1258-1647 Mtoe (IEA, 2013),

The domestic production of crude oils is less which satisfies 18 per cent of the demand, the balance of the crude oil requirement is met through imports. India imports 4543645 barrels a day of crude oil to meet the energy requirements, expected to reach more than 8 million barrels per day in 2035. The imports cost close to the US$140 billion (2) for the Country. Hence, India look for alternate sources for crude oils considering their high fluctuation of cost in the global market and political turbulences at Middle East Countries who are the primary supplier of crude oil to India. Demirbas (2009) reported that the wells of fossil fuels will be exhausted by 2050 considering the existing production rate. Similarly, the use of fossil fuels results in the production of greenhouses gases contributing to Global warming.

Biofuel are considered among the most promising candidate in this context, which are produced locally to use as substitute for diesel and petrol. Biofuel are derived from recently grown agricultural biomass, which includes agricultural crops and by-products, wood and poles, aquatic plants and algae, organic solid wastes and bio solids. Biofuel from agricultural biomass is considered as carbon neutral, because the carbon dioxide from atmosphere is assimilated to produce agricultural biomass through photosynthesis, which in turn will be emitted back to atmosphere while using the biomass as fuel. Biodiesel is derived from vegetable oil and animal fats through transesterification process. Oil crops such as soybean, canola, sunflower, mustard and tree borne oilseeds such as Jatropha, Pongamia, *Basia latifolia*, animal fats and lipids stored in algae are the sources widely used to produce biodiesel. Biodiesel consist of long-chain alkyl (methyl, ethyl, or propyl) esters, which is produced reacting vegetable oils with an alcohol to produce fatty acid esters. The use of biodiesel in conventional engines will reduce the emission of carbon dioxide (CO_2) and carbon monoxide (CO). The presence of traces of hydrogen and oxygen in biodiesel compared to petro diesel results in complete combustion of fuel in the engines thereby reducing hydrocarbons in exhaust gases. Cetane number is the indicator of diesel quality, which refers to inverse function of fuels ignition delay, the time interval between fuel injection and auto ignition in engines. The fuel with higher cetane number will have short ignition delay, where ignition occurs in less time after the injection of fuel, whereas the fuel with less cetane number will have longer ignition delays (Cataluña and da Silva., 2012). Generally, biodiesel have higher cetane number compared to petro diesel (Bazooyar *et al.*, 2015)

Bioethanol is an alcohol made through fermentation and distillation of sugars, Bioethanol is obtained from renewable feedstock such as maize, sorghum, sugar beet, grasses (switch grass and miscanthus) and sugarcane and woody biomass. Cellulose, starch and carbohydrates in the plant biomass are converted into ethanol through hydrolysis, fermentation and distillation. Hydrolysis converts plant components into simple sugars which up on anaerobic fermentation is converted into alcohol in the presence of yeast *Saccharomyces ceveresiae*. The efficiency of converting cellulose into sugars through hydrolysis depends on the pretreatment of plant biomass, to deconstruct the lignin to release cellulose and hemicellulose from the biomass. Generally, cellulose and hemicellulose are wrapped in lignin matrix where pretreatment is to remove lignin and enhance the availability of cellulose and hemicellulose for hydrolysis and fermentation. Octane rating of bioethanol is higher than gasoline, where octane rating is a measure of resistance of fuels to detonate in spark-ignition internal combustion engines. Higher performance engines with high compression ratios are prone for detonation where it requires higher-octane fuel to resist the detonation in the spark engines

Biofuel are classified as conventional and advanced biofuel, based on the feedstock used for the production of biofuel. Conventional biofuel are referred to first generation biofuel derived from edible parts of the plant such grains, tubes and canes. This first generation biofuel promptly evokes debate as it competes with crops grown for food/feed purpose. Bioethanol is produced from maize grains in United States, which affected the export of maize to Mexico resulting in doubling the price of corn tortillas staple food of Mexicans. Advanced biofuel are referred to second to fourth generation biofuel produced from non-food/feed crops sustainably grown feedstock and agricultural (municipal) wastes (Table 25.1)

Table 25.1. Classification of Biofuel

Classification	*Sub Classification*	*Feedstock Used for Biofuel Production*
Conventional biofuel	First Generation (Food/Feed stock)	Cereals, sugar beet/sugarcane for making bioethanol
		Soybean, rapeseed, palm oil, animal fats, waste oils for producing biodiesel
Advanced Biofuel	Second Generation (Plant Biomass and Tree borne oilseeds)	Cellulosic ethanol from switch grass, miscanthus, wheat straw, poplar, willow, corn stover and stalks and stubble (stems), leaves, seed pods, as well as forest/park residues)
		Biodiesel from non-food crops *Jatropha curcus, Derris indica, Calophyllum inophylum* grown in non-arable lands (waste lands)
	Third Generation (Biodiesel and Bioethanol)	Algal systems
	Fourth Generation (Biodiesel and Bioethanol)	Genetically Modified Crops

Scope of Nanotechnology for Biofuel Production

Nanoscience is a fascinating field that involves atom-by-atom manipulation resulting novel properties for wider application in engineering, bio imaging, diagnostics, drug delivery, energy and environment applications. Nanoparticles of 1 and 100 nm have unique properties from bulk materials with large specific surface area, high surface energy, and quantum confinement. This nanomaterial has high-end applications in biofuel production. Nanotechnology plays significant role for the optimization of biodiesel production. The nanocatalyst-mediated processes are of more efficient, economic, durable, and stable nano-catalysts holding the potential to achieve higher product quality and yields. Titanium dioxide (Gardy *et al.*, 2017), calcium oxide (Liu *et al.*, 2008), magnesium oxide (Verziu *et al.*, 2008), and strontium oxide (Liu *et al.*, 2007) are used for biodiesel production. The use of magnetic nanoparticle-based catalysts is advantageous as this nanoparticles are separated easily from reaction media through magnetic field and reused, leading to more economical, industrial-scale biodiesel production

Enzyme immobilization using nanomaterials is another promising application in the biofuel industry. The nanostructures offers high enzyme loading, higher enzymatic stability, and possibility of enzyme reusability. The process of immobilization reduce the operational cost of large-scale biofuel production.

Nanotechnology for Pre-treatment of Lignocellulosic Biomass

Second-generation biofuel are produced from non-food agricultural residue and waste feedstock. It is estimated that agricultural residues will be available for biofuel production @ 796 and 887 MT per annum in 2020 and 2030 respectively (Purohit and Dhar. 2018). The agricultural residues are lignocellulosic material, which are ready to use feedstock for biofuel production. Lignocellulosic biomass comprise of primarily cellulose (38–50 per cent), hemicellulose (23–32 per cent), and lignin (15–25 per cent), including other minor components, such as proteins, pectin, and extractives (Fengel and Wegener 1989). The cellulose is a homopolymer of glucose molecules linked by the glycolic bond β-1,4. The hydrogen bonds form to link the cellulose inside (intramolecular) and outside the chains (intermolecular). The hemicelluloses are pentose (xylose, arabinose) and hexose (glycose, mannose, and galactose) heteropolysaccharides. Lignin is a complex biomolecule composed of three phenolic compounds: p-cumaril, coniferyl, and alcohol sinapil (Wang 2012). Greatest challenge lies in converting lignocellulosic biomass into bioethanol, and pretreatment which facilitates the removal of lignin matrix from the plant biomass to expose cellulose and hemi cellulose for hydrolytic and fermentation. The pretreatment is the crucial step, which disrupts hydrogen bonds in crystalline cellulose; breaks down the cross-linked matrix of hemicelluloses and lignin, and finally, the raising porosity and surface area of cellulose for hydrolysis (Figure 25.1)

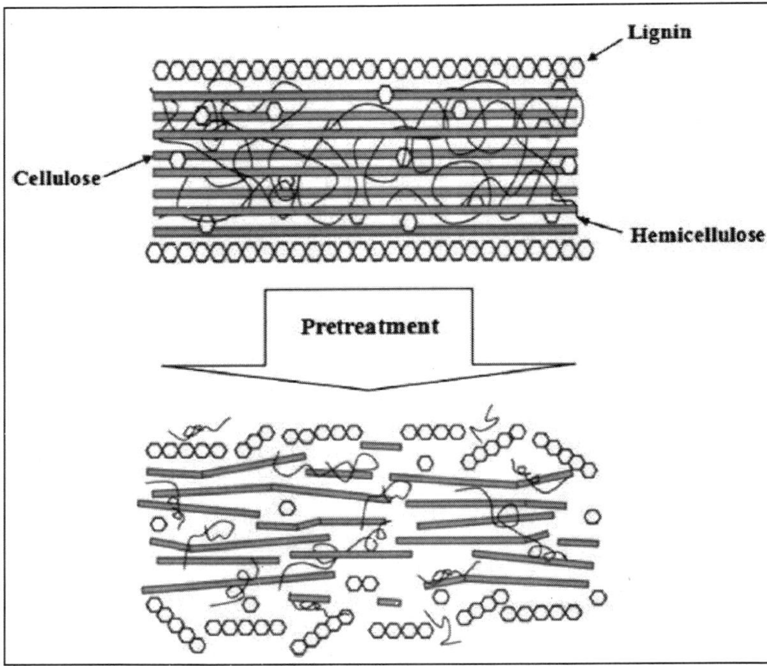

Figure 25.1. Illustration of Pre-treatment of Lignocellulosic Biomass to Liberalize Cellulose and Hemicellulose for Microbial Hydrolysis and Subsequent Fermentation.

Electron Microscopy to Evaluate Pre-treatment Technologies

Electron microscopy is a high-end instrumentation to capture images of samples at a greater resolution than with a light microscope. There are scanning and transmission electron microscopy, where scanning electron microscopy focuses the beam of electrons into a small spot that scans across the surface of a sample. The signals that are emitted back from the sample due to specimen-electron beam interaction are captured to produce sample image. The image provides in depth surface information of the sample. Transmission electron microscopy transmits the beam of electrons through a thin sample onto a screen to provide image of ultrastructure inside the sample. Hence, this electron microscopy will be useful to evaluate pre- treatment technologies by comparing structural changes in plant tissues before and after the pretreatment step. Similarly, electron microscopy images are used to observe the changes in the sample due to enzymatic hydrolysis. Donohoe *et al.* (2011) studied the surface and ultrastructural changes of raw and pretreated switch grass samples using electron microscopy (Figure 25.2). The pretreatment technologies comprise of Ammonia fiber expansion pretreatment (AFEX), lime pretreatment, soaking in aqueous ammonia pretreatment (SAA), liquid hot water (LHW) and dilute sulfuric acid pretreatment (DA)

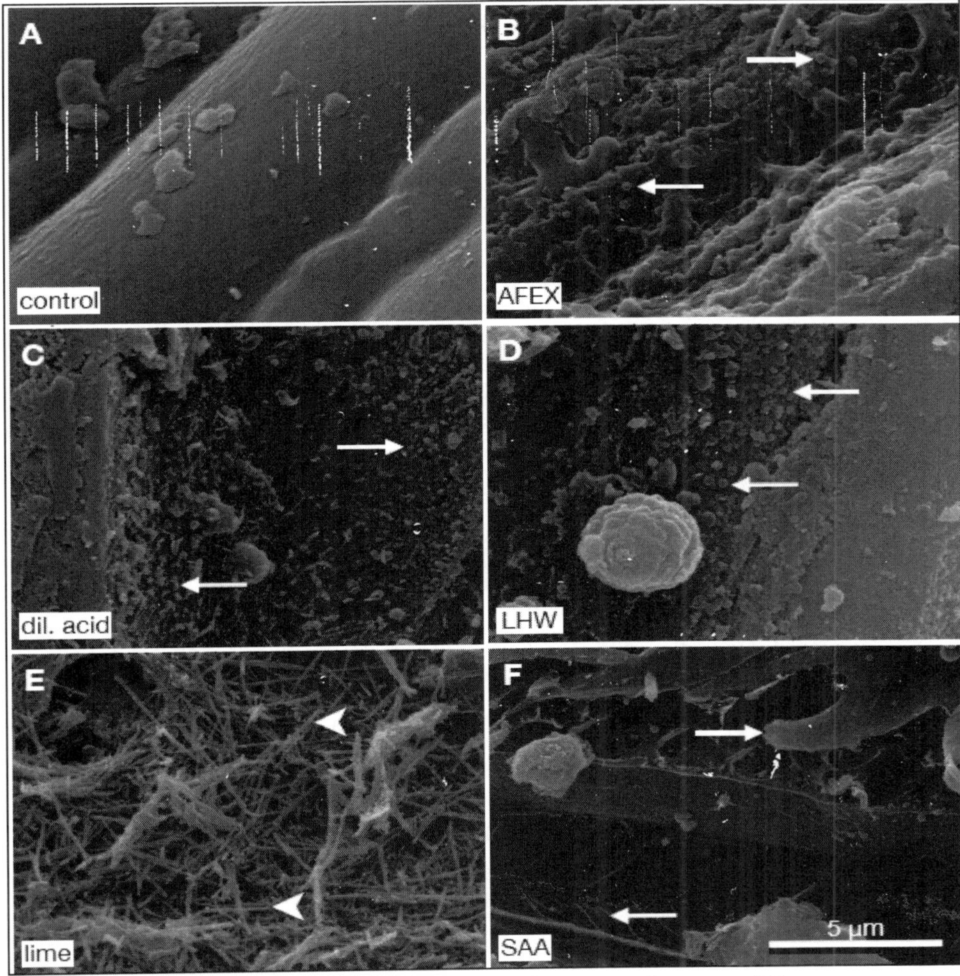

Figure 25.2. Scanning Electron Microscopy Images of Raw and Pretreated Switch Grasses.
Courtesy: Donohoe *et al.*, 2011.

Electron microscopy images revealed that relatively smooth surface of the cell in control, while pretreated samples showed *etc.*hing or micro pitting of cell wall. Ammonia fiber expansion pretreatment caused surface erosion cell wall matrix exhibiting highly irregular surfaces. Dilute acid treatment and liquid hot water treatment showed mild surface erosion, however lignin-rich globules because of re localization of lignin were notices over the surface of the sample. Lime treated samples showed complete removal of cell wall matrix leaving behind cellulose micro fibrils, while soaking in aqueous ammonia showed more uniformly eroded surface and fractured walls. Transmission electron microscopy images of pretreated switch grass are shown in Figure 25.3. The transmission electron microscopy images revealed the delamination and porous nature of samples due to pretreatment technologies. The sample was stained to highlight the presence of lignin globules. Ammonia fiber expansion treated samples showed new pore structure containing coalesced lignin. The dilute acid treated samples exhibited delamination interspaced with re-localized lignin, while the lime treated samples displayed extraction of the middle lamella. Samples soaked in aqueous ammonia showed that cell walls seems to be loosened from wall thickness.

Similarly, atomic force microscopy is low cost imaging techniques carried out at ambient environment conditions without any artefacts to samples. The atomic force microscopy scans the sample with probe measuring the interactive forces between the probe and sample surface, converting into images. The images are not true to type, provides surface topography information about the images, however, 3D images are obtained with atomic force microscope to assess the surface roughness of the sample. Atomic force microscope images cell wall boundaries, middle lamellas, and arrangement of the micro fibrils to the native state of the sample. X-ray diffraction measures the crystallinity of samples before and after the pretreatment where the removal of hemicellulose and lignin results in improving crystallinity index. Crystallinity Index is calculated by dividing the intensity at I_{200} and the height of the minimum among the (2 0 0) and (1 1 0) peaks. I_{100} peak refers to peak obtained @$2\theta = 18^{o}$.

Crystallinity Index = $(I_{200}–I_{100}/I_{200})$ * 100

Where I_{200} refers to intensity of crystalline peak, while I_{100} refers to peak intensity of amorphous regions

Nano Shear Hybrid Alkaline Pre-treatment for Lignocellulosic Biomass

Nano-scale shear hybrid alkaline (NSHA) pretreatment method of lignocellulosic biomass was designed by Wang *et al.* (2013) to extract cellulose from the feedstock. The reaction was carried out in a modified Taylor Couette reactor (nano mixer) where lignocellulosic biomass were fed into mixer along with mild alkali solution (Figure 25.4). The shear rate of 12500 s^{-1} was maintained in the vessel to reach the temperature of 100°C under ambient pressure. The shearing generated by the turbine in the vessel cause the twist of corn stover. This facilitates the degradation of the nano-structured cell wall complex in the corn stover, while alkali effect and simultaneous thermal effect due to the severe friction between the

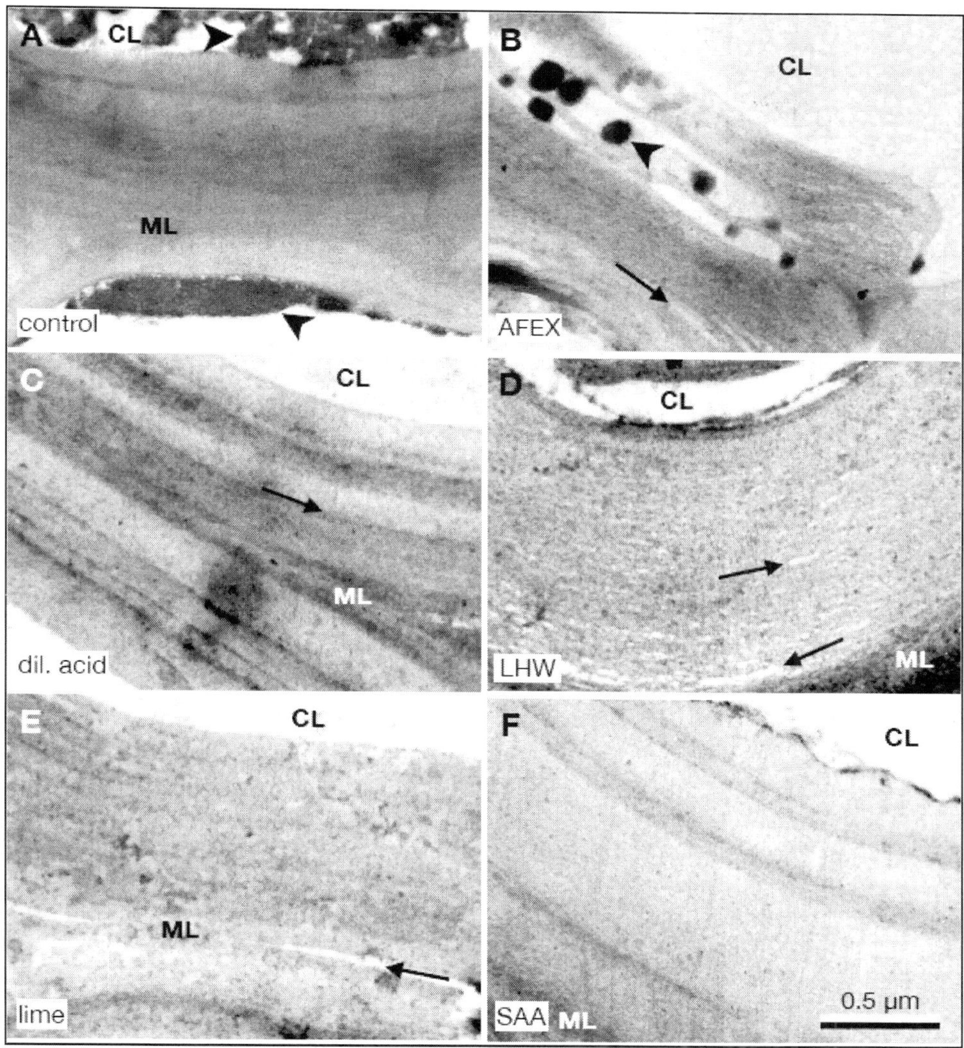

Figure 25.3. Transmission Electron Microscopy Images of Raw and Pretreated Switch Grasses
Courtesy: Donohoe *et al.*, 2011.

turbine and heterogeneous biomass results in faster and efficient elimination of lignin. Pretreatment of corn stover through NSHA technique resulted in the recovery of cellulose @ 72.7 per cent, hemicellulose of 9.9 per cent and lignin with 9.7 per cent compared to that of cellulose @ 64.7 per cent, hemicellulose 24.6 percent and lignin of 8.6 per cent with the pretreatment of corn stover with 2 per cent sodium hydroxide alone. Generally, sodium hydroxide is used as an alkali in the NSHA pretreatment, where sodium hydroxide has the ability to delignify the biomass and solubilize hemicellulose. The duration required for delignification and solubilization of hemicellulose ranged to several hours, while high shearing force under alkaline conditions in NSHA technology improves the efficiency of lignin removal and solubilization of hemicellulose in couple of minutes

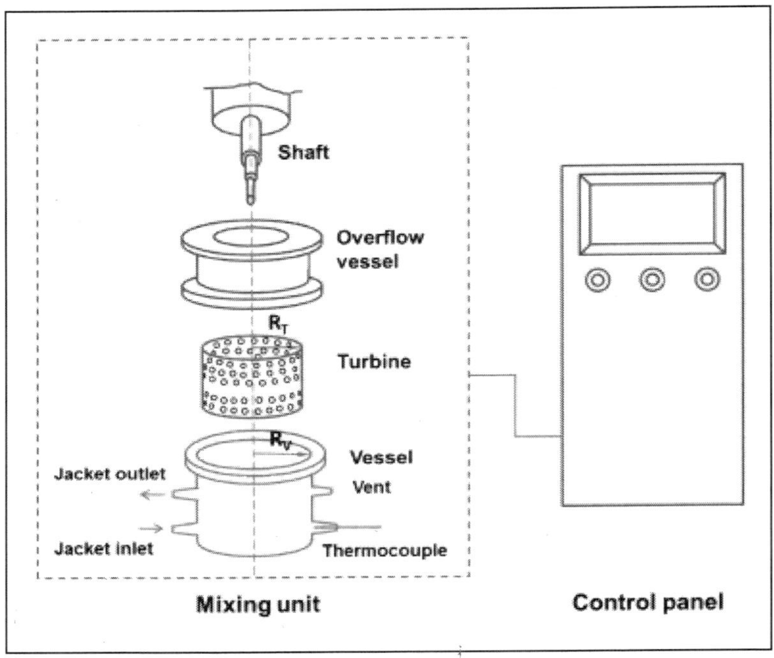

Figure 25.4. Modified Taylor Couette reactor (nano mixer).
Courtesy: Wang *et al.*, 2013.

Nanomaterials for Enzymes Immobilization to Produce Biofuel

Ethanol production from lignocellulosic biomass is a complex process, which involves pretreatment of lignocelluloses, hydrolysis of cellulose and hemicellulose, sugar fermentation, and distillation of ethanol (Figure 25.5). The feedstock used as substrate for ethanol production and enzyme complex used for fermentation of cellulose into simple sugars, are the major factors deciding the economic feasibility of second generation bioethanol (Gomez *et al.*, 2016). Cellulase enzyme complex comprise of endoglucanases, exoglucanases and b-glucosidases that is capable of breaking insoluble crystalline cellulose into free sugars. The simple sugars up on action by ethanologens, coverts into ethanol (Wen *et al.*, 2004). However, cellulase based hydrolysis of lignocellulosic biomass is less efficient process, and remains as costly process. Reuse of enzymes in multiple times will reduce the cost on enzyme technology.

Enzyme immobilization is fixing an enzyme with a biocompatible and inert supports. The supports are referred as substrate over which enzymes are immobilized. The technique of immobilization improves and sustains catalytic

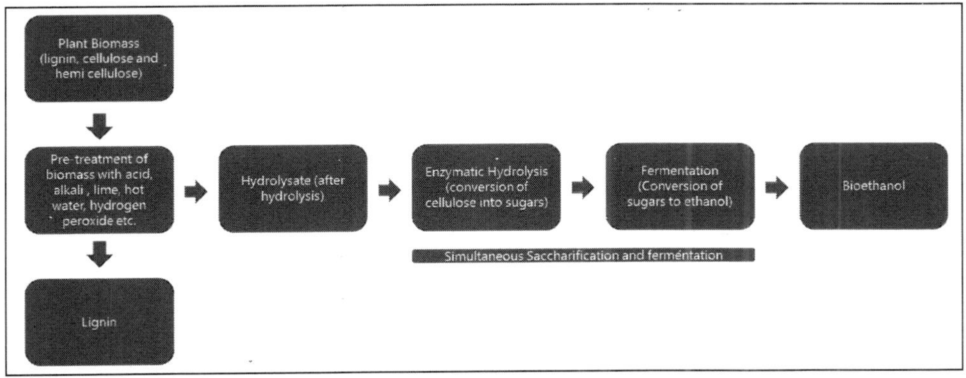

Figure 25.5. The Process of Converting Lignocellulosic Biomass into Bioethanol.

activity, offers thermal and pH stability and facilitates the reuse of enzymes. Nano immobilization is fixing of biocatalysts in nanomaterial support taking the higher surface area per unit mass and superior physical properties (like strength, chemical reactivity, and conductivity) than conventional bulk materials. The nanomaterial improves the diffusion limitations of enzymes from the support to the reaction media and achieve higher loading of enzymes exploiting higher surface area of nanomaterials. Physical retention and chemical bonding are the common methods used for immobilizing enzymes over the supports. Binding of enzymes over the support may be of physical adsorption, ionic binding, covalent binding with support, and entrapment into support. Adsorption is a physical process of reversible surface interaction between support and enzyme achieved through Van der Waals forces, ionic bond and hydrogen bonding interactions (Figure 25.6)

Crosslinking is widely used for enzyme immobilization for industrial commercial applications. The technique involves formation of number of covalent bonds between enzyme and the support using multi-functional reagents (*e.g.*, glutardialdehyde, glutaraldehyde, glyoxal, diisocyanates, hexamethylene diisocyanate and toluene diisocyanate). Generally, amino groups of lysine, sulfhydryl groups of cysteine, phenolic OH groups of tyrosine, or imidazol group of histidine are used for enzyme binding over the supports. Cross-linking between biocatalysts and supports, results in irreversible intermolecular bonds withstanding pH and temperature extremes during the catalytic processes. Generally magnetic and silica nanoparticles are widely employed for the immobilization of cellulase enzyme for bioethanol production. The extend of reuse of enzyme is critical for the economic operation, where the most of literature demonstrated that immobilized enzymes retains catalytic activities even after five or six cycles of use. The adsorption of enzymes on the support is crucial for assessing the immobilization efficiency, which decides the enzyme loading in nanomaterials. More the enzyme loading in the supports results in more catalytic activity of getting better conversion of complex substances into sugars (Table 25.2)

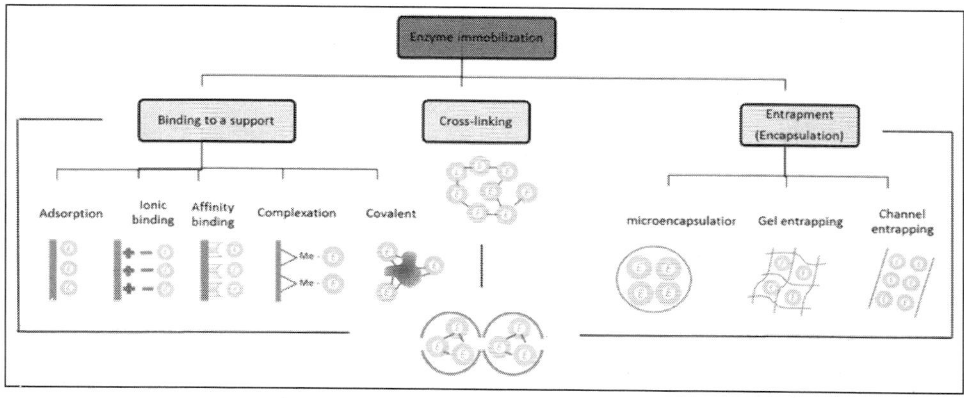

Figure 25.6. The Various Methods of Enzyme Immobilization with Supports.
Courtesy: Vaghari *et al.*, 2015.

Nanomaterials for Laccase and Lipase Immobilization in Biodiesel Production

Biodiesel consists of monoalkyl esters of long-chain fatty acids derived from renewable vegetable oils or animal fats. The acid composition, high viscosity and free fatty acids of vegetable oils decides the quality of biodiesel. The free fatty acids of less than 2 per cent in the vegetable oil is suitable for transesterification. The higher content of free fatty acid results in gum formation and carbon dots in the engines.

Transesterification involves triacylglycerols reacting with short-chain monohydric alcohols (methanol) in the presence of catalyst (alkali, acid or enzyme) at elevated temperatures to form fatty acid alkyl esters (FAAE) and glycerol. Transesterification route is the easiest and most cost-effective process to produce biodiesel. The first step in transesterification is to convert triglycerides to diglycerides, which then are converted to monoglycerides, while third step is the conversion of monoglycerides to glycerol. To sum up the process, 3 mole of alcohol and 1 mole of triglyceride are combined to produce 1 mole of glycerol and 3 mole of fatty acid esters, *i.e.* biodiesel. The catalysts for the transesterification includes alkali, acid, enzyme or heterogeneous catalysts. Alkali catalysts such as sodium hydroxide, sodium methoxide, potassium hydroxide and potassium methoxide are more effective, while acid catalysts are sulfuric, hydrochloric and sulfonic acids are used in transesterification. Lipase or laccase are used for transesterification. Lipases are mostly of microbial origin such as *Candida antarctica, Candida rugosa, Candida sp. Burkholderia cepacia,* and *Rhizopus oryzae.* Lipases catalyze specific types of oil or fat to produce certain amount of fatty acids. Hence, combi-lipases are used in enzymatic transesterification. Immobilization of lipases are crucial for improving the catalytic activity and reusability. In this context, nanomaterials are used as supports for immobilizing lipases (Table 25.3)

Nanotechnology in Algae Biofuel Production

Algal biofuel production is of immense interest owing its fast growing nature

Table 25.2. Immobilization of Enzymes for Hydrolysis of Cellulosic Biomass

Sl.No.	*Enzyme Immobilized*	*Support Used*	*Results*	*Authors*
1	Cellulase from *Trichoderma reesei*	Hematite and Ferrite nanoparticles	94 per cent of cellulase loading with support. Thermal stability upto 80°C, 50 per cent of the initial activity after five cycles, storage ability for 45 days and hydrolysis of 93 per cent with hemp hurd biomass	Abraham *et al.* (2014)
2	Cellulase from *Aspergillus fumigatus*	MnO_2 nanoparticles	Immobilized cellulase were thermally stable upto 72°C and maintains enzyme activity for 2h and retains 60 per cent of its original activity after four cycles	Cherian *et al.* (2015)
3	Cellobiohydrolase D (CelD) and dippa-glucosidase A (BglA) from *Aspergillus nidulans*	Magnetic supports (Fe_3O_4) functionalized with silica	Immobilized dippa glucosidase A retained 43.3 per cent of the original activity, while Cellobiohydrolase D maintains 43.4 per cent after 10 cycles of reuse. Co-immobilization of BglA and CelD on functionized magnetic nanoparticles promote synergistic action of cellulases while lowering enzyme consumption	Song *et al.* (2016)
4	Cellulase from *Aspergillus fumigatus*	Zinc oxide nanoparticles	Thermal stability at 65°C for 10h and pH stability upto 10.5 and retains 53 per cent of relative activity	Srivastava *et al.* (2016)
5	Halocellulase from *Aspergilus niger*	Magnetic nanoparticles (Fe_2O_3)	Sugar yield of 375.39 mg gram dry substrate $^{-1}$ with immobilized enzyme using alkali treated paddy straw compared with yield of 339.99 mg gram dry substrate $^{-1}$ using free enzyme for accaharification and 55 per cent recovery after two cycles	Kumar *et al.* (2016)
6	Cellulase	Zinc Ferrite nanoparticles	The immobilized enzyme was stable at 60°C, while retaining its activity up to 3 recycles. Exhibited 53 per cent hydrolysis yield on pretreated sun hemp biomass.	Monasa *et al.* (2017)
7	Cellulase from *Aspergillus fumigatus*	Cobalt nanoparticle (CO_3O_4)	Sustained 75 per cent of the catalytic activity after 4 cycles with ethanol yield of 21 g l^{-1} of hydrolysates	Cherian *et al.* (2017)
8	Pectinase, Xylanase and cellulaase	Magnetic nanoparticles activated with 3, Amino propyl Triethoxy silane (APTS)	Thermo stability at 70° C and retained its original activity upto five cycles of reuse. Immobilized enzyme produce ethanol of 0.5g l^{-1} of hydrolysate compared to 0.37g l^{-1} with free enzyme	Perwez *et al.* (2017)

Sl.No.	*Enzyme Immobilized*	*Support Used*	*Results*	*Authors*
9	Cellulase from *Trichoderma reesei*	Silica nanoparticles	Immobilized enzymes lost 1.6 per cent of catalytic activity upto five cycles 43.6 per cent conversion of glucose into 5 hydroxy methyl furfural	Song *et al.*, (2017)
10	Cellulase from *Trichoderma reesei*	3 Amino propyl Trimethoxy Silane activated Magnetic nanoparticles	Retained 74 per cent of the initial activity while using carboxy methyl cellulose as substrate after six cycles; 21 per cent yield of sugars from bamboo biomass with four cycles retaining 38 per cent of initial catalytic activity	Jia *et al.*, (2017)
11	Cellulase	Magnetic supports (Fe_3O_4) with ionic and epoxy dongles	Higher loading @115mg g^{-1} with enzyme immobilization; higher catalytic activity @ 60°C compared to free enzyme @ 50°C; Free enzymes loses its activity at 12h compared to retaining 50 per cent activity after 12h; immobilized cellulase retained 60 per cent of its initial activity after 6 cycles	Hosseini *et al.*, (2018)
12	Cellulase from *Trichoderma longibrachiatum*	Maghemite nanoparticles (magnetic nanoparticles)	Recovery of 95 per cent of the magnetite nanocatalysts and moderate loss of catalytic activity over 5 cycles	Schnell *et al.*, (2019)
13	Cellulase	Magnetic nanoparticles functionized with glutaraldehyde	Free enzymes converts 13.7±0.25 gm l^{-1} of glucose in 48 h, while immobilized enzyme produce 19.5±0.15 gm l^{-1} of glucose and 6.15±0.15 gm l^{-1} and 3.3±0.25 gm l^{-1} of glucose in second and third cycles	Gaikwad *et al.*, (2019)
14	Cellulase from *Trichoderma sps.*	Low Temperature Hydrothermal Carbonization produced charcoal cross-linked enzyme	High residual enzyme activity	Primozic *et al.*, (2019)

Table 25.3. Immobilization of Laccase and Lipase Enzymes for Biodiesel Production

Sl.No.	*Enzyme immobilized*	*Support used*	*Results*	*Authors*
1	Laccase	Chitosan coated Magnetic Nanoparticles activated with carbodiimide	Immobilization efficiency of laccase enzyme was 79.2 per cent and retaining 90 per cent of the catalytic activity	Das *et al.* (2016)
2	Lipase from *Pseudomonas cepacia*	Polydopamine coated magnetite nanoparticles	Immobilized enzyme produced biodiesel yield of 93 per cent in soybean compared to yield of 86 per cent with free enzyme within 12h. Immobilized enzymes retains its activity upto 3 cycles	Andrade *et al.* (2016)
3	Lipase from *Candida regusa*	Silica nanoparticles activated with 3 amino propyl triethoxy silane	Immobilized lipase maintains 81 per cent of the activity for 28 days while it retains 80 percent of the activity after 8 cycles of use. Sustains 55 per cent of the catalytic activity @ 70°C	Ali *et al.* (2016)
4	Lipase from *Burkholderia cepacia*	Amino grafted magnetic nanoparticle coated with silica	Immobilized lipase achieved 91 per cent conversion from waste cooking oil to biodiesel and slight activity loss while repeated use	Karimi (2016)
5	Lipase from *Aspergillus niger*	Magnetic nanoparticles	Immobilized enzyme in magnetic nanoparticles retains catalytic activity for 5 cycles with maximum yield of 88 per cent of biodiesel in *Chlorococcum sphacosum.*	Donatus *et al.* (2017)
6	Lipase from *Burkholderia cepacia*	2,3-epoxy propyl trimethyl-ammonium chloride with epoxy and quaternary ammonium group and glutaraldehyde were grafted onto aminated magnetic nanoparticles	96.8 per cent biodiesel conversion in soybean oil with immobilized enzyme with in 12h. Retains 75 per cent of the activity after seven cycles of use	Li *et al.* (2017)
7	Lipase from *Candida rugosa*	Interfacial activated magnetic nanoparticle with polyvinyl benzyl chloride	Immobilized enzyme retains more than 69.8 per cent of its initial activities after 10 times reuse. Catalytic activity of the immobilized lipase retained about 99.6 ± 3.3 per cent of the free enzyme activity	Chen *et al.* (2017)
8	Laccase (Novozym 51003)	Ferromagnetic (iron and nickel particles) nanoparticles encapsulated in silica (Tetraethyl Ortho silicate)	Pretreatment of Oil palm empty fruit bunches with immobilized laccase degrades 58 per cent of lignin after 24 hours, increasing saccharification by 161 per cent	Rashid *et al.* (2019)

Sl.No.	*Enzyme immobilized*	*Support used*	*Results*	*Authors*
9	Laccase from *Trametes versicolor*	Ferrite and Copper ferrite nanoparticles functionalized with 3- amino propyl triethoxy silane	Immobilized laccase exhibited 18 per cent more activity than free enzyme, 70 per cent of initial activity after six cycles with the *Ipomoea carnea* Immobilized laccase in copper magnetic nanoparticles showed lignin removal of 43.28± 1.46 per cent compared to 38.16±0.78 per cent after 20h of incubation	Muthuvelu *et al.* (2019)
10	Lipase from *Candida antarctica*	Chitosan coated magnetic nanoparticles cross-linked with glutaraldehyde	Half-life of the immobilized enzymes was 8-11 times higher than that of soluble enzymes. Immobilized enzyme in magnetic nanoparticles showed highest activity at pH 10, while free enzyme recorded the maximum activity at pH 7	Monteiro *et al.* (2019)
11	Lipase	Multi walled Carbon Nano Tubes (MWCNT)	Highest immobilization efficiency of enzyme (loading) was 79 per cent	Khan *et al.* (2019)

and their huge oil producing capacity throughout the year. Algae serve as an alternative to petroleum derived biodiesel. The efficiency of biofuel relies on downstream refinement for the extraction of lipids from biomass. The recovery of oil per unit acre is estimated to be 30 times higher than the palm oil (Demirbas and Faith Demirbas 2011). However, many techniques are followed to improve the lipid accumulation in algae. Nanotechnology is a potential strategy in the field of algal biofuel production (Hassanuddin *et al.*, 2018; Kim *et al.*, 2006, Safarik *et al.*, 2016). Metallic nanoparticles with size range of 5-100 nm facilitate the mass transfer coefficient at gas liquid interface, improving CO_2 concentration, thereby influencing algal photosynthesis and accumulation of various valuable bio products (He *et al.*, 2017). The algal biomass and lipid production are enhanced with suitable nanomaterials at appropriate concentration. Nanotechnology renders solution in algal growth, harvest, extraction and processing of biofuels (Pattarkine and Paterkine 2012).

The response of nanoparticles in algae systems is the production of reactive oxygen species in cells causing cell damage and oxidative stress (Manke *et al.*, 2013; He *et al.*, 2017). The interruption of electron transport chain of light reaction and distortion of enzymes involved in Calvin cycle of algae cells, and inhibition of cell division are reported in algae (Navarro *et al.*, 2015) for the challenge of nanomaterials. Similarly, application of nanoparticles stimulates anti-oxidant response (He *et al.*, 2012) and enhances the total lipid production (Pham 2019) in algal systems. Sarma *et al.* (2014) have hypothesized that induction of photosynthesis by nanoparticles improves the lipid production. The nanoparticles readily react with cell surface, easily penetrate into cytosol and markedly increase the growth behavior of algae by serving as a source of trace element (Kadar *et al.*, 2012). Addition of nanomaterials in the culture medium of algae helps in accumulation of lipids. For instance, silica or iron oxide nanoparticles creates shear between the cells and particles, which renders the competition among the cells for the uptake of nutrients and leads to lipid accumulation in algal cells (Zhang *et al.*, 2013). The lipid productivity improve the growth rate of heterotrophic microalgae (Gao *et al.*, 2010). However nanoparticles have great impact towards the lipid buildup, which depends on the nature of nanomaterial, use limit and its effect on the strain are to be optimized. The biomass and the lipid production of algae are the prime criteria for its biofuel production, where the application of nanomaterials alters suitable in the recent research. Application of Zero-valent iron nanoparticles (nZVI) enhances the lipid accumulation and decreased the content of saturated and monounsaturated fatty acids in green algae and cyanobacterium (Padrova *et al.*, 2015). The oxidative stress triggered by contact of nZVI with algal cells upregulates the lipid accumulation. Several other nanoparticles like TiO_2 induced reactive oxygen species generation and the accumulation of lipids in algal cells

Lipid extraction from algae is critical for assessing the yield and the efficiency of lipid extraction determine the economics of algal biofuel (Davis *et al.*, 2014). However, the lipid recovery is higher from dried biomass, drying the biomass is energy intensive process (Lardon *et al.*, 2009; Xu *et al.*, 2011). Nanotechnology are explored for the application of nanoparticles for lipid extraction. Diverse nanoparticles like dielectric, hard, magnetic and enzymatic type nanomaterials are used in the lipid extraction process. Organic nano clay for lipid extraction from algal biomass has resulted in low lipid extraction efficiencies, but conversion of lipids to biodiesel was more than 98 per cent (Lee *et al.*, 2013). Cetyl trimethylammonium bromide aided foam floatation in *Chlorella* has increased the extractable lipid by its adsorption into the cell surface during harvesting (Coward *et al.*, 2014). Amino clay-conjugated TiO_2 in algal harvest and cell disruption resulted in higher harvesting efficiency of 85 per cent (Lee *et al.*, 2014). The use of nanomaterials augments lipid extraction accomplishing no harm to the microalgae used.

Immobilization of lipase in the nanomaterial provides enormous surface area and enhances the diffusion rate of lipase at active sites of reaction. Shah *et al.* (2007) showed highest activity of lipase (97 per cent) immobilized into carbon nano tubes. Similarly, lipase encapsulated polymer gel exhibited 80 per cent activity, while the free lipase displayed less than 10 per cent of the catalytic activity, suggesting the high stability of lipase in nanogel environment or nanosized silica (Kwon *et al.*, 2007). The advantage of using solid nanomaterial in transesterification process offers high quality biodiesel and by-products besides their ease in recovery. The use of solid base nanocatalyst used in transesterification of *Nannochloropsis oculata* with Al_2O_3 supported CaO/MgO, increase the biodiesel yield from 23 to 97.5 per cent (Teo *et al.*, 2014). Nickel supported zeolites facilitates hydrogenolysis of triacyl glycerols, hydrogenation of carboxylic acids, aldehydes and alkenes, thereby supporting the conversion of micro algal oil to diesel (Peng *et al.*, 2012). The qualitative and quantitative analysis of algae treated with nanoparticles revealed the increased methyl esters and fatty acids. Use of 0.2 per cent silica nanoparticles ($SiO_2 - CH_3$) resulted in increased cell dry weight (1.49g L^{-1}) and fatty acid methyl ester productivity (Jeon *et al.*, 2017). The biodiesel conversion efficiency of about 91 per cent was achieved using calcium oxide nanoparticles (CaO) during catalytic transesterification (Patternike and Patterni, 2012). The use of magnetic particles at low pH for harvesting oleaginous *Chlorella* sp has resulted in higher lipid content 0.35-0.41 g fatty acid methyl esters g^{-1} cell biomass (Lee *et al.*, 2015). Merging nanotechnology with bioprocess of algae has led to several breakthrough in enhancing efficiency, extraction yield, conversion lipids to biodiesel. Nanoparticles are also explored to find solutions in down streaming process.

Conclusion

Fossil fuels are exhausting at faster rate besides the damage to the environment due to the emission of carbon monoxide, carbon dioxide, sulfur and particulate matter. Hence, biofuel is considered as right candidate to satisfy the energy requirement. Biofuel sector has leaped from conventional to advanced biofuel based on the feedstock used for the extraction of biofuel. However, the progress of biofuel is facing several key challenges, especially for advanced biofuels: cellulosic ethanol and algae-based fuels to make economically viable enterprises. Nanotechnology provide solutions to unresolved challenges in the biofuel industries to take forward for commercialization of advanced fuels at a large scale. Nanotechnology is regarded as a technology of the future.

REFERENCES

1. Ali, Z., Tian, L., Zhao, P., Zhang, B., Ali, N., Khan, M., and Zhang, Q. (2016). Immobilization of lipase on mesoporous silica nanoparticles with hierarchical fibrous pore. *Journal of Molecular Catalysis B: Enzymatic, 134,* 129-135.

2. Andrade, M. F., Parussulo, A. L., Netto, C. G. C. M., Andrade, L. H., and Toma, H. E. (2016). Lipase immobilized on polydopamine-coated magnetite nanoparticles for biodiesel production from soybean oil. *Biofuel Research Journal, 3*(2), 403-409.

3. Bazooyar, B., Shariati, A., and Hashemabadi, S. H. (2015). Economy of a utility boiler power plant fueled with vegetable oil, biodiesel, petrodiesel and their prevalent blends. *Sustainable Production and Consumption, 3,* 1-7.

4. Bechara, R., Gomez, A., Saint-Antonin, V., Schweitzer, J. M., and Maréchal, F. (2016). Methodology for the design and comparison of optimal production configurations of first and first and second generation ethanol with power. *Applied energy, 184,* 247-265.

5. Cataluña, R., and Da Silva, R. (2012). Effect of cetane number on specific fuel consumption and particulate matter and unburned hydrocarbon emissions from diesel engines. *Journal of Combustion, 2012.*

6. Chen, Z., Liu, L., and Yang, R. (2017). Improved performance of immobilized lipase by interfacial activation on Fe_3O_4 @ PVBC nanoparticles. *RSC Advances, 7*(56), 35169-35174.

7. Das, A., Singh, J., and Yogalakshmi, K. N. (2016). Comparative Analysis of Laccase Immobilization on Magnetic Iron Nanoparticles using Two Activating Agents: EDAC and Cyanuric Chloride.

8. Davis R, Kinchin, C., Markham, J., Tan, E., Laurens, L., Sexton, D., Knorr, D., Schoen, P., and Lukas, J. (2014) Process design and economics for the conversion of algal biomass to biofuels: algal biomass fractionation to lipid-products process design and economics for the conversion of algal biomass to biofuels: algal biomass fractionation to lipid- and carbohyd.

9. Demirbas A, Fatih Demirbas M (2011) Importance of algae oil as a source of biodiesel Energy Conversion and Management 52: 163-170

10. Donohoe, B. S., Vinzant, T. B., Elander, R. T., Pallapolu, V. R., Lee, Y. Y., Garlock, R. J., and Ladisch, M. R. (2011). Surface and ultrastructural characterization of raw and pretreated switchgrass. *Bioresource technology, 102*(24), 11097-11104.

11. Fengel, D., and Wegener, G. (Eds.). (2011). Wood: *chemistry, ultrastructure, reactions.* Walter de Gruyter.

12. Gao C, Zhai Y, Ding Y, Wu Q (2010) Application of sweet sorghum for biodiesel production by heterotrophic microalga *Chlorella protothecoides* Applied Energy 87: 756-761 doi: https: //doi.org/10.1016/j.apenergy.2009.09.006

13. Gardy, J., Hassanpour, A., Lai, X., Ahmed, M. H., and Rehan, M. (2017). Biodiesel production from used cooking oil using a novel surface functionalised TiO_2 nano-catalyst. *Applied Catalysis B: Environmental,* 207, 297-310.

14. Hasannuddin AK *et al.* (2018) Nano-additives incorporated water in diesel emulsion fuel: Fuel properties, performance and emission characteristics assessment Energy Conversion and Management 169: 291-314 doi: https: //doi.org/10.1016/j.enconman.2018.05.070

15. He D, Dorantes-Aranda JJ, Waite TD (2012) Silver Nanoparticle–Algae Interactions: Oxidative Dissolution, Reactive Oxygen Species Generation and Synergistic Toxic Effects Environmental Science and Technology 46: 8731-8738 doi: 10.1021/es300588a

16. He M, Yan Y, Pei F, Wu M, Gebreluel T, Zou S, Wang C (2017) Improvement on lipid production by *Scenedesmus obliquus* triggered by low dose exposure to nanoparticles Scientific reports 7: 15526-15526 doi: 10.1038/s41598-017-15667-0

17. https: //www.ceicdata.com/en/india/petroleum-overview/petroleum-consumption-crude-oil. accessed on 21.11.2019

18. International Energy Outlook 2013. With Projections to 2040 U.S. Energy Information, Administration Office of Energy Analysis, U.S. Department of Energy, Washington, DC 20585, 300 pages

19. Jeon H-S, Park SE, Ahn B, Kim Y-K (2017) Enhancement of biodiesel production *in Chlorella vulgaris* cultivation using silica nanoparticles Biotechnology and Bioprocess Engineering 22: 136-141 doi: 10.1007/s12257-016-0657-8

20. Kadar E, Rooks P, Lakey C, White DA (2012) The effect of engineered iron nanoparticles on growth and metabolic status of marine microalgae cultures Science of The Total Environment 439: 8-17 doi: https: //doi.org/10.1016/j.scitotenv.2012.09.010

21. Karimi, M. (2016). Immobilization of lipase onto mesoporous magnetic nanoparticles for enzymatic synthesis of biodiesel. *Biocatalysis and Agricultural Biotechnology, 8,* 182-188.

22. Khan, A. K., Mubarak, N. M., Abdullah, E. C., Khalid, M., Nizamuddin, S., Baloch, H. A., and Siddiqui, M. T. H. (2019, April). Immobilization of Lipase Enzyme Carbon Nanotubes via Adsorption. In *IOP Conference Series: Materials Science and Engineering* (Vol. 495, No. 1, p. 012055). IOP Publishing.

23. Kim J, Jia H, Wang P (2006) Challenges in biocatalysis for enzyme-based biofuel cells Biotechnology Advances 24: 296-308 doi: https: //doi.org/10.1016/j.biotechadv.2005.11.006

24. Kwon SS, Jeon SH, Shon JK, Kim DH, Chang IS, Kim JM (2007) Preparation and Stabilization of Chitosan-Lipase Composite within Mesoporous Silica Material Solid State Phenomena 124-126: 1717-1720 doi: 10.4028/www.scientific.net/SSP.124-126.1717

25. Lardon L, Hélias A, Sialve B, Steyer J-P, Bernard O (2009) Life-Cycle Assessment of Biodiesel Production from Microalgae Environmental Science and Technology 43: 6475-6481 doi: 10.1021/es900705j

26. Lee Y-C *et al.* (2013) Lipid extractions from docosahexaenoic acid (DHA)-rich and oleaginous Chlorella sp. biomasses by organic-nanoclays Bioresource Technology 137: 74-81 doi: https: //doi.org/10.1016/j.biortech.2013.03.090

27. Lee Y-C *et al.* (2014) Aminoclay-conjugated TiO2 synthesis for simultaneous harvesting and wet-disruption of oleaginous *Chlorella* sp Chemical Engineering Journal 245: 143-149 doi: https: //doi.org/10.1016/j.cej.2014.02.009

28. Lee Y-C, Lee K, Oh Y-K (2015) Recent nanoparticle engineering advances in microalgal cultivation and harvesting processes of biodiesel production: A review Bioresource Technology 184: 63-72 doi: https: //doi.org/10.1016/j.biortech.2014.10.145

29. Li, K., Fan, Y., He, Y. *et al.*, Burkholderia cepacia lipase immobilized on heterofunctional magnetic nanoparticles and its application in biodiesel synthesis. *Sci Rep* 7, 16473 (2017) doi: 10.1038/s41598-017-16626-5

30. Li, K., Fan, Y., He, Y. *et al.*, *Burkholderia cepacia* lipase immobilized on heterofunctional magnetic nanoparticles and its application in biodiesel synthesis. *Sci Rep* 7, 16473 (2017) doi: 10.1038/s41598-017-16626-5

31. Liu, X., He, H., Wang, Y., and Zhu, S. (2007). Transesterification of soybean oil to biodiesel using SrO as a solid base catalyst. *Catalysis Communications, 8*(7), 1107-1111.

32. Liu, X., He, H., Wang, Y., Zhu, S., and Piao, X. (2008). Transesterification of soybean oil to biodiesel using CaO as a solid base catalyst. *Fuel, 87*(2), 216-221.

33. Manke A, Wang L, Rojanasakul Y (2013) Mechanisms of Nanoparticle-Induced Oxidative Stress and Toxicity BioMed Research International 2013: 15 doi: 10.1155/2013/942916

34. Micheal Donatus, S., Vijayalakshmi, S., and Ranjitha, J. (2017). Sustainable Biodiesel Production from a Newly Isolated Micro-Algal Species Chlorococcum Sphacosum Using a Novel Lipase Immobilised Functionalized Nanobiocataylst. *Int J Chem Sci, 15*(3), 157

35. Ministry of Petroleum and Natural Gas. 2012. Basic Statistics on Indian Petroleum and Natural Gas 2011-12. Ministry of Petroleum and Natural Gas (MPNG), Government of India, New Delhi.

36. Monteiro, R. R., Lima, P. J., Pinheiro, B. B., Freire, T. M., Dutra, L. M., Fechine, P. and Fernandez-Lafuente, R. (2019). Immobilization of lipase A from Candida antarctica onto chitosan-coated magnetic nanoparticles. *International journal of molecular sciences, 20*(16), 4018.

37. Muthuvelu, K. S., Rajarathinam, R., Selvaraj, R. N., and Rajendren, V. B. (2019). A Novel Method for Improving Laccase Activity by Immobilization onto Copper Ferrite Nanoparticles for Lignin Degradation. *International journal of biological macromolecules*.

38. Navarro E, Wagner B, Odzak N, Sigg L, Behra R (2015) Effects of Differently Coated Silver Nanoparticles on the Photosynthesis of *Chlamydomonas reinhardtii*

Environmental Science and Technology 49: 8041-8047 doi: 10.1021/acs.est.5b01089

39. Pádrová K *et al.* (2015) Trace concentrations of iron nanoparticles cause overproduction of biomass and lipids during cultivation of cyanobacteria and microalgae Journal of Applied Phycology 27: 1443-1451 doi: 10.1007/s10811-014-0477-1
40. Pattarkine MV, Pattarkine VM (2012) Nanotechnology for Algal Biofuels. In: Gordon R, Seckbach J (eds) The Science of Algal Fuels: Phycology, Geology, Biophotonics, Genomics and Nanotechnology. Springer Netherlands, Dordrecht, pp 147-163. doi: 10.1007/978-94-007-5110-1_8
41. Peng B, Yao Y, Zhao C, Lercher JA (2012) Towards Quantitative Conversion of Microalgae Oil to Diesel-Range Alkanes with Bifunctional Catalysts Angewandte Chemie International Edition 51: 2072-2075 doi: 10.1002/anie.201106243
42. Pham T-L (2019) Effect of Silver Nanoparticles on Tropical Freshwater and Marine Microalgae Journal of Chemistry 2019: 7 doi: 10.1155/2019/9658386
43. Purohit, P., and Dhar, S. (2018). Lignocellulosic biofuels in India: current perspectives, potential issues and future prospects. *Aims Energy, 6*(3), 453-486.
44. Rashid, S. S., Huan, H. S., Ab Rahim, M. H. B., Sikder, M. B. H., and Sasi, A. A. H. (2019, October). Laccase Enzyme Immobilization on Ferromagnetic Support Material for Oil Palm Empty Fruit Bunches (OPEFB) Pretreatment. In *IOP Conference Series: Materials Science and Engineering* (Vol. 627, No. 1, p. 012025). IOP Publishing.
45. Safarik I, Prochazkova G, Pospiskova K, Branyik T (2016) Magnetically modified microalgae and their applications Critical Reviews in Biotechnology 36: 931-941 doi: 10.3109/07388551.2015.1064085
46. Sarma SJ, Das RK, Brar SK, Le Bihan Y, Buelna G, Verma M, Soccol CR (2014) Application of magnesium sulfate and its nanoparticles for enhanced lipid production by mixotrophic cultivation of algae using biodiesel waste Energy 78: 16-22 doi: https: //doi.org/10.1016/j.energy.2014.04.112
47. Sebastian Patrick Schwaminger, Priya Anand, Monika Borkowska-Panek, Silvia Angela Blank-Shim, Paula Fraga-García, Karin Fink, Sonja Berensmeier and Wolfgang Wenzel, Rational Design of Iron Oxide Binding Peptide Tags, *Langmuir*, 10.1021/acs.langmuir.9b00729, (2019).
48. Shah S, Solanki K, Gupta MN (2007) Enhancement of lipase activity in non-aqueous media upon immobilization on multi-walled carbon nanotubes Chemistry Central Journal 1: 30 doi: 10.1186/1752-153X-1-30
49. Teo SH, Islam A, Yusaf T, Taufiq-Yap YH (2014) Transesterification of *Nannochloropsis oculata* microalga's oil to biodiesel using calcium methoxide catalyst Energy 78: 63-71 doi: https: //doi.org/10.1016/j.energy.2014.07.045

50. Vaghari, H., Jafarizadeh-Malmiri, H., Mohammadlou, M., Berenjian, A., Anarjan, N., Jafari, N., and Nasiri, S. (2016). Application of magnetic nanoparticles in smart enzyme immobilization. *Biotechnology letters, 38*(2), 223-233.

51. Verziu, M., Cojocaru, B., Hu, J., Richards, R., Ciuculescu, C., Filip, P., and Parvulescu, V. I. (2008). Sunflower and rapeseed oil transesterification to biodiesel over different nanocrystalline MgO catalysts. *Green Chemistry, 10*(4), 373-381.

52. Wang, H., Hashimoto, S., Moriguchi, Y., Yue, Q., and Lu, Z. (2012). Resource use in growing China: Past trends, influence factors, and future demand. *Journal of Industrial Ecology, 16*(4), 481-492.

53. Wang, W., Ji, S., and Lee, I. (2013). Fast and efficient nanoshear hybrid alkaline pretreatment of corn stover for biofuel and materials production. *Biomass and bioenergy, 51*, 35-42

54. Wang, W., Ji, S., and Lee, I. (2013). Fast and efficient nanoshear hybrid alkaline pretreatment of corn stover for biofuel and materials production. *Biomass and bioenergy, 51*, 35-42.

55. Wen, Z., Liao, W., and Chen, S. (2004). Hydrolysis of animal manure lignocellulosics for reducing sugar production. *Bioresource Technology*, 91(1), 31-39.

56. Xu L, Brilman DWF, Withag JAM, Brem G, Kersten S (2011) Assessment of a dry and a wet route for the production of biofuels from microalgae: Energy balance analysis Bioresource Technology 102: 5113-5122

57. Zhang XL, Yan S, Tyagi RD, Surampalli RY (2013) Biodiesel production from heterotrophic microalgae through transesterification and nanotechnology application in the production Renewable and Sustainable Energy Reviews 26: 216-223

Chapter 26

Organic and Inorganic Nanoparticles in Pollution Management and Ecosystem Health

*V. Karthika, M. Praghadeesh and A. Lakshmanan**

Department of Nano Science and Technology,
Tamil Nadu Agricultural University, Coimbatore – 641 003, Tamil Nadu
**e-mail: microlaxman@gmail.com*

Introduction

Nanomaterials are defined as regular organic or inorganic frameworks in the size of nanometer. Nanomaterials are of great interest because of their excellent pore structure, surface to volume ratio and high electrical conductivity. Nanomaterials have unique physical and chemical properties such as thermal stability, chemical resistance, hardness, large surface area, electron transfer, conductance and impedance. Nanomaterials include organic materials and their derivatives and inorganic materials like metals/metal oxides. Nanomaterials have wide application in various fields of environmental sciences like wastewater treatment, pollution reduction in soil and air. Environmental nanotechnology is the sub-division of nanotechnology intended to create ecofriendly products for cleaner environment, provide green methods in manufacturing to reduce pollution and produce more cost effective alternative energies (Sivasubramanian, 2016). There are variety of nanostructured materials explored including metal organic frameworks (MOFs), Nanoporous carbonaceous materials, Nano silica, Nano zeolites, functionalized nanomaterials and nanocomposites which are discussed in detail for its efficiency in preserving the environment. Role of nanotechnology in environmental implications are enormous and also has great potential. This review highlights the different nanomaterials used in the process of environmental remediation.

Classification of Nanomaterials

Nanomaterials are broadly classified into two broad categories based on the composition. They are organic nanomaterials which include CNT's, Graphene, Fullerenes, Carbon fibers, Chitosan based nanoparticles and inorganic nanomaterials include metal based, ceramics based, silica based and zeolite based nanomaterials (Figure 26.1).

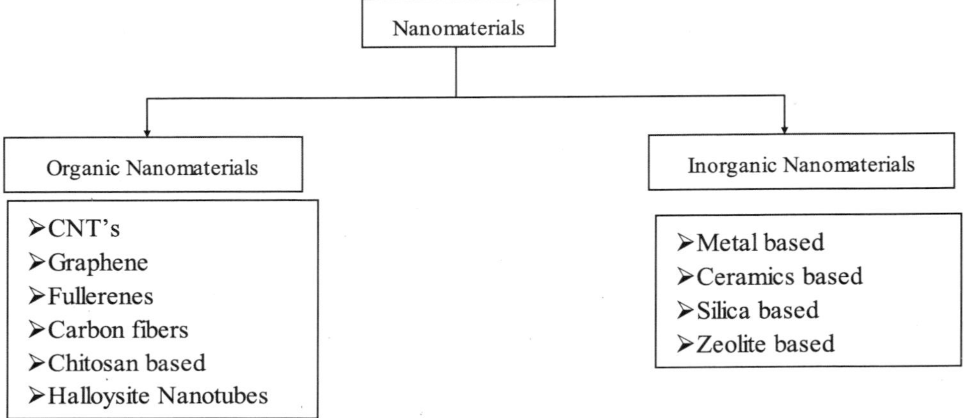

Figure 26.1. Classification of Nanomaterials.

Organic Nanomaterials

Organic Nanomaterials have attracted increasing attention in the past few years. Organic Nanomaterials are formed by the self-organization or chemical binding of the several organic molecules. Zwitterions are key components in the fabrication of organic nanoparticles with polar and non-polar regions aids in the encapsulation process. Interactions among the organic nanoparticles are weaker in nature compared to inorganic nanoparticles, in which micelles and vesicles are formed. Encapsulation of pollutants is relatively simple in using the organic nanoparticles for the environmental remediation or cleanup process. Organic nanomaterials like Carbon nanotubes (CNTs), Fullerenes and Graphene nanomaterials are efficient in the remediation of persistent pollutants in the environment.

Carbon Nanotubes

Carbon nanotubes are allotropes of carbon with a cylindrical nanostructure which has length-to-diameter ratio of up to 132,000,000:1. They are made up of sp^2-bonded carbon atoms assembled into a tube like nanostructures (Iijima, 1991). CNTs are preferred over the other conventional carbonaceous materials in the pollutant removal as a result of its unique physicochemical properties, including extremely high mechanical strength, high electrical conductivity, high ductility, high specific heat and thermal conductivity and high adsorptive behavior. There are two types of carbon nanotube: multi-wall carbon nanotubes (MWCNTs) and single-wall carbon nanotube (SWCNTs) classified based on the layer of graphene sheets present in the nanotubes. Based on the two-dimensional form of the graphene sheets, CNTs are further grouped into three types. Zigzag form which has a hexagonal pattern moving around the body of the tubule, Armchair form that contains one or two cyclohexane conformers, with the carbon atom describing a hexagonal pattern as it moves around the body of the tubule body and Chiral forms indicates handedness and the tube can be twisted in any direction resembles zigzag and armchair forms (Baughman*et al.*, 2002). Carbon nanotubes are efficiently used in many environmental applications and the mechanisms for pollutant removal are represented in Figure 26.2. CNTs are used as adsorbents to remove heavy metals and various organic and inorganic pollutants from the aqueous solutions and water bodies (Fiyadh *et al.*, 2019).Multi-walled carbon nanotubes (MWCNTs) promote heavy metal sedimentation and immobilization in water–sediment interface and also alleviate the bioavailability and transformation ability of the metals. It is also noted that the MWCNTs reduced the aromaticity, hydrophobicity and humification of fulvic acid (FA) and humic acid (HA) and microbial activity in the sediments (Xu*et al.*, 2019). The hydrophobic parts and ï-ï interactions are the mechanisms that create strong affinity in the binding of CNTS and the pollutants (Tang*et al.*, 2014). Surface modification of CNTs using magnetic nanoparticles, pollutant binding and degrading receptors and enzymes provides novel properties to the CNTS like fibrous shape, high aspect ratio, large surface to volume ratio and accessible mesopores in environmental applications (Ali *et al.*, 2014).

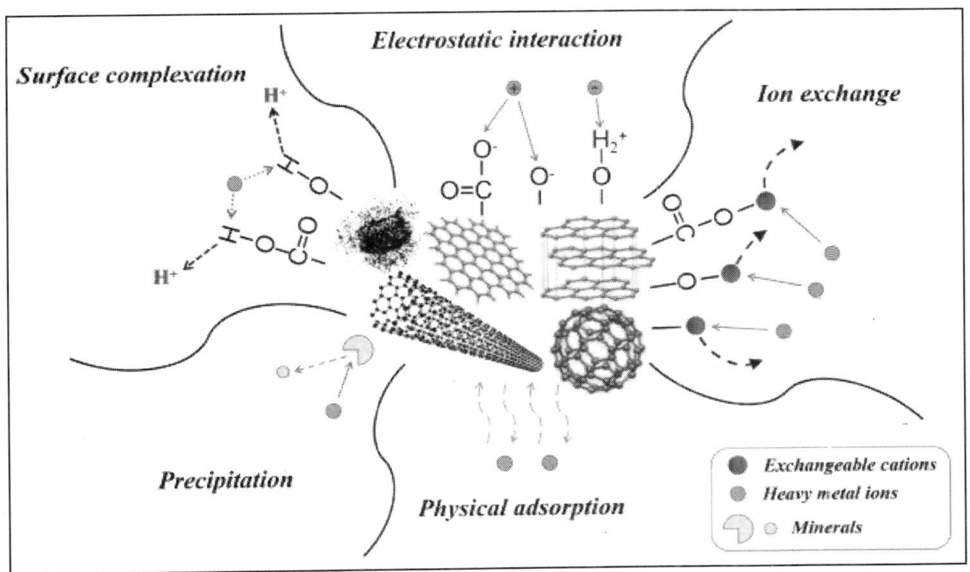

Figure 26.2. Schematics of Pollutants Removal Mechanisms by Carbon Based Adsorbents.
Courtesy: Yang *et al.*, 2019.

Graphene

Single atomic layer of graphite is termed as "Graphene" and introduced by International Union of Pure and Applied Chemistry (IUPAC). Graphene is a single sheet of 2D hexagonal network of carbon accessible in variety of forms including graphene, graphene oxide, reduced graphene oxide and functionalized graphene in the pollution reduction process. Graphene based materials have high specific surface area, electrical, thermal and mechanical properties, outstanding chemical steadiness and charge mobility, flexibility, negligible thickness and rich functional groups, ideal for the removal of different kinds of pollutants from the aqueous solutions and polluted habitats (Novoselov, 2012). Graphene with its high surface area, geometry and chemical properties used as promising sorbent for organic and inorganic pollutants in the wastewater treatment (Ersan *et al.*, 2017). Heavy metals in the water bodies and aqueous solutions are eliminated through the surface complexation mediated as result of binding with oxygen functional groups and the aromatic framework present in the surface of graphene (Gao *et al.*, 2018). Toxic organic pollutants like dyes, brominated flame retardants, pharmaceuticals, endocrine disrupting chemicals and polycyclic aromatic hydrocarbons in the wastewater are removed efficiently with the graphene-based adsorbents (Baig *et al.*, 2019).

In addition to the adsorption process, a good adsorbent should also possess high desorption ratio for the process of cyclic regeneration. Graphene has excellent adsorption-desorption rate, which is an economically important attribute in reusing the nanomaterial in the wastewater treatment. HCl, EDTA and thiourea-HCl were studied for elution efficiency in the elution of adsorbed pollutants and in the most of the studies it is reported that graphene has maintained 90 per cent adsorption efficiency even after 10 cycles (Wu *et al.*, 2013). These parameters support the use of graphene as adsorbent in the real time application of wastewater treatment in the near future.

Fullerenes

Fullerenes are enclosed cage-like structures made up of twelve 5-member rings and an unspecified number of 6-memberrings. The bonding rate, reactive carbon sites and strain energies of fullerenes are determined by the number of hexagons, preferably fewer present in the structure. The C-60 (Buckminsterfullerene) and C-20 fullerenes differs in the chemical characters based on the aromatic molecule and the straight chained alkene in the structure which is critical for the environmental applications. The supramolecular pathways of the fullerenes are further altered for exploiting in the field of environmental remediation (Mauter and Elimelech, 2008). Fullerenes are non-toxic and has wide spectrum of positive biological activity which could be useful in environmental remediation process without affecting the indigenous organisms (Andrievsky *et al.*, 2005). Fullerenes are potential antioxidants has the ability of both quenching and generating reactive oxygen species (Markovic and Trajkovic, 2008) without any acute or subacute toxicity (Gharbi *et al.*, 2005). Fullerene and its composites have been studied widely for sorption of heavy metals like copper, nickel, zinc, and cadmium ions and the sorption efficiency of the fullerenes depending on the heavy metal cations (Alekseeva *et al.*, 2016). Fullerenes have been extensively used in the advanced wastewater treatment systems to remove various organic and inorganic pollutants because of its adsorption capacity and the regeneration capacity to reuse it (Gupta and Saleh, 2013). The structure and function of the soil microbial community upon the treatment with fullerenes are not affected and it shows that the land application of treated waste along with little fullerene are not toxic to microbes or soil (Tong *et al.*, 2007).

Carbon Nanofibers

Nanofibers fabricated by electrospinning gained much attention in the recent days and are utilized widely for environmental applications. Nanofibers are lengthy nanomaterials, could be embedded in the carrier material easily than any other nanostructures (Yoon *et al.*, 2008). Multiwalled carbon nanotube-reinforced electrospun polymer nanofibers containing zero-valent iron nanoparticles has excellent capability to decolorize the dyes and also found to be effective in degrading the hydrocarbons (Xiao *et al.*, 2010). Nanofibers/mesoporous carbon composite harvested from the pyrolysis of waste biomasscould be effectively used to store electrochemical energy (Liu *et al.*, 2014). One dimensional hollow carbon nanofibers with excellent structural advantages and N-doped composition exhibits a remarkable catalytic degradation of tetracycline in the advanced oxidation process (Wang *et al.*, 2019). Electrospun carbon nanofiber/graphene oxide composite aerogels possess enhanced mechanical compressive strength and recycling capacity performed high adsorption capacity for diverse oils (Lin *et al.*, 2019). Carbon nanofibers incorporated with MoS_2/CdS/TiO_2 ternary composite acquired great potential in the degradation of organic pollutants and the methylene blue in the wastewater (Pant *et al.*, 2019). Carbon nanofibers film prepared from the coal liquefaction residue by the electrospinning process found to be effective in coal waste management and also used as super capacitor electrodes (Li *et al.*, 2019).

Chitosan based Nanomaterials

Chitosan, a biopolymer and natural polysaccharide attracted the attention of researchers in recent years due to its low cost and its chemical structure that holds the amino and hydroxyl groups which can serve as active sites for the removal of pollutants from wastewater (Yue *et al.*, 2014). Chitosan has high affinity toward pollutants compared to other biomaterials because of its unique characteristics such as high reactivity with pollutants, exceptional chelation behavior and notable chemical stability. Silicate-chitosan nanocomposites prepared by layer-by-layer deposition is an excellent biosorbent and has the stability and capability for the removal of Cd(II), Cr(III), and Cr(VI) (Copello *et al.*, 2008). Chitosan/graphene oxide composite adsorbent possess pH-tunable surface charge and morphology that aids in the removal of multiple types of pollutants from water, including various anionic and cationinc dyes (methylene blue and methyl orange) and metal ions like, Cu(II) and Cr(VI) (Yan *et al.*, 2016). Multifunctional nanocomposites of chitosan with silver nanoparticles, copper nanoparticles and carbon nanotubes have a higher antimicrobial activity, in relative short contact times and low concentration, thus act as a promising integral material for multi-contaminant water treatment (Morsi *et al.*, 2017). Chitosan based nanomaterials such as nanogels, nanocomposites and nanoparticles attracted researchers in the field of agriculture due to the extraordinary water holding capacity and thus it serves as moisture retention agent in the dryland farming. It has been demonstrated by the *Department of Nano Science and Technology*, Tamil Nadu Agricultural University, that one gram of chitosan based nanomaterial has the capability to adsorb 99ml of water (Gowtham, 2019). This ability of high water holding capacity of chitosan-based nanomaterial could be used in many areas like dryland farming, water conservation and water transport.

HNTs

Halloysite is an aluminosilicate clay mineral (Al, Si, O... (OH)„) made up of aluminium, silicon and hydrogen. These are the naturally occurring nanoporous materials with low cost, adequate availability, environmental compatibility, good adsorption efficiency and environmental-friendly in nature. Halloysites are nanotubes of 500–1000 nm in length and 10–100 nm in inner-diameter. Halloysites are modified to expand the specific surface area by thinning the walls of the nanotube or by making holes on the walls. Nanoporous-walled silica and alumina nanotubes derived from halloysite has excellent adsorption capacity up to 427 mg/g for methylene blue (Shu *et al.*, 2015). Dual surface amidoximated halloysite nanotube, an efficient adsorbent showed high uranium adsorption capacity of 456.24 mg/g^{-1} in 32 ppm uranium spiked simulated seawater. It suggests that amidoximated halloysite nanotube could be used for economical extraction of uranium from the oceans over the current extraction prices that are equivalent the uranium cost (Zhao *et al.*, 2019). Halloysite nanotubes possess a negatively charged exterior and a positively charged inner lumen and this provides sites for adsorptive removal of anions like phosphates. HNTs modified with nano-scale iron oxide are proved to be effective in distribution of these oxide surfaces, which are known to adsorb phosphate via ligand exchange and by forming inner-sphere complexes (Almasri *et al.*, 2019). Highly porous HNTs sponges with skeletons made of electrospun nanofibers showed high performances as dye adsorbent and catalyst support (Xu *et al.*, 2019).

Inorganic Nanomaterials

Inorganic Nanomaterials include metals and metal oxide based nanomaterials, metal organic frameworks, ceramic based nanomaterials, silica based nanomaterials and polymer based nanomaterials. Inorganic materials work on the pollutants based on the principle of ion exchange and electrostatic interactions to make it less harmful or in toxic in the environment. Beyond the advantages of using inorganic nanomaterials in the pollutant degradation process, it is necessary to ensure the biocompatibility and environment benevolence of the materials used (Kabir *et al.*, 2018).

Metal based Nanomaterials

Diverse range of metal based nanomaterials are synthesized for the remediation of the environmental pollutants, but most of them are intended to study the removal of heavy metals and chlorinated organic pollutants from wastewater. Metal and metal oxide based nanomaterials are ideal candidates for the adsorption of recalcitrant pollutants because of the fast kinetics mechanism and high adsorption capacity (Santhosh *et al.*, 2016). Some of the different metal-based nanomaterials that have been studied for removal of pollutants in the environment are summarized in the Table 26.1.

Table 26.1. Metal-based Nanomaterials for Removal of Pollutants in the Environment

Metal based Nanomaterials	*Environmental Applications*
TiO_2	Photocatalysts, Water disinfectant, Solar hydrogen production, Dye degradation (methylene blue), Air Pollutants degradation such as Nitric Oxide, Trichloroethylene (Choi *et al.*, 2012, Puddu *et al.*, 2010, Wen *et al.*, 2005)
Ag	Antimicrobial agents, Water disinfectant, Dye degradation (bromophenol blue), Organic pollutants degradation, Solar Fuel Production(Parmar *et al.*, 2019, Galante *et al.*, 2019, Otari *et al.*, 2019)
Fe/nZVI	Photocatalysts, Chlorinated solvents remediation, Dye degradation (bromophenol blue), Heavy metals, Polyhalogenated hydrocarbons (Thangavel *et al.*, 2019, Su *et al.*, 2020, Gallo *et al.*, 2019, Berger (2019))
Zn	Dye degradation (methyl orange), Organosulfur polluatnts, Antibacterial activity, Water treatment, Photocatalysts, Cyanide degradation (Baruah *et al.*, 2012, Bagabas *et al.*, 2013, Deb *et al.*, 2019, Khalafi *et al.*, 2019, Liu *et al.*, 2019)
Cu	Degradation of aromatic compounds (2-nitrophenol (2NP) and 4-nitrophenol (4NP)), Organic dye (cresyl blue) degradation, Antimicrobial activity (Zain *et al.*, 2014, Haider *et al.*, 2016)
Pd	Wastewater treatment, Diatrizoate removal, Chromate reduction and immobilization, Dechlorination of Trichloroethylene (Hennebel *et al.*, 2010, Tuo *et al.*, 2013, Ma *et al.*, 2012)
Au	Nitroaromatics reduction in water, Water analysis, Biosensors (Qin *et al.*, 2019, Mahajan *et al.*, 2019, Ribovski *et al.*, 2019)

Besides magnetic metallic nanoadsorbents are exceptionally good in wastewater treatment due to reproducible nature and easily separated from treated water. Iron and iron oxides nanoparticles are comprehensively described in many studies for the removal of heavy metals like Ni^{2+}, Co^{2+}, Cd^{2+}, Cr^{6+} and Cu^{2+} removal of chlorinated organic solvents and recalcitrant inorganic pollutants.

Zero Valent Iron Nanoparticles

Borohydride reduction method to synthesize nZVI-kaol was reported by Üzüm *et al.* (2009) in which iron nanoparticles had low aggregation and enhanced activity to remove Copper and Cobalt ions from aqueous solutions. The excellent removal mechanism of Cu^{2+} and Co^{2+} ions from wastewater was achieved using nZVI by adsorption and redox reactions respectively. The clay mineral supported nZVIs have lower aggregation, higher sorption capacity, increased stability and higher removal efficiency than bare nZVI particles. Kanel *et al.* (2007) used pristine nZVI in removal of As (III) from ground water and laser light scattering analysis demonstrated the inner-sphere surface complexes of nZVI- As(III). Cd (II) adsorption with possible interaction mechanism was further proved by Li *et al.* (2017) using nZVI rather than redox process. Using the three mechanisms such as reduction, adsorption and precipitation, nZVI supported with a novel composite of (Mg $(OH)_2$) showed possible reduction of Pb(II) to PbO resulting in 47 per cent PbO was formed as result of reduction mechanism (Wang *et al.*, 2015). Zhou *et al.* (2015) proved that zeolite supported nZVI particles could restore 80 per cent Sb (III) into Sb (O) by reduction process which was confirmed by XPS analysis.

Metal Organic Frameworks

Metal Organic Frameworks (MOFs) are the emerging field with plenty of scope in decontamination and remediation. MOFs are the crystalline polymers engrossing structural design which is capable of assembling organic bridging linkers or organic ligands and metal ions or metal clusters such as lanthanide metals or transition metals. Organic ligands can be amines, carboxylates, sulfonates and phosphates. MOFs are versatile novel adsorbent with high surface area than other adsorbents and having an ability to adopt any coordination geometry. MOFs can be utilized in various fields such as gas storage, catalysis, sensors, drug delivery, adsorption and separation. There are several methods for synthesizing MOFs, *viz.*, slow evaporation, solvothermal, sonochemical, electrochemical, mechanochemical and microwave assisted methods. Some of its application showed promising results that attracts more of research work to be done on improving the functional characteristics of MOFs, to explore the interaction towards any pollutant and mechanism involved behind it.

Recent developments illustrate the use of functionalized MOFs in removal of heavy metals from wastewater such as Cu based MOF for Pb (II) and Cd (II) which showed higher adsorption compared to graphene and zeolite (Rahimi and Mohaghegh, 2016). Thiol functionalized MOFs demonstrate more affinity towards Hg (II) and adsorption performance (Ke *et al.*, 2011). In addition, MOFs like carbomethoxy-functionalized Cd-MOFs are utilized as robust Pb (II) adsorbents because of its flexible coordination with lead and EDTA grafted MOF-88(Zr) via formic acid substitution also reported the removal efficiency (>99 per cent) for a total of 22 kinds of metal ions. Metal-organic framework (MOF-5) nanomaterial synthesized by solvothermal method contains abundant oxygen-containing functional groups, efficiently adsorb U(VI) from aqueous solutions based on the surface complexation and electrostatic interaction mechanisms (Wu *et al.*, 2018). Review shows that there is a need of exploration in MOFs in remediating or removing hazardous metals from wastewater. Some areas are to be concerned while developing MOFs such as evolving methods with less reaction time and high yield, developing composite with faster removal efficiency and high adsorption capacity, synthesizing green MOFs possessing ultrahigh porosity, environmentally friendly and low-cost method, stable mechanical structure and its reusability. These areas are yet to be explored for practical applicability of MOFs in the environmental remediation.

Ceramics based Nanoparticles

Ceramic nanocrystalline materials are solids with grain size of lower than

100 nanometers and are characterized by higher mechanical strength, tensile strength and hardness, which are utilized in many fields. These materials have low ductility and cracks grow easily in the brittle nanocrystallines limits their practical utility. Ceramic nanocrystallines are used mainly in the design of materials for the development of sensors, analytical tools and the semiconductors (Piticescu *et al.*, 2003). Ag doped WO_3 ceramic nanomaterials are used in the development of humidity sensors due to the high surface area available for the adsorption of the water molecules (Pandey *et al.*, 2013). Nanostructured bioceramics, cements and coatings are emerging as promising candidates for specific interactions with biological entities in the medical field like drug delivery, suturing and biocoatings (Traykova *et al.*, 2006).

Silica based Nanoparticles

Mesoporous silica nanoparticles possess many desirable properties like very high surface to volume ratio, tunable pore size and shape selectivity that are useful in the environmental applications like adsorption, bioenergy production and catalysis reactions. Mesoporous organosilica adsorbents are organically-modified nanoengineered silica materials for the removal of inorganic and organic pollutants from aqueous solutions (Walcarius and Mercier, 2010). Silica-based organic–inorganic hybrid nanomaterials used as chemosensors in environmental analytics as well as adsorbents for inorganic molecules and in biological applications like drug delivery (Jung *et al.*, 2010). Thiol-functionalized silica-coated magnetite nanoparticles has uniform pore size of <"2.1 nm has the higher adsorption capacity for the removal of Hg (II) (Hakami *et al.*, 2012). Carbon electrode modified with functionalized mesoporous silica nanoparticles used in anodic stripping square wave voltammetry for the detection of Pb(II) ions in the non-pretreated natural waters (Sánchez *et al.*, 2010). A selective fluoroionophore based on BODIPY-functionalized magnetic silica nanoparticles are used in the removal of Pb(II) ions from the human blood (Lee *et al.*, 2009). Bisindole anchored mesoporous silica nanoparticles produces spectroscopic changes that would be used as indicator for sensing cyanide ions in the aqueous solutions (Kim *et al.*, 2010).

Zeolites based Nanoparticles

Zeolites are renowned catalysts used in the petroleum refining industry and chemical production industries. Environmental catalysts are designed in the recent days to deal with the arising environmental issues such as reduction in automobile emissions, green chemistry in the manufacturing sectors and renewable energy sources (Grassian, 2005). Production of lead-free octane enhancers in petroleum industry and ion-exchangers in powdered laundry detergents to remove calcium and magnesium ions for softening the hard water are some of the notable environmental applications of the zeolites (Marcus and Cormier, 1999). Natural nanozeolites are utilized as potential adsorbents for the removal of petroleum monoaromatics including benzene, toluene, ethylbenzene, and xylenes (Seifi *et al.*, 2011). Phytoestrogens are plant derived compounds that are similar to the chemical structure of human oestrogen. These are effectively removed from the aqueous solutions by the tailored nanozeolites (Goyal *et al.*, 2019). Electrospun cellulose acetate/iron-modified nanozeolite could be effectively used as potential adsorbent for the removal of nickel from the aqueous solution (Tolentino *et al.*, 2019).

Concluding Remarks and Future Perspectives

Nanomaterials have drawn the attention of researchers in the recent days in the field of environmental sciences. Investigations are still going to explore the possible outcomes in widening the applications of nanoporous materials as well as in developing methodologies for the fabrication process of nanostructures in the application of environment clean up and energy storage. Environmental science is a multidisciplinary in nature and requires contributions from physicists, chemists and engineers to provide sustainable solutions to the problems existing in the globe. Application of nanomaterials in remediation process in commercial scale and producing cost effective nanomaterials are still challenging tasks.

Nanomaterials acts as platform for the development of environmental monitoring sensors, membranes for water purification, carrier materials for gas storage and transport, photocatalysts for pollutant degradation and adsorbents for pollutant removal. Research is mandatory in preventing the agglomeration of nanoparticles upon aging to enhance the shelf life of the nanoparticles applied in the environmental remediation. Non-magnetic nanoporous materials should be analyzed in a better way to increase their reception potential in the remediating sites besides exploring new nanoporous materials for the environmental applications. In the same way attention should be given to promotion of already existing nanomaterials to make them resourceful and efficient. Importantly any nano material before their use in the environment, the biosafety concerns need to be addressed.

REFERENCES

1. Alekseeva, O. V., Bagrovskaya, N. A., and Noskov, A. V. (2016). Sorption of heavy metal ions by fullerene and polystyrene/fullerene film compositions. *Protection of Metals and Physical Chemistry of Surfaces*, 52(3), 443-447.
2. Ali, M. E., Das, R., Maamor, A., and Hamid, S. B. A. (2014). Multifunctional carbon nanotubes (CNTs): a new dimension in environmental remediation. *In: Advanced Materials Research.*, 832: 328-332. Trans Tech Publications.
3. Almasri, D. A., Saleh, N. B., Atieh, M. A., McKay, G., and Ahzi, S. (2019). Adsorption of phosphate on iron oxide doped halloysite nanotubes. *Scientific reports*, 9(1), 3232.
4. Andrievsky G, Klochkov V, Derevyanchenko L. Is the C60 fullerene molecule toxic? *Fullerenes Nanotub Car N.* 2005;13: 363–376
5. Bagabas, A., Alshammari, A., Aboud, M. F., and Kosslick, H. (2013). Room-temperature synthesis of zinc oxide nanoparticles in different media and their application in cyanide photodegradation. *Nanoscale research letters*, 8(1), 516.
6. Baig, N., Sajid, M., and Saleh, T. A. (2019). Graphene-based adsorbents for the removal of toxic organic pollutants: A review. *Journal of environmental management*, 244, 370-382.

7. Baruah, S., K Pal, S., and Dutta, J. (2012). Nanostructured zinc oxide for water treatment. *Nanoscience and Nanotechnology-Asia*, 2(2), 90-102.

8. Baughman, R.H., Zakhidov, A.A., De Heer, W.A. 2002. Carbon nanotubes-the route toward applications. *Science* 297 (5582), 787e792.

9. Berger, A. (2019). Sulfidation of Zero-valent Iron Nanoparticles for Environmental Remediation.

10. Choi, S. K., Yang, H. S., Kim, J. H., and Park, H. (2012). Organic dye-sensitized TiO_2 as a versatile photocatalyst for solar hydrogen and environmental remediation. *Applied Catalysis B: Environmental*, 121, 206-213.

11. Copello, F. Varela, R. Vivot, L. Diaz Immobilized chitosan as biosorbent for the removal of Cd(II), Cr(III) and Cr(VI) from aqueous solutions. *Bioresour.* Technol., 99 (2008), pp. 6538-6544.

12. Deb, A., Kanmani, M., Debnath, A., Bhowmik, K. L., and Saha, B. (2019). Ultrasonic assisted enhanced adsorption of methyl orange dye onto polyaniline impregnated zinc oxide nanoparticles: kinetic, isotherm and optimization of process parameters. *Ultrasonics sonochemistry*, 54, 290-301.

13. Diallo, M. S., Fromer, N. A., and Jhon, M. S. (2013). Nanotechnology for sustainable development: retrospective and outlook. *In:* Nanotechnology for Sustainable Development (pp. 1-16). Springer, Cham.

14. Ersan, O.G. Apul, F. Perreault, T. Karanfil. (2017).Adsorption of organic contaminants by graphene nanosheets, a review. *Water Res.*, 126: 385-398

15. Fiyadh, S. S., AlSaadi, M. A., Jaafar, W. Z. B., AlOmar, M. K., Fayaed, S. S., Mohd, N. S. B., and El-Shafie, A. 2019. Review on heavy metal adsorption processes by carbon nanotubes. Journal of Cleaner Production.

16. Galante, M. T., Sotelo, P., Hossain, M. K., Vali, A., Raamann, A., Longo, C and Rajeshwar, K. (2019). Silver Oxide Based Semiconductors for Solar Fuels Production and Environmental Remediation: a Solid State Chemistry Approach. ChemElectroChem, 6(1), 87-96.

17. Gallo, A., Bianco, C., Tosco, T., Tiraferri, A., and Sethi, R. (2019). Synthesis of eco-compatible bimetallic silver/iron nanoparticles for water remediation and reactivity assessment on bromophenol blue. *Journal of cleaner production*, 211, 1367-1374.

18. Gharbi, N., Pressac, M., Hadchouel, M., Szwarc, H., Wilson, S. R., and Moussa, F. (2005). [60] Fullerene is a powerful antioxidant *in vivo* with no acute or subacute toxicity. *Nano Letters*, *5*(12), 2578-2585.

19. Goyal, N., Bulasara, V. K., and Barman, S. (2019). Tailoring of nanozeoliteNaX for enhanced removal of a phytoestrogen from its aqueous solutions. *Separation Science and Technology*, 54(2), 224-232.

20. Grassian, V. H. (2005). Environmental catalysis. CRC press.

21. Gupta, V. K., and Saleh, T. A. (2013). Sorption of pollutants by porous carbon, carbon nanotubes and fullerene-an overview. *Environmental science and pollution research*, *20*(5), 2828-2843.

22. Haider, S., Kamal, T., Khan, S. B., Omer, M., Haider, A., Khan, F. U., and Asiri, A. M. (2016). Natural polymers supported copper nanoparticles for pollutants degradation. *Applied Surface Science*, 387, 1154-1161.

23. Hakami, O., Zhang, Y., and Banks, C. J. (2012). Thiol-functionalised mesoporous silica-coated magnetite nanoparticles for high efficiency removal and recovery of Hg from water. *Water research*, 46(12), 3913-3922.

24. Hennebel, T., De Corte, S., Vanhaecke, L., Vanherck, K., Forrez, I., De Gusseme, B. and Boon, N. (2010). Removal of diatrizoate with catalytically active membranes incorporating microbially produced palladium nanoparticles. *Water research*, 44(5), 1498-1506.

25. Iijima S. 1991. Nature 354: 56.

26. Jung, J. H., Park, M., and Shinkai, S. (2010). Fabrication of silica nanotubes by using self-assembled gels and their applications in environmental and biological fields. *Chemical Society Reviews*, *39*(11), 4286-4302.

27. K.S. Novoselov, V.I. Falko, L. Colombo, P.R. Gellert, M.G. Schwab, K. Kim. 2012. A roadmap for graphene. Nat., 490 pp. 192-200.

28. Kabir, E., Kumar, V., Kim, K. H., Yip, A. C., and Sohn, J. R. (2018). Environmental impacts of nanomaterials. *Journal of environmental management*, *225*, 261-271.

29. Kanel, SR, RR Goswami, TP Clement, MO Barnett, and D Zhao. 2007. "Two dimensional transport characteristics of surface stabilized zero-valent iron nanoparticles in porous media." *Environmental science and technology* 42 (3): 896-900.

30. Ke, F., Qiu, L. G., Yuan, Y. P., Peng, F. M., Jiang, X., Xie, A. J., and Zhu, J. F. (2011). Thiol-functionalization of metal-organic framework by a facile coordination-based postsynthetic strategy and enhanced removal of Hg^{2+} from water. Journal of hazardous materials, 196, 36-43.

31. Khalafi, T., Buazar, F., and Ghanemi, K. (2019). phycosynthesis and enhanced photocatalytic Activity of Zinc oxide Nanoparticles toward organosulfur pollutants. Scientific reports, 9(1), 6866.

32. Kim, H. J., Lee, H., Lee, J. H., Choi, D. H., Jung, J. H., and Kim, J. S. (2011). Bisindole anchored mesoporous silica nanoparticles for cyanide sensing in aqueous media. *Chemical Communications*, 47(39), 10918-10920.

33. Lee, H. Y., Bae, D. R., Park, J. C., Song, H., Han, W. S., and Jung, J. H. (2009). A selective fluoroionophore based on BODIPY functionalized magnetic silica nanoparticles: removal of Pb2+ from human blood. *AngewandteChemie International Edition*, 48(7), 1239-1243.

34. Li, Shaolin, Wei Wang, Feipeng Liang, and Wei-xian Zhang. 2017. "Heavy metal removal using nanoscale zero-valent iron (nZVI): Theory and application." *Journal of hazardous materials* 322: 163-171.

35. Li, X., Tian, X., Yang, T., He, Y., Liu, W., Song, Y., and Liu, Z. (2019). Coal Liquefaction Residues Based Carbon Nanofibers Film Prepared by

Electrospinning: An Effective Approach to Coal Waste Management. *ACS Sustainable Chemistry and Engineering, 7*(6), 5742-5750.

36. Lin, Y. Z., Zhong, L. B., Dou, S., Shao, Z. D., Liu, Q., and Zheng, Y. M. (2019). Facile synthesis of electrospun carbon nanofiber/graphene oxide composite aerogels for high efficiency oils absorption. *Environment international, 128*, 37-45.
37. Liu, W. J., Tian, K., He, Y. R., Jiang, H., and Yu, H. Q. (2014). High-yield harvest of nanofibers/mesoporous carbon composite by pyrolysis of waste biomass and its application for high durability electrochemical energy storage. *Environmental science and technology, 48*(23), 13951-13959.
38. Liu, Y., He, L., Mustapha, A., Li, H., Hu, Z. Q., and Lin, M. (2009). Antibacterial activities of zinc oxide nanoparticles against Escherichia coli O157: H7. Journal *of applied microbiology*, 107(4), 1193-1201.
39. Ma, H., Huang, Y., Shen, M., Guo, R., Cao, X., and Shi, X. (2012). Enhanced dechlorination of trichloroethylene using electrospun polymer nanofibrous mats immobilized with iron/palladium bimetallic nanoparticles. Journal of *hazardous materials*, 211, 349-356.
40. Mahajan, P. G., Dige, N. C., Vanjare, B. D., Hong, S. K., and Lee, K. H. (2019). Gallotannin functionalized gold nanoparticles for rapid, selective and sensitive colorimetric detection of Ag+ in aqueous medium: Approach to eco-friendly water analysis. Sensors and Actuators B: Chemical, 281, 720-729.
41. Marcus, B. K., and Cormier, W. E. (1999). Going green with zeolites. Chem. *Eng. Prog, 95*(6), 47-53.
42. Markovic, Z., and Trajkovic, V. (2008). Biomedical potential of the reactive oxygen species generation and quenching by fullerenes (C60). Biomaterials, 29(26), 3561-3573.
43. Mauter, M. S., and Elimelech, M. (2008). Environmental applications of carbon-based nanomaterials. *Environmental science and technology, 42*(16), 5843-5859.
44. Morsi, R. E., Alsabagh, A. M., Nasr, S. A., and Zaki, M. M. (2017). Multifunctional nanocomposites of chitosan, silver nanoparticles, copper nanoparticles and carbon nanotubes for water treatment: antimicrobial characteristics. International journal of biological macromolecules, 97, 264-269.
45. Otari, S. V., and Yadav, H. M. (2019). Photocatalytic Active Silver Phosphate for Photoremediation of Organic Pollutants. Photocatalytic Functional Materials for Environmental Remediation, 163-189.
46. Pandey, N. K., Tiwari, K., Roy, A., Mishra, A., and Govindan, A. (2013). Ag Loaded WO3 Ceramic Nanomaterials: Characterization and Moisture Sensing Studies. *International Journal of Applied Ceramic Technology*, 10(1), 150-159.
47. Pant, B., Park, M., and Park, S. J. (2019). MoS_2/CdS/TiO2 ternary composite incorporated into carbon nanofibers for the removal of organic pollutants from water. *Inorganic Chemistry Communications, 102*, 113-119.

48. Parmar, A., Kaur, G., Kapil, S., Sharma, V., Choudhary, M. K., and Sharma, S. (2019). Novel biogenic silver nanoparticles as invigorated catalytic and antibacterial tool: A cleaner approach towards environmental remediation and combating bacterial invasion. Materials Chemistry and Physics, 238, 121861.

49. Piticescu, R. M., Piticescu, R. R., Taloi, D., and Badilita, V. (2003). Hydrothermal synthesis of ceramic nanomaterials for functional applications. Nanotechnology, 14(2), 312.

50. Puddu, V., Choi, H., Dionysiou, D. D., and Puma, G. L. (2010). TiO2 photocatalyst for indoor air remediation: Influence of crystallinity, crystal phase, and UV radiation intensity on trichloroethylene degradation. Applied Catalysis B: Environmental, 94(3-4), 211-218.

51. Qin, L., Huang, D., Xu, P., Zeng, G., Lai, C., Fu, Y. and Zhou, C. (2019). In-situ deposition of gold nanoparticles onto polydopamine-decorated g-C3N4 for highly efficient reduction of nitroaromatics in environmental water purification. Journal of colloid and interface science, 534, 357-369.

52. Rahimi, E., and Mohaghegh, N. (2016). Removal of toxic metal ions from Sungun acid rock drainage using mordenite zeolite, graphene nanosheets, and a novel metal–organic framework. Mine water and the environment, 35(1), 18-28.

53. Ribovski, L., dos Santos, F. A., Zucolotto, V., and Janegitz, B. C. (2019). Gold nanorods and poly (amido amine) dendrimer thin film for biosensing. Journal of Solid State Electrochemistry, 23(5), 1581-1591.

54. Sánchez, A., Morante-Zarcero, S., Pérez-Quintanilla, D., Sierra, I., and Del Hierro, I. (2010). Development of screen-printed carbon electrodes modified with functionalized mesoporous silica nanoparticles: Application to voltammetric stripping determination of Pb (II) in non-pretreated natural waters. Electrochimica Acta, 55(23), 6983-6990.

55. Santhosh, C.; Velmurugan, V.; Jacob, G.; Jeong, S.K.; Grace, A.N.; Bhatnagar, A. Role of nanomaterials in water treatment applications: A review. Chem. Eng. J. 2016, 306, 1116–1137.

56. Seifi, L., Torabian, A., Kazemian, H., Bidhendi, G. N., Azimi, A. A., and Charkhi, A. (2011). Adsorption of petroleum monoaromatics from aqueous solutions using granulated surface modified natural nanozeolites: systematic study of equilibrium isotherms. Water, Air, and Soil Pollution, 217(1-4), 611-625.

57. Shu, Z., Chen, Y., Zhou, J., Li, T., Yu, D., and Wang, Y. (2015). Nanoporous-walled silica and alumina nanotubes derived from halloysite: controllable preparation and their dye adsorption applications. Applied Clay Science, 112, 17-24.

58. Sivasubramanian, V. (2016). Nanotechnology in Environmental Applications. In Environmental Sustainability Using Green Technologies (pp. 75-116). CRC Press.

59. Su, C., Puls, R. W., Krug, T. A., Watling, M. T., O'Hara, S. K., Quinn, J. W., and Ruiz, N. E. (2020). Long-Term Performance Evaluation of Groundwater

Chlorinated Solvents Remediation Using Nanoscale Emulsified Zerovalent Iron at a Superfund Site. In Waste Management: Concepts, Methodologies, Tools, and Applications (pp. 1352-1371). IGI Global.

60. Tang, G.M. Zeng, J.L. Gong, J. Liang, P. Xu, C. Zhang, B.B. Huang. 2014. Impact of humic/fulvic acid on the removal of heavy metals from aqueous solutions using nanomaterials: a review. Sci. Total Environ., 468: 1014-1027.

61. Thangavel, S., Raghavan, N., and Venugopal, G. (2019). Magnetically Separable Iron Oxide Based Nanocomposite Photocatalytic Materials for Environmental Remediation. Photocatalytic Functional Materials for Environmental Remediation, 243-265.

62. Tolentino, M. S., Aquino, R. R., Tuazon, M. R. C., Basilia, B. A., Liana, M. J., and Cosico, J. A. M. C. (2019, October). Adsorptive removal of Ni2+ ions in wastewater using electrospun cellulose acetate/iron-modified nanozeolite nanostructured membrane. In IOP Conference Series: Earth and Environmental Science (Vol. 344, No. 1, p. 012044). IOP Publishing.

63. Tong, Z., Bischoff, M., Nies, L., Applegate, B., and Turco, R. F. (2007). Impact of fullerene (C60) on a soil microbial community. *Environmental science and technology*, *41*(8), 2985-2991.

64. Traykova, T., Aparicio, C., Ginebra, M. P., and Planell, J. A. (2006). Bioceramics as nanomaterials.

65. Tuo, Y., Liu, G., Zhou, J., Wang, A., Wang, J., Jin, R., and Lv, H. (2013). Microbial formation of palladium nanoparticles by Geobactersulfurreducens for chromate reduction. Bioresource technology, 133, 606-611.

66. Üzüm, Çaðrý, Talal Shahwan, Ahmet E Eroðlu, Keith R Hallam, Thomas B Scott, and Ingo Lieberwirth. 2009. "Synthesis and characterization of kaolinite-supported zero-valent iron nanoparticles and their application for the removal of aqueous Cu2+ and Co2+ ions." Applied Clay Science 43 (2): 172-181.

67. Walcarius, A., and Mercier, L. (2010). Mesoporous organosilica adsorbents: nanoengineered materials for removal of organic and inorganic pollutants. Journal of Materials Chemistry, 20(22), 4478-4511.

68. Wang, C., Kim, J., Kim, M., Lim, H., Zhang, M., You, J. and Yamauchi, Y. (2019). Nanoarchitectured metal–organic framework-derived hollow carbon nanofiber filters for advanced oxidation processes. *Journal of Materials Chemistry A*, *7*(22), 13743-13750.

69. Wang, Zifeng, Na Liu, Feng Feng, and Zhanfang Ma. 2015. "Synthesis of cadmium, lead and copper alginate nanobeads as immunosensing probes for the detection of AFP, CEA and PSA." Biosensors and Bioelectronics 70: 98-105.

70. Wen, M., Zhang, S., Dai, W., Li, G., and Zhang, D. (2015). In situ synthesis of Ti3+ self-doped mesoporous TiO2 as a durable photocatalyst for environmental remediation. Chinese Journal of Catalysis, 36(12), 2095-2102.

71. Wu, Y. Yang, H. Zhou, T. Ye, Z. Huang, R. Liu, Y. Kuang. Highly efficient removal of Cu(II) from aqueous solution by using graphene oxide Water. Air. Soil. Poll., 224 (2013), p. 1372.

72. Wu, Y., Pang, H., Yao, W., Wang, X., Yu, S., Yu, Z., and Wang, X. (2018). Synthesis of rod-like metal-organic framework (MOF-5) nanomaterial for efficient removal of U (VI): batch experiments and spectroscopy study. Science bulletin, 63(13), 831-839.

73. Xiao, S., Shen, M., Guo, R., Huang, Q., Wang, S., and Shi, X. (2010). Fabrication of multiwalled carbon nanotube-reinforced electrospun polymer nanofibers containing zero-valent iron nanoparticles for environmental applications. *Journal of Materials Chemistry, 20*(27), 5700-5708.

74. Xu, P., Chen, M., Zeng, G., Huang, D., Lai, C., Wang, Z., Yan, M., Huang, Z., Gong, X., Song, B. and Li, T. 2019. Effects of multi-walled carbon nanotubes on metal transformation and natural organic matters in riverine sediment. Journal of hazardous materials, 374, pp.459-468.

75. Xu, T., Zheng, F., Chen, Z., Ding, Y., Liang, Z., Liu, Y., and Fong, H. (2019). Halloysite nanotubes sponges with skeletons made of electrospun nanofibers as innovative dye adsorbent and catalyst support. Chemical Engineering Journal, 360, 280-288.

76. Y. Gao, X. Ren, J. Wu, T. Hayat, A. Alsaedi, C. Cheng, C. Chen. Graphene oxide interactions with co-existing heavy metal cations: adsorption, colloidal properties and joint toxicity Environ. Sci. Nano., 5 (2018), pp. 362-371.

77. Yan, H., Yang, H., Li, A., and Cheng, R. (2016). pH-tunable surface charge of chitosan/graphene oxide composite adsorbent for efficient removal of multiple pollutants from water. Chemical Engineering Journal, 284, 1397-1405.

78. Yoon, K., Hsiao, B. S., and Chu, B. (2008). Functional nanofibers for environmental applications. *Journal of Materials Chemistry, 18*(44), 5326-5334.

79. Yue, L. Zhang, H. Zhong. 2014. Carboxymethyl chitosan: a new water soluble binder for Si anode of Li-ion batteries. *J. Power Sources*, 247: 327-331.

80. Zain, N. M., Stapley, A. G. F., and Shama, G. (2014). Green synthesis of silver and copper nanoparticles using ascorbic acid and chitosan for antimicrobial applications. Carbohydrate polymers, 112, 195-202.

81. Zhao, S., Yuan, Y., Yu, Q., Niu, B., Liao, J., Guo, Z., and Wang, N. (2019). A Dual Surface Amidoximated Halloysite Nanotube for High Efficiency Economical Uranium Extraction from Seawater. Angewandte Chemie International Edition, 58(42), 14979-14985.

82. Zhou, Zhe, Chaomeng Dai, Xuefei Zhou, Jianfu Zhao, and Yalei Zhang. 2015. "The removal of antimony by novel NZVI-zeolite: the role of iron transformation." Water, Air, and Soil Pollution 226 (3): 76.

Chapter 27

Nanomembranes in Water Treatment

*V. Karthika, M. Praghadeesh and A. Lakshmanan**

Department of Nano Science and Technology,
Tamil Nadu Agricultural University, Coimbatore – 641 003, Tamil Nadu
**e-mail: microlaxman@gmail.com*

Introduction

Globally, around two billion people use fecal contaminated water for the drinking purpose (WHO, 2019). This single fact implies the need of development of cost effective novel technology for the proper treatment of drinking water. Contaminated water can transmit many diseases such as diarrhea, cholera, dysentery, typhoid, and polio and it is estimated to cause 485000 diarrheal deaths every year. Water bodies are polluted heavily with too many contaminants especially after the industrial revolution. Protection of water sources from the pollution from various sources and controlling is a challenge in the recent days. Detecting and detoxifying existing contaminants and preventing the forthcoming pollution are among the challenges. There are several technologies involved in the remediation of contaminated water bodies and nanotechnology plays vital role in water purification and decontamination.

Nanotechnology is the revolutionary science and art of manipulating matter at the atomic or molecular scale that has cut across such disciplines as chemistry, physics, biology, and engineering. Nanoparticles have very high surface area to volume ratio and reactivity, which successfully remove pollutants in the water. Nanosorbents, Nanocatalysts, Nanostructured catalytic membranes, Nanofilms, Bioactive nanoparticles and nanoparticle enhanced filtration are some of the nano based products developed for the water purification. This review focuses on the use of nanomaterials for the development of water purification.

Conventional Water Treatment Technologies

Most of the water treatment plants use a combination of coagulation, sedimentation, flocculation, filtration and disinfection to get potable water. In the other words physical treatment, chemical treatment and biological treatment are applied for the treatment of water. Physical treatment involves the processes like coagulation, sedimentation, flocculation and filtration. In the chemical treatment, chemicals are applied in the water to remove the contaminants or pathogens. For example, application of chlorine powder to remove the pathogenic microorganisms is a chemical treatment. In the biological treatment of water, elite microbes are engaged to remove or detoxify the contaminants in the water.

Nanotechnology in Water Purification

The above said conventional methods are inefficient to meet the specific guideline for the drinking water or produces adverse effects upon the usage (Abbaszadegan *et al.*, 2008). The common water purification method includes electrodialysis, electro-deionization, microfiltration, ultrafiltration and nanofiltration for the water and wastewater and the nanofiltration and reverse osmosis for the desalination and water reclamation and reuse (Hoek *et al.*, 2014). Potential advantages of membrane filtration includes continuous process without any hazardous chemicals, comparatively low-energy consumption, easy scale-up and hybridization with other processes, high process intensity (*i.e.*, small land area per unit volume of water processed) and highly automated process control. The major disadvantages in using membrane based filtration are short span of life, limited

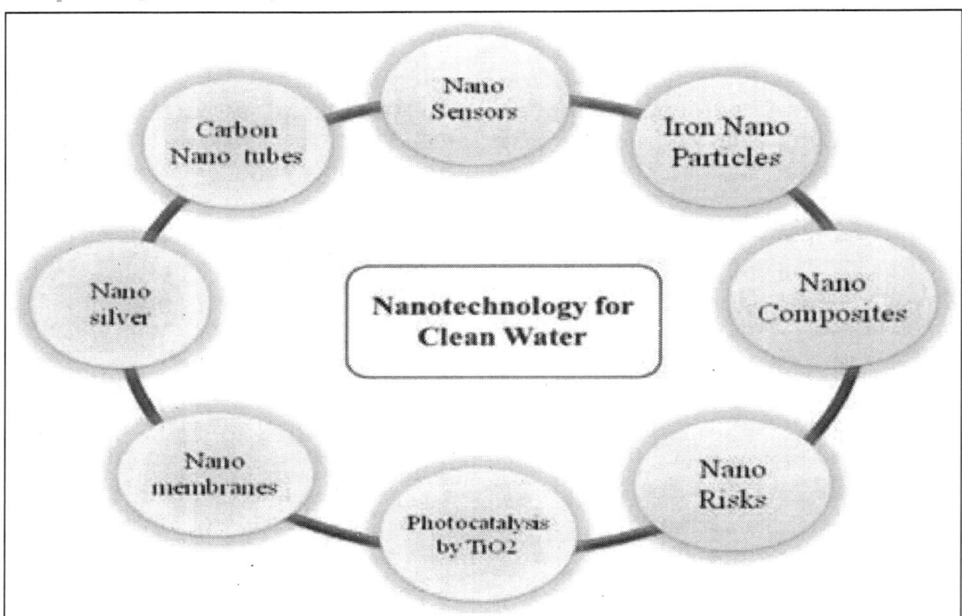

Figure 27.1. Summary of Nano-technologies Used for Water Purification.
Courtesy: Ahmed *et al.*, 2014.

selectivity towards chemicals and biological materials, concentration polarization, and fouling of membranes (Noble and Stern, 1995). Membranes are considered as one of the key components in the development of advanced methods of water purification. There is a demand and search for new materials for the membrane development. In this regard, Nanomaterials are considered as one of the emerging materials for the development of membranes for water purification.

Nanomembranes for Water Purification

Membrane is a selective barrier between two phases, which allows the transport of particular species which might be a liquid or gaseous phase. Membrane based separation process is a new technology that offers effective and efficient separation of two phases. Nanofiltration is a promising approach in water purification that immobilizes nano-materials on or within a solid matrix called "Nanomembrane". These membranes exhibit excellent separation performance, high chemical stability, thermal stability, mechanical stability, interfacial properties and anti-fouling ability depending on the nano-material selected. The nanomaterials based membranes used in the water treatment are zeolite-coated ceramic membranes, CNTs based membranes, polymer based biomimetic membranes, organic-inorganic nanocomposites based membranes and graphene-based membranes. Nanofiltration is widely used in many treatment processes like hard water softening, desalination of sea water and brackish water and removal of pathogens, micro-pollutants from contaminated water.

Mixed Matrix Membranes (MMM)

Mixed matrix membranes consist of continuous polymer phase in which the nanomaterials like zeolites, carbon nanotubes and carbon molecular sieve are incorporated (Figure 27.2).

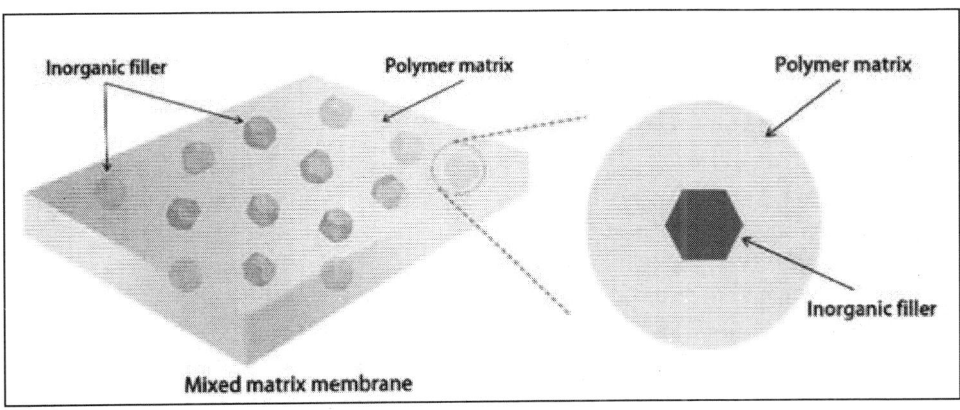

Figure 27.2. Schematic Diagram of Mixed Matrix Membrane with Dispersed Phase and Polymer Matrix

Courtesy: Reproduced from Maghami and Abdelrasoul, 2018.

Mixed matrix membranes provide stability of the ceramic material along with desired membrane characters like high permeability, resistance to foul, temperature and pH, chemical and mechanical stability. Mixed matrix membranes are divided into three broad categories based on the filler materials used in the dispersed phase. They are inorganic filler based MMMs, organic filler based MMMs and hybrid filler based MMMs. Inorganic filler based MMMs contains zeolites, silica, TiO_2 and silver nanoparticles in dispersed phase whereas the organic MMMs are confined with polymers like chitosan, cellulose acetate, clay minerals like halloysites and carbon nanoparticles. Hybrid MMMs possess both organic and inorganic nanoparticles as filler in their composition.

Inorganic Filler Based Mixed Matrix Membranes

Zeolite based Nanomembranes

Zeolites are naturally occurring aluminosilicate minerals with highly uniform subnanometer and nanometer scale crystalline structures. Zeolite membranes are intrinsically inert amorphous silicate, aluminosilicate, or aluminophosphate crystalline structures with extreme thermal and chemical stability formed via hydrothermal synthesis or layer-by-layer crystallization and dry gel conversion in the presence of a template-water vapor (Kazemimoghadam, 2010). Zeolite based nanomembranes are currently developing in the water purification due to the water permeability on the range of ultrafiltration with the solute selectivity of nano or RO membranes and the shape selective catalytic properties. Zeolite nano-particles impregnated polysulfone membranes enhanced water hydraulic permeability and metal exchange capacity of the membrane and the maximum sorption capacity of the hybrid membrane for the lead and nickel ions was measured as 682 and 122 mg/g respectively at the end of 60 min of filtration, under 1 bar of transmembrane pressure (Yurekli, 2016). Zeolites/chitosan nanocomposite membranes with 1.0 and 1.25 wt per cent zeolite designed for removal of arsenic showed the highest rejection (>94 per cent) for 1000 μg/L concentration of arsenic trioxide aqueous solution (Mukhopadhyay *et al.*, 2019).

Silica based Nanomembranes

Silica is eliciting a growing interest in the membrane development for the water treatment due to its various remarkable properties including high specific surface area, chemical inertness, good thermal and mechanical stability and low cost of manufacture. Silica is a reservoir of abundant functional groups in the surface, which serves as active sites of binding the pollutants and also that can be tailored for excellent selectivity towards specific pollutant. Monodisperse Stober silica increased the hydrophilicity membranes due to the high water affinity of nano-SiO_2, resulting in higher water permeability and the fouling resistance increased by 70 per cent (Lin *et al.*, 2016). It has tbeen proved that Synergistic effect of UV irradiation and thermal annealing improves the separation performance *i.e.* permeability and selectivity of the Polyethersulfone-nano silica membrane used in the water treatment (Kusworo, *et al.*, 2017). Thin-film nanocomposite (TFN) membrane containing porous MCM-41 silica nanoparticles (NPs) was prepared by the in situ interfacial polymerization (IP) process. By increasing concentration of MCM-41 NPs, hydrophilicity, roughness and zeta potential of TFN membranes increased with the permeate water flux increased from 28.5±1.0 to 46.6 ±1.1 L/m^2/h, while maintaining high rejections of NaCl (97.9±0.3 per cent) and Na_2SO_4 (98.5±0.2 per cent) (Yin *et al.*, 2012). Green silica based low cost ceramic hollow fibre membrane prepared from waste rice husk achieved a good mechanical strength (71.2/ MPa) and excellent porosity (50.2 per cent) hold promise for water filtration application (Hubadillah *et al.*, 2018). Polyamide thin-film nanocomposite membranes prepared on nanosilica-modified polysulfone showed that the wettability of substrates was improved with increasing concentrations of SiO_2 and also it increased the water permeability in the forward osmosis applications (Huang *et al.*, 2018).

Metals/Metal Oxide based Nanomembranes

Metal and metal oxide based nanomembranes are attracting more attention because of their superior thermal, chemical and mechanical stability, antifouling property and chemical inertness towards various pollutants. Inorganic nanofiltration consists of thin inorganic nanomembrane which are mixed oxides from aluminium (Al), zirconium (Zr) or titanium (Ti) over a porous ceramic support layer which provides good mechanical stability (Mottern *et al.*, 2007). Inorganic metal based membranes consumes low energy for the operation while maintain good water flux (Nangmenyi *et al.*, 2013). Cost effective Ag-doped fly ash/polyurethane spider-web-like nanofibrous membrane enhanced the absorption capacity to remove carcinogenic heavy metal arsenic and toxic organic dyes, and antibacterial properties with reduce bio-fouling for membrane filter application (Pant *et al.*, 2014). Multifunctional nanomembrane prepared with carbon nanofibers decorated by silver nanoparticles had enhanced photocatalytic activity attributed to the highly accessible surface areas, good light absorption capability, and high separation efficiency of photogenerated electron-hole pairs and exhibited excellent dye degradation and bacteria disinfection (Taha, 2015). Cellulose coated magnetite-silver nanoparticles antimicrobial properties that can be used against *E. coli* and it could be used effectively for water purification (Namdeo *et al.*, 2009).

Titania-γ-alumina multilayer nanomembranes showed different rejections for ions up to 25 per cent with complete separation of the microorganisms and higher than 10 bars of pressure, the permeability stayed constant due to the increase of flux (Shayesteh *et al.*, 2016). Plasma-induced poly (acrylic acid)-TiO_2 coated polyvinylidene fluoride membrane fabricated based on the mechanism of nano-TiO_2 immobilization had the permeation flux more than four times compared unmodified membrane, and the oil rejection rate was higher than 92 per cent (Chen *et al.*, 2019). Flexible and heterostructured silver/bismuth oxyiodide/titania nanofibrous membrane exhibited superior photocatalytic disinfection activity (7.5 log inactivation of *E. coli* within 1 h) under visible light illumination for high volume contaminated water and maintained good structural integrity and mechanical flexibility (Song *et al.*, 2019).

Zirconia membranes functionalized by various alkyl- and fluoroalkylsilanes showed increased wettability and decreased surface roughness and adhesion while removing the hazardous volatile organic compounds (VOCs) butanol, methyl-tert-butyl ether, and ethyl acetate (Kujawa *et al.*, 2019). Engineered amyloid fibril/ZrO_2 hybrid membranes exhibited superior selectivity for F^- against various competitive ions at both low and high (ca. 200 mg L–1) F– concentrations, the membrane efficiency exceeds 99.5 per cent removal, exceeding commercial ion exchange resins (IRA 900) by 180 times and outdoing the performance of most commercial carbon activated aluminum membranes (Zhang *et al.*, 2019). Modified vapor grown carbon nanofibers and zirconia in cellulose acetate blend membranes with improved antifouling and hydrophilic character showed 96 per cent, 99 per cent and 93 per cent direct dye rejection for Red, blue and orange respectively (Wasim *et al.*, 2019).

Alumina NF membranes exhibit a high mechanical, chemical stability and anti-corrosion and fouling resistance regarded as an important material in the development of nanofilters. Flux-enhanced α-alumina tight ultrafiltration membranes showed acid and alkali resistance at high temperatures with high permeability with high surface porosity to treat dye/salt waste water (Zou *et al.*, 2019).

Organic Filler based Mixed Matrix Membranes

CNTs based Nanomembranes

Carbon nanotubes are emerging as the novel materials for the synthesis of nano membranes with attractive features for water purification due to their large surface area, ease of functionalization, high aspect ratio and fast water transport (Lee *et al.*, 2016). CNTs can be used either as directly as filters or as a supporting matrial to improve the membrane performance. CNTs created remarkable revolution in the membrane development due to the improved antifouling behavior, strength, disinfection, rejection and permeability (Lee, 2019). Figure 27.3 describes Illustration of possible pathways for water transportation in a CNT/polymer blend membrane due to (1) hydrophobic effect enhanced transport, (2) nano-confinement enhanced flux, (3) ultrafast transport through the CNT pores, and (4) direct transport through the membrane matrix.

Polymer/carbon nanotubes mixed matrix membranes are specially designed to improve the separation performance of the membranes, such as water flux, ion selectivity, and antifouling characteristics (Farahani *et al.*, 2019). Superoleophobic CNTs/nanoparticles multifunctional hybrid membranes constructed separated separate oil-in-water emulsion and also catalytically decomposed different organic pollutants in water (Yan *et al.*, 2019).

Graphene based Nanomembranes

Graphene oxide (GO) membranes have been demonstrated for high salt rejection and water flux provides an alternate for the existing aromatic polyamide membranes in the water purification process. *etc.*hing holes into graphene sheets forms holey graphene membranes with improved permeability upto four times through the additional routes/holes generated for water transport. Ethylene glycol-modified graphene oxide membranes designed by a vacuum filtration method have interlayer

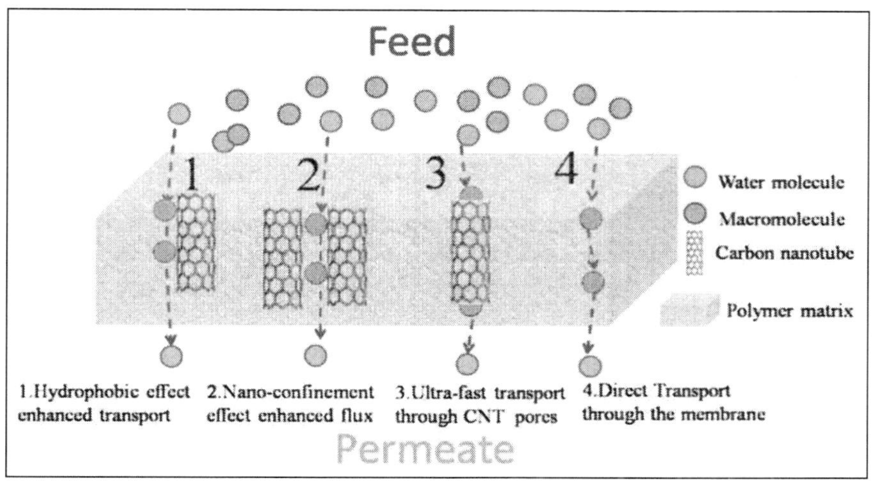

Figure 27.3. Illustration of Possible Pathways for Water Transportation in a CNT/Polymer Blend Membrane

Courtesy: Li *et al.*, 2014.

spacing of GO membranes 0.825 nm achieved high water flux (94.56 per cent) and high rejection of Rhodamine B (97.13 per cent) (Zhang *et al.*, 2019). Photocatalytic self-cleaning carbon nitride nanotube intercalated reduced graphene oxide membranes enlarged the rGO interlayer spacing for improved water permeance and also had synergetic removal more than 98 per cent of organic dye Rhodamine B (RhB) (Wei *et al.*, 2019). Graphene oxide/polyacrylamide composite membranes with interlayer spacing of 0.68 nm had the best comprehensive separation performance as well as the high water flux and higher percentage (95.43 per cent) of retention of rhodamine B compared to unmodified graphene oxide membrane (85.03 per cent) (Cheng *et al.*, 2019). The challenges in adopting graphene based membranes in the commercial sector are the limited availability of the materials and the availability of the cheaper alternatives for the graphene.

Chitosan based Nanomembranes

Chitosan-based membrane in the water treatment is is a fast-emerging field of interest to environmental scientists and research sector due to the reliablility, efficiency, and low-cost for removing the challenging persistent pollutants like heavy metals through the adsorption in the aqueous solutions. Metal oxides, nanocomposites, and polymer mixed chitosan membranes. Are some of the examples in the recent developments in chitosan-based membranes. Nanofiltration membranes prepared by laccase-triggered coatings of tea catechins/chitosan showed 99 per cent rejection of Na_2SO_4 with a high water permeation flux of 45 L m^{-2} h^{-1} under 0.6 MPa (Qiu *et al.*, 2016). Nano-hybrid membrane containing carbon nanotube based on chitosan modified with perfluorooctanesulfonyl fluoride exhibited super superhydrophobicity, high mechanical strength, low energy surface and excellent antibacterial property (Song *et al.*, 2015). Nano-graphene oxide assisted hydrotalcite/chitosan biocomposite displayed high mechanical strength and high sorption capacity of 42.64/ mg of Cr (VI)/g of composite within 50/ min contact time (Periyasamy *et al.*, 2019).

Cellulose Acetate based Nanomembranes

Cellulose acetate is the acetate ester of cellulose used in designing nanomembranes for water and wastewater treatment due to the excellent properties like mechanical strength, stability and most importantly degradable in nature. High flux nanocellulose-cellulose acetate nanocomposite membrane created hydrophilic nanochannels that reduced membrane resistance and facilitated the water transport while the BSA rejection was found to be 85 per cent (Li *et al.*, 2019). Biodegradable cellulose acetate ultrafiltration membranes incombination with graphene oxide and HKUST-1 (a metal–organic framework) exhibited higher hydrophilicity, antifouling property and highest water permeate flux of 183.51/ L/ m^{-2} h^{-1} (Yang *et al.*, 2019). Cellulose acetate added inorganic additives like $Ni(NO_3)_2.6H_2O$ showed the separation efficiency of 75 per cent for 200 ppm sodium alginate at 3to 10 bar and 89 per cent salt rejection at 7 bar and the the flux of distilled water was about 50 LMH at a maximum pressure of 8 bar (Kim *et al.*, 2019). Phosphonate cellulose acetate membranes used for the adsorptive removal of cationic methylene blue at different concentrations and pH values at 25°C. Novel polymer additive synthesized from cellulose acetate improved the surface hydrophilicity, polymer compatibility, mechanical strength.and oleophobicity with 92.5 per cent of oil rejections and 90 per cent relative flux recovery (Shen *et al.*, 2019).

Biomimetic Membranes

Biomimetic membranes are also called bio-inspired membranes are the synthetic membranes imitate the bimolecules like aquaporins, vesicles, liposomes, polymerosomes in their structure. The problems associated with the conventional membranes such as limited permeability, low selectivity, and excessive energy consumption should be easily tackled by the biomimetic approach. Nanoparticles are also employed in fabricating the biomimetic membranes. Bio-inspired superhydrophobic and superoleophilic nanofibrous membranes made with carbon black nanoparticles with hydrophobic nanofibrous support had high flux and selectivity for separation of oil from water (Shahabadi and Brant, 2019). Aquaporin-inspired graphene-based nano membrane simulated showed that water flow through the channel decreases by increasing the charge, while the ion rejection rate of the channel is enhanced (Lohrasebi and Koslowski, 2019).

Nanomembranes in Desalination of Seawater

Hybrid zeolite-polyamide thin film nanocomposite reverse osmosis membranes incorporated with Linde type A type zeolite molecular sieve nanocrystals in the interfacial polymerization reaction exhibited higher water permeability and salt rejection and also the hydrophilic and smooth interfaces, which appeared to inhibit adhesion of bacteria cells onto the membranes (Lind *et al.*, 2009). Polyamide and nano-NaX zeolites thin film nanocomposite membranes have higher thermal stability, more water permeability and lower salt rejection in a reverse osmosis process (Fathizadeh *et al.*, 2011). Zeolite nano-membrane for seawater desalination studied using comprehensive molecular dynamics simulation showed that there is relationship between the permeability and thickness of the zeolite membrane and the pressure drop (Liu and Chen, 2013). NaY zeolite nanoparticles dispersed in the dense nodular polyamide on the polysulfone high-flux reverse osmosis membranes for brackish water desalination permitted the water flux of 0.95 to 1.78 $m^3/m^2/day$ along with 98.8 per cent salt rejection (Dong *et al.*, 2015). Nano zeolite-Y incorporated poly (vinyl) alcohol- networked cellulose membrane exhibited a smooth surface and registered high salt rejection of 99.9 per cent for 25000 mg/L NaCl feed (Anis *et al.*, 2019). Figure 27.4 represents the cross-sectional image of zeolite nanocomposite reverse osmosis membrane for water purification through desalination process.

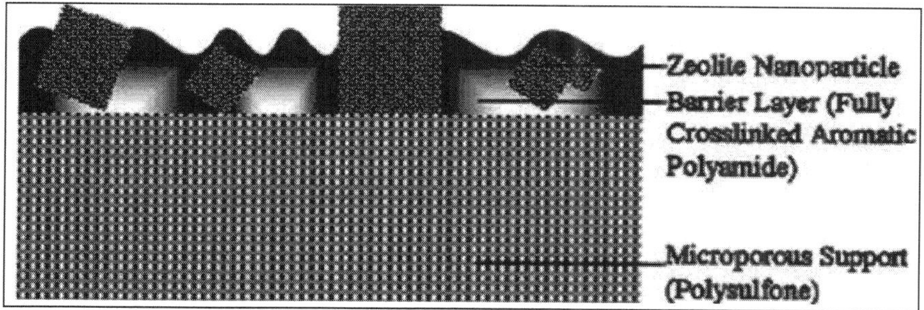

Figure 27.4. Schematic Diagram of Cross Section of Zeolite Nanocomposite Membrane.
Courtsey: Reproduced from Maghami and Abdelrasoul, 2018.

Graphene is one of the most promising capacitive deionization electrode materials with good salt removal ability due to its high specific surface area, good electrical conductivity, and controllable morphology structure. Reduced graphene oxide has tremendous application in desalination through the interlayer spacing (0.34-0.37/ nm) in the structure that confine the passage of sodium ions and the hydroxyl and carboxyl functional groups are largely responsible for the formation of uniform rGO membranes under the treatment of 160°C for 2 hours (Huang *et al.*, 2019). Cellulose acetate/polyethylene glycol membranes in the presence of candle soot carbon nanoparticles contributed good percentage of salt rejection with a slight reduction in the water flux behavior could be used for water desalination.

Capacitive Deionization

Capacitive deionization (CDI) based one electrosorption is an environmental friendly desalination technique compared to other conventional technologies of membrane separation and thermal distillation. CDI is an electrochemically controlled method based on electric double-layer capacitors. The charged ions can be adsorbed onto the surface or the pores of porous electrodes to form electrostatic double-layer capacitors when an external voltage is applied; ions can be released from the electrodes when the charge is removed or reversed (Figure 27.5).

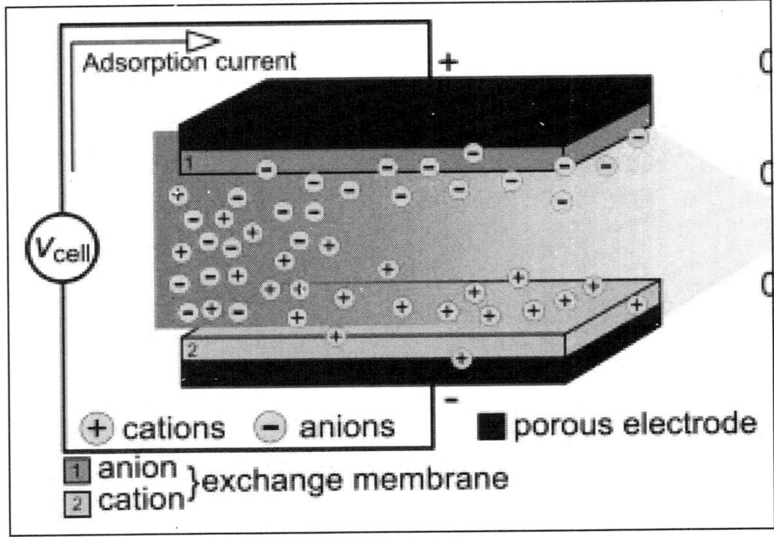

Figure 27.5. Schematic Diagram of Capacitative Deionization
Courtsey: Zhao *et al.*, 2013.

Capacitive deionization of saline water using nanocomposites is a emerging trend in the desalination process. Nanoparticles are embedded in to the electrodes or the nanocomposites used as electrodes in the desalination using capacitive deionization. Graphene/metal oxide nanoparticle hybrid, Ag-Cu bimetallic nanoparticle decorated graphene nanocomposites (Vengatesan *et al.*, 2019), Zirconia nanoparticles incorporated carbon electrodes (Yasin *et al.*, 2019) are some of the examples for the use of nanotechnology in the capacitive deionization.

Conclusion

In this chapter different types of nanomembranes and the recent updates in the field of development membranes for nanofiltration are discussed. This provides information about the possible nanomaterials available for the water purification/desalination process to replace the existing membranes with limitations like high cost, short life span, low salt rejection and water flux. Nanomaterials used in the membrane development are also having drawbacks like high cost, lesser selectivity, and lower water flux and lacking stability. However these problems could be solved with further research such as using nanomaterials in combination for the desired characteristics and improving the performance of the membrane. Environmental fate, chemical interaction and toxicity of a nanomaterials used in membranes should be further studied to avoid cross contamination and side effects.

REFERENCES

1. Abbaszadegan, M., Monteiro, P., Nwachuku, N., Alum, A., and Ryu, H. (2008). Removal of adenovirus, calicivirus, and bacteriophages by conventional

drinking water treatment. Journal of Environmental Science and Health, Part A, 43(2), 171-177.

2. Anis, S. F., Hashaikeh, R., and Hilal, N. (2019). Flux and salt rejection enhancement of polyvinyl (alcohol) reverse osmosis membranes using nano-zeolite. *Desalination, 470*, 114104.

3. Chen, X., Huang, G., An, C., Feng, R., Yao, Y., Zhao, S. and Wu, Y. (2019). Plasma-induced poly (acrylic acid)-TiO2 coated polyvinylidene fluoride membrane for produced water treatment: Synchrotron X-Ray, optimization, and insight studies. Journal of Cleaner Production, 227, 772-783.

4. Cheng, M. M., Huang, L. J., Wang, Y. X., Zhao, Y. C., Tang, J. G., Wang, Y. and Wickramasinghe, S. R. (2019). Synthesis of graphene oxide/polyacrylamide composite membranes for organic dyes/water separation in water purification. Journal of materials science, 54(1), 252-264.

5. Dong, H., Zhao, L., Zhang, L., Chen, H., Gao, C., and Ho, W. W. (2015). High-flux reverse osmosis membranes incorporated with NaY zeolite nanoparticles for brackish water desalination. Journal of Membrane Science, 476, 373-383.

6. Farahani, M. H. D. A., and Vatanpour, V. (2019). Polymer/carbon nanotubes mixed matrix membranes for water purification. *In:* Nanoscale Materials in Water Purification (pp. 87-110). Elsevier.

7. Fathizadeh, M., Aroujalian, A., and Raisi, A. (2011). Effect of added NaX nano-zeolite into polyamide as a top thin layer of membrane on water flux and salt rejection in a reverse osmosis process. Journal of membrane science, 375(1-2), 88-95.

8. Hoek, E. M., Pendergast, M. T. M., and Ghosh, A. K. (2014). Nanotechnology-based membranes for water purification. In Nanotechnology applications for clean water (pp. 133-154). William Andrew Publishing.

9. Huang, H. H., Joshi, R. K., De Silva, K. K. H., Badam, R., and Yoshimura, M. (2019). Fabrication of reduced graphene oxide membranes for water desalination. Journal of membrane science, 572, 12-19.

10. Huang, Y., Jin, H., Yu, P., and Luo, Y. (2016). Polyamide thin-film composite membrane based on nano-silica modified polysulfone microporous support layer for forward osmosis. Desalination and Water Treatment, 57(43), 20177-20187.

11. Hubadillah, S. K., Othman, M. H. D., Ismail, A. F., Rahman, M. A., Jaafar, J., Iwamoto, Y and Yusop, M. Z. M. (2018). Fabrication of low cost, green silica based ceramic hollow fibre membrane prepared from waste rice husk for water filtration application. Ceramics International, 44(9), 10498-10509.

12. Kim, H. Y., Cho, Y., and Kang, S. W. (2019). Porous cellulose acetate membranes prepared by water pressure-assisted process for water-treatment. Journal of Industrial and Engineering Chemistry.

13. Kujawa, J., Kujawski, W., Cyganiuk, A., Dumée, L. F., and Al-Gharabli, S. (2019). Upgrading of zirconia membrane performance in removal of hazardous VOCs

from water by surface functionalization. Chemical Engineering Journal, 374, 155-169.

14. Kusworo, T. D., Utomo, D. P., and Aryanti, N. (2017). Synergistic effect of UV irradiation and thermal annealing to develop high performance polyethersulfone-nano silica membrane for produced water treatment. Journal of environmental chemical engineering, 5(4), 3290-3301.
15. Lee, J. (2019). Carbon Nanotube-Based Membranes for Water Purification. In Nanoscale Materials in Water Purification (pp. 309-331). Elsevier.
16. Lee, S. Jeong, Z. Liu Progress and challenges of carbon nanotube membrane in water treatment Crit. Rev. Environ. Sci. Technol., 46 (2016), pp. 999-1046.
17. Li, N., Zheng, J., Hadi, P., Yang, M., Huang, X., Ma, H. and Hsiao, B. S. (2019). Synthesis and Characterization of a High Flux Nanocellulose–Cellulose Acetate Nanocomposite Membrane. Membranes, 9(6), 70.
18. Lin, J., Ye, W., Zhong, K., Shen, J., Jullok, N., Sotto, A., and Van der Bruggen, B. (2016). Enhancement of polyethersulfone (PES) membrane doped by monodisperse Stöber silica for water treatment. Chemical Engineering and Processing: Process Intensification, 107, 194-205.
19. Lind, M. L., Jeong, B. H., Subramani, A., Huang, X., and Hoek, E. M. (2009). Effect of mobile cation on zeolite-polyamide thin film nanocomposite membranes. Journal of Materials Research, 24(5), 1624-1631.
20. Liu, Y., and Chen, X. (2013). High permeability and salt rejection reverse osmosis by a zeolite nano-membrane. Physical Chemistry Chemical Physics, 15(18), 6817-6824.
21. Lohrasebi, A., and Koslowski, T. (2019). Modeling water purification by an aquaporin-inspired graphene-based nano-channel. Journal of molecular modeling, 25(9), 280.
22. M. Kazemimoghadam, New nanopore zeolite membranes for water treatment, Desalination 251 (2010) 176180.
23. Maghami, M., and Abdelrasoul, A. (2018). Zeolites-Mixed-Matrix Nanofiltration Membranes for the Next Generation of Water Purification. In Nanofiltration. IntechOpen.
24. Mottern, M.L., Shqau, K., Shi, J.Y., Yu, D. and Verweij, H. (2007) 'Thin supported inorganic membranes for energy-related gas and water purification', International Journal of Hydrogen Energy, Vol. 32, No. 16, pp.3713–3723.
25. Mukhopadhyay, M., Lakhotia, S. R., Ghosh, A. K., and Bindal, R. C. (2019). Removal of arsenic from aqueous media using zeolite/chitosan nanocomposite membrane. Separation Science and Technology, 54(2), 282-288.
26. Namdeo, M., Hintz, W., and Tomas, J. (2009). Synthesis of Cellulose Coated Magnetite–Silver Nanoparticles (CCMS) and Their Application in Water Purification. Journal of Bionanoscience, 3(2), 105-109.

27. Nangmenyi, G. N., Lin, X., Lee, E., and Yuan, C. (2013). Experimental investigation of inorganic nanofiltration membranes prepared by atomic layer deposition and sol-gel methods for saltless water softening. International Journal of Environmental Technology and Management, 16(1-2), 21-33.

28. Noble, R. D., and Stern, S. A. (Eds.). (1995). Membrane separations technology: principles and applications (Vol. 2). Elsevier.

29. Pant, H. R., Kim, H. J., Joshi, M. K., Pant, B., Park, C. H., Kim, J. I and Kim, C. S. (2014). One-step fabrication of multifunctional composite polyurethane spider-web-like nanofibrous membrane for water purification. Journal of hazardous materials, 264, 25-33.

30. Periyasamy, S., Manivasakan, P., Jeyaprabha, C., Meenakshi, S., and Viswanathan, N. (2019). Fabrication of nano-graphene oxide assisted hydrotalcite/chitosan biocomposite: An efficient adsorbent for chromium removal from water. International journal of biological macromolecules, 132, 1068-1078.

31. Qiu, W. Z., Zhong, Q. Z., Du, Y., Lv, Y., and Xu, Z. K. (2016). Enzyme-triggered coatings of tea catechins/chitosan for nanofiltration membranes with high performance. Green Chemistry, 18(23), 6205-6208.

32. S. Li, G. Liao, Z. Liu, Y. Pan, Q. Wu, Y. Weng, X. Zhang, Z. Yang, O.K.C. Tsui. 2014. Enhanced water flux in vertically aligned carbon nanotube arrays and polyethersulfone composite membranes J. Mater. Chem. A., 2: 12171-12176,

33. Shahabadi, S. M. S., and Brant, J. A. (2019). Bio-inspired superhydrophobic and superoleophilic nanofibrous membranes for non-aqueous solvent and oil separation from water. Separation and Purification Technology, 210, 587-599.

34. Shayesteh, M., Samimi, A., Shafiee Afarani, M., and Khorram, M. (2016). Synthesis of titania–γ-alumina multilayer nanomembranes on performance-improved alumina supports for wastewater treatment. Desalination and Water Treatment, 57(20), 9115-9122.

35. Shen, S. S., Chen, H., Wang, R. H., Ji, W., Zhang, Y., and Bai, R. (2019). Preparation of antifouling cellulose acetate membranes with good hydrophilic and oleophobic surface properties. Materials Letters, 252, 1-4.

36. Song, J., Yu, J., Sun, G., Si, Y., and Ding, B. (2019). Visible-light-driven, hierarchically heterostructured, and flexible silver/bismuth oxyiodide/titania nanofibrous membranes for highly efficient water disinfection. Journal of colloid and interface science, 555, 636-646.

37. Song, K., Gao, A., Cheng, X., and Xie, K. (2015). Preparation of the superhydrophobic nano-hybrid membrane containing carbon nanotube based on chitosan and its antibacterial activity. Carbohydrate polymers, 130, 381-387.

38. Taha, A. A. (2015). Direct synthesis of mesostructured carbon nanofibers decorated with silver-nanoparticles as a multifunctional membrane for water treatment. Advances in Natural Sciences: Nanoscience and Nanotechnology, 6(4), 045003.

39. Vengatesan, M. R., Darawsheh, I. F. F., Govindan, B., Alhseinat, E., and Banat, F. (2019). Ag-Cu bimetallic nanoparticle decorated graphene nanocomposite as an effective anode material for hybrid capacitive deionization (HCDI) system. Electrochimica Acta, 297, 1052-1062.

40. Wasim, M., Sabir, A., Shafiq, M., and Khan, R. U. (2019). Fractionation of direct dyes using modified vapor grown carbon nanofibers and zirconia in cellulose acetate blend membranes. Science of The Total Environment, 677, 194-204.

41. Wei, Y., Zhu, Y., and Jiang, Y. (2019). Photocatalytic self-cleaning carbon nitride nanotube intercalated reduced graphene oxide membranes for enhanced water purification. Chemical Engineering Journal, 356, 915-925.

42. World Health Oragnisation. Fact sheet. June 2019.

43. Yan, L., Zhang, G., Zhang, L., Zhang, W., Gu, J., Huang, Y and Chen, T. (2019). Robust construction of underwater superoleophobic CNTs/nanoparticles multifunctional hybrid membranes via interception effect for oily wastewater purification. Journal of membrane science, 569, 32-40.

44. Yang, S., Zou, Q., Wang, T., and Zhang, L. (2019). Effects of GO and MOF@ GO on the permeation and antifouling properties of cellulose acetate ultrafiltration membrane. Journal of membrane science, 569, 48-59.

45. Yasin, A. S., Mohamed, I. M., Amen, M. T., Barakat, N. A., Park, C. H., and Kim, C. S. (2019). Incorporating zirconia nanoparticles into activated carbon as electrode material for capacitive deionization. Journal of Alloys and Compounds, 772, 1079-1087.

46. Yin, J., Kim, E. S., Yang, J., and Deng, B. (2012). Fabrication of a novel thin-film nanocomposite (TFN) membrane containing MCM-41 silica nanoparticles (NPs) for water purification. Journal of membrane science, 423, 238-246.

47. Yurekli, Y. (2016). Removal of heavy metals in wastewater by using zeolite nano-particles impregnated polysulfone membranes. Journal of hazardous materials, 309, 53-64.

48. Zhang, Q., Bolisetty, S., Cao, Y., Handschin, S., Adamcik, J., Peng, Q., and Mezzenga, R. (2019). Selective and efficient removal of fluoride from water: in situ engineered amyloid fibril/ZrO2 hybrid membranes. Angewandte Chemie International Edition, 58(18), 6012-6016.

49. Zhang, Y., Huang, L. J., Wang, Y. X., Tang, J. G., Wang, Y., Cheng, M. M. and Hedayati, M. (2019). The preparation and study of ethylene glycol-modified graphene oxide membranes for water purification. Polymers, 11(2), 188.

50. Zou, D., Chen, X., Qiu, M., Drioli, E., and Fan, Y. (2019). Flux-enhanced α-alumina tight ultrafiltration membranes for effective treatment of dye/salt wastewater at high temperatures. Separation and Purification Technology, 215, 143-154.

Chapter 28

Harnessing Nanotechnology to Manage Greenhouse Gases

A. Lakshmanan[1], R. Gowtham[2], L. Sivashankari[1] and V. Geethalakshmi[3]

[1]Department of Nano Science and Technology, [2]Agro-Climatic Research Center
[3]Director (Crop Management)
Tamil Nadu Agricultural University, Coimbatore – 641 003, Tamil Nadu

Climate Change and Greenhouse Gases

Climate change due to increased atmospheric concentration of greenhouse gases (GHGs) is a cause for concern worldwide. Agriculture impacts climate change by the way of releasing greenhouse gases like methane and nitrous oxide. Irrigated rice farming is an important anthropogenic source of CH_4 and nitrous oxide and anaerobic decomposition of organic material in flooded rice fields produces methane (CH_4), which escapes to the atmosphere primarily by diffusive transport through the rice plants during the growing season. The Inter-governmental Panel on Climate Change (IPCC, 1996) estimated the global emission rate of methane from paddy fields at 60 Tg/yr, with a range of 20 to 100 Tg/yr. This is about 5-20 per cent of the total emission from all anthropogenic sources. Nitrous oxide is emitted when nitrogen is added to the soil through the use of synthetic fertilizers. Nitrous oxide is also emitted during the breakdown of nitrogen in livestock manure and urine, which contributed to 5 per cent of N_2O emissions. As efforts to mitigate climate change increase, there is a need to identify cost effective ways to avoid emissions of Green House Gases (GHGs) from agro eco systems. Nanotechnology has become a rapidly growing field with potential applications ranging from advanced application to daily used products. Nanotechnology has been exploited widely in engineering, health, electronics and material sciences and agricultural scientists have began to use it as a tool to improve the input use efficiencies by integrating nanotechnological approaches in conventional production system. If Indian agriculture is to attain its broad national goal of sustainable agriculture growth of over 4 per cent, it is important that the nanotechnology research is extended to the agricultural total production – consumption system, that is, across the entire agricultural value chain (Subramanian, Paulraj, and Natarajan, 2008). This chapter summarizes the use of various nano particles and nano systems in mitigating climate change by minimizing the emission of greenhouse gases, in particular methane and carbon dioxide, from agricultural eco systems and other sources.

Nano Clays and Nano Zeolites as Adsorbents of Methane from Rice Soil

Methane is a substantial driver of global climate change, contributing 30 per cent of current net global warming and is emitted at a wide range of concentrations from a variety of sources, including natural gas systems, livestock, landfills, coal mining, manure management, wastewater treatment, rice cultivation and a few combustion processes. Methane is far more potent as a greenhouse gas than CO_2 and its capture poses a challenge that can only be addressed through extensive material screening and ingenious molecular-level designs. Methane is generally produced in soil-water interface in rice fields and later released in the environment through ebullition, diffusion and plant mediated transfer. The water management system under which rice is grown is one of the most important factors affecting CH_4 emissions. Upland rice fields are not flooded, and therefore are not believed to produce CH_4. Certain nanomaterials such as nano clays and nano zeolites can adsorb methane and helps in oxidation of methane. Nanoclays are nanoparticles of layered mineral silicates. Depending on chemical composition and nanoparticle morphology, nanoclays are organized into several classes such as montmorillonite, bentonite, kaolinite, hectorite and halloysite and generally affect the mechanical properties of the soil in addition to some modifications in soil hydraulic conductivities. It depends on soil chemistry, soil physics or soil mechanics.

Among the potential adsorbent nanoporous candidates, zeolites are very promising due to their environmentally friendly character and their strong affinity for carbon dioxide and methane gases. These materials also offer attractive choices as sorbents because they have a number of parameters that are easily tuned in the laboratory: pore size, Si/Al ratio, and choice of extra-framework cation. Zeolite membranes in combination with noble metals can also be used as catalysts for reforming carbon dioxide into fuels. Faujasite zeolites (*e.g.* NaY) are considered promising environmental catalysts because of their cation exchange capacity and acid–base properties, and nanocrystalline NaY zeolites have shown enhanced performance for the selective catalytic reduction of NOx (NOx= NO + NO_2) with urea (Yin *et al.*, 2018).

The nano zeolite crystal consists of cages linked by double six membered rings, forming columns in the C-direction. Connection of these columns give rise to 12-membered rings with a free diameter varying from 0.71 nm to 1.26 nm (Agger, Meza, Cundy, Plaisted, and Anderson, 2005). Significant change in properties of zeolite nanocrystals can be observed in comparison with conventional micro sized crystals. The decrease in particle size causes a strong increase in the surface to volume ratios and enhances the catalytic reactions.

Nano zeolites exhibit a molecular sieving effect for lighter hydrocarbons; hence, the hydrocarbon molecules with sizes larger than the pore size are mainly adsorbed on the external surface. Accordingly, an increase in the external surface area is effective for increasing catalytic activity. Zeolites with crystal sizes less than 100 nm have a higher adsorption capacity (Song *et al.*, 2004).

Nano zeolite can play a role as hosts for functional molecules and green house gases. Adsorption of gases on zeolites is mainly determined by two main factors, structure and physiochemical properties of adsorbent. The pore size, which depends on the structure of zeolite, determines the accessibility of methane molecules into the zeolite framework. However, in certain cases, it can adapt the aperture upon adsorption because of the flexibility of the framework. The degree of flexibility is a function of the framework structures in the presence of extra-framework cations and molecules (Coker *et al.*, 1998). The availability of internal space volume is another interesting characteristic of zeolites for separation/purification applications.

Nano clays and nano zeolite increase the water use efficiency as well as alter CH_4 producing (Methanogens) and consuming (Methanotrophs) microbes, and thereby produce complex effects on CH_4 emissions. Nano clay and nano zeolite usage may be used as a strategy to achieve the water use efficiency and methane emission reduction in the paddy field. The Nano clay and Nano zeolite applied field generally requires less amount of water than conventional irrigation system, so the quantities of CH_4 released from fields are believed to be significantly less.

Application of nano clay and nano zeolite minimized the methane flux of experimental rice soil significantly. The nano clay applied soil @ 50 g m^{-2} of soil recorded low CH_4 flux of 13.4 mg m^{-2} day^{-1} followed by nano zeolite application @ 50 g m^{-2} of soil (13.8 mg m^{-2} day^{-1}). Structure and physiochemical properties of zeolites influence methane adsorption. In terms of structure, zeolite X and Y with faujasite structure have pore openings that permit the diffusion of methane. With open cage type framework, zeolite X and Y are more capable to adsorb methane in rice ecosystem. The study results indicate that nano clay and nano zeolite enhance the water holding capacity, rice crop growth and yield in addition to minimizing the emission of methane from rice soils. Due to the combined effect of growth enhancement, moisture retention and lesser methane flux, nano clay and nano zeolite might be potential and eco friendly ingredients in rice cultivation (Mohanraj, Lakshmanan, and Subramanian, 2017; Ragaventhiran, 2015).

New Functionalized Nanomaterials for CO_2 Capture

Carbon dioxide is a minor but very important component of the atmosphere, and is released through natural processes such as respiration and volcanic eruptions and through human activities such as deforestation, land use changes, and burning fossil fuels. Humans have increased atmospheric CO_2 concentration by more than a three since the Industrial Revolution began. This is the most important long-lived "forcing" of climate change as CO_2 is one among the prime green house gases.

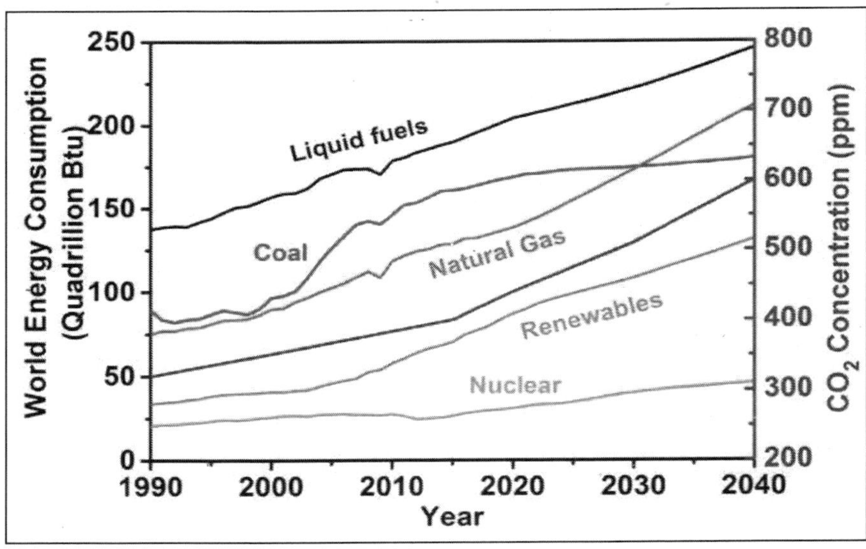

Figure 28.1: Global Energy Consumption by Energy Sources and Resultant Changes in Atmospheric CO_2 Concentration.
Courtsey: U.S. Department of Energy, 2016.

Most methods of fighting climate change are about reducing greenhouse gas emissions by inventing cleaner power plants, engineering greener cars. But many researchers focus on drawing gases from the atmosphere once they have already been released. The concept of "carbon dioxide capture" has been controversial, often dismissed as impractical or inadequate. Yet as global efforts to reduce emissions have proven difficult and sometimes disappointing, the approach seems increasingly alluring. Mitigating rising CO_2 levels is of prime importance. In a new development, scientists at the Tata Institute of Fundamental Research, Mumbai, have developed a novel design of CO_2 sorbents that show superior CO_2 capture capacity and stability over conventional materials. The immobilization of functional amines on a porous solid support can result in stable and efficient CO_2 sorbent materials compared to similar liquid sorbents. A critical disadvantage, however, is a drastic decrease in the textural properties of these supports (their surface area and pore volume), leading to a decrease in the CO_2 capture capability. To overcome this challenge, novel functionalized nanomaterials have been designed that allows higher amine loading with a minimal decrease in surface area (Singh and Polshettiwar, 2016).

Sarker *et al.* (2018) compared the effectiveness of different nanoparticles in minimizing carbon dioxide, methane, and hydrogen sulfide emissions from liquid dairy manure. The three nanoparticles studied were zinc silica nanogel liquid, copper silica nanogel liquid, and n-acetylcystein liquid coated zinc oxide quantum dots. The researchers found that all three types of nanoparticles significantly reduced gas production by over 90 per cent. Additionally, the scientists also found that there were less bacteria in nanoparticle treatments indicating that the nanoparticles might be inhibiting the microorganisms that produce the greenhouse gases.

A new invention, from scientists at University of California, Berkeley, offers a novel take on carbon capture. The researchers have created a nanomaterial that destroys carbon dioxide by splitting it into oxygen and carbon monoxide. Scientists have long tried to get rid of carbon dioxide by splitting its molecule. These splitting attempts can be energy-intensive, which defeats the environmental purpose. So researchers have used various catalysts to speed up the reaction, reducing the amount of electricity needed to split the molecules. Many scientists have focused on porphyrins, ring-shaped organic molecules, to make these reactions happen. Though porphyrins can have various atoms at their centers, the ones used for this purpose are cobalt porphyrins, which are especially catalytically active. When these porphyrins are added to a solution with two electrodes, an electrolyte and some dissolved carbon dioxide, the porphyrins are attracted to the electrolyte. This causes the electrons to move to the carbon dioxide, splitting it into carbon monoxide and oxygen. But this approach has not been perfect. The porphyrins clump together and lose effectiveness over time, and the solutions used to make the process happen are environmentally questionable themselves. The Berkeley researchers seem to have found a new way to deal with this by creating a porous nanomaterial linking porphyrins together into a mesh-like substance. This is called a covalent organic framework (COF). The carbon dioxide percolates through the COF, splitting into carbon monoxide and oxygen with very little added energy. It works about 60 times more efficiently than splitting the carbon dioxide using free-floating porphyrins (Lin *et al.*, 2015).

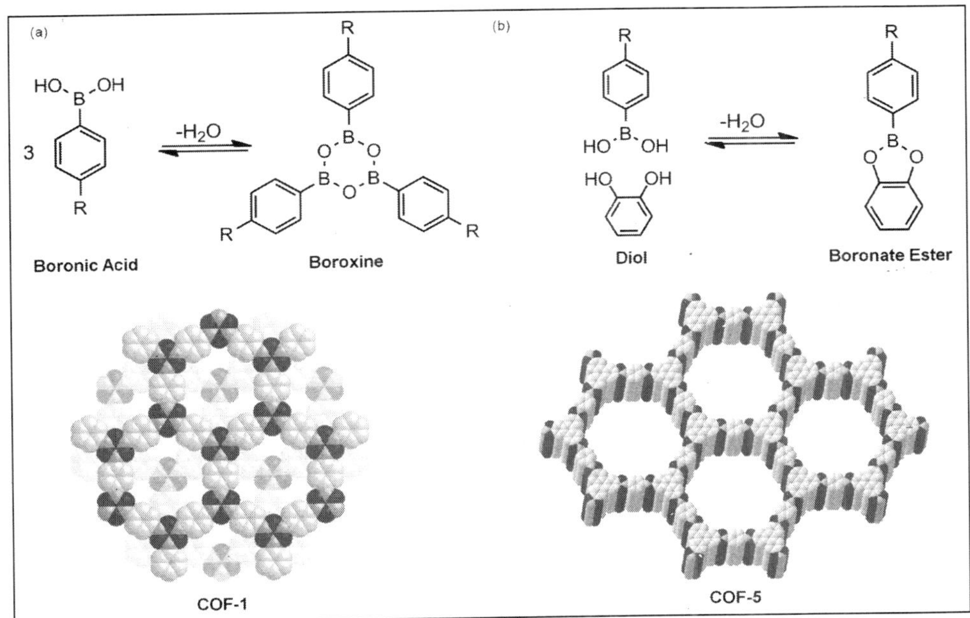

Figure 28.2: Covalent Organic Framework.
Courtesy: Dogru and Bein, 2014.

Green Nanotechnology in Minimizing Carbon Dioxide Emissions

Nanotechnology has the potential of reducing greenhouse gases emissions significantly and thereby mitigating global warming. The rate at which global warming is increasing could endanger this planet in the future. The last two decades of the 20th century were the hottest in 400 years. The Earth has experienced climate changes in the past but the current changes in climate are more rapid and destructive. Our addiction to fossil fuels is most likely the major reason for global warming. There has been a dramatic increase in greenhouse gases, particularly carbon dioxide in recent times from automobiles and industries. Nanotechnology addresses the global warming issue by minimizing/eliminating fossil fuels, decreasing energy consumption, and increasing the efficiency of existing technologies (Ashley *et al.*, 2012; Mamalis, Ramsden, Holt, Vortselas, and Mamali, 2011).

There are various approaches in many major areas to reducing energy consumption and thereby reducing greenhouse gas emissions. Nanotechnology has greatly impacted our lives by providing a possible alternative to fossil fuels and thereby reducing greenhouse gases. Transportation and industry are the two major contributors of greenhouse emissions in the atmosphere. According to the U.S. Environmental Protection Agency, the transportation sector contributes 28 per cent of carbon dioxide emissions. By reducing the weight of vehicles using nanocomposites instead of metal can reduce fuel consumption. It has been determined that a 10 per cent reduction in fuel consumption could be achieved with a 10 per cent weight reduction in vehicles, thereby leading to a proportionate reduction in greenhouse gas emissions.

Nanotechnology has the potential to harvest solar power, just one of many nano applications that may be used to combat greenhouse gas emissions (Ahmadi *et al.*, 2019). Scientists and researchers are designing nanoscale sieves that can filter out toxic agents in fuels and fuel byproducts. Newly designed porous nanoparticles can extract more clean gas from every barrel of oil. Engineers and scientists are designing nanosieves that can capture carbon dioxide before it can enter the atmosphere. Certain nano additives can be incorporated into premium commercial diesel fuel, which reduces fuel consumption by up to 10 per cent and reduces carbon dioxide emissions by up to 15 per cent.

Nanotechnology is also playing an important role in increasing the efficiency of existing technology. A new concept of green nanotechnology involves the development of clean technologies to minimize potential environmental and health hazards associated with the use of nano manufacturing and nano products, and also the design and development of new products or the replacement of existing nanoproducts that are friendly to the environment and humans throughout their life cycle. Green nanotechnology has two main purposes: producing engineered nanomaterials without causing damage to the environment or any hazardous conditions to human health, and producing nanomaterials that provide a solution to environmental problems. This concept uses the principles of green chemistry and green engineering to fabricate nanomaterials without toxicity, using less energy and renewable resources and using lifecycle thinking at all stages (Gadekar and Kadam, 2014; Rognmo, Horjen, and Fernø, 2017).

Metal-Organic Frameworks (MOFs) in Carbon Capture

Metal-organic frameworks (MOFs) are versatile compounds hosting nano-sized pores in their crystal structure. Because of their nanopores, MOFs are now used in a wide range of applications, including separating petrochemicals, mimicking DNA, and removing heavy metals, fluoride anions, hydrogen, and even gold from water. Gas separation in particular is of great interest to a number of industries, such as biogas production, enriching air in metal working, purifying natural gas, and recovering hydrogen from ammonia plants and oil refineries. "The flexible 'lattice' structure of metal-organic frameworks soaks up gas molecules that are even larger than its pore window making it difficult to carry out efficient membrane-based separation" (Khan, Hasan, and Jhung, 2013).

Scientists have greatly improved the gas separation by making the MOF lattice structure rigid. A novel "post-synthetic rapid heat treatment" is applied to MOF –ZIF:8 (zeolitic imidazolate framework 8) at 360°C for a few seconds. This treatment drastically improves ZIF-8's gas-separation performance, specifically in 'carbon capture', a process that captures carbon dioxide emissions produced from the use of fossil fuels, preventing it from entering the atmosphere. The scientists attribute the improvement to a shrinkage of the lattice parameters which makes the chemical bonds of MOF more rigid. The essential chemical composition, bonding environment, and crystallinity of the material is unaffected by the new procedure. Rapid heat treatment is an easy and versatile technique that can vastly improve the gas separation performance of the MOF membranes (Agarwal and Gupta, 2017; Al-Maythalony *et al.*, 2015).

Scientists from the Department of Energy's SLAC National Accelerator Laboratory and Stanford University have taken the first images of carbon dioxide molecules within a **molecular cage** – part of the **highly porous metal-organic**

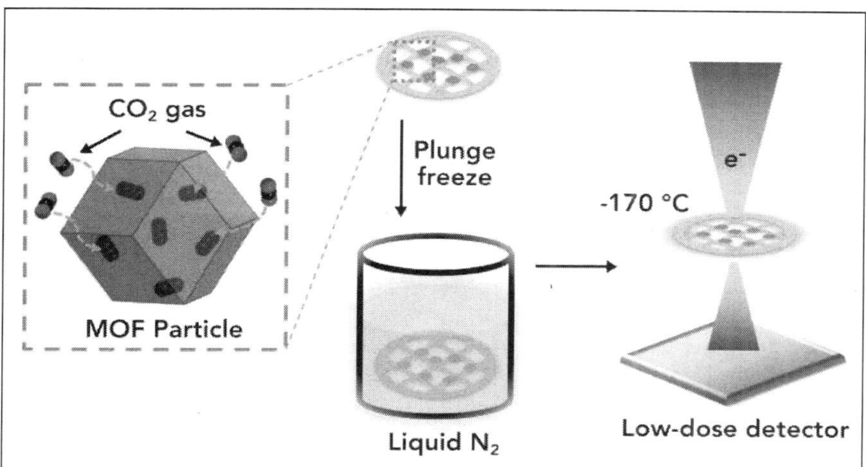

Figure 28.3. Process of Trapping Carbon Dioxide Molecules in Highly Porous Manoparticles – MOFs.
Courtsey: X. Li *et al.*, 2019.

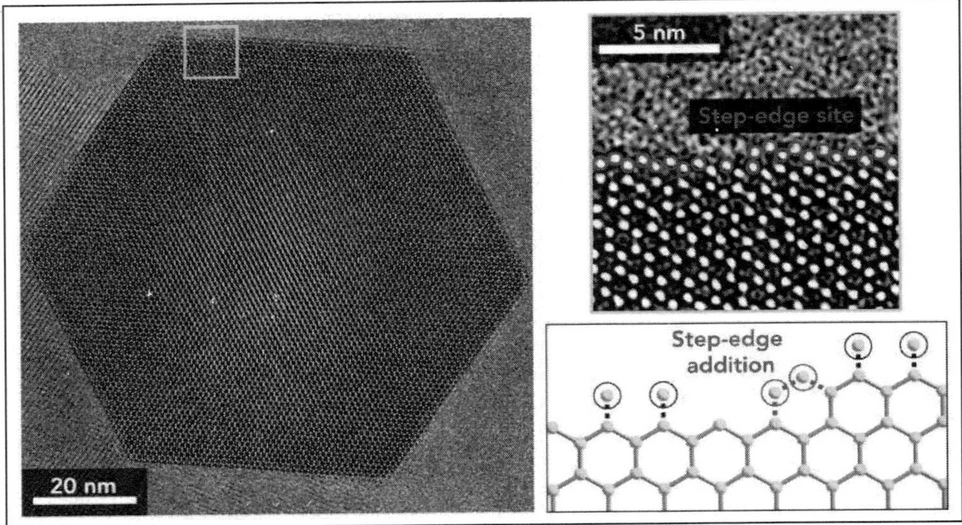

Figure 28.4: The New Cryo-EM Images that Reveal Step-like Features at the Edges of MOF Particles (Upper right) where Scientists think New Cages may Form as the Particle Grows (Bottom right).
Courtesy: Y. Li *et al.*, 2019.

framework that has great potential for separating and storing gases and liquids. The images, made at the **Stanford-SLAC Cryo-EM Facilities**, show two configurations of the CO_2 molecule in its cage, in what scientists call a guest-host relationship; reveal that the **cage expands slightly as the CO_2 enters**; and zoom in on jagged edges where MOF particles may grow by adding more cages.

Carbon Sequestration through Nano Sieves

Carbon sequestration is the capture and secure storage of carbon that would otherwise be emitted to, or remain in the atmosphere. Its aim is to mitigate climate change and its effects caused by greenhouse gas emissions. Silicon and Magnesium are the 8th and 9th most abundant elements on Earth respectively. They are readily available and their use as a compound for carbon sequestration is effective and safer as there is minimal pollution of underground water sources and the atmosphere due to storage related impact. The mode of application would be as molecular sieves for adsorption and as natural sinks (Hatori, Takagi, and Yamada, 2004). Hydrous Magnesium Silicates when reacted with Carbon dioxide form Magnesium carbonate, nanotechnology would ensure more surface area per volume for the molecular sieves.

Nanotechnology is the Science, engineering and technology conducted at the at the Nano scale (about 1 to 100 Nano meters). It involves the ability to see and control individual atoms and molecules. Around each atom a 'sphere of impermeability' can be described beyond which no other matter can penetrate under usual conditions. Using layered molecular sieves of Magnesium Silicate would allow creation of Nano scale junctions between materials where reactions can take place. Carbon dioxide captured through other processes such as absorption or cryogenic fractionation can be stored in Magnesium Silicate rich rock formations as a way of enhancing natural sinks. Micro porous membranes engineered at the Nano scale using Magnesium Silicates can also be used to capture and store carbon at Carbon dioxide sources.

Future opportunities for Carbon dioxide capture may arise from producing hydrogen fuels from carbon rich feed stocks such as natural gas, coal and biomass. The molecular sieves and micro porous membranes should be used to sequester carbon in large stationary sources of Carbon dioxide which include fossil fueled power plants, refineries, cement manufacturing plants, ammonia production plants and in natural gas wells.

Carbon Nanotubes from Carbon Dioxide

The concentration of atmospheric carbon dioxide is at its highest levels in hundreds of thousands of years and a novel idea is to turn it into something useful, specifically, high-quality carbon nanotubes. Scientists at Vanderbilt University in Tennessee, have developed a non-traditional way to construct small-diameter carbon nanotubes, called CNTs, while actively using carbon dioxide.

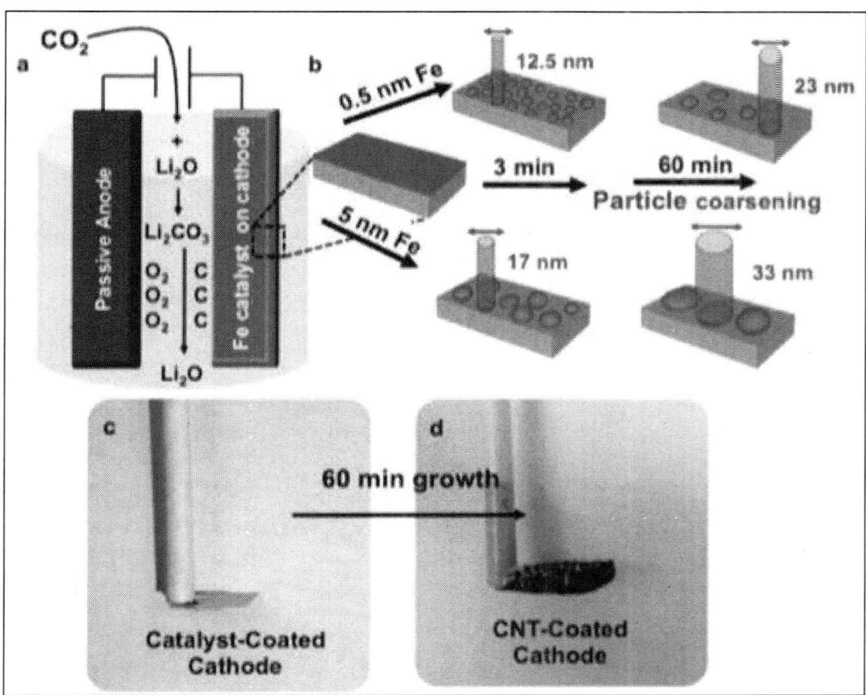

Figure 28.5: Schematic Illustrations: (a) Electrolysis setup (b) Catalyst of varying thickness and growth time controlling the CNT diameter; (c) Catalyst-coated stainless steel cathode before electrochemical CNT growth and (d) with CNT coating after 60 min of growth.

The process relies on a type of molten salt known as lithium carbonate, or Li_2CO_3. More commonly used for health purposes, lithium carbonate is often used in tandem with treatment of mental health disorders. But the scientists melt it by heating it up to 723°C. This process creates dilithium oxide (Li_2O) and releases both oxygen and carbon into the air. Scientists are able to capture this liberated carbon and place it on an iron catalyst. When that connection occurs, it creates carbon nanotubes. The Li_2O reacts with carbon dioxide already in the air and becomes lithium once again. CNTs have a wide variety of uses. They are 100 times stronger than steel but with only one-sixth the weight, as well as remarkable thermal conductivity, and small amounts are used in things like high-end baseball bats and golf clubs. The iron catalyst is not that stable and depending on how long it is in use is creating irregularly size CNTs and this needs further validation. Once the electricity issue is figured out, the entire endeavor would only cost half of current CNT synthesis (Douglas, Carter, Li, and Pint, 2018).

Conclusion

Methane and carbon dioxide levels have to be brought under control in the coming years amidst industrialization, intensive agriculture and population increases. In order to meet GHG emissions reduction target, a complimentary range of technological approaches, including methane and nitrous oxide emission reduction at source in agriculture, improving energy efficiency and conservation, adopting clean fuels and clean coal technologies, developing renewable energy, and implementing CCS (carbon capture and sequestration), have been considered by various countries according to their own circumstances. Nano particles and nano systems have been found highly effective in greenhouse gases emission reduction and carbon capture and thereby mitigating global warming. Nano zeolites role in paddy ecosystem in methane capture and oxidation is a subject that needs to be field validated further to promote as a technology. Nano systems mediated absorption is the most mature CO_2 separation process, due to its high efficiency and lower cost; although issues related to environmental impact have to be fully understood. Although technologies regarding the capture and storage of CO_2 exist, the overall cost of using current CCS procedures using nano systems is still high and must be substantially reduced before it can be widely deployed.

REFERENCES

1. Agarwal, R. A., and Gupta, N. K. (2017). CO2 sorption behavior of imidazole, benzimidazole and benzoic acid based coordination polymers. *Coordination Chemistry Reviews, 332*, 100-121.
2. Agger, J., Meza, L., Cundy, C., Plaisted, R., and Anderson, M. (2005). Control of zeolite surface nucleation. In *Studies in Surface Science and Catalysis* (Vol. 158, pp. 35-42): Elsevier.
3. Ahmadi, M. H., Ghazvini, M., Alhuyi Nazari, M., Ahmadi, M. A., Pourfayaz, F., Lorenzini, G., and Ming, T. (2019). Renewable energy harvesting with the application of nanotechnology: A review. *International Journal of Energy Research, 43*(4), 1387-1410.

4. Al-Maythalony, B. A., Shekhah, O., Swaidan, R., Belmabkhout, Y., Pinnau, I., and Eddaoudi, M. (2015). Quest for anionic MOF membranes: continuous sod-ZMOF membrane with CO2 adsorption-driven selectivity. *Journal of the American Chemical Society, 137*(5), 1754-1757.

5. Ashley, M., Magiera, C., Ramidi, P., Blackburn, G., Scott, T. G., Gupta, R. Biswas, A. (2012). Nanomaterials and processes for carbon capture and conversion into useful by products for a sustainable energy future. *Greenhouse gases: science and technology, 2*(6), 419-444.

6. Coker, E. N., Roelofsen, D. P., Barrer, R. M., Jansen, J. C., and van Bekkum, H. (1998). Sorption of bulky aromatic molecules into zeolite NaX. *Microporous and mesoporous materials, 22*(1-3), 261-268.

7. Dogru, M., and Bein, T. (2014). On the road towards electroactive covalent organic frameworks. *Chemical Communications, 50*(42), 5531-5546.

8. Douglas, A., Carter, R., Li, M., and Pint, C. L. (2018). Toward small-diameter carbon nanotubes synthesized from captured carbon dioxide: critical role of catalyst coarsening. *ACS Applied Materials and Interfaces, 10*(22), 19010-19018.

9. Gadekar, J. M., and Kadam, K. D. (2014). Nanotechnology: green innovation. *International Journal of Science and Research, 3*(1), 225-228.

10. Hatori, H., Takagi, H., and Yamada, Y. (2004). Gas separation properties of molecular sieving carbon membranes with nanopore channels. *Carbon, 42*(5-6), 1169-1173.

11. Khan, N. A., Hasan, Z., and Jhung, S. H. (2013). Adsorptive removal of hazardous materials using metal-organic frameworks (MOFs): a review. *Journal of hazardous materials, 244*, 444-456.

12. Li, X., Sun, Y., Zhang, T., Bai, Y., Lyu, X., Cai, W., and Li, Y. (2019). N-doping nanoporous carbon microspheres derived from MOFs for highly efficient removal of formaldehyde. *Nanotechnology, 30*(10), 105702.

13. Li, Y., Wang, K., Zhou, W., Li, Y., Vila, R., Huang, W. Tsao, Y. (2019). Cryo-EM structures of atomic surfaces and host-guest chemistry in metal-organic frameworks. *Matter*.

14. Lin, S., Diercks, C. S., Zhang, Y.-B., Kornienko, N., Nichols, E. M., Zhao, Y. Yaghi, O. M. (2015). Covalent organic frameworks comprising cobalt porphyrins for catalytic CO2 reduction in water. *science, 349*(6253), 1208-1213.

15. Mamalis, A., Ramsden, J., Holt, G., Vortselas, A., and Mamali, A. (2011). The effect of nanotechnology on mitigation and adaptation strategies in response to climate change. *Collegium, 159*.

16. Mohanraj, J., Lakshmanan, A., and Subramanian, K. (2017). Nano-Zeolite Amendment to Minimize Greenhouse Gas Emission in Rice Soil. *J. Environ. Nanotechnol, 6*(3), 73-76.

17. Ragaventhiran, M. (2015). *Assessing the impact of nanoclay and nano zeolite on water use efficiency, productivity, and methane flux in rice cultivation.* (B.tech (Agricultural Information Technology) Masters Thesis), Tamil Nadu Agricultural University,

18. Rognmo, A., Horjen, H., and Fernø, M. (2017). Nanotechnology for improved CO2 utilization in CCS: Laboratory study of CO2-foam flow and silica nanoparticle retention in porous media. *International Journal of Greenhouse Gas Control, 64,* 113-118.

19. Sarker, N. C., Keomanivong, F., Borhan, M., Rahman, S., and Swanson, K. (2018). In vitro evaluation of nano zinc oxide (nZnO) on mitigation of gaseous emissions. *Journal of animal science and technology, 60*(1), 27.

20. Singh, B., and Polshettiwar, V. (2016). Design of CO2 sorbents using functionalized fibrous nanosilica (KCC-1): insights into the effect of the silica morphology (KCC-1 vs. MCM-41). *Journal of Materials Chemistry A, 4*(18), 7005-7019.

21. Song, W., Justice, R., Jones, C., Grassian, V., and Larsen, S. (2004). Size-dependent properties of nanocrystalline silicalite synthesized with systematically varied crystal sizes. *Langmuir, 20*(11), 4696-4702.

22. Subramanian, K., Paulraj, C., and Natarajan, S. (2008). Nanotechnological approaches in nutrient management. *Nanotechnology applications in agriculture. TNAU technical bulletin, Coimbatore, India,* 37-42.

23. Yin, D., Li, J., Wang, J., Ling, L., and Qiao, W. (2018). Low-Temperature Selective Catalytic Reduction of NO x with Urea Supported on Carbon Xerogels. *Industrial and Engineering Chemistry Research, 57*(20), 6842-6852.

Chapter 29

Impact of Nanoparticles on Soil Microbial Communities

A. Lakshmanan[1], L. Sivashankari[1] and S.K. Rajkishore[2]

[1]Department of Nano Science and Technology, [2]O/o The Dean (SPGS), Tamil Nadu Agricultural University, Coimbatore – 641 003, Tamil Nadu

Introduction

Nanotechnology involves research of materials at atomic or molecular levels in the range of 1-100 nm (10^{-9}). Any matter at these range exhibit unusual, unique physical and chemical properties due to their increased surface to mass ratio when compared to their bulk counterpart. This characteristic of nanomaterials makes it possible to utilize them in wide range of applications which includes computing, electronics, medicine, agriculture, packaging and many more including space research. Thus, this tiny science begun to creep upon us and our future. Nanoparticles (NPs) exhibit exceptional properties that draw the interests of scientists. In general, NPs consists of three layers i) the outer/the surface layer – that can be functionalized or manipulated based on the desired function/application, ii) the shell, the layer above the core usually differ chemically from the core and iii) the core, which itself is referred as NPs in general (Shin, Cho, Kannan, Lee, and Kim, 2016). In case of agriculture, application of nanotechnology has rooted deeper during last decade. Nanoparticles can serve as a magic potion in increasing the productivity of the crops and can also play a major role in disease control in agricultural crops. Many engineered Nanoparticles like Au, SiO_2, ZnO, TiO_2 and CNTs have been reported to contribute in plant growth and development. Though the NPs have number of reports stating their positive impact on plant growth and in improving soil conditions, the actual impact depends on the size, dose, composition and the interaction of NPs with the ecosystems they are released in. Ultimately any nanoparticle released to soil environment gets absorbed by the microbial community in the soil if size of the NP is ~50nm. When compared to air and water ecosystems, soil serves as the major sink of NPs and as Exposure modeling suggests (Sun *et al.*, 2014), the soil microbial communities are the most sensitive and pertinent indicators of soil perturbation. Terrestrial microbial community plays a major role in balancing the soil ecosystem and the protection of such organisms becomes crucial. As the science of nanotechnology grows, more nano inputs and nano wastes would be released in to the soil ecosystem. Hence it is important to understand the impact of nanoparticles on soil microbial population by investigating the toxic effects of NPs under *in vitro* and *in vivo* conditions. In this chapter, the impact of various types of NPs on the growth and metabolism of various soil microbial communities is summarized along with the methodology to assess the level of toxicity.

Types of Nanoparticles

The nanoparticles are divided usually into organic and inorganic and carbon-based NPs. The organic group includes micelles, dendrimers, liposomes and compact polymers. The second group has metal and metal oxide NPs and the carbon based NPs includes Carbon Nanotubes (CNT), Carbon dots, Fullerenes, Graphene and Carbon Nanofibers (Cartaxo, 2015).

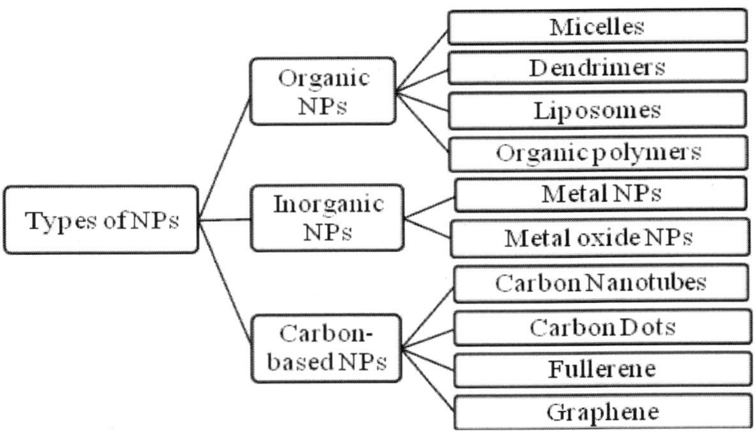

Figure 29.1: Different Types of Nanoparticles.

Effect of Organic NPs on Soil Microbial Communities

The organic nanoparticles include micelles, dendrimers, liposomes and organic polymers that are eco-friendly and has potential applications in biomedicine and bio-imaging *etc.* Liposomes have an aqueous part as its core, surrounded by one or more concentric bilayers of phospholipid structure. The diameter of liposomes ranges from 1 nm to several microns. (Alavi and Hamidi, 2019).

Lipid-polymer hybrid nanoparticle forms to be a novel polymeric structure that has a core and outer shell, utilized for its beneficial characteristics in delivery system. It consists of two major parts, namely, Polymer and the lipid layer. The former one serves as a core in loading diagnostic agents and small drug molecules, whereas, lipid shell is used to provide stability and biocompatibility. Poly(lactic-co-glycolic acid) (PLGA), known for its excellent biodegradability and bio-compatibility, is the most commonly used polymer layer. (Wakaskar, 2018).

Dendrimers are molecules that are built layer wise, by repetitive branching from the core. The dendrimer have an interior and the surface, that are essentially monodispersing and its size are usually controllable during synthesis The presence of large number of surface groups is an added advantage, which are utilized for multivalent presentation of ligands bound to the surface. Depending on the branch-cell unit, a dendrimer can have cavities inside the molecule, which can accommodate small molecules, but small molecules can also be non-covalently bound to the surface groups of the dendrimer giving rise to endo- or exo-complexation depending on where the guest molecule is located (Christensen, 2018).

The above discussed organic nanosystems are usually biodegradable and ecofriendly and they are used commonly in medicinal application. Thus, they do not often find a place in soil ecosystem and hence it does not or very rarely cause an impact on soil microbes.

Effect of Inorganic NPs on Soil Microbial Population

The inorganic NPs are generally termed as Engineered Nanoparticles (ENPs). ENPs have impact on soil microorganisms through direct contact (toxic effects) or by altering the bioavailability of nutrients or toxins in soil and it also causes indirect effects through interactions with toxic organic compounds that could alleviate their toxicity (Simonet and Valcarcel, 2009). The major mechanisms by which the metal NPs influence soil microorganisms are given below.

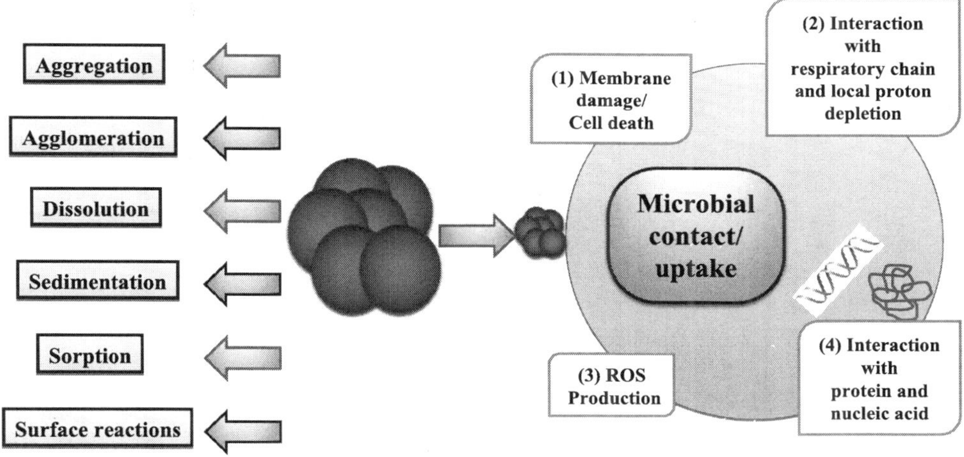

Figure 29.2: Dynamic Transformations of ENMs in Terrestrial Environment and their Impact on Microbial Biomass.
Courtesy: Hegde, Brar, Verma, and Surampalli, 2016.

Generation of Reactive Oxygen Species

More often, the toxic nature of engineered Nanoparticles are attributed to formation of ROS or ROS induced damage (Applerot *et al.*, 2012; Karlsson, Di Bucchianico, Collins, and Dusinska, 2015; Neal, Kabengi, Grider, and Bertsch, 2012). There are different types of Reactive oxygen species (ROS) that decides the bactericidal action of engineered Nanoparticles; OH radicals, hydrogen peroxide and superoxide ions are the major type of ROS associated with nanotoxicity. ROS are produced either by UV illumination or visible illumination. The UV radiation causes a variety of DNA damage due to absorption of UV-B by DNA molecules or it causes oxidative stress in bacterial cells due to formation of free radicals and ROS. Generation of ROS is linked with the presence of defects in microorganisms such as Oxygen vacancies (Krishnamoorthy, Manivannan, Kim, Jeyasubramanian, and Premanathan, 2012).

Metal Ion Release

The other mechanism of metal nanoparticle toxicity is the solubility or the release of metal ions from metal oxide in aqueous solution. This mechanism of toxicity are tested, comparing metal oxide nanoparticle with soluble metal salt toxicity with different metal salts like chlorides, sulfates *etc.* (Heinlaan, Ivask, Blinova, Dubourguier, and Kahru, 2008). The release of metal ions are time-dependent and the initial response are due to ROS production while the secondary toxic response are due to the metal ions dissolved (Misra *et al.*, 2014). This is contributed to the fact that after particular time, most of the ions from metal NPs dissolve.

Other Mechanism of Nanotoxicity

Accumulation of nanomaterials on membrane surfaces and internalization of nanomaterials are the other mechanisms of nano toxicity in addition to the production of ROS and metal ion release. The contact of nanoparticles with the cell wall may contribute to the toxicity on microorganisms (Applerot *et al.*, 2012; Applerot *et al.*, 2009). The electrostatic interaction between the positively charged nanoparticles that attracts the negatively charged bacterial cell wall can play a significant role in nanomaterial toxicity (Feris *et al.*, 2009; Thill and Zeyons, 2006). The presence of electrolytes and alteration in pH may affect the electrostatic interaction between microorganisms and nanoparticle. Thus, the surface charge has a major effect on interaction toxicity (positive charge = toxicity, negative charge = little or no effect). Surface charge is also a key in affecting internalization, subcellular localization, and toxicity of many nanoparticles. The particles ones, if attached onto the cell surface, the redox reactions takes place on the surface of adsorbed nanoparticles that then results in oxidative stress, thus causing cytotoxicity (Neal *et al.*, 2012; Thill and Zeyons, 2006).

Impact of Metal-based NPs

Silver nanoparticles (AgNPs) that is known for its anti-microbial properties are the most exploited nanomaterials into a wide variety of commercial goods and are also used for water treatment, bone prostheses, drug and gene delivery,

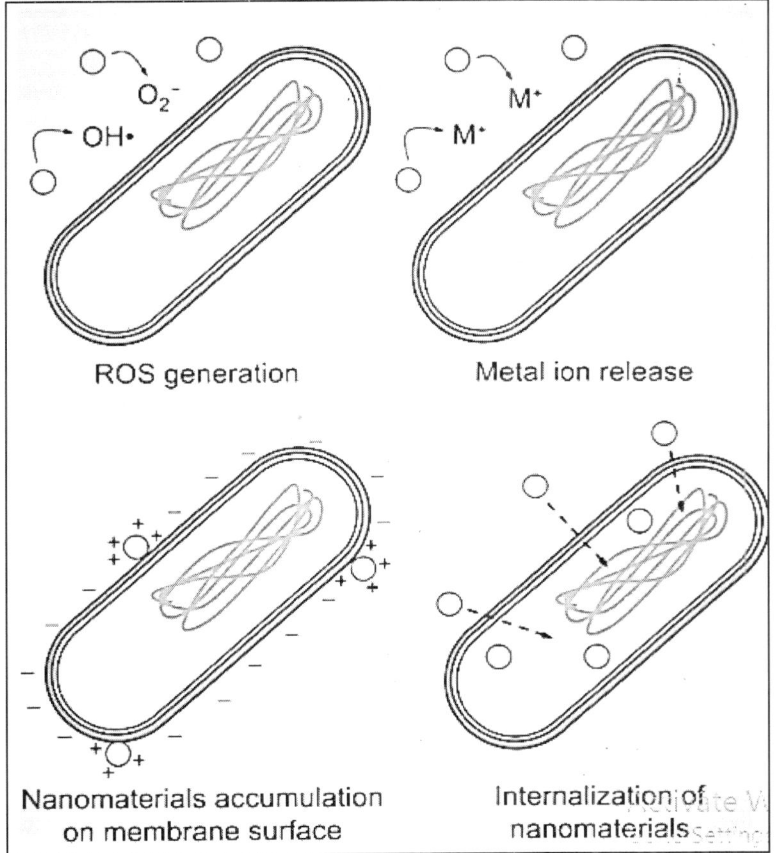

Figure 29.3: The Image Illustrates the Possible Toxicity Mechanisms towards Bacteria Cells.
Courtesy: Djurišiæ *et al.*, 2015.

implantable materials, bioimaging devices, and in biosensors(Ahamed, AlSalhi, and Siddiqui, 2010).The accidental or incidental environmental releases of AgNP-containing products could impact biogeochemical cycling. Many studies have been conducted worldwide to assess the impact of silver nanoparticles on soil microflora and their mode of entry in to various eco systems. In United states, the probabilistic material flow analysis predicted the AgNP concentrations to be as follows: 0.0001 mg/L in surface waters; 9 mg/L in wastewater treatment plants (WWTPs); 1550 mg/L in sewage sludge. The great concern is that such higher concentrations of silver NPs could pose a threat to nitrogen cycling in natural or engineered systems (Radniecki and Lauchnor, 2011). Exposure of AgNPs at 100 mg/L to 1000 mg/L of various types has reported to inhibit the process of nitrification(Choi and Hu, 2009). Arnaout and Gunsch investigated the toxicity of AgNPs by observing the heavy metal stress response through different surface coatings to the nitrifier, *Nitrosomonas europaea*. The results from the investigations on the interactions of AgNPs and Ag+

with 3 model N-cycling bacteria: the denitrifier *Pseudomonas stutzeri*, the nitrogen fixer *Azotobacter vinelandii*, and the nitrifier *N. europaea* reported a decrease in nitrification activity and that it hinders nitrogen removal and treatment efficiency in waste water treatment plants. Overall, localized impacts of AgNPs in soil could affect soil fertility, water quality, and ecosystem productivity, which underscores the importance to mitigate and intercept incidental and accidental releases of AgNPs and their associated silver ions. Nevertheless, fewer research has been conducted on how soil microbes respond to AgNPs at the molecular level and the resulting impact on the associated metabolic pathways.

The Copper (Cu) Nanoparticles are the next most extensively used ones for its antimicrobial activity. The unintentional release of nano-Cu in soil increases its concentration that causes major distractions in metabolism of soil microbial population (Trujillo-Reyes, Majumdar, Botez, Peralta-Videa, and Gardea-Torresdey, 2014). VandeVoort and Arai (2018) investigated the impact of commercially-available copper NPs on soil nitrification kinetics, and the results from X-ray absorption near edge structure spectroscopy showed that the NPs readily oxidize to Cu(II) and are strongly retained in soils with minimum dissolution; Vmax suggested a negative impact by CuNP at 1 to 100 mg/L, the suppression of nitrification kinetics was observed at 1 to 100 mg/kg and the effect was concluded to be dose- dependent. The Cu NPs of 20 nm exhibits significant decrease in substrate utilization in arctic soils (Kumar, Shah, and Walker, 2011). The mechanism for toxicity for CuNPs could be due to ROS production via Fenton reaction ($Cu^+ + H_2O_2 \rightarrow Cu^{2+} + OH + OH^-$) or by Haber-Weiss ($H_2O_2 + O_2^- \rightarrow OH + OH^- + O_2$) reactions (Letelier, Sanchez-Jofre, Peredo-Silva, Cortes-Troncoso, and Aracena-Parks, 2010).

The nanoscale zero-valent iron particles (nZVI) being the most promising remediation for the contaminated soils also increases the concern of ecotoxicity on soil microbial communities and, hence, soil quality. The mechanisms responsible for nZVI toxicity are: (i) direct contact that causes physical damage, thereby disrupts cell membrane and the membrane permeability; (ii) indirect effects by oxidative stress that causes molecular and biochemical destruction (Xue *et al.*, 2018). An inhibitory impact of nZVI on soil microorganisms is observed in sandy-loam soil compared to clay-loam soil which is probably due to the "protective" effect of a high clay and organic matter content, that renders nZVI inactive for interaction with microbial cells (Gomez-Sagasti *et al.*, 2019).

Gold (Au) nanoparticles (GNPs) are the noble nanoparticle that has gained a lot of attention due to their potential applications in diverse areas from electronics to biomedicine. These properties of Au NPs comes from the specific physical and chemical properties that they exhibit at nanoscale (Sreeprasad and Pradeep, 2013) and their chemical inertness and resistance to surface oxidation makes Au, the most notable material in making technologies at nano-scale. The Au(I/III) complexes are usually precipitated by archaea, fungi, actinomycetes and bacteria under a wide range of experimental conditions (Karthikeyan and Beveridge, 2002; Lengke and Southam, 2005). The active intracellular precipitation of gold particles by *Acidothiobacillus thiooxidans* from $Au(S_2O_3)$ takes place even after several months the bacterial growth had stopped, they formed gold particles of round and irregular shape and octahedral gold crystals (Lengke and Southam, 2005). Further experimental evidences suggests that gold toxicity on *R. metallidurans* has led to death of more than 90 per cent of viable cells in the medium containing AuCl4 (50mM). GNPs are combined with near infra-red (NIR) light or laser pulses to potentially use it for antimicrobial activities (Letfullin, Joenathan, George, and Zharov, 2006). GNPs of 30-40 nm diameter are not toxic to bacterial cells and do not accumulate in bacteria (Simon-Deckers *et al.*, 2008). Though few studies suggest the toxic nature of GNPs, most experimental studies state that AuNPs are comparatively safe.

Impact of Metal Oxide-based NPs

The Metal oxide nanomaterials represent a class of nanomaterials with the highest global annual production and are used widely in practical applications. Thus, in the past decade the toxicity of the metal oxide nanoparticles has been studied intensively but still there remains the question of "from where does the toxicity arises: from the NPs themselves or from the metal ions released". Many metal oxide nanoparticles though considered non-toxic keeping in account about their bulk counterpart, these NPs typically exhibit toxicity to bacteria and fungi in terrestrial ecosystem. The metal oxide NPs potentially dissolve in aqueous media, which results in toxic metal ions released into their surrounding media. Metal and metal oxide NPs exhibit detrimental effects on the microbial count, metabolic activities and diversity even at very low concentrations (<1mg/kg).

Investigations conducted on microbial respiration, ammonification, dehydrogenase (DH) activity, and fluorescent diacetate hydrolase (FDAH) activity in the soil in order to demonstrate the toxicological effect of ZnO-NPs on soil microorganisms showed that ZnO-Nanoparticles suppressed thermogenic metabolism, and it caused reduction in colonies of *Azotobacter*, phosphorous-solubilizing and potassium-solubilizing bacteria (Shen, Chen, Hou, Li, and Lu, 2015). The effect of ZnO-NP also inhibited enzymatic activities in diacetate hydrolysis activity (Chai *et al.*, 2015). The reports also showcased the inhibition of ammonification process by up to 37.8 per cent in the soil treated with ZnO NPs (1mg/gram of soil) over a period of 3 months. The ZnO NPs of 30-200 nm expressed pH-dependent effects strongly between the NPs and soil microbial diversity that altered the composition of microbial cells at pH levels of 4.8 and 5.9 and at pH level of 7.2 microbial community in soil were less susceptible to the effect of zinc NPs (Read *et al.*, 2016).

Iron Oxide Metal Nanoparticles (IOMNP) potentially stimulate growth in some bacteria and it could also alter the structure of soil bacterial community to some extent, without altering the bacterial abundance (He *et al.*, 2011). Several potential IOMNPs stimulated bacteria are related to Actinobacteria (*Duganella, Streptomycetaceae* or *Nocardioides*) based on molecular fingerprinting and sequencing analysis. Meanwhile, many studies state that the soil invertase and urease activities increase significantly with IOMNPs amendment and this is considered to be the consequence of changes in the bacterial community.

The four metal oxide nanoparticles, namely, zinc oxide nanoparticles (nZnO), cerium dioxide nanoparticles (nCeO_2), titanium dioxide nanoparticles (nTiO_2)

and magnetite nanoparticles (nFe_3O_4), were taken to evaluate their impact on soil enzymatic activities *i.e.*, invertase, urease, catalase, and phosphatase and bacterial communities in black soil and saline-alkali soil. The soils exposed to these metal oxide nanoparticles at 0.5, 1.0, and 2.0 mg g^{-1} for 15 and 30 days exhibited change in soil enzyme activities and soil bacterial community. Among the four Nanoparticles assessed nZnO had a stronger effect on soil enzymatic activities and out of two variable soils tested the susceptible ones was found to be saline-alkali soil than black soil (He *et al.*, 2011).

The pyrosequencing analysis also revealed high variance in composition of bacterial community, from the samples of saline-alkali soil exposed to nZnO. *e.g.*, *α-proteobacteria*, and *γ-proteobacteria* class. Thus, the soil type is a key component in dictating the effect of metal oxide nanoparticles on the bacterial community composition and size. The antibacterial effects of Co_3O_4, Fe_2O_3, NiO and Cr_2O_3 tested, showed moderate antibacterial effects, with EC_{50} values of about 200 mg/L, 130 mg/L, and 260 mg/L, in Fe_2O_3, Co_3O_4, Cr_2O_3 NPs respectively; whereas relatively, NiO NPs has lower antibacterial effects, with an EC_{50} value of up to 870 mg/L and Fe^{3+} found to have a significant increase in antibacterial effect (Wang, Liu, Liu, and Jiang, 2014).

Titanium dioxide nanoparticles (TiO_2-NPs) is one another metal oxide NPs that are widely used in commercial products such as toothpastes and sunscreens. It is also used in industrial products like paper, lacquers and paint, and in water treatment (Wang *et al.*, 2014; Wen *et al.*, 2015 and Mitrano, Motellier, Clavaguera, and Nowack, 2015). Eventually, TiO2-NPs are discharged indirectly into agro ecosystem through sewage-sludge application or via irrigation and directly as nano pesticides or Nanofertilizers (Servin and White, 2016). Simonin and Richaume (2015) reported that the Titanium dioxide NPs even at lower concentration (1 mg kg^{-1} dry soil) impose a major concern on nitrogen cycle in soil as it modifies the structure of bacterial population in agricultural soil. The aged TiO_2 –NPs after a long exposure to soil microbial biomass expresses greater effects on microorganisms even at low concentrations. This could be due to low mobility and long residence time of TiO_2 –NPs in ecosystem (Niazi and Gu, 2009), but still the studies lack the incubation period not exceeding 60 days (Simonin and Richaume, 2015). Toxicity of TiO_2-NPs are associated with the production of reactive oxygen species (ROS) causing an oxidative stress and osmotic stress in pure cultures of bacteria after being adsorbed to their cell membrane.

Impact of Carbon-based NPs

Carbon nanomaterials (CNMs) are a class of engineered nanomaterials (ENMs) that have been significantly the research of interest over last two decades due to their exceptional optical, electrical, mechanical, and thermal properties. This evolution dated back when Buckminster (C_{60}) fullerenes were discovered in 1985, followed by CNTs in 1991, and the graphene (Bergmann and Machado, 2015). To produce structures such as graphene sheets and nano-diamonds, Carbon obtains thermodynamically favorable configurations from sp^3 to sp^2, depending on the pressure conditions and heat of formation and they combine quantum effects, stable resonance structures and tunable physico-chemical properties due sp^3 character of C-C bond nano scale. The CNPs are extensively used in wide range of applications resulted in fatal increase in concentration of CNPs in soil ecosystem/landfills estimated to be 80 per cent, according to life cycle release modeling studies. Carbon Nanotubes has number of reports suggesting its positive impact on plants (Lahiani *et al.*, 2013) but the evidences on the effects of CNPs on soil microbes are bit contrary.

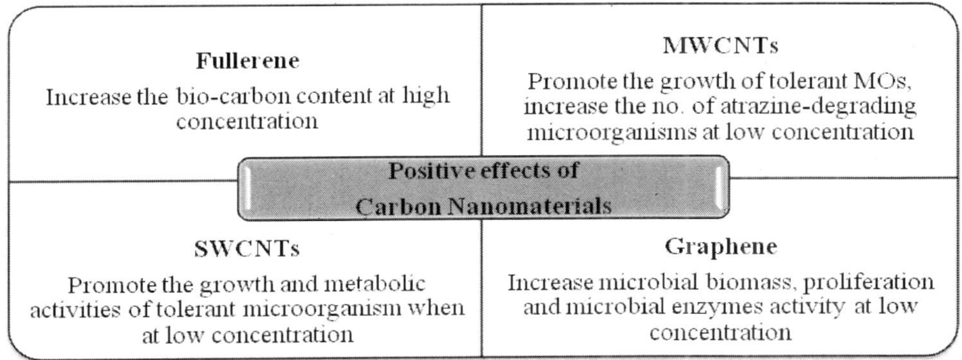

Figure 29.4. Positive effects of Carbon Nanomaterials

CNTs are cylindrical molecules that are the rolled-up sheets of graphene and can further be categorized into single-wall (SWCNTs) and multi-wall nanotubes (MWCNTs) depending on the number of concentric layers of rolled graphene sheets (Yang *et al.*, 2011; De Volder, Tawfick, Baughman, and Hart, 2013). This widely used carbon nano structure exhibits a strong interaction with bacterial cell membrane and eventually disrupts the membrane structure by oxidative stress and/or physical damage (Jackson *et al.*, 2013).

The investigations on the toxic influence of fullerene on microbial communities in fertile soils for 30 days showed that the increase in concentration of fullerene (upto 500mg/kg) reduced the amount of *Planctomyces, Nitrospira* and *Lysobacter* (Hao *et al.*, 2018). A decline in growth trend in number of fast-growing bacteria by 3–4 times with exposure of fullerene at 25 and 50/ mg/kg concentration for 3/ days were also reported.

Toxicity Assessment of Nanoparticles on Soil Microbial Communities and other Biosystems

Various Nanoparticles exhibit different effects on soil microorganisms depending on the size, shape, dose and their interactions with the microbes and the type of environment. Assessment of nanoparticle toxicity become inevitably necessary as NPs are widely used in biomedical applications, cosmetics and in agriculture as nano agricultural inputs. The higher reactive nature of NPs that comes from its greater surface area to volume ratio many times contributes its toxicity to non-targeted organisms. It is important to understand the mechanism of toxicity and the methods to assess them. The higher reactivity of NPs with biological system

Table 29.1: Negative Effects of Carbon Nanomaterials on Various Mmicroorganisms (*Courtesy*: Mukherjee *et al.*, 2014)

Reference	CNM	Organism	Treatment	Effect
Kang et al., 2007	SWNTs	*Escherichia coli*	1–50 μg/mL	Strong antimicrobial properties.
Fang et al., 2007	C_{60} aggregates	*Bacillus subtilis* (Gram-positive) and *Pseudomonas putida* (Gram-negative)	0.01 mg/L	Significantly impacted levels of iso-and anteiso-branched fatty acids in *Bacillus subtilis* (Gram-positive).
Johansen et al., 2008	Fullerenes C_{60}	Microbial communities and protozoans	0–50 mg/kg	Results after 14 days showed a threefold decrease in the number of fast-growing bacteria. No significant changes in protozoan population.
Liu et al., 2009	SWCNT dispersed and SWCNT agglomerates in saline solution	Gram-negative *Escherichia coli*, *Pseudomonas aeruginosa*, and gram-positive *Staphylococcus aureus*, *Bacillus subtilis*	5 g/mL	Higher antibacterial activity of dispersed SCNT to gram-positive bacteria in comparison with the agglomerates.
Chung et al., 2011	MWCNT	Microbial communities	0, 50, 500, and 5000 μg/g	Enzyme activity decreased after 30 min of the incubation. Microbial biomass and extracellular enzyme activity decreased (up to ~50%).
Rodrigues et al., 2013	Carboxyl-functionalized SWCNTs	Bacterial and fungal communities	0.5 mg/L	Alteration on *Pseudomonas putida* (Gram-negative) phase transition temperatures and levels of unsaturated fatty acids. Higher doses had a maximum biomass loss at 3 days and the fungal community was unable to recover even after 14 days.
Jin et al., 2014	SWCNT	Gram- positive, Gram-negative bacteria, and fungal populations	0.03 to 1 mg/g	Decreased biomass of microbial groups and fungal populations.
Chung et al., 2015	Graphene oxide	Microbial communities/soil enzymes	0.5–1 mg/kg	Decreases of up to 50% in the enzyme activities after 21 days of incubation.
Kerfahi et al., 2015	Raw and acid treated or functionalized MWCNTs	Microbial communities	0–5000 mg/kg	Bacterial community composition was affected but recovered after 8 weeks.

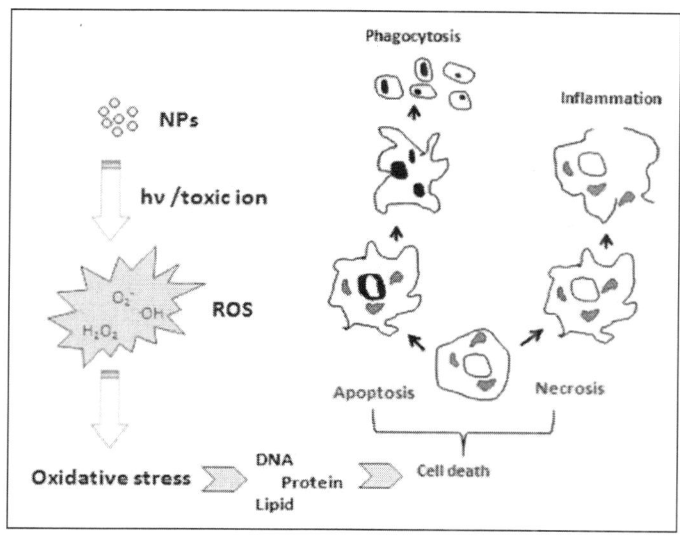

Figure 29.5: Cytotoxicity Mechanism and ROS Generation Induced by NPs
Courtesy: Fard, Jafari, and Eghbal, 2015.

results in the increased production of ROS. One of the mechanisms involving in the toxicity of NPs is the ROS production that in turn causes oxidative stress, consequent damages to the proteins, inflammation and damages/punctures cell membrane and DNA. The chemicals that induce oxidative stress by forming reactive oxygen species (ROS) or through inhibiting the antioxidants are called Prooxidants. NPs react with such prooxidants and cause effects via intracellular generation of ROS involving activation of NADPH-dependent enzyme systems and mitochondrial respiration.

Table 29.2: Different Types of *in vitro* Assays to Evaluate Nanotoxicity (Fard *et al.*, 2015)

Assay	*Type of NPs*	*Type of Cells (system)*
MTT assay	QDs	Human embryonic kidney cells
	TiO_2	Human erythrocyte/lymphocyte cells
Natural red	TiO_2 NPs	Zebrafish embryos
LDH test	TiO_2 NPs	Human kidney cells
	CNTs	human pneumocytes cells
MTS assay	Ag NPs	mouse embryonic fibroblasts
	Gold NPs	Mammalian cells
Trypan blue	Gold NPs	mouse fibroblast
	TiO_2 NPs	human lung epithelial cells

Cytotoxicity assays are used in basic assessment of cell damage in microorganisms. They are classified as *in vivo* and *in vitro* tests. The former ones (cell-based assay) are time-consuming, expensive and involves many ethical issues but *in vitro* toxicity tests are cell cultured-based and are faster, convenient, less expensive method that are devoid of any ethical issues. These advantages of *in vitro* assays make them a better choice for toxicity assessment of most nanomaterials. The common and widely used in vitro methods include lactate dehydrogenase (LDH) assay, MTT and MTS assays. Cell membrane integrity and the metabolic activity of viable cells are assessed using *in vitro* techniques.

Lactate dehydrogenase (LDH) is based on the leakage of substances/based on measuring the cytoplasmic enzyme and the measurement of LDH activity in the extracellular media. Lactate dehydrogenase is a cytoplasmic enzyme that is released rapidly into the cell culture supernatant when the cell membrane is damaged. The principle of LDH assay, involves quantifying NADH produced during the conversion of lactate to pyruvate and the practical procedure measures the reduction of tetrazolium salt (yellow) by NADH into water soluble formazan (red) and absorbed at 492 nm.

Apart from LDH, the membrane integrity can also be determined by penetration of dyes such as tryptan blue and neutral red into the damaged cells and staining intracellular components. These dyes cannot enter living cells. Metabolic activity of viable cells could be determined through colorimetric assays, such as the MTT and MTS assays.

To assess the genotoxicity, the comet assay and micronucleus (MN) test are the most commonly employed methods for assessing NP-induced damage at DNA and chromosome level, respectively. The comet assay is the most sensitive method to detect DNA strand breaks and as well as oxidative damage in DNA at the level of single cells. COMET Assay is performed to assess the geno toxicity effects of NPs on microbes by the action of mobility of nano particles into the cell/nuclear membrane resulting in irreversible DNA damage. The COMET assay was conducted to assess the impact of nano zeolites on *Rhizobium* and the micro slides image depicted no quantifiable DNA damage in nano-zeolite treated cells in comparison to the control cultures. Further, the tail formation (Comet shape) which is an outcome of DNA repair was not observed in nano-zeolite treated bacterium entrapped in micro slides. (Sivashankari *et al.*, 2019).

Conclusion

Though utilization of novel Nanoparticles in wide range of applications is being reported every day, their potential impact on environment is equally raising up. The increased availability of nanoparticle-based products would expose the environment to these materials and NPs are finding their way into the soil environment through nano products, industrial wastes and other anthropogenic sources. The fate of nanoparticles released in environment is mostly bound to soil ecosystem than in aquatic ecosystem or atmosphere. As the soil microbial communities are the major service providers for the ecosystem and humankind, it is critical to study the effects of NPs exposure on soil microbes. It is very difficult to make a general statement on the effect of NPs on soil microbial communities as the intensity of impact is decided by many factors such as type and doze of NPs, soil conditions and the type of microbial community. Any nanoparticle before the release into soil should be tested for its biosafety under laboratory conditions or in controlled conditions as soil serves as a sink of microbial community and it is important to maintain its balance as to maintain the biogeochemical cycle.

REFERENCES

1. Ahamed, M., AlSalhi, M. S., and Siddiqui, M. (2010). Silver nanoparticle applications and human health. *Clinica chimica acta, 411*(23-24), 1841-1848.
2. Alavi, M., and Hamidi, M. (2019). Passive and active targeting in cancer therapy by liposomes and lipid nanoparticles. *Drug metabolism and personalized therapy, 34*(1).
3. Applerot, G., Lellouche, J., Lipovsky, A., Nitzan, Y., Lubart, R., Gedanken, A., and Banin, E. (2012). Understanding the antibacterial mechanism of CuO nanoparticles: revealing the route of induced oxidative stress. *Small, 8*(21), 3326-3337.
4. Applerot, G., Lipovsky, A., Dror, R., Perkas, N., Nitzan, Y., Lubart, R., and Gedanken, A. (2009). Enhanced antibacterial activity of nanocrystalline ZnO due to increased ROS mediated cell injury. *Advanced Functional Materials, 19*(6), 842-852.
5. Bergmann, C. P., and Machado, F. M. (2015). *Carbon nanomaterials as adsorbents for environmental and biological applications*: Springer.
6. Cartaxo, A. L. P. (2015). Optimization of the size of biodegradable polymeric nanoparticles as intracellular drug delivery devices.
7. Choi, O., and Hu, Z. (2009). Role of reactive oxygen species in determining nitrification inhibition by metallic/oxide nanoparticles. *Journal of Environmental Engineering, 135*(12), 1365-1370.
8. Christensen, J. B. (2018). Dendrimers-Based Nanoparticles for Cancer Therapy and Bioimaging *Nanooncology* (pp. 281-304): Springer.
9. De Volder, M. F., Tawfick, S. H., Baughman, R. H., and Hart, A. J. (2013). Carbon nanotubes: present and future commercial applications. *science, 339*(6119), 535-539.
10. Djurišiæ, A. B., Leung, Y. H., Ng, A. M., Xu, X. Y., Lee, P. K., Degger, N., and Wu, R. (2015). Toxicity of metal oxide nanoparticles: mechanisms, characterization, and avoiding experimental artefacts. *Small, 11*(1), 26-44.
11. Fard, J. K., Jafari, S., and Eghbal, M. A. (2015). A review of molecular mechanisms involved in toxicity of nanoparticles. *Advanced pharmaceutical bulletin, 5*(4), 447.
12. Feris, K., Otto, C., Tinker, J., Wingett, D., Punnoose, A., Thurber, A. Hanna, C. (2009). Electrostatic interactions affect nanoparticle-mediated toxicity to gram-negative bacterium Pseudomonas aeruginosa PAO1. *Langmuir, 26*(6), 4429-4436.

13. Gómez-Sagasti, M. T., Epelde, L., Anza, M., Urra, J., Alkorta, I., and Garbisu, C. (2019). The impact of nanoscale zero-valent iron particles on soil microbial communities is soil dependent. *Journal of hazardous materials, 364*, 591-599.

14. He, S., Feng, Y., Ren, H., Zhang, Y., Gu, N., and Lin, X. (2011). The impact of iron oxide magnetic nanoparticles on the soil bacterial community. *Journal of Soils and Sediments, 11*(8), 1408-1417.

15. Hegde, K., Brar, S. K., Verma, M., and Surampalli, R. Y. (2016). Current understandings of toxicity, risks and regulations of engineered nanoparticles with respect to environmental microorganisms. *Nanotechnology for Environmental Engineering, 1*(1), 5.

16. Heinlaan, M., Ivask, A., Blinova, I., Dubourguier, H.-C., and Kahru, A. (2008). Toxicity of nanosized and bulk ZnO, CuO and TiO2 to bacteria Vibrio fischeri and crustaceans Daphnia magna and Thamnocephalus platyurus. *Chemosphere, 71*(7), 1308-1316.

17. Jackson, P., Jacobsen, N. R., Baun, A., Birkedal, R., Kühnel, D., Jensen, K. A. Wallin, H. (2013). Bioaccumulation and ecotoxicity of carbon nanotubes. *Chemistry Central Journal, 7*(1), 154.

18. Karlsson, H. L., Di Bucchianico, S., Collins, A. R., and Dusinska, M. (2015). Can the comet assay be used reliably to detect nanoparticle induced genotoxicity? *Environmental and molecular mutagenesis, 56*(2), 82-96.

19. Karthikeyan, S., and Beveridge, T. (2002). Pseudomonas aeruginosa biofilms react with and precipitate toxic soluble gold. *Environmental Microbiology, 4*(11), 667-675.

20. Krishnamoorthy, K., Manivannan, G., Kim, S. J., Jeyasubramanian, K., and Premanathan, M. (2012). Antibacterial activity of MgO nanoparticles based on lipid peroxidation by oxygen vacancy. *Journal of Nanoparticle Research, 14*(9), 1063.

21. Kumar, N., Shah, V., and Walker, V. K. (2011). Perturbation of an arctic soil microbial community by metal nanoparticles. *Journal of hazardous materials, 190*(1-3), 816-822.

22. Lahiani, M. H., Dervishi, E., Chen, J., Nima, Z., Gaume, A., Biris, A. S., and Khodakovskaya, M. V. (2013). Impact of carbon nanotube exposure to seeds of valuable crops. *ACS Applied Materials and Interfaces, 5*(16), 7965-7973.

23. Lengke, M. F., and Southam, G. (2005). The effect of thiosulfate-oxidizing bacteria on the stability of the gold-thiosulfate complex. *Geochimica et Cosmochimica Acta, 69*(15), 3759-3772.

24. Letelier, M. E., Sánchez-Jofré, S., Peredo-Silva, L., Cortés-Troncoso, J., and Aracena-Parks, P. (2010). Mechanisms underlying iron and copper ions toxicity in biological systems: Pro-oxidant activity and protein-binding effects. *Chemico-biological interactions, 188*(1), 220-227.

25. Letfullin, R. R., Joenathan, C., George, T. F., and Zharov, V. P. (2006). Laser-induced explosion of gold nanoparticles: potential role for nanophotothermolysis of cancer.

26. Misra, S. K., Nuseibeh, S., Dybowska, A., Berhanu, D., Tetley, T. D., and Valsami-Jones, E. (2014). Comparative study using spheres, rods and spindle-shaped nanoplatelets on dispersion stability, dissolution and toxicity of CuO nanomaterials. *Nanotoxicology, 8*(4), 422-432.

27. Mitrano, D. M., Motellier, S., Clavaguera, S., and Nowack, B. (2015). Review of nanomaterial aging and transformations through the life cycle of nano-enhanced products. *Environment international, 77*, 132-147.

28. Neal, A. L., Kabengi, N., Grider, A., and Bertsch, P. M. (2012). Can the soil bacterium Cupriavidus necator sense ZnO nanomaterials and aqueous Zn2+ differentially? *Nanotoxicology, 6*(4), 371-380.

29. Niazi, J. H., and Gu, M. B. (2009). Toxicity of metallic nanoparticles in microorganisms-a review *Atmospheric and biological environmental monitoring* (pp. 193-206): Springer.

30. Radniecki, T. S., and Lauchnor, E. G. (2011). Investigating Nitrosomonas europaea stress biomarkers in batch, continuous culture, and biofilm reactors *Methods in enzymology* (Vol. 496, pp. 217-246): Elsevier.

31. Read, D. S., Matzke, M., Gweon, H. S., Newbold, L. K., Heggelund, L., Ortiz, M. D. Svendsen, C. (2016). Soil pH effects on the interactions between dissolved zinc, non-nano-and nano-ZnO with soil bacterial communities. *Environmental Science and Pollution Research, 23*(5), 4120-4128.

32. Servin, A. D., and White, J. C. (2016). Nanotechnology in agriculture: next steps for understanding engineered nanoparticle exposure and risk. *NanoImpact, 1*, 9-12.

33. Shen, Z., Chen, Z., Hou, Z., Li, T., and Lu, X. (2015). Ecotoxicological effect of zinc oxide nanoparticles on soil microorganisms. *Frontiers of Environmental Science and Engineering, 9*(5), 912-918.

34. Shin, W.-K., Cho, J., Kannan, A. G., Lee, Y.-S., and Kim, D.-W. (2016). Cross-linked composite gel polymer electrolyte using mesoporous methacrylate-functionalized SiO 2 nanoparticles for lithium-ion polymer batteries. *Scientific reports, 6*, 26332.

35. Simon-Deckers, A., Gouget, B., Mayne-L'Hermite, M., Herlin-Boime, N., Reynaud, C., and Carriere, M. (2008). In vitro investigation of oxide nanoparticle and carbon nanotube toxicity and intracellular accumulation in A549 human pneumocytes. *Toxicology, 253*(1-3), 137-146.

36. Simonet, B. M., and Valcárcel, M. (2009). Monitoring nanoparticles in the environment. *Analytical and bioanalytical chemistry, 393*(1), 17.

37. Simonin, M., and Richaume, A. (2015). Impact of engineered nanoparticles on the activity, abundance, and diversity of soil microbial communities: a review. *Environmental Science and Pollution Research, 22*(18), 13710-13723.

38. Sreeprasad, T. S., and Pradeep, T. (2013). Noble metal nanoparticles *Springer Handbook of Nanomaterials* (pp. 303-388): Springer.

39. Sun, M., Qu, S., Hao, Z., Ji, W., Jing, P., Zhang, H. Shen, D. (2014). Towards efficient solid-state photoluminescence based on carbon-nanodots and starch composites. *Nanoscale, 6*(21), 13076-13081.

40. Thill, A., and Zeyons, O. (2006). spalla O, Chauvat f, Rose J, Auffan M, flank AM. Cytotoxicity of CeO2 nanoparticles for Escherichia coli: Physico-chemical insight of the toxicity mechanism. *Environ. sci. Technol, 40,* 6151-6156.

41. Trujillo-Reyes, J., Majumdar, S., Botez, C., Peralta-Videa, J., and Gardea-Torresdey, J. (2014). Exposure studies of core–shell Fe/Fe3O4 and Cu/CuO NPs to lettuce (Lactuca sativa) plants: are they a potential physiological and nutritional hazard? *Journal of hazardous materials, 267,* 255-263.

42. VandeVoort, A., and Arai, Y. (2018). Macroscopic Observation of Soil Nitrification Kinetics Impacted by Copper Nanoparticles: Implications for Micronutrient Nanofertilizer. *Nanomaterials, 8*(11), 927.

43. Wakaskar, R. R. (2018). General overview of lipid–polymer hybrid nanoparticles, dendrimers, micelles, liposomes, spongosomes and cubosomes. *Journal of Drug Targeting, 26*(4), 311-318. doi: 10.1080/1061186X.2017.1367006

44. Wang, C., Liu, H., Liu, Y., and Jiang, C. (2014). Comparative activity of TiO2 microspheres and P25 powder for organic degradation: Implicative importance of structural defects and organic adsorption. *Applied Surface Science, 319,* 2-7.

45. Wen, J., Li, X., Liu, W., Fang, Y., Xie, J., and Xu, Y. (2015). Photocatalysis fundamentals and surface modification of TiO2 nanomaterials. *Chinese Journal of Catalysis, 36*(12), 2049-2070.

46. Xue, W., Huang, D., Zeng, G., Wan, J., Zhang, C., Xu, R. Deng, R. (2018). Nanoscale zero-valent iron coated with rhamnolipid as an effective stabilizer for immobilization of Cd and Pb in river sediments. *Journal of hazardous materials, 341,* 381-389.

47. Yang, S.-Y., Chang, K.-H., Tien, H.-W., Lee, Y.-F., Li, S.-M., Wang, Y.-S. Hu, C.-C. (2011). Design and tailoring of a hierarchical graphene-carbon nanotube architecture for supercapacitors. *Journal of Materials Chemistry, 21*(7), 2374-2380.

Chapter 30

Toxicity of Nanoparticles on Natural Enemies and Pollinators

Malaichamy Kannan[1] *, *Kolanthasamy Elango*[2] and *Govindaraju Kasivelu*[3]

[1]*Department of Plant protection, HC&RI, Periyakulam, Tamil Nadu*
[2]*Department of Entomology, Tamil Nadu Agricultural University, Coimbatore – 641 003, Tamil Nadu*
[3]*Centre for Ocean Research, Sathyabama Institute of Science and Technology, Chennai – 600 119, Tamil Nadu*
e-mail: kanento@gmail.com

Introduction

The fast emergence of nanotechnology in various fields of science and technology necessitates to understand their toxic effects on the development and physiology of non-target organisms and environment. The application of nanoscience and technology in plant pest and disease management is in infant stage. Nanomaterial can be produced by natural, incidental or engineered; their forms could be amorphous, crystalline, polymeric or composites with different in shapes, *viz.*, spheres, tubes, rods, cones and fibres. It can be non-metal (*e.g.*, carbon, fullerene), metallic (*e.g.*, Au, Ag), semiconductor (*e.g.*, Cd, Se), or a combination (Huang *et al.*, 2007). Nanomaterials are formed atom by atom by building process or from macro, micro sized material to nano size by breaking process. Using the above methods, scientific communities developed inorganic, organic, composites nanomaterials and effectively utilize in almost all the science sector to social sector. Even then, the applicability of nanomaterials in agriculture is still under explored, especially in crop protection and production (Salata, 2004). Nanomaterial has been widely used as conductor, semiconductor, medical device, sensors, coating, catalytic agents and also as pesticides in the recent.

Nanotechnology set to offer a platform to transform agriculture sector from production, protection, processing and storage. The key contributions due to application of nanotechnology in agriculture are delivery systems that aid in slow release with more use efficiency in agro-inputs usage. In India, the post green revolution demands to reduce the use of higher quantities of fertilizers and pesticides to safe guard the future agriculture and environment. The increasing use of pesticides has a negative impact on farmers, consumers, non-target organism and the environment. A challenge to balance between crop production and environmental protection can be achieved by adopting nanotechnology. Several questions such as the way by which the nanoparticles enter in the cell and arise physiological events inside the cell are raised on the toxic effects of nanoparticles. The main aim of this chapter is accounting a comprehensive knowledge on the impacts of nanomaterials on the growth, development, parasitism or predatory efficiency, adult emergence of effective beneficial insects and also identifying the way to conserve parasitoids, predators and pollinators from nanotoxicity in the cropping system.

Nanocides for Insect Control

Various engineered nanoparticles are classified based on their physico-chemical composition and their properties. Further, the activity of nanoparticles changes based on their morphology, size, shape, composition, uniformity, distribution and agglomeration (Buzea *et al.*, 2007; Nam *et al.*, 2008). In spite of how these nanomaterials are classified, a variety of nanoparticles even within a single chemical due to variation in size, shape, specific surface area, would result in different chemical reactivity, bio-availability and eco-toxicity (Handy *et al.*, 2008).

Nanomaterials are effective against diversified insect pests of crops. Various kinds of nanomaterials *viz.*, titanium dioxide, silver, silica, zinc oxide, and aluminium oxide nanoemulsion can be developed as novel insecticide against insect pests in field and storage. The modes of action of nanoparticles reveals that the applied nanomaterials on the plant get adsorbed in insect body when they crawl over the surface and absorbed into the cuticular lipids (Chandrashekharaiah *et al.*, 2015). Further, the peristalic movement of the insect facilitates the nanoparticle enter in to the body tissue and cause physical death to cell organelle, followed by oozing out of body fluid lead to desiccational death. Ulriches *et al.* (2005) stated that the surface charge modified hydrophobic nanoparticles can be successfully advocated for management of ectoparasites in animals and insect pests of agriculture.

Many scientific workers have reported the growth and opportunity in nanotechnology as potent tool in changing agriculture. Several nanoparticles like aluminium oxide, silver, gold, silicon dioxide, zinc oxide, titanium dioxide and silica nanoparticles biologically derived from plant have proved its insecticidal activity, larval mortality, larval, pupal, adult developmental periods and fecundity against rice weevil, *Sitophilus oryzae* L.; red flour beetle, *Tribolium castaneum* H., larvae of *Spodoptera litura* L., oleander aphid, *Aphis nerii* B., bruchid beetle, *Callosobruchus maculatus* F. and diamond backmoth, *Plutella xylostella* L., respectively (Wang *et al.*, 2007; Barik *et al.*, 2008; Yang *et al.*, 2009; Bhattacharyya *et al.*, 2010; Goswami *et al.*, 2010; Chakravarthy *et al.*, 2012; Rouhani *et al.*, 2012 and Sahayaraj *et al.*, 2016).

These studies can expand the application of nanoparticle based technologies in pest management. Hence, nanomaterials associated insecticides formulations can change the activity by improved formulation, target specific easy application, improved efficiency, reduced dose and less environmental threats over conventional insecticide.

Entomophagous Insects in Agriculture

Beneficial insects are important component in the food chain and they afford amenity for maintaining the ecological imbalance in the agro ecosystem. The members in the group of beneficial insects are pollinators, natural enemies, weed killers, productive insects (honey bees, silkworms, and lac insects), scavengers and decomposers. Pollinators like honey bees, solitary bees, bumblebees, pollen wasps, ants, hoverflies, mosquitoes, butterflies, moths and flower beetles play vital role in pollinating the agricultural crop species for seed and fruit set for enhancing agricultural productivity (Capinera 2005). Similarly, weeds feeding insects, *viz.*, *Cochnieal* insect, *Aristalochia* butterfly, *Caotropis* butterfly, AK grasshopper, Water hyacinth weevil and Parthenium weed killer feed on prickly pear, aristalochia, calotropis, eichhornia and parthenium, respectively.

In addition, insects namely scavenger beetles, soil ants, larvae of cut-worms, flies, crickets, termites and wasps live in soil make tunnels. Insect's activities in soil improve soil aeration, activity of beneficial soil microorganisms bring the subsoil surface which enhances the earthworm activity indeed improves the soil properties such as texture, structure and pore space. The adults and larvae insects like rove beetles, chafer beetles, darkling beetles, nitidulids, water scavenger beetle, daddy long legs, muscid flies, termites, ants feed on decaying matter and convert the complex material into simple substances through biochemical cycling which prevent the environment hazards. The adults and larvae insects like rove beetles, chafer beetles, darkling beetles, nitidulids, water scavenger beetle, daddy long legs, muscid flies, termites and ants feed on decaying matter.

The parasitoids and predators devour the agricultural insect pests at egg, larval/nymphal, pupal and adult stages and bring down the pest load below economic threshold level. The entomophagous insects have the ability of reducing the pest population below 30 per cent under field condition even without insecticide application. The biological control was achieved mainly by parasitoids and predators of invasive insect pests like Pink hibiscus mealybug, *Maconellicoccus hirsutus* G., Cassava mealy bug, *Phenacoccus manihoti* M., sugarcane woolly aphid, *Ceratovacuna lanigera* Z. and papaya mealy bugs, *Paracoccus marginatus* W. were managed by *Cryptolaemus montrouzieri* M., *Apoanagyrus* (*Epidinocarsis*) *lopezi* D., *Dipha aphidivora* M., *Acerophagus papaya* N., respectively are some of the best examples for natural enemies role in pest management (Ballal and verghese 2015) (Tables 30.1 and 30.2).

Toxicity of Chemical Pesticides and Nanoparticles on Natural Enemies

The growth of Indian agriculture was positively progressed through green revolution by application broad-spectrum insecticides and fertilizer responsive

Table 30.1. Important Insect Entomophages

Order	*Predator*	*Prey insect*	*Host Stage*
Odonata - Dragonflies	*Sympetrum frequens* *S. infuscatum*	Mosquito	Larva - wriggler
Coleoptera - Lady bird beetles	*Coccinella septumpunctata*	Aphids	Nymphs and Adults
	Cryptolaemus montrouzieri	Papaya mealy bug, Cushiony Scales	Nymphs and Adults
	C. transversalis	Aphids	Nymphs and Adults
	Cheilomenes sexmaculatus	Aphids	Nymphs and Adults
	Scymnus coccivora		
	Chilochorus nigritus	Scales	Nymphs and Adults
	Rodalia cardinalis	Cottony cushion scale	Nymphs
Neuroptera	*Chrysoperla* spp.	All soft body insects, lepidopterans	Nymphs and Adults; Eggs stages in lepidopterans
	Micromus igrotus	Sugarcane wooly aphid	Nymphs
Lepidoptera	*Dipha aphidivora*	Sugarcane wooly aphid	Nymphs and Adults
Hemiptera	*Cyrtorhinus lividipennis*	Rice hoppers	Nymphs
	Platymeris laevicollis	Coconut rhinoceros beetle	Grub
	Eucantheconidia furcellata	Red hairy caterpillar	Larva
	Rhynocoris sp	Soft body insects	All stages
	Geocoris sp	Insect eggs, aphids, mealybugs, spider mites, leafhoppers, plant bugs, whiteflies, caterpillars, and beetle larvae.	Eggs, Nymphs and grub stages
	Belostomatid sp	Variety of aquatic insects	Nymphs and Adults
Predatory Spider	*Lycosa pseudoannulata*	Rice BPH	Nymphs
Predatory Mite	*Phytoseiulus persimilis*	Red spider mite	All stages
Mantids	*Mantis religiosa*	Flies, grasshopper and many caterpillars	All statges
Hymenoptera	*Vespidae*	caterpillars	Larva

Source: Ballal and Verghese (2015).

varieties have created impact on the reduction of natural enemies' populations which increased the pest resurgence, resistance and residues. Further, indiscriminate use of fertilizers and pesticides are threatening the sustainability of agriculture around the globe (Bueno and Freitas, 2004). Several broad spectrum pesticides and their metabolites are litigated unsafe to natural enemies when they exposed directly. Further, studies conducted by Lord *et al.* (1968) and Stapel *et al.* (2000) evidenced that systemic novel group of insecticides caused ill effects to pollinators, predators and parasitoids (Gill and Garg, 2014).

Table 2. Important insect parasitoids

Crops	*Pest*	*Family*	*Parasitoids*	*Category*
		HYMENOPTERA		
Rice, Sugarcane, Vegetables	Many lepidopterous pests	Trichogrammatidae	*Trichogramma chilonis*	Egg parasitoid
Cotton	*Spodoptera litura*	Scelionidae	*Telenomus remus*	Egg parasitoid
Coconut	*Opisina arenosella*	Braconidae	*Bracon brevicornis*	Larval parasitoid
Cotton	Spotted bollworms, *Earias* spp.		*Chelonus blackburni*	Egg larval parasitoid
Potato	Tuber moth	Encyrtidae	*Capidosoma koehleri*	Egg larval parasitoid
Rice	Gall midge	Platygasteridae	*Platygaster oryzae*	Larval parasitoid
Coconut	*Opisina arenosella*	Chalcididae	Brachymeria nephantidis	Larval parasitoid
		Bethylidae	*Goniozus nephantidis*	Larval parasitoid
Sugarcane	Top borer	Ichneumonidae	*Isotima javensis*	Larval pupal parasitoid
Papaya	Mealy bug	Encyrtidae	*Acerophagous papaya*	Nymphal adult parasitoid
Cotton	Whitefly	Aphelinidae	*Encarsia formosa*	Nymphal adult parasitoid
Coconut	*Opisina arenosella*	Eulophidae	*Trichospilus pupivora*	Pupal parasitoids
			Tetrastichus israeli	Pupal parasitoids
		DIPTERA		
Sugarcane	Shoot borer *Chilo infuscatellus*	Tachinidae	*Sturmiopsis inferens*	Larval parasitoid
Coconut	*Opisina arenosella*		*Spoggosia bezziana*	Larval parasitoid
Cotton, Sun flower and fruits	*H. armigera*		*Eucelatoria bryani*	Larval pupal parasitoid
		LEPIDOPTERA		
Sugarcane	Leafhopper, *Pyrilla perpusilla*	Epiricanidae	*Epiricania melanoleuca*	Nymphal adult parasitoid

Source: Ballal and Verghese (2015).

The possibility for utilization of nanotechnological tools in pest management are enormous. Nanotechnology enables techniques for slow release of fertilizer, efficient dosage of water through nanoporous zeolites, nanoencapsulates, nanomaterials for delivery of plant protection inputs in pest management, nano sensors for pest detection and monitoring. Further studies are required to assess the effects of these nano inputs on the beneficials to reduce the adverse effects and conserve them to maintain the ecological balance (Figure 30.1).

Toxicity of Nanoparticles on Entomophagous

The dragon flies, damsel flies, preying mantis, giant water bug, robber flies, hover flies and wasps are the common entomophagous of insect pests. The exposure of low-concentration silver nanoparticles altered the behaviour, survival, and reproduction in common natural predatory dragonfly (Green Darner, *Anax junius* Drury: Odonata) naiads (Pokhrel and Dubey 2012). It clearly evident that Ag nanoparticle (NP) exposures could influence the life history and behavioural traits in Odonata (Lovern *et al.*, 2007; Brausch *et al.*, 2011). Murugan *et al.* (2015) also documented that AgNPs synthesized from *Datura metal* increased the predator potential of young instars of dragonfly (blue Darner, *Anax immaculifrons* Rambur) against malarial vector *Anopheles stephensi* Liston from 75.5 to 95.5 and 53.5 to 78 per cent in second and third instars of dragonfly naiads. At lower doses the thorn apple, *Datura metel* L. synthesized Ag NPs reduce the larval population of malaria vectors and have no adverse effect on behaviour of dragonfly naiads.

Nanomaterials use as a soil amendment affect not only the soil environment but also the entire soil microbial community and only little effects are known and many on the interaction of nanomaterials with soil microbial biota are unknown. McKee and Filser (2016) concluded that metal nanomaterials decrease the nitrogen and increase carbon emissions. Because of the hindrance of nanomaterials on the population ratio of microorganisms lead the trophic transfer from bacteria to fungi. Further, nanomaterial affecting the food web through biomagnification is also possible. Similarly, the short-term exposure of applied chemical insecticides namely imidacloprid, fipronil and its metabolites fipronil sulfone in rice ecosystem caused significant harm as acute toxicity, feeding inhibition and delayed toxicity to dragon flies (Autumn Darter, *Sympetrum frequens* S. and *Sympetrum infuscatum* S.) naiads (Jinguji *et al.*, 2018). Accordingly, paralysis due to fipronil and fipronil sulfone sublethal effects may cause feeding inhibition which leads to death in dragon flies species and realistic application fipronil may eliminate dragonflies in paddy ecosystem. The findings of the above studies suggested that the chemical insecticides imidacloprid, fipronil and fipronil sulfone caused mortality to dragonflies naiads. On the contrary, the Ag NPs synthesized from *Datura metal* increased the predator potential of young instars of dragonfly *A. immaculifrons*.

The green lacewings are efficient and successful predators in the management of insect pests. Karthika *et al.* (2015) tested the safety of hexanal nanoemulsion to the *Chrysoperla carnea* Stephens grubs and stated that nanoemulsion was not toxic to the predator grub, 100 per cent pupation and adult emergence was also noticed. Similarly, Kannan and Elango (2019) studied the effect of silica nano particles at

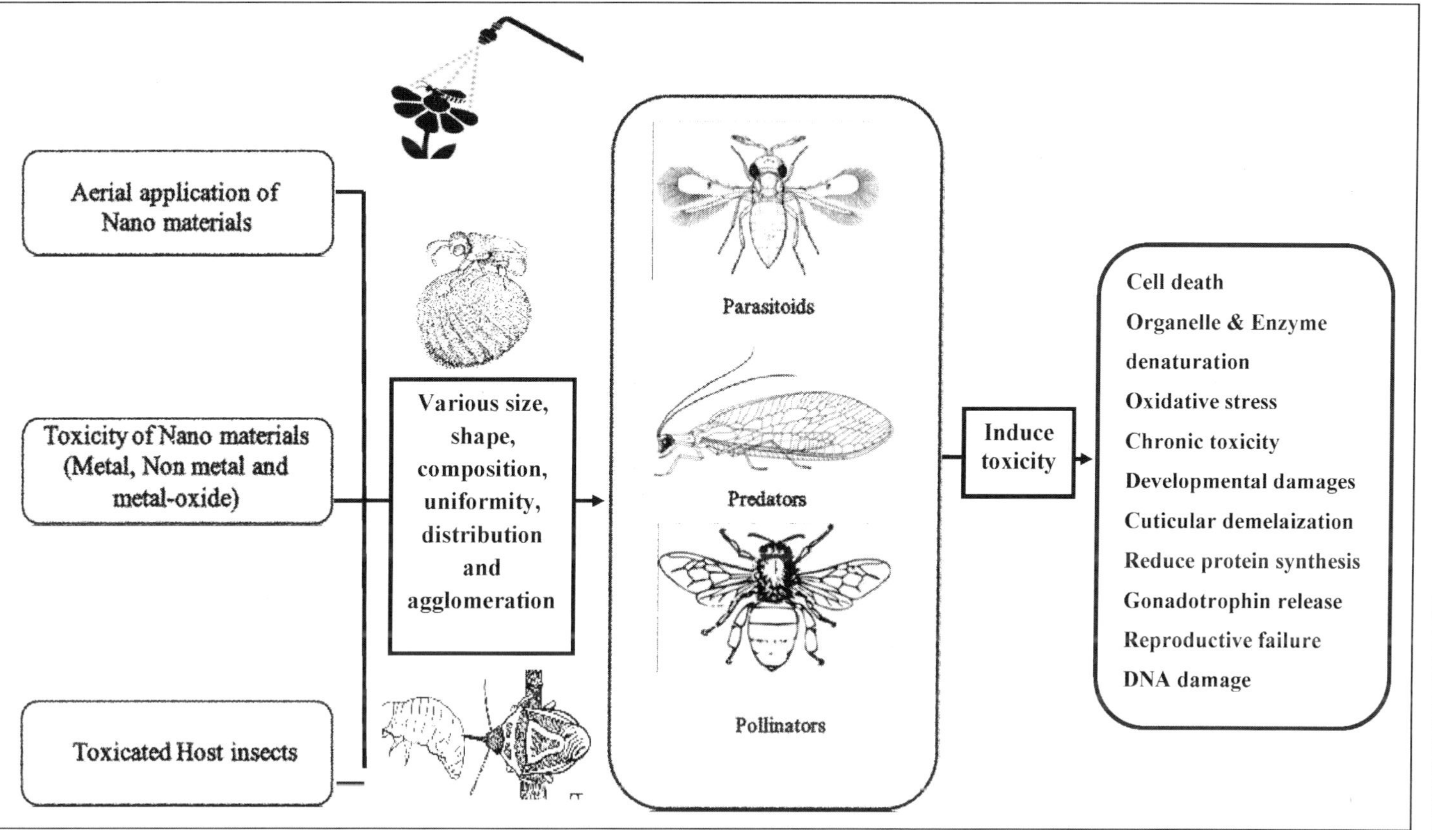

Figure 30.1. Scheme for Toxicity Mechanisms of Nanoparticles against Beneficial Insects

different doses (20000, 15000, 10000, 5000, 4500, 1000, 500, 100 and 50 ppm) on the grubs of green lacewing, *Chrysoperla zastrowi* sillemi (Esben-Peterson) and revealed that the emergence was only 80.00, 60.00 per cent in 20000 and 15000 ppm, respectively compared to the untreated check (Table 30.3 and Figures 30.2 and 30.3).

Table 30.3. Toxicity of Silica Nanoparticles to Grubs of Green Lacewing, *Chrysoperla zastrowi sillemi* **(Esben-Peterson)**

Treatments PPM	*Per cent Develop-ment of Pupa**	*Per cent Mortality**	*Pre pupal Duration* (Days ± SD*)*	*Pupal Duration (Days± SD*)*	*Per cent Adult Emergence**	*Category*
20000	20.00	80.00	3.00±0.71	8.00±0.00	0.00	Moderately harmful
15000	40.00	60.00	2.50±0.00	7.75±0.71	20.00	Slightly harmful
10000	40.00	60.00	2.00±0.58	7.50±0.71	20.00	Slightly harmful
5000	60.00	40.00	2.33± 0.57	7.33±0.58	40.00	Harmless
2500	60.00	40.00	1.67±0.50	7.33±0.65	40.00	Harmless
1000	80.00	20.00	1.75± 0.71	7.00±0.00	60.00	Harmless
500	100.0	0.00	2.00± 0.84	6.40±0.89	80.00	Harmless
100	100.0	0.00	1.80±0.45	6.40±0.71	100.00	Harmless
50	100.0	0.00	1.80±0.55	6.00±0.55	100.00	Harmless
Control	100.0	0.00	1.40±0.71	6.00±0.45	100.00	Harmless

* <50 mortality - harmless; 50-79 - Slightly harmful; 80-89 - Moderately harmful; >90 - Harmful

Courtesy: Kannan and Elango, 2019.

Quite the reverse, compared to nanomaterials, even at low doses, chemical insecticides caused negative impact on the growth and development of the common green lacewing, *Chrysoperla carnea* (Stephens) (Neuroptera: Chrysopidae). The fenoxycarb has been reported to cause the prolonged development time in all the stages of the predator *Chysoperla rufilabris* Burmeister (Liu and Chen, 2001) and imidacloprid, propargite and pymetrozine also caused significant reductions in fecundity (Rezaei *et al.*, 2007). Similarly, the direct contact of spinosad caused reduced progeny production and malformation of biological stages in *C. carnea* (Medina *et al.* (2001, 2003; Desneux *et al.*, 2007). Mandour (2009) and Maia and Moore (2011) also stated that direct spray of spinosad on immature stages of *C. carnea* affected the reproduction and survival of adult.

Toxicity of Nanoparticles on Parasitoids

Trichogrammatidae are effective egg parasitoids and parasitize the eggs of several insects belonging to more than eight orders in aquatic and terrestrial habitats. Karthika *et al.* (2015) observed lesser toxicity by nanoemulsion of hexanal @ 0.02 per cent to the immature stages of *Trichogramma japonicum*

Control

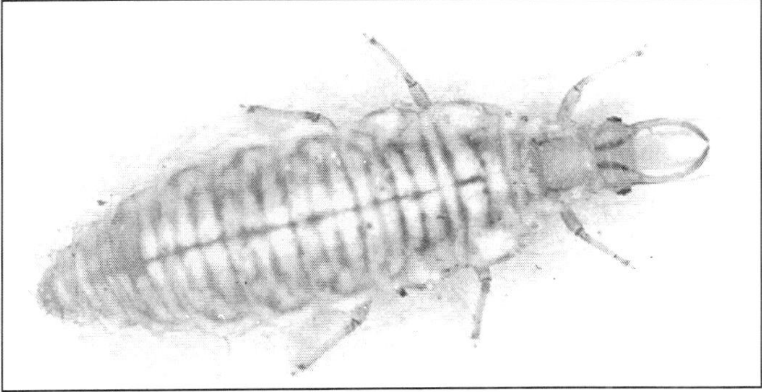

Healthy grub (Greenish yellow colour)

Treated with 20000 ppm of SiO_2

Died grub (Body parts turned into black colour

Figure 30.2. Effect of Silica Nano Particles to Grubs of Green Lacewing, *Chrysoperla zastrowi sillemi* (Esben-Peterson).

Ashmead. The mean adult emergence of two experiments ranged from 97.15 to 93.05 per cent in different doses. The hexanal treatments had little impact on the parasitization of *Trichogramma chilonisi* Ishii. Further, Mohan *et al.* (2017) revealed that nanoemulsion of hexanal was safer to *Trichogramma pretiosum* Riley and *T. chilonis* with 98.53 and 97.88 per cent parasitization and 97.57 and 96.60 adult emergence respectively. In addition, after the treatment the respective II generation *T. pretiosum* and *T. chilonis* adults also showed 96.12 and 97.65 per cent parasitization with 94.29 and 96.40 per cent adult emergence, respectively. Finally they decided that hexanal being

Control

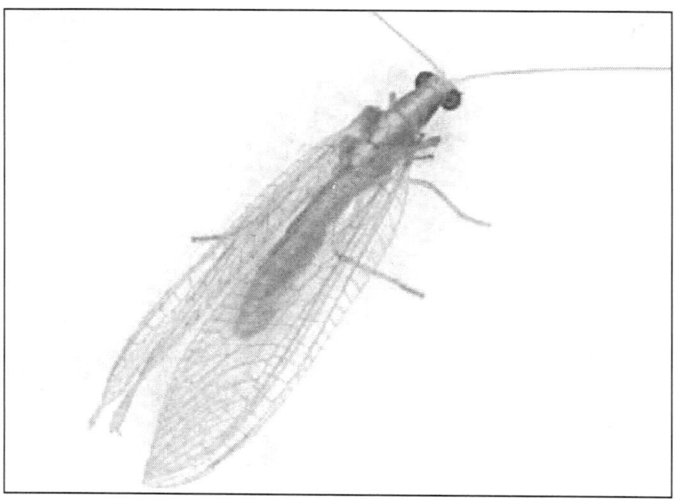

Healthy Adult

Treated with 20000 ppm of SiO_2

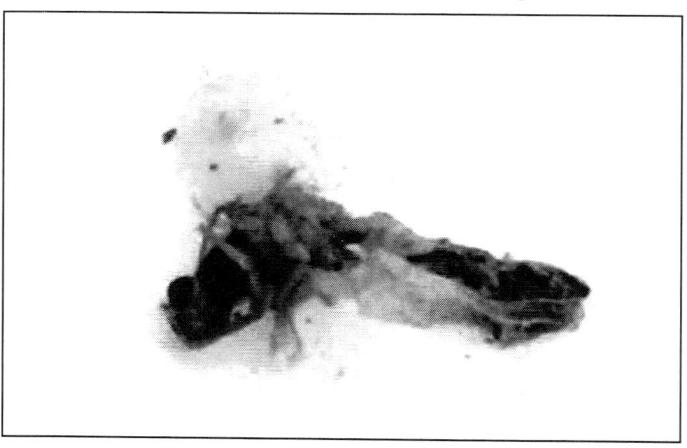

Malformed Adult

Figure 30.3. Effect of Silica Nano Particles to Adults of Green Lacewing, *Chrysoperla zastrowi sillemi* (Esben-Peterson).

a naturally occurring alkyl aldehyde compound in plants, hence the nano hexanal formulation was safer to egg parasitoids.

Likewise, Preetha *et al.* (2018) reported that the neem oil based nanoemulsion was safer to egg parasitoid, *T. chilonis* than macro-emulsion. The parasitism (86.00 per cent) and adult emergence (79.98 per cent) was significantly higher in neem oil nanoemulsion; the lowest per cent adult emergence (48.45 per cent) and parasitism

(66.78 per cent) was recorded in the highest concentration of neem oil macro emulsion. The adult emergence was found on increasing rate by decreasing the concentrations of neem nano and macroemulsions. Equally, Kannan and Elango (2019) also studied the effect of silica nano particles at different doses (20000, 15000, 10000, 5000, 4500, 1000, 500, 100 and 50 ppm) on the parasitization potential of treated *Trichogramma chilonis* adults and revealed that maximum parasitization in untreated check (94.30 per cent). At 20000 and 15000 ppm concentration only silica nano particles showed maximum effect on parasitization of *T. chilonis* (42.78 and 46.10 per cent respectively) (Table 30.4 and Figure 30.4).

Table 30.4. Toxicity of Silica Nano Particles to Adults, *Trichogramma chilonis* Ishii

Treatments PPM	*Total No of Eggs*	*Total No of Parasitized Eggs*	*Per cent Prasiti-zation**	*Per cent un parasitized Eggs**	*Per cent Reduction Over Control*	*Category*
20000	180.00	77.00	42.78	57.22	54.64	Slightly harmful
15000	154.00	71.00	46.10	53.90	51.11	Slightly harmful
10000	163.00	83.00	50.92	49.08	46.00	Harmless
5000	158.00	96.00	60.76	39.24	35.57	Harmless
2500	172.00	108.00	62.79	37.21	33.42	Harmless
1000	166.00	118.00	71.08	28.92	24.62	Harmless
500	154.00	126.00	81.82	18.18	13.24	Harmless
100	179.00	158.00	88.27	11.73	6.40	Harmless
50	166.00	150.00	90.36	9.64	4.18	Harmless
Control	158.00	149.00	94.30	5.70	0.00	Harmless

*<50 mortality - harmless; 50-79 - Slightly harmful; 80-89 - Moderately harmful; >90 - Harmful.

Courtesy: Kannan and Elango, 2019.

In contrast, except methoxyfenozide and tebufenozide, lambda cyhalothrin, cypermethrin, thiodicarb, profenophos and spinosad, adversely affected the emergence, adult survival, life span and fitness parameters of *Trichogramma exiguum* (Pinto and Platner) in *Helicoverpa zea* (Boddie) host eggs (Suh *et al.*, 2000). Similarly, Shoeb (2010) also observed the effects, when they treated the immature stages of *Trichogramma evanescens* (Westwood) with Lambda-cyhalothrin, Spinosad, and Fenitrothion. Subsequently, they concluded that negative effects on parasitism and emergence of parasitoids were observed. But the neem products based bio insecticides achieved nil serious side effect on parasitism and emergence rates of *T. pretiosum* (Riley) and *T. minutum* (Riley) and have given good control against *Helicoverpa armigera* (Hübner) under greenhouse condition (El-Wakeil *et al.*, 2006).

Toxicity of Nanoparticles on Pollinators (Honey Bees)

Plant pollination through insects is the natural event happens in reproduction of the common flowering plants of domesticated species. Many plant species

Control

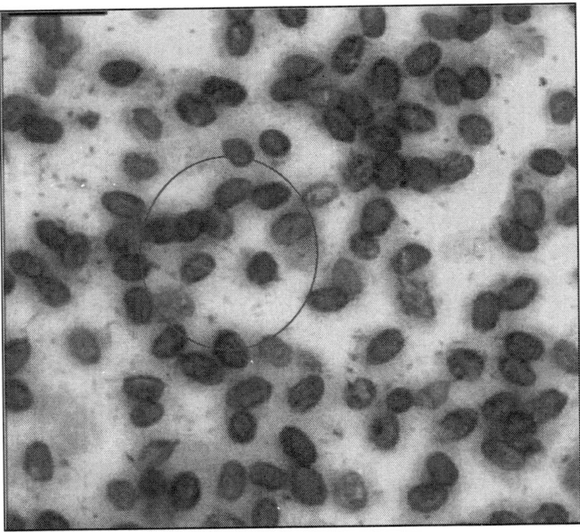

Parasitized eggs

Treated with 20000 ppm of SiO_2

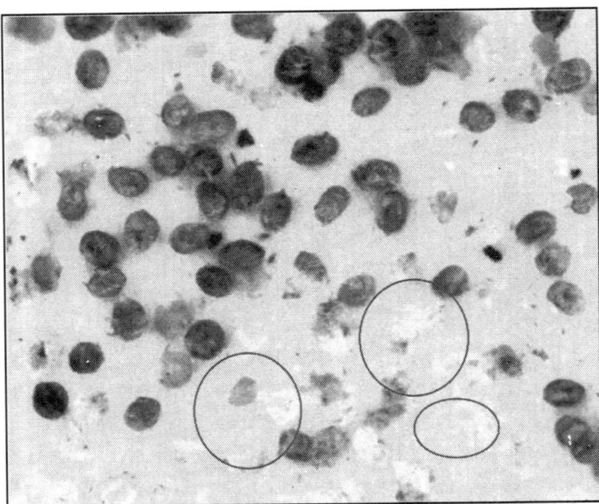

Unparasitized eggs

Figure 30.4. Parasitization Potential of Silica Treated *Trichogramma chilonis* Adults on Corcyra Eggs.

insects are the only source of pollinating agent (Table 30.3) and their contribution in pollination was around 36 per cent which enhances the fruit set and yield of the cultivated crops. Near future, nanomaterial based agricultural inputs application may yield effect on the beneficial aspects of crop cultivation but also permits the possibility of adverse effects on the beneficials in the agro ecosystem. Some of the laboratory studies summarized below enlist the effects of nanomaterials on the pollinators. Karthika *et al.* (2015) concluded that no mortality was found in any of the nano hexanal formulation treatments even after 48 hours after treatment. The results are in comparable with Flesar *et al.* (2010) who reported that acute oral toxicity of the most active natural products was determined on adult honey bees, showing them as non-toxic at concentrations as high as 100 μg/bee. Likewise Mohan *et al.* (2017) reported that all the concentration of nano formulation of hexanal was harmless to worker honey bees (maximum mortality of 7.65 per cent in *Apis cerana indica* (Fabricius), 5.80 per cent in *Apis mellifera* (Linnaeus) and 5.20 per cent in *Apis florea* (Fabricius) at 24 hours after treatment) which recorded a mortality of < 30 percent as per the threshold prescribed by IOBC (Table 30.5).

Table 30.5. Effect of Hexanal Nano Emulsion to Honey Bees

Treatments	*Italian bees Apis mellifera L.*	*Indian bees Apis cerana indica F.*	*Little bees Apis florea F.*
	Mortality of Worker Bees (per cent)		
	12 Hours After Treatment		
T1 - Nano emulsion of hexanal @ 0.02 per cent	3.33	4.25	3.33
T2 - Nano emulsion of hexanal @ 0.04 per cent	4.05	4.55	3.45
T3 - Nano emulsion of hexanal @ 0.06 per cent	4.90	5.00	4.30
T4 - Pure hexanal @ 0.02 per cent	5.89	6.40	6.20
T5 - Pure hexanal @ 0.04 per cent	6.67	7.65	6.67
T6 - Pure hexanal @ 0.06 per cent	7.95	7.90	8.90
T7 - Tween 20 @ 0.2 per cent	3.30	3.33	3.30
T8 - Ethanol @ 0.2 per cent	2.20	1.10	2.20
T9 - Control	0.00	0.00	0.00

Courtesy: Mohan *et al.*, 2017.

On the contrary, compared to nanoformulations, chemical insecticides caused considerable adverse effect on the pollinators. Bees exposed to imidaclopride insecticides altered the foraging behavior in *Apis millifera* and bees are not returned to the hive then control bees (Yang *et al.*, 2008). In the same way, bees intoxicated with the sublethal doses of neonicotinoid insecticides affected the navigation, foraging efficiency of bees and adult of natural enemies. The persistence toxicity of insecticides *viz.*, imidacloprid, flipronil, indoxacarb, dimethoate, cypermenthrin and azadirachtin to honey bees assert that fipronil followed by imidacloprid recorded highest residual toxicity than other insecticides Morandin and Winston (2003).

Conclusions

Nanomaterials are effective nano-cides against insects and pests. The use of different type of available metallic nanomaterials and nanoemulsion can be effectively used as a novel insecticide in the management applications. Development of appropriate nano formulations could open up novel plant protection tool in insect pest management than chemical insecticides to conserve, enhance the toxicity, persistence to UV radiation, safer to environment and extended self-life may reduce the inconsistence in field performance of many potential biomolecules and bioagents for sustainable agriculture. The knowledge on the impacts of nanomaterials on growth, development, parasitism or predatory efficiency and emergence capacity as well as to safeguard beneficial insects are essential to conserve parasitoids, predators and pollinators from nanotoxicity, at the same time to suppress the insect pests in the cropping system.

Based on the recent literatures demonstrated the impacts of nanomaterials against beneficial insects especially predators, parasitoids and pollinators, results clearly indicated that nanomaterials are not causing deleterious effects to the beneficial insects. There was a little harmfulness was observed in 500 fold (20000ppm) increased LC_{50} dose of host insect pests. Compared to nanomaterials the synthetic chemical insecticides, even at very low doses caused toxicity and impaired the development of natural enemies from host insects. Hence, the knowledge on the effects of nanomaterials on biological parameters, parasitism or predatory efficiency and emergence capacity of beneficial insects are essential to conserve parasitoids, predators and pollinators from nanotoxicity to suppress the insect pests in the cropping system. This chapter clearly briefed that nanomaterials are not causing harmful effects to the beneficial insects but synthetic chemical insecticides, even at very low doses caused acute toxicity, feeding inhibition, growth impairment and delayed toxicity to beneficial insects.

Nanotechnological principles and concepts can be exploited for early detection, monitoring and management of insect pests. It is set to offer a platform to transform agriculture sector from production, protection, processing and storage. The key contributions due to application of nanotechnology in agriculture are delivery systems that aid in slow release and efficient delivery of agro-inputs. The growing population around the world demands more use of agricultural inputs to meet the profuse food stock with low cost of production. The increasing use of pesticides have a negative impact on farmers, consumers, non-target organism and environment. A challenge to balance between crop production and environmental protection can be achieved by adopting nanotechnology.

REFERENCES

1. Ballal, C. R and Verghese, A. (2015). Role of Parasitoids and Predators in the Management of Insect Pests. In *New Horizons in Insect Science: Towards Sustainable Pest Management* (pp. 307-326). Springer, New Delhi.

2. Barik, T. K., Sahu, B and Swain, V. (2008). Nanosilica–from medicine to pest control. *Parasitology research, 103*(2), 253.

3. Bhattacharyya, A., Bhaumik, A., Rani, P. U., Mandal, S and Epidi, T. T. 2010. Nano-particles-A recent approach to insect pest control. *African Journal of Biotechnology, 9*(24), 3489-3493.

4. Brausch, K. A., Anderson, T. A., Smith, P. N and Maul, J. D. 2011. The effect of fullerenes and functionalized fullerenes on Daphnia magna phototaxis and swimming behavior. *Environmental toxicology and chemistry, 30*(4), 878-884.

5. Bueno, A. F and Freitas, S. 2004. Effect of the insecticides abamectin and lufenuron on eggs and larvae of Chrysoperla externa under laboratory conditions. *BioControl, 49*(3), 277-283.

6. Buzea, C., Pacheco, I. I and Robbie, K. 2007. Nanomaterials and nanoparticles: sources and toxicity. *Biointerphases, 2*(4), MR17-MR71.

7. Capinera, J. L. 2005. Relationships between insect pests and weeds: an evolutionary perspective. *Weed science, 53*(6), 892-901.

8. Chakravarthy, A. K., Kandakoor, S. B., Atanu, B., Dhanabala, K., Gurunatha, K and Ramesh, P. 2012. Bio efficacy of inorganic nanoparticles CdS, Nano-Ag and Nano-TiO2 against Spodoptera litura (Fabricius) (Lepidoptera: Noctuidae). *Current Biotica, 6*(3), 271-281.

9. Chandrashekharaiah, M., Kandakoor, S. B., Gowda, G. B., Kammar, V and Chakravarthy, A. K. 2015. Nanomaterials: A Review of Their Action and Application in Pest Management and Evaluation of DNA-Tagged Particles. In *New Horizons in Insect Science: Towards Sustainable Pest Management* (pp. 113-126). Springer, New Delhi.

10. Desneux, N., Decourtye, A., and Delpuech, J. M. 2007. The sublethal effects of pesticides on beneficial arthropods. *Annu. Rev. Entomol., 52*, 81-106.

11. El-Wakeil, N. E., Gaafar, N. M and Vidal, S. 2006. Side effect of some neem products on natural enemies of Helicoverpa (Trichogramma spp.) and Chrysoperla carnea. *Archives of Phytopathology and Plant Protection, 39*(6), 445-455.

12. Flesar, J., Havlik, J., Kloucek, P., Rada, V., Titera, D., Bednar, M and Kokoska, L. 2010. In vitro growth-inhibitory effect of plant-derived extracts and compounds against Paenibacillus larvae and their acute oral toxicity to adult honey bees. *Veterinary microbiology, 145*(1-2), 129-133.

13. Gill,K and Garg H. 2014. Pesticides: Environmental Impacts and Management Strategies, Pesticides - Toxic Aspects, Marcelo L. Larramendy and Sonia Soloneski, IntechOpen, DOI: 10.5772/57399

14. Goswami, A., Roy, I., Sengupta, S and Debnath, N.2010. Novel applications of solid and liquid formulations of nanoparticles against insect pests and pathogens. *Thin solid films, 519*(3), 1252-1257.

15. Handy, R. D., Von der Kammer, F., Lead, J. R., Hassellöv, M., Owen, R and Crane, M. (2008). The ecotoxicology and chemistry of manufactured nanoparticles. *Ecotoxicology, 17*(4), 287-314.

16. Huang, J., Li, Q., Sun, D., Lu, Y., Su, Y., Yang, X and Hong, J. 2007. Biosynthesis of silver and gold nanoparticles by novel sundried Cinnamomum camphora leaf. *Nanotechnology, 18*(10), 105-104.

17. Jinguji, H., Ohtsu, K., Ueda, T and Goka, K. 2018. Effects of short-term, sub lethal fipronil and its metabolite on dragonfly feeding activity. *PloS one, 13*(7), e0200299.

18. Ju-Nam, Y and Lead, J. R. (2008). Manufactured nanoparticles: an overview of their chemistry, interactions and potential environmental implications. *Science of the total environment, 400*(1-3), 396-414.

19. Kannan, M and Elango K.2019. Effect of Silica nano particles to egg parasitoids, *Trichogramma chilonis* Ishii and green lacewing, *Chrysoperla zastrowi sillemi* (Esben-Peterson). Unpublished data

20. Karthika, S., Kumar, N. B. N., Gunasekaran, K and Subramanian, K. S. 2015. Biosafety of nanoemulsion of hexanal to honey bees and natural enemies. *Indian J Sci Technol, 8*, 1-7.

21. Liu, T. X and Chen, T. Y. 2001. Effects of the insect growth regulator fenoxycarb on immature Chrysoperla rufilabris (Neuroptera: Chrysopidae). *Florida entomologist, 84*(4), 628.

22. Lord, K. A., MAY, M. A and Stevenson, J. H. 1968. The secretion of the systemic insecticides dimethoate and phorate into nectar. *Annals of Applied Biology, 61*(1), 19-27.

23. Lovern, S. B., Strickler, J. R and Klaper, R. 2007. Behavioral and physiological changes in Daphnia magna when exposed to nanoparticle suspensions (titanium dioxide, nano-C60, and C60HxC70Hx). *Environmental science and technology, 41*(12), 4465-4470.

24. Maia, M. F and Moore, S. J. 2011. Plant-based insect repellents: a review of their efficacy, development and testing. *Malaria journal, 10*(1), 511.

25. Mandour, N. S. 2009. Influence of spinosad on immature and adult stages of Chrysoperla carnea (Stephens)(Neuroptera: Chrysopidae). *BioControl, 54*(1), 93.

26. Medina P, Budia F, Del Estal PE, Viñuela E. 2003. Effects of three modern insecticides, pyriproxyfen, spinosad and tebufenozide, on survival and reproduction of *Chrysoperla carnea* adults. Annals of Applied Biology 142 (1): 55-61.

27. Medina, P., Budia, F., Tirry, L., Smagghe, G and Vinuela, E. 2001. Compatibility of spinosad, tebufenozide and azadirachtin with eggs and pupae of the predator Chrysoperla carnea (Stephens) under laboratory conditions. *Biocontrol Science and Technology, 11*(5), 597-610.

28. Mohan, C., S. Sridharan, K. S. Subramanian, N. Natarajan and S. Nakkeeran.2017. Effect of nanoemulsion of hexanal on honey bees (Hymenoptera; Apidae). Journal of Entomology and Zoology Studies 5(3): 1415-1418.

29. Mohan, C., Sridharan, S., Gunasekaran, K., Subramanian, K. S and Natarajan, N. 2017. Biosafety of hexanal as nanoemulsion on egg parasitoid Trichogramma Spp. *Journal of Entomology and Zoology Studies, 5*(2), 1541-1544.

30. Morandin, L. A and Winston, M. L. 2003. Effects of novel pesticides on bumble bee (Hymenoptera: Apidae) colony health and foraging ability. *Environmental Entomology, 32*(3), 555-563.

31. Murugan, K., Dinesh, D., Kumar, P. J., Panneerselvam, C., Subramaniam, J., Madhiyazhagan, P and Higuchi, A. 2015. Datura metel-synthesized silver nanoparticles magnify predation of dragonfly nymphs against the malaria vector Anopheles stephensi. *Parasitology research, 114*(12), 4645-4654.

32. Pokhrel, L. R. And Dubey, B. 2012. Potential impact of low-concentration silver nanoparticles on predator–prey interactions between predatory dragonfly nymphs and Daphnia magna as a prey. *Environmental science and technology, 46*(14), 7755-7762.

33. Preetha S, Kannan M, Lokesh S, Gowtham V.2018. Effect of neem oil based nanoemulsion on egg parasitoid, *Trichogramma chilonis* (Ishii) (Hymenoptera: Trichogrammatidae). Journal of Biological Control 2 (2): 103-107

34. Rezaei, M., Talebi, K., Naveh, V. H and Kavousi, A. 2007. Impacts of the pesticides imidacloprid, propargite, and pymetrozine on Chrysoperla carnea (Stephens)(Neuroptera: Chrysopidae): IOBC and life table assays. *BioControl, 52*(3), 385-398.

35. Rouhani, M., Samih, M. A and Kalantari, S. 2012. Insecticide effect of silver and zinc nanoparticles against Aphis nerii Boyer De Fonscolombe (Hemiptera: Aphididae). *Chilean journal of agricultural research, 72*(4), 590.

36. Sahayaraj, K., Madasamy, M and Radhika, S. A. 2016. Insecticidal activity of bio-silver and gold nanoparticles against Pericallia ricini Fab.(Lepidaptera: Archidae). *Journal of Biopesticides, 9*(1), 63.

37. Salata, O. V.2004. Applications of nanoparticles in biology and medicine. *Journal of nanobiotechnology, 2*(1), 3.

38. Shoeb, M. A. 2010. Effect of some insecticides on the immature stages of the egg parasitoid Trichogramma evanescens West. (Hym., Trichogrammatidae). *Egyptian Academic Journal of Biological Sciences. A, Entomology, 3*(1), 31-38.

39. Stapel, J. O., Cortesero, A. M. and Lewis, W. J. 2000. Disruptive sublethal effects of insecticides on biological control: altered foraging ability and life span of a parasitoid after feeding on extrafloral nectar of cotton treated with systemic insecticides. *Biological Control, 17*(3), 243-249.

40. Suh, C. P. C., Orr, D. B and Van Duyn, J. W. 2000. Effect of insecticides on Trichogramma exiguum (Trichogrammatidae: Hymenoptera) preimaginal development and adult survival. *Journal of Economic Entomology, 93*(3), 577-583.

41. Ulrichs, C., Mewis, I. And Goswami, A. 2005. Crop diversification aiming nutritional security in West Bengal: biotechnology of stinging capsules in nature's water-blooms. *Ann Tech Issue of State Agri Technologists Service Assoc,* 1-18.

42. Wang, L., Li, X., Zhang, G., Dong, J and Eastoe, J. 2007. Oil-in-water nanoemulsions for pesticide formulations. *Journal of colloid and interface science, 314*(1), 230-235.

43. Yang, E. C., Chuang, Y. C., Chen, Y. L and Chang, L. H. (2008). Abnormal foraging behavior induced by sublethal dosage of imidacloprid in the honey bee (Hymenoptera: Apidae). *Journal of economic entomology, 101*(6), 1743-1748.

44. Yang, F. L., Li, X. G., Zhu, F and Lei, C. L. (2009). Structural characterization of nanoparticles loaded with garlic essential oil and their insecticidal activity against Tribolium castaneum (Herbst) (Coleoptera: Tenebrionidae). *Journal of agricultural and food chemistry, 57*(21), 10156-10162

Chapter 31

Environmental Concerns and Toxicology of Engineered Nanoparticles

S.K. Rajkishore, A. Lakshmanan and S. Shanmugapriya

Department of Nano Science and Technology, Tamil Nadu Agricultural University, Coimbatore – 641 003, Tamil Nadu

e-mail: rajkishoresk@gmail.com

Introduction

Nanoparticles are not new, as they have been around in the atmosphere since the dawn of the universe. However, interest in nanoparticles, in particular "engineered nanoparticles," has increased significantly owing to its unique property of very large surface area to mass ratio, which is considered to be a boon to create novel products. But this distinctive property is a cause for concern as these nanoparticles challenge the way we identify, understand and address potential risks (Rajkishore *et al.*, 2013). The global nanotechnology market is expected to exceed US$ 125 Billion by 2024 (https://nano-magazine.com/news/2019/7/16/investing-in-the-nanotechnology-market). According to the current statistics (November, 2019) published by nanotechnology products database, there are 8871 nanoproducts in the global market, which are produced by 2343 companies across 60 countries (https://statnano.com/). The sector wise nanoproducts that are visible in the global level are agriculture (226), automotive (629), construction (767), cosmetics (844), electronics (1925), environment (533), food (345), home appliance (313), medicine (1000), petroleum (297), printing (152), renewable energies (469), sports and fitness (140) and textiles (743) and others (498). Such outpouring of nanoproducts from diverse sectors are likely to cause biosafety issues. Such anxiety is due to the fact that the production of engineered nanomaterials was 2000 tons in 2004 and it is expected to increase to 58,000 tons in 2011 - 2020 (Maynard, 2006). Thus, nanotoxicity is a serious concern to be addressed sooner than later as nano-based processes and products in various fields are being extensively exploited. In this chapter, the term nanoparticles refers mainly to engineered nanoparticles.

Potential Ecological Risk

Nanotoxicology is the science of engineered nanodevices and nanostructures that deals with their effects in living organisms (Oberdörster *et al.*, 2005). Kahru and Dubourguier (2010) assessed the literature on nanotoxicity to main food chain levels (bacteria, algae, crustaceans, ciliates, fish, yeasts and nematodes) and classified the NPs of Ag and ZnO as "extremely toxic", C60 fullerenes and CuO NP as "very toxic", SWCNTs and MWCNTs as "toxic" and TiO_2 NP as "harmful". In order to understand the biosafety issues of nanoparticles so as to safely and effectively employ these novel chemicals in agriculture, an extensive review was carried out in the Dept. of Nano Science and Technology, TNAU as early in 2013 and such assessment in a short span of time was sufficient to alert us to the fact that some engineered

Table 31.1. Nanotoxicity Observed in Various Ecological Levels

Model System	*Nanoparticle*	*Effects*	*References*
Algae	TiO_2	Affected photosynthetic activity	(Navarro *et al.*, 2008; Aruoja *et al.*, 2009)
	CeO_2	Inhibited growth	(Hoecke *et al.*, 2009)
Microbes	ZVI	Cell disruption	(Lee *et al.*, 2008)
	CuO	Antibacterial activity	(Mahapatra *et al.*, 2008; Dimkpa *et al.*, 2011)
	Ag	Increased cell permeability and ultimately cell death	(Sondi and Salopek-Sondi, 2004)
	ZnO	100 per cent mortality	(Jiang *et al.*, 2009)
Plants	CNT	Programmed cell death	(Shen *et al.*, 2010)
	ZnO	Inhibited seed germination and root growth	(Lin and Xing, 2007)
Fish	Se	Hyper-accumulation	(Li *et al.*, 2008)
	ZVI	Morphologic changes	(Li *et al.*, 2009a)
Rodents	Al_2O_3	Inflammation, crossed blood brain barrier	(Li *et al.*, 2009b)
	TiO_2	Inflammation	(Oberdörster *et al.*, 1992)and (Baggs *et al.*, 1997)
	ZVI	Toxic to nerve cells	(Phenrat *et al.*, 2008)
	CNT	Death, necrosis, inflammation and cell injury	(Lam *et al.*, 2004) (Shvedova *et al.*, 2010)
	Au	Moved from mother's placenta to fetus	(Warheit *et al.*, 2004)
Humans	Magnetite	Neurodegenerative diseases	(Kirschvink *et al.*, 1992; Dunn *et al.*, 1995)
	CNT	Mesothelioma	(Donaldson *et al.*, 2006; Fisher *et al.*, 2012)
	Au	Penetrated into the sperm head and tails causing fragmentation.	(Wiwanitkit *et al.*, 2009)

nanoparticles do indeed behave differently to their conventional counterparts and may present new and unusual risks (Rajkishore *et al.*, 2013). Our study concluded that most of the nanoparticles which find its application in a wide spectrum of disciplines have tremendous potential to cause toxicity at various trophic levels in ecological pyramids (Rajkishore *et al.*, 2013). Numerous investigations have also documented that particle at nanoscale exhibit toxicity effects in a wide array of biological systems (Table 31.1).

Literature survey reveals that even primitive organisms such as algae which have proved to tolerate adverse and fluctuating environmental conditions are prone to nanotoxicity.

Further, there are pouring evidences from across the globe that demonstrates toxicity to microbes, plants, fishes and rodents as a result of nanoparticle interaction. Nevertheless, many of these studies are *in vitro* experiments and even in the case of *in vivo* studies, the dose administered have been reported to be very high, which rarely coincides with the expected concentrations in the environmental media. This warrants much more ecotoxicological studies in general and biokinetics of nanoparticles with reference to human system in particular.

Toxicodynamics

In order to assess the toxic potential of engineered nanoparticles, experiments on toxicodynamics are highly essential. Toxicodynamics is the study of assessing the interaction of nanoparticles either at cellular or organ level to understand its ability to cause damage to the specific organs or even to lethality.

Toxicokinetics

The potent harmful effects of nanoparticles can be well understood only when its fate, distribution, persistence and bioaccumulation in biological system is investigated thoroughly. Hence, the study of absorption, distribution, metabolism, accumulation and excretion (ADME) of nanoparticles from the organism is pivotal and this dataset put together is called as toxicokinetics (TK). This biological process is the application of pharmacokinetic (PK) principles to characterize the fate of xenobiotics in the biological system. While testing the biosafety of nanoparticles, the dose in terms of particle surface area per unit weight or surface activity per unit body weight is more relevant than a mass-based dose metric. Though, key dose-metric-related difference exists between bulk and nanoparticles, a mass based dose metric has been used in most biosafety studies with nanoparticles. This has blurred the distinction between PK and TK. Hence, many have started to use a new term "biokinetics" that is devised to include both PK and TK. Biokinetics is also applied in cases where the data cannot be clearly defined for one or the other.

Bio-kinetics of Nanoparticles in Human System

The knowledge on biokinetics of nanoparticles is very much essential to design effective and safe nanoproducts. In kinetic point of view, the Figure 31.1 gives an overview of the ADME processes in the body. It is understandable that nanoparticles can be distributed to the same organ by several routes of exposure. For instance,

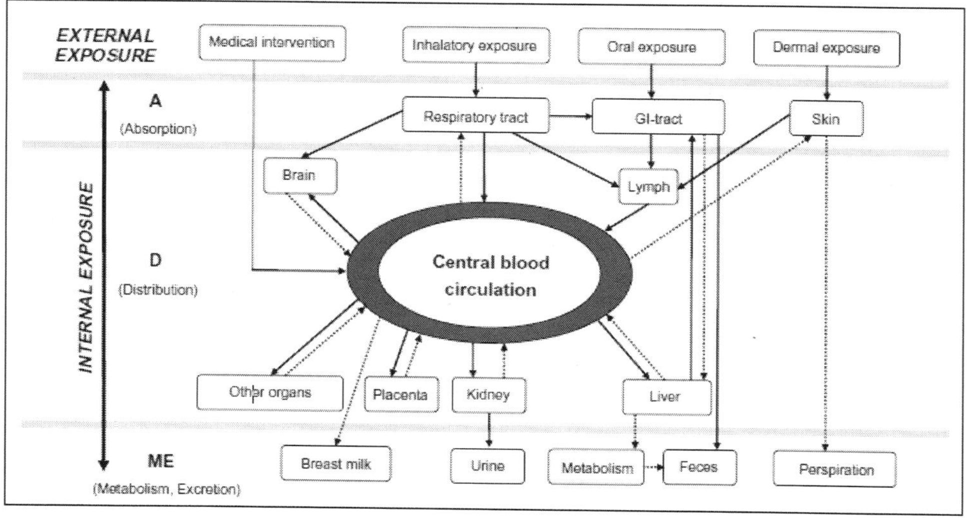

Figure 31.1. Kinetics of Nanoparticles.
Courtesy: Hagens *et al.*, 2007.

nanoparticles found in the gastrointestinal tract can originate from ingested products. In other cases, nanoparticles can also reach the gastrointestinal tract *via* an indirect route such as respiratory tract or the skin, where these nanoparticles can be absorbed in the systemic circulation. From there, nanoparticles can be distributed to the liver, taken up by hepatocytes and excreted in the bile to the gastrointestinal tract. In the Figure 31.1 external exposure refers to the total ingested, inhaled or dermally applied dose of nanoparticles. The internal exposure is the part of the external dose of nanoparticles that follows the systemic circulation and other organs and tissues. This shows clearly that it is necessary to conduct bio-kinetic experiments if the exact route (exposure, absorption, distribution, metabolism and excretion) of engineered nanoparticles in the body has to be revealed.

Mode of Entry and Fate of Nanoparticles in Human System

In order to assess nanoparticle toxicity, *in-vitro* studies alone cannot completely predict possible hazards to humans. Thus, *in-vivo* studies are required to elucidate mechanisms, pathways and entry routes of nanoparticles in a complex multicellular organism. The most common routes of entry of nanoparticles in human system includes,

1. Dermal
2. Inhalation (Pulmonary)
3. Ingestion (Gastrointestinal tract)
4. Injection (Circulatory)

Dermal Route of Entry

The skin which accounts for 16 per cent of the total body weight present all over the body, act as a barrier for the mobilisation of nanoparticles into the internal organs and organelles. Hence, understanding of the different layers in human skin (epidermis, dermis and hypodermis) is pertinent to study the kinetics of nanoparticles.

Epidermis, the outermost element of the skin which acts as a primary barrier contains five layers such as basal cell layer (Startum basale), Prickle cell layer (Stratum spinosum), granular cell layer (Stratum lucidum and stratum granulosum) and horny layer (Stratum corneum). Keratinocytes (dead cells) and corneocytes (damaged cells) accounts for 90 – 95 per cent of the skin cells and other cells such as melanocytes (Pigment cells), langerhans cells and markel cells (Sensory receptor cells) are present in these layers.

Dermis, present between the epidermis and hypodermis is made of three types of fibres namely elastin, collagen and reticular fibres containing fibroblast, histocyte cells, mast cells, lymphocyte cells, plasma cells and granulocytes. Dermis layer carries sweat glands, sebaceous glands, nerve endings, lymphatic vessels and other skin appendages such as hair follicles.

Hypodermis (Subcutaneous layer) forms the inner layer of the skin which contains connective tissues and adipose tissues. Thus, the nanoparticles entry through the multifaceted structures in human skin is very difficult. But still, because of the unique physiochemical characteristics of the nanoparticles there are three routes of entries namely intracellular, trans appendageal route and intercellular and this is illustrated in the Figure 31.2.

Intracellular Route

This route favours the entry of nanoparticles mainly through passive diffusion process because of the large amount of dead cells (hydrated keratenocytes). Usually, this pathway facilitates the movement of hydrophilic nanoparticles.

Trans Appendageal Route

This pathway forms the simplest and shortest route for the nanoparticles to easily reach the circulatory system and its subsequent distribution. Lymphatic and blood vessels which are having contact with the hair follicles are the major destination points of nanoparticles.

Inter Cellular Route

It is the third pathway in which the nanoparticles move in between the lipophillic space present in the cells and hence it is the most short route for lipophillic nanoparticles.

Overall, the nanoparticles of size more than 45 nm can't pass through the dermal and particles between 20-45 nm passes through damaged skin alone. It is clear from several studies that 4 – 20 nm sized particles pass through both intact and damaged cells whereas particles below 4 nm internalise through the intact cells.

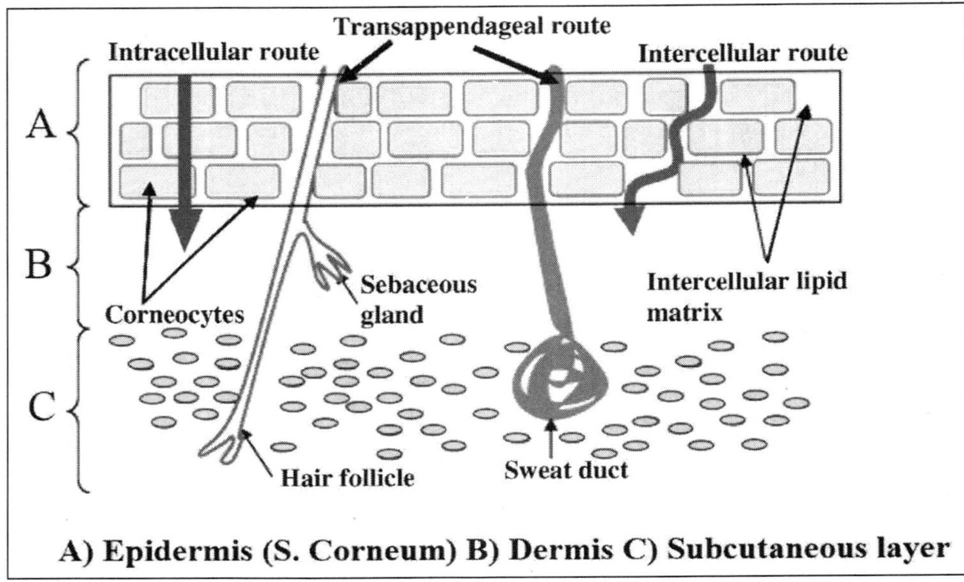

Figure 31.2. Routes of Entry of Nanoparticles in Skin.
Courtesy: Nafisi and Maibach, 2018.

Inhalation (Pulmonary System)

This route of entry is considered to be one of the pathway through which nanoparticles enter the human system. Several biokinetics studies with ultrafine particles have demonstrated that inhalation is the main entry ports and this holds good with our particles of interest. In inhalation toxicology, the term translocation refers to particle uptake from the lung and deposition in target organs. However, in a bio-kinetic point of view, translocation represents both the absorption and distribution process. Though reports indicate that more than 80 per cent of the nanoparticles inhaled (<100 nm) may be deposited in the respiratory tract, it also necessary to understand how these particles get eliminated or obtain clearance as the probability of entry of nanoparticles to different parts in the respiratory system cannot be ignored. The respiratory region starts at nasal airway and ends at alveolar sac with the intermediates as pharynx, larynx, trachea and bronchitis. The nanoparticles once inhaled along with the air, first interact with the host defence system mainly the Respiratory Lining Fluid (RLF). It includes mucus and surfactant present in proximal and distal part respectively. Mucus and surfactant forms a layer over the nanomaterial called corona. The interaction and internalisation of nanoparticles through the mucous layer depends on the physiochemical properties of the nanoparticles and the chemical nature of the mucus layer is depicted in the Figure 31.3.

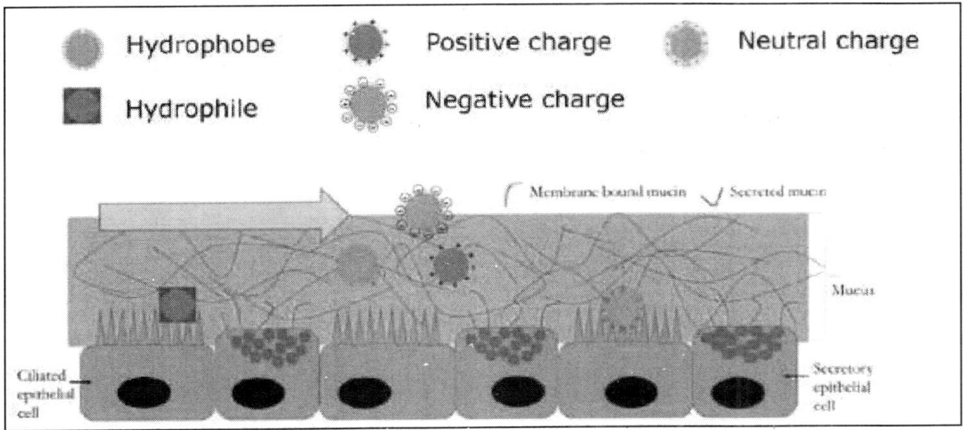

Figure 31.3. Movement of Nanoparticles in Mucus
Courtesy: Gustafsson and Osterlund, 2016.

To reach the endothelial cells (inner lining of lungs), the nanoparticles should cross the mucus. Thus, the nanoparticles should have steric hindrance (repulsion) toward mucin. If the nanoparticles are attached to the mucin, then they will be moved out of the respiratory tract through the ciliary movement. Since 90 per cent of mucin is hydrophilic in nature, hydrophilic nanoparticles may be cleared easily, whereas the lipophillic nanoparticles are likely to be internalised easily. Very importantly, the surface charge of nanoparticles plays a key role for movement in the respiratory tract. Positively and neutral charged nanoparticles are likely to penetrate easily whereas negatively charged nanoparticles are repelled because of negatively charged oligosaccharides domains in mucin. Those nanoparticles that makes its way through mucus, reaches the lungs. The deposition of these nanoparticles (corona coated) in lungs occurs through five different mechanism such as sedimentation, inertial impaction, interception, electrostatic deposition and diffusion. Finally, the respiratory tract ends at alvelor region and hence, nanoparticle of size 1nm – 0.1μm can easily reach the alveolar sac.

Clearance of Nanoparticles from Alveolar Sac

Alveoli consist of two type of epithelial cells such as type I which is responsible for gas exchange and type II cells that plays a role in secretion of surfactant. Apart from epithelial cell it also contains antigen presenting cells such as denntritic cells and macrophages. In the pulmonary system, the alveoli is mainly responsible for initiating the clearance mechanism aganist nanoparticles. The five underlying mechanisms that pave way for nanoparticle clearance from lungs are furnished below along with an illustration in Figure 31.4.

1. Lysosomal destruction of nanoparticles within the macrophages
2. Upward movement of nanoparticles engulfed macrophages towards larynx through the mucociliary movement.

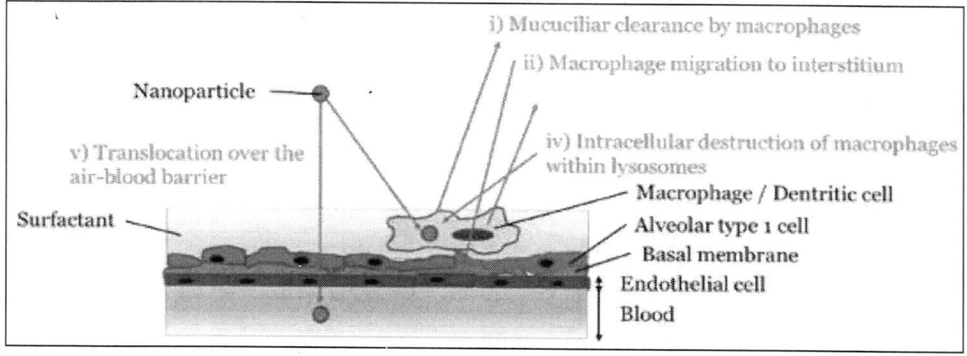

Figure 31.4. Clearance of Nanoparticle through Alveolar Sac
Courtesy: Gustafsson and Osterlund, 2016.

3. Mobilisation of macrophages through lymphatic fluid and its deposition lymphatic vessels.
4. Migration of macrophages to interstitial tissue
5. Migration to secondary organs through circulatory system.

Ingestion (Gastrointestinal Delivery)

This is an easiest way through which nanoparticles of any size and physiochemical characteristics can enter the human system. Nanoparticles can be taken as additives or supplements of food or in some cases where water is contaminated with engineered nanoparticles.

The gastrointestinal tract of human beings starts from mouth and ends at large intestine comprising oesophagus, stomach, small intestine (Duodenum, ileum and jejunum), large intestine, liver, gall bladder, pancreas and renal system. Each portion of gastrointestinal tract have varying pH such as mouth (slightly acidic pH 5.0), stomach (highly acidic pH: 1.2 - 2) and small intestine (neutral pH 6 – 7). This sudden change in pH may cause dissolution of nanoparticles. Apart from rapid change in pH, peristaltic movement of stomach also creates a pressure of 15 mm/Hg imposing a change in properties of nonmaterials. Finally, the nanoparticles from gastrointestinal tract can be distributed to blood through the following four pathways and each of this route is also depicted in the Figure 31.5.

1. Absorption of small nanoparticles (<50 nm) through enterocytes (Nutrient absorptive cells) of villi present in small intestine.
2. Payers patch region which possess M cells helps in translocation of intact materials from lumen to circulation
3. Persorbtion is the process by which nanoparticles moves through damage or opening present on enterocytes caused by other smaller nanoparticles taken up by villi.

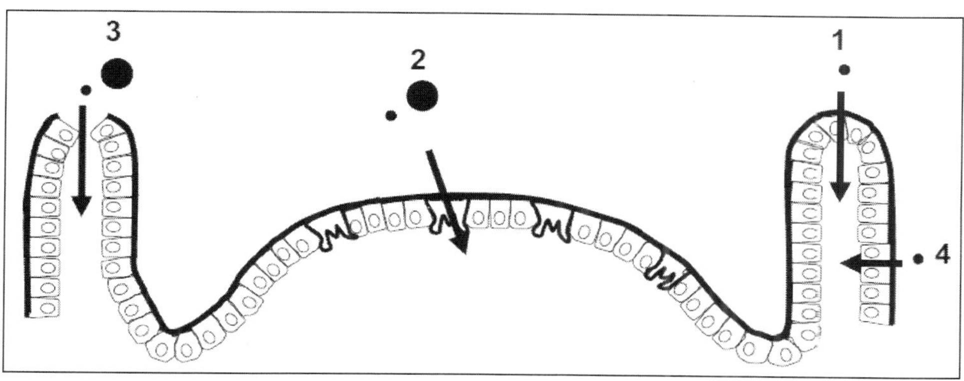

Figure 31.5. Nanoparticle Translocation Across Gastrointestinal Tract
Courtesy: Powell *et al.*, 2010.

4. Paracellular transcytosis is another pathway through which nanoparticles move through the tight junction of the epithelial cells. This route mainly occurs during diseased or damaged conditions.

Injection (Circulatory System)

The circulatory system plays a key role in collection of nanoparticles from dermal, pulmonary and gastrointestinal tract followed by their distribution to hepatobiliary excretory system, renal system, central nervous system and reproductive system. The basics of human physiology makes to understand that superior and inferior vena cava which receives deoxygenated blood and waste materials from different organelles shall carry the nanoparticles to the heart from where aorta which supplies oxygenate blood to the different organelles may distribute the nanoparticles. Nanoparticles after reaching the systemic circulation, potentially, interact with plasma-proteins, coagulation factors, platelets and other blood cells. Above all, the binding of nanoparticles to plasma components may further lead to distribution and excretion of nanoparticles. Furthermore, one of the key biosafety issues of nanoparticles is that it has the ability to cross the blood–brain barrier (BBB) which controls the passage of substances from the blood into the central nervous system. Though permeability of this physical barrier is highly restricted to molecules which are either lipophillic, actively transported compounds or are small soluble molecules (<500 Da), it is still a cause of concern and many experiments need to be carried out to understand better.

Overall, this suggests that heart is the major exposure of nanoparticles and it triggers oxidative stress, signal cascade, change in heart rhythm, inflammation due to interaction with blood cells and vascular dysfunction.

Elimination of Nanoparticles

The most common routes of elimination of nanoparticles from the human system are,

1. Expired air (Pulmonary)
2. Feces (Hepatobillary elimination)
3. Urine (Renal system)

In case of clearance of nanoparticles in the respiratory system, as explained in previous section, the nanoparticles are either expelled through mucus or after reaching alveoli sac, different mechanisms lead to nanoparticle elimination to circulatory system.

Hepatobilary Excretion

Liver is the main site for metabolism (biotransformation) of all xenobiotics including nanoparticles. Liver consist of two lobes as small left hepatic lobe and large right hepatic lobe separated by hepatic vein. It consist of specialised epithelial cells called hepatocytes and other cells such as kupffer cells, liver sinusoidal endothelial cells, stellate cells and macrophages. There are two main source of blood supply to the liver which includes hepatic artery and portal vein. Hepatic artery supplies the oxygenated blood from heart which accounts only for 30 per cent of total blood that liver receives whereas portal vein supplies nutrient rich blood from small intestine accounting for 70 per cent of total blood. Once the nanoparticles enters the liver sinusoid endothelial layer through hepatic artery and portal vein, particles of size of 150 – 200 nm or above will be sequestrated by the kupffer cell and those below them moves to the space of diss and make direct contact with hepatocytes which engulf them using endocytosis mechanism. Those particles engulfed by hepatocytes moves to bile duct through bile canaliculi. From bile it moves to duodenum of small intestine through sphincter of odii and slowly it will be excreted through feces (Figure 31.6).

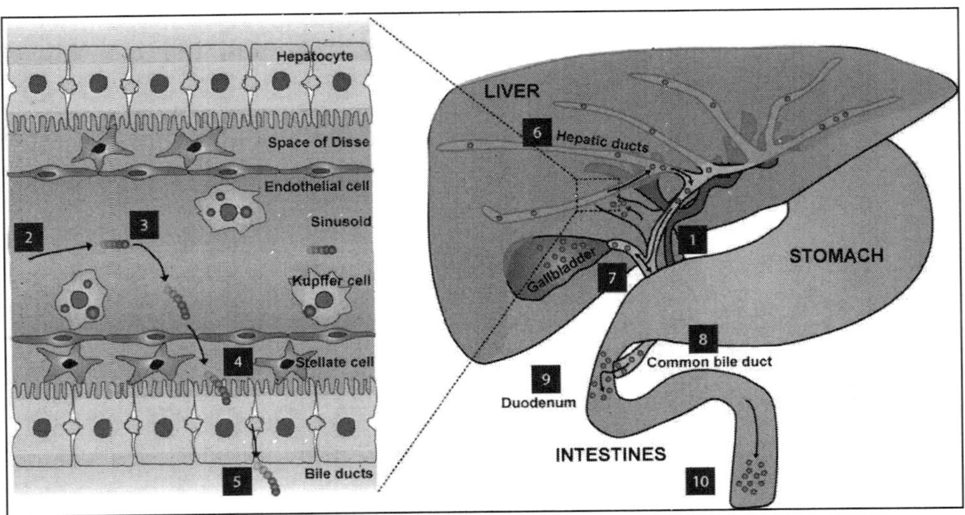

Figure 31.6. Elimination by Hepatobiliary.
Courtesy: Zhang *et al.*, 2016.

Processing of Nanoparticles in Hepatocytes

It is quite interesting that those nanoparticles within size regime of 100 nm are reported to be processed by hepatocytes. Thus, nanoparticles from different organelles entering the hepatocytes may either be functionalised, defunctionalised, hydrophilic or hydrophobic in nature as indicated in the Figure 31.7.

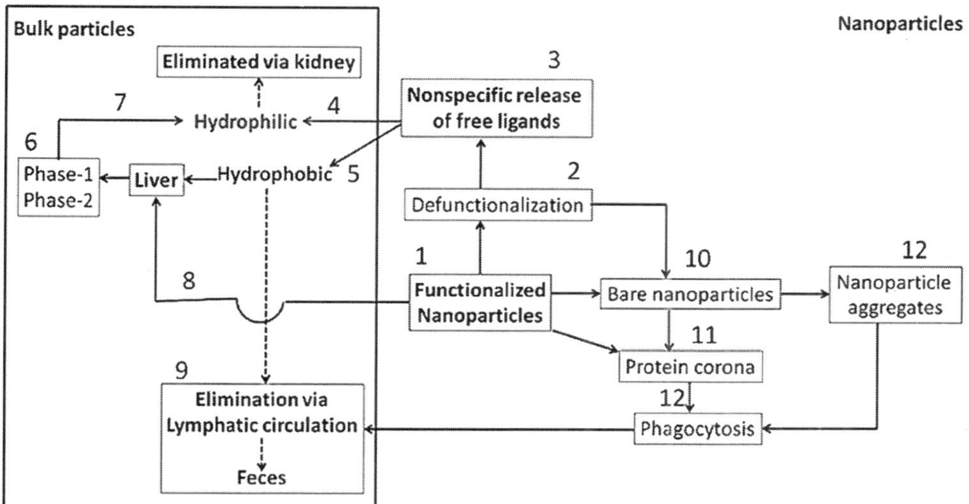

Figure 31.7. Metabolism and Elimination of Nanoparticles.
Courtesy: Singh, 2015.

Specific Metabolism in Hepatocytes

Nanoparticles with hydrophilic nature are reported to taken by the kidney for its elimination through urine whereas nanoparticles with hydrophobic character will undergo phase I and phase II processing for converting them to hydrophilic nature and subsequent elimination in kidney as outlined below (Figure 31.8).

Phase I : The process in which the hydrophobic molecules will be functionalised such that during the phase II process it will be converted to hydrophillic molecule.

Phase II: During this process, the end metabolite of Phase I will be converted to polar groups using some transferase enzymes such as aminoacid transferase, N-acetyl transferase *etc.*

Renal System

The third most important clearance mechanism for elimination of nanoparticles is undertaken by renal system. Hence to better understand the nanoparticles interaction in this system, the basic anatomy of renal is explained below.

Anatomy of Nephron and Bowmn's Capsule (Filteration unit)

The basic structural and functional unit of kidney is called nephron (Figure 31.9). The nephron consist of Glomerulus carposcule surrounded by glomerulus capsule.

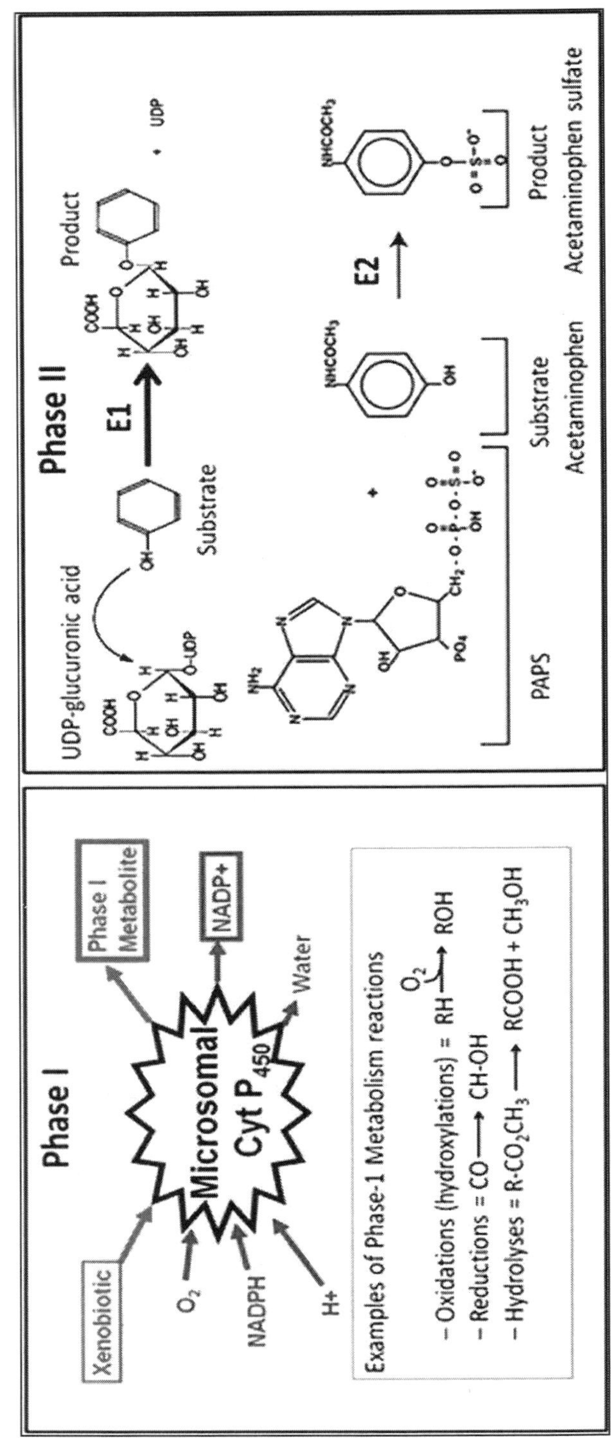

Figure 31.8. Phase I and Phase II Reaction in Hepatocytes
Courtesy: Singh, 2015.

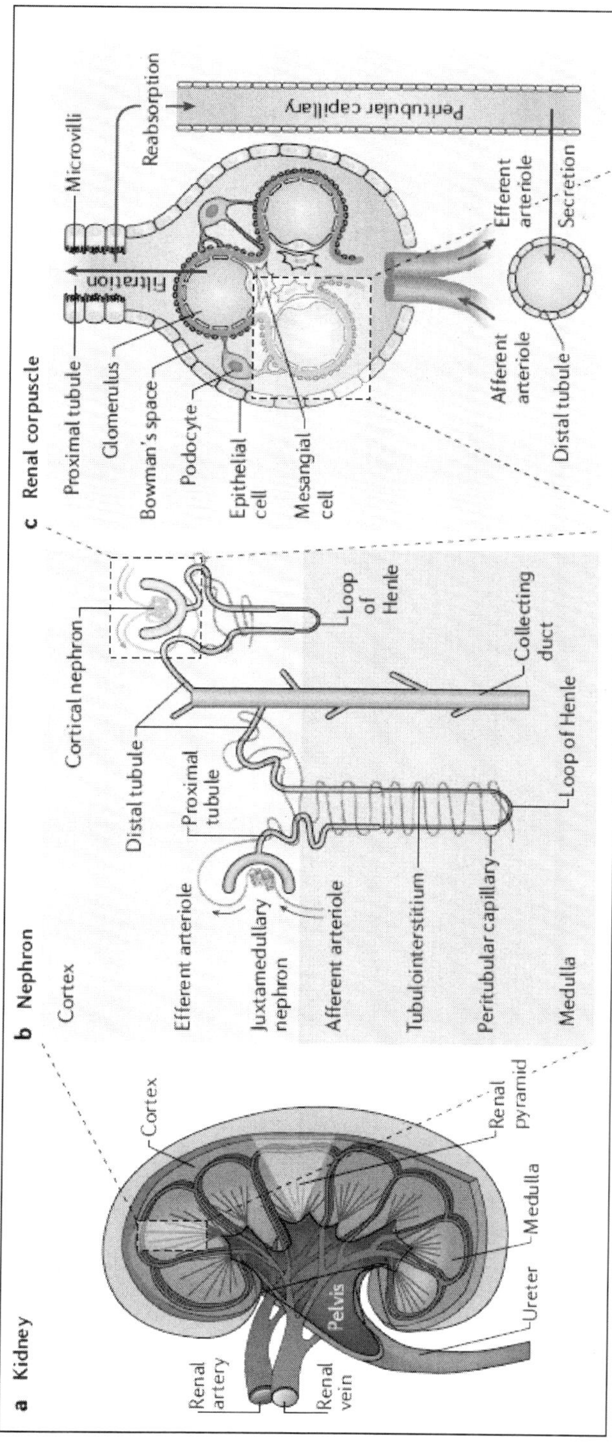

Figure 31.9. Kidney Anatomy.
Courtesy: Du *et al.*, 2018.

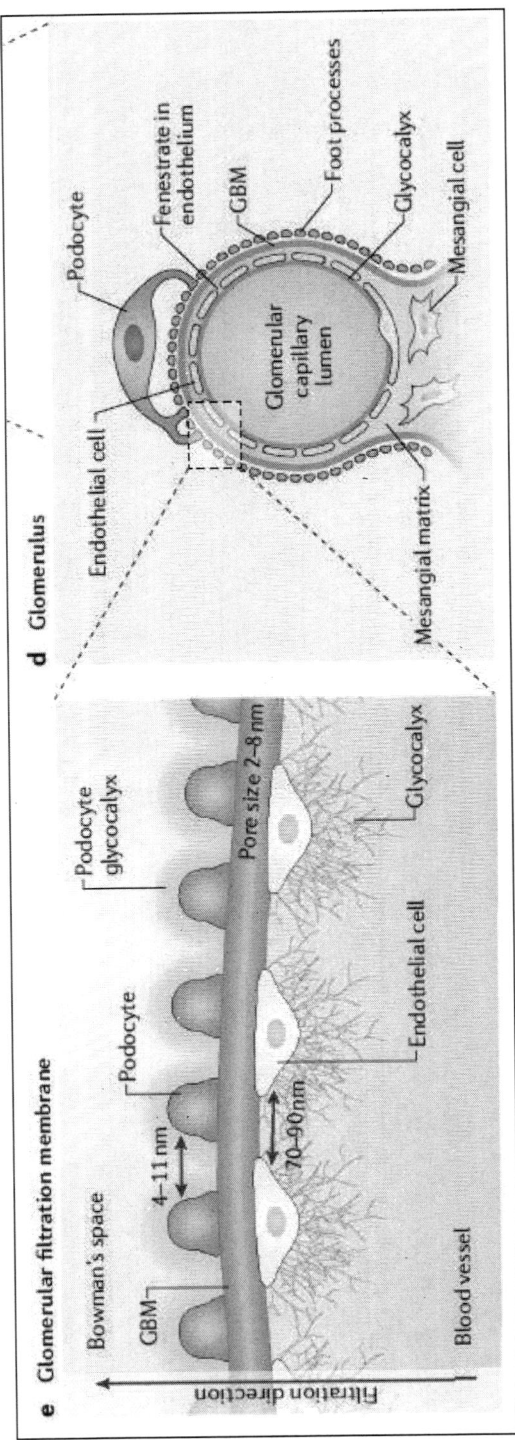

Figure 31.10. Glomerular Filtration Membrane.
Courtesy: Du *et al.*, 2018.

The proximal convoluted tubule from glomerulus capsule extend and connect with loop of henle. From loop of henle distal convoluted tubule extend and connect to collecting duct. Each glomerular capsule consist of three glomerular carposcule. And each glomerular carposcule have glomerular filteration membrane (GFM) which consists of Bowman's space, podocytes, glomerular basement membrane (GBM), Endothelial cells and glycoclayx.

One side of GBM consist of endothelial cells placed at a distance of 70 – 90 nm while the other side is arranged with podocytes at a distance of 4 -11 nm (Figure 31.10). These arrangements of cells makes filtration an efficient process and it depends on the size, shape, charge and composition of nanoparticles.

Size Dependent Filteration of Nanoparticles

The each component of glomerular filtration membrane will restrict the movement of different categories of nanoparticles. Endothelial cells will restrict movement of nanoparticles of size > 100 nm and Glomerular basement membrane restrict the movement of nanoparticles of size 6- 100 nm. On the other hand, the particles of size below 6 nm seem to bypasses the filtration as shown in Figure 31.11. Research in this line suggests that nanoparticles less than 6 nm may be subjected to renal clearance. However, there exists a research gap on the fate and toxicity issues of nanoparticles above the size regime of 6 nm.

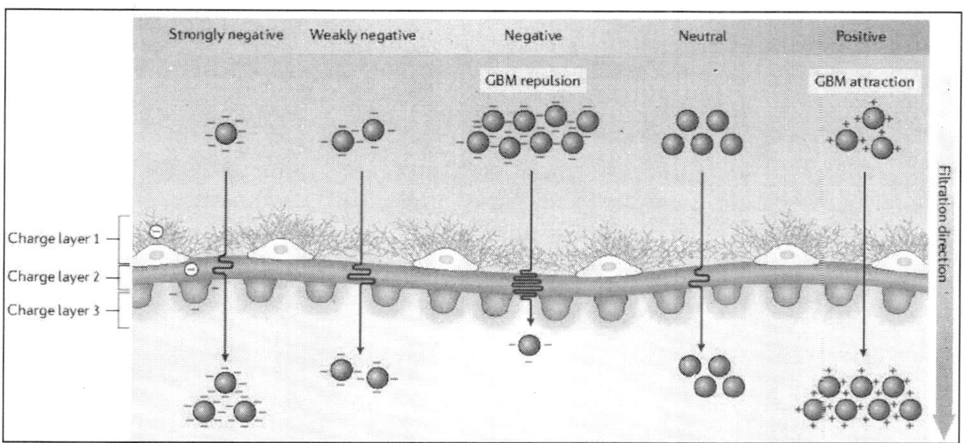

Figure 31.11. Size Dependent Filtration of Nanomaterial in GFM.
Courtesy: Du *et al.*, 2018.

Charge Dependent Filteration of Nanoparticles

In the context of nanochemistry, the Glomerular filteration membrane is likely to facilitae easy movement of positive and neutrally charged nanoparticles particles because of its attraction and diffusion toward the negatively charged glomerular filteration membrane. When the negative potential of nanoparticle increases, filteration slow down (Figure 31.12). The cleaarance will be more faster for positively and neutrally charged particles.

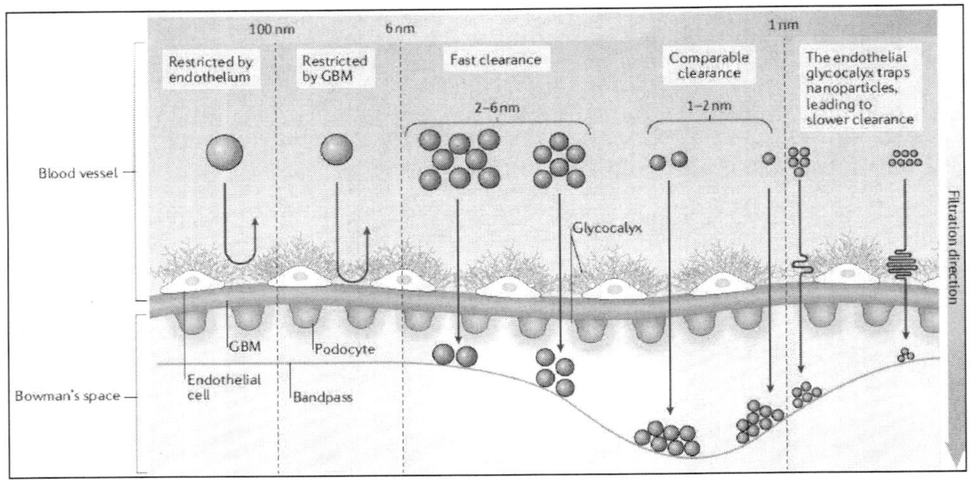

Figure 31.12. Charge Dependent Filtration of Nanomaterial in GFM.
Courtesy: Du *et al.*, 2018.

Nano-Bio Interface

As discussed in the previous section, nanoparticles due to its unique physico-chemical properties, enter, interact and or gets eliminated in the biological system. In order to throw insights into the interactions of nanoparticles at cellular or sub cellular level, the following mechanisms are explained so as to get a clear picture of how cells uptake nanoparticles and cause toxicity for some of the reported nanoparticles. In general, cell membrane maintains cell haemostasis and thus, the nanoparticles should cross the membrane to interact with the cellular organelles. It includes four different mechanisms such as:

1. Phagocytosis
2. Caveolae dependent Endocytosis
3. Clathrin mediated Endocytosis
4. Macropinocytosis

Phagocytosis

Phagocytosis is performed by phagocytic cells such as dentritic cells, macrophages, and monocytes. Once the nanoparticles comes near the extracellular fluid, the protein or immunglobulins are adsorbed to the outer surface of nanoparticles forming a protein corona. This process of formation of protein corona is called opsonisation. Depending on the adsorbed proteins, through ligand receptor interaction, signal cascade process will be activated which initiates the reassembly of actin filaments of plasma membrane making an invagination. This invagination will slowly forms a vesicle called phagosomes (Figure 31.13).

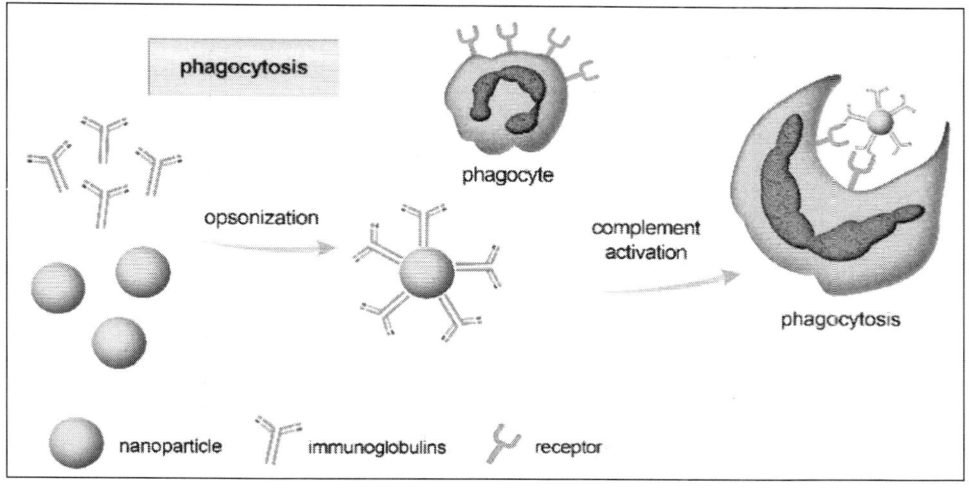

Figure 31.13. Phagocytosis
Courtesy: Behzadi *et al.*, 2017.

Clathrin Mediated Endocytosis

This is the process by which the cell receives its nutrient, cholesterol for plasma membrane and iron. CME can occur either as receptor dependent or receptor independent mechanism. In receptor dependent process ligand specific reaction will occur whereas receptor independent process utilises non specific electrostatic interactions.

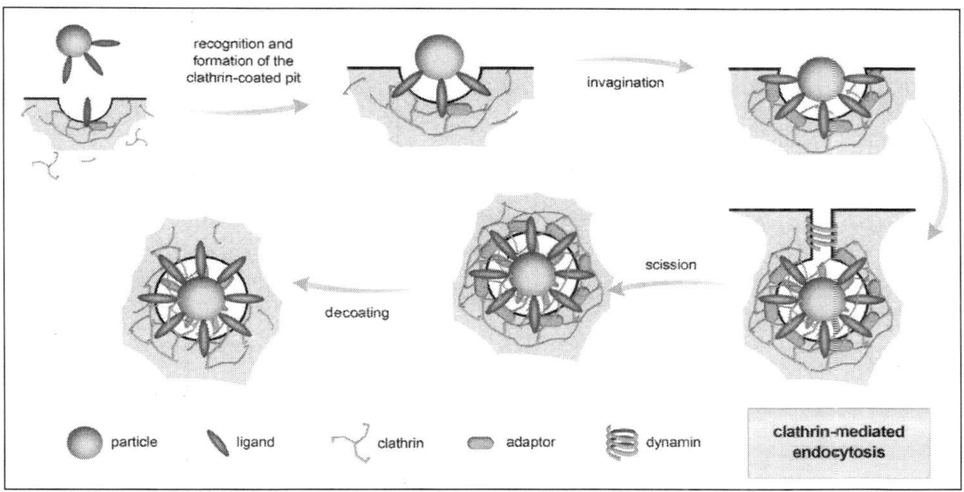

Figure 31.14. Clathrin Mediated Endocytosis.
Courtesy: Behzadi *et al.*, 2017.

Clathrin protein accounts for 0.5 – 2 per cent of cell surface and the individual unit of clathrin protein is called triskelion which contains 3 heavy and light chains. They will spontaneously co-assemble forming an alternative pentagons and hexagons forming a vesicle. This will be facilitated by the accessory proteins resulting in vesicle formation engulfing foreign (nanomaterial) particle. A scissor protein called dynamin will cleave the vesicle from the plasma membrane (Figure 31.14).

Caveolae Dependent Endocytosis

It is the process through which facilitate the movement of cells structural components such as proteins and lipids. It contain two caveolae protein, caveolae 1 will form the flask shaped vesicular structure at the plasma membrane surface and caveolae protein is involved in stabilisation of the formed structure. A scissor protein named Bin amphiphysin rvs (BAR) protein involved in cleavage of vesicle from the plasma membrane (Figure 31.15).

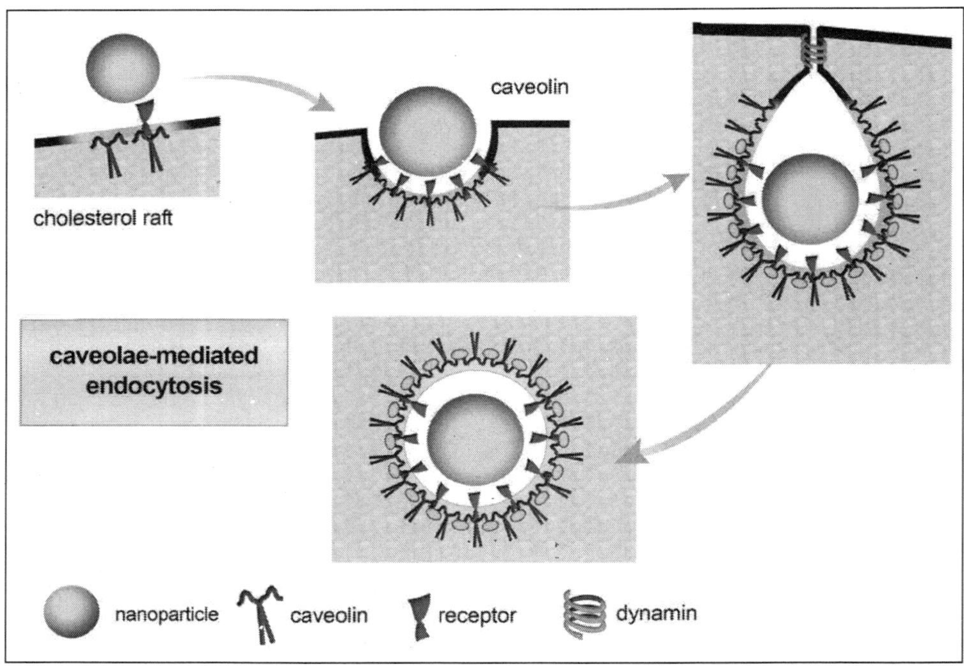

Figure 31.15. Caveolae Dependent Endocytosis
Courtesy: Behzadi *et al.*, 2017.

Macropinocytosis

It is the process through extracellular fluids mostly dissolved ions will be removed off. It is different from other endocytosis process, because it won't involve vesicle formation, instead cytoskeleton rearrangement will takes place forming a protrusion and engulf the extracellular fluid (Figure 31.16).

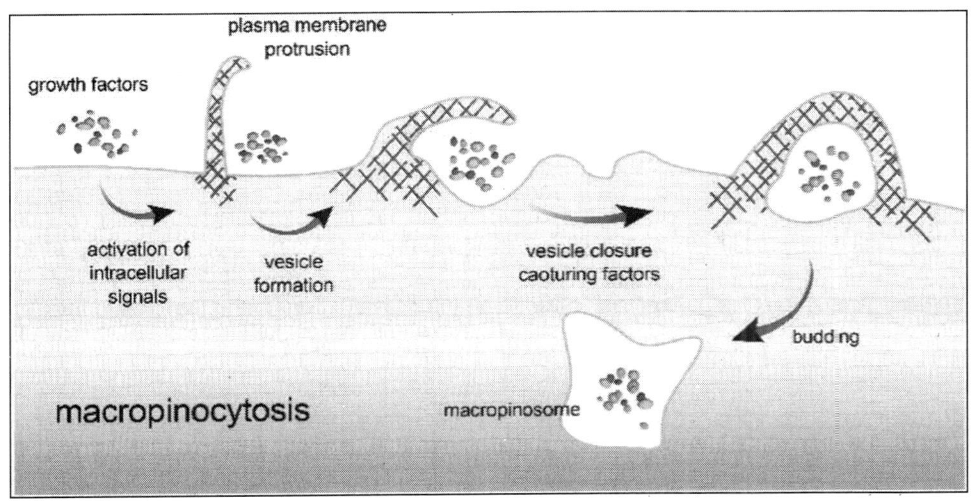

Figure 31.16. Micropinocytosis.
Courtesy: Behzadi *et al.*, 2017.

Cellular Trafficking of Nanoparticles

Those nanoparticles that successfully cross the cell membrane and forms vesicle in the cell by any of the four said mechanisms, finally interact with endosomes. In general endosomes are classified as early endosomes, recycling endosomes and late endosomes (lysosomes). The vesicle formed will fuse with the early endosomes and will be analysed for their requirement in the cell and will be classified as essential, toxic and unknown. The essential molecules will be mobilised to their specific location whereas unknown and toxic will be fused with the recycle endosome and late endosomes, respectively. Recycle endosomes will release the unknown compound out of the cell whereas the late endosome will either degrade the nanoparticles using lysosomal enzymes and excrete them through vacuoles or lysis of lysosomes may happen and the nanoparticles will be released in the cytosol and prone to cascade of reactions leading to cyto and or genotoxicity.

Oxidative Stress as the Cause for Nanotoxicity

The key reason for nanotoxicity in most of the living organisms is the excessive generation of reactive oxygen species (ROS) leading to oxidative stress as a result of nanoparticle interaction at the sub cellular level (Oberdörster *et al.*, 2005); (Shvedova *et al.*, 2010); (Rajkishore *et al.*, 2013). The chemistry of nanoparticles trigger the production of ROS are represented in Figure 31.17 as well as discussed below.

Electron Hole Pair Theory

This is one of the mechanisms in which TiO_2 nanoparticles during UV exposure, generates ROS through the formation of electron-hole pairs as result of photoactivation effect resulting in oxidative stress and inflammation. To explain in detail, upon irradiation, the electrons in the valence band of nanoparticles are promoted to conduction band, leaving a hole with an oxidation potential of +2.6 V when compared to normal hydrogen electrode and ultimately oxidising water or hydroxide in the biological medium into hydroxyl radicals.

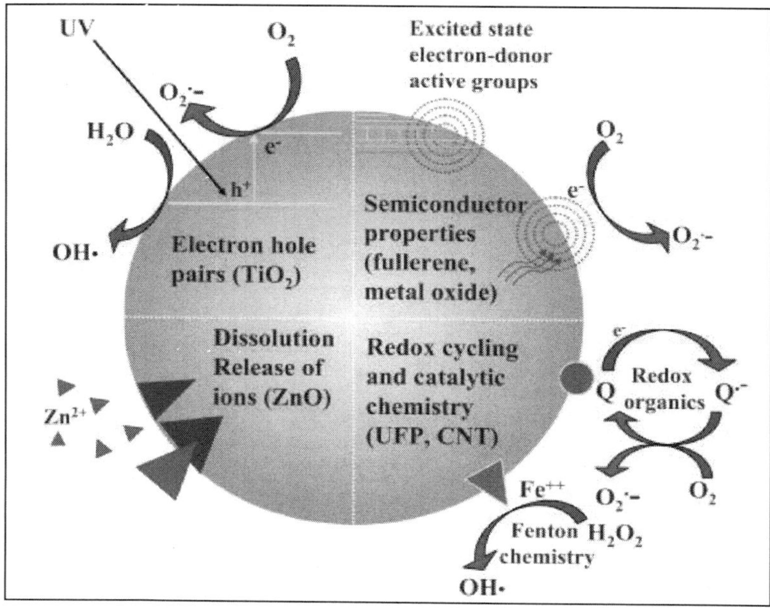

Figure 31.17. Nanoparticle and ROS Production
Courtesy: Li *et al.*, 2008 and Xia *et al.*, 2009.

Fenton Chemistry

This is the phenomena in which metal impurities like ferrous iron on the CNT surface can react with hydrogen peroxide that are produced during normal conditions and finally producing large quantities of hydroxyl radical. Thus transition metals or redox cycling organic chemicals on the nanoparticle surfaces as contaminants/surface modification can produce excess ROS.

Active Electronic State

It is reported that the surface of nanoparticles that possess discontinuous crystal planes or material defects have the potential to create active electronic state thereby facilitating in the generation of ROS.

Particle Dissolution

Toxicity is also seems to be produced as a result of particle dissolution (*e.g.*, ZnO, CdSe, Cu) as the disassociated free ions may trigger variety of ROS (Derfus *et al.*, 2004); (Meng *et al.*, 2007). However, this mechanisms is not fully understood and needs further confirmation studies.

Nanoparticle Induced Molecular Interactions

It is quite clear that nanoparticles generate ROS ultimately resulting in cellular and tissue injury responses such as inflammation, apoptosis, necrosis, fibrosis, hypertrophy, metaplasia, and carcinogenesis ((Nel *et al.*, 2006). Thus, ROS produced by nanoparticle interactions play key role in the initiation of numerous signal transduction pathways that are linked to apoptosis, inflammation and proliferation (Shukla *et al.*, 2003).

The molecular interaction of nanoparticles causes direct genotoxicity and indirect genotoxicty. In case of direct genotoxicity, nanoparticles right away interact with the genetic information and in the category of indirect genotoxicity, protein responsible for DNA replication will be inactivated. However, in both the cases cell death regulation P53 pathway gets activated causing toxicity.

The nanoparticles that produces excessive ROS triggers the signalling pathway by upregulating P38 proapoptotic kinase. This molecule phosphorylates BCl_2 family proteins (responsible for membrane hoemostasis) and destabilise mitochondrial membrane. Once haemostasis is altered, all the respiratory components will be released into cytosol. Especially cytochrome C, when released will bind to apaf -1 (apoptosis protease activating factor -1) gene and get activated. When protein associated with apaf-1 gene increases, caspases-9 will also get increased thereby initiates a cascading activity to produce caspases – 3. In such circumstances, caspases – 3 reduces the activity of poly ADP ribo polymerase enzyme, an enzyme involved in replication process and ultimately halting the DNA replication. Overall, all this P53 signalling pathway destroys all energy sources of cell as well causes genetic damage.

There is another signalling pathway named Nrf-2 and it is observed that ROS produced by nanoparticles causes downregulation. In normal conditions, Nrf-2 gene produces HO -1 enzyme which act as a carrier of oxygen for haem molecule of blood. Therefore, when Nrf -2 is inactivated/downregulated, oxygen will be depleted for all the regular functions causing stress to cell. These are some of the examples that suggest how nanoparticles causes toxicity by either upregulating or downregulating gene and much more pathways do exist that can also be altered by nanoparticles at genetic level.

Conclusion

Many research findings confirm the fact that engineered nano particles have to be assessed thoroughly for their toxicity before commercialization as they have immense potential to cause toxicity. Bio nano systems, nano systems from natural polymers, green synthesized nano particles and nanomaterials developed from natural minerals such as zeolites have been found to exhibit very low or no toxicity to non targeted organisms.Hence these nano systems need to be promoted as carrier or functional systems in agriculture for eco safety without any major bio safety concerns. The toxicological results of nano particles cannot be generalized as several physio-chemical properties of nanoparticles play a major role in causing such unwarranted biological effects. Hence, it is pertinent to judge the biosafety of nanoparticles on case to case basis with appropriate assays or analytical methods that have the capacity to detect effectively. In addition, nanoparticles released into the environment interact with air, water and soil and thus the surface chemistry and concentration of nanoparticles is most likely to be different when compared to pristine nanoparticles that are studied in the *invivo* or *invitro* studies. Thus, it is more necessary to understand nanoparticles and their environmental setting as a "complex" to better predict their fate and toxicity.

REFERENCES

1. Aruoja, V., H.-C. Dubourguier, K. Kasemets, and A. Kahru. 2009. "Toxicity of nanoparticles of CuO, ZnO and TiO2 to microalgae Pseudokirchneriella subcapitata." *Science of the total environment* 407 (4): 1461-1468.
2. Baggs, R., J. Ferin, and G. Oberdörster. 1997. "Regression of pulmonary lesions produced by inhaled titanium dioxide in rats." *Veterinary pathology* 34 (6): 592-597.
3. Behzadi, S., V. Serpooshan, W. Tao, M.A. Hamaly, M.Y. Alkawareek, E.C. Dreaden, D. Brown, A.M. Alkilany, O.C. Farokhzad, and M. Mahmoudi. 2017. "Cellular uptake of nanoparticles: journey inside the cell." *Chemical Society Reviews* 46 (14): 4218-4244.
4. Derfus, A.M., W.C. Chan, and S.N. Bhatia. 2004. "Probing the cytotoxicity of semiconductor quantum dots." *Nano letters* 4 (1): 11-18.
5. Dimkpa, C.O., A. Calder, D.W. Britt, J.E. McLean, and A.J. Anderson. 2011. "Responses of a soil bacterium, Pseudomonas chlororaphis O6 to commercial metal oxide nanoparticles compared with responses to metal ions." *Environmental pollution* 159 (7): 1749-1756.
6. Donaldson, K., R. Aitken, L. Tran, V. Stone, R. Duffin, G. Forrest, and A. Alexander. 2006. "Carbon nanotubes: a review of their properties in relation to pulmonary toxicology and workplace safety." *Toxicological sciences* 92 (1): 5-22.
7. Du, B., M. Yu, and J. Zheng. 2018. "Transport and interactions of nanoparticles in the kidneys." *Nature Reviews Materials* 3 (10): 358-374.
8. Dunn, J., M. Fuller, J. Zoeger, J. Dobson, F. Heller, J. Hammann, E. Caine, and B.M. Moskowitz. 1995. "Magnetic material in the human hippocampus." *Brain research bulletin* 36 (2): 149-153.
9. Fisher, C., A.E. Rider, Z.J. Han, S. Kumar, I. Levchenko, and K. Ostrikov. 2012. "Applications and nanotoxicity of carbon nanotubes and graphene in biomedicine." *Journal of Nanomaterials* 2012: 3.
10. Hagens, W.I., A.G. Oomen, W.H. de Jong, F.R. Cassee, and A.J. Sips. 2007. "What do we (need to) know about the kinetic properties of nanoparticles in the body?" *Regulatory toxicology and pharmacology* 49 (3): 217-229.
11. Hoecke, K.V., J.T. Quik, J. Mankiewicz-Boczek, K.A.D. Schamphelaere, A. Elsaesser, P.V.d. Meeren, C. Barnes, G. McKerr, C.V. Howard, and D.V.D.

Meent. 2009. "Fate and effects of CeO2 nanoparticles in aquatic ecotoxicity tests." *Environmental science and technology* 43 (12): 4537-4546.

12. Jiang, W., H. Mashayekhi, and B. Xing. 2009. "Bacterial toxicity comparison between nano-and micro-scaled oxide particles." *Environmental pollution* 157 (5): 1619-1625.
13. Kahru, A., and H.-C. Dubourguier. 2010. "From ecotoxicology to nanoecotoxicology." *Toxicology* 269 (2-3): 105-119.
14. Kirschvink, J.L., A. Kobayashi-Kirschvink, and B.J. Woodford. 1992. "Magnetite biomineralization in the human brain." *Proceedings of the National Academy of Sciences* 89 (16): 7683-7687.
15. Lam, C.-W., J.T. James, R. McCluskey, and R.L. Hunter. 2004. "Pulmonary toxicity of single-wall carbon nanotubes in mice 7 and 90 days after intratracheal instillation." *Toxicological sciences* 77 (1): 126-134.
16. Lee, W.M., Y.J. An, H. Yoon, and H.S. Kweon. 2008. "Toxicity and bioavailability of copper nanoparticles to the terrestrial plants mung bean (Phaseolus radiatus) and wheat (Triticum aestivum): plant agar test for water-insoluble nanoparticles." *Environmental Toxicology and Chemistry: An International Journal* 27 (9): 1915-1921.
17. Li, H., J. Zhang, T. Wang, W. Luo, Q. Zhou, and G. Jiang. 2008. "Elemental selenium particles at nano-size (Nano-Se) are more toxic to Medaka (Oryzias latipes) as a consequence of hyper-accumulation of selenium: a comparison with sodium selenite." *Aquatic toxicology* 89 (4): 251-256.
18. Li, H., Q. Zhou, Y. Wu, J. Fu, T. Wang, and G. Jiang. 2009a. "Effects of waterborne nano-iron on medaka (Oryzias latipes): antioxidant enzymatic activity, lipid peroxidation and histopathology." *Ecotoxicology and environmental safety* 72 (3): 684-692.
19. Li, X.-b., H. Zheng, Z.-r. Zhang, M. Li, Z.-y. Huang, H.J. Schluesener, Y.-y. Li, and S.-q. Xu. 2009b. "Glia activation induced by peripheral administration of aluminum oxide nanoparticles in rat brains." *Nanomedicine: Nanotechnology, Biology and Medicine* 5 (4): 473-479.
20. Lin, D., and B. Xing. 2007. "Phytotoxicity of nanoparticles: inhibition of seed germination and root growth." *Environmental pollution* 150 (2): 243-250.
21. Mahapatra, O., M. Bhagat, C. Gopalakrishnan, and K.D. Arunachalam. 2008. "Ultrafine dispersed CuO nanoparticles and their antibacterial activity." *Journal of Experimental Nanoscience* 3 (3): 185-193.
22. Maynard, A.D. 2006. "Nanotechnology: assessing the risks." *Nano Today* 1 (2): 22-33.
23. Meng, H., Z. Chen, G. Xing, H. Yuan, C. Chen, F. Zhao, C. Zhang, and Y. Zhao. 2007. "Ultrahigh reactivity provokes nanotoxicity: explanation of oral toxicity of nano-copper particles." *Toxicology letters* 175 (1-3): 102-110.

24. Nafisi, S., and H.I. Maibach. 2018. "Skin penetration of nanoparticles." In *Emerging Nanotechnologies in Immunology*, 47-88. Elsevier.

25. Navarro, E., A. Baun, R. Behra, N.B. Hartmann, J. Filser, A.-J. Miao, A. Quigg, P.H. Santschi, and L. Sigg. 2008. "Environmental behavior and ecotoxicity of engineered nanoparticles to algae, plants, and fungi." *Ecotoxicology* 17 (5): 372-386.

26. Nel, A., T. Xia, L. Mädler, and N. Li. 2006. "Toxic potential of materials at the nanolevel." *science* 311 (5761): 622-627.

27. Oberdörster, G., J. Ferin, R. Gelein, S.C. Soderholm, and J. Finkelstein. 1992. "Role of the alveolar macrophage in lung injury: studies with ultrafine particles." *Environmental health perspectives* 97: 193-199.

28. Oberdörster, G., E. Oberdörster, and J. Oberdörster. 2005. "Nanotoxicology: an emerging discipline evolving from studies of ultrafine particles." *Environmental health perspectives* 113 (7): 823-839.

29. Phenrat, T., T.C. Long, G.V. Lowry, and B. Veronesi. 2008. "Partial oxidation ("aging") and surface modification decrease the toxicity of nanosized zerovalent iron." *Environmental science and technology* 43 (1): 195-200.

30. Powell, J.J., N. Faria, E. Thomas-McKay, and L.C. Pele. 2010. "Origin and fate of dietary nanoparticles and microparticles in the gastrointestinal tract." *Journal of autoimmunity* 34 (3): J226-J233.

31. Rajkishore, S., K. Subramanian, N. Natarajan, and K. Gunasekaran. 2013. "Nanotoxicity at various trophic levels: a review." *Bioscan* 8 (3): 975-982.

32. Shen, C.X., Q.F. Zhang, J. Li, F.C. Bi, and N. Yao. 2010. "Induction of programmed cell death in Arabidopsis and rice by single-wall carbon nanotubes." *American Journal of Botany* 97 (10): 1602-1609.

33. Shukla, A., M. Gulumian, T.K. Hei, D. Kamp, Q. Rahman, and B.T. Mossman. 2003. "Multiple roles of oxidants in the pathogenesis of asbestos-induced diseases." *Free Radical Biology and Medicine* 34 (9): 1117-1129.

34. Shvedova, A.A., V.E. Kagan, and B. Fadeel. 2010. "Close encounters of the small kind: adverse effects of man-made materials interfacing with the nano-cosmos of biological systems." *Annual review of pharmacology and toxicology* 50: 63-88.

35. Singh, A.K. 2015. *Engineered nanoparticles: structure, properties and mechanisms of toxicity*: Academic Press.

36. Sondi, I., and B. Salopek-Sondi. 2004. "Silver nanoparticles as antimicrobial agent: a case study on E. coli as a model for Gram-negative bacteria." *Journal of colloid and interface science* 275 (1): 177-182.

37. Warheit, D.B., B.R. Laurence, K.L. Reed, D.H. Roach, G.A. Reynolds, and T.R. Webb. 2004. "Comparative pulmonary toxicity assessment of single-wall carbon nanotubes in rats." *Toxicological sciences* 77 (1): 117-125.

38. Wiwanitkit, V., A. Sereemaspun, and R. Rojanathanes. 2009. "Effect of gold nanoparticles on spermatozoa: the first world report." *Fertility and Sterility* 91 (1): e7.

39. Xia, T., N. Li, and A.E. Nel. 2009. "Potential health impact of nanoparticles." *Annual review of public health* 30: 137-150.

40. Zhang, Y.-N., W. Poon, A.J. Tavares, I.D. McGilvray, and W.C. Chan. 2016. "Nanoparticle–liver interactions: Cellular uptake and hepatobiliary elimination." *Journal of controlled release* 240: 332-348.

Chapter 32

Biosafety Guidelines for Nano Agri Inputs and Nano Agri Products: Global and Indian Scenario

S.K. Rajkishore[1], *A. Lakshmanan*[1], *K.S. Subramanian*[2] *and S. Shanmugapriya*[1]

[1]*Department of Nano Science and Technology,* [2]*Directorate of Research, Tamil Nadu Agricultural University, Coimbatore – 641 003, Tamil Nadu*

Introduction

Nanotechnology is widely believed to be one of the promising cross-cutting technologies for improving the farming system (Subramanian *et al.,* 2018). Nano-fertilizers with improved nutrient use efficiency, nano-composites for slow and controlled release of pesticides or herbicides, nano-materials for food packaging to enhance shelf-life and nanobiosensors to diagnose plant disease, disorder or nutrient deficiency are some of the prominent areas where nanotechnology research is in progress. Despite the fact that nanotechnology applications in the sphere of agricultural sciences begin to increase, biosafety concerns are also mounting up and warrant better and transparent testing (Rajkishore *et al.,* 2013). This is one of the main reasons for the slow progress of nano research and commercialization of nano products in agricultural domain when compared with other sectors. The scarcity of commercial nanoproducts in agriculture is mainly due to limited existence of biosafety guidelines (Arts *et al.,* 2014). This chapter is designed to provide an outline of the biosafety guidelines for nanomaterials employed in agricultural sciences besides highlighting a comparative assessment of regulatory frameworks for agricultural sector in different countries.

Need for Risk Assessment

The nanotoxicity dataset suggests that engineered nanoparticles in many cases regardless of biological organisms in different trophic levels have showed deleterious effects. Hence, the risk assessment which focuses on management of a toxicological hazard in a specific exposure scenario is considered to be highly relevant for nanoproducts. In such context, regulatory bodies have now started to understand the importance of systematically carryout the risk assessment of nanoproducts for developing robust regulatory framework besides promoting nanobased agricultural technologies (Amenta *et al.*, 2015).

In general, risk assessment comprises four steps namely hazard characterization, dose-response determination, exposure assessment and risk characterization. In the first step, hazard identification is quite pertinent to quantitatively measure the hazard potential where the NMs are subjected to hazard screening. Secondly, establishing dose-response relationship is important as this assessment quantitatively determines the relationship between adverse effects (*i.e.*, hazards) and a concentration of a substance in a given environment. This exercise helps to determine no-observed-adverse-effect levels (NOAELs) and occupational exposure limits (OELs). Thirdly, the exposure assessment provides exposure scenarios describing how a chemical is used or how it is released into the environment. The exposure will be assessed through one or more routes (*i.e.* inhalation, ingestion or dermal) or environmental compartments (*e.g.* water, sediment, soil) under varying exposure scenarios. Finally, the risk characterization deals with the estimation of the potential for adverse health effects to occur, evaluation of uncertainty and summarization of risk information. Hence, all these four steps are prime elements to design a foolproof regulatory framework for risk assessment of nano agri inputs and nano agri products. At the same time, employing appropriate tools for risk assessment is also important to predict the risks with utmost scientific caliber. More than forty eight risk assessment tools specific for nanomaterials have been identified and these can be chosen to the nanoproduct or nonmaterial of interest (Hristozov *et al.*, 2016). Risk assessment tools are categorized into control banding, risk screening, occupational and consumer exposure, environmental fate and exposure, hazard assessment, physicochemical characterization, and decision support tools. Despite the large number of tools that are being identified for risk assessment, no regulatory framework is well structured for nanoproducts and there exists a large scope to focus to develop a framework for engineered nanomaterials which can fulfill all evaluation criteria that are stipulated by experts for a robust regulatory framework.

Prioritizing Life Cycle Assessment (LCA)

The Life Cycle Assessment (LCA) is considered as the best methodology for assessing the environmental impacts throughout a product's life from raw materials and till it gets recycled in the ecosystem (*i.e.* cradle-to-grave). The ISO framework for LCA (ISO 14040:2006) is reported be completely applicable for engineered nanomaterials or nanoproducts, even in the cases where data on the elementary flows and impacts are uncertain and scarce (Klöpffer *et al.*, 2007). Despite these facts, publications or reports on LCA for nano-products are scanty. Furthermore, few papers dealing with LCA for nanoproducts concentrate only cradle-to-gate assessment instead of cradle to grave. Therefore, while undertaking LCA studies for nano-agri products or nano-agri inputs, the use phase and the after-use phase (recycling, disposal) need be focused without omitting.

Regulatory Guidelines for Nano Agriculture: Global Scenario

In order to assess the strengths and limitations of the existing regulatory frameworks pertaining to nanotechnology in general and nano agri products in particular, a theoretical assessment was undertaken in the Department of Nano Science and Technology, TNAU (Subramanian and Rajkishore, 2018). This assessment which was carried out to examine current regulatory frameworks across the spectrum of countries highlighted that at the moment the European Union along with Switzerland is the only geographical region in the globe that has legislated specific provisions of "nano" (Amenta *et al.*, 2015). Europe also insists on mandatory labeling for nano products. Thus, it is very clear that European Union has taken a lead step to formulate regulatory policy for nanoproducts (Justo-Hanania and Dayan 2014). In United States, the FDA has issued specific guidance for application of nanotechnology in food system. However, it is yet to frame a regulatory definition for nano-materials. The Health Canada and Canadian Food Inspection Agency (CFIA) has jointly stipulated the guidelines for the use of nano-materials in general whereas the Public Health Agency of Canada (PHAC) is entrusted with a responsibility for regulating food products. Though specific regulation for nano-products is not available but such products are brought under the ambit of existing regulatory frameworks. With respect to India, Food Safety and Standards Act (2006) is not yet amended to include provisions to cover nanomaterials. However, recently, Department of Biotechnology has taken initiatives to draft the guidelines for evaluation of nano-agri input products and nano-agri products and the details are discussed separately in this chapter.

Overall, it is observed that regulatory frameworks for agri-food sector across globe still lacks clarity. More specifically, there is no consensus for the definition to refer the size of nanoproducts or nanomaterials. This is very important because without regulatory definition for nano-materials, the framework cannot be made legally binding. As defined by International Organization for Standardization (ISO), "nanotechnology" is the "application of scientific knowledge to manipulate and control matter predominantly in the nanoscale (*i.e.*, approximately from 1 to 100 nm) to make use of size- and structure-dependent properties and phenomena distinct from those associated with individual atoms or molecules, or extrapolation from larger sizes of the same material". Though several such definitions for nanotechnology or nano-material exists, the European Commission (EC) definition of the term "nanomaterial it is greatly agreed by several experts at this moment. Accordingly, "a material, regardless of its origin (*i.e.*, natural, incidental or manufactured), is considered as nanomaterial if it contains 50 per cent or more of unbound, agglomerated or aggregated particles with one or more external dimensions in the size range between 1 and 100 nm". However, this rule is exempted for fullerenes, graphene flakes and single-wall carbon nanotubes as the lower size limit remains ineffectual, but the upper limit of 100 nm still applies. In addition, it is also stated, "the 50 per cent threshold referring to the number of particles with a size between 1 and 100 nm may be lowered down to 1 per cent in specific cases and whenever warranted due to concerns triggered by environmental, health, and safety concerns". The definition of nanomaterial provided by various regulatory bodies in select countries are presented in Table 32.1. Furthermore, to provide the global scenario in regulatory framework, a comparative analysis made by Amenta *et al.* (2015) and Adholeya *et al.* (2017) is presented in the Table 32.2 which has been adopted from Subramanian and Rajkishore (2018) for better understanding.

Salient Features in India's Draft Guidelines

- ✰ DBT proposed draft guidelines for evaluation of nano agri inputs and nano agri products clearly defines these two types. Accordingly, nano-agriinproduct (NAIP) is defined as "an agricultural input preparation containing nanomaterials intended for external and internal applications (through soil, seed, foliar, and drip in crops as well as by other means) on crop for the purpose of agricultural farming". In case of nano agriproduct (NAP) it is defined as "an agricultural preparation containing nanomaterials intended for consumption or applications in food/feed and their supplements as well as nutraceutical delivery" (DBT, 2019).
- ✰ Since, physico-chemical characterization is very important for assessing the safety evaluation, the guidelines categorizes nanoagri input products under four types according to its degradation nature (biodegradable or non biodegradable), chemical nature (organic, inorganic and composite), nanoform of the active ingredient (nanocarriers loaded with active ingredient (AI), Active ingredient converted to nano form), synthesis (biological, chemical or physical means).
- ✰ The guidelines also give impetus to understand the nature of excipients (stabilizers, surfactants and coupling agents *etc.*,) used in nanoproducts for safety evaluation.
- ✰ It has guided to adopt four principal types of storage stability studies as per OECD TG 318, FAO/WHO, AVMPA protocols.
- ✰ In order to obtain information for evaluation of nanoproducts, it has provided a detailed check list.
- ✰ Most importantly, a list of cytotoxicity and genotoxicity assays as well as ecotoxicity tests has been given for adoption and the insights of these protocols are discussed separately hereunder.

Importance of Physicochemical Characterization

The key factors that affect nanotoxicity are dose, surface area, size, crystalline structure, surface coating and opsonization. The notion of "Dose makes the poison" is most debated in the field of nanotoxicology and it observed that in many a times the toxicity of nanoparticles does not always correlate with particle mass dose. This is reasoned out that high concentration of nanoparticles may promote particle aggregation and could therefore reduce toxic responses compared to lower

Table 32.1. Comparison of Nanomaterial Definition for Agri-food Sector in Regulatory Frameworks of Various Countries

Country	*Regulatory Body/ Responsible Organization*	*Definition of Nanomaterial*			*Relevant Legislation*
		Size	*Distribution Threshold*	*Nanoscale Dimension*	
European Union	European Commission (European Chemical agency)	1 -100 nm	50 per cent by number	Atleast one	EC No. 1907/2006; Regulation, Evaluation and Authorization of Chemicals (REACH) for plant protection products and food contact materials
	Specific to agri sector European Parliament and the Council of the European Union on the Provision of Food Information to Consumers	1 -100 nm and extend* upto 1 μm	Not mentioned	Atleast one	Regulation on the Provision of Food Information to Consumers (1169/2011), EC Regulation No 258/97 concerning "novel foods" and "novel food ingredients"
United States	United States Food and Drug Adminisiration	1 -100 nm and extends* upto 1 μm	Not mentioned	Atleast one	Federal Food, Drug, and Cosmetic Act (FFDCA)
Canada	Health Canada	1 -100 nm and larger*	Not mentioned	Atleast one	Food and Drugs Act
India	Guidelines proposed by DBT	1-100 nm and also 100-1000nm if#	Not mentioned	Atleast one	Yet to make it legislate

* Larger size particles (up to 1 μm are considered as nanomaterials if it exhibits nanoscale properties); # it has altered the agri-input or agri product characteristics associated with the application of nanotechnology compared with active ingredient

Courtesy: Adopted and modified from Subramanian and Rajkishore, 2018.

Table 32.2. Global Scenario of Regulatory Guidelines for Nanoproducts in Agri-food Systems

Country	*Regulatory Body/ Responsible Organization*	*Legislation*	*Provisions in the available Legislation*	*Remarks*
USA	Food and Drug Administration (FDA) and US-Environmental Protection Agency (US-EPA)	Federal Food, Drug and Cosmetic Act (FFDCA), Federal Insecticide, Fungicide and Rodenticide Act (FIFRA),	No specifications on nanoproducts in FFDCA or FIFRA	FDA issued a specific guidance that recommends preliminary safety assessments for food/food contact substances produced at nanoscale FDA yet to issue a regulatory definition of NMs
Canada	Canadian Food Inspection Agency (CFIA) and Public Health Agency of Canada (PHAC)	Food and Drugs Act	Nanoproducts are regulated under existing legislative	No nano specific regulatory framework
EU	European commission (EC), European Parliament and Council	1. Regulation on the provision of food information to consumers (1169/20119) 2. Regulation on plastic food contact materials and articles (10/2011) 3. Regulation on active and intelligent materials and articles (450/2009) 4. The biocidal products regulation (528/2012) 5. The cosmetic products regulation (1223/2009) 6. Registration, Evaluation, Authorization and Restriction of Chemicals (REACH) regulation (1907/2006)	These legislations have nano specific provisions. But no exclusive NM legislation	Taking a lead step in regulating nanoproducts

Country	*Regulatory Body/ Responsible Organization*	*Legislation*	*Provisions in the available Legislation*	*Remarks*
Switzerland	Federal Office for the Environment (FOEN) Swiss Federal office of public health (FOPH)		Existing regulations are ensured for safety of NMs	Information on physico-chemical parameters of NMs is mandatory while application submission
Australia and New Zealand	Food standards Australia New Zealand (FSANZ)	Australian New Zealand Food Standards Code	Amended FSANZ Application Handbook to support new food regulations to manage risks from nanoproducts	First regulatory program for "industrial nanomaterials was intruducted by "National Industrial Chemicals Notification and Assessment Scheme (NICNAS) Tasmanian Institute of Agriculture has published a comprehensive review on Nanotechnology in Agricultural Products Logistics Management
China	Ministry of Agriculture Ministry of Health	Food Safety Law of China	National Centre for Nanoscience and Technology (NCNST) and the commission on nanotechnology standardization	Nanoproducts in agri-food systems is not yet approved by the authority
South Korea	Ministry of Food and Drug safety (MFDS), Korean Food and Drug Administration (KFDA) Korean Agency for Technology and Science (KATS)	National Nano-Safety Strategic Plan (2012/2016) and Food Sanitation Act	No NM specifications	Ministry of Knowledge and Economy (MKE) and the Korean Agency for Technology and Standards published Guidance on safety management of nano-based products
Japan	Ministry of Health, Labour and Welfare	Food Sanitation Law	No NM specific legislation	Basic survey on safety of nanoproducts was done
Iran	Iran Nanotechnology Initiative Council (INIC) Nanotechnology Committee of Food and Drug Organization			Food and Drug Organization (FDO) constituted guidelines for nanoproducts in food, beverages, pharmaceutical, medical equipment but agriculture is not yet included.

Country	*Regulatory Body/ Responsible Organization*	*Legislation*	*Provisions in the available Legislation*	*Remarks*
Taiwan	Taiwan Nanotechnology Industry Development Association (TANIDA)		Nanoproducts are certified	TANIDA established nanoMark system to certify the nanoproducts
Thailand	Food and Drug Administration of the Ministry of Public Health			NanoQ label has been introduced for nano products (but not for agri/food) that are certified by the Nanotechnology Association of Thailand
India	Food Safety and Standards Authority of India	Food Safety and Standards Act (2006)		DST Nano mission has come up with safe handling of NMs in workplace and industry. Though no legislation/ regulation available for agriculture, DBT has recently drafted the guidelines for evaluation of nano agri inputs and nano-agri products. Hence, these guidelines are developed in provisions of Insecticides Act, FCO, BIS and FSSAI, with specific requirements and adaptations for NAIPs and NAPs wherever considered necessary

Courtesy: Presented with modification from Subramanian and Rajkishore, 2018.

concentrations of the same particles (Buzea *et al.*, 2007). In the aspect of surface area, the relative portion of surface atoms to bulk atoms is considerably different in nano-sized when compared to microsized particles of same chemistry. For instance, less than 1 per cent of atoms of a microparticle occupy surface positions, while 10 per cent of the atoms in a 10 nm particle reside on its surface. In the case of size as the factor, even within the nano size regime (1-100 nm), different intensities of toxicity are exhibited. For example, gold particles with a size of 13 nm and above, are considered as non-toxic (Jahnen-Dechent and Simon, 2008) and the gold particles below 2 nm have demonstrated unexpected degree of toxicity in varied cell types (Schmid, 2008). The other important factor is type of crystalline nature and therefore different crystal forms are considered to cause different levels of toxicity. While it is known that surface chemistry is the key for exhibiting different properties by a same chemical compound, in many instances, the contaminants on the surface of nanoparticles have been the reason for toxicity. Finally, opsonization is very important factor in nanotoxicity, as nanoparticles bind to proteins in biological fluids and form protein corona, giving a different identity to the nanoparticle and thereby changing its toxicity potential. Therefore, a comprehensive characterization of the test materials (nano-material of interest), media and dose are most essential elements establishing realistic exposure scenarios.

Details of Select Classic Bioassays

The biosafety concerns of nano agri inputs or nano agri products can be effectively addressed by employing appropriate assays. In general, nanotoxicity studies include *in vitro, in vivo* tests and *in silico* models. Cytotoxic, genotoxic and alterations in gene expression assays are commonly suggested by regulatory bodies across the globe to evaluate the toxicity of nanomaterials in *in vitro* system (Rajkishore *et al.*, 2013). A total of 59 Endpoints were identified by Working Party on Manufactured Nanomaterials (WPMN) of OECD for the testing of nano-materials and grouped into the following 6 domains:

1. Nanomaterial information/identification,
2. Physical-chemical properties and material characterization,
3. Environmental fate,
4. Environmental toxicology
5. Mammalian toxicology
6. Material safety

Among the six, physical-chemical data are very unique for characterizing nanomaterials. The ten years of OECD's work has concluded that the approaches for the testing and assessment of traditional chemicals are in general, appropriate for assessing the safety of nanomaterials (OECD 2009), but may have to be adapted to the specificities of nanomaterials. It is suggested to evaluate nano agri inputs and nanoagri products as proposed in the DBT draft guidelines (DBT, 2019). However for better understanding an outline of some classic bioassays are discussed below.

Cytotoxicity Assays	*Genotoxicity Assays*
Cell titer- Glo luminescent Cell Viability assay	Bacterial Reverse Mutation test
Nuetral red uptake	Chromosomal aberration test
LDH Assay	Micronuclei test
MTT Assay	Cell gene mutation Test
XTT Assay	Unscheduled DNA Synthesis
WTS – 8 Assay	COMET assay
Tryphan Blue assay	
Live and dead cell Counting	

Cytotoxicity Assays

Cell Titer-Glo Luminescent Cell Viability Assay

It is used to determine the number of viable cells present in the culture by measuring the **ATP,** the major source of energy of active cells.

Beetle Luciferin $+ ATP + O_2$ —(Ultra-Glo™ Recombinant Luciferase, Mg^{2+})→ Oxyluciferin $+ AMP + PP_i + CO_2 +$ Light

In this test, the luciferin substrate is converted by luciferase enzyme in the presence of ATP and Mg^{2+} into a luminescent oxyluciferin molecule. Thus, Metabolically active cell produces luminescence whereas dead cells doesn't produce it (Hannah *et al.*, 2001).

Neutral Red Uptake

It is employed to measure the viable cells because of the ability of neutral red dye to move into lysosome at the optimum physiological pH of the cell.

Neutral red is a weekly cationic dye, moves to the lysozome of cell through nonionic passive diffusion mode and binds to phosphate group of anionic charge present in lysozomal matrix through electrostatic hydrophobic bond. After entry of dye, proton gradient will occur in lysozome to maintain the pH gradient with cytoplasm during which dye completely become cationic and retained within the lysosome. If the cell is not alive, there won't be a balance in pH gradient such that the dye retention in lysozome will be reduced.

Amount of dye retained in the cell is measured at 540nm. Intensity of Absorption increases with increase in no. of. viable cells (Repetto *et al.*, 2008).

Lactate Dehydrogenase Assay

This assay provides the number of non – viable cells by measuring the lactate dehydrogenase enzyme produced during the unfavorable condition of cell.

Lactate dehydrogenase is an enzyme used to convert the stored form of energy (lactate) into a active energy (Pyruvate) during this conversion the co – factor NAD+ is reduced to NADH.

Damaged cells

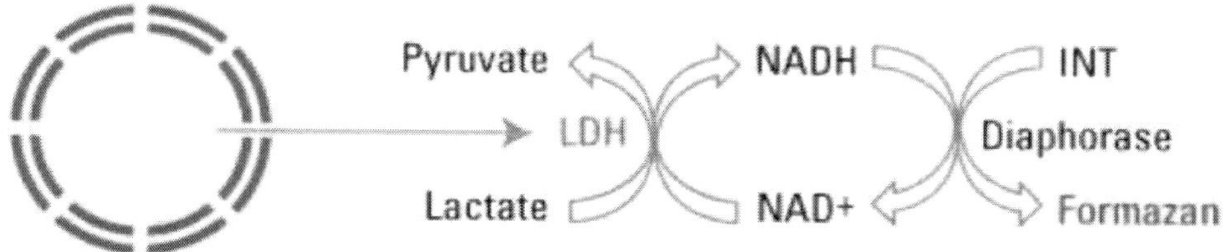

The diopharase enzyme (added externally) is capable of converting INT tetrazolium salt (added externally) into a red coloured formazon product by utilizing the proton released during the oxidation of NADH to NAD+ (Parhamifar *et al.*, 2013).

The red coloured formazon product was measured at 490nm. The bsorbance increases with increase in number of non – viable cells and viceversa.

MTT Assay

MTT denotes the chemical name (3-(4,5-Dimethylthiazol-2-yl)-2,5-Diphenyltetrazolium Bromide of the substrate molecule.

It is used to determine the number of viable cells by measuring the mitochondrial succinate dehydrogense enzyme. An enzyme which converts succinate to fumarate in kreb's cycle of respiration.

Mitochondria

Dehydrogenase

MTT

MTT Formazan

The enzyme is capable of converting MTT tetrazolium salt to insoluble formazon crystals whose absorbance is at 540 nm.

Viable cells produces more enzyme than the non viable cells, thus more absorbance in viable cells.

The coloured formazon crystals won't give appropriate absorbance reading Hence, to solubilize it needs further procedures. To overcome this disadvantage, XTT assay was developed.

XTT Assay

XTT denotes the chemical name of the substrate sodium, 3′-[1-{phenylamino}-carbonyl]-3,4-tetrazolium]-bis(4-methoxy-6-nitro)benzene –sulfonic acid hydrate.

XTT, a slight yellow colored oxidized form of XTT compound will be reduced to bright yellow or orange colour compound by the dehydrogenase enzyme. The breakdown of positively charged tetrazolium ring causes the colour change.

The dehydrogenase enzyme is produced in the viable cells to maintain the pyridine nucleotide redox state.

The absorbance was measured at 630 -690nm, and the absorbance increases with increase in number of viable cells.

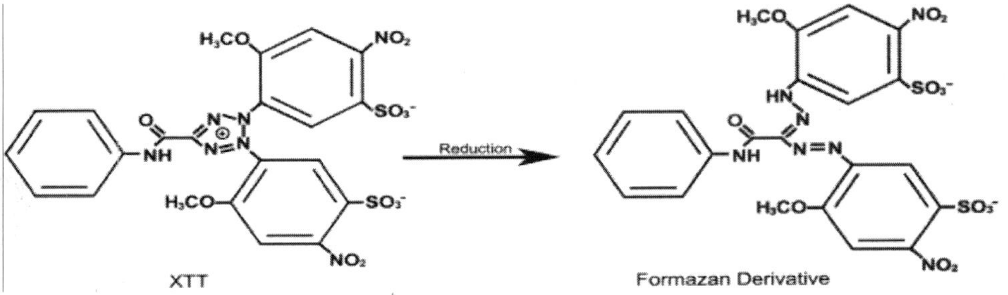

Since XTT is negatively charged it can't permeate the cell membrane, hence the reduction occurs on the outer surface of the cell with the help of trans membrane electron transport chain. This electron transport can further be facilitated by the chemical PMS (N- methyl dibenzine methyl sulphate). The reduced XTT than taken in to the plasma membrane.

WST (Water Soluble Tetrazolium) Assay

This test is very similar to XTT assay and it determine the viable cells by measuring the dehydrogenase enzyme which is essential to maintain redox state of the cell (Stockert *et al.*, 2018).

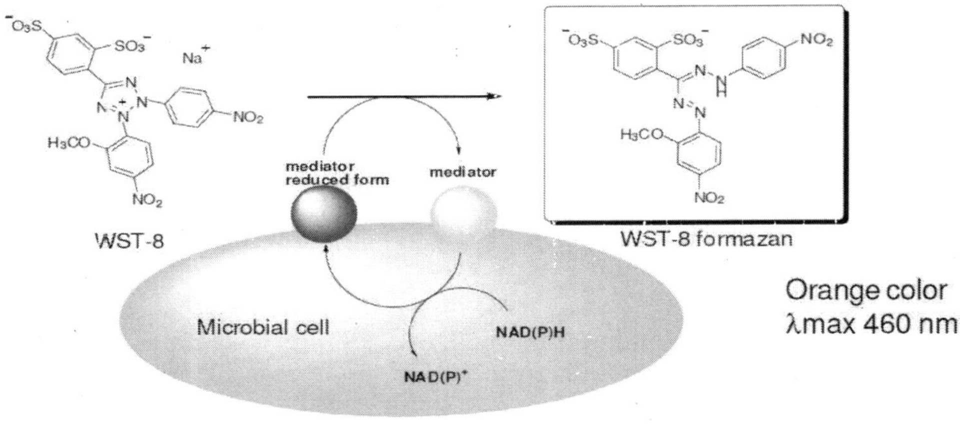

Dehydrgenase enzyme in presence of NADPH converts colourless WST -8 to a water soluble orange coloured formazon compound whose absorbance can be measured at 460nm.

Thus, absorbance increases with increase in the number of viable cells.

Tryphan Blue Assay

It is used to determine the number of non – viable cells present in the culture.

The basic principle behind this technique is that the intact cells have plasma membrane such that, certain type of dyes can not be internalised inside the cell which includes tryptaphan blue.

Thus, viable cells containing intact plasma membrane doesn't localise the dye and remain colourless whereas dead cell which don't have the intact plasma membrane takes the dye and become coloured (Strober, 2015).

Live or Dead Cell counting

There are two type of test used to measure or count the viable cells either spectrophotometricaly or using haemocytometer respectively.

- Fluorescent based detection
- Non- fluorescent based detection

Fluorescent Based Detection

It is used to determine the number of viable cells present in the cell culture by measuring amount of esterase enzyme of the cell. The esterase enzyme convert the non fluorescent substrate into an fluorescent compound. Thus fluorescence increases with increase in the number of viable cells.

Example

- Resazurin Reduction Assay Concept
- Protease viability marker assay concept
- Cell titre- glo Assay test

Resazurin Reduction Assay

In this test, Non fluorescent resazurin is reduced to resorufin by accepting the electron released due to the oxidation to NADH to NAD+ with the help of esterase enzyme produced in viable cells. Thus, viable cell fluoresce as pink colour whose absorbance is measured at 560 nm.

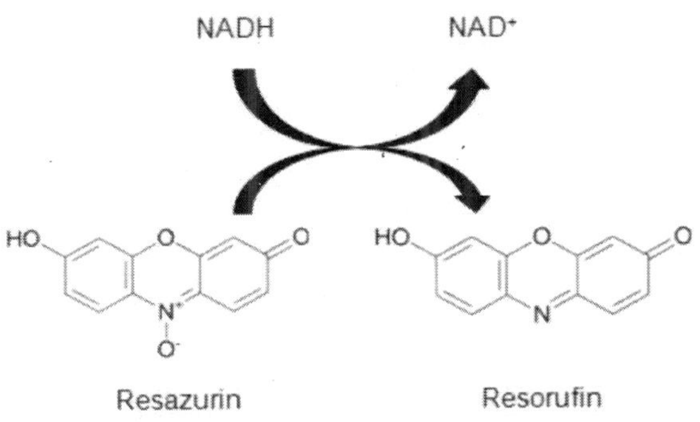

Protease Viability Marker Assay

It is also used to verify the presence of viable cells by indirectly measuring the aminopeptidase enzyme, an active enzyme involved in protein synthesis.

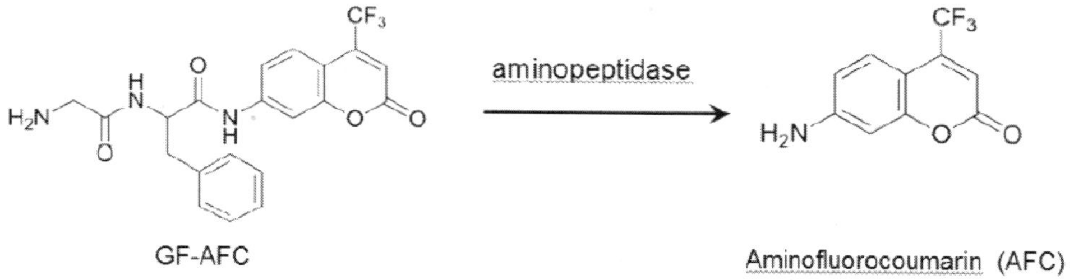

aminopeptidase

GF-AFC

Aminofluorocoumarin (AFC)

Aminopeptidase converts glycyl phenyl alanile amino fluoro coumarin into a aminofluorocoumarin whose fluorescent intensity is measured at 380 to 400nm

Non-Fluorescent Based Detection

It includes,

- Tryphan Blue Test
- MTT Assay

Genotoxicity Assay

Reverse Mutation Test

It is also called as Ames test. A general test used for preliminary assessment of genotoxicity or point mutation. It utilises different mutant starins of *Solmonella typhimurium* and *Escherichia coli* which can't grow in the absence of some essential compounds (*i.e.*, minimal media) Such as histidine for *S.typhimurium* and tryptophan for *E.coli.* The bacterial cultures used for the test are called mutants.

When these mutant culture are treated with the nanoparticles, and if they were found to be grown in minimal medium, then there is possibilities for the reversion of the mutation. Thus, cell culture those grown after the nanoparticles treatment are called as reverse mutants. If reverse mutants present, nanoparticles can be a genotoxic compound. The assay test can be performed according to the protocol of OECD Tg 471.

Chromosomal Aberration Test

It is used to determine the chromosomal damage like break, deletion, addition, inversion and transposition. It can be easily visualised during the metaphase stage of cell of division during which the chromatids will be arranged in centre of the cell before daughter cell formation.

The overnight grown cultures are treated with nanoparticles, and the cells are harvested during the metaphase stage of cell cycle. The cell cycle can be arrested at metaphase stage by adding colcemid chemical that helps in arresting the spindle fibre formation which pulls the chromatids to opposite poles for cell division. Thus, metaphase stage easily visualise the chromatids.

The harvested cells are then incubated in dyes and observed under microscope. The assay protocol can be followed as per the guidelines of OECD TG 473.

Micronuclei Test

It is used to determine the presence of micronucleus in the cell formed as a result of clastogenic event *i.e.*, chromosomal damage.

It can be easily visualised during the interphase stage of cell division. The cells can be arrested at interphase stage using cytchalasin B chemical. The harvested chemicals then be stained and observed using microscope. The assay protocol can be followed as per the guidelines of OECD TG 487

Cell Gene Mutation Test

It is used to determine the mutation at HPRT locus which codes for the enzyme hypoxanthine guanine phosphoribosyl transferase gene and XPRT locus which codes for Xanthine guanine phosphoribosyl transferase.

The enzyme HPRTase converts guanosine monophosphate to guanosine an active enzyme essential in salvage pathway of nucleotide synthesis.

In *invitro* studies, the enzyme HPRTase converts 6 – Thioguanine to 6 – thioguanine monophosphate, 6 – thioguanisine is a toxic compound which inhibit the cell growth.

If there is any mutation in HPRT gene due to nanoparticles, it can't convert 6- thioguanine to 6- thio guanine monophosphate and colonies will easily grow. Thus, presence of colonies indicates the genotoxicity of nanoparticles. The assay protocol can be followed as per the guidelines of OECD TG 476.

Unscheduled DNA Synthesis

This determines the mutation or mismatches in the cell by utilising the cells natural nucleotide excision repair (NER).

Nucleotide excision repair is a DNA repair mechanism of normal dividing cells in which the damaged DNA or mismatched nucleotide if present will be replaced and ligated by the DNA polymerase I and ligases enzyme during replication.

The cells treated with nanoparticles will be incubated in radioactively labelled thyamidine. If there is any damage in DNA due to nanoparticles, the cell's DNA repair mechanism utilises radioactively labelled thyamidine to replace the DNA damage or mismatches. The amount of radioactive thyamidine consumed indicates the presence of mutation which can be found using autoradiograph. The assay protocol can be followed as per the guidelines of OECD TG 482.

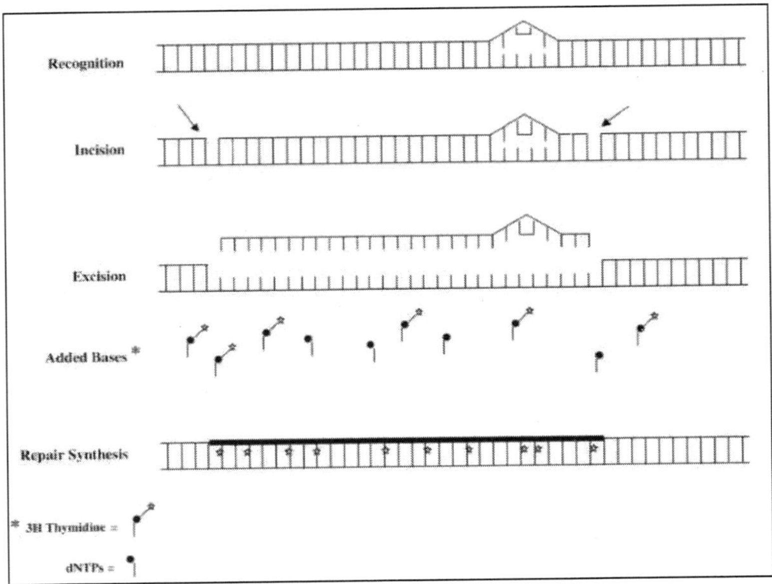

Nucleotide Excision Repair.
Courtesy: Latimer and Kelly, 2016.

COMET Assay

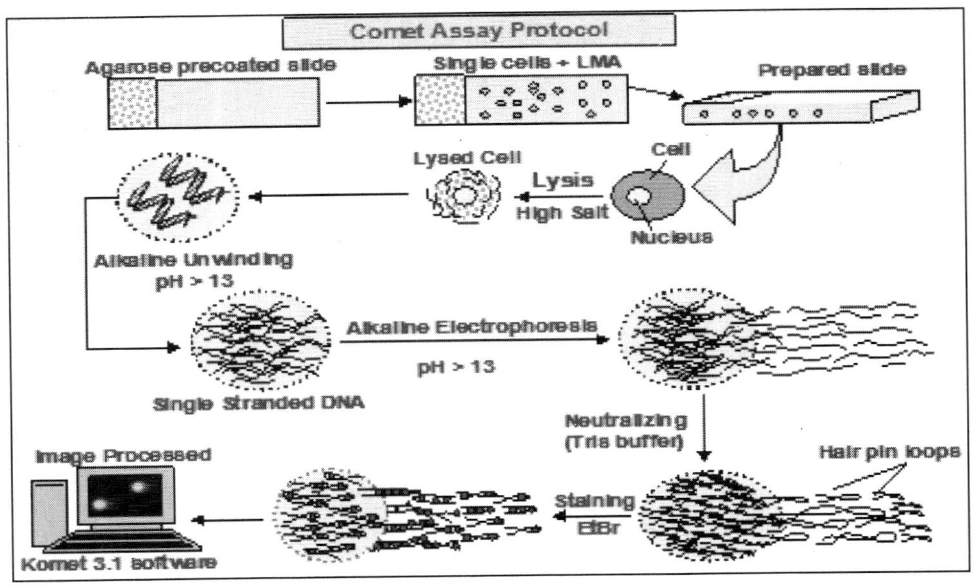

COMET Assay.
Courtesy: Bajpaye *et al.*, 2015.

It is also called as single cell gel electrophoresis Assay. Till now, most reliable test to identify the DNA break in individual eukaryotic cell.

In this test, cells treated with nanoparticles and without nanoparticles were immobilised over the glass slide using agarose. The glass slide will be incubated in lysis buffer to break the cell wall and plasma membrane without disturbing the double stranded DNA. Now, cells were treated alkaline buffer solution at which double strand DNA convert to single strand. Now the glass slide is stained with EtBr and subjected to gel electrophoresis and visualised under autoradiogram. If the DNA is damage due to nanoparticles, the autoradiograph will show a tail behind the spherical spot indicating the presence of small broken single strand DNA.

Conclusion

The regulatory science considers nano-materials as enigmatic matter and its associated ill-effects are found to vary with size, shape, structure, dose and environmental conditions. This warrants a clear cut risk assessment strategy for developing a robust regulatory framework. As of now, it is observed that the European Union is taking a lead step in formulating a nano-regulatory framework. In general, it is observed that there is a lack of emphasis on risk related research in nanotechnology and it is vital to undertake the biosafety evaluation at an early stage when applicability of specific nano-applications is ascertained, especially when the prototype of the product has been developed. Since, the pace of nanomaterial innovation far surpasses acquisition of environmentally relevant data, it is highly essential to prioritize the biosafety research to develop validated protocols that are reproducible across laboratories.

REFERENCES

1. Adholeya A, Das RK, Dubey MK, Kochar M, Singh R (2017) Zero draft policy on regulation of nanoproducts in agriculture. The Energy and Resources Institute (TERI), New Delhi
2. Amenta V, Aschberger K, Arenaa M, Bouwmeester H, Moniz FB, Brandhoff P, Gottardo S, Marvin HJP, Mech A, Pesudo Q, Rauscher H, Schoonjans R, Vettori MV, Weigel S, Peters RJ (2015) Regulatory aspects of nanotechnology in the agri feed food sector in EU and non-EU countries. Regulatory Toxicology and Pharmacology 73: 463-476
3. Arts JHE, Hadi M, Irfan AM, Keene AM, Kreiling R, Lyon D, Maier M, Michel K, Petry T, Sauer UG, Warheit D, Weinch K, Wohlleben W, Landseidel R (2014) A critical appraisal of existing concepts for the grouping of nanomaterials. Regul. Toxicol. Pharmacol. 70, 492–506
4. Bajpayee, M., Pandey, A. K., Parmar, D., Mathur, N., Seth, P. K., and Dhawan, A. (2015). Comet assay responses in human lymphocytes are not influenced by the menstrual cycle: a study in healthy Indian females. Mutation Research/ Genetic Toxicology and Environmental Mutagenesis, 565(2), 163-172.
5. DBT Department of Biotechnology (2019) Guidelines for guidelines for evaluation of nanoagriinput products and nanoagriproducts in India,

Department of Biotechnology, Ministry of Science and Technology, Govt. of India, pp1-26

6. DST Dept of Science and Technology (2016) Guidelines and best practices for safe handling of nano-materials in research laboratories and industries. Centre for Knowledge Management of Nanoscience and Technology, Dept. of Science and Technology, Govt. of India, pp1-24.
7. Hannah, R., Beck, M., Moravec, R., and Riss, T. (2001). CellTiter-Glo™ Luminescent cell viability assay: a sensitive and rapid method for determining cell viability. *Promega Cell Notes, 2*, 11-13.
8. Hristozova D, Gottardo S, Semenzin E, Oomen A, Bos P, Peijnenburg W, Tongeren MV, Nowack B, Hunt N, Brunelli A, Scott FJ.J, Tran L, Marcomini A (2016) Environment International 95: 36-53.
9. Justo HR, Dayan T (2014) European risk governance of nanotechnology: Explaining the emerging regulatory policy. Research Policy 44: 1527-153
10. Klöpffer W, Curran MA, Frankl P, Heijungs R, Köhler A, Olsen SI (2007). Nanotechnology and Life Cycle Assessment. A systems approach to Nanotechnology and the environment: Synthesis of Results Obtained at a Workshop Washington, DC 2–3 October 2006. European Commission, DG Research, jointly with the Woodrow Wilson International Center for Scholars.
11. Latimer, J. J., and Kelly, C. M. (2014). Unscheduled DNA synthesis: the clinical and functional assay for global genomic DNA nucleotide excision repair. In Molecular Toxicology Protocols (pp. 511-532). Humana Press, Totowa, NJ.
12. Parhamifar, L., Andersen, H., and Moghimi, S. M. (2013). Lactate dehydrogenase assay for assessment of polycation cytotoxicity *Nanotechnology for Nucleic Acid Delivery* (pp. 13-22): Springer.
13. Rajkishore SK, Subramanian KS, Natarajan N, Gunasekaran K (2013) Nanotoxicity at various trophic levels. A review The Bioscan 8(3): 975-982
14. Repetto, G., Del Peso, A., and Zurita, J. L. (2008). Neutral red uptake assay for the estimation of cell viability/cytotoxicity. *Nature protocols, 3*(7), 1125.
15. Stockert, J. C., Horobin, R. W., Colombo, L. L., and Blazquez-Castro, A. (2018). Tetrazolium salts and formazan products in Cell Biology: Viability assessment, fluorescence imaging, and labeling perspectives. *Acta histochemica, 120*(3), 159-167.
16. Strober, W. (2015). Trypan blue exclusion test of cell viability. *Current protocols in immunology, 111*(1), A3. B. 1-A3. B. 3.
17. Subramanian KS, Manikandan A, Thirunavukkarasu M, Rahale CS (2015) In: Rai M et al.(eds.), Nanotechnologies in Food and Agriculture. Springer International Publishing, Switzerland, pp 69-80
18. Subramanian KS, Raja K, Marimuthu S (2018) Multifarious applications of nanotechnology for enhanced productivity in agriculture. Emerging trends in agri-nanotechnology: Fundamentals and applied aspects (Eds: Singh H.B, Mishra S, Fraceto LF, Lima Rd. CAB International, pp56-77

19. Subramanian KS, Tarafdar JC (2009) Nanotechnology in soil science. In: Proceedings of the Indian society of soil science-platinum jubilee celebration, December 22-25, IARI Campus, New Delhi, pp199

20. Subramanian K.S and Rajkishore S K (2018). Regulatory framework for Nanomaterials in Agri Food Systems. In: M Rai, J K Biswas (eds) Nanomaterials: Ecotoxicity, Safety and Public Perception,pp 319-342: Springer.

Chapter 33

Nanotechnology in Agriculture: Emerging Opportunities and Future Challenges

A. Lakshmanan[1], K.S. Subramanian[2], R. Abinayaa[1] and L. Sivashankari[1]

[1]Department of Nano Science and Technology, [2]Directorate of Research, Tamil Nadu Agricultural University, Coimbatore – 641 003, Tamil Nadu

Introduction

The ever increasing human population is exerting serious pressure in food, water and energy sectors and human pressure over natural resources has already reached critical levels. Soliciting scientific research in order to identify innovative solutions to support the primary sectors is the need of the hour. Nanotechnology is a rapidly evolving field with innovative tools to ensure food and nutritional security. The potentials of nano systems have just started showing the results and better understanding of nanomaterials can contribute hugely to agriculture. Many unresolved problems in agriculture such as over dependency of chemicals, loss in soil fertility, yield plateau, pest and disease management and post-harvest losses can be better handled with the intervention of nanotechnology. Implying nanotechnology in agriculture is aimed at reducing the amount of active ingredients by developing "smart delivery" of nutrients and plant protective agents into the plant system and thereby increasing crop yield and income to farmers. Natural and synthetically prepared nanomaterials used in agriculture are expected to be a potential alternative to the existing less efficient conventional methods and therefore, the application of nanomaterials is directed towards the development of sustainable agriculture. The efficiency of the nanomaterials is mainly based on their physico-chemical properties like chemical composition, crystalline/amorphous nature, particle size, shape, porous structure, surface structure, surface charge, hydrophilic/hydrophobic ratio, particle agglomeration and dispersion. The physico-chemical properties of nanomaterials can be further tuned to achieve higher efficiency and to face the future challenges in agriculture and allied sectors. Promising nanomaterials to be used in agriculture, plant nanobionics to enhance basic functions in plant systems, Internet of NanoThings in agriculture, nanotechnology and artificial intelligence, nanotechnology in precision farming, nanotechnology to augment drones in agriculture and nano barcoding are some of the fascinating areas that could transform the agriculture in the coming years and these topics are discussed in this chapter.

Nanomaterials with Prospective Applications in Agriculture

Halloysite Nanotubes

Halloysite Nanotube (HNT) is a naturally found 1:1 aluminosilicate clay mineral with an empirical formula of $Al_2Si_2O_5(OH)_4.2H_2O$. The beginning of the research on HNT as a mineral goes long ago. HNT was discovered in 18th century by Berthier and was named after Omalius d'Halloy, who analyzed the mineral at first. In recent times, HNT as natural nanomaterial is gaining attention in the field of nanotechnology for various eco-friendly applications. They have many appealing characteristics like 1 dimensional (1D) nanotube morphology, two layered sheets (rolled up to nanotube) with the outer negatively charged siloxane nanotube layer and the inner positively charged aluminol nanotube layer. The outer diameter of HNT is about 50–60 nm, an inner diameter of 12–15 nm with a length of 0.5–1 µm (Lvov *et al.*, 2008).

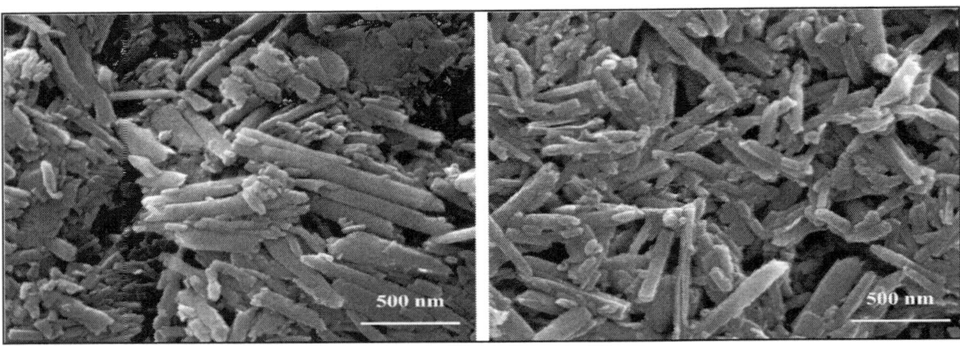

Figure 33.1: FESEM Micrograph of HNT.
Courtesy: Abhinayaa, Jeevitha, Mangalaraj, Ponpandian, and Meena, 2019.

In agriculture, HNT with hollow inner lumen structure can be used as a nano-carrier to loaded fertilizer, herbicides, pesticides, and other micro-nutrients to facilitate sustainable release for a longer duration and thereby can reduce the higher amount of fertilizer and pesticide application in the agriculture soil. HNT with differential surface charges can be surface modified with different functional molecules to enhance their reactivity. HNT with porous structure with gap of about

7 Å or 10Å between the double nanotube *i.e.*, the dehydrated and the hydrated form and with hollow inner lumen can be used as an adsorbent material. To promote rain fed agriculture, a superabsorbent material is needed to facilitate water retention capacity in the soil. Many earlier works confirm that the addition of HNT into the polymer matrix could increase the water sorption capacity of the polymer and HNT/ polymer nano-composite can be further studied to design potential superabsorbent nanomaterial for moisture adsorption.

Graphene Oxide

Graphene oxide sheets or graphene oxide (GO) are 2 D monolayer or few-layers of carbon atoms. GO sheets are chemically synthesized using modified Hummers method. GO sheets has one dimension in nanometer range and can be functionalized to load various agrochemicals and the functional groups present in the outer surface of GO can interact with microbial cells and can kill phytopathogens.

Figure 33.2: SEM Micrograph of Graphene Oxide Sheets
Courtesy: Jang, Zhamu, and Guo, 2016.

GO based core-shell type fertilizer has been developed by encapsulating KNO_3 pellets and the GO is reduced to rGO⁻ reduced graphene oxide after heat treatment. After the core- shell structure is formed, KNO_3 pellets from the rGO core exhibits the slow-release of the fertilizer. This new coating technology could hold great promise for the development of environmentally benign controlled-release fertilizer for crop production (Zhang *et al.*, 2008). GO nanocomposites is effective in inhibiting phytopathogen, *Xanthomonas perforans,* which causes bacterial spot a major disease in tomatoes, leading 10-50 per cent reduction in the crop production. The DNA-directed silver (Ag) nanoparticles grown on (GO) effectively decreases *Xanthomonas perforans,* cell viability in culture and on plants. This indicates that GO can be utilized for disease control in agriculture and can be exploited to other applications like water disinfection, site specific delivery of functional molecules, detection of chemical residues in farm produces *etc.*

Chitosan Nanoparticles

Chitosan, a biopolymer and deacetylated derivative of chitin that is naturally present in the fungal cell wall and in crustacean shells from which it can be easily extracted. Chitosan has distinct features such as biodegradability, non-toxicity, biocompatibility *etc*. The chitosan nanoparticle based systems have a range of applications in crop protection that works against phytopathogens.

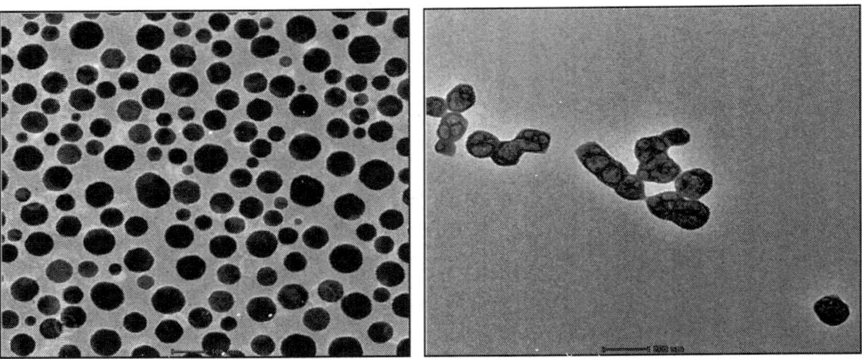

Figure 33.3: SEM Micrograph of Chitosan Nanospheres.

Many studies confirm that chitosan nanoparticle has drastically reduced the cell viability of fungal phytopathogens which cause diseases like wilt and rot. The chitosan nanoparticle behaves as a resistance elicitor inducing both local and systemic plant defense responses even when applied to the seeds. The chitosan used as soil amendment has shown many benefits to different plant species by reducing the pathogen attack and infection. Chitosan nanoparticle also stimulates the growth and increases yield of plants as well as induces the immune system of plants. Seed soaking and foliar spraying with chitosan solution shows significant increase in many crops. Chitosan can also be employed as carrier to deliver functional molecules in soil and plant systems. As chitosan is a natural polymer, any product developed out of chitosan nano systems are having fewer biosafety concerns and hence commercialization would be easy. The chitosan nanomaterials have the potentials for multiple application in agriculture and food preservation and needs to be studied further.

Plant Nanobionics to Enhance Basic Functions in Plant Systems

Scientists have been deliberating for many years on enhancing the photosynthetic efficiency of plants using different strategies such as engineering C_3 crops to use C_4 photosynthesis pathway, improving the efficiency of Rubisco, modifying the chlorophyll antenna size of chloroplast photosystems *etc*. Nanotechnology could provide us with the means to engineer changes to photosynthesis to enhance the productivity and to overcome the yield plateau. Understanding the relationship between plants and nanomaterials is an important area and more research works need to be devoted and plant nanobionics would provide more insight into plant - nanomaterials interaction. Many studies are reporting the positive effects of nanomaterials on photosynthesis. Titanium oxide nanoparticles (nTiO$_2$) is one good example and the high photocatalytic activity of anatase crystal nTiO$_2$ has been hypothesized to have a role in the improvement of light absorbance by plant leaves, thus sustaining an increase in photosynthesis. nTiO$_2$ protects the chloroplast from aging due to photochemical stress, activates Rubisco carboxylation, promoting an enhancement of the photosynthetic rate and positively influences biophysics traits of photosynthesis, such as electron chain transport and Chl-photophosphorylation activity besides improving leaf water conductance and transpiration rate. More recently, the original idea to merge nanomaterials with living plants to enhance their native functions and to give them non-native functions has been more accurately focused. This approach assumed the name of "plant nanobionics" and potentially allowed to engineer faster-growing plants and become the key factor to design and develop artificial photosynthetic systems, a potential source of clean energy (Marchiol, 2018).

Employing Carbon Nano Tubes (CNTs) to deliver DNA to plant cells has been an upcoming and most prominent nanotechnological invention and CNT mediated genome editing has been found more effective compared to *Agrobacterium* or gene gun based techniques. Carbon nanotubes are tiny, hollow cylinders, just one nano meter in diameter, made entirely of carbon. Their unusual molecular properties make them valuable tool for genome editing. Carbon nanotubes are so narrow and can slip through the cell wall and deliver the DNA with higher accuracy with no harmful side effects. Earlier attempts confirmed the fact that carbon nanotube based genome editing had 85–95 per cent efficiency. CNT editing tools are present just long enough to edit the plant's DNA before being broken down by the plant, leaving no footprint. The plant's DNA would only be changed at the target site and unlike with some current plant genome editing methods, no foreign DNA would be inserted into the plant's genome.

Another interesting fact is that chloroplasts have their own genome that codes for many essential photosynthesis genes. But editing chloroplast genomes is even more challenging than editing the main genome of a plant cell. Amazingly the results of CNT tracking studies confirm that the carbon nanotubes end up in over 90 per cent of chloroplasts suggesting the availability of new powerful tool to deliver genes to chloroplasts that would far exceed the capabilities of current technology. Proof of concepts are available on the enhancement of a basic plant function in response to incorporation of nanomaterials and plant nanobionics can open up a new avenue in genome editing and this area could be exploited to breed new varieties with many desirable traits.

Internet of NanoThings in Agriculture

Site specific crop management is a form of Precision Agriculture in which decisions on resource application and agronomic practices are improved to better match soil and crop requirements as they vary in the field. Applications that are derived from convergence between Internet of Things (IoT) and nanotechnologies are developing very rapidly in industrial, information and communication technologies, and biomedical sectors. The future interactions between the Internet and nanodevices introduce a new perspective which has been referred to as the "Internet of NanoThings" (IoNT). A properly designed network of nano sensors would allow the optimization of crop yield and the most efficient agronomic management of factors, such as fertilizers, water, herbicides and pesticides. Large numbers of nanoscale-sensing devices could be positioned on the plant leaves through suspension in a spray treatment. Nanonetworks for monitoring plant conditions can alert automatically suggesting a more efficient usage of crop inputs. The real time monitoring of the crop growth through nano sensors helps to enhance the use efficiency of various inputs that reduces cost and environmental pollution besides enhancing the yield (Marchiol, 2018). Agriculture sector can be benefitted hugely by Internet of NanoThings due to its potential impact on smart agriculture and this area could be concentrated more in the coming years.

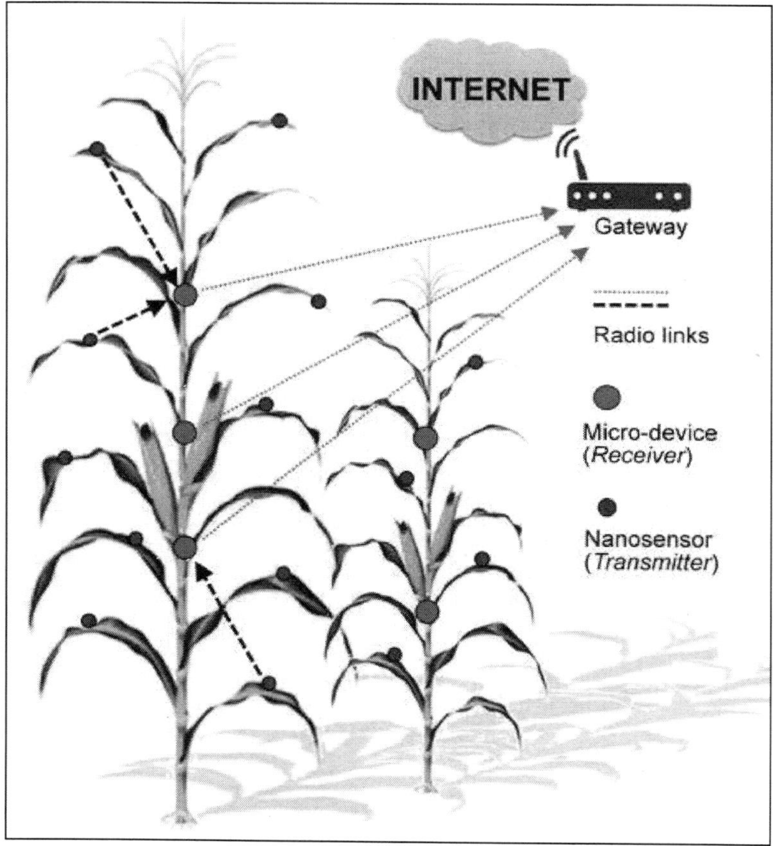

Figure 33.4: Structure and Components of a Nanonetwork Designed for Plant Monitoring Applications.

Courtesy: Marchiol, 2018.

Nanotechnology and Artificial Intelligence: Prospects in Agriculture

"To eradicate poverty and hunger" as stated in Millennium Development Goal, UN Summit 2000, it is important to adapt latest technologies in agriculture to bring a balance between the demand from growing world population and agricultural production despite the bitter fact of depleting land resources and lack of labour availability. In this context, agriculture is seeing rapid adoption of Nanotechnology in Articial Intelligence (AI) and Machine Learning (ML) to obtain better insights on field conditions, disease infestations and other process like fertilizer and pesticide applications using drones. Nano sensors and more specifically, nanobiosensors is the dedicated miniature sensors with highly miniaturization based on the ideas of nanotechnology. The wide range of application of Nano sensors in agriculture and food industry is represented below.

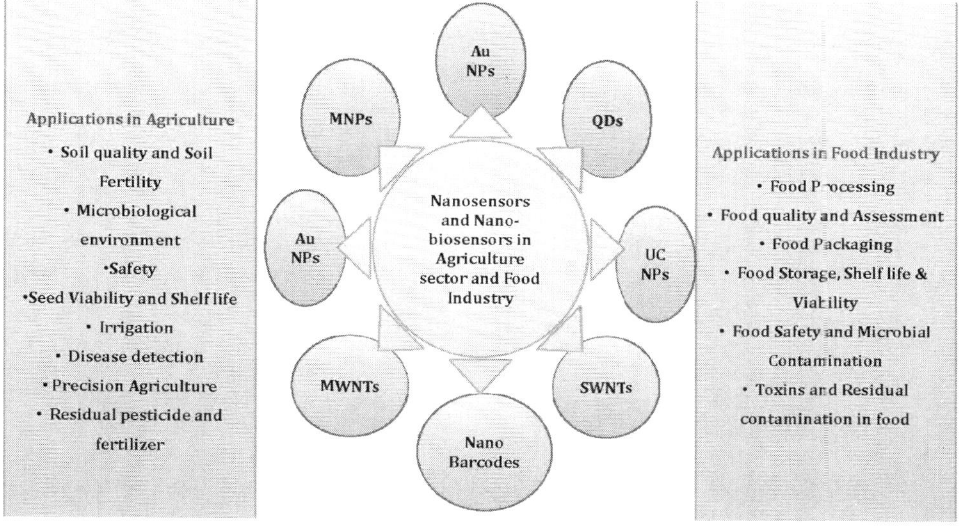

Figure 33.5: Schematics Depicting the Major Nanostructures Used in the Area of Food and Agriculture Sector.
Courtesy: Srivastava, Dev, and Karmakar, 2018.

The biological recognition molecule interacts with target compound, and the physical transducer converts the biological response to a detectable signal, quantitates as redox changes, and detects electrochemically, optically, acoustically, mechanically, calorimetrically, or electronically, which can be correlated with the analyte concentration (Sharma and Rogers, 1994). The nanobiosensors constitute transducer based on nanomaterials, having physical nanoscale confinement or nanofabrication of nanoparticles or nanostructured surfaces.

Nanotechnology can also help in cognitive computing, that plays a major role in agriculture services in terms of understanding, learning, and responding to different field conditions and has high efficiency.

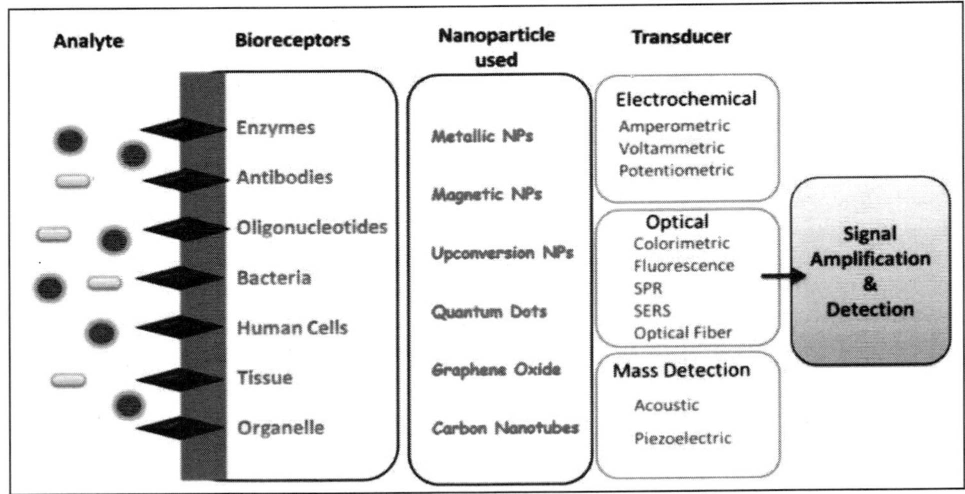

Figure 33.6: Essential Components of a Typical Nanobiosensor from Sample Analyte to Bioreceptor, Transducer with Integrated Nanostructures, and Finally Detectors.
Courtesy: Srivastava *et al.*, 2018.

The most popular applications of Artificial Intelligence in the coming years are categorized under three major divisions (Sennaar, 2017).

I. Agricultural Robots

Companies are developing and programming autonomous robots to handle essential agricultural tasks such as harvesting crops at a higher volume and faster pace than human laborers.

II. Crop and Soil Monitoring

Companies are leveraging computer vision and deep-learning algorithms to process data captured by drones and/or software-based technology to monitor crop and soil health.

III. Predictive Analytics

Machine learning models are being developed to track and predict various environmental impacts on crop yield such as weather changes.

In all the above mentioned tasks nano sensors could be an integral compound and engineered nano particles can enhance the sensitivity and selectivity of the sensing process

Nanotechnology in Precision Farming

Precision farming is all about "Right Place, Right Time, Right Product". This system replaces the repetitive and labor-intensive and is more accurate and has

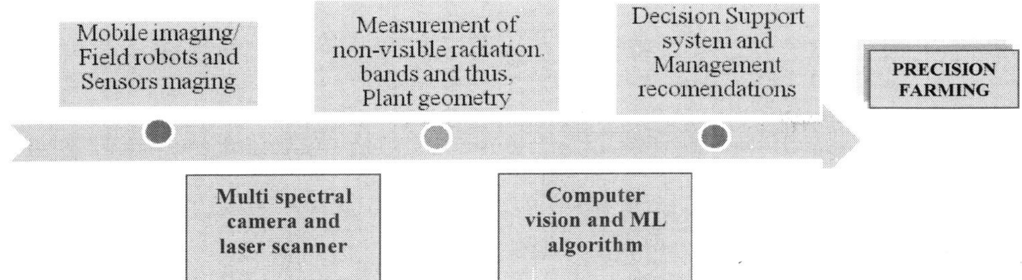

Figure 33.7: Robotics and Machine Learning in Precision Farming.

controlled mechanisms. Precision farming provide complete guidance to smart farmers on planting and harvesting time, pest and disease management, crop rotation, optimum water and nutrient management and so on.

In precision farming encapsulated nanostructures could help in the controlled release of nutrients or any functional molecules. The outer shell of nanocapsules is engineered and programmed to open when stimulated by environmental factors or man-induced pulses. Here are some examples of possible control mechanisms that can be exploited in precision farming to achieve smart and sustained release of nutrients and other agro inputs.

- **Slow release:** The capsule releases its payload slowly over a longer period of time so as to synchronize plant assimilation and limit leaching.
- **Quick-release:** The capsule shell breaks upon contact with a leaf surface.
- **Specific release:** The nutrient release occurs through a recognition mechanism between a receptor (molecule or functional group) bound to the shell and a target molecule.
- **Moisture release:** The shell breaks down and releases nutrients in the presence of water.
- **pH release:** The shell breaks up only in specific alkaline/acidic environment (*e.g.*, within plant tissues or inside a cell).
- **Magnetic/ultrasonic pulses:** The shell opens in response to a magnetic or ultrasonic pulse emitted by a man-controlled system (precision agriculture).

Nanotechnology to Augment Drones in Agriculture

In agriculture, the drone-based applications has lots of significance in terms of productivity, managing adverse weather conditions and precision farming. The major advantages of drones in agriculture are listed below:

1. Drones are used to produce a 3-Dimentional field mapping that provides detailed information on terrain, drainage, soil viability and irrigation. Before the crop cycle, nitrogen-levels in soil and their management can also be done using drones.

2. Drones provide greater facility in aerial spraying of plant nutrients into the soil to provide necessary supplements for crops.
3. They also help in monitoring crop growth and assessment of plant health conditions that remains one of the most signicant areas of drones in agriculture. Drone-based technology in combination with Articial Intelligence and machine learning/computer vision technology provides solutions for plant health issues.
4. The drones installed with high-resolution cameras can be programmed to pass through nook and corner of the field for better identification of weed infested areas, plant stress levels in various growth stages *etc.*, In case of plant infestation, it scans the crops in both RGB and near-infra red light, that makes it possible to generate multispectral images. With these images, it is possible to specify the exact infested plants and their location in a vast eld for further disease management, instantly.

Engineered nanoparticles could be employed to design efficient nano sensors that can transmit ground information to drones to carryout operations accordingly. This would help in the optimization of crop yield and the most efficient agronomic management of factors.

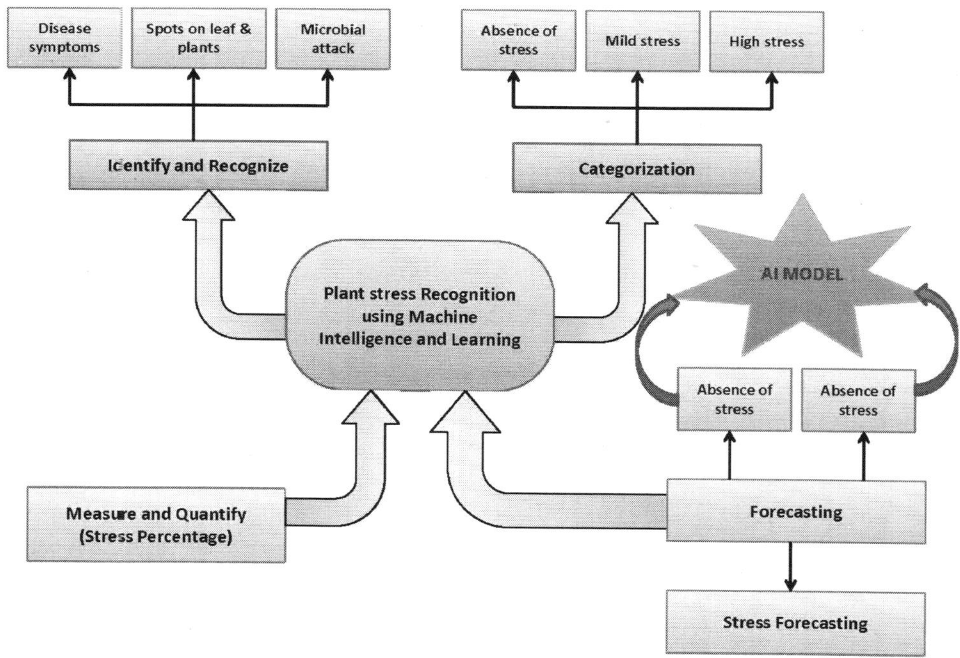

Figure 33.8: Machine Learning and Intelligence in Recognizing Plant Stress.
Courtesy: Dharmaraj and Vijayanand, 2018.

Nanobarcode Technology

Barcode technology utilizes the principle of symbology by interpreting the encoded information or data in the form of a map. This technology is formally termed as "biobarcode" or "DNA barcode" technology. It serves as novel tool for identifying single species of animal, plant and microbes by taking a small stretch of a single gene (Ferri *et al.*, 2009). Micron-sized barcodes are the most established ones. They are the structures encoded with different elements with varied properties including different sizes, shapes, colour, *etc.*, Micro-barcodes exhibit efficient binding, and they are easy to handle and detect, thus it is widely used in multiplex bioassays. Luminex's xMAP technology using dye encoded microbeads is one commercially established micro-barcode systems. However, the size of micro-barcodes becomes a limiting factor in certain applications like imaging and labelling of cells and biomolecules. Nano-barcodes due to their advantages pertaining to nanoscale dimension, finds a place for various applications. The nano-barcodes identify smaller sized targets due to their confined size range for imaging applications. The nanostructures are encoded by unique signatures to prepare distinct codes, as encoding carriers. They are referred as encoding elements and they could be magnetic, optical (fluorescent and non-fluorescent) and phase change nanoparticles of different shapes and sizes. These encoding elements can be decorated on the surface of nanostructures or embedded inside or can be self-assembled to prepare the nano-barcodes.

A wide varieties of nano-barcodes such as encoded nano-beads, nano-wires, nano-clusters, nano-rods, nano-tubes, nano-sheets, nanoscale metal organic frameworks (NMOF), and bio-barcodes have been developed. With the availability of surplus of nano-barcodes, detection of biomolecules (nucleic acid and proteins), imaging, security, drug delivery *etc.* (Hung *et al.*, 2017; Liu *et al.*, 2017; Sun *et al.*, 2017) has a huge scope in the coming years in agriculture and allied sectors.

Challenges

Nanotechnological interventions could certainly make huge positive impacts in agriculture in the coming years and we have already started realizing the potentials. However, there are few concerns on the commercialization and field applications of nano products which have to be addressed clearly. As the proof of concepts related to nano products are highly fascinating, the industries are showing great expectations to market the products. Hence the first and major concern is the biosafety of nano products. Nanotechnology is a fast growing field and it needs support from all sides from academic to policy interventions. At the same time a technology with missing links cannot be promoted in a hasty way. Hence biosafety protocols and guidelines to promote nano products have to be drafted very carefully so that unbiased, yet prudent strategy prevails that takes care of safe technology outreach. Application of nanomaterials in agriculture still needs more extensive research to understand the fate of nanomaterials in the agricultural environment. Life cycle assessment (LCA) of nano products is very important to get the complete information of a product from cradle to grave. Due to its holistic and comprehensive assessment, LCA is an essential tool to analyse, evaluate, understand and manage

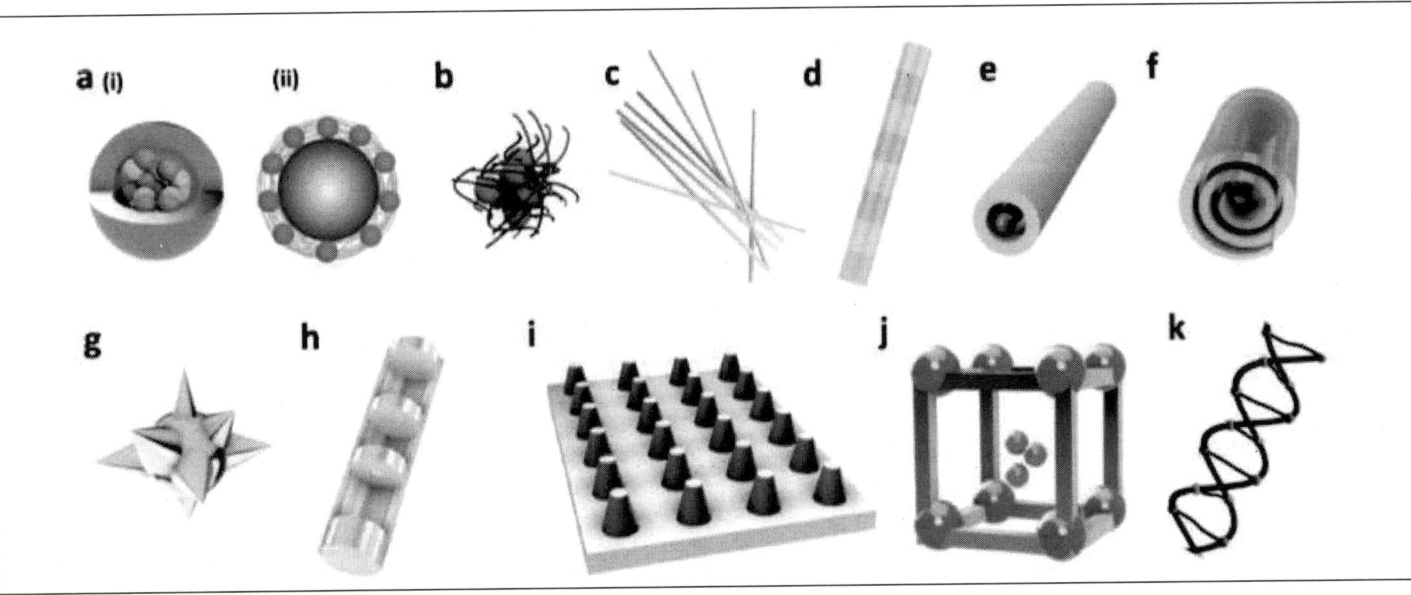

Figure 33.9: Schemas for different barcoded nanostructures of different geometries. Encoded (a) nanobeads with encoding elements; (b) nano-clusters, where nanoparticles are assembled as super-particles and stabilized by polymers; (c) nano-wire array which has nanoscale diameter and length up to several microns; (d) nanorods with different metal compositions in a striped pattern; (e) nano-tube crosssection showing nanoparticles encapsulated inside; (f) nano-sheets with nanoparticles loaded inside; (g) nano-stars that are branched nanostructures with small cores prepared from metal nanoparticles; (h) nano-disks with disk-shaped structures; (i) nano-pillars with pointed top ends and tapered bases; (j) NMOFs consisting of coordination bonds between metal ions and organic linkers with fluorescent nanoparticles encapsulated inside and (k) sequence barcodes made of oligonucleotide (or peptides) sequences as encoding elements. Note that the encoding elements on each nanostructure depicted are only for representation and can be changed to other elements (*Courtesy*: Shikha, Salafi, Cheng, and Zhang, 2017).

the environmental and health effects of nanotechnology. The evaluation of health and environmental impacts of nanotechnologies, throughout the whole of their life-cycle by using LCA methodology, has been mentioned in a number of EU policy documents. Currently, only few LCA's studies on nanotechnology are carried out, and only fewer studies assess the aspects relating to ecotoxicity. There are a number of challenges in undertaking LCA of nano-based products such as the lack of inventory input/output data; absence of toxicological tests results; wide process-to-process variation; lack of understanding in certain areas. Even though LCA has challenges and involves more time and cost, it should be encouraged to ensure safe commercialization of nano products.

REFERENCES

1. Abhinayaa, R., Jeevitha, G., Mangalaraj, D., Ponpandian, N., and Meena, P. (2019). Toxic influence of pristine and surfactant modified halloysite nanotubes on phytopathogenic bacteria. *Applied Clay Science, 174*, 57-68.
2. Dharmaraj, V., and Vijayanand, C. (2018). Artificial Intelligence (AI) in Agriculture. *Int. J. Curr. Microbiol. App. Sci, 7*(12), 2122-2128.
3. Ferri, G., Alu, M., Corradini, B., Licata, M., and Beduschi, G. (2009). Species identification through DNA "barcodes". *Genetic testing and molecular biomarkers, 13*(3), 421-426.
4. Hung, A. H., Lilley, L. M., Hu, F., Harrison, V. S., and Meade, T. J. (2017). Magnetic barcode imaging for contrast agents. *Magnetic resonance in medicine, 77*(3), 970-978.
5. Jang, B. Z., Zhamu, A., and Guo, J. (2016). Nano-scaled graphene plate films and articles. In: Google Patents.
6. Liu, H., Jayakumar, M. K., Huang, K., Wang, Z., Zheng, X., Ågren, H., and Zhang, Y. (2017). Phase angle encoded upconversion luminescent nanocrystals for multiplexing applications. *Nanoscale, 9*(4), 1676-1686.
7. Marchiol, L. (2018). Nanotechnology in Agriculture: New Opportunities and Perspectives. *New Visions in Plant Science*, 121.
8. Sennaar, K. (2017). AI in Agriculture–present applications and Impact. *TechEmergence, October, 16.*
9. Sharma, A., and Rogers, K. R. (1994). Biosensors. *Measurement Science and Technology, 5*(5), 461.
10. Shikha, S., Salafi, T., Cheng, J., and Zhang, Y. (2017). Versatile design and synthesis of nano-barcodes. *Chemical Society Reviews, 46*(22), 7054-7093.
11. Srivastava, A. K., Dev, A., and Karmakar, S. (2018). Nano sensors and nanobiosensors in food and agriculture. *Environmental chemistry letters, 16*(1), 161-182.
12. Sun, T., Xu, B., Chen, B., Chen, X., Li, M., Shi, P., and Wang, F. (2017). Anti-counterfeiting patterns encrypted with multi-mode luminescent nanotaggants. *Nanoscale, 9*(8), 2701-2705.

Index

D

E

F

G

N

O

P

Q

R

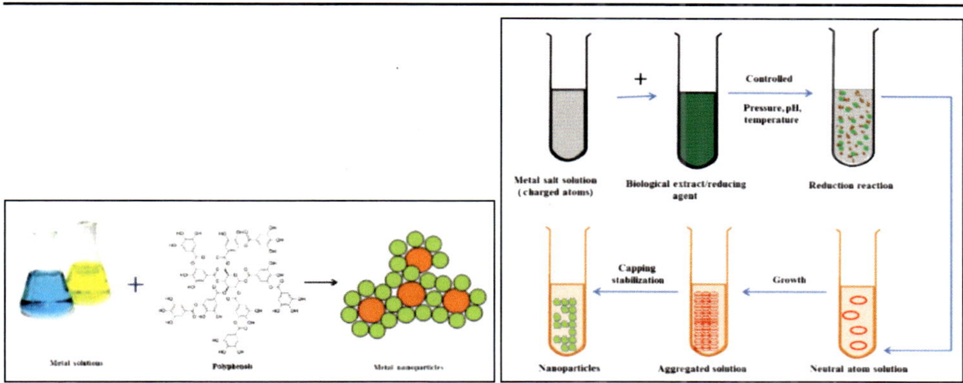

Figure 11.1. Process of Synthesis of Metal Nanoparticles from Plant Extract
Courtesy: Devatha and Thala, 2018. P. 253

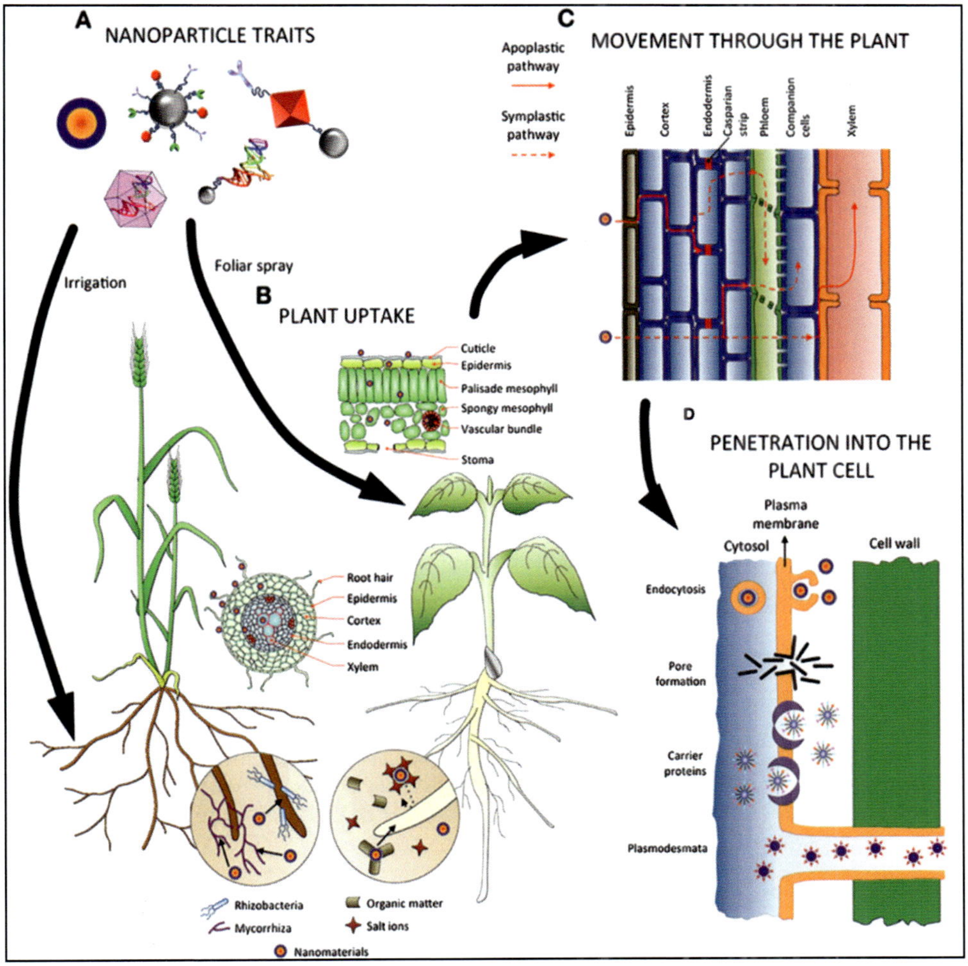

Figure 13.2. Entry of Encapsulated Herbicide into Plant System.
Courtesy: Anna *et al.*, 2019. **P. 290**

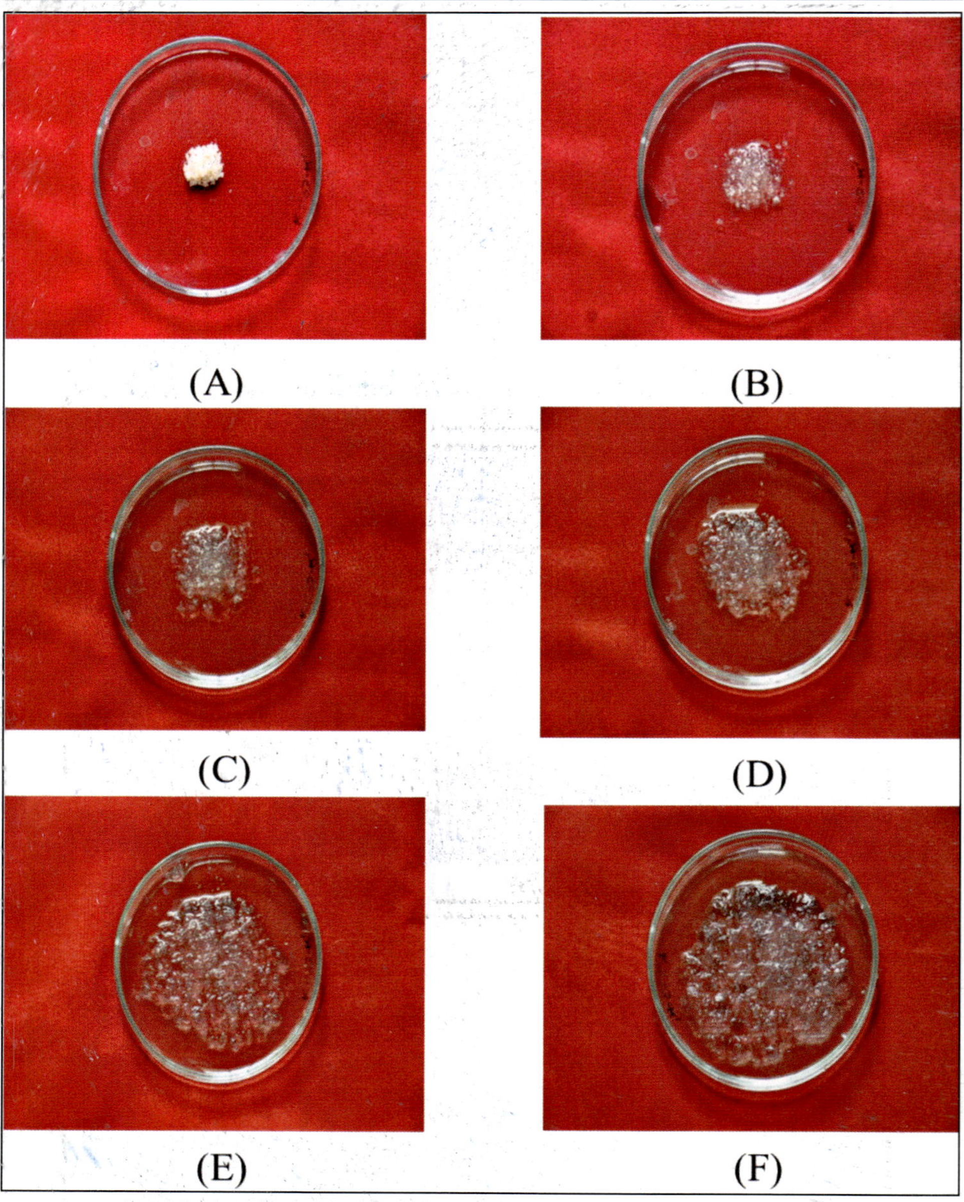

Figure 14.2. Moisture Absorbed by Cross-Linked Biopolymer at different Time in Interval. A: Suberabsorbent polymer without moisture; B: Superabsorbent polymer in water after 5 minutes; C: Superabsorbent polymer in water after 10 minutres; D: Superabsorbent polymer in water after 15 minutres; E: Superabsorbent polymer in water after 20 minutres; F: Superabsorbent polymer in water after 25 minutres. p. 299

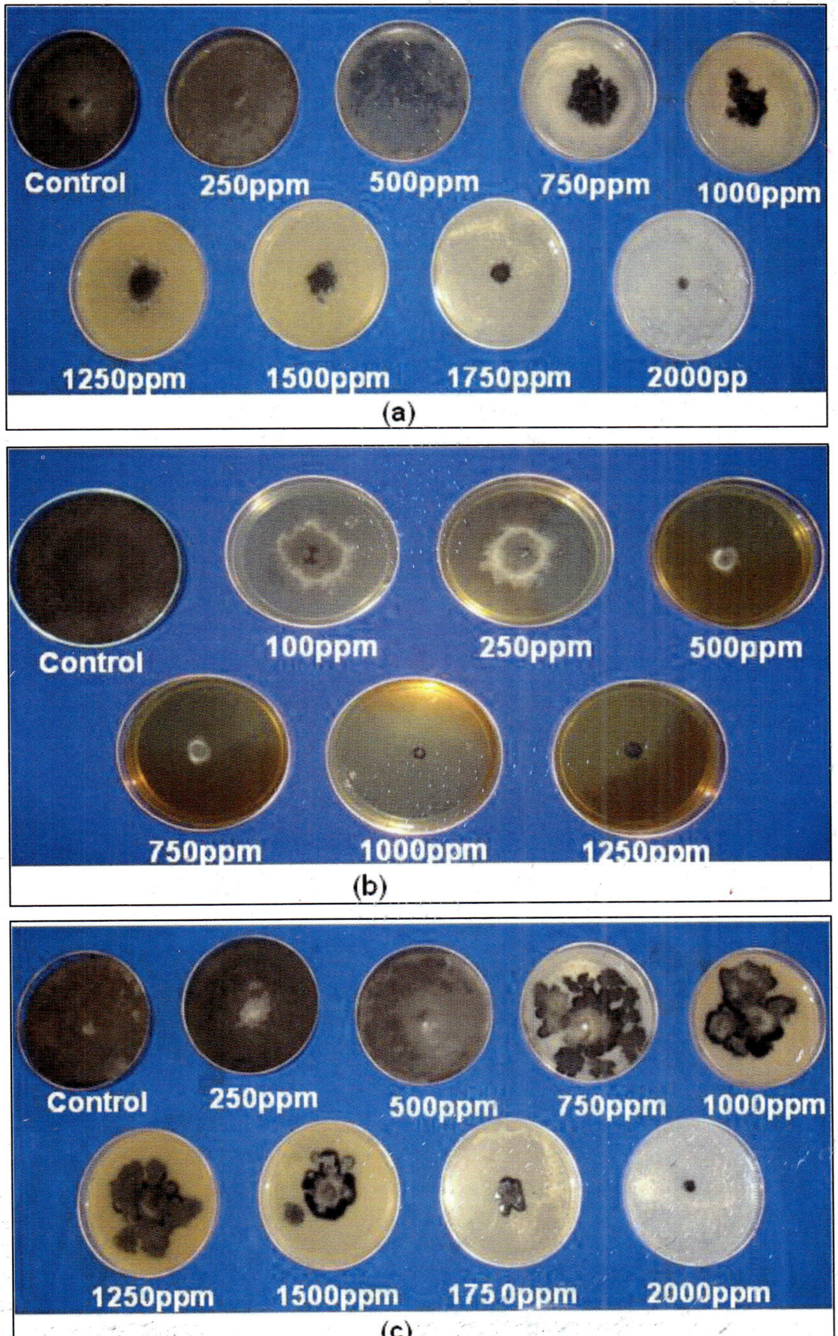

Figure 15.9. Inhibition Zones Caused by different Concentrations of Nanoparticles of (a) ZnO, (b) Ag and (c) TiO_2 against *Macrophomina phaseolina*. P. 328

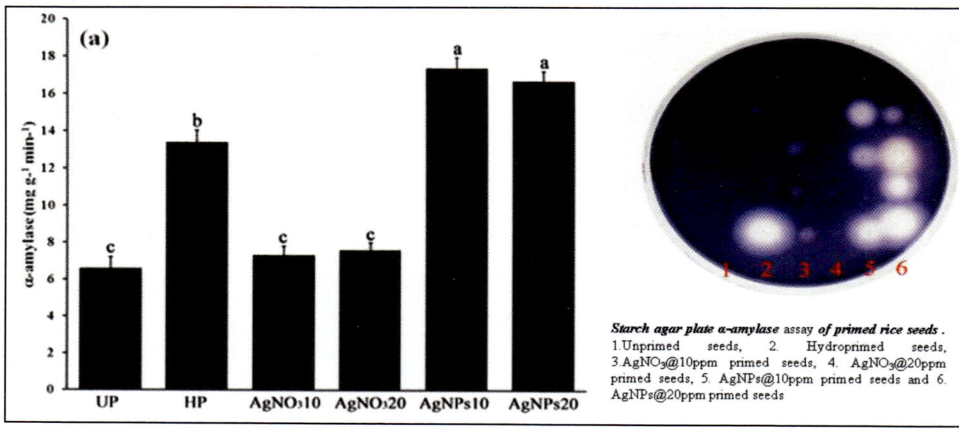

Figure 15.12. Nanopriming on Alpha Amylase Activity in Rice Seeds.
Courtesy: Wuttipong Mahakham *et al.*, 2017. **P. 331**

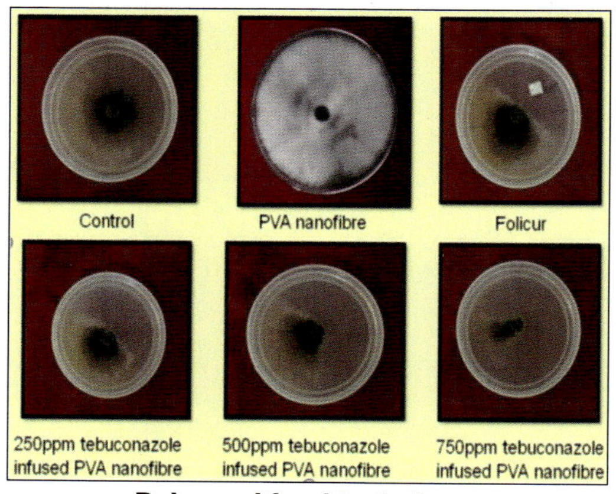

Poisoned food technique.

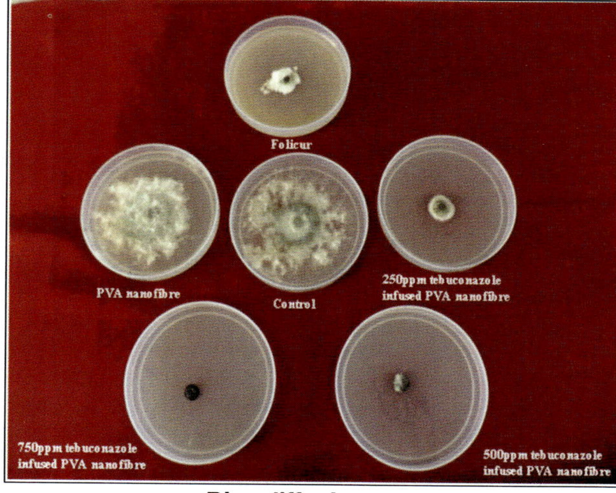

Disc diffusion assay.

Figure 15.22. Bio-efficacy of Tebuconazole Infused PVA Nanofibres on *Macrophomina phaseolina*. P. 341

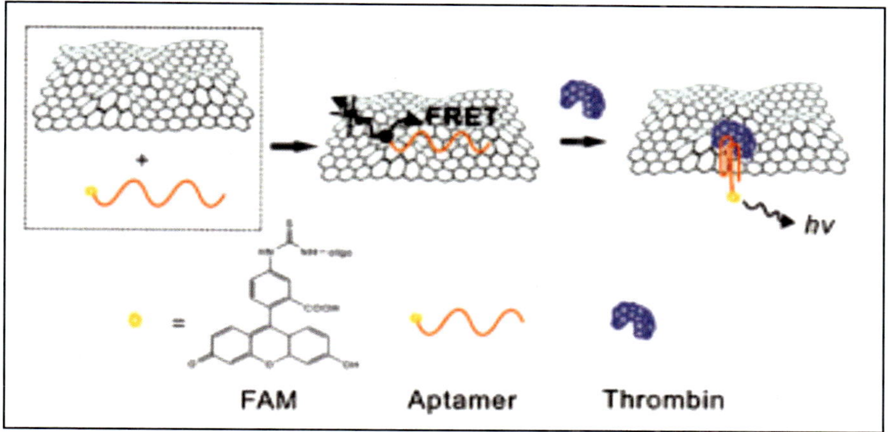

Figure 17.4. Graphene Based Probe.
Courtesy: Chang *et al.*, 2010. **P. 371**

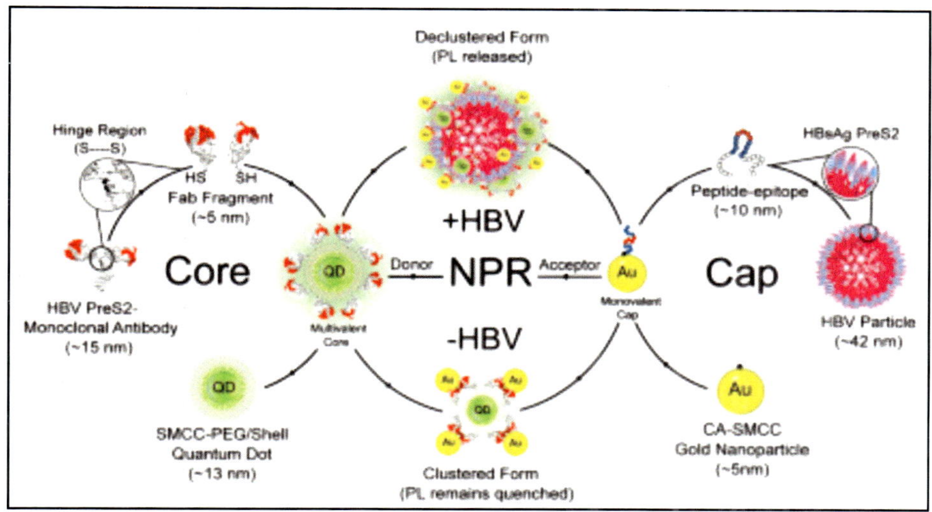

Figure 17.6. Detection Hepatitis B Virus by Quenching Nanoprobes.
Courtesy: Draz *et al.*, 2012. **P. 372**

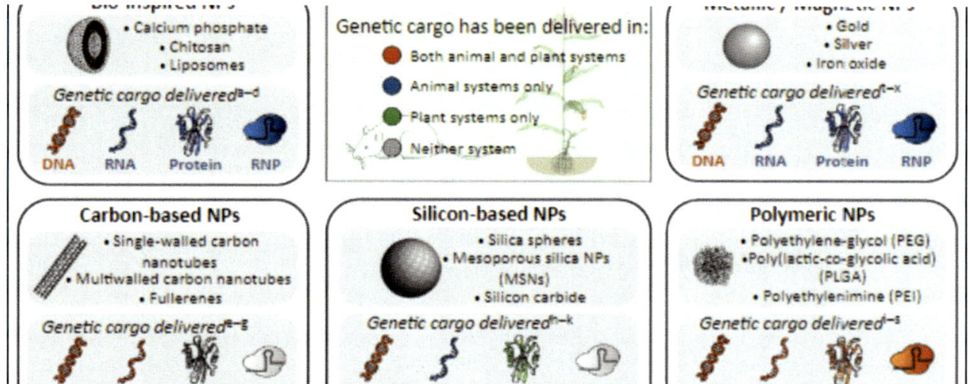

Figure 18.2. Nanoparticle (NP) Mediated Genetic Cargo Delivery to Animals and Plants.
Courtesy: Cunningham *et al.*, 2018. **P. 389**

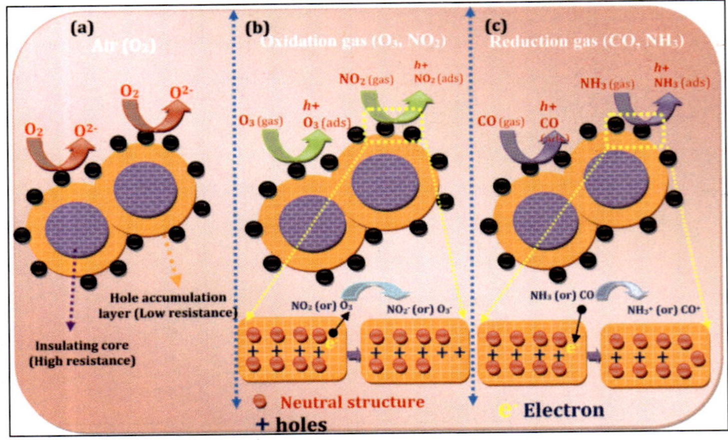

Figure 19.1. Schematic Diagrams for Oxidation (O_3 and NO_2) and Reducing (CO and NH_3) Type Gas Sensing Operating Mechanism on *p*-type $CuAlO_2$ Microspheres Constituted of Nanoparticles (S. Thirumalairajan *et al.*, 2015). P. 405

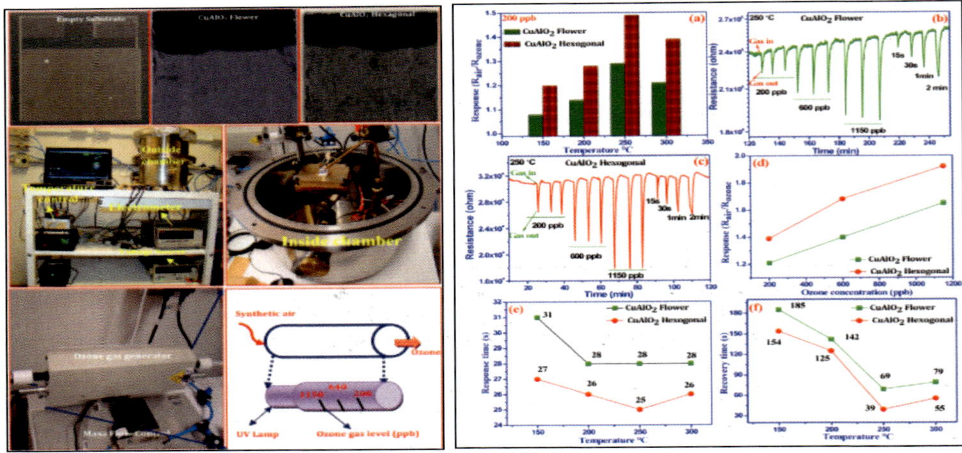

Figure 19.4. (a) Photograph of sensing films and chemical sensor measurement setup and (b) analysed chemical sensor parameter.
Courtesy: S. Thirumalairajan *et al.*, 2014. **P. 411**

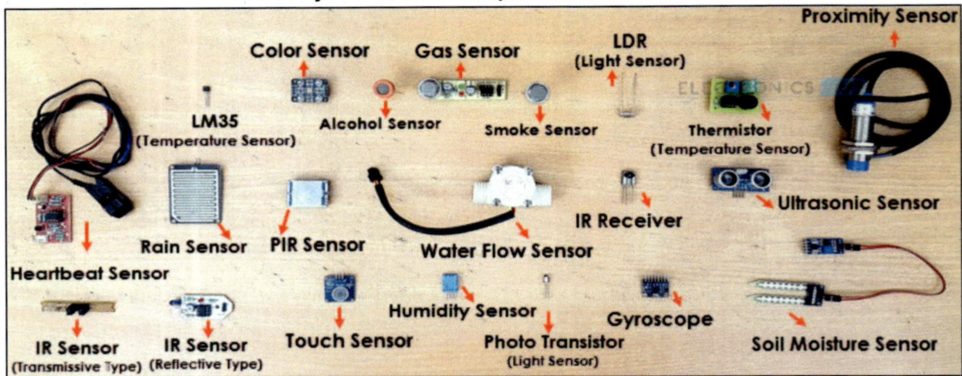

Figure 19.5 P. 411

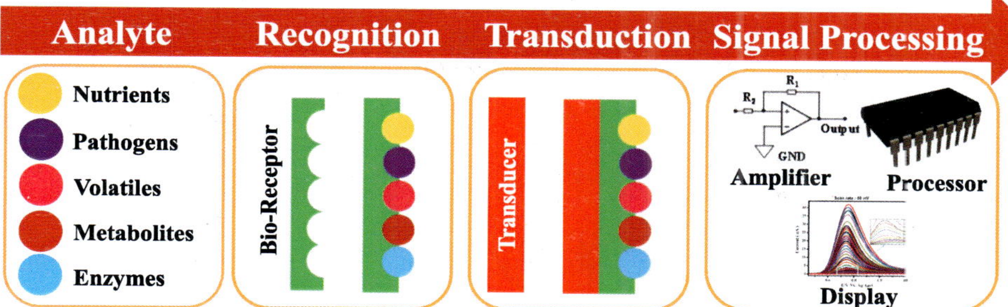

Figure 20.1. Schematic Workflow and Components of Typical Biosensor. P. 422

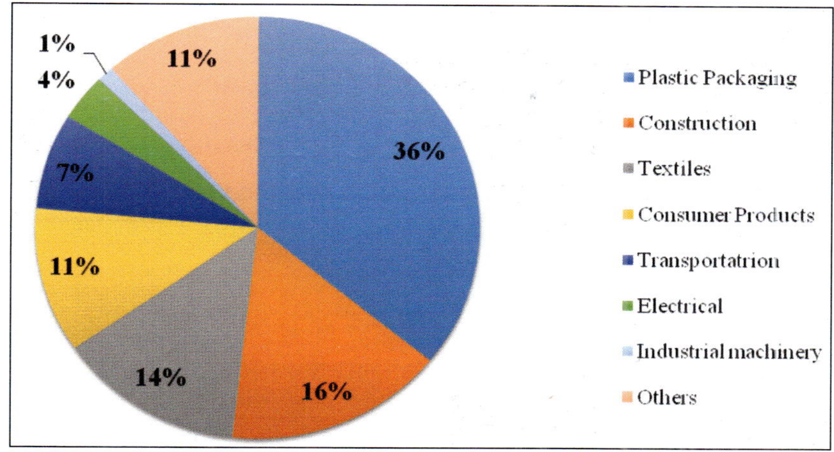

Figure 21.1. Use of Plastics in Various Fields of Applications
Courtesy: https://www.bloombergquint.com/charts/worlds-plastic-burden-weight-of-a-billion-african-elephants **P. 211**

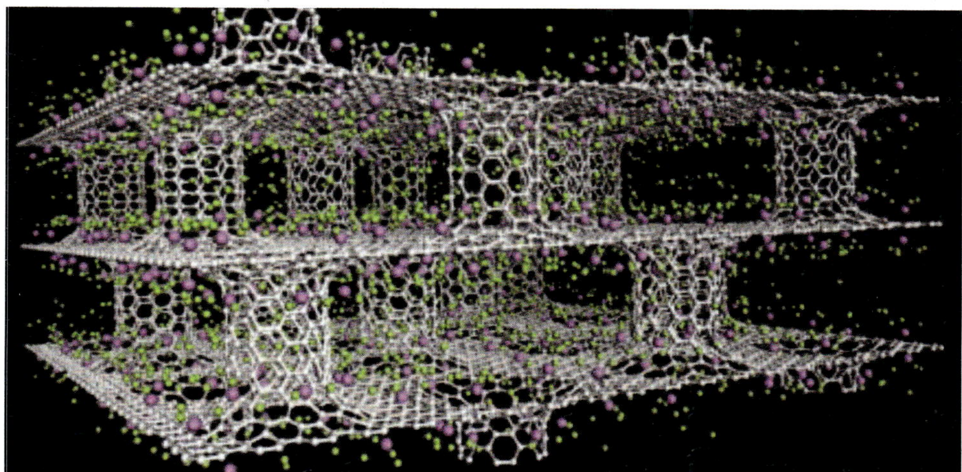

Figure 23.11: Cartoon Representation of CNT and Carbon Nanomaterials for Storage of Hydrogen. P. 494

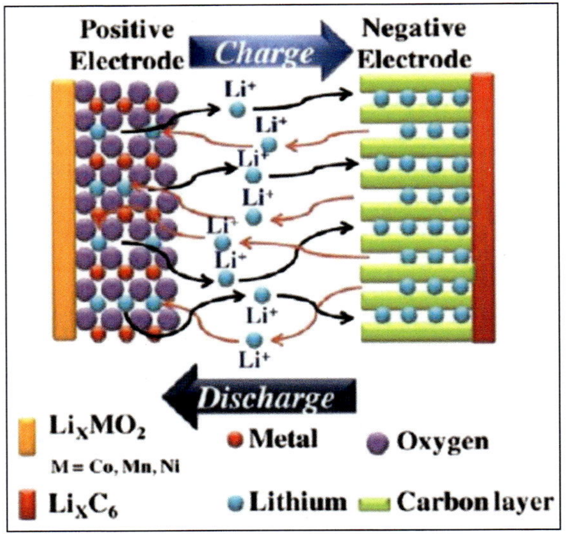

Figure 24.1: Depiction of the Li-ion Battery.
Courtesy: Tiwari *et al.*, 2012.**P. 499**

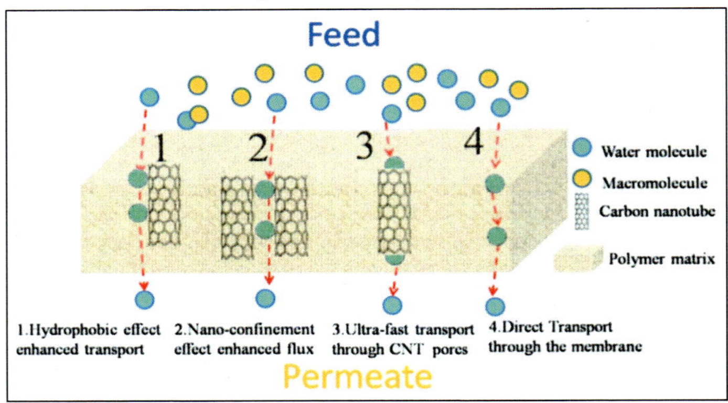

Figure 27.3. Illustration of Possible Pathways for Water Transportation in a CNT/Polymer Blend Membrane
Courtesy: Li *et al.*, 2014. P. 557

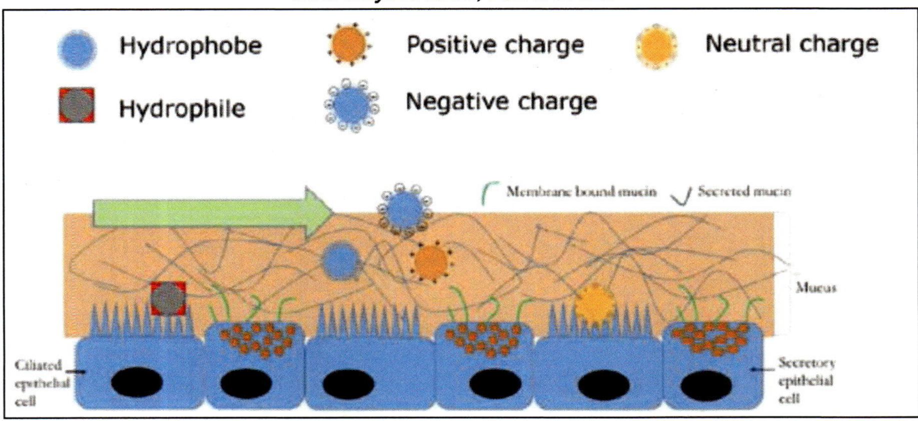

Figure 31.3. Movement of Nanoparticles in Mucus
Courtesy: Gustafsson and Osterlund, 2016. **P. 617**